**Probability and Its Applications**

René Carmona Jean Lacroix

# Spectral Theory of Random Schrödinger Operators

Birkhäuser
Boston • Basel • Berlin

René Carmona
Department of Mathematics
University of California at Irvine
Irvine, CA 92717, USA

Jean Lacroix
Département de Mathématiques
Université de Paris XIII
93400 Villetaneuse
Fessy 74
France

Library of Congress Cataloging-in-Publication Data
Carmona, R. (René)
Spectral theory of random Schrödinger operators / René Carmona, Jean Lacroix.
p. cm. — (Probability and its applications)
Includes bibliographical references and index.
ISBN 0-8176-3486-X (alk. paper)
1. Schrödinger operator. 2. Spectral theory (Mathematics)
I. Lacroix, Jean, 1944- . II. Title. III. Series.
QA329.C37 1990
515'.7242—dc20 90-39801

Printed on acid-free paper.



ISBN 0-8176-3486-X
ISBN 3-7643-3486-X

Camera-ready copy provided by the authors using LaTEX.
Printed and bound by Edwards Brothers, Inc., Ann Arbor, Michigan.
Printed in the U.S.A.

9 8 7 6 5 4 3 2 1

# PREFACE

Since the seminal work of P. Anderson in 1958, localization in disordered systems has been the object of intense investigations. Mathematically speaking, the phenomenon can be described as follows: the self-adjoint operators which are used as Hamiltonians for these systems have a tendency to have pure point spectrum, especially in low dimension or for large disorder.

A lot of effort has been devoted to the mathematical study of the random self-adjoint operators relevant to the theory of localization for disordered systems. It is fair to say that progress has been made and that the understanding of the phenomenon has improved. This does not mean that the subject is closed. Indeed, the number of important problems actually solved is not larger than the number of those remaining. Let us mention some of the latter: .

- A proof of localization at all energies is still missing for two dimensional systems, though it should be within reachable range. In the case of the two dimensional lattice, this problem has been approached by the investigation of a finite discrete band, but the limiting procedure necessary to reach the full two-dimensional lattice has never been controlled.
- The smoothness properties of the density of states seem to escape all attempts in dimension larger than one. This problem is particularly serious in the continuous case where one does not even know if it is continuous.
- The structure of the absolutely continuous spectrum is very nebulous and still at the stage of conjectures. It is a common belief that the higher the dimension the easier it should be to show existence of this type of spectrum. In fact several announcements have been made on the existence of absolutely continuous spectrum and of a mobility edge (i.e. sharp transition between pure point spectrum and absolute continuous spectrum) in the case of the Bethe lattice (or Caley tree).
- The inverse theory has made interesting progress and it is about to

give a characterization of almost periodic potentials in one dimension. But more needs to be done, in particular in higher dimensions.

We hope that the present book will provide the necessary background to newcomers trying to get into this exciting field.

One of the beauties of the subject is the intimate interaction between sophisticated mathematical concepts, probabilistic objects and physical arguments. Unfortunately, their intricate involvement makes it less accessible and sometimes quite painful to learn.

Both authors worked for a long time (and they still do) on some of the challenging mathematical problems which attracted so many mathematicians. What made us more likely to end up writing a book like the present one is that we both tried very early to make the subject more popular to a wide audience. Indeed, the first named author published a set of Lecture Notes from a course he gave in 1984 in a Probability Summer School. The purpose of the notes was to offer to probabilists a concise exposition of the spectral theory of self-adjoint operators in general, to Schrödinger operators in particular, and an introduction to the spectral theory of their random counterparts. Except for some results on the density of states, most of the set of notes was devoted to the *one dimensional continuous case*. On the other hand, the second named author wrote the second half of a book devoted to the theory of random matrices and their applications. More than a text on Schrödinger operators, the chapters were designed to illustrate how the ergodic theory of random matrices could be used to investigate the spectral properties of the *one-dimensional discrete Schrödinger operators* and of their analogs on the strip.

Several excellent survey articles and book chapters have been written and published since then. Some of them give a good account of results which appeared in the meantime. Among them are the multidimensional results which followed the fundamental contribution of Frölich and Spencer, and the inverse theory initiated by Kotani. But still no book devoted entirely to the spectral theory of random Schrödinger operators has yet appeared.

The decision to write this book was made during the Spring quarter of 1988 while the second named author (J.L.) was visiting Irvine. We thought that it would not take too much effort to put together our respective monographs. It did not seem very hard to add recent results on the multidimensional case and the inverse theory which were not touched upon in our previous writeups. We were wrong. It seemed at the time that the enterprise would be easier than what it ended up being. We are glad it is over and we hope that probabilists as well as functional analysts and mathematical physicists will benefit from our efforts to bring this exciting material to the level of a graduate text. We originally aimed at presenting the best features of the subject with a minimum of prerequisites and technicalities. But our desire to give a fair account of all the results and approaches to the problems led us to expand the text into an encyclopaedic treatise. We hope that the reading guide proposed as Chapter 0 will help the reader to find the tree he is looking for in the dense forest of the book.

The bulk of the actual writing of the book was done during the Spring and the Summer of 1989 while the second named author was visiting the Department of Mathematics at U.C. Irvine. The latter should be thanked for its hospitality and its financial support. Both authors benefited over the years from very useful discussions with many colleagues and friends on the subject of the book. The list of those to thank for this kind of help would be too long for this preface. Nevertheless, we would like to thank the few individuals who took a quick look at the first drafts of some of the chapters and who gave us some constructive feedback. I. Goldsheid, A. Klein, S. Kotani and F. Perez are among them. We would also like to thank Mary Moore and Angie Velez for typing part of an early draft of the manuscript, and M. Dowd, W.C. Masters and especially P. Dugan for helping us proofread part of the final version.

The first named author (R.C.) would like to acknowledge partial support from N.S.F. under Grant DMS-8701320. He wishes to add a dedication to the above list of acknowledgments. He would like to express his deep gratitude to an exceptional mathematical physicist who proved to be over the years an enjoyable co-worker, a reliable friend but most of all an enlightening mentor. Thank you Barry Simon. The second named author (J.L.) would like to thank Yves Guivarc'h for his interest and his generous help during the preparation of the book.

The last word of this preface should be an apology. English is not the native language of either one of the authors. As a consequence, and despite strong efforts on our part, many sentence constructions will be awkward if not syntactically incorrect. We simply hope that this will not be too frustrating and that it will not prevent the reader from appreciating the beauty of the mathematics involved.

# READER'S GUIDE

The purpose of this short introductory chapter is to help the reader find her/his way through the dense forest of results which comprise the present book. By reading this outline, the reader will understand the scope of the book and find a clear path to the special results she/he is interested in.

## 1. Background and Motivation

This book is about conservative linear wave propagation. The effects of interest are ultimately scattering effects for there is no contribution from friction or dissipation.
The focus of study is propagation in a random environment. This is perhaps a good start toward modelling the complexity of the real world. The prototype equation for such propagation is the Schrödinger equation of quantum mechanics. This is, up to a few variations, the only equation that is explicitly considered in this book.

Let us first describe a nonrandom model for the sake of illustration. We choose the example of a simple quantum system, an atom with $N$ electrons. We assume that the nucleus is infinitely heavier that the electrons so that it will stay fixed, say at the origin of our space $\mathbb{R}^3$. The Hamilton function of classical mechanics can be written as:

$$h = \sum_{j=1}^{N} \frac{1}{2m_j} |p_j|^2 - \sum_{j=1}^{N} \frac{Ne^2}{|x_j|} + \sum_{1 \le j < k \le N} \frac{e^2}{|x_j - x_k|}$$

where $m_j$ denotes the mass of the $j$-th electron, $x_j \in \mathbb{R}^3$ its position, $p_j \in \mathbb{R}^3$ its momentum (i.e. $m_j$ times its velocity) and $e$ the charge of an electron. The first term represents the total kinetic energy of the system (recall that the nucleus is regarded as being at rest), the second term represents the attraction of the electrons by the nucleus and the last term represents the mutual repulsion of the electrons.

Quantum theory requires each $x_j$ to become the operator of multiplication by the function $x_j$ on the appropriate Hilbert space, and the momentum $p_j$ to become $i$ times the operator of partial derivation with respect to the variable $x_j$ (here and in the following, $i$ denotes the complex number $\sqrt{-1}$ and the constant $h$ is set equal to 1). Consequently the Hamilton function for the energy of the system becomes the formal partial differential operator

$$H = H_0 + V(x) \tag{1}$$

where:

$$H_0 = -\sum_{j=1}^{N} \frac{1}{2m_j} \Delta_j \tag{2}$$

with $\Delta_j$ denoting the Laplacian operator in the variable $x_j$, and where $V(x)$ denotes the operator of multiplication by the function:

$$V(x) = \sum_{j=1}^{N} \frac{Ne^2}{|x_j|} - \sum_{1 \leq j < k \leq N} \frac{e^2}{|x_j - x_k|} \tag{3}$$

where $x = (x_1, x_2, \cdots, x_N)$ is now an element of $\mathbb{R}^{3N}$. We will make the convention $m_j = 1/2$ so that our operator $H_0$ representing the quantum mechanical kinetic energy will simply be minus the usual Laplacian operator in $\mathbb{R}^{3N}$. The Hilbert space on which these operators are acting is chosen to be $\mathcal{H} = L^2(\mathbb{R}^{3N}, dx)$. A quantum mechanical state is an element of the Hilbert space $\mathcal{H}$ instead of being an element of $\mathbb{R}^{3N}$. More precisely it is a unit vector, say $f$, in this Hilbert space, hence a function on $\mathbb{R}^{3N}$ the integral of the square of the modulus of which is equal to one. The square of this modulus is interpreted as a density of probability: If $A$ is any Borel subset of $\mathbb{R}^{3N}$ the quantity $\int_A |f(x)|^2 dx$ represents the probability that the system described by $f$ is in $A$.

To describe the time evolution of the system we need to define its dynamics. It is given by the Schrödinger equation:

$$i\frac{\partial \psi}{\partial t} = H\psi. \tag{4}$$

Hence, if the state of the system is $\psi_0$ at time $t = 0$, the system at any time $t$ is given by the solution $\psi(t, x)$ of (4) satisfying the initial condition $\psi(0, \cdot) = \psi_0(\cdot)$.

Let us assume that some existence and uniqueness theorem holds for equation (4). Consequently, for any real numbers $s$ and $t$ there exists an operator $U(s, t)$ on $\mathcal{H}$ such that $\psi(t, \cdot) = U(t, s)\psi(s, \cdot)$ is the solution at time

$t$ if $\psi(s,\cdot)$ is the solution at time $s$. $\{U(t,s);\, s,t \in \mathbb{R}\}$ is usually called the propagator of equation (4). The existence and uniqueness assumption forces the propagator to be a family of bounded (one-to-one onto) operators on $\mathcal{H}$ which satisfy:

$$U(s,t)U(t,u) = U(s,u)$$

for all real numbers $s, t$ and $u$. Obviously $U(s,s) = I$ for each real number $s$. Moreover, saying that a state at a given time will remain a state at all times simply means that these operators are norm preserving, hence unitary. Finally, if we assume that this propagator is strongly continuous (in fact, weakly measurable is enough) then Stone's theorem tells us that

$$U(t,s) = e^{-i(t-s)\tilde{H}}$$

for some self-adjoint operator $\tilde{H}$ on $\mathcal{H}$. This self-adjoint operator $\tilde{H}$ has to be equal to (or at least be an extension of) the Schrödinger operator $H$ given in (1), (2) and (3). Consequently, we can solve the Schrödinger equation in an $L^2$-sense if the initial state $\psi_0 \in \mathcal{H}$ is in the domain of the self-adjoint operator $H$. We are interested in understanding the time evolution of the states and especially in the asymptotic behavior of these states for large times. For example, we would like to find out if:

$$\lim_{T\to\infty} \frac{1}{2T} \int_{-T}^{+T} \left( \int_{|x|\le r} |\psi_t(x)|^2 dx \right) dt = 0 \tag{5}$$

for each real $r > 0$ or if:

$$\lim_{r\to\infty} \sup_{-\infty<t<+\infty} \int_{|x|\ge r} |\psi_t(x)|^2 dx = 0. \tag{6}$$

Relation (5) says that for any given ball, the probability that the state is in this ball for large time is, at least in a Cesaro average sense, going to 0. The state $\psi_0$ is said to be an *unbound state* because of this wandering out to infinity. The unbound states form a subspace of the Hilbert space $\mathcal{H}$. Relation (6) says that, up to small controllable errors, we can find a large enough ball in which the system stays at all times. Such an initial state $\psi_0$ is naturally called a *bound state.* The bound states form a subspace of the original Hilbert space $\mathcal{H}$ and this subspace is orthogonal to the space of unbound states. It is comforting to realize that, in most of the cases, the Hilbert space $\mathcal{H}$ is actually equal to the direct sum of these subspaces. In fact, this physical decomposition of the space coincides with a decomposition made on purely mathematical grounds, by considering the elements of $\mathcal{H}$ having an absolutely continuous spectral measure and those having a point spectral measure.

A time honored method to study equations of this form is to look first for solutions of a particular form, the so called stationary solutions. This leads to an equivalent form which is now time independent:

$$H\psi = \lambda\psi$$

where both the (possibly generalized) eigenfunction $\psi$ and the eigenvalue $\lambda$ are to be found. As pointed out by Schrödinger himself in his original work, this form of the problem is fundamental. Indeed, the knowledge of all the eigenfunctions of the operator $H$ can be used to reconstruct all the spectral quantities which are relevant to the problems at hand. In particular, the propagator and all the other functions of the operator $H$ can be recovered by eigenfunction expansions.

The functional analysis we just outlined is described in Chapters I and III. Namely, we gather the main fact of the theory of (possibly unbounded) self-adjoint operators on a Hilbert space. We also show how Schrödinger operators can be defined as self-adjoint operators, or at least as symmetric operators having self-adjoint extensions. If more than one such extension exists, we also show how to find and control those which are relevant to the physics of the problem. We discuss in great detail the typical spectral characteristics of self-adjoint operators and the decomposition of the Hilbert space according to the various components of the spectrum (i.e. the so-called Lebesgue decomposition of the Hilbert space).

The above example was borrowed from the quantum theory of $N$-body systems. The difficult analytic problems which are inherent to the interactions between the electrons are far beyond what can be analyzed when randomness is added to the potential function $V$. The problems which we shall consider are not very realistic. They typically concern only the properties of one electron. Let us describe another example to illustrate that fact. Let us assume that an atom $x_j$ is sitting at each site of the lattice $\mathbb{Z}^d$, and it creates a potential $v_j$. A particle, whose position we denote by $x$, will feel a total potential:

$$V(x) = \sum g_j v_j(x-j)$$

where the "coupling constants" $g_j$ are real numbers. If we assume that the particles in the lattice are identical, as in the study of crystals (which are perfectly periodic structures), then all the functions $v_j$ are equal and so are all the constants $g_j$, and the potential function $V$ is periodic. Ordinary differential operators and partial differential operators with periodic coefficients have attracted the attention of mathematicians for a long time. Moreover, the spectral properties of periodic Schrödinger operators are of crucial importance for applications in solid state physics and the study of conductor and insulator materials. Such a spectrum is a union of intervals $[a_n, b_n]$ with $b_n \leq a_{n+1}$. The intervals $[a_n, b_n]$ are called bands of allowed energies while the intervals $(b_n, a_{n+1})$ are called gaps of forbidden energies. These properties are mentioned in Chapter II and in Chapter III and detailed proofs are given in Chapter VII for one dimensional systems.

We finally come to the discussion of some of the ideas of P.W. Anderson. Let us imagine for example that the coupling constants are random and form a sequence of independent identically distributed (i.i.d. for short) random variables. Or, let us imagine that these constants are equal but that the particles $x_j$ are randomly located (according to a Poisson point process in $\mathbb{R}^d$ for example) instead of sitting at the very sites of the lattice $\mathbb{Z}^d$. In both cases the potential function is random. It is a stationary random process, and the Schrödinger equation involves an operator which is now a random self-adjoint operator. Its spectral properties are expected to be drastically different from those of the deterministic counterparts. Indeed, according to the predictions of the physicists, one expects dense pure point spectrum with exponentially decaying eigenfunctions in one and two dimensions (so, no scattering states, only bound states) and, in higher dimensions, dense pure point in some energy range and purely absolutely continuous spectrum in the rest of the spectrum. The transition between these two regimes is expected to be sharp. It is called the *mobility edge*.

Most of these claims are still at the stage of mathematical conjectures. The purpose of this book is to present the mathematical results which have been proved so far and the mathematical theories involved in the proofs.

As we already explained, the first three chapters are devoted to the functional analysis of self-adjoint operators in general, of Schrödinger operators in particular, and of one (or quasi-one) dimensional systems. The last four chapters deal with the actual spectral characteristics of *random* Schrödinger operators. Except for some results on the integrated density of states which are used in each of the last three chapters, Chapters VI, VII, VIII and IX are dealing with different facets of the field and they can be read independently.

## 2. Random Schrödinger Operators

We consider self-adjoint operators $H$ of the form $H = H_0 + V$ where $H_0$ is a deterministic quantization of the kinetic energy, and $V$ is the operator of multiplication by a potential function. We call $H_0$ the free Hamiltonian and we regard $V$ as a perturbation of $H_0$. We assume that $V$ is random. More precisely we assume that $V(\cdot,\omega)$ is a sample from a stationary ergodic stochastic process (or random field) defined on some probability space. The stationarity means that all the averages over the random parameter $\omega$ will be translation invariant, and this is a very natural assumption. Ergodicity is not less natural for experience shows that most of the physical systems one encounters are actually ergodic. Consequently, the full Hamiltonian $H$ becomes a function of the random parameter $\omega$ and we are facing a *random self-adjoint operator* $H(\omega)$ which should be stationary and ergodic in some sense. The first part of Chapter V is devoted to the rigorous definition of these operators and to the treatment of the various measurability problems. We also take full advantage of the ergodicity property to prove that essentially all the spectral sets which one usually considers (spectrum, absolutely continuous spectrum, singular continuous spectrum, ...) are nonrandom. This has to be interpreted in the following sense: for almost every $\omega$ with respect to the probability measure giving the stochastic characteristics of the potential function $V$, the operators $H(\omega)$ have the same spectrum, say $\Sigma$, the same absolutely continuous spectrum, say $\Sigma_{ac}$, ....

We should emphasize that regarding the potential $V$ as a perturbation of the free Hamiltonian $H_0$ is not always the right thing to do. Indeed, $H_0$ is bounded in some cases (the kinetic energy of a system on the lattice $\mathbb{Z}^d$ is such a case) and the potential energy may be much larger (systems with high disorder) and it is more convenient in these situations to regard the operator $H_0$ as a perturbation of the potential operator $V$! This is the approach to localization used in Chapter VII when we discuss almost periodic potentials in one dimension, and in Chapter IX for multidimensional lattice systems.

## 3. The Integrated Density of States

One of the earliest successes of the mathematical theory of random operators was the construction and the estimation of the so-called integrated density of states. This quantity measures in some sense the number of eigenvalues (energy levels) per unit volume. Its definition depends on the approximation by bounded regions $\Lambda$ to which we restrict the operator $H(\omega)$. A particular choice of a set of boundary conditions is necessary but we will ignore this technical detail for the purpose of the present discussion. As explained in Chapter II, the resolvent operators of these restrictions $H_\Lambda(\omega)$ are compact and this implies that the spectrum of each $H_\Lambda(\omega)$ is discrete and that its eigenvalues can be enumerated. Repeating each eigenvalue according to its multiplicity, the eigenvalue counting function is then defined as:

$$N_{\Lambda,\omega}(\lambda) = \#\{\text{eigenvalues of } H_\Lambda(\omega) \leq \lambda\}.$$

This counting function is almost additive when we consider the union of disjoint domains and the ergodic theorem can be used to show that:

$$\frac{1}{|\Lambda|} N_{\Lambda,\omega}(\lambda) \longrightarrow N(\lambda),$$

where $|\Lambda|$ denotes the volume of the domain $\Lambda$. The limit $N(\Lambda)$ is not random. It is called the ***integrated density of states***. In fact, because of the ergodicity assumption many quantities associated with $H_\Lambda(\omega)$ will converge to a nonrandom limit. This phenomenon was known to many physicists. It goes very often under the name of ***self-averaging***. The function $N(\lambda)$ is nonnegative and nondecreasing. As such, it can be viewed as the distribution function of a nonnegative measure on the real line. The topological support of this measure is equal to the closed set $\Sigma$ singled out as the almost sure spectrum of the operators $H(\omega)$. This implies that the function $N(\lambda)$ is zero outside the spectrum $\Sigma$ and an interesting question is to determine the behavior of $N(\lambda)$ when $\lambda$ approaches one of the edges of the spectrum from inside.

Let us consider for example the case of the infimum $\lambda_0$ of $\Sigma$ when the spectrum is bounded below. As early as 1965, I.M. Lifschitz gave heuristic arguments to demonstrate that the decay of $N(\lambda)$ had to be exponential in a specific power of $(\lambda - \lambda_0)$. This asymptotic behavior, appropriately named Lifschitz tail, can be investigated via Tauberian theorems once the Laplace transform of $N(\lambda)$ is computed. This study had a significant impact on several developments in the asymptotic theory of stochastic processes because this Laplace transform has a simple expression in terms of exponential functions of the conditioned Brownian motion.

It was believed for quite some time that the integrated density of states could be used to distinguish between parts of the spectrum $\Sigma$ where the operator $H(\omega)$ had different spectral characteristics. In particular it was conjectured that the function $N(\lambda)$ should have a singularity of some kind at the mobility edge because of the transition from absolutely continuous

to dense pure point spectrum. We know now that this is not the case because these subtle properties of the operators $H(\omega)$ are washed out in the averaging involved in the very construction of $N(\lambda)$. In fact, there are examples of random Schrödinger operators having the same integrated density of states $N(\lambda)$ and very different spectral characteristics.
These results are presented in Chapter VI. The existence and the asymptotic results are used to illustrate the two possible approaches: the probabilistic approach and the functional analytic one. According to the reader's taste and/or background some sections may be skipped to avoid duplication.
The integrated density of states is a very rich and colorful mathematical object which deserves to be investigated for its own sake. We tried to give as faithful a review as possible of the various possible approaches because there are still many challenging problems remaining. The absolute lack of knowledge of any kind of regularity property of the function $N(\lambda)$ in the multidimensional continuous case is obviously the most striking one.
Despite all that, the integrated density of states cannot be reduced to the level of a mere mathematical curiosity. Indeed some of its properties play a crucial role in the fine analysis of the spectral types of the operators $H(\omega)$. Let us mention here two examples from the theory of the discrete Schrödinger operator on the lattice $\mathbb{Z}^d$ when the values of the potential at different sites are independent.

1. In the one dimensional case, the integrated density of states is related to the Lyapunov exponent $\gamma(\lambda)$ defined below via the formula:

$$\gamma(\lambda) = \int \log |\lambda' - \lambda| dN(\lambda')$$

   known as the Thouless formula. The latter can be used to prove the local Hölder continuity of the integrated density of states from the corresponding property of $\gamma(\lambda)$. This regularity property of the function $N(\lambda)$ turns out to be instrumental in a crucial estimate for the proof of localization for potentials taking only finitely many values.

2. When the distribution of the potential at one site has a bounded density, then the function $N(\lambda)$ is absolutely continuous and its derivative is bounded. This is the only available tool to control the probability that a resonance occurs at an energy $\lambda$ (in the sense that the spectrum of the restriction to a bounded domain of the operator $H(\omega)$ becomes too close to the energy $\lambda$). It is done by a simple application of Chebychev's inequality. It is known as Wegner's estimate. Such

an estimate is not known in the multidimensional continuous case and this is one of the main obstacles to a proof of localization in this setting.

These properties of the integrated density of states, together with many other regularity results, are proved in Chapter VI. Most of this chapter can be skipped in a first reading by someone mostly interested in studying the almost sure spectral characteristics of the operators $H(\omega)$.

# 4. The Absolutely Continuous Spectrum

There is essentially no mathematical result concerning the absolutely continuous spectrum of stationary random Schrödinger operators in more than one dimension. This lack of knowledge is certainly the most challenging of the remaining open problem. Consequently, we restrict the present review to the one dimensional case. Moreover, we further concentrate on the continuous case of the ordinary differential operator for the sake of clarity. All the results that we discuss below, once appropriately reformulated, also hold in the case of the one dimensional finite difference analogue and its extension to the strip.
As we already mentioned, the asymptotic behavior of the solutions of the eigenvalue problem:

$$-\ddot{y} + Vy = \lambda y \tag{7}$$

can be used to determine the nature of the spectrum of the operator $H$. This asymptotic behavior is best understood from the asymptotics of the propagator $U_\lambda(t,s)$ of the ordinary differential equation:

$$\dot{Y}(t) = A_\lambda(t)Y(t) \tag{8}$$

where

$$Y(t) = \begin{pmatrix} \dot{y}(t) \\ y(t) \end{pmatrix} \qquad \text{and} \qquad A_\lambda(t) = \begin{pmatrix} 0 & V(t) - \lambda \\ 1 & 0 \end{pmatrix}$$

are chosen so that (7) is equivalent to (8). In the case of a random potential, the matrices $U_\lambda(t,s)$ are random and the subadditive ergodic theorem gives the existence of the limit:

$$\gamma(\lambda) = \lim_{|t|\to\infty} \frac{1}{|t|} \log ||U_\lambda(t,0)||.$$

The number $\gamma(\lambda)$ so-defined is called the Lyapunov exponent of the problem. Its positivity implies that there exists an eigenfunction $\psi_+$ (resp. $\psi_-$) which decays exponentially when $t \to +\infty$ (resp. $t \to -\infty$) and such that any eigenfunction which is linearly independent of $\psi_+$ (resp. $\psi_-$) explodes exponentially. This exponential dichotomy plays a crucial role.

It is possible to view a periodic function (and more generally any almost periodic function) as a sample from a stationary ergodic random process. The idea is simply to consider a shift of the function chosen at random from a period interval. Consequently, ergodic Schrödinger operators can have absolutely continuous spectrum. This is not the typical behavior which one expects from "random" operators, at least in one dimension. Indeed, periodic processes stand at one end of the spectrum of random processes. In some sense they are *deterministic* because their knowledge in an arbitrary neighborhood of $-\infty$ is sufficient to determine them completely over the whole line. It is shown in Chapter VII that nondeterministic potentials do not have any absolutely continuous spectrum. The reason is that the Lyapunov exponent $\gamma(\lambda)$ is positive for almost every $\lambda \in \mathbb{R}$, and the exponential dichotomy of the solutions which follows, does rule out the existence of absolutely continuous spectrum. In other words, *enough randomness makes it impossible to have absolutely continuous spectrum.* But the question is then: what are the spectral characteristics of random operators with moderate randomness? In particular, what is the answer for almost periodic potentials. The answer is very complex. These operators have a tendency to have Cantor spectra. To get an intuitive idea of why this should be the case, one should think of an almost periodic potential as a superposition of periodic ones. Since each periodic component opens gaps all over the spectrum, it is not surprising to see that the union of these gaps becomes everywhere dense!! Not only is the spectrum unusual as a set, but the spectral types are also remarkable. Any kind of spectrum is possible: dense pure point, purely absolutely continuous, singular continuous. What is even more remarkable is that there exist many families of potentials depending on one or two parameters for which the three spectral types do occur when one varies these parameters. In chapter VII we review without proofs the main results concerning almost periodic potentials. A complete treatment of this class would require another 600 page book.

We already mentioned that nondeterministic potentials did not have any absolutely continuous spectrum. But there are many random deterministic potentials (sorry for the barbarism) which do not have absolutely continuous spectrum either. A section of Chapter VII is devoted to the discussion of some models of this type. Another section is devoted to the inverse theory. It is a very natural generalization of the classical theory for periodic potentials, but it is still in the infancy for random potentials. This chapter depends very much on the classical theory presented in Chapter III. It is not possible to read it without a good knowledge of the notation and the results of Chapter III. Also, several facts from the general theory of products of random matrices presented in Chapter IV and of ergodic self-adjoint operators of Chapter V are necessary.

# 5. Dense Point Spectrum and Localization

The spectral properties of periodic potentials illustrate the fact that the stationarity and ergodicity of a random potential are far from enough to guarantee the absence of diffusion which is expected in most of the disordered systems. A weak form of localization occurs when the wave packets (in other words the solutions to the Schrödinger equation) are trapped and cannot travel to infinity. A strong form of localization occurs (in an energy range $\mathcal{I}$) when the spectrum of the operators $H(\omega)$ is pure point in $\mathcal{I}$ for almost all the $\omega$'s and when the corresponding eigenfunctions are exponentially decaying. It is possible that the *dense pure point spectrum* so typical of random systems does occur without the exponential decay of the eigenfunctions. We will mention some cases where this phenomenon has been observed. Actually, the belief that this could be the case in two dimensions for small disorder is widespread.

Exponential localization is well understood in one dimension. As we already explained, the positivity of the Lyapunov exponent guarantees some sort of exponential dichotomy of the solutions. Moreover, if the randomness of the potential is mathematically tractable, it should be possible to match the solution which is exponentially decaying at $-\infty$ to the solution which is exponentially decaying at $+\infty$ often enough to see the dense pure point appearing. This approach is very classical in spirit and it was the first one to be successfully pursued. The mathematical details are very involved even though the idea is relatively simple: controlling the exponential decay of all the generalized eigenfunctions corresponding to energies on which the spectral measures put some weight. The third part of Chapter VIII is devoted to a thorough exposé of this proof scheme.

Soft analysis methods have been designed to avoid the technicalities of the estimates controlling the decay of the eigenfunctions. The strategy depends on some prior knowledge of the decay. Let us assume for example that we already know that the Lyapunov exponent is positive (in the one-dimensional case) or that the Green's function decays exponentially (in higher dimension). Then tricks can be used to get, in a very elegant way, the dense pure point nature of the spectrum. This approach to localization is presented in the first two sections of Chapter VIII for one dimensional systems and in the first part of Chapter IX for multidimensional systems.

These soft analysis techniques are very elegant but they cannot be used without the results of fine estimates on the decay of the eigenfunctions. As we already pointed out, this decay can be obtained in the one-dimensional case from the strict positivity of the Lyapunov exponent. This is most clearly understood as a by-product of the theory of random matrices. The latter is a beautiful mathematical theory and we included a long Chapter IV to present all the elements which are relevant to the properties of the random systems which we are considering in this book. It is important to notice that this long chapter can be skipped in a first reading. Its results are only used in the proof of localization in one-dimension. And even in this case there are shortcuts to get to the exponential decay of the eigenfunctions without appealing to this heavy machinery. On the other hand, the proofs of localization for quasi-one dimensional systems (such as the strip) require a heavy use of the results on the Lyapunov spectrum and its simplicity, and no simple shortcut seems possible. Moreover, the intuition provided by recasting the search for the asymptotic properties of the solution of the eigenvalue problem into the study of the asymptotics of products of random matrices is very helpful. Finally, once this theory is presented in Chapter IV, the regularity properties of the Lyapunov exponents as functions of the energy can be advantageously investigated. This is done in the second half of Chapter V and the results are used in Chapter VI to obtain regularity properties of the integrated density of states.

From the mathematical point of view, the soft analysis proofs of localization are neat and quite satisfactory. Unfortunately, they fail to address the following important practical problem: when is it possible to see the exponential localization in a finite or bounded region. After all, real life problems and experiments are by nature of a finite character and one should be able to recognize or predict those for which the mathematical results proved in an asymptotic regime apply. More precisely, it seems to us that answers to the following questions are of prime importance.

- given an energy $\lambda$, or more generally an energy range $\mathcal{I}$, how large should a domain $\Lambda$ be so that the eigenfunctions of the operators $H_\Lambda(\omega)$ show some exponential localization in a small subregion of $\Lambda$, (which should of course depend on $\Lambda$, $\omega$, and the eigenvalue)
- given a fixed (large) bounded or finite region $\Lambda$, can one tell if an

energy range $\mathcal{I}$ is still delocalized in the sense that the eigenfunctions with energy in $\mathcal{I}$ are spread out over $\Lambda$, or is it already localized in the sense given above.

Even though we did not tackle these problems as such in one dimension, the approach based on the approximation of the operators $H(\omega)$ and its spectral measures by the corresponding objects restricted to bounded or finite domains, is in this spirit and it could be pushed to provide some answers to these important questions.

In fact, the estimation of the decay of the Green's function (which is the object of Chapter IX) is very much in this spirit. One first defines the exponential decay of the Green's function in a bounded domain as

$$|G_\Lambda(\lambda, x, y)| \leq e^{-\gamma|x-y|}$$

whenever $x, y \in \Lambda$ and $|x-y| > \delta(\Lambda)$ where $\delta(\Lambda) > 0$ depends on the size of the domain $\Lambda$. Then the probability that such an estimate holds is estimated for a large enough domain $\Lambda_0$. Finally a multiscale analysis is used to control the probability that the estimate holds for larger and larger boxes extending to the whole space.

The exponential decay of the Green's function for a fixed energy is very similar to the positivity of the upper Lyapunov exponent. In fact, we show in Chapter IX that the soft analysis tricks which we used to conclude the proof of the one dimensional exponential localization can also be used in the multidimensional discrete case of the lattice $\mathbb{Z}^d$. These tricks require some smoothness of the distribution of the potential at one site. They rely on the existence of a density. This hypothesis excludes simple models such as the Bernouilli potential which takes only two values. There is a way around this difficulty. One has to control the probability that the Green function decays simultaneously in energies restricted to an interval. Indeed, in this case the generalized eigenfunctions can be shown to decay exponentially. This is slightly more technical than the proof of the exponential decay of the Green function for fixed energy, but a similar multiscale argument can be given to prove it.

The multiscale analysis is dimension independent. It depends on an initial estimate. Finding an initial domain such that the probability of exponential decay of the Green's function is large enough can be done by considering that the potential is multiplied by a coupling constant and by choosing this constant large. This is usually called *large disorder regime.* In one dimension, the positivity of the upper Lyapunov exponent can be used to show that this initial step of the proof can be done at all disorders. It is possible that this is also true in dimension two but no proof is known.

But of course the most important problem remaining open in the mathematical theory of random Schrödinger operators is still the existence of absolutely continuous spectrum in dimensions larger than or equal to 3 and for small disorder, and no complete understanding of the disordered systems can be claimed until this outstanding mathematical problem is solved.

# CONTENTS

# I

# SPECTRAL THEORY OF SELF-ADJOINT OPERATORS

Contents:



The present chapter is devoted to the introduction of the notation, the definitions and most of the results from functional analysis which will be needed in the sequel. Because of lack of space, we refrain from explaining the motivations behind the numerous definitions we introduce. We merely illustrate them with examples of Schrödinger operators and we postpone a more detailed study to Chapter II. Rather than a serious introduction to the spectral theory of self-adjoint operators, this chapter should be understood as a glossary and a summary of the terms and results to be used in the sequel.

Our goal is to present the functional calculus for self-adjoint operators on an infinite dimensional (complex) Hilbert space. This is done in Sections 1.2 and 1.5 where we state the spectral theorem and where we discuss some questions of integration with respect to projection valued measures, of multiplicity and of canonical forms. We use the classic approach based on the representation theorems recalled in Section 1.3. Unfortunately the operators we want to study will not always be given as self-adjoint operators but merely as formally symmetric partial differential operators. The usefulness of the spectral theory of self-adjoint operators will depend on our being able to construct self-adjoint extensions for these symmetric operators. We tackle this problem in Section 1.6. Among other things, we illustrate the use of boundary conditions and of quadratic form techniques in the study of self-adjoint operators.

## I.1 Domains, Adjoints, Resolvents and Spectra

An operator $H$ on a Hilbert space $\mathcal{H}$ is a linear map from a vector subspace $\mathcal{D}(H)$ of $\mathcal{H}$ into $\mathcal{H}$. $\mathcal{D}(H)$ is called the *domain* of $H$.

- All the abstract Hilbert spaces $\mathcal{H}$ to be considered below will be implicitly assumed separable and unless stated otherwise, all the operators will be assumed to be densely defined.

Consequently, to be given an operator is to be given a dense vector subspace and an action on the elements of this vector space. For example $H = \Delta$ does not define an operator on $\mathcal{H} = L^2(\mathbb{R}^d, dx)$ until one chooses a domain. The *graph* of the operator $H$ is the subset $\{(f, Hf) : f \in \mathcal{D}(H)\}$ of $\mathcal{H} \times \mathcal{H}$. It will be denoted by $gr(H)$. $\mathcal{H} \times \mathcal{H}$ is a Hilbert space for the inner product:

$$< (f_1, g_1), (f_2, g_2) >=< f_1, f_2 > + < g_1, g_2 >$$

and the operator $H$ is said to be *closed* if $gr(H)$ is a closed subset of $\mathcal{H} \times \mathcal{H}$. Another operator, say $H_1$, is said to be an *extension* of $H$ if $gr(H) \subset gr(H_1)$, that is if $\mathcal{D}(H) \subset \mathcal{D}(H_1)$, and if $Hf = H_1 f$ whenever $f \in \mathcal{D}(H)$. We will use the notation $H \subset H_1$. $H$ is said to be *closable* if it has at least one closed extension. When this is the case, the smallest of these extensions is denoted by $\overline{H}$ and is called the *closure* of $H$. A given operator $H$ does not always have closed extensions. Indeed, the closure $\overline{gr(H)}$ of the graph of $H$ need not be the graph of an operator. Nevertheless we will see later that symmetric operators always have such closed extensions. Also it is easy to see that the graph of the closure of $H$ is always the closure of the graph of $H$, i.e., $gr(\overline{H}) = \overline{gr(H)}$, whenever $H$ is closable. Recall that an operator with domain the whole space is said to be bounded if it is continuous (or equivalently if the image of the unit ball is bounded). By the closed graph theorem this is equivalent to being closed.

We now come to the important concept of ***adjoint*** of an operator $H$. This is the operator $H^*$ defined by:

$$\mathcal{D}(H^*) = \{f \in \mathcal{H};\ \exists g \in \mathcal{H}, \forall h \in \mathcal{D}(H), < Hh, f >=< h, g >\}$$

and

$$H^* f = g \tag{I.1}$$

if $f$ and $g$ are as in the definition of $\mathcal{D}(H^*)$. Note that $H^*$ is well defined on $\mathcal{D}(H^*)$. Indeed, $\mathcal{D}(H)$ is implicitely assumed to be dense. In fact, it is easy to see that $H^*$ is always closed and that $H$ is closable if and only if $\mathcal{D}(H^*)$ is dense in which case $\overline{H} = H^{**}$ and $(\overline{H})^* = H^*$.

The following example plays a crucial role. Its simplicity makes the fundamental result of spectral theory even more remarkable: every self-adjoint operator will be shown to be of this form, at least up to a unitary equivalence.

**Example I.1.1** Let $\mu$ be a nonnegative $\sigma$-finite measure on a separable measurable space $(\Omega, \mathcal{B})$ and let $\mathcal{H} = L^2(\Omega, \mathcal{B}, d\mu)$. For each measurable function $\varphi : \Omega \to \mathbb{C}$ we define the ***multiplication operator*** by $\varphi$, which we denote $M_\varphi$, by:

$$\mathcal{D}(M_\varphi) = \{f \in \mathcal{H}; \varphi f \in \mathcal{H}\}$$

and:

$$M_\varphi f = \varphi f, \qquad f \in \mathcal{D}(M_\varphi).$$

One easily checks that $\mathcal{D}(M_\varphi)$ is dense, that $M_\varphi$ is closed and that $M_\varphi^* = M_{\overline{\varphi}}$. When $\Omega$ is a subset of $\mathbb{C}$ and when $\mathcal{B}$ is the trace $\sigma$-field, the multiplication operator by the identity function will be simply denoted by $M$.

If $H$ is an operator on $\mathcal{H}$ and if $z \in \mathbb{C}$, the operator $zH$ is defined by $\mathcal{D}(zH) = \mathcal{D}(H)$ and $[zH]f = z(Hf)$ whenever $f$ is in $\mathcal{D}(H)$. One easily checks that $[zH]^* = \bar{z}H^*$.

Now, if $H$ and $K$ are operators on $\mathcal{H}$, the operator $H + K$ is defined by $\mathcal{D}(H + K) = \mathcal{D}(H) \cap \mathcal{D}(K)$ and $[H + K]f = Hf + Kf$ whenever $f$ is in $\mathcal{D}(H) \cap \mathcal{D}(K)$. Note that we may have $\mathcal{D}(H) \cap \mathcal{D}(K) = \{0\}$ eventhough $\mathcal{D}(H)$ and $\mathcal{D}(K)$ are dense. Hence the sum of densely defined operators is not necessarily densely defined. Nevertheless we always have $[H + K]^* = H^* + K^*$ if either $H$ or $K$ is bounded. Also, the operator $HK$ is defined by

$[HK]f = H(Kf)$ whenever $f$ is in $\mathcal{D}(HK) = \{f \in \mathcal{D}(K);\ Kf \in \mathcal{D}(H)\}$, and one can check that one always has $K^*H^* \subset (HK)^*$ while the equality $K^*H^* = (HK)^*$ holds whenever $H$ is bounded.
We will use the notation $I$ for the *identity operator* of $\mathcal{H}$ and $L(\mathcal{H})$ for the Banach space (for the uniform norm) of bounded operators on $\mathcal{H}$.
If the operator $H$ is one-to-one, its *inverse* is the operator $H^{-1}$ defined by $\mathcal{D}(H^{-1}) = \{Hf;\ f \in \mathcal{D}(H)\}$ and $H^{-1}g = f$ if $g = Hf$. Note that $H^{-1}$ is closed if $H$ is so (but of course, $\mathcal{D}(H^{-1})$ is not necessarily dense any more).
Let us now assume that the operator $H$ is closed. The set of complex numbers $z$ for which the operator $zI - H$ is one-to-one from $\mathcal{D}(H)$ onto $\mathcal{H}$ (in which case $[zI - H]^{-1}$ is bounded by the closed graph theorem) is denoted by $\rho(H)$ and is called the *resolvent set* of $H$. The bounded operator $[zI - H]^{-1}$ is often denoted by $R(z, H)$ and is called the *resolvent operator* of $H$ at $z$. The resolvent set is an open subset of $\mathbb{C}$ and the function $z \hookrightarrow R(z, H)$ is a strongly analytic function on $\rho(H)$ with values in $L(\mathcal{H})$. Moreover, for all $z$ and $z'$ in $\rho(H)$, the operators $R(z, H)$ and $R(z', H)$ commute and satisfy the so-called *first resolvent identity:*

$$R(z, H) - R(z', H) = (z' - z)R(z', H)R(z, H) \tag{I.2}$$

The complement $\mathbb{C} \setminus \rho(H)$ of the resolvent set is called the *spectrum* of $H$ and will be denoted $\Sigma(H)$. The operator $H$ is said to be *symmetric* if $H \subset H^*$, i.e., $\mathcal{D}(H) \subset \mathcal{D}(H^*)$ and $Hf = H^*f$ whenever $f$ is in $\mathcal{D}(H)$, or equivalently if

$$< Hf, g >=< f, Hg >$$

for all $f$ and $g$ in $\mathcal{D}(H)$. It is said to be *self-adjoint* if $H^* = H$, that is if $H$ is symmetric and if $\mathcal{D}(H) = \mathcal{D}(H^*)$. A symmetric operator is always closable since $H \subset H^*$ and $H^*$ is closed. Moreover, since $H^*$ is a closed extension of $H$, the smallest one of them, namely $H^{**}$, has to be smaller than $H^*$. Hence, we have:

$$H \subset H^{**} \subset H^*,$$

whenever $H$ is symmetric and:

$$H = H^{**} \subset H^*,$$

whenever $H$ is closed and symmetric and:

$$H = H^{**} = H^*,$$

whenever $H$ is self-adjoint.

**Example I.1.2** Let us consider the case of Example I.1.1 above and let us assume that $\varphi$ is real valued. For such a function $\varphi$ we define the *essential range* as the set of $\lambda \in \mathbb{R}$ for which:

$$\mu\{\omega \in \Omega;\ \lambda - \epsilon < \varphi(\omega) < \lambda + \epsilon\} > 0,$$

for all $\epsilon > 0$. It is easy to see that the operator $M_\varphi$ is self-adjoint and that its spectrum is the essential range of the function $\varphi$. See Problem I.7.2 for details.

The following example is easier in some sense. It deals with the so-called *discrete Laplacian*. It is the finite difference analog of the usual (negative) Laplacian. It is an example of a bounded self-adjoint operator. Its definition is very straightforward and the investigation of its spectral properties is very simple.

**Example I.1.3** Let $\mathcal{H}$ be the space $\ell^2(\mathbb{Z}^d)$ of square summable sequences of complex numbers endowed with the Hilbert space structure giving the norm:

$$||f||^2 = \sum_{\mathbf{n} \in \mathbb{Z}^d} |f_{\mathbf{n}}|^2.$$

For each $f \in \mathcal{H}$ we set:

$$[H_0 f]_{\mathbf{n}} = 2d\, f_{\mathbf{n}} - \sum_{|\mathbf{m}-\mathbf{n}|=1} f_{\mathbf{m}}, \qquad \mathbf{n} \in \mathbb{Z}^d.$$

It is easy to see that the operator $H_0$ so defined is a bounded self-adjoint operator and that its spectrum is the interval $[0, 4d]$. This last claim can be easily proved by checking that the Fourier transform gives a unitary equivalence between $H_0$ and a multiplication operator.

A symmetric operator is said to be *essentially self-adjoint* if its closure is self-adjoint. Hence, it has a unique self-adjoint extension. In fact, the converse statement is also true. Moreover, $H$ is essentially self-adjoint if and only if $H^*$ is symmetric in which case one has $\bar{H} = H^*$.
If $H$ is symmetric, if $z \in \mathbb{C}$ and if $f \in \mathcal{D}(H)$, then

$$||[zI - H]f|| \geq |\Im m\, z|\ ||f|| \tag{I.3}$$

because:

$$||[zI - H]f||^2 = < [zI - H]f, [zI - H]f > = ||[H - xI]f||^2 + y^2||f||^2 \geq y^2||f||^2,$$

if $z = x + iy$ with $x$ and $y$ real. It follows from this estimate that the range of $zI - H$ is closed whenever $H$ is closed and $z$ is not real.

We now define the usual Laplacian operator as an essentially self-adjoint partial differential operator. But first we need to introduce some notation.

If $f$ is an element of the Schwartz space $\mathcal{S}(\mathbb{R}^d)$ of rapidly decreasing test functions, its Fourier transform is the function $\hat{f}$ defined on $\mathbb{R}^d$ by:

$$\hat{f}(p) = \frac{1}{(2\pi)^{d/2}} \int_{\mathbb{R}^d} e^{-ip\cdot x} f(x)\, dx$$

The Fourier transform of distributions is defined by duality. If $T$ is a tempered distribution, its Fourier transform $\hat{T}$ is characterized by:

$$\hat{T}(f) = T(\hat{f}) \qquad f \in \mathcal{S}(\mathbb{R}^d).$$

The Fourier transform is a bijection of $\mathcal{S}(\mathbb{R}^d)$ onto itself. Its inverse transform is given by:

$$\check{f}(x) = \frac{1}{(2\pi)^{d/2}} \int_{\mathbb{R}^d} e^{ix\cdot p} f(p)\, dp$$

and similarly for the distributions. The Fourier transform is without question the most efficient tool in the study of partial differential operators with constant coefficients. The following discussion is a clear illustration of this fact.

**Example I.1.4** There are several candidates for the definition of the Laplacian as a self-adjoint operator on $L^2(\mathbb{R}^d, dx)$. The first one is the operator $H_0^{min}$ defined by $\mathcal{D}(H_0^{min}) = C_c^\infty(\mathbb{R}^d)$ the space of $C^\infty$-functions with compact supports in $\mathbb{R}^d$ and by:

$$H_0^{min} f = -\Delta f$$

whenever $f \in C_c^\infty(\mathbb{R}^d)$. Some readers might be surprised by the minus sign in the above definition. We will find a justification in the next chapter when we introduce the so-called free Hamiltonians and the Schrödinger operators. So we apologize for calling very often Laplacian what is actually the negative of the usual Laplacian operator. At the present stage, the explanation for the minus sign is that the qualitative properties of the operator in question do not depend on its sign and since the operator defined above will have a nonnegative spectrum, this should be regarded as a plus.

If instead of trying a minimal domain we want the largest possible domain, we will consider the operator $H_0^{max}$ whose domain $\mathcal{D}(H_0^{max})$ is the set of all elements of $\mathcal{H} = L^2(\mathbb{R}^d, dx)$ whose Laplacian in the sense of distributions is still a function in $L^2(\mathbb{R}^d, dx)$ and we set:

$$H_0^{max} f = -\Delta f$$

whenever $f \in \mathcal{D}(H_0^{max})$. It is then easy to see that:

i) $H_0^{min}$ is essentially self-adjoint,
ii) $H_0^{max}$ is self-adjoint and $\overline{H_0^{min}} = H_0^{max}$.

The proof relies on the following simple facts:

- for each tempered distribution $T$ we have $\widehat{-\Delta T} = |p|^2 \hat{T}$.
- $f \in \mathcal{D}(H_0^{max})$ if and only if the function $p \hookrightarrow |p|^2 \hat{f}(p)$ is square integrable and in this case, $[H_0^{max} \hat{f}](p) = |p|^2 \hat{f}(p)$

The reader is refered to the Section IX.7 of [274] for details.

The following is an easy to prove sufficient condition for a real number to be in the spectrum of a self-adjoint operator. It goes very often under the name of *Weyl criterion*. We leave the proof as an exercise to the reader.

**Proposition I.1.5** *$\lambda \in \mathbb{R}$ is in the spectrum of the self-adjoint operator $H$ whenever there exists a sequence $\{f_n;\, n \geq 1\}$ of unit vectors in the domain of $H$ satisfying* $\lim_{n\to\infty} \|[H - \lambda I]f_n\| = 0$.

We close this section by introducing the concept of convergence which we will use in the sequel. Let $\{H_k; k \geq 1\}$ be a sequence of self-adjoint operators on $\mathcal{H}$. We would like to define the convergence of $H_k$ to some operator $H$ on $\mathcal{H}$ in such a way that the important properties of $H$ could be infered from the corresponding properties of the approximants $H_k$. The first idea which comes to mind is to define this convergence by the convergence in $\mathcal{H}$ of the $H_k f$'s to $Hf$. Though natural, this idea cannot be implemented because the set of $f$'s for which $H_k f$ and $Hf$ make sense may be very small. Indeed, even though the domains $\mathcal{D}(H_k)$ are dense, their intersection may well reduce to $\{0\}$. This last point is the main reason why the use of the convergence of unbounded operators requires a modicum of care. Since it is obviously easier to deal with bounded operators we choose the following definition:

**Definition I.1.6** *The sequence $\{H_k; k \geq 1\}$ of self-adjoint operators is said to converge to the self-adjoint operator $H$, in the sense of strong resolvent convergence, if, for each $z$ in $\mathbb{C}\setminus\mathbb{R}$, the resolvent operator $[H_k - zI]^{-1}$ converges strongly to $[H - zI]^{-1}$, that is if we have:*

$$\lim_{k\to\infty} [H_k - zI]^{-1}f = [H - zI]^{-1}f \tag{I.4}$$

*for all $f$ in $\mathcal{H}$.*

Our first trial to define the concept of convergence was too naive. Nevertheless it makes perfectly good sense in some cases in which it turns out to be equivalent to the more sophisticated definition which we formulated above. This is particularly the case when all the domains have in common a set of elements $f$ for which the convergence of $H_k f$ to $Hf$ occurs, this set being rich enough to determine $H$. This justifies the following definition:

**Definition I.1.7** *If $H$ is a closed operator and if $D$ is a subset of its domain, we say that $D$ is a core for $H$ if the closure of the restriction of $H$ to $D$, which we will denote $H_D$, is equal to $H$ itself.*

The complete knowledge of an operator is given by its knowledge on a core. Now it is easy to show that:

**Proposition I.1.8** *If $H$ and $H_k$ are self-adjoint operators on $\mathcal{H}$ and if $D$ is a core for $H$ which is contained in all the $\mathcal{D}(H_k)$ and for which we have:*

$$\lim_{k\to\infty} H_k f = Hf \tag{I.5}$$

*for all $f$ in $D$, then $H_k$ converges to $H$ in the strong resolvent sense.*

The convergence of self-adjoint operators plays a crucial role in the important study of perturbations. Approximation arguments are very appealing but unfortunately, checking that a sequence of operators actually converges is not always easy. We give in Problem I.7.5 a convenient characterization of the concept of convergence.

## I.2 Resolutions of the Identity

Throughout this section $\mathcal{H}$ will be a fixed (separable) complex Hilbert space and $(\Omega, \mathcal{B})$ a fixed measurable space and we will say projection in $\mathcal{H}$ for orthogonal (i.e. self-adjoint) projection in $\mathcal{H}$.

**Definition I.2.1** *A function $E$ on $\mathcal{B}$ with values in the space of projections in $H$ is called a resolution of the identity (resp. subresolution of the identity) of $\mathcal{H}$ on $(\Omega, \mathcal{B})$ if:*

*i)* $E(\Omega) = I$ *(resp. i')* $E(\emptyset) = 0)$
*ii)* $E(\cup_{n\geq 1} A_n) = \Sigma_{n\geq 1} E(A_n)$ *whenever* $\{A_n; n \geq 1\}$ *is a sequence in* $\mathcal{B}$ *whose elements are disjoint.*

The convergence in *ii)* has to be understood in the sense of the strong convergence of operators. This means that for each fixed $f \in \mathcal{H}$, the series $\Sigma_{n\geq 1} E(A_n) f$ converges in $\mathcal{H}$ to $E(\cup_{n\geq 1} A_n) f$. Hence, for each fixed $f \in \mathcal{H}$ the function $\mathcal{B} \ni A \hookrightarrow E(A) f \in \mathcal{H}$ is countably additive. This is what is usually called a *vector valued measure*. Since the norm of a projection is 0 or 1, the series in *ii)* cannot converge in the norm of $L(\mathcal{H})$, unless all but finitely many $E(A_n)$ are zero. Consequently, except for some trivial cases, $E$ is not countably additive as a function in $L(\mathcal{H})$. In other words, $E$ is not an $L(\mathcal{H})$-valued measure. Nevertheless one easily checks that:

**Proposition I.2.2** *A projection valued function $E$ on $\mathcal{B}$ is a resolution (resp. subresolution) of the identity of $H$ if and only if:*

*i)* $E(\Omega) = I$ *(resp. i')* $E(\emptyset) = 0)$
*ii) for all $f$ and $g$ in $\mathcal{H}$ the complex function $E_{f,g}$ defined on $\mathcal{B}$ by* $E_{f,g}(A) = \langle E(A) f, g \rangle$ *is a complex measure.*

**Remark I.2.3** If $E$ is a subresolution of the identity of $\mathcal{H}$ and if $\mathcal{H}'$ denotes the range of the projection $E(\Omega)$, then the function $E$ defined on $\mathcal{B}$ by setting $E(A)$ equal to the restriction of $E(A)$ to $\mathcal{H}'$ is a resolution of the identity of $\mathcal{H}'$.

**Example I.2.4** Let $\mu$ be a nonnegative measure on $(\Omega, \mathcal{B})$, let us assume that the Hilbert space $\mathcal{H} = L^2(\Omega, \mathcal{B}, \mu)$ is separable, and let us set $E(A) f =$

$\mathbf{1}_A f$ for all $f \in \mathcal{H}$ and $A \in \mathcal{B}$ where the notation $\mathbf{1}_A$ stands for the indicator function (also called the characteristic function) of the set $A$. Then, $E$ is a resolution of the identity of $\mathcal{H}$. We will see later that every resolution of the identity on $(\mathbb{R}, \mathcal{B})$ is unitarily equivalent to a resolution of the identity of this form or to a direct sum of them.

More properties of vector valued measures and of resolutions of the identity are given as problems.

The following theorem is the main result of this section. It gives the definition and the main properties of functions $\phi(E)$ of a resolution of the identity $E$. These properties are easy to check when the functions $\phi$ are simple, more involved when the functions $\phi$ are merely bounded and measurable and rather subtle in the general case of measurable functions $\phi$. We merely make the following remarks as a preparation for the proof.

If $\phi = \sum_{1 \leq j \leq n} a_j \mathbf{1}_{A_j}$ is a simple function with $a_j \in \mathbb{C}$ and $A_j \in \mathcal{B}$ for $j = 1, \cdots, n$, and if $g \in \mathcal{H}$ we set:

$$\int_\Omega \phi(\omega)\ E(d\omega)g = \sum_{1 \leq j \leq n} a_j E(A_j)g.$$

Note that $\int_\Omega \phi(\omega)\ E(d\omega)g$ is the only element of $f \in \mathcal{H}$ satisfying:

$$\forall h \in \mathcal{H}, \qquad < f, h >= \int_\Omega \phi(\omega)\ E_{g,h}(d\omega),$$

where the integral in the right hand side is understood in the usual sense of integration with respect to a scalar valued measure. Moreover:

$$\|\int_\Omega \phi(\omega)\ E(d\omega)g\|^2 = \int_\Omega |\phi(\omega)|^2\ E_{g,g}(d\omega).$$

So, if $g \in \mathcal{H}$ is fixed, for each measurable function $\phi$ on $\Omega$ which is square integrable with respect to the nonnegative measure $E_{g,g}$ we define the integral $\int_\Omega \phi(\omega)\ E(d\omega)g$ in the following way. First, we approximate the function $\phi$ by a sequence $\{\phi_n;\ n \geq 1\}$ in $L^2(\Omega, E_{g,g}(d\omega))$. Then we notice that:

$$\|\int_\Omega \phi_n(\omega)E(d\omega)g - \int_\Omega \phi_m(\omega)E(d\omega)g\|^2 = \int_\Omega |\phi_n(\omega) - \phi_m(\omega)|^2\ E_{g,g}(d\omega),$$

and finally we define the integral $\int_\Omega \phi(\omega)\ E(d\omega)g$ as the limit in $\mathcal{H}$ of the Cauchy sequence $\{\int_\Omega \phi_n(\omega)\ E(d\omega)g;\ n \geq 1\}$. It is easy to check that this definition does not depend on the particular choice of the approximating sequence $\{\phi_n;\, n \geq 1\}$ and that the integral $\int_\Omega \phi(\omega)\, E(d\omega)g$ so defined has the properties we stated for the integrals of simple functions.

**Theorem I.2.5** *Let $E$ be a resolution of the identity of $\mathcal{H}$ on $(\Omega, \mathcal{B})$*
*i) for each measurable function $\phi : \Omega \to \mathbb{C}$, the subspace:*

$$\mathcal{D}_\phi = \{f \in H;\, \int_\Omega |\phi(w)|^2 E_{f,f}(dw) < \infty\} \tag{I.6}$$

*is dense in $\mathcal{H}$ and there is a unique operator $\phi(E)$ with domain $\mathcal{D}_\phi$ and such that:*

$$< \phi(E)f, g >= \int_\Omega \phi(w) E_{f,g}(dw) \tag{I.7}$$

*for all $f \in \mathcal{D}_\phi$ and $g \in \mathcal{H}$. Moreover, for each $f \in \mathcal{D}_\phi$ we have:*

$$||\phi(E)f||^2 = \int_\Omega |\phi(w)|^2 E_{f,f}(dw) \tag{I.8}$$

*ii) if $\phi$ and $\psi$ are measurable functions on $\Omega$ we have:*

$$\mathcal{D}(\phi(E)\psi(E)) = \mathcal{D}_\psi \cap \mathcal{D}_\phi \text{ and } \phi(E)\psi(E) \subset [\phi\psi](E), \tag{I.9}$$

*and consequently $\phi(E)\psi(E) = [\phi\psi](E)$ if and only if $\mathcal{D}_{\phi\psi} = \mathcal{D}_\psi$.*
*iii) for each measurable function $\phi$ on $\Omega$ we have:*

$$\phi(E)^* = \bar{\phi}(E) \quad \text{and} \quad \phi(E)\phi(E)^* = |\phi|^2(E) = \phi(E)^*\phi(E) \tag{I.10}$$

The proof is left as an exercise. See Problem I.7.7.

**Example I.2.6** If $\phi$ is a measurable function on $(\Omega, \mathcal{B})$ and if $E$ is the resolution of the identity introduced in Example I.2.4 above one easily checks that $\phi(E) = M_\phi$.

The above example shows that the operator $\phi(E)$ can be viewed as the function $\phi$ of the operator $H = \phi_o(E)$ whenever $\Omega$ is a Borel subset of $\mathbb{C}$ and $\mathcal{B}$ is the trace $\sigma$-field (recall that $\phi_o(w) = w$ by definition). Indeed, (I.9) shows that this interpretation is accurate for polynomial functions $\phi$. Hence Theorem I.2.5 provides us with a functional calculus for the self-adjoint operators $H$ of the form $\phi_o(E)$. We will see later in this chapter that every self-adjoint operator is of this form.

Numerous concepts from classical scalar valued measure theory remain meaningful for projection valued measures. For example, if $E$ is a resolution of the identity of $H$ we define the *E-essential range* of a measurable function $\phi$ as the complement of the largest open subset $U$ of $\mathbb{C}$ which satisfies $E(\phi^{-1}(U)) = 0$ and the function $\phi$ is said to be *E-essentially bounded* if its $E$-essential range is bounded, the *E-essential supremum* of $\phi$, which we will denote $||\phi||_\infty$ being defined as the supremum of the $|z|$ for $z$ in the $E$-essential range. The reader is asked to prove as an exercise that the spectrum of the operator $\phi(E)$, i.e. $\Sigma(\phi(E))$, is nothing but the $E$-essential range of $\phi$.

**Definition I.2.7** *If $\mu$ is a nonnegative $\sigma$-finite measure on $(\Omega, \mathcal{B})$, if $E$ is a subresolution of the identity of $\mathcal{H}$ on $(\Omega, \mathcal{B})$, by the $\mu$-essential support of $E$ we will refer to any set $A \in \mathcal{B}$ satisfying:*

$$\exists A' \in B,\ \mu(A') = 0 \text{ and } E((A \cup A')^c) = 0$$
$$\forall A'' \in B, \quad E(A'' \cap A) = 0 \Leftrightarrow \mu(A \cap A'') = 0.$$

Note that all the $\mu$-essential supports differ only by $\mu$-negligible sets and we will freely talk of the $\mu$-essential support even though the latter is not uniquely defined as a set. Note also that this support does not change if one replaces $\mu$ by an equivalent measure, finite for example. Finally, we will simply talk about the essential support without even mentioning $\mu$ whenever $\Omega$ is the real line and $\mu$ is the Lebesgue measure.

**Definition I.2.8** *If $\Omega$ is a topological space and if $\mathcal{B}$ is its Borel $\sigma$-field, the topological support of $E$ is defined in the usual way as the complement of the largest open set $U$ for which $E(U) = 0$. It will be denoted supp$E$.*

We end this section with a discussion of the *Lebesgue decomposition* of a subresolution of the identity and of the associated operators. But first, we make the following important remark.

**Remark I.2.9** Let $E_1$ and $E_2$ be subresolutions of the identity of $\mathcal{H}$ on $(\Omega, \mathcal{B})$ and let us assume that $\mathcal{H}$ is the orthogonal direct sum of $\mathcal{H}_1 = E_1(\Omega)\mathcal{H}$ and $\mathcal{H}_2 = E_2(\Omega)\mathcal{H}$. Then, the function $E$ defined by $E(A) = E_1(A) + E_2(A)$ for all $A \in \mathcal{B}$ is a resolution of the identity of $\mathcal{H}$ on $(\Omega, \mathcal{B})$ and the functional calculus of Theorem I.2.5 gives, for each measurable function $\phi$ on $\Omega$ and for $j = 1, 2$:

$$E_j(\Omega)\mathcal{D}(\phi(E)) \subset \mathcal{D}(\phi(E)) \quad \text{and} \quad \phi(E)\mathcal{H}_j \subset \mathcal{H}_j$$

which is summarized by saying that the decomposition reduces the operator since $\phi(E)$ commutes with the projections $E_j(\Omega)$ on the $\mathcal{H}_j$ and:

$$\mathcal{D}(\phi(E_j)) = \mathcal{D}(\phi(E)) \cap \mathcal{H}_j \quad \text{and} \quad \phi(E_j)f = \phi(E)f \quad \text{if} \quad f \in \mathcal{H}_j$$

which is summarized by saying that the components of $\phi(E)$ on $\mathcal{H}_1$ and $\mathcal{H}_2$ are the operators $\phi(E_1)$ and $\phi(E_2)$.

Recall that each complex measure $\mu$ on $(\mathbb{R}, \mathcal{B}_{\mathbb{R}})$ has a decomposition:

$$\mu = \mu_{pp} + \mu_{ac} + \mu_{sc} \tag{I.11}$$

where $\mu_{pp}$ is a ***pure point measure*** (i.e., a sum of Dirac point masses), where $\mu_{ac}$ is ***absolutely continuous*** with respect to Lebesgue measure and where $\mu_{sc}$ is a ***singular continuous*** measure (***continuous*** means that it has no point masses, i.e., $\mu_{sc}\{x\} = 0$ for all $x \in \mathbb{R}$ and ***singular*** means that it is carried by a set of zero Lebesgue measure). This Lebesgue decomposition of the scalar measures $\mu$ suggests that, given a resolution of the identity of $H$ on $(\mathbb{R}, \mathcal{B}_{\mathbb{R}})$, say $E$, we set:

$$\begin{aligned}
\mathcal{H}_{pp}(E) &= \mathcal{H}_{pp} = \{f \in \mathcal{H};\ E_{f,f} \text{ is pure point}\} \\
\mathcal{H}_{ac}(E) &= \mathcal{H}_{ac} = \{f \in \mathcal{H};\ E_{f,f} \text{ is absolutely continuous }\} \\
\mathcal{H}_{sc}(E) &= \mathcal{H}_{sc} = \{f \in \mathcal{H};\ E_{f,f} \text{ is singular continuous }\}
\end{aligned}$$

It is easy to see that the $\mathcal{H}_\alpha$'s are orthogonal closed subspaces of $\mathcal{H}$ (note that here and in the sequel, $\alpha$ stands for *pp*, *ac* or *sc*) and that:

$$\mathcal{H} = \mathcal{H}_{pp} \oplus \mathcal{H}_{ac} \oplus \mathcal{H}_{sc}. \tag{I.12}$$

In fact, if $f \in \mathcal{H}$ and if $f = g + h + k$ is the decomposition of $f$ in the above direct sum, and if $E_{f,f} = \mu_1 + \mu_2 + \mu_3$ is the Lebesgue decomposition of $E_{f,f}$ then we necessarily have $\mu_1 = E_{g,g}$, $\mu_2 = E_{h,h}$, and $\mu_3 = E_{k,k}$.

If, $\alpha$ being again either *pp*, or *ac* or *sc*, we denote by $\pi_\alpha$ the projection of $\mathcal{H}$ onto $\mathcal{H}_\alpha$ and if for $f$ and $g$ in $\mathcal{H}$ and $A \in \mathcal{B}_{\mathbb{R}}$ we set:

$$< E_\alpha(A)f, g >= [E_{f,g}]_\alpha(A)$$

where the left hand side is defined as the right hand side, then one easily checks that the $\{E_\alpha(A);\ A \in \mathcal{B}_{\mathbb{R}}\}$'s are resolutions of the identity of:

$$E_{pp}(\mathbb{R})\mathcal{H} = \mathcal{H}_{pp}, \quad E_{ac}(\mathbb{R})\mathcal{H} = \mathcal{H}_{ac} \quad \text{and} \quad E_{sc}(\mathbb{R})\mathcal{H} = \mathcal{H}_{sc}$$

respectively, and that:

$$E = E_{pp} + E_{ac} + E_{sc}. \tag{I.13}$$

$E_{pp}$ (resp. $E_{ac}, E_{sc}$) is a pure point (resp. absolutely continuous, singular continuous) subresolution of the identity of $\mathcal{H}$ in the sense that as $f$ varies in $\mathcal{H}$, all the scalar measures $< E_{pp}(\cdot)f, f >$ (resp. $< E_{ac}(\cdot)f, f >$ and $< E_{sc}(\cdot)f, f >$) are pure point (resp. absolutely continuous and singular continuous). For this reason the decomposition (1.13) is called the *Lebesgue decomposition of the resolution of the identity* $E$.

Consequently, for each measurable function $\phi$, the operator $\phi(E)$ can be decomposed in three parts $\phi(E_{pp})$, $\phi(E_{ac})$ and $\phi(E_{sc})$. In the particular case of $H = \phi_0(E)$ we have three self-adjoint operators $H_{pp} = \phi_0(E_{pp})$, $H_{ac} = \phi_0(E_{ac})$ and $H_{sc} = \phi_0(E_{sc})$ on the Hilbert spaces $\mathcal{H}_{pp}$, $\mathcal{H}_{ac}$ and $\mathcal{H}_{sc}$ respectively. They are called the *pure point part,* the *absolute continuous part* and the *singular continuous part* of the operator $H$. Their spectra $\Sigma(H_{pp}), \Sigma(H_{ac})$ and $\Sigma(H_{sc})$ are denoted $\Sigma_{pp}(H)$, $\Sigma_{ac}(H)$ and $\Sigma_{sc}(H)$ and are called the *pure point spectrum,* the *absolutely continuous spectrum* and the *singular continuous spectrum* of $H$. These spectra are the topological supports of the resolutions of the identity $E_{pp}, E_{ac}$ and $E_{sc}$. Notice that:

$$\Sigma(H) = \Sigma(H_{pp}) \cup \Sigma(H_{ac}) \cup \Sigma(H_{sc})$$

but these sets may not be disjoint. Also, $\Sigma(H_{ac}) \cup \Sigma(H_{sc})$ is denoted $\Sigma_c(H)$ and called the continuous spectrum of $H$. Finally we warn the reader that in many texts, $\Sigma_{pp}(H)$ stands for the set of eigenvalues of $H$ while for us, it is actually the closure of this set.

**Example I.2.10** Let $\{f_k;\ k \geq 1\}$ be an orthonormal basis in the Hilbert space $\mathcal{H}$, let $\{\lambda_k;\ k \geq 1\}$ be a sequence in $\mathbb{R}$ and let us set:

$$E(A) = \sum_{\lambda_k \in A} f_k \otimes f_k \qquad A \in \mathcal{B}_{\mathbb{R}}.$$

This defines a resolution of the identity of $\mathcal{H}$ on $(\mathbb{R}, \mathcal{B}_{\mathbb{R}})$ such that $E = E_{pp}$ and hence $\mathcal{H} = \mathcal{H}_{pp}$. The operator $H = \phi_0(E)$ is such that:

$$\mathcal{D}(H) = \{f \in \mathcal{H};\ \sum_{k \geq 1} \lambda_k^2 \,|< f, f_k >|^2 < \infty\},$$

and:

$$Hf = \sum_{k\geq 1} \lambda_k < f, f_k > f_k,$$

whenever $f \in \mathcal{D}(H)$. Moreover $\Sigma(H) = \Sigma_{pp}(H) = \overline{\{\lambda_1, \lambda_2, \cdots\}}$ and $\Sigma_c(H) = \emptyset$.

At some point in what follows we will consider the *essential spectrum* of $H$. It will be denoted by $\Sigma_{ess}(H)$. It is defined as the (closed) set whose elements are the real numbers $\lambda$ for which the rank of the spectral projection $E((\lambda - \epsilon, \lambda + \epsilon))$ is infinite for all $\epsilon > 0$. We will also say a few words about the *discrete spectrum* $\Sigma_{disc}(H)$ which is defined as the set of isolated eigenvalues of finite multiplicity. Notice that $\Sigma_{ess}(H) = \Sigma(H) \setminus \Sigma_{disc}(H)$.

## I.3 Representation Theorems

We review for later purposes, classical results concerning some specific classes of harmonic and analytic functions. We use the following notation:

$$\mathbb{D} = \{z \in \mathbb{C};\ |z| < 1\} \qquad \text{(I.14)}$$
$$\Pi_+ = \{z \in \mathbb{C};\ \Im m\, z > 0\} \qquad \text{(I.15)}$$
$$\mathbb{T} = \{z \in \mathbb{C};\ |z| = 1\} \qquad \text{(I.16)}$$

and:

$$p_r(\theta) = \frac{1 - r^2}{1 - 2r cos\theta + r^2} \qquad r \in [0, 1),\ \theta \in \mathbb{R}$$

for the Poisson kernel. $\mathbb{T}$ will be often identified with intervals in $\mathbb{R}$ and an integral over $\mathbb{T}$ with respect to a measure $\mu \in \mathcal{M}(\mathbb{T})$ will be denoted $\int_{[a,a+2\pi)} - d\mu$ while the notation $\int_a^{a+2\pi} -$ will only be used for Lebesgue integrals (i.e., with respect to the Haar measure). We use the notation $\mathcal{M}(S)$ for the space of finite (signed) measures on the Borel $\sigma$-field of the topological space $S$. For $\mu \in \mathcal{M}(\mathbb{T})$ we set:

$$P_\mu(z) = \frac{1}{2\pi} \int_{[-\pi,+\pi)} p_r(\theta - t)\mu(dt) \qquad z = re^{i\theta} \in \mathbb{D}$$

for the Poisson integral of $\mu$. $P_\mu(z)$ is harmonic in $\mathbb{D}$ and the absolutely continuous measures $p_\mu(re^{i\theta})d\theta$ converge weakly to $\mu$ as $r \to 1$. In fact we have:

- a nonnegative function $f$ on $\mathbb{D}$ is harmonic if and only if it is the Poisson integral of some nonnegative measure $\mu$ on $\mathbb{T}$ (i.e., $f = P_\mu$ for some $\mu \in \mathcal{M}_+(\mathbb{T})$) and $\mu$ is uniquely determined by $f$,

where we use the obvious notation $\mathcal{M}_+(S)$ for the subset of nonnegative elements of $\mathcal{M}(S)$. Whether or not the measure $\mu$ is nonnegative we have the famous Fatou's theorem:

- any Poisson integral $f = P_\mu$ has nontangential limits at almost every point of $\mathbb{T}$, the limit is equal to the Radon-Nikodym derivative of the absolute continuous part of $\mu$ with respect to the normalized Lebesgue measure of the circle, and its singular part is carried by the set of points where the limit exists and is $\infty$.

The last statement is often refered to as the "de La Vallée Poussin theorem." Using the fact that the real and imaginary parts of analytic functions are harmonic we obtain:

- a function $f$ on $\mathbb{D}$ is analytic and has a nonnegative real part if and only if it has a representation:

$$f(z) = i\beta + \int_{[-\pi,+\pi)} \frac{e^{it}+z}{e^{it}-z}\mu(dt) \qquad z \in \mathbb{D},$$

  for some real number $\beta$ and some nonnegative measure $\mu$ on $\mathbb{T}$ which are uniquely determined by the function $f$.

This representation theorem, together with Fatou's theorem makes possible the study of the fine structure of the measure $\mu$ from the investigation of the boundary behavior of the analytic function $f$. The classical results which we just recalled are not exactly in the form we will need them. Indeed the measures we intend to study are on $\mathbb{R}$ instead of $\mathbb{T}$. Using the one-to-one map:

$$z \hookrightarrow \lambda = i\frac{1+z}{1-z},$$

from $\mathbb{D}$ onto $\Pi_+$ the above representation theorem is transformed in the more convenient form:

- a function $f$ on $\Pi_+$ is analytic and has a nonnegative imaginary part if and only if it has a representation:

$$f(\lambda) = \alpha + \beta\lambda + \int_{\mathbb{R}} \left[\frac{1}{\xi-\lambda} - \frac{\xi}{1+\xi^2}\right] d\sigma(\xi) \qquad \lambda \in \Pi_+$$

for some $\alpha \in \mathbb{R}$, some $\beta \geq 0$ and some nonnegative measure $\sigma$ on $\mathbb{R}$ which satisfies:

$$\int_{\mathbb{R}} \frac{d\sigma(\xi)}{1+\xi^2} < \infty .$$

The measure $\sigma$ in the above representation is sometimes called the spectral measure of the function $f$. The functions having such a representation will be called Herglotz functions and we will denote by $\mathcal{H}erg$ the set of such functions. This representation can still be improved modulo an extra assumption. More precisely,

- a function $f$ on $\mathbf{\Pi}_+$ has a representation:

$$f(w) = \int_{\mathbb{R}} \frac{1}{\lambda - w} \sigma(d\lambda) \qquad w \in \mathbf{\Pi}_+ \tag{I.17}$$

for a nonnegative finite measure $\sigma$ on $\mathbb{R}$ if and only if $f$ is analytic, its imaginary part is nonnegative and

$$\sup_{y>0} y|f(iy)| < \infty. \tag{I.18}$$

The continuous part of the spectral measure can easily be recovered from the Herglotz function itself. Indeed, if $\mathcal{I}$ is an interval whose endpoints are not charged by the measure $\sigma$ one has:

$$\sigma(\mathcal{I}) = \frac{1}{\pi} \lim_{\epsilon \searrow 0} \int_{\mathcal{I}} \Im m\, f(\xi + i\epsilon)\, d\xi .$$

Moreover, Fatou's and de la Vallée Poussin's theorems are transported to the present setting so that the function $\Im m\, f$ has nontangential limits at almost every point of $\mathbb{R}$ (which is the boundary of $\mathbf{\Pi}_+$) and these limits are equal to $\pi$ times the density of the absolutely continuous component $\sigma_{ac}$ of $\sigma$. In other words:

$$\frac{d\sigma_{ac}(\xi)}{d\xi} = \frac{1}{\pi} \lim_{\epsilon \searrow 0} \Im m\, f(\xi + i\epsilon)$$

for almost every $\xi \in \mathbb{R}$ with respect to the Lebesgue measure of $\mathbb{R}$. Also, the singular component $\sigma_s$ of $\sigma$ is carried by the set of points where the limit exists and equals $+\infty$. We will make intensive use of these two facts in the sequel. In fact we will need more sophisticated properties of the functions of the Herglotz class and we will devote an entire section to the proofs of these properties in Chapter VI.

## I.4 The Spectral Theorem

We saw that we could associate a self-adjoint operator $\phi_0(E)$ to each resolution of the identity $E$ of $\mathcal{H}$ on $(\mathbf{R}, \mathcal{B}_{\mathbb{R}})$. The present section is devoted to the converse of this result. It goes under the name of the *spectral theorem for self-adjoint operators.*

**Theorem I.4.1** *A densely defined operator $H$ on $\mathcal{H}$ is self-adjoint if and only if there exists a resolution of the identity of $\mathcal{H}$ on $(\mathbb{R}, \mathcal{B}_{\mathbb{R}})$, say $E$, such that $H = \phi_0(E)$.*

*Proof:*
The proof of this converse is as follows. For each $f \in \mathcal{H}$, the function $\Phi_f(z) = -< [H - zI]^{-1} f, f >$ is defined and analytic outside the spectrum and in particular on $\mathbf{\Pi}_+$. The resolvent identity gives:

$$-2\Im m\, z \|[H - zI]^{-1} f\|^2 = 2i\Im m < [H - zI]^{-1} f, f >$$

which shows that the imaginary part of $\Phi_f$ is nonnegative in the upper half plane and consequently that $\Phi_f$ is a Herglotz function (i.e. it belongs to the class $\mathcal{H}erg$). Then the representation theorem gives the existence of a finite nonnegative measure $\mu_f$ on $\mathbb{R}$ such that:

$$< [H - zI]^{-1} f, f > = \int_{\mathbb{R}} \frac{1}{\lambda - z}\, d\mu_f(\lambda) \qquad z \in \mathbf{\Pi}_+ \qquad \text{(I.19)}$$

Now, for $f$ and $g$ in $\mathcal{H}$ we set:

$$\mu_{f,g} = \frac{1}{4}[\mu_{f+g} - \mu_{f-g} + i(\mu_{f+ig} - \mu_{f-ig})]$$

By polarization we get:

$$< [H - zI]^{-1} f, g > = \int_{\mathbb{R}} \frac{1}{\lambda - z}\, d\mu_{f,g}(\lambda) \qquad z \in \mathbf{\Pi}_+ \qquad \text{(I.20)}$$

The uniqueness part of the representation theorem gives that for each Borel set $A$ in $\mathbb{R}$ the function $(f, g) \hookrightarrow \mu_{f,g}(A)$ is linear in $f$ and antilinear in $g$ and hence there is a linear map $E(A)$ from $\mathcal{H}$ into itself such that $< E(A)f, g > = \mu_{f,g}(A)$ for all $f$ and $g$ in $\mathcal{H}$. It is then a routine exercise to show that $E(A)$ is an orthogonal projection and that $\{E(A); A \in \mathcal{B}_{\mathbb{R}}\}$ is a resolution of the identity of $\mathcal{H}$ such that $\phi_0(E) = H$. ∎

An easy consequence of the spectral theorem is the weak convergence of the spectral resolutions of the identity of a convergent sequence (in the sense of the strong resolvent convergence of operators) of self-adjoint operators. See Problem I.7.17 for a complete statement and a proof.

We now come to the important problems of the *canonical form* and of the *spectral multiplicity* of a self-adjoint operator.

A subspace $\mathcal{K}$ of $\mathcal{H}$ is said to be a *generating subspace* for $E$ if the set $\{E(A)f;\ A \in \mathcal{B}_{\mathbb{R}},\ f \in \mathcal{K}\}$ is total in $\mathcal{H}$. The *multiplicity* of $E$ is then defined as the infimum of the dimensions of the generating subspaces of $E$. If $E$ is the resolution of a self-adjoint operator $H$, the *spectral multiplicity* of $H$ is defined as the multiplicity of $E$. This definition can be localized to the components of $H$ in its Lebesgue decomposition or to a part of the spectrum of $H$ by considering the restriction of $E$ to the Borel subset in question. If the spectral multiplicity of $H$ is one, any nonzero element of a generating subspace is called a *cyclic vector* for $H$. Our interest in cyclic vectors can be justified by the following:

**Proposition I.4.2** *If $g$ is a cyclic vector for $H$, the formula:*

$$Vf = \int_{\mathbb{R}} f(\lambda)E(d\lambda)g \tag{I.21}$$

*for $f \in L^2(\mathbb{R}, \mu_g)$, defines a unitary isomorphism of the Hilbert spaces $\mathcal{H}$ and $L^2(\mathbb{R}, \mu_g)$. $V$ maps the domain of the operator $M_{\phi_0}$ in $L^2(\mathbb{R}, \mu_g)$ onto $\mathcal{D}(H)$ and $H(Vf) = V(\phi_0 f)$ for $f$ in $\mathcal{D}(M_{\phi_0})$.*

*Proof:*
By definition of the integral in (I.21) with respect to a measure with values in $\mathcal{H}$, we have, for $A \in \mathcal{B}_{\mathbb{R}}$ and $f \in L^2(\mathbb{R}, \mu_g)$:

$$< E(A)g, Vf >= \int_{\mathbb{R}} f(\lambda) < E(d\lambda)g, E(A)g >$$

and consequently:

$$||Vf||^2 = \int_{\mathbb{R}} |f(\lambda)|^2 < E(d\lambda)g, g >$$

which shows that $V$ preserves the norm and hence the inner product by polarization. $V$ is onto because $g$ is cyclic and the end of the proof is straightforward. ∎

See Problem I.7.13 for a surprising consequence of this result. Proposition I.4.2 easily generalizes to the finite multiplicity case if one uses the theory of $L^2$-spaces with respect to matrix measures.

**Definition I.4.3** *A matrix valued measure $M$ on a locally compact metric space $\mathcal{X}$ is an array $M = [\mu_{i,j}]_{i=1,\cdots,m,\, j=1,\cdots,n}$ of complex measures $\mu_{i,j}$ on $\mathcal{X}$. Furthermore, $M$ is said to be nonnegative if $m = n$ and if for every $f \in C_c(\mathcal{X})$ the space of continuous functions with compact support the matrix $[\int_{\mathcal{X}} f\, d\mu_{i,j}]_{i,j=1,\cdots,n}$ is a nonnegative definite Hermitian matrix.*

It is easy to see that the matrix measure is nonnegative if and only if $m = n$ and for every Borel subset $A$ of $\mathcal{X}$ the matrix $[\mu_{i,j}(A)]_{i,j=1,\cdots,n}$ is nonnegative definite. If $M$ is an $n \times n$ nonnegative matrix measure we denote by $\mathcal{L}^2(\mathcal{X}, dM)$ the vector space of $n$-tuples $F = (f_1, \cdots, f_n)$ of measurable functions for which:

$$||F||^2 = \int_{\mathcal{X}} < F, dM\, F >= \sum_{i,j=1}^{n} \int_{\mathcal{X}} f_i(x)\overline{f_j(x)}\, d\mu_{i,j}(x) < \infty, \qquad \text{(I.22)}$$

and we define $L^2(\mathcal{X}, dM)$ as the set of all equivalence classes in $\mathcal{L}^2(\mathcal{X}, dM)$ modulo the space of $M$-null functions, i.e. those $F$'s for which $||F|| = 0$. As usual we will not distinguish between equivalence classes and their representatives. The square root of (I.22) defines a norm arising from the inner product:

$$< F, G >= \int_{\mathcal{X}} < F, dM\, G >= \sum_{i,j=1}^{n} \int_{\mathcal{X}} f_i(x)\overline{g_j(x)}\, d\mu_{i,j}(x). \qquad \text{(I.23)}$$

Before proving that the space $L^2(\mathcal{X}, dM)$ is complete for this norm we give an example to illustrate some of the properties of these spaces.

**Example I.4.4** Let us set $M = \text{diag}[\mu_1, \cdots, \mu_n]$ where $\mu_1, \cdots, \mu_n$ are $n$ nonnegative finite measures on $\mathcal{X}$. $M$ is a nonnegative matrix measure and:

$$||F||^2 = \sum_{j=1}^{n} \int_{\mathcal{X}} |f_j(x)|^2\, d\mu_j(x), \qquad \text{(I.24)}$$

which shows that in this case the space $L^2(\mathcal{X}, dM)$ can be identified with the direct sum $L^2(\mathcal{X}, d\mu_1) \oplus \cdots \oplus L^2(\mathcal{X}, d\mu_n)$.

From now on we assume that $M = [\mu_{i,j}]_{i,j=1,\cdots,n}$ is a nonnegative matrix measure on the metrizable locally compact space $\mathcal{X}$. Then:

$$\mu = \text{tr}M = \sum_{j=1}^{n} \mu_{j,j}$$

is a nonnegative measure which satisfies:

$$\mu(A) = 0 \qquad \Leftrightarrow \qquad M(A) = 0,$$

and hence, all the measures $\mu_{i,j}$ are absolutely continuous with respect to $\mu$. Let us denote by $m_{i,j}(x)$ the corresponding Radon-Nikodym derivative. Then one can easily prove that the matrix $M(x) = [\mu_{i,j}(x)]_{i,j=1,\cdots,n}$ is defined and is nonnegative definite for $\mu$-almost every $x \in \mathcal{X}$. Moreover, $M(x) \leq I$ in the sense that $< M(x)\xi, \xi > \leq ||\xi||^2$ for all $\xi \in \mathbb{C}^n$. We can now proceed to the proof of our main result concerning the space $L^2(\mathcal{X}, dM)$.

**Proposition I.4.5** *The space $L^2(\mathcal{X}, dM)$ is a separable Hilbert space.*

*Proof:*
To prove completeness we pick a Cauchy sequence $\{F_k;\, k \geq 1\}$ in the sense of the $L^2(\mathcal{X}, dM)$ norm. The result of Problem I.7.15 shows that the sequence $\{F_k M^{1/2};\, k \geq 1\}$ is a Cauchy sequence in the Hilbert space $L^2(\mathcal{X}, d\mu; \mathbb{C}^n)$ of $\mu$-equivalence classes of $\mathbb{C}^n$-valued square integrable functions on $\mathcal{X}$. Consequently there exists a $\mathbb{C}^n$-valued function $G$ whose norm is square integrable with respect to $\mu$ and such that:

$$\lim_{k\to\infty} \int_{\mathcal{X}} ||F_k(x)M(x)^{1/2} - G(x)||^2 \, d\mu(x) = 0.$$

In order to complete the proof we need only to show that the limit function $G$ is of the form $G(x) = F(x)M(x)^{1/2}$ for some $\mathbb{C}^n$-valued measurable function. We know from classical linear algebra that, since $M(x)$ is nonnegative definite, we must have $J(x)M(x)^{1/2} = M(x)^{1/2} = M(x)^{1/2}J(x)$ if $J(x)$ denotes the matrix of the orthogonal projection on the range of $M(x)$. Moreover, there exists an invertible Hermitian matrix $H(x)$ satisfying $H(x)M(x)^{1/2} = J(x) = M(x)^{1/2}H(x)$. A close look at the construction of the matrices $J(x)$ and $H(x)$ show that they are actually measurable functions of the variable $x$. Then for $\mu$- almost every $x \in \mathcal{X}$ we have:

$$\begin{aligned} F(x)M(x)^{1/2} &= G(x)H(x)M(x)^{1/2} \\ &= \lim_{k\to\infty} F_k(x)M(x)^{1/2}J(x) \end{aligned}$$

$$
\begin{aligned}
&= \lim_{k\to\infty} F_k(x)M(x)^{1/2} \\
&= G(x).
\end{aligned}
$$

The separability follows from the properties proven in Problem I.7.16 which the $L^2$-spaces constructed over matrix measures share with the usual $L^2$-spaces. ∎

The proof of Proposition I.4.2 can be easily adapted to the situation of the $L^2$-spaces constructed from matrix measures and we obtain:

**Proposition I.4.6** *If $H$ is of finite multiplicity, if $\{g_1, \cdots, g_n\}$ is an orthonormal basis of a generating subspace and if $M$ is the matrix measure $[\mu_{g_j,g_k}]_{j,k=1,\cdots n}$ then the formula:*

$$VF = \sum_{1\le j\le n} \int_{\mathbb{R}} f_j(\lambda)E(d\lambda)g_j, \qquad F = (f_1, \cdots, f_n) \in L^2(\mathbb{R}, dM)$$

*defines a unitary isomorphism of the Hilbert spaces $L^2(\mathbb{R}, dM)$ and $H$ which also satisfies $Y\mathcal{D}(M_{\phi_0}) = \mathcal{D}(H)$ and $H(VF) = V(\phi_0 F)$ for $F$ in $\mathcal{D}(M_{\phi_0})$.*

The isomorphism $V$ is said to realize the *canonical form* of $H$. In the case of finite multiplicity, many properties of the operator $H$ can be obtained from the knowledge of the nonnegative measure $\mu = \Sigma_{1\le j\le n} M_{jj}$, i.e. the trace of the matrix measure $M$. In fact it is easy to see that:

- $\Sigma(H)$ is the topological support of the measure $\mu$. Moreover, $\mu$ is the least upper bound (for the order given by the absolute continuity of measures) of the family $\{E_{f,f}; f \in \mathcal{H}\}$.

In the general case an induction argument leads to:

**Proposition I.4.7** *For each self-adjoint operator $H$ there exist a (possibly infinite) integer $N$ and nonnegative measures $\mu_1, \cdots, \mu_N$ on $\mathbb{R}$ and a unitary isomorphism $V$ of $\otimes_{1\le j\le N} L^2(\mathbb{R}, \mu_j)$ onto $\mathcal{H}$ such that $V\mathcal{D}(M_{\phi_0}) = \mathcal{D}(H)$ and $H(VF) = V(\phi_0 F)$ for $F = (f_1, \cdots, f_N) \in \mathcal{D}(M_{\phi_0})$ and with $[\phi_0 F](\lambda) = (\lambda f_1(\lambda), \cdots, \lambda f_N(\lambda))$.*

The above result is another form of the spectral theorem. See the Notes for references. Its content is strikingly simple: any self-adjoint operator is unitarily equivalent to a multiplication operator by the identity function $\lambda$ on some $L^2$-space. The measure on which the $L^2$-space is constructed shows the spectral types and the spectral multiplicities of the operator in an obvious way.

# I.5 Quadratic Forms and Self-adjoint Operators

If $Q$ is a dense subspace of the Hilbert space $\mathcal{H}$ and if $q : Q \times Q \to \mathbb{C}$ is a sesquilinear form, the function $Q \ni f \to q(f) = q(f,f) \in \mathbb{C}$ is called a *quadratic form.* We use the same notation for the sesquilinear form and the corresponding quadratic form for we think that there is no confusion possible. $Q$ is called the *domain* of the form $q$. It is denoted by $\mathcal{Q}(q)$. $q$ is said to be symmetric if $q(f,g) = \overline{q(g,f)}$ for all $f$ and $g$ in $\mathcal{Q}(q)$ and nonnegative if $q(f) \geq 0$ for all $f$ in $\mathcal{Q}(q)$. More generally we will say that the sesquilinear form (or the corresponding quadratic form) is *bounded below* if:

$$q(f,f) \geq m||f||^2$$

for some finite constant $m$ and all $f$ in $\mathcal{Q}(q)$. We will then say that $q$ is closed if furthermore the vector space $\mathcal{Q}(q)$ is closed for the norm

$$||f||_{(q)} = [q(f,f) + (-m+1)||f||^2]^{1/2}.$$

If $H$ is a self-adjoint operator and if $H$ is nonnegative (i.e. if $\Sigma(H) \subset \mathbb{R}_+$), then the quadratic form $q$ defined by $\mathcal{Q}(q) = \mathcal{D}(H^{1/2})$ and

$$q(f,g) = < H^{1/2}f, H^{1/2}g > \tag{I.25}$$

for $f$ and $g$ in $\mathcal{D}(H^{1/2})$ is a closed nonnegative form, and we say that it is the form of the operator $H$. The above right hand side will very often be denoted $< Hf, g >$ even though $f$ may not belong to the domain of $H$. Conversely, and this is a very important result, every closed nonnegative quadratic form is associated with a unique self-adjoint operator whose spectrum is contained in $\mathbb{R}_+$.

**Example I.5.1** The Neumann and Dirichlet Laplacians.

We now discuss the case of the (negative) Laplacian in an open domain instead of the whole $\mathbb{R}^d$. Such "an operator" is formally symmetric and the theory of boundary conditions discussed in the next section could be used to find self-adjoint extensions. This is possible and usually done for domains with smooth boundaries. We will concentrate on Dirichlet and Neumann boundary conditions and we rely on the theory of quadratic forms presented in this section. More general boundary conditions will be considered in the one dimensional case in the next chapter.

Let $\Lambda$ be a bounded open subset of $\mathbb{R}^d$ and let us set $\mathcal{H} = L^2(\Lambda, dx)$ for a while.

i) Let $q_\Lambda^{(N)}$ be the sesquilinear form defined by:

$$q_\Lambda^{(N)}(f,g) = \int_\Lambda \nabla f(x) \cdot \overline{\nabla g(x)}\, dx$$

on its domain:

$$Q(q_\Lambda^{(N)}) = \{f \in \mathcal{H};\ \nabla f \text{ is square integrable in } \Lambda\}.$$

where the gradient $\nabla$ is understood in the sense of distributions. $q_\Lambda^{(N)}$ is closed and nonnegative. The corresponding self-adjoint operator is called the Neumann Laplacian in $\Lambda$ and is denoted $H_\Lambda^{(N)}$. The terminology is justified for one can prove that if the boundary of $\Lambda$ is smooth and if $f$ is $C^\infty$ in a neighborhood of the closure of $\Lambda$, then the solution of the classical Neumann problem is given by the function

$$u = [zI - H_\Lambda^{(N)}]^{-1} f$$

in the sense that $u$ satisfies $\Delta u + zu = f$ in $\Lambda$, and the normal derivative of $u$ vanishes on the boundary of $\Lambda$.

ii) The sesquilinear form defined by:

$$(f,g) \hookrightarrow \int_\Lambda \nabla f(x) \cdot \overline{\nabla g(x)}\, dx$$

with domain the space $C_c^\infty(\Lambda)$ of $C^\infty$-functions with compact supports contained in $\Lambda$ is closable. Its closure $q_\Lambda^{(D)}$ is a nonnagative closed sesquilinear form and the corresponding self-adjoint operator is denoted by $H_\Lambda^{(D)}$ and called the Dirichlet Laplacian in $\Lambda$ because its resolvent operator gives the solution of the classical Dirichlet problem in $\Lambda$.

We will investigate the spectral properties of these operators in Chapter II and use them in Chapter VI to define the integrated density of states.

Notice that the operator $H_\Lambda^{(D)}$ is the Friedrichs extension of the operator defined on the space of $C^\infty$-functions with compact support contained in $\Lambda$ as the usual negative Laplacian.

The first important property of the Dirichlet and Neumann boundary conditions is that they *decouple* the space in the following sense:

**Proposition I.5.2** *If $\Lambda_1$ and $\Lambda_2$ are disjoint open subsets of $\mathbb{R}^d$ (in which case one has $L^2(\Lambda_1 \cup \Lambda_2, dx) = L^2(\Lambda_1, dx) \oplus L^2(\Lambda_2, dx)$), then we have:*

$$H^{(\cdot)}_{\Lambda_1\cup\Lambda_2} = H^{(\cdot)}_{\Lambda_1} \oplus H^{(\cdot)}_{\Lambda_2},$$

*where the dot $\cdot$ can be replaced by $D$ in the case of Dirichlet boundary conditions as well as $N$ in the case of Neumann boundary conditions.*

The definition and some of the properties of direct sums of operators are given in Problem I.7.18.

*Proof:*
If $f \in C_c^\infty(\Lambda_1 \cup \Lambda_2)$ then obviously $f_{|\Lambda_1} \in C_c^\infty(\Lambda_1)$ and $f_{|\Lambda_2} \in C_c^\infty(\Lambda_2)$ and moreover we have the relation:

$$\int_{\Lambda_1\cup\Lambda_2} |\nabla f(x)|^2\, dx = \int_{\Lambda_1} |\nabla f_{|\Lambda_1}(x)|^2\, dx + \int_{\Lambda_2} |\nabla f_{|\Lambda_2}(x)|^2\, dx,$$

which can easily be extended to all the functions of the quadratic form domains in question. This completes the proof of the Dirichlet case. The proof of the Neumann case is similar and is left as an exercise to the reader.
■

Let us introduce a notation which will be convenient whenever we have to deal with several self-adjoint operators simultaneously. Usually, we denote by $E^{(H)}(\,\cdot\,)$ the spectral resolution of the identity of the self-adjoint operator $H$ and by $N^{(H)}(\lambda)$ the dimension of the range of the spectral projection $E^{(H)}((-\infty, \lambda])$. For typographical convenience we will use the short notation $N_\Lambda^{(\cdot)}(\lambda)$ for $N^{(H_\Lambda^{(\cdot)})}(\lambda)$.
The above result implies immediately the following useful corollary.

**Corollary I.5.3** *If $\Lambda_1, \dots, \Lambda_N$ are disjoint open domains for all $\lambda \in \mathbb{R}$ we have:*

$$N^{(\cdot)}_{\cup_{i=1}^N \Lambda_i}(\lambda) = \sum_{i=1}^N N^{(\cdot)}_{\Lambda_i}(\lambda)$$

*where the dot $\cdot$ can again be replaced by $D$ in the case of Dirichlet boundary conditions as well as $N$ in the case of Neumann boundary conditions.*

The reader is invited to look at Problem I.7.19 for a similar decoupling result concerning mixed boundary conditions.

The second important property of the Dirichlet and Neumann boudary conditions is the fact that they move the original spectrum in opposite directions. More precisely, we are about to show that the introduction of a Dirichlet boundary condition raises the energy spectrum while a Neumann boundary condition lowers it. Combined with Corollary I.5.3 above, this will give us a very convenient tool to estimate the energy levels. But to make these facts rigorous we need some sort of order on operators or quadratic forms. In fact, quadratic forms are very convenient to define an order on the bounded below self-adjoint operators. More precisely:

**Definition I.5.4** *If $H$ and $K$ are bounded below self-adjoint operators on $\mathcal{H}$ we say that $H \leq K$ if and only if:*
*i) $\mathcal{Q}(K) \subset \mathcal{Q}(H)$,*
*ii) for all $f \in \mathcal{Q}(K)$, one has $< Hf, f > \leq < Kf, f >$.*

The typical example involves the Dirichlet and Neumann Laplacians in bounded domains. In fact, from the definitions we gave one immediately gets:

**Proposition I.5.5** *If $\Lambda$ and $\Lambda'$ are bounded domains such that $\Lambda \subset \Lambda'$ then one has:*

$$0 \leq H_{\Lambda'}^{(D)} \leq H_{\Lambda}^{(D)} \tag{I.26}$$

*and for each bounded domain $\Lambda$:*

$$0 \leq H_{\Lambda}^{(N)} \leq H_{\Lambda}^{(D)} \tag{I.27}$$

*which goes under the name of Dirichlet-Neumann bracketing. Finally, if $\Lambda_1$ and $\Lambda_2$ are disjoint open domains such that the complement of $\Lambda_1 \cup \Lambda_2$ in the interior $\Lambda$ of the closure of $\Lambda_1 \cup \Lambda_2$ is of measure 0, then one also has:*

$$0 \leq H_{\Lambda}^{(D)} \leq H_{\Lambda_1 \cup \Lambda_2}^{(D)} \tag{I.28}$$

*and:*

$$0 \leq H_{\Lambda_1 \cup \Lambda_2}^{(N)} \leq H_{\Lambda}^{(N)} \tag{I.29}$$

As a useful application of this notion of order we state the so-called *min-max principle* which we present in a form which will be convenient later.

If $H$ is a bounded below self-adjoint operator on $\mathcal{H}$ we set:

$$\mu_n(H) = \sup_{f_1,\cdots,f_{n-1}} \inf_{g\in\mathcal{Q}_1(H,f_1,\ldots,f_{n-1})} < Hg, g >,$$

where $\mathcal{Q}_1(H, f_1, \ldots, f_{n-1})$ is the set of unit vectors of the form domain of $H$ which are orthogonal to $f_1, \ldots, f_{n-1}$. Then either:

- there are at least $n$ eigenvalues (counting multiplicity) below the infimum of the essential spectrum of $H$ and $\mu_n(H)$ is the $n$-th eigenvalue (counting multiplicity),

or:

- $\mu_n(H)$ is the infimum of the essential spectrum, there are at most $n-1$ eigenvalues (counting multiplicity) below $\mu_n(H)$ and $\mu_n(H) = \mu_{n+1}(H) = \mu_{n+2}(H) = \cdots$.

If $H$ and $K$ are bounded below self-adjoint operators satisfying $H \leq K$ then a simple application of the min-max principle gives:

i) for all $n \geq 1$ one has $\mu_n(H) \leq \mu_n(K)$,

and:

ii) for all $\lambda \in \mathbb{R}$ one has $\dim E^{(H)}((-\infty, \lambda]) \leq \dim E^{(K)}((-\infty, \lambda])$

if we denote by $E^{(H)}$ and $E^{(K)}$ their respective resolutions of the identity.

**Example I.5.6** The discrete Neumann and Dirichlet Laplacians:

The corresponding decoupling problem for the *finite difference operators* on the lattice $\mathbb{Z}^d$ can be achieved in various way. But it is more difficult to find natural analogues of the Dirichlet and Neumann boundary conditions which have the same decoupling properties as in the continuous case.

Notice that, in the lattice case, any operator $H$ is determined by its matrix elements $H(\mathbf{m}, \mathbf{n})$ in the sense that:

$$[Hf](\mathbf{m}) = \sum_{\mathbf{n}\in\mathbb{Z}^d} H(\mathbf{m}, \mathbf{n}) f(\mathbf{n}),$$

whenever this expression makes sense. At least formally, all the operators on $\mathcal{H} = \ell^2(\mathbb{Z}^d)$ have a kernel and the values of this kernel are given by the matrix elements $H(\mathbf{m}, \mathbf{n}) =< H\delta_{\mathbf{m}}, \delta_{\mathbf{n}} >$ where $\delta_{\mathbf{m}}$ denotes the function on $\mathbb{Z}^d$ whose value at $\mathbf{n}$ is 1 if $\mathbf{m} = \mathbf{n}$ and 0 otherwise.

With this in mind there is an obvious way to define the operator on a subset of the whole space. Indeed, if $\Lambda$ is a subset of the lattice $\mathbb{Z}^d$, one can define the operator $H_\Lambda$ as the operator given by its matrix elements:

$$H_\Lambda(\mathbf{m},\mathbf{n}) = \begin{cases} H(\mathbf{m},\mathbf{n}), & \text{if } \mathbf{m} \in \Lambda \text{ and } \mathbf{n} \in \Lambda \\ 0, & \text{otherwise.} \end{cases} \tag{I.30}$$

In the case of the Laplacian $H_0$ one can easily see that:

$$< H_0 f, f > = < (H_0)_\Lambda f, f > \ . \tag{I.31}$$

whenever the function $f$ vanishes outside $\Lambda$. We now follow Simon [300] to define the discrete analogues of the Dirichlet and Neumann Laplacians in a domain $\Lambda$. We present them as curiosities more than anything else for we will not use them in the sequel. First we define, for each $\mathbf{m},\mathbf{n} \in \mathbb{Z}^d$ the operators $L^{\mathbf{m},\mathbf{n}}$ and $S^{\mathbf{m},\mathbf{n}}$ on $\ell^2(\mathbb{Z}^d)$ by their action on an element $f$ of $\ell^2(\mathbb{Z}^d)$:

$$[L^{\mathbf{m},\mathbf{n}} f](\mathbf{k}) = \begin{cases} 0, & \text{if } \mathbf{k} \neq \mathbf{m} \text{ or } \mathbf{k} \neq \mathbf{n} \\ f(\mathbf{m}) - f(\mathbf{n}), & \text{if } \mathbf{k} = \mathbf{m} \\ f(\mathbf{n}) - f(\mathbf{m}), & \text{if } \mathbf{k} = \mathbf{n} \end{cases} \tag{I.32}$$

and:

$$[S^{\mathbf{m},\mathbf{n}} f](\mathbf{k}) = \begin{cases} 0, & \text{if } \mathbf{k} \neq \mathbf{m} \text{ or } \mathbf{k} \neq \mathbf{n} \\ f(\mathbf{m}) + f(\mathbf{n}), & \text{if } \mathbf{k} = \mathbf{m} \\ f(\mathbf{n}) + f(\mathbf{m}), & \text{if } \mathbf{k} = \mathbf{n} \end{cases} \tag{I.33}$$

Notice that $< L^{\mathbf{m},\mathbf{n}} f, f > = |f(\mathbf{m}) - f(\mathbf{n})|^2$ and $< S^{\mathbf{m},\mathbf{n}} f, f > = |f(\mathbf{m}) + f(\mathbf{n})|^2$ which show that $L^{\mathbf{m},\mathbf{n}} \geq 0$ and $S^{\mathbf{m},\mathbf{n}} \geq 0$. Notice also that:

$$H_0 = \sum_{|\mathbf{m}-\mathbf{n}|=1} L^{\mathbf{m},\mathbf{n}}$$

where each pair is counted only once in the above sum. Now, given a set $B$ of bonds (i.e. pairs $(\mathbf{m},\mathbf{n})$ such that $|\mathbf{m}-\mathbf{n}| = 1$) the Dirichlet and Neumann Laplacians are defined as:

$$H_0^{B,(D)} = H_0 + \sum_{(\mathbf{m},\mathbf{n})\in B} S^{\mathbf{m},\mathbf{n}}$$

and:

$$H_0^{B,(N)} = H_0 - \sum_{(\mathbf{m},\mathbf{n})\in B} L^{\mathbf{m},\mathbf{n}}$$

respectively. This definition implies immediately the bracketing:

$$H_0^{B,(N)} \leq H_0 \leq H_0^{B,(D)}$$

and it is easy to check that if $\mathbb{Z}^d$ is disconnected into the union of disjoint components $S_i$ by removing the bonds of $B$, then both $H_0^{B,(N)}$ and $H_0^{B,(D)}$ are direct sum operators on $\ell^2(\mathbb{Z}^d) = \oplus \ell^2(S_i)$.

# I.6 Self-adjoint Extensions of Symmetric Operators

Most of the unbounded operators to be considered below are differential or partial differential operators or finite difference operators which are only given formally. It is usually easy to choose domains (made of test functions) on which these operators become symmetric, but it will be sometimes difficult to see if they are self-adjoint, or if they have self-adjoint extensions. And even if this is the case, we would like to be able to classify the various self-adjoint extensions in order to choose among them those which are relevant to the physics of the problem. The theory of boundary conditions which we now outline gives a satisfactory answer. We proceed to the discussion of a few fundamental examples.

**Proposition I.6.1** *If the operator $H$ is closed and symmetric, the function $z \hookrightarrow \mathrm{dimKer}[zI - H^*]$ is equal to a constant $n_+(H)$ in the upper half plane $\mathbf{\Pi}_+$ and to a constant $n_-(H)$ in the lower half plane $\mathbf{\Pi}_-$. Moreover $\Sigma(H)$ is:*

- *either the closed upper half plane $\overline{\mathbf{\Pi}}_+$*
- *or the closed lower half plane $\overline{\mathbf{\Pi}}_-$*
- *or the whole complex plane $\mathbb{C}$*
- *or a closed subset of $\mathbb{R}$.*

The integers $n_+(H)$ and $n_-(H)$ are called the *deficiency indices* of $H$.

It follows easily from the above result that $H$ is self-adjoint if and only if $\Sigma(H) \subset \mathbb{R}$, or equivalently, $n_+(H) = n_-(H) = 0$. Consequently, extending closed symmetric operators to self-adjoint ones amounts to finding extensions with smaller and smaller deficiency indices until they finally vanish. This is the purpose of the theory of boundary conditions which we outline now.

Let $H$ be a symmetric operator on $\mathcal{H}$. We define the inner product $< .,. >_H$ on $\mathcal{D}(H^*)$ by:

$$< f,g >_H = < f,g > + < H^*f, H^*g > \qquad f,g \in \mathcal{D}(H^*).$$

$\mathcal{D}(H^*)$ is a Hilbert space for this inner product and "$H$-closure", "$H$-orthogonality", "$H$-continuity", $\cdots$ will be understood in the sense of this inner product. An $H$-continuous linear form $B$ on $\mathcal{D}(H^*)$ which vanishes on $\mathcal{D}(H)$ will be called a boundary value for $H$ and the equation

$$B(f) = 0 \qquad f \in \mathcal{D}(H)$$

will be called a boundary condition for $H$. A set $\{B_j(f) = 0; 1 \leq j \leq k\}$ of boundary conditions is said to be symmetric if:

$$(B_j(f) = B_j(g) \quad j = 1, \cdots, k) \quad \Rightarrow \quad < H^*f, g > = < f, H^*g > .$$

The importance of these definitions lies in the following important result:

- if $H$ is a closed symmetric operator with equal deficiency indices $n = n_+(H) = n_-(H)$, the self-adjoint extensions of $H$ are the restrictions of $H^*$ to the subspaces of $\mathcal{D}(H^*)$ determined by symmetric families of $n$ linearly independent boundary conditions (i.e., to subspaces of the form $\{f \in \mathcal{D}(H^*);\ B_j(f) = 0 \quad 1 \leq j \leq n\}$ where $\{B_j;\ 1 \leq j \leq n\}$ is a symmetric family of $n$ linearly independent boundary conditions).

We close this section with a short discussion of the so-called *Friedrichs extension* which is still another method to construct self-adjoint extensions of formally symmetric operators.

Let $H$ be a nonnegative operator (i.e. $< Hf, f > \geq 0$ for all $f \in \mathcal{D}(H)$). Then the quadratic form $q$ defined on the domain $\mathcal{Q}(q) = \mathcal{D}(H)$ by $q(f) = < Hf, f >$ is closable. Its closure is a nonnegative closed quadratic form so, according to the discussion of the previous section, it corresponds to a unique nonnegative self-adjoint operator. The latter is denoted by $H_F$ and is called the *Friedrichs extension* of $H$. It is maximal among all the bounded below self-adjoint extensions of $H$ for one easily shows that if $\tilde{H}$ is any other bounded below self-adjoint extension of $H$, then the quadratic form of $\tilde{H}$ is an extension of $q$ and $\tilde{H} \subset H_F$.

See Problem I.7.23 for an example of an application.

# I.7 Problems

**Problem I.7.1** *Let $H$ be a self adjoint operator and $\{f_i\ ;\ i = 1\ldots k\}$ be an orthonormal set of vectors in a Hilbert space $\mathcal{H}$. One assumes that for some real number $\lambda$ and some positive $\epsilon$ one has:*

1. $\|\,(H-\lambda)f_i\| \leq \epsilon$ *for* $i = 1\ldots k$
2. $< f_i\,, Hf_i > = < Hf_i\,, Hf_j >$ *for* $i \neq j$

*Let $P = E_{[\lambda-\epsilon,\lambda+\epsilon]}$ be the spectral projector on the interval $[\lambda-\epsilon, \lambda+\epsilon]$ and $V$ be the linear span of the set of vectors $\{f_i\ ;\ i = 1\ldots k\}$. Prove that:*

1. *For any $f \in V$ one has $\|\,(H-\lambda)f\| \leq \epsilon\|f\|$*
2. *Prove that if the dimension of Range($P$) is no greater than $k-1$ then there exists a nonzero vector $f \in V$ which is orthogonal to Range($P$), hence such that $\|\,(H-\lambda)f\| > \epsilon\|f\|$.*

*Conclude that the dimension of Range($P$) is no less than $k$.*

**Problem I.7.2** *Prove the claim of Example I.1.2, namely that any multiplication operator by a real valued function is self-adjoint and that its spectrum is the essential range of the function in question.*

**Problem I.7.3** *Let $K_1$ be a bounded operator and let $K_2$ be a self-adjoint operator. Prove that if $H = K_1 + K_2$ is defined as the operator sum, then any $\lambda \in \Sigma(H)$ satisfies $d(\lambda, \Sigma(K_2)) \leq \|K_1\|$. A proof can be found in Theorem 4.10 of Chapter V of [167].*

**Problem I.7.4** *Let $\lambda \in \mathbb{R}$ and let $H$ be a self-adjoint operator. Prove that $\lambda \in \Sigma(H)$ whenever there exists a sequence $\{f_n;\ n \geq 1\}$ of unit vectors in $\mathcal{D}(H)$ such that* $\lim_{n\to\infty} \|[H-\lambda I]f_n\| = 0$.

**Problem I.7.5** *Let $\{H_n; n \geq 1\}$ be a sequence of self-adjoint operators on $\mathcal{H}$ and let $\{E_n; n \geq 0\}$ be the corresponding sequence of resolutions of the identity. Show that the convergence of the sequence $\{H_n; n \geq 1\}$ to a self-adjoint operator $H$ (whose resolution of the identity will be denoted by $E$) in the strong resolvent sense is equivalent to any of the following three properties:*

**i)** *for each $z \in \mathbb{C}\setminus\mathbb{R}$, $[zI - H_n]^{-1}$ converges strongly to $[zI - H]^{-1}$,*

**ii)** *for each* $t \in \mathbb{R}$, $e^{itH_n}$ *converges strongly to* $e^{itH}$,
**iii)** *for each* $f \in \mathcal{H}$, *the measure* $< E_n(\cdot)f, f >$ *converges vaguely to the measure* $< E(\cdot)f, f >$.

*One can also consider the narrow (or weak) convergence of measures since all the measures in iii) are nonnegative and have the same total mass* $||f||^2$. *In fact, one can restate iii) in the following equivalent form obtained by polarization:*

**iii)'** *for each* $f$ *and* $g$ *in* $\mathcal{H}$, *the measure* $< E_n(\cdot)f, g >$ *converges vaguely to the measure* $< E(\cdot)f, g >$.

*Also, when all the spectra of the* $H_n$ *are contained in the same half line* $[a, \infty)$ *(i.e. when the operators are uniformly bounded below) the Laplace transform can be used instead of the Fourier transform in the characterization of the weak convergence of measures and ii) becomes equivalent to:*

**ii)'** *for each* $t > 0$, $e^{-tH_n}$ *converges strongly to* $e^{-tH}$.

**Problem I.7.6** *Let* $E$ *be a finitely additive function (in the sense that:*

$$E(\cup_{1 \leq j \leq n} A_n) = \Sigma_{1 \leq j \leq n} E(A_j)$$

*whenever* $\{A_n; n \geq 1\}$ *is a finite disjoint sequence in* $\mathcal{B}$*) from the* $\sigma$*-field* $\mathcal{B}$ *into the space of orthogonal projections in* $\mathcal{H}$. *Prove that:*

**i)** $E$ *is monotone in the sense that:* $E(A) \leq E(B)$ *whenever* $A \subset B$.
**ii)** $E$ *is subtractive in the sense that:* $E(B \setminus A) = E(B) - E(A)$ *whenever* $A \subset B$.
**iii)** $E$ *is modular in the sense that:* $E(A \cup B) + E(A \cap B) = E(A) + E(B)$ *whenever* $A, B \in \mathcal{B}$.
**iv)** $E$ *is multiplicative in the sense that:* $E(A \cap B) = E(A)E(B)$ *whenever* $A, B \in \mathcal{B}$.

**Problem I.7.7** *Provide the necessary approximation and monotone class arguments to pursue the discussion preceeding Theorem I.2.5 and make it into a complete proof of this theorem.*

**Problem I.7.8** *If* $E$ *is a resolution of the identity of* $\mathcal{H}$ *and if* $\phi$ *is a measurable function, then prove that the spectrum of the operator* $\phi(E)$ *is the* $E$*-essential range of the function* $\phi$.

**Problem I.7.9** *Let $A$ be an essential support for the spectral resolution of the identity $E$ of a self-adjoint operator $H$ and let $A' = A \cap \Sigma(H)$. Prove that $A'$ is also an essential support for $E$ and that $\overline{A'} = \Sigma(H)$. Check that the existence of an essential support whose closure is equal to the spectrum is a result which extends to the other kinds of (closed) spectra.*

**Problem I.7.10** *Let $\mathcal{H} = L^2(\Lambda, dx)$ for some (not necessarily bounded) region $\Lambda$ in $\mathbb{R}^d$ and let $E$ be the resolution of the identity of the multiplication operator $H = M_\phi$ by some real-valued measurable function $\phi$ on $\Lambda$. Show that for each $\lambda \in \mathbb{R}$ the range of $E((-\infty, \lambda])$ is the set of $f \in \mathcal{H}$ satisfying $f(x) = 0$ whenever $\phi(x) > \lambda$ and that:*

$$||E(A)f||^2 = \int_{\phi^{-1}(A)} |f(x)|^2 \, dx,$$

*for all $A \in \mathcal{B}_{\mathbb{R}}$ and $f \in \mathcal{H}$. Finally show that the spectrum of $H$ is purely absolutely continuous (i.e. $H = H_{ac}$ and $E = E_{ac}$) if $\phi$ is continuously differentiable and its gradient is nonzero almost everywhere.*

**Problem I.7.11** *Let $\{f_k; k \geq 1\}$ be an orthonormal basis in $\mathcal{H}$, $E$ a subresolution of the identity of $\mathcal{H}$ and let us define the measure $\sigma$ by:*

$$\sigma(A) = \sum_{k \geq 1} 2^{-k} < E(A) f_k, f_k > \qquad A \in \mathcal{B}$$

*a) Check that $\sigma(A) = 0$ if and only if $E(A) = 0$*
*b) Conclude that the $\mu$-essential supports of $\sigma$ and of the projection valued measure $E$ coincide.*

**Problem I.7.12** *Show that a subspace $\mathcal{K}$ of the $C^\infty$-vectors of a self-adjoint operator $H$, i.e. of the subspace $\cap_{n \geq 1} \mathcal{D}(H^n)$, is generator for $H$ if the set $\{H^n f;\, n \geq 1,\, f \in \mathcal{K}\}$ is total in $\mathcal{H}$.*

**Problem I.7.13** *Let $E$ be the resolution of the identity of a self-adjoint operator $H$ with simple spectrum (i.e. with spectrum of multiplicity one) and let $g \in \mathcal{H}$ be a cyclic vector for $H$. Prove that a finite nonnegative measure $\sigma$ on $\mathbb{R}$ is absolutely continuous with respect to $\sigma_g =< E(\cdot)g, g >$ if and only if $\sigma = \sigma_f$ for some $f \in \mathcal{H}$.*

**Problem I.7.14** *Let us assume that for each bounded Borel subset $A$ of $\mathbb{R}$, $E(A)$ is a self-adjoint trace class operator on $\mathcal{H}$ such that one has the relation $E(A) = \lim_{n\to\infty}\sum_{1\leq j\leq n} E(A_j)$ in the sense of the strong convergence of operators whenever $A = \cup_{n\geq 1} A_n$ with $A_i \cap A_j = \emptyset$ for $i \neq j$.*

**i)** *Prove that the formula $\sigma(A) = \mathrm{tr}E(A)$ defines a nonnegative Radon (i.e. finite on compact sets) measure $\sigma$ on the real line.*

**ii)** *Let us choose an orthonormal basis $\{f_j;\ j \geq 1\}$ of $\mathcal{H}$ and for each $i, j \geq 1$ let us define the measure $\sigma_{i,j}$ by the formula $\sigma_{i,j}(A) = < E(A)f_i, f_j >$. Check that:*

**a.** *for each finite set $\{n_1, \cdots, n_N\}$ of integers, $[\sigma_{n_i,n_j}]_{i,j=1,\cdots,N}$ is a nonnegative matrix measure (in the sense given in Section I.4) and $\sum_{j\geq 1}\sigma_{j,j} = \sigma$.*

**b.** *each measure $\sigma_{i,j}$ is absolutely continuous with respect to $\sigma$ and all the Radon-Nikodym densities $e_{i,j}(\lambda) = d\sigma_{i,j}(\lambda)/d\sigma$ can be simultaneously defined on the same set of full $\sigma$-measure.*

**c.** *for $\sigma$-almost every $\lambda \in \mathbb{R}$ the formula:*

$$E(\lambda) = \sum_{j\geq 1} e_{i,j}(\lambda) < f_i, \cdot > f_j,$$

*defines a nonnegative trace class operator $E(\lambda)$ on $\mathcal{H}$ satisfying $trE(\lambda) = 1$ and such that:*

$$E(A) = \int_A E(\lambda)\, d\sigma(\lambda),$$

*for every bounded Borel subset $A$ of $\mathbb{R}$.*

*Recall that a trace class operator on $\mathcal{H}$ is an operator $H$ for which one has:*

$$\sum_{j\geq 1} | < Hf_j, f_j > | < \infty,$$

*for one (and hence for all) orthonormal basis $\{f_j;\ j \geq 1\}$ of $\mathcal{H}$. Recall also that the operator is said to be Hilbert-Schmidt if the quantities $| < Hf_j, f_j > |^2$ are summable.*

**Problem I.7.15** *i) Show that if $F = (f_1, \cdots, f_n)$ is an $n$-tuple of measurable functions then we have:*

$$F \in L^2(\mathcal{X}, dM) \qquad \Leftrightarrow \qquad x \hookrightarrow \|F(x)M(x)^{1/2}\|^2 \in L^1(\mathcal{X}, d\mu),$$

*where we used the notation of Section I.4.*
*ii) Show also that if* $F, G \in L^2(\mathcal{X}, dM)$ *then the function*

$$x \hookrightarrow F(x)M(x)G(x)^*$$

*is in* $L^1(\mathcal{X}, d\mu)$ *and that:*

$$< F, G >= \int_{\mathcal{X}} F(x)M(x)G(x)^* \, d\mu(x).$$

*iii) Prove that if* $\tilde{\mu}$ *is a nonnegative measure on* $\mathcal{X}$ *with respect to which all the measures* $\mu_{i,j}$ *are absolutely continuous and if* $\tilde{M}(x) = [\tilde{m}_{i,j}(x)]_{i,j=1,\cdots,n}$ *denotes the corresponding matrix of Radon-Nikodym densities, then part ii) above remains true with* $M$ *and* $\mu$ *replaced by* $\tilde{M}$ *and* $\tilde{\mu}$ *respectively.*

**Problem I.7.16** *Let* $\{\mathcal{X}_k;\ k \geq 1\}$ *be an increasing sequence of measurable subsets of* $\mathcal{X}$ *whose union equals the whole space* $\mathcal{X}$ *and let* $F = (f_1, \cdots, f_n) \in L^2(\mathcal{X}, dM)$. *Show that the sequence* $\{F_k;\ k \geq 1\}$ *defined by* $F_k = \mathbf{1}_{\mathcal{X}_k} F$ *converges to* $F$ *in the* $L^2(\mathcal{X}, dM)$*-sense.*
*ii)* $F$ *is said to be simple if it can be written in the form:*

$$F = \sum_{j=1}^{N} \mathbf{1}_{\mathcal{X}_j} \xi^{(j)},$$

*for some family* $\{\mathcal{X}_j;\ j = 1, \cdots, N\}$ *of disjoint measurable subsets of* $\mathcal{X}$ *and a family* $\{\xi^{(j)};\ j = 1, \cdots, N\}$ *of elements of* $\mathbb{C}^n$ *with rational components. Show that the set of simple elements is dense in* $L^2(\mathcal{X}, dM)$.

**Problem I.7.17** *Let* $\{H_n;\ n \geq 1\}$ *be a sequence of self-adjoint operators converging (in the sense of the strong resolvent convergence of operators) to a self-adjoint operator* $H$. *Show that the corresponding resolutions of the identity* $E_n$ *converge weakly to the resolution of the identity of* $H$, *say* $E$, *in the sense that for every* $f$ *and* $g$ *in* $\mathcal{H}$ *we have:*

$$\lim_{n\to\infty} < E_n(\cdot)f, g >=< E(\cdot)f, g >,$$

*in the sense of the weak convergence of measures on* $\mathbb{R}$. *Recall that a sequence* $\{\mu_n;\ n \geq 1\}$ *of finite (possibly signed) measures on* $\mathbb{R}$ *(and more generally on a Polish space) is said to converge weakly to a measure* $\mu$ *whenever:*

$$\lim_{n\to\infty} \int f(x)\, d\mu_n(x) = \int f(x)\, d\mu(x),$$

*for every bounded continuous function $f$ on $\mathbb{R}$. Hint: use the characterization of the weak convergence in terms of the Fourier transform and the spectral theorem.*

**Problem I.7.18** *Let us assume that the Hilbert space $\mathcal{H}$ is equal to the direct sum of two subspaces $\mathcal{H}_1$ and $\mathcal{H}_2$, i.e. $\mathcal{H} = \mathcal{H}_1 \oplus \mathcal{H}_2$, and let $H_1$ and $H_2$ be self-adjoint operators on $\mathcal{H}_1$ and $\mathcal{H}_2$ respectively. Let $H$ be the operator defined on its domain $\mathcal{D}(H) = \{f = f_1 + f_2;\ f_1 \in \mathcal{D}(H_1),\ f_2 \in \mathcal{D}(H_2)\}$ by $Hf = H_1f_1+H_2f_2$. We use the standard notation $H = H_1 \oplus H_2$. Prove the following facts:*

1. *The operator $H_1 \oplus H_2$ is self-adjoint.*
2. *If $D_1$ is a core for $H_1$ and $D_2$ is a core for $H_2$, then the set:*

$$D_1 \oplus D_2 = \{f+ = f_1 + f_2;\ f_1 \in D_1,\ f_2 \in D_2\}$$

   *is a core for $H_1 \oplus H_2$.*
3. *$\mathcal{Q}(H_1 \oplus H_2) = \mathcal{Q}(H_1) \oplus \mathcal{Q}(H_2)$ and:*

$$< [H_1 \oplus H_2](f_1 + f_2), f_1 + f_2 >=< H_1 f_1, f_1 > + < H_2 f_2, f_2 >,$$

   *whenever $f_1 + f_2 \in \mathcal{Q}(H_1) \oplus \mathcal{Q}(H_2)$.*
4. *$E^{H_1 \oplus H_2}(A) = E^{H_1}(A) \oplus E^{H_2}(A)$ for all Borel subsets $A \subset \mathbb{R}$.*
5. *$N^{H_1 \oplus H_2}(\lambda) = N^{H_1}(\lambda) + N^{H_2}(\lambda)$ for all $\lambda \in \mathbb{R}$.*

*where we used the notation introduced in the text.*

**Problem I.7.19** *The purpose of this problem is to introduce Lapacian operators with mixed boundary conditions and to prove a corresponding decoupling result.*
*Let $\Lambda$ be a convex polyhedron in $\mathbb{R}^d$, let us denote by $\partial\Lambda$ its boundary and let us fix a function $\chi$ in the space $L^\infty(\partial\Lambda, d\sigma)$ where $d\sigma$ denotes the surface measure on $\partial\Lambda$. Recall that functions $f$ in the Sobolev space $\mathcal{H}^1(\Lambda)$ (i.e. the domain of the Neumann Laplacian in $\Lambda$) have a trace on $\partial\Lambda$ so that one can define the quadratic form $q_\Lambda^{(\chi)}$ by:*

$$q_\Lambda^{(\chi)}(f) = \int_\Lambda |\nabla f(x)|^2\, dx + \int_{\partial\Lambda} \chi(x)|f(x)|^2\, d\sigma(x)$$

*with the obvious choice of a domain. This quadratic form is bounded below and the (unique) corresponding self-adjoint operator is denoted by $H_{0,\Lambda}^{(\chi)}$ and it is called the negative Laplacian on $\Lambda$ with mixed boundary condition given by $\chi$.*

**a.** *Show that the Neumann Laplacian* $H_{0,\Lambda}^{(N)}$ *corresponds to the choice* $\chi \equiv 0$ *and that the Dirichlet Laplacian* $H_{0,\Lambda}^{(D)}$ *corresponds to the limit* $\chi \nearrow \infty$.

**b.** *Let us assume that the plane* $S$ *divides the polyhedron* $\Lambda$ *into two convex polyhedra* $\Lambda_1$ *and* $\Lambda_2$ *in such a way that* $S = \overline{\Lambda_1} \cap \overline{\Lambda_2}$. *Then for any bounded function* $\chi$ *on* $\partial\Lambda$ *and any bounded function* $\varphi$ *on* $S$ *prove that we have:*

$$H_{0,\Lambda}^{(\chi)} \geq H_{0,\Lambda_1}^{(\chi_1)} \oplus H_{0,\Lambda_2}^{(\chi_2)}$$

*where* $\chi_i(x) = \chi(x)$ *if* $x \in \partial\Lambda_i \cap \partial\Lambda$ *and* $\chi_i(x) = (-1)^i\varphi(x)$ *if* $x \in S$.

**Problem I.7.20** *We investigate the discrete analogue of the decoupling result proved above for Laplacians with mixed boundary conditions.*
*Let* $H = [H(\mathbf{m},\mathbf{n})]_{\mathbf{m},\mathbf{n}\in \mathbb{Z}^d}$ *be a self-adjoint operator on* $\mathcal{H} = \ell^2(\mathbb{Z}^d)$ *and let* $\Lambda$ *be a subset of* $\mathbb{Z}^d$. *We define the inner boundary of* $\Lambda$ *relative to* $H$ *as the set:*

$$\partial^{in}\Lambda = \{\mathbf{m} \in \Lambda;\ H(\mathbf{m},\mathbf{n}) \neq 0 \text{ for some } \mathbf{n} \in \Lambda^c\}$$

*and for each function* $\chi$ *which vanishes outside* $\partial^{in}\Lambda$ *we define the operator* $H$ *in* $\Lambda$ *with boundary condition* $\chi$ *on* $\partial^{in}\Lambda$, *say* $H_\Lambda^{(\chi)}$, *by:*

$$H_\Lambda^{(\chi)} = H_\Lambda + M_\chi$$

**a.** *Prove that if* $\Lambda = \Lambda_1 \cup \Lambda_2$ *with* $\Lambda_1 \cap \Lambda_2 = \emptyset$, *then, for any nonnegative function* $\varphi$ *which vanishes outside* $\partial^{in}\Lambda_1 \cup \partial^{in}\Lambda_2$ *we have the bracketing:*

$$H_{\Lambda_1}^{(\chi_1^-)} \oplus H_{\Lambda_2}^{(\chi_2^-)} \leq H_\Lambda^{(\chi)} \leq H_{\Lambda_1}^{(\chi_1^+)} \oplus H_{\Lambda_2}^{(\chi_2^+)}$$

*where the functions* $\chi_i^\pm$ *vanish outside* $\Lambda_i$ *and satisfy:*

$$\chi_i^\pm(\mathbf{m}) = \chi(\mathbf{m}) \pm \sum_{\mathbf{n}\in\partial^{in}\Lambda_{3-i}} |H(\mathbf{m},\mathbf{n})|\varphi(\mathbf{n})^{1-2i}$$

*for* $\mathbf{m} \in \partial^{in}\Lambda_i$ *and* $i = 1,2$.

**b.** *We now restrict ourselves to finite* $\Lambda$*'s and for each* $\lambda \in \mathbb{R}$ *we set:*

$$N_\Lambda^{(\chi)}(\lambda) = \#\{\lambda_i^{(\chi)} \leq \lambda\}$$

*where the* $\lambda_i^{(\chi)}$*'s denote the eigenvalues (counting multiplicity) of the operator* $H_\Lambda^{(\chi)}$. *If one uses the superscript* $(0)$ *for the function* $\chi \equiv 0$ *and* $(\infty)$ *for* $\varphi \equiv \infty$ *(with the convention* $0 \times \infty = 0$*) prove that:*

*i)* $N_\Lambda^{(\infty)}(\lambda) \le N_\Lambda^{(\chi)}(\lambda) \le N_\Lambda^{(0)}(\lambda)$ *for all* $\chi$, $\Lambda$ *and* $\lambda$.
*ii)* $N_{\Lambda_1\cup\Lambda_2}^{(0)}(\lambda) \le N_{\Lambda_1}^{(0)}(\lambda) + N_{\Lambda_2}^{(0)}(\lambda)$ *for all* $\lambda$ *provided* $\Lambda_1 \cap \Lambda_2 = \emptyset$.
*iii)* $N_{\Lambda_1}^{(\infty)}(\lambda) + N_{\Lambda_2}^{(\infty)}(\lambda) \le N_{\Lambda_1\cup\Lambda_2}^{(\infty)}(\lambda)$ *for all* $\lambda$ *provided* $\Lambda_1 \cap \Lambda_2 = \emptyset$.
*iv)* $0 \le N_{\Lambda_k}^{(0)}(\lambda) - N_{\Lambda_k}^{(\infty)}(\lambda) \le 2dk^{d-1}$ *for all* $\lambda$ *where* $\Lambda_k$ *denotes the hypercube with sides of length* $k$ *and the origin as a vertex and the coordinate axes as sides.*

**Problem I.7.21** *Let* $L \ge 1$ *be an integer and for each* $\mathbf{m} \in \mathbb{Z}^d$ *let us set* $\Lambda_L(\mathbf{m}) = \{\mathbf{n} \in \mathbb{Z}^d;\ Lm_j + 1 \le n_j \le (L+1)m_j,\ j = 1,\cdots,d\}$. *We denote by* $B(L)$ *the set of bonds coupling distinct* $\Lambda_L(i)$*'s and by* $H_0^{B(L),(N)}$ *and* $H_0^{B(L),(D)}$ *the corresponding Neumann and Dirichlet Laplacians in the direct sum form.*

**a.** *Show that the eigenvalues of* $H_0^{B(L),(D)}$ *are the set of numbers of the form:*

$$\sum_{j=1}^{d}(2 - 2\cos\frac{k_j\pi}{L})$$

*with* $k_j \in \{1,\cdots,L\}$ *and all the d-tuples* $(k_1,\cdots,k_d)$ *occur once. Notice that in particular, the lowest eigenvalue is* $d(2-2\cos(\pi/L))$.

**b.** *Show that the eigenvalues of* $H_0^{B(L),(N)}$ *are given by the set of numbers given by the same formula but with* $k_j \in \{0,1,\cdots,L-1\}$ *instead of* $\{1,\cdots,L\}$. *In particular the lowest eigenvalue is* 0.

**Problem I.7.22** *Let* $E$ *be the spectral resolution of the identity of a self-adjoint operator* $H$ *on the separable Hilbert space* $\mathcal{H}$.

**1.** *Use the spectral theorem to show that for each* $\lambda \in \mathbb{R}$, *each* $\epsilon > 0$ *and each* $f \in \mathcal{H}$ *one has:*

$$\Im m < [H - (\lambda + i\epsilon)I]^{-1} f, f >= 2\epsilon < [H - (\lambda + i\epsilon)I]^{-2} f, f > .$$

**2.** *We now assume that* $\mathcal{H} = \ell^2(\mathbb{Z}^d)$ *and we denote by* $G(z,\mathbf{m},\mathbf{n})$ *the quantity* $< [H - zI]^{-1}\delta_{\mathbf{m}}, \delta_{\mathbf{n}} >$. *Prove that the result of part 1. can be rewritten as:*

$$\Im m\, G(\lambda + i\epsilon, \mathbf{n}, \mathbf{n}) = 2\epsilon \sum_{\mathbf{m}\in\mathbb{Z}^d} |G(\lambda + i\epsilon, \mathbf{n}, \mathbf{m})|^2$$

*for all* $\mathbf{n} \in \mathbb{Z}^d$ *and use the de la Vallée Poussin theorem to show that the condition:*

$$\lim_{\epsilon \searrow 0} \sum_{\mathbf{m} \in \mathbb{Z}^d} |G(\lambda + i\epsilon, \mathbf{n}, \mathbf{m})|^2 = 0, \text{ for all } \mathbf{n} \in \mathbb{Z}^d \text{ and almost all } \lambda \in \mathcal{I}$$

*for some Borel set* $\mathcal{I} \subset \mathbb{R}$ *implies that the operator* $H$ *has no absolutely continuous spectrum in* $\mathcal{I}$.

**3.** *Generalize the above result to the case of* $\mathcal{H} = L^2(\mathbb{R}^d, dx)$ *or more generally any* $L^2$*-space over a locally compact measure space, whenever the resolvent operators* $[H - zI]^{-1}$ *have jointly continuous kernels* $G(z, x, y)$ *in the sense that:*

$$[H - zI]^{-1} f(x) = \int G(z, x, y) f(y)\, dy$$

*for all* $f \in \mathcal{H}$.

**Problem I.7.23** *Let* $\mathcal{H} = L^2([0, \infty), dx)$ *and let us define the operator* $H$ *by* $Hf = -f''$ *for* $f \in \mathcal{D}(H) = \{f \in \mathcal{H};\ f' \in \mathcal{H},\ f'' \in \mathcal{H},\ f(0) = 0,\ f'(0) = 0\}$. *Check that* $H \geq 0$ *and that its Friedrichs extension is given by* $\mathcal{D}(H_F) = \{f \in \mathcal{H};\ f' \in \mathcal{H},\ f'' \in \mathcal{H},\ f(0) = 0\}$ *and* $H_F f = -f''$ *whenever* $f \in \mathcal{D}(H_F)$. *In other words,* $H_F$ *is the Dirichlet Laplacian on* $[0, \infty)$.
*Check that the operator* $H_N$ *associated to the quadratic form* $q_N$ *defined by* $q_N(f) = \langle f', f' \rangle$ *for* $f \in \mathcal{Q}(q_N) = \{f \in \mathcal{H};\ f' \in \mathcal{H}\}$ *is given by* $H_N f = -f''$ *for* $f \in \mathcal{D}(H_N) = \{f \in \mathcal{H};\ f' \in \mathcal{H},\ f'' \in \mathcal{H},\ f'(0) = 0\}$. *This is the Neumann Laplacian on* $[0, \infty)$.

**Problem I.7.24** *Let us denote by* $\mathbb{N}$ *the set of nonnegative integers and by* $\mathbb{N}^*$ *the set of strictly positive integers. For each* $\alpha > 0$ *we denote by* $H_0^{(D)}(\alpha)$ *and by* $H_0^{(N)}(\alpha)$ *the Dirichlet and Neumann negative Laplacians on the domain* $\Lambda(\alpha) = [-\alpha, +\alpha]^d$.

1. *Show that the eigenvalues and the eigenfunctions of* $H_0^{(D)}(\alpha)$ *are labeled by* $(\mathbb{N}^*)^d$ *and are given by the formulas:*

$$\lambda_n(\alpha) = \left(\frac{\pi}{2\alpha}\right)^2 \sum_{i=1}^{d} n_i^2 \tag{I.34}$$

*and:*

$$\Phi_{\mathbf{n},\alpha}(x) = \alpha^{-d/2} \prod_{i=1}^{d} \varphi_{n_i}(\alpha^{-1}x_i)$$

*where:*

$$\begin{aligned} \varphi_k(x) &= \cos\frac{k\pi x}{2}, \qquad k = 1,3,5,\ldots \\ \varphi_k(x) &= \sin\frac{k\pi x}{2}, \qquad k = 2,4,6,\ldots \end{aligned}$$

2. *Show that the eigenvalues and the eigenfunctions of $H_0^{(N)}(\alpha)$ are labeled by $\mathbb{N}^d$, that the eigenvalues are given by the same formula (I.34) and that the eigenfunctions are given by:*

$$\begin{aligned} \psi_0(x) &\equiv \sqrt{2}/2 \\ \psi_k(x) &= \cos\frac{k\pi x}{2}, \qquad k = 1,3,5,\ldots \\ \psi_k(x) &= \sin\frac{k\pi x}{2}, \qquad k = 2,4,6,\ldots \end{aligned}$$

3. *Prove that:*

$$|N^{(\cdot)}_{\Lambda(\alpha)}(\lambda) - \tau_d(\frac{2\alpha}{2\pi})^d\lambda^{d/2}| \leq c(1 + (\alpha^2\lambda)^{(d-1)/2})$$

*for some constant $c > 0$ and all $\lambda \geq 0$ where $\cdot$ stands for $D$ in the case of Dirichlet boundary condition and for $N$ in the case of Neumann boundary condition and where $\tau_d$ denotes the volume of the unit ball in $\mathbb{R}^d$.*

*Hint: Use bracketing to prove that:*

$$N^{(D)}_{\Lambda(\alpha)}(\lambda) \leq \tau_d(\frac{2\alpha}{2\pi})^d\lambda^{d/2} \leq N^{(N)}_{\Lambda(\alpha)}(\lambda)$$

*and then that:*

$$N^{(N)}_{\Lambda(\alpha)}(\lambda) - N^{(D)}_{\Lambda(\alpha)}(\lambda) \leq c(1 + (\alpha^2\lambda)^{(d-1)/2})$$

*by noticing first that:*

$$N^{(N)}_{\Lambda(\alpha)}(\lambda) - N^{(D)}_{\Lambda(\alpha)}(\lambda) = \#S_\lambda \cap (\mathbb{N}^d \backslash (\mathbb{N}^*)^d$$

*where $S_\lambda$ denotes the ball of radius $\frac{2}{\pi}\sqrt{\lambda}$ centered at the origin.*

*In fact, the results and the hints can be reformulated in the following way: the explicit formula giving the eigenvalues of $H_0^{(D)}(\alpha)$ and $H_0^{(N)}(\alpha)$ shows that $N_{\Lambda(\alpha)}^{(D)}(\lambda)$ and $N_{\Lambda(\alpha)}^{(N)}(\lambda)$ are respectively equal to the numbers of points of $(\mathbb{N}^*)^d$ and $\mathbb{N}^d$ in the ball of radius $\frac{2\alpha}{\pi}\sqrt{\lambda}$ centered at the origin. This is equal to the volume of an octant of this ball (i.e. $\tau_d(\frac{2\alpha}{2\pi})^d\lambda^{d/2}$) up to errors of the order of the surface of the boundary (i.e. of the order of $(\alpha^2\lambda)^{(d-1)/2}$).*

## I.8 Notes and Complements

The notation and the terminology of this chapter are standard. They are used in the basic text books on the subject. Our presentation was strongly influenced by the book of Kato [167] and the four volumes of the series of Reed and Simon [274], [271], [272] and [273]. Our discussion of the resolutions of the identity, of the projection valued measures and of the associated functional calculus is patterned after the books of Halmos [139] and Rudin [279]. We payed special attention to the Lebesgue decomposition for it plays a crucial role in the problems we are interested in. Also our choice of the illustrative examples was strongly biased by the applications to Schrödinger operators which we have in mind.

The representation theorems for analytic and harmonic functions are also classical. They can be found in Hoffman's book [149] or in the book of Donoghue [85]. The de La Vallé Poussin's Theorem is certainly less well known than all the other ones. It can be found for example in Rudin's book [280] where the interested reader will have to put together Theorem 11.12 and Theorem 8.10.

The details of the proof of the spectral theorem can be found for example in Fuhrmann's book [109] where the interested reader will also find some complements on the canonical representations. The theory of multiplicity is very subtle. A nice account can be found in Halmos' book [139] while the intuitive ideas are enlightened in the case of Schrödinger operators in Section C.5 of Simon's article [294].

The results on quadratic forms which we discuss are borrowed from Kato's book [167], Faris' set of lectures [97] and most importantly Reed and Simon's series. Our discussion of the Dirichlet and Neumann boundary conditions is patterned after the material of Section XIII.5 of the fourth volume of Reed and Simon's series to which we refer the reader for proofs and complements.

Finally, the reader interested in the theory of the self-adjoint extensions of symmetric operators may consult the volume I of the Reed and Simon collection [274], and the book of Dunford and Schwarz [92] and Faris' monograph [98] for the approach based on the concepts of boundary conditions.

# II

# SCHRÖDINGER OPERATORS

Contents:



This chapter can be viewed as a crash course on Schrödinger operators. We discuss their definitions and some of their spectral properties. We illustrate the abstract theory of Chapter I by the concrete examples of the generalized Laplacian operators in Section II.1 and their perturbations by multiplication operators in Section II.2. The latter are of special importance since they are the Schrödinger operators we want to study. We consider the problem of the essential self-adjointness of these perturbations. We devote a subsection to the properties of the trace class perturbation theory in connection with the stability of the absolutely continuous spectrum. Obviously, the choice of the models we consider and of the techniques we use is strongly biased by the applications we contemplate. We consider only the so-called two-body Hamiltonians and we choose to give a probabilistic proof whenever such a proof is available. It turns out that, once properly defined as self-adjoint operators, Schrödinger operators generate semigroups (when they are bounded below) and unitary groups (when they are defined on the lattice $\mathbb{Z}^d$) which can be written using path integrals. We discuss the relevant formulas in Section II.3 of this chapter. They will be very useful in the investigation of the so-called integrated density of states in Chapter VI. Finally, we use these path integral methods to show the regularity properties of the semigroups and the decay of the $L^2$-eigenfunctions. These properties are important in the investigation of the so-called generalized eigenfunction expansions. This problem was not very popular before the wave of interest for the spectral properties of random Schrödinger operators. Indeed, the result on the existence of polynomially bounded solutions which we prove at the end of this chapter will play a crucial role in most of the proofs of localization which we shall discuss later.

This chapter is mostly concerned with the general theory of Schrödinger operators. Moreover, we postpone to the next chapter the discussion of the results and the techniques which are specific to the one-dimensional cases.

## II.1 The Free Hamiltonians

In classical mechanics, observables are measurable functions on the state space of the system. For example, the momentum is represented by a vector $\vec{p} = (p_1, p_2, p_3)$ in $\mathbb{R}^3$ and the kinetic energy is then given by $e = m|\vec{p}|^2/2$. Throughout the rest of the book we assume that an appropriate system of units has been chosen so that all the physical constants (masses, Planck constant, ...) magically become equal to 1. This convention does not change the mathematics and will make the exposition easier.

In quantum mechanics, the observables are operators (hopefully self-adjoint) on some infinite dimensional Hilbert space. For example, the components $p_j$ of the momentum are transformed by the quantization procedure into the partial differential operators $i\partial/\partial x_j$ on $L^2(\mathbb{R}^3, dx)$ and consequently, the kinetic energy which was:

$$e = p_1^2 + p_2^2 + p_3^2 \tag{II.1}$$

is transformed into the operator:

$$H_0 = -\frac{\partial^2}{\partial x_1^2} - \frac{\partial^2}{\partial x_2^2} - \frac{\partial^2}{\partial x_3^2}. \tag{II.2}$$

$H_0$ is thus equal to the negative Laplacian in $\mathbb{R}^3$. It is usually called the *free Hamiltonian* because it does not take into account any interaction energy. We can use the theory presented in Chapter I to make sure that $H_0$ is defined as a self-adjoint operator on the Hilbert space $L^2(\mathbb{R}^3, dx)$.

The kinetic energy (II.1) is adequate for one three-dimensional particle if one neglects relativistic effects. The kinetic energy for an ensemble of $N$ particles in three dimensions can still be written as $|p|^2$ provided that $p$ is regarded as a generic element of the momentum space $\mathbb{R}^{3N}$. This space is the dual of the configuration space $\mathbb{R}^{3N}$. This Euclidean space is sometimes replaced by $\mathbb{R}^{3(N-1)}$ when the center of mass motion is removed. We shall not pursue the discussion of the formalism of the $N$-body quantum problems for the mathematics we will be concerned with will be those of the 2-body problems. In any case we use the letter $d$ to denote the dimension of our sytstem. In order to take into account relativistic corrections to the kinetic energy (II.1), the following formula is often substituted for (II.1):

$$e = \sqrt{|p|^2 + m^2} - m \tag{II.3}$$

for some parameter $m > 0$ which can be given the interpretation of a mass. The corresponding self-adjoint operator is then:

$$H_0^{(rel)} = \sqrt{-\Delta + m^2} - m. \tag{II.4}$$

The latter is often traded for the operator:

$$H_0^{(1)} = \sqrt{-\Delta}. \tag{II.5}$$

(which has nicer scaling properties) in situations where they have the same behavior. In any case, the operator giving a mathematical account of the quantum kinetic energy is an operator of the form $H_0 = F(p)$ for some real valued function $F$ on $\mathbb{R}^d$ where we use the notation $p$ for the partial differential operator $-i\nabla$.

In addition to the examples discussed above the mathematical literature is full of Schrödinger operators of interest with a free part $H_0$ different from the usual quadratic function:

$$F^{(2)}(p) = |p|^2.$$

The free Hamiltonian:

$$H_0 = \cos(p) = \cos(-i\nabla),$$

is for example crucial in the investigation of the so-called Harper's equation.

Let us emphasize once more that, in all of the examples discussed above the free part of the Schrödinger Hamiltonian is of the form $H_0 = F(-i\nabla) =$

$F(p)$ for some function $F(p)$. A natural requirement one should have is that it generates a semigroup $\{e^{-tH_0}; t \geq 0\}$ of positivity preserving transformations of $L^2(\mathbb{R}^d, dx)$. It is possible to prove that $F(p)$ is such a function if and only if for each cut–off function $h : \mathbb{R}^d \to \mathbb{R}^d$ with compact support satisfying $h(x) = x$ in a neighborhood of the origin one has the representation:

$$F(p) = a + i\, b \cdot p + \; p \cdot Cp - \int_{\mathbb{R}^d} \left[e^{ip\cdot x} - 1 - i\, p \cdot h(x)\right] \; \nu(dx), \qquad \text{(II.6)}$$

for some real constant $a$, some vector $b \in \mathbb{R}^d$, some nonnegative definite matrix $C$ and some nonnegative measure $\nu$ satisfying $\int_{\mathbb{R}^d} 1 \wedge |x|^2 \; \nu(dx) < \infty$. We use the notation $a \wedge b$ to denote the minimum of the two real numbers $a$ and $b$. Note that if $F(p)$ has a representation of the above type for some cut–off function $h$, it has a similar representation for any other cut–off function. The choice of the cut–off function only affects the constant $a$ and the vector $b$. The latter is called *the drift*, C is called *the covariance* and the measure $\nu$ is called *the Lévy measure*. Formula (II.6) is the famous *Lévy–Kintchine formula* and the function $F$ is called *the exponent function* because of (II.7) below. Well known examples are the relativistic function $F^{(rel)}(p) = \sqrt{|p|^2 + m^2} - m$ and the trigonometric function $F^{(trig)}(p) = \cos(p)$ which we already encountered, and the stable one:

$$F^{(\alpha)}(p) = |p|^\alpha,$$

for $0 < \alpha \leq 2$ which is a generalization of the quadratic one. The self-adjoint operators $H_0 = F(p)$ defined in this way for exponent functions $F(p)$ as in the Lévy-Kintchine formula will be called *generalized Laplacians* in the sequel. The additive constant $a$ appearing in the Lévy–Kintchine formula will be chosen to be equal to zero. This normalization is very convenient for it implies the existence of a convolution semigroup $\{\mu_t; t \geq 0\}$ of (infinitely divisible) probability measures on $\mathbb{R}^d$ such that:

$$\hat{\mu}_t(p) = e^{-tF(p)}. \qquad \text{(II.7)}$$

for all $t > 0$ and $p \in \mathbb{R}^d$. In other words, the free semigroup $\{e^{-tH_0}; t \geq 0\}$ is a convolution semigroup given by:

$$[e^{-tH_0} f](x) = \int_{\mathbb{R}^d} f(x + y) \; \mu_t(dy). \qquad \text{(II.8)}$$

Note also that if $z = -\beta$ with $\beta > 0$, then the resolvent operator $[H_0 - zI]^{-1}$ is also a convolution operator since:

$$[H_0 - zI]^{-1} f(x) = \int_{\mathbb{R}^d} f(x + y) \; g_\beta(dy). \qquad \text{(II.9)}$$

with:

$$g_\beta = \int_0^\infty e^{-\beta t}\mu_t\,dt.$$

Convolution semigroups are generated by ***stochastic processes with stationary independent increments*** also known as ***Lévy processes.*** We will dwell more on this point in Section II.3.

It is relatively easy to investigate the *smoothing* properties of these semigroups. We are mostly interested in their action on the various classical $L^p(\mathbb{R}^d, dx)$ spaces and for this reason, it will be convenient to know that the measures $\mu_t$'s have densities which we can control. Note that this rules out the case of the trigonometric function of the Harper's equation and more generally all the *lattice* Lévy processes.

In the classical case of the negative Laplacian, it is well known that the measure $\mu_t$ has a density $p_t$ given by the formula:

$$p_t(x) = \frac{1}{\sqrt{4\pi t}^d} e^{-|x|^2/(4t)}, \qquad \text{(II.10)}$$

For more general Lévy processes the assumption:

$$\int_{\mathbb{R}^d} e^{-tF(p)}\,dp < \infty, \qquad t > 0, \qquad \text{(II.11)}$$

implies that, for each $t > 0$ the measure $\mu_t$ has a density given by its inverse Fourier transform:

$$p_t(x) = \frac{1}{(2\pi)^d}\int_{\mathbb{R}^d} e^{-tF(p)}e^{ip\cdot x}\,dx, \qquad x \in \mathbb{R}^d.$$

Note that $p_t$ is a bounded continuous function such that $\lim_{|x|\to\infty} p_t(x) = 0$. Moreover, it is possible to prove that, for each fixed $x \in \mathbb{R}^d$, $p_t(x)$ is an analytic function of $t$ on $(0, \infty)$ and that $0 < p_t(x) \le p_t(0) < \infty$ for all $t > 0$ and $x \in \mathbb{R}^n$. Moreover, using Young's inequality we get:

$$||e^{-tH_0}f||_s = ||p_t * f||_s \le ||p_t||_q ||f||_p,$$

provided $1 + 1/s = 1/p + 1/q$. We use the notation $||\cdot||_p$ for the norm of the Lebesgue space $L^p(\mathbb{R}^d, dx)$.

**Lemma II.1.1** *If* $1 \le s \le p \le \infty$ *and* $t > 0$*, then* $e^{-tH_0}$ *is a bounded operator from* $L^s(\mathbb{R}^d, dx)$ *into* $L^p(\mathbb{R}^d, dx)$ *and its norm is bounded above by* $||p_t||_{L^q(\mathbb{R}^d)}$ *where* $1 + 1/s = 1/p + 1/q$.

It is sometimes possible to obtain explicit bounds for the time dependence of the norm of this operator using scaling arguments. In the particular case of the stable exponent function $F^{(\alpha)}$, one has:

$$\|p_t^{(\alpha)}\|_p = t^{-d/\alpha p'}\|p_1^{(\alpha)}\|_p,$$

where $p'$ is the conjugate exponent of $p$. Note that all the above examples satisfying assumption (II.11) also satisfy:

$$\int_{\mathbb{R}^d} |p|^m e^{-tF(p)}\, dp < \infty,$$

for all $t > 0$ and all integers $m \geq 0$. This implies that the density $p_t$ is actually $C^\infty$, and by convolution $e^{-tH_0}f$ inherits this smoothness. In any case our assumption (II.11) implies that the density $p_t$ is continuous and consequently all the ranges of the operators $e^{-tH_0}$ are formed by continuous functions.

**Remark II.1.2** *Except for a handfull of papers, the theory of random Schrödinger operators with free Hamiltonian $H_0$ given by a generalized Laplacian has not been worked out. One may want to look at some of the problems for which the usual Laplacian plays a crucial role and try to extend them in a form or another. One then will need sharp estimates for the transition probability $p_t(x)$. We briefly review some of the estimates which can be helpful in this respect. In the stable case corresponding to $F^{(\alpha)}(p) = |p|^\alpha$ one can use the usual scaling argument to reduce the problem to estimating only $p_1(x)$, and then use the existence of positive constants $c_1$ and $c_2$ such that:*

$$\frac{c_1}{|x|^{d+\alpha}} \leq p_1^{(\alpha)}(x) \leq \frac{c_2}{|x|^{d+\alpha}}, \qquad |x| \geq 1,$$

*if $\alpha \neq 2$. We will not discuss the well known exponential bounds of the Brownian motion case corresponding to $\alpha = 2$. In the case of the relativistic Hamiltonian $F^{(rel)}(p) = \sqrt{p^2 + m^2} - m$, the transition density can be given by:*

$$p_t^{(rel)}(x) = (2\pi)^{-n}\frac{t}{\sqrt{|x|^2+t^2}}\int_{\mathbb{R}^d} e^{mt}e^{-\sqrt{(|x|^2+t^2)(p^2+m^2)}}\, dp$$

*which we borrowed from [146] and from which one gets:*

$$p_t^{(rel)}(x) \geq c_d t\, e^{-m|x|}(|x|^2+t^2)^{-(n+1)/2},$$

*for some positive constant $c_d$. The density $p_t^{(rel)}(x)$ has also a corresponding exponential upper bound which one can derive in a similar way.*

Let us now discuss briefly the spectral type of the generalized Laplacians. The latter are unitarily equivalent to the operator of multiplication by the exponent function $F(p)$. Consequently the spectrum, as a set, is the essential range of the function $F$. Recall Example I.1.1 of Chapter I. For example, in the case of the usual Laplacian given by the exponent function $F(p) = |p|^2$, as well as in the relativistic case given by the function $F^{(rel)}$ or any of the stable cases given by the functions $F^{(\alpha)}$, the spectrum is the nonnegative half axis $[0, \infty)$. Moreover, we know from Problem I.7.10 of Chapter I that this spectrum is purely absolutely continuous (i.e. $H = H_{ac}$ and $E = E_{ac}$) whenever the exponent function $F$ is differentiable almost everywhere which is the case for all the examples which we just mentioned.

We close this section with a short discussion of the free Hamiltonians $H_0$ in bounded open domains $\Lambda$ of $\mathbb{R}^d$. We consider only the cases of the Dirichlet and Neumann boundary conditions. In either case the resolvent operators are compact operators on $L^2(\Lambda,\ dx)$. Indeed, we show in Section II.3 below that the semigroup and the resolvent operators have a bounded kernel. Consequently, the spectra of the operators $H_{0,\Lambda}^{(D)}$ and $H_{0,\Lambda}^{(N)}$ are of the form $\{\lambda_0^{(D)}(\Lambda), \lambda_1^{(D)}(\Lambda), \lambda_2^{(D)}(\Lambda), \ldots\}$ and $\{\lambda_0^{(N)}(\Lambda), \lambda_1^{(N)}(\Lambda), \lambda_2^{(N)}(\Lambda), \ldots\}$ where $\{\lambda_n^{(D)}(\Lambda);\ n \geq 0\}$ and $\{\lambda_n^{(N)}(\Lambda);\ n \geq 0\}$ are unbounded nondecreasing sequences of real numbers. These spectra are actually discrete since all the eigenvalues are of finite multiplicity. See Section II.3 for complements and path integral proofs. The Dirichlet-Neumann bracketing discussed in Section I.5 of Chapter I gives:

$$\lambda_n^{(N)}(\Lambda) \leq \lambda_n^{(D)}(\Lambda), \qquad n = 0, 1, \ldots. \tag{II.12}$$

We discuss the discrete case later in this chapter.

## II.2 Schrödinger Operators as Perturbations

This section is concerned with the definition of the sum of self-adjoint operators $H_0$ and $V$ as a self-adjoint operator and to the investigation of some of its spectral characteristics. We saw in Chapter I that this problem could be very delicate. In any case, we want to investigate the properties of the operator $H = H_0 + V$. We can think of $H$ as the energy Hamiltonian of a quantum system, $H_0$ being the kinetic energy and the operator $V$ (which is the notation we will use for the operator of multiplication $M_V$ by the measurable function V) representing the potential energy.

The lattice case is not discussed in this section for the free Hamiltonian $H_0$ is a bounded self-adjoint operator and the sum $H = H_0+V$ is always a self-adjoint operator on its domain $\mathcal{D}(H) = \mathcal{D}(V) = \{f \in \mathcal{H};\ Vf \in \mathcal{H}\}$. Note also that in this case it is appealing from a mathematical point of view to view $H_0$ as a perturbation of the multiplication operator $V$. In particular, any core for the latter will also be a core for the operator sum $H$. It will be convenient to use from time to time the subspace $\mathcal{D}$ of $\mathcal{H} = \ell^2(\mathbb{Z}^d)$ formed by the sequences $f = (f(\mathbf{n}))_{\mathbf{n}\in\mathbb{Z}^d}$ for which $f(\mathbf{n}) = 0$ except for at most finitely many $\mathbf{n} \in \mathbb{Z}^d$. This space is the analog in the discrete case of the space $C_c^\infty(\mathbb{R}^d)$ of $C^\infty$-functions with compact support in $\mathbb{R}^d$.

### II.2.1 Self-adjointness

When the operator $V$ is bounded (in our case when the function $V$ is bounded) the operator $H$ is self-adjoint as we already noticed in Chapter I. Its spectral properties can be investigated from the characteristics of the unitary group $\{e^{itH};\ t \in \mathbb{R}\}$ generated by $H$. The latter can be computed thanks to the famous Trotter product formula which we now recall in the particular case we are interested in.

**Proposition II.2.1** *Let $H = H_0 + V$ with $H_0$ self-adjoint and $V$ self-adjoint and bounded. Then:*

$$e^{itH}f = \lim_{n\to\infty} [e^{i(t/n)H_0}e^{i(t/n)V}]^n f \tag{II.13}$$

*for all $t \in \mathbb{R}$ and $f \in \mathcal{H}$.*

The proof of this result can be found in many standard text books. We refer the interested reader to Section VIII.9 of the first volume of the Reed and Simon series [274]. The boundedness assumption on the potential $V$ is not necessary. It is convenient because it simplifies the proof and after all, it is all that we will need here. But the result is still true in more general situations. In particular, formula (II.13) still holds when $V$ is a general self-adjoint operator such that the operator sum $H = H_0 + V$ is essentially self-adjoint on $\mathcal{D}(H_0) \cap \mathcal{D}(V)$. Moreover, if $H_0$ and $V$ are bounded from below, then a semigroup version of formula (II.13)holds, namely:

$$e^{-tH}f = \lim_{n\to\infty} [e^{-(t/n)H_0}e^{-(t/n)V}]^n f \tag{II.14}$$

for all $t \geq 0$ and all $f \in \mathcal{H}$.

The potential functions we will have to deal with are most of the time unbounded and the self-adjointness problem is usually difficult.

The situation is simpler when we consider the case of the Schrödinger operator in a bounded open subset $\Lambda$ of $\mathbb{R}^d$, in which case the Hilbert space $\mathcal{H}$ is chosen to be $L^2(\Lambda, dx)$. If the measurable function $x \hookrightarrow V(x)$ is locally bounded or even locally square integrable, then the multiplication operator by the function $V$ defines a bounded operator on $\mathcal{H}$ and the perturbed operator $H = H_0^{(\cdot)} + V$ is self-adjoint whether the dot stands for $D$ in the case of Dirichlet boundary condtions or $N$ in the case of Neumann boundary conditions.

The situation is more delicate when the local singularities of the potential function are not so easily controllable or when we consider Schrödinger operators on the whole space. The basic perturbation result is the so-called Kato-Rellich theorem which says that:

> If $H_0$ is a self-adjoint operator on a Hilbert space $\mathcal{H}$ and if $V$ is a symmetric operator such that $\mathcal{D}(H_0) \subset \mathcal{D}(V)$ and:
>
> $$||Vf|| \leq a||H_0 f|| + b||f||, \tag{II.15}$$
>
> for positive real numbers $a$ and $b$ satisfying $a < 1$ and for all $f \in \mathcal{D}(H_0)$, then the operator $H_0 + V$ is self-adjoint on $\mathcal{D}(H_0)$ and any core for $H_0$ is a core for $H_0 + V$.

The interested reader will find a proof of this result in Section X of [271]. When $\mathcal{D}(H_0) \subset \mathcal{D}(V)$ and (II.15) holds, then the symmetric operator $V$ is said to be *relatively (operator) $H_0$-bounded* and the *relative bound* is the infimum of the positive numbers $a$ for which (II.15) holds. Checking the assumptions of the Kato-Rellich theorem involves Sobolev's type estimates. This has been a very successful way to solve the self-adjointness problem of Schrödinger operators. Nevertheless, we will find more practical to use a type of perturbation result based on quadratic form techniques. Here is our main tool.

**Theorem II.2.2** *Let $H_0$ be a nonnegative self-adjoint operator on a Hilbert space $\mathcal{H}$ and let $q$ be a symmetric quadratic form such that $\mathcal{Q}(H_0) \subset \mathcal{Q}(q)$ and:*

$$|q(f)| \leq a < H_0 f, f > +b||f||^2, \tag{II.16}$$

*for positive real numbers $a$ and $b$ satisfying $a < 1$ and for all $f \in \mathcal{Q}(H_0)$. Then the form sum $f \hookrightarrow < H_0 f, f > +q(f)$ is the quadratic form of a* *unique self-adjoint operator $H$ which is bounded below by the constant $-b$. This operator will be denoted $H = H_0 + q$, $\mathcal{Q}(H) = \mathcal{Q}(H_0)$ and each core for $H_0$ is a form core for $H$.*

When $\mathcal{Q}(H_0) \subset \mathcal{Q}(q)$ and (II.16) holds, then the symmetric quadratic form $q$ (resp. the self-adjoint operator whose quadratic form is $q$) is said to be *form $H_0$-bounded* and the *relative bound* is the infimum of the positive numbers $a$ for which (II.16) holds. It is possible (though not completely obvious) to show that a self-adjoint perturbation is relatively form bounded whenever it is relatively operator bounded.

*Proof:*
Let us define the quadratic form $\tilde{q}$ by $\mathcal{Q}(\tilde{q}) = \mathcal{Q}(H_0)$ and:

$$\tilde{q}(f) = < H_0 f, f > + q(f),$$

for all $f \in \mathcal{Q}(H_0)$. Our assumptions imply that $\tilde{q}$ is bounded below by $-b$. Moreover:

$$\begin{aligned}(1-a) < H_0 f, f > + \|f\|^2 &\leq \|f\|^2_{(\tilde{q})} = < H_0 f, f > + (b+1)\|f\|^2 \\ &\leq (1+a) < H_0 f, f > + (2b+1)\|f\|^2\end{aligned}$$

which shows that the norms $\|\cdot\|_{(H_0)}$ and $\|\cdot\|_{(\tilde{q})}$ are equivalent on $\mathcal{Q}(H_0)$. Since $\mathcal{Q}(H_0)$ is closed for the norm $\|\cdot\|_{(H_0)}$, we conclude that it is also closed for the norm $\|\cdot\|_{(\tilde{q})}$. Thus $\tilde{q}$ is a closed bounded below quadratic form on its domain $\mathcal{Q}(H_0)$. We conclude by using the fundamental representation result of Section I.5 of Chapter I. ∎

**Remark II.2.3** *There is a version of this result without the assumption that $H_0$ is nonnegative (or more generally bounded below). We did not give it for the proof is slightly more technical and we did not want to obscure its very nature.*

The class of potential functions which is most often used, especially for the use of path integral techniques, seems to have been singled out by Aizenman and Simon in [3]. It is called the class $K_d$ or Kato's class. A function $V$ is said to be in the class $K_d$ if:

$$\lim_{\beta \nearrow \infty} \sup_{x \in \mathbb{R}^d} [(H_0 + \beta)^{-1}|V|](x) = 0. \qquad \text{(II.17)}$$

As we already pointed out, the semigroup operators $e^{-tH_0}$ corresponding to generalized Laplacians are convolutions with probability measures $\mu_t$.

These measures have a density, say $p_t(x)$, with respect to the Lebesgue measure in most of the models. If for each $\delta > 0$, these densities satisfy:

$$\sup_{t>0} \|\mathbf{1}_{\{|y|>\delta\}} p_t\|_{\ell^1(L^\infty)} < \infty, \tag{II.18}$$

where for each measurable function $f$ on $\mathbb{R}^d$ we set:

$$\|f\|_{\ell^1(L^\infty)} = \sum_{\mathbf{n}\in \mathbb{Z}^d} \sup_{x\in C_{\mathbf{n}}} |f(x)| < \infty,$$

where $C_{\mathbf{n}}$ denotes the cube centered at $\mathbf{n} \in \mathbb{Z}^d$ with sides of lengths 1, i.e. $C_{\mathbf{n}} = \mathbf{n}+C_0$ with $C_0 = \{x = (x_1, \cdots, x_d) \in \mathbb{R}^d;\ |x_j| < 1/2 \quad j = 1, \cdots, d\}$, then the Kato's condition (II.17) is in fact equivalent to the condition:

$$\lim_{\delta\searrow 0} \sup_{x\in \mathbb{R}^d} \int_{|x-y|<\delta} g_1(x-y)V(y)\, dy = 0.$$

See for example [49] for a proof. In the particular (original) case of the usual Laplacian, the Kato's condition can be written as:

$$\lim_{a\searrow 0} \sup_{x\in \mathbb{R}^d} \int_{|x-y|\le a} \frac{|V(y)|}{|x-y|^{d-2}} dy = 0$$

when $d \ge 3$,

$$\lim_{a\searrow 0} \sup_{x\in \mathbb{R}^2} \int_{|x-y|\le a} |V(y)| \log \frac{1}{|x-y|} dy = 0$$

when $d = 2$, and:

$$\sup_{x\in \mathbb{R}} \int_{|x-y|\le 1} |V(y)| dy < \infty$$

when $d = 1$. The function $V$ will be said to be in $K_{d,loc}$ when $V|_\Lambda$ is in $K_d$ for all bounded domains $\Lambda$.

Let us assume that the potential function $V$ has a decomposition $V = V_1 - V_2$ with $V_1$ bounded below and in $K_{d,loc}$ and $V_2 \ge 0$ in $K_d$. By the definition of the class $K_d$, the norm of $V_2(H_0+\lambda)^{-1}$ as an operator from $L^1(\mathbb{R}^d, dx)$ into itself goes to 0 as $\lambda \to \infty$. By duality, the norm of $(H_0+\lambda)^{-1}V_2$ as an operator from $L^\infty(\mathbb{R}^d, dx)$ into itself also goes to 0 as $\lambda \to \infty$. Since the norm of $V_2^{1/2}(H_0+\lambda)^{-1/2}$ as an operator on $L^2(\mathbb{R}^d, dx)$ is equal to the norm of the operator $V_2^{1/2}(H_0+\lambda)^{-1}V_2^{1/2}$, it has to go to 0 as $\lambda \to \infty$. This proves that $V_2$ is $H_0$-form bounded with relative bound 0,

so that $H_0 + V$ defines a bounded below closed quadratic form on the form domain $\mathcal{Q}(H_0) \cap \mathcal{Q}(V_1)$, and according to the above discussion the operator $H = H_0 + V$ can be defined in the sense of quadratic forms as a bounded below self-adjoint operator.
This argument was borrowed from [294] where it is also proven that $C_c^\infty$ is a form core for $H$ and that it is an operator core whenever $V$ is also locally square integrable.

This result is very satisfactory for most of the applications. Unfortunately, the study of random Schrödinger operators brings about very naturally a lot of models for which the Hamiltonian should not lead to a bounded below operator. This will be typically the case when the random potential is Gaussian.
When the potential function $V$ is locally square integrable, for every function $f \in C_c^\infty(\mathbb{R}^d)$ the Laplacian $H_0 f$ is still smooth with compact support while $Vf$ is square integrable so that:

$$Hf = H_0 f + Vf$$

is an element of $\mathcal{H}$. One can thus define the operator $H$ on the natural domain $C_c^\infty(\mathbb{R}^d)$. Moreover this operator is easily seen to be symmetric. Whenever it is essentially self-adjoint it has a unique self-adjoint extension which we will still denote by $H$, and in many cases it is possible to show that its domain $\mathcal{D}(H)$ is actually the set of elements $f \in L^2(\mathbb{R}^d, dx)$ for which the quantity $H_0 f + Vf$ is a square integrable function (which is the value of $Hf$). This discussion screams for essential self-adjointness criteria which would apply to potential functions which are not bounded below and which give rise to unbounded below operators. The following one turns out to be very convenient. Indeed, it is capable of handling most of the cases where one wants to deal with an unbounded below Hamiltonian. It is due to Faris and Lavine (the curious reader may consult Section X.5 in the second volume of the Reed and Simon series [271] for a proof and complements).
We need to introduce some more notation. $L^p_{loc,unif}(\mathbb{R}^d, dx)$ will denote the space of measurable functions on $\mathbb{R}^d$ which satisfy:

$$\sup_{\mathbf{n} \in \mathbb{Z}^d} \int_{C_\mathbf{n}} |f(x)|^p \, dx < \infty,$$

and $\ell^1(L^p(\mathbb{R}^d, dx))$ the space of measurable functions on $\mathbb{R}^d$ which satisfy:

$$\sum_{\mathbf{n} \in \mathbb{Z}^d} [\int_{C_\mathbf{n}} |f(x)|^p \, dx]^{1/p} < \infty.$$

**Proposition II.2.4** *Let $V_1$ and $V_2$ be real valued measurable functions on $\mathbb{R}^d$ such that:*

*i)* $V_1 \in L^2_{loc}(\mathbb{R}^d, dx)$ *and* $V_1(x) \geq -a(|x|^2+1)$ *for some* $a > 0$,
*ii)* $V_2 \in L^p_{loc,unif}(\mathbb{R}^d, dx)$ *with* $p \geq 2$ *if* $d \leq 3$, $p > 2$ *if* $d = 4$ *and* $p \geq d/2$ *if* $d \geq 5$.

*Then* $H = H_0 + V_1 + V_2$ *is essentially self-adjoint on* $C_c^\infty(\mathbb{R}^d)$.

Some of the problems discussed above are easier when the potential function has some of the symmetries of the Laplacian operator. In particular, the essential self-adjointness, and in fact the complete analysis, of Schrödinger operators with spherically symmetric potentials reduce to one-dimensional problems as we are about to demonstrate.

**Remark II.2.5 Spherically Symmetric Potentials**
A real valued measurable function $V$ on $\mathbb{R}^d$ is said to be spherically symmetric if $V(x) = v(|x|)$ if $x \neq 0$ for some function $v$ on $(0,\infty)$. We use the fact that the Hilbert space $\mathcal{H} = L^2(\mathbb{R}^d, dx)$ can be considered as $L^2((0,\infty) \times S^{d-1}, r^{d-1}dr\,d\vartheta)$ where $S^{d-1}$ denotes the unit sphere of $\mathbb{R}^d$ and $d\vartheta$ the usual surface measure on $S^{d-1}$. Consequently we have (up to a unitary isomorphism):

$$\mathcal{H} = L^2((0,\infty), r^{d-1}dr) \otimes L^2(S^{d-1}, d\vartheta)$$

and we try to take advantage of this representation to study Schrödinger operators with spherically symmetric potentials. Let us first consider the action of the operator $H$ on functions of the form $f(r)g(\vartheta)$ for a smooth function $f$ with compact support in $(0,\infty)$ and a smooth function $g$ on $S^{d-1}$. Computing the Laplacian in spherical coordinate we get:

$$[-\Delta+V]f\otimes g(r,\vartheta) = \left(-\frac{d^2}{dr^2} + v(r) - \frac{d-1}{r}\frac{d}{dr}\right) f(r)\,g(\vartheta) - \frac{1}{r^2}f(r)\,[Lg](\vartheta)$$

where $L$ is the Laplace-Beltrami operator on $L^2(S^{d-1}, d\vartheta)$. It is well known, and easy to check in dimensions two and three, that the operator $L$ is actually essentially self-adjoint and nonpositive. In fact, it has discrete spectrum and its eigenvalues will be denoted by $-\kappa_k$ for $k = 0, 1, \ldots$. One can show that $\kappa_k = k(k+d-2)$. If we denote by $K_k$ the eigenspace of the eigenvalue $\kappa_k$, then we have:

$$\mathcal{H} = \bigoplus_{k=0}^{\infty} \mathcal{K}_k$$

with $\mathcal{K}_k = L^2((0,\infty), r^{d-1}dr) \otimes K_k$. Let us denote by $\tilde{D}$ the space of finite linear combinations of functions of the form $f \otimes g$ used above. Then, on $\tilde{D}_k = \tilde{D} \cap \mathcal{K}_k$ the operator $-\Delta + V$ coincides with the operator:

$$\left(-\frac{d^2}{dr^2} + v(r) - \frac{d-1}{r}\frac{d}{dr} + \frac{\kappa_k}{r^2}\right) \otimes I$$

and $-\Delta + V$ will be essentially self-adjoint on $\tilde{D}$ if and only if the operator

$$-\frac{d^2}{dr^2} + v(r) - \frac{d-1}{r}\frac{d}{dr} + \frac{\kappa_k}{r^2}$$

is essentially self-adjoint on $C_c^\infty((0,\infty)) \subset L^2((0,\infty), r^{d-1}dr)$ for each $k \geq 0$. To rewrite the problem on a Lebesgue space without weight, one introduces the unitary operator $U : L^2((0,\infty), r^{d-1}dr) \ni \varphi(r) \to [U\varphi](r) = r^{(d-1)/2}\varphi(r) \in L^2((0,\infty), r^{d-1}dr)$ and if one sets:

$$\tilde{H}_k = U\left(-\frac{d^2}{dr^2} + v(r) - \frac{d-1}{r}\frac{d}{dr} + \frac{\kappa_k}{r^2}\right)U^{-1}$$

we have:

$$\tilde{H}_k = -\frac{d^2}{dr^2} + v(r) + p_k(r) \qquad \text{where} \qquad p_k(r) = \left[\frac{(d-1)(d-3)}{4} + \kappa_k\right]\frac{1}{r^2}.$$

We shall study the essential self-adjointness of the operators $\tilde{H}_k$, and many other one dimensional Schrödinger operators, in the next chapter. See Remark III.1.15.

We close this subsection with a discussion of some of the conventions and notation of the discrete case.

Recall from Chapter I that the free Hamiltonian $H_0$ was defined on $\mathcal{H} = \ell^2(\mathbb{Z}^d)$ as the discrete analogue of the negative Laplacian, i.e. by:

$$[H_0 f](\mathbf{n}) = -2d\, f(\mathbf{n}) + \sum_{|\mathbf{m}-\mathbf{n}|=1} f(\mathbf{m}) = \sum_{|\mathbf{n}-\mathbf{m}|=1} (f(\mathbf{n}) - f(\mathbf{m})) \qquad \mathbf{n} \in \mathbb{Z}^d,$$

for all $f \in \mathcal{H}$. We already noticed that:

$$< H_0 f, f >= \sum_{|\mathbf{m}-\mathbf{n}|=1} |f(\mathbf{m}) - f(\mathbf{n})|^2,$$

which is the discrete analogue of the classical expression of the quadratic form of the negative Laplacian. The potential is again a multiplication operator by a function $V$ on the lattice $\mathbb{Z}^d$. As we already pointed out, the self-adjointness problem is irrelevant for the full Hamiltonian $H = H_0+V$ is always self-adjoint on $\mathcal{D}(H) = \mathcal{D}(V)$. However, it is common in the physics and the mathematical physics literature to absorb the diagonal part in the potential. In other words, to consider the free Hamiltonian $H_0$ defined by:

$$[H_0 f](\mathbf{n}) = -\sum_{|\mathbf{m}-\mathbf{n}|=1} f(\mathbf{m}), \qquad \mathbf{n} \in \mathbb{Z}, \tag{II.19}$$

and the new perturbed Hamiltonian $H$ defined by:

$$[Hf](\mathbf{n}) = [H_0 f](\mathbf{n})+V(\mathbf{n})f(\mathbf{n}) = -\sum_{|\mathbf{m}-\mathbf{n}|=1} f(\mathbf{m})+V(\mathbf{n})f(\mathbf{n}), \qquad \mathbf{n} \in \mathbb{Z}, \tag{II.20}$$

It is even frequent to see studies where the sign of the free Hamiltonian $H_0$ has been changed. The effect of these changes are minor and none of the spectral properties are qualitatively changed for the new and the old Hamiltonians are unitarily equivalent up to additive constants. We will nevertheless try to avoid this last change.

If we look at these new free Hamiltonians as *infinite* matrices, all the entries are zero except those of the secondary diagonals where they are equal to $-1$. It is of some interest, especially for the inverse spectral theory to be discussed later, to consider generalizations of these Hamiltonians where the ones of the secondary diagonals are replaced by real numbers. See Problem III.6.5 of Chapter III for a discussion of the self-adjointness problem in one dimension.

## II.2.2 Perturbation of the Absolutely Continuous Spectrum

We will see numerous examples of spectral types of Schrödinger operators. For the time being, we present a set of perturbation results borrowed from scattering theory. They are presented in a very abstract form and as such, they could have been included in Chapter I.

They state that if two self-adjoint operators $H_0$ and $H$ differ by a compact operator, or more generally if their resolvents differ by a compact operator, then they have the same essential spectra and the same absolutely continuous spectra.

The first spectral stability result is part of the so-called Weyl's essential spectral theory while the second one is part of a theory which goes under the names of Kato and Birman. Both can be found in many books. We include the proof of the second result only. Indeed, this second result seems to open promising avenues to a better understanding of the positivity of the Lyapunov exponent and of the localization phenomenon. Moreover, the first result is used in a very marginal way in the sequel. We follow closely Simon's discussion [288] of the Kato-Birman's theorem. We will apply this result to simple examples of one-dimensional lattice Schrödinger operators in the way it was done in [305]. But first we state the Weyl's essential spectrum theorem in a form appropriate to our future use.

**Theorem II.2.6** *Let $H_0$ and $H$ be two self-adjoint operators on a (separable) Hilbert space $\mathcal{H}$ for which the resolvent difference:*

$$[H_0 + zI]^{-1} - [H - zI]^{-1}$$

*is a compact operator. Then we have $\Sigma_{ess}(H_0) = \Sigma_{ess}(H)$.*

If the operators $H$ and $H_0$ satisfy the assumption of the above theorem, then $H$ is said to be a *compact perturbation* of $H_0$. There exists a similar notion of *form compact perturbation* leading to the same stability of the essential spectra. The reader is refered to Section XIII.4 of Volume IV of the Reed and Simon series [274] for proofs and complements. See also Problem II.5.1 for a typical example.

If $H_0$ and $H$ are two self-adjoint operators on a (separable) Hilbert space $\mathcal{H}$, then the time evolution of the states of the systems governed by the Schrödinger equations:

$$i\frac{\partial f}{\partial t} = H_0 f \qquad \text{and} \qquad i\frac{\partial f}{\partial t} = H f \tag{II.21}$$

are given by the unitary groups, namely by $e^{-itH_0}\phi$ and $e^{-itH}\psi$ where the functions $\phi$ and $\psi$ are the initial conditions at time $t = 0$. One of the goals of scattering theory is to compare the asymptotics of the free and the perturbed dynamics and since:

$$||e^{-itH_0}\phi - e^{-itH}\psi|| = ||\phi - e^{itH_0}e^{-itH}\psi||,$$

the results will depend on the existence of the limits:

$$W^{\pm}(H_0, H) = \lim_{t\to\mp\infty} e^{itH_0}e^{-itH}E^H_{ac}, \tag{II.22}$$

where $E^H$ denotes the spectral resolution of the identity of $H$. The spectral projection is inserted for the pure point and the singular continuous spectra are unstable under perturbations and because the limit is mostly interesting for the initial conditions in the absolutely continuous part of

$\mathcal{H}$ for $H$. The operators $W^{\pm}(H_0, H)$ are the central object of scattering theory. They are called the *wave operators*. They are said to be *complete* when $W^{\pm}(H_0, H)\mathcal{H} = \mathcal{H}_{ac}^{H_0}$, i.e. the range of the absolutely continuous part of the resolution of the identity of $H_0$. Completeness implies that $H_0|_{\mathcal{H}_{ac}^{H_0}}$ and $H|_{\mathcal{H}_{ac}^{H}}$ are unitarily equivalent, and in particular, that the operators $H_0$ and $H$ have the same absolutely continuous spectrum. The next result will shorten some of the proofs.

**Lemma II.2.7** *Let us assume that the wave operators $W^{\pm}(H_0, H)$ exist. Then they are complete if and only if the wave operators $W^{\pm}(H, H_0)$ also exist.*

*Proof:*
For an element $\psi \in \mathcal{H}_{ac}^{H}$, $\lim_{t\to-\infty} e^{itH_0}e^{-itH}\psi$ exists and is equal to $\phi$ if and only if $\lim_{t\to-\infty} \|e^{-itH_0}\phi - e^{-itH}\psi\| = 0$ or equivalently if and only if:

$$\psi = \lim_{t\to-\infty} e^{itH}e^{-itH_0}\phi.$$

Notice that:

$$\begin{aligned} < e^{-itH_0}\phi, \phi > &= \lim_{t\to\infty} < e^{-isH_0}e^{itH_0}e^{-itH}\psi, e^{i(t-s)H_0}e^{-i(t-s)H}\psi > \\ &= < e^{-itH}\psi, \psi > \end{aligned}$$

which shows that $\phi$ must be in $\mathcal{H}_{ac}^{H_0}$, and consequently one concludes that, for $\psi \in \mathcal{H}_{ac}^{H}$, $\lim_{t\to-\infty} e^{itH_0}e^{-itH}\psi$ exists if and only if $\psi$ is in the range of the wave operator $W^{+}(H_0, H)$. The conclusion follows for this range is necessarily contained in $\mathcal{H}_{ac}^{H}$ as the above calculation shows. ∎

The main result of the Birman-Kato theory is the following. It gives a sufficient condition for the existence and the completeness of the wave operators.

**Theorem II.2.8** *Let $H_0$ and $H$ be self-adjoint operators on $\mathcal{H}$ and let us assume the existence of a bounded operator $J$ such that $H_0J - JH$ is trace class. Then the generalized wave operators:*

$$W_J^{\pm}(H_0, H) = \lim_{t\to\mp\infty} e^{itH_0}Je^{-itH}E_{ac}^{H},$$

*exist.*

*Proof:*
We reproduce the proof given in [288] by Simon. For each $t \in \mathbb{R}$ we set $W(t) = e^{itH_0}Je^{-itH}$ and we prove that $W(t)\phi$ has a limit for all $\phi \in \mathcal{H}^H_{ac}$. Since $\|W(t)\| = \|J\|$ for all $t$, the operators $W(t)$ are uniformly bounded and it is enough to prove the existence of the limit for a dense subspace of $\mathcal{H}^H_{ac}$, namely the space $\mathcal{AC}_\infty$ of the vectors $\phi$ for which the spectral measure $\sigma_\phi$ has a bounded density. See part a. of Problem II.5.4. Then we notice that:

$$\begin{aligned}&\|[W(t) - W(s)]\phi\|^2 \\ &\quad =< [W(t) - W(s)]\phi, W(t)\phi > - < [W(t) - W(s)]\phi, W(s)\phi >\end{aligned}$$

which shows that we need only to prove that $< W(t)^*[W(t) - W(s)]\phi, \phi >$ goes to 0 and we consider only the case $s \leq t$ going to $\infty$ since the other cases can be handled similarly. If we set:

$$G(u) = e^{iuH}W(t)^*W(s)e^{-iuH},$$

then we have:

$$\frac{dG(u)}{du} = e^{iuH}Y(t,s)e^{-iuH}$$

with:

$$Y(t,s) = e^{itH}C^*e^{-i(t-s)H_0}Je^{-isH} - e^{itH}J^*e^{-i(t-s)H_0}Ce^{-isH}$$

where we set $C = H_0J - JH$. Then:

$$W(t)^*W(s) = e^{iTH}W(t)^*W(s)e^{-iTH} - \int_0^T e^{iuH}Y(t,s)e^{-iuH}\,du$$

so that:

$$\begin{aligned}&< W(t)^*[W(t) - W(s)]\phi, \phi > \\ &\quad =< [W(t) - W(s)]e^{-iTH}\phi, W(t)e^{-iTH}\phi > \\ &\qquad + < \int_0^T e^{iuH}[Y(t,s) - Y(t,t)]e^{-iuH}\,du\,\phi, \phi >\end{aligned}$$

and taking the limit $T \to \infty$ one gets:

$$\begin{aligned}&< W(t)^*[W(t) - W(s)]\phi, \phi > \\ &\quad =< [W(t) - W(s)]e^{-iTH}\phi, W(t)e^{-iTH}\phi > \\ &\qquad + < \int_0^\infty e^{iuH}[Y(t,s) - Y(t,t)]e^{-iuH}\,du\,\phi, \phi > \quad \text{(II.23)}\end{aligned}$$

since the first term goes to 0 because $W(t) - W(s) = i\int_s^t e^{iuH_0}Ce^{-iuH}\,du$ is compact since $C$ is, and because of part b. of Problem II.5.4. Now, the right hand side of (II.23) is a sum of terms of the form::

$$< \int_0^\infty e^{iuH}Q(t,s)e^{-iuH}\,du\,\phi, \phi >$$

and of their complex conjugates for some uniformly bounded family of operators $Q(t,s)$. One then concludes using part d. of Problem II.5.4. ∎

We will use this theory in the form of the following corollary which is credited to Birman and Kuroda.

**Corollary II.2.9** *Let $H_0$ and $H$ be self-adjoint operators such that the difference of their resolvents, say $(H_0 + i)^{-1} - (H + i)^{-1}$, is a trace class operator. Then the wave operators $W^\pm(H_0, H)$ exist and are complete, and in particular the operators $H_0$ and $H$ have the same absolutely continuous spectra.*

*Proof:*
According to Lemma II.2.7, it suffices to prove that the wave operators denoted by $W^\pm(H_0, H)$ exist. Let us set $J = [H_0 + iI]^{-1} - [H + iI]^{-1}$. Then our assumption implies that the operator $H_0J - JH$ is of trace class so that we can apply Theorem II.2.8. Notice that for $\phi \in \mathcal{D}(H)$ we have:

$$W_J^\pm(H_0, H)[H + iI]\phi = W^\pm_{[H_0+iI]^{-1}}(H_0, H)\phi,$$

which shows that the generalized wave operator $W^\pm_{[H_0+iI]^{-1}}(H_0, H)$ exists if one uses the density of $\mathcal{D}(H)$. From the definition of the generalized wave operator $W_J^\pm(H_0, H)$ and our trace class assumption one easily sees that, for $J = (H_0 + i)^{-1} - (H + i)^{-1}$, the generalized wave operator $W_J^\pm(H_0, H)$ exists and is equal to 0 because of the compactness of $J$. The linearity in $J$ of the definition of the generalized wave operators implies the existence of $W_J^\pm(H_0, H)$ for $J = [H + iI]^{-1}$ and we can then prove that $W^\pm(H_0, H)$ exist in the same way we proved that $W^\pm_{[H_0+iI]^{-1}}(H_0, H)$ exists. ∎

We will give in SectionIII.4 of Chapter III several interesting applications of this abstract result. The Birman-Kato theory is used to show existence or nonexistence of absolutely continuous spectrum. It needs the comparison of the resolvent with the resolvent of an operator the absolutely continuous spectrum of which is known to us.

### II.2.3 An Approximation Argument

We close this subsection with a discussion of an approximation argument which will be used very often in the sequel. Typically, the spectral properties of the Schrödinger operators will be infered from the properties of the corresponding operators restricted to bounded domains and the following limiting argument.

Let $\{\mathcal{H}_n; n \geq 1\}$ be an increasing sequence of subspaces of the Hilbert space $\mathcal{H}$, let $\mathcal{D}$ be a core for a self-adjoint operator $H$ on $\mathcal{H}$ and for each $n \geq 1$ let $H_n$ be a self-adjoint operator on $\mathcal{H}_n$. The reader should think of $\mathcal{H}_n = L^2(\Lambda_n, dx)$ for some increasing sequence $\{\Lambda_n; n \geq 1\}$ of bounded domains whose union is the whole $\mathbb{R}^d$ (or the corresponding situation in $\mathbb{Z}^d$) and of $H_n$ as the Schrödinger operator on $\Lambda_n$ constructed from the Dirichlet or Neumann Laplacian.

Let us make the crucial assumption that, for each $f \in \mathcal{D}$ there exists an integer $n_f$ such that $f \in \mathcal{D}(H_n)$ and $Hf = H_n f$ whenever $n \geq n_f$. Then we have the same conclusions as in the case of the strong resolvent convergence of the operators $H_n$ toward $H$, and even better, we have convergence of the corresponding integral kernels.

Indeed, for each $n \geq 1$ one can use the direct sum decomposition of $\mathcal{H}$ as $\mathcal{H}_n \oplus \mathcal{H}_n^\perp$ to define the operator $\tilde{H}_n$ as the direct sum of the operator $H_n$ on $\mathcal{H}_n$ and the identity (but anything would do) on $\mathcal{H}_n^\perp$, for in these conditions one has:

$$\lim_{n\to\infty} \tilde{H}_n = H,$$

in the sense of the strong resolvent convergence presented in Chapter I.

Indeed, if $\psi \in \mathcal{H}$ is of the form $\psi = [H + iI]\varphi$ for some $\varphi \in \mathcal{D}$ one has:

$$[\tilde{H}_n + iI]^{-1}\psi - [H + iI]^{-1}\psi = [\tilde{H}_n + iI]^{-1}(H - \tilde{H}_n)\varphi,$$

which is 0 if $n \geq n_\varphi$, and the conclusion follows from the density of this set of $\psi$ (recall that $\mathcal{D}$ is assumed to be a core for $H$).

Since the spectral resolution of the identity of the operator $\tilde{H}_n$ is equal to the direct sum of the resolution of the identity of $H_n$ and of the identity on $\mathcal{H}_n^\perp$ (recall Problem I.7.18 of Chapter I), one gets the convergence of the corresponding spectral measures.

## II.3 Path Integral Formulas

This section is devoted to the discussion of probabilistic formulas which will be very useful in the sequel. Roughly speaking, they give path integral representations for the exponential functions of the Schrödinger operators and they make possible the use of the machinery developed to study the process of Brownian motion and more general Lévy processes.

## II.3.1 Brownian Motions and the Free Hamiltonians

Our brief discussion of the possible use of Lévy processes is enough to make possible the simultaneous treatment of the continuous and the lattice case. Nevertheless we will still present parallel discussions to help the casual reader find appropriate information quickly with notation he is used to. The systematic reader of this text is nevertheless invited to understand the results and the proofs given in the two cases as two facets of the same object.

### The Continuous Case

We first introduce the new notation we will need. We let $\mathbf{W}$ be the space $C([0,\infty),\mathbb{R}^d)$ of continuous functions on $[0,\infty)$ with values in $\mathbb{R}^d$. This will be our path space and its elements will be denoted by $w$ as opposed to the usual $\omega$'s which run through the generic probability spaces $\Omega$'s. For each $t \geq 0$, we denote by $C([0,\infty),\mathbb{R}^d)$ the $t$-th coordinate function $\mathbf{W} \ni w \to X_t(w) = w(t) \in \mathbb{R}^d$. $\mathcal{W}$ denotes the smallest $\sigma$-field of subsets of $\mathbf{W}$ for which all $X_t' s$ are measurable. Finally, for each $x \in \mathbb{R}^d$ we use the notation $\mathbb{P}_x$ for the Wiener measure on $(\mathbf{W},\mathcal{W})$ for the Brownian motion starting from $x$ at time 0. The corresponding expectation is denoted by $\mathbb{E}_x$. Recall that the measure $\mathbb{P}_x$ is uniquely determined by the fact that for every choice of $0 = t_0 < t_1 < \cdots < t_n < \infty$ the distribution of the random vector $(X_{t_0}, X_{t_1}, \cdots, X_{t_n})$, i.e. the measure $\mathbb{P}_x\{X_{t_0} \in dx_0, X_{t_1} \in dx_1, \cdots, X_{t_n} \in dx_n\}$, is equal to the convolution $\delta_x * \mu_{t_1} * \mu_{t_2-t_1} * \cdots * \mu_{t_n-t_{n-1}}$ where the measure $\mu_t$ is the mean zero Gaussian measure with variance $tI$ the density of which is given in formula (II.10).

Let $H_0$ be the free Hamiltonian defined in Section II.1 as the natural realization as a self-adjoint operator of the negative Laplacian on $\mathbb{R}^d$. Using the above notation we have:

$$[e^{-tH_0}f](x) = \mathbb{E}_x\{f(X_t)\}, \tag{II.24}$$

for each $f \in L^2(\mathbb{R}^d, dx)$.

**Remark II.3.1** *It would be foolish to believe that the above discussion applies only to the negative Laplacian and the process of Brownian motion.*

*Indeed, there exist stochastic processes which bear to the generalized Laplacians of Section II.1 the same relation as the process of Brownian motion does to the negative Laplacian. They are called Lévy processes by the probabilists. For example, the process generated by the operator $H_0 = \sqrt{-\Delta}^{\alpha}$ is known under the name of symmetric stable process of order $\alpha$. The process governed by the operator $H_0 = \sqrt{-\Delta + m^2} - m$ has been neglected by the probabilists but used by mathematical physicists. These Lévy processes have the independence and the stationarity of the increments in common with the process of Brownian motion. But their properties are very poor in general. This is partly due to the fact that the sample paths are not continuous but merely right continuous with left limits. In particular the path space* **W** *is not the same any more. See for example [49] and [47] for details.*

Formula II.24 makes sense for much more general functions than those in $L^2(\mathbb{R}^d, dx)$. The above representation of the semigroup depends only on the position of the sample paths at a single instant $t$. We should not expect anything spectacular from this formula since it involves only simple quantities. In this respect, the representation of the semigroup generated by the Dirichlet Hamiltonian $H^D_{0,\Lambda}$ of a bounded domain $\Lambda$ is more interesting. Let $T_\Lambda$ be the first exit time of $\Lambda$ for the Brownian path, i.e.

$$T_\Lambda(w) = \inf\{t > 0;\ X_t(w) \notin \Lambda\}.$$

One then has:

$$[e^{-tH^{(D)}_{0,\Lambda}} f](x) = \mathbb{E}_x\{f(X_t);\ t < T_\Lambda\} \tag{II.25}$$

for $f \in L^2(\Lambda, dx)$, $t \geq 0$ and $x \in \Lambda$. This formula is well known. A proof can be found for example in [267] or [289] where the definition of the Dirichlet operator $H^{(D)}_{0,\Lambda}$ by the method of the quadratic forms is used.

**Remark II.3.2** *The semigroup generated by the Neumann Hamiltonian $H^{(N)}_{0,\Lambda}$ has also a path integral representation, at least when the boundary of the domain is smooth. The stochastic process involved is a reflected Brownian motion instead of a killed Brownian motion. We will not discuss it for it is not always easy to construct this process and to identify the corresponding semigroup and the infinitesimal generator.*

**Remark II.3.3** *The first exit time of the domain $\Lambda$ can be defined in the same way for general Lévy processes. Moreover, the right hand side of formula (II.25) still defines a strongly continuous semigroup of contractions on $L^2(\Lambda, dx)$. Unfortunately, it is more difficult to identify the infinitesimal generator. It is nevertheless possible to use the theory of Dirichlet forms as presented for example in Fukushima's book [111] to define the Dirichlet*

*Hamiltonian of a domain $\Lambda$ in the case of the genralized Laplacians and then to extend the correspondence we considered in the case of the usual Laplacian.*

We end this subsection with a discussion of the lattice case. Not only will we express the semigroup generated by $H_0$ via a path integral formula, but we will do the same for the unitary group $\{e^{-itH_0}; t \in \mathbb{R}\}$. Such a formula, if any, should be named after Feynman since he proposed such a formula for the continuous case of $\mathbb{R}^d$. But we also know that, in this case, it cannot be given by an integral with respect to a regular measure on path space. Surprisingly, it can be shown that the Feynman formula makes perfectly good mathematical sense whenever the Lévy process generated by the generalized Laplacian $H_0$ is a pure jump process. See Problem II.5.5 for details. In fact, there is no loss of generality in considering only processes with values in the lattice $\mathbb{Z}^d$. This forces the processes to be of pure jump type and the Feynman formula will be proven rigorously, both for the free Hamiltonian and for the perturbed ones.

### The Discrete Case

As before, we first introduce the necessary notation. We let $\mathbf{W}^{(d)}$ be the space of right continuous functions with left limits from $\mathbb{R}$ into $\mathbb{Z}^d$. For each $t \geq 0$, we still denote by $X_t$, the coordinate function $\mathbf{W}^{(d)} \ni w \to X_t(w) = w(t) \in \mathbb{Z}^d$. We will use the notation $\mathcal{W}_t^{(d)}$ for the smallest $\sigma$-field of subsets of $\mathbf{W}^{(d)}$ for which all the the $X_s$'s for $0 \leq s \leq t$ are measurable. The simpler notation $\mathcal{W}^{(d)}$ will be used in the case $t = \infty$. For each $s \geq 0$ and $\mathbf{n} \in \mathbb{Z}^d$ we denote by $\mathbb{P}_\mathbf{n}^{(d)}$ the probability measure on $(\mathbf{W}^{(d)}, \mathcal{W}^{(d)})$ of the *Brownian motion* on $\mathbb{Z}^d$ starting from $\mathbf{n}$ at time 0. The corresponding expectation will be denoted by $\mathbb{E}_\mathbf{n}^{(d)}$. Remember that under $\mathbb{P}_\mathbf{n}^{(d)}$, the process $\{X_t; t \geq s\}$ is at $\mathbf{n}$ at time 0, stays there for an exponential holding time, and then jumps to one of the $2d$ neighbors of $\mathbf{n}$ in $\mathbb{Z}^d$ with equal probabilities. It stays there an exponential holding time independent of everything else, and jumps again to one of the $2d$ neighbors with equal probabilities, .... As before we will use the short notation $\mathbb{P}_\mathbf{n}^{(d)}$ and $\mathbb{E}_\mathbf{n}^{(d)}$ in the case $s = 0$.

We did not talk about the value $\alpha$ of the parameter of the exponential distribution of the time spent at each site of the lattice. A convenient choice will be $\alpha = 2d$.

For each $t \geq 0$ and $w \in \mathbf{W}^{(d)}$ we denote by $N(t, w)$ the number of jumps of the path $w$ before time $t$. $\{N(t); t \geq 0\}$ is a Poisson process with parameter $\alpha = 2d$ independent of the sites of $\mathbb{Z}^d$ visited by the path.

**Proposition II.3.4** *If $H_0$ is the free Hamiltonian in the discrete case, then for each $f \in \mathcal{H} = \ell^2(\mathbb{Z}^d)$, $t \geq 0$ and $\mathbf{n} \in \mathbb{Z}^d$ we have:*

$$[e^{-itH_0} f](\mathbf{n}) = e^{2dt} \mathbb{E}_{\mathbf{n}}^{(d)} \{ f(X_t) i^{N(t)} \} \tag{II.26}$$

*and:*

$$[e^{-tH_0} f](\mathbf{n}) = e^{2dt} \mathbb{E}_{\mathbf{n}}^{(d)} \{ f(X_t) \} \tag{II.27}$$

*Proof:*
The operator $H_0$ is self-adjoint and bounded. Consequently, in order to prove formula (II.26) we need only to show that the quantity $u(t, \mathbf{n})$ defined as the right hand side of (II.26) satisfies:

$$\frac{\partial u}{\partial t} = -i\, H_0 u \tag{II.28}$$

and $u(0, \mathbf{n}) = f(\mathbf{n})$ for all $\mathbf{n} \in \mathbb{Z}^d$. The latter is obvious. Moreover:

$$\begin{aligned} u(t + \delta t, \mathbf{n}) &= e^{2dt} e^{2d\,\delta t} \mathbb{E}_{\mathbf{n}}^{(d)} \{ f(X_{t+\delta t}) i^{N(t+\delta t)} \} \\ &= e^{2\,\delta t} \mathbb{E}_{\mathbf{n}}^{(d)} \{ u(t, X_{\delta t}) i^{N(\delta t)} \} \end{aligned}$$

where we used the Markov property at time $\delta t$. Moreover:

$$\begin{aligned} \mathbb{E}_{\mathbf{n}}^{(d)} \{ u(t, X_{\delta t}) i^{N(\delta t)} \} = {} & u(t, \mathbf{n}) \mathbb{P}_{\mathbf{n}}^{(d)} \{ N(\delta t) = 0 \} \\ & + i\, \mathbb{E}_{\mathbf{n}}^{(d)} \{ u(t, X_{\delta t}); N(\delta t) = 1 \} \\ & + \mathbb{E}_{\mathbf{n}}^{(d)} \{ u(t, X_{\delta t}) i^{N(\delta t)}; N(\delta t) > 1 \} \end{aligned} \tag{II.29}$$

The first term of the right hand side becomes $u(t, \mathbf{n})$ after multiplication by $e^{2d\,\delta t}$ and the second term gives:

$$\begin{aligned} i\mathbb{E}_{\mathbf{n}}^{(d)} \{ u(t, X_{\delta t}); N(\delta t) = 1 \} &= i\mathbb{P}_{\mathbf{n}}^{(d)} \{ N(\delta t) = 1 \} \frac{1}{2d} \sum_{|\mathbf{m}-\mathbf{n}|=1} u(t, \mathbf{m}) \\ &= i\, \delta t \sum_{|\mathbf{m}-\mathbf{n}|=1} u(t, \mathbf{m}) + 0(\delta t^2). \end{aligned}$$

Finally, we notice that the function $u$ is uniformly bounded because $f$ is so and consequently the last expectation in (II.29) is of the order of the probability of more than two jumps in less than $\delta t$, which is of the order of $(\delta t)^2$. Hence, its product with $e^{2d\,\delta t}$ converges to 0 as $\delta t \to 0$, even after dividing by $\delta t$. Collecting all the terms we obtain:

$$\frac{u(t + \delta t, \mathbf{n}) - u(t, \mathbf{n})}{\delta t} = i \sum_{|\mathbf{m}-\mathbf{n}|=1} u(t, \mathbf{m}) + 0(\delta t)$$

which is exactly the desired conclusion (II.28) provided we notice that the convergence is not only pointwise but also in the $\ell^2(\mathbb{Z}^d)$ sense. In order to prove formula (II.27) we need only to show that the function $u(t, \mathbf{n})$ defined as the right hand side of (II.27) satisfies:

$$\frac{\partial u}{\partial t} = -H_0 u \tag{II.30}$$

and $u(0, \mathbf{n}) = f(\mathbf{n})$ for all $\mathbf{n} \in \mathbb{Z}^d$. The proof is omitted for it is similar and even simpler than the above one. ∎

**Remark II.3.5** *The above formulas deal with our version of the free Hamiltonian $H_0$. The latter involves only a nearest neighbor interaction due to the fact that the process jumps only to one of the 2d nearest points of the lattice. It is possible to consider more general Hamiltonians and prove for them the analogues of the above Feynman and Feynman-Kac formulas. This is done in Problem II.5.5.*

We close this subsection with a discussion of the path integral version of the semigroup generated by the restriction of the free Hamiltonian to a box. Since every subset of the lattice $\mathbb{Z}^d$ is closed as well as open, the notion of boundary is somewhat meaningless. We will introduce and use two kinds of boundaries. Let $\Lambda$ be a subset of the lattice $\mathbb{Z}^d$ and let us denote by $\partial\Lambda$ its boundary defined by:

$$\partial\Lambda = \{(\mathbf{m}, \mathbf{n}) \in \mathbb{Z}^d \times \mathbb{Z}^d;\ \mathbf{m} \in \Lambda,\ \mathbf{n} \notin \Lambda,\ |\mathbf{m} - \mathbf{n}| = 1\}, \tag{II.31}$$

by $\partial^{in}\Lambda$ its inner boundary defined by:

$$\partial^{in}\Lambda = \{\mathbf{n} \in \Lambda;\ |\mathbf{n} - \mathbf{m}| = 1 \text{ for some } \mathbf{m} \in \Lambda^c\} \tag{II.32}$$

and $\partial^{out}\Lambda$ its outer boundary defined by:

$$\partial^{out}\Lambda = \{\mathbf{m} \in \Lambda^c;\ |\mathbf{n} - \mathbf{m}| = 1 \text{ for some } \mathbf{n} \in \Lambda\}. \tag{II.33}$$

Now, if $\Gamma$ is any subset of $\mathbb{Z}^d$, we defined the operator $H_{\Gamma^c}$ as the negative infinitesimal generator of the semigroup defined by:

$$[e^{-tH_{\Gamma^c}} f](\mathbf{n}) = \mathbb{E}_{\mathbf{n}}\{f(X_t);\ t < T\}$$

where the killing time $T$ is the first hitting time of $\Gamma$ defined by:

$$T = \inf\{t > 0;\ X_t \in \Gamma\}.$$

It is easy to see that our notation for the operator $H_{\Gamma^c}$ is consistent with our discussion of operators restricted to regions of $\mathbb{Z}^d$ which we had in SectionI.5 for the operator $H_{\Gamma^c}$ is actually the restriction of the operator $H$ to the set $\Gamma^c$. In the applications we have in mind, $\Gamma$ is some sort of hypersurface separating the lattice into two connected components $\Lambda_1$ and $\Lambda_2$, the first of which is finite, in such a way that $\mathbb{Z}^d = \Lambda_1 \cup \Gamma \cup \Lambda_2$ where the sets involved in this union are disjoint. The operator $H_\Gamma$ is defined on $\ell^2(\Lambda_1 \cup \Lambda_2)$ which can be identified with the subspace of $\ell^2(\mathbb{Z}^d)$ of the sequences vanishing on $\Gamma$. Notice that we have $\ell^2(\Lambda_1 \cup \Lambda_2) = \ell^2(\Lambda_1) \oplus \ell^2(\Lambda_2)$ and that the operator $H_\Gamma$ preserves this direct sum decomposition. In particular, the operator $H_\Gamma$ can be regarded as an operator on $\ell^2(\Lambda_1)$ as well as an operator on $\ell^2(\Lambda_2)$

## II.3.2 The Feynman-Kac Formula

The first formula we consider deals with the continuous case of $\mathbb{R}^d$. It is the famous Feynman-Kac formula. Let us assume that $H_0$ is the negative Laplacian in $\mathbb{R}^d$ and $V$ a real-valued bounded and measurable function on $\mathbb{R}^d$. The Schrödinger operator $H = H_0 + V$ is well defined on the domain of $H_0$ and is self-adjoint. The corresponding semigroup can be expressed with the following *path integral* formula:

$$[e^{-tH} f](x) = \mathbb{E}_x\{f(X_t)e^{-\int_0^t V(X_s)ds}\} \tag{II.34}$$

We use the convention $e^{-\infty} = 0$ throughout. There are many ways to prove this formula. A functional analytic approach consists in expressing the left hand side by the semigroup version of the Trotter product formula and identifying the approximants so-obtained as the approximants of the right hand side when one approximates the Riemman integral $\int_0^t V(X_s)ds$ by Riemman sums. The formula is proved in this way for all the functions $f \in L^2(\mathbb{R}^d, dx)$ and almost every $x \in \mathbb{R}^d$. It is possible to extend this formula to more general functions $f$ and to more general potential functions $V$ by various approximation arguments. See for example Section X.11 of [271]. This formula holds most of the time for every $x \in \mathbb{R}^d$ rather than merely for almost every $x$. This can be proven by checking that both sides of the formula (II.34) are actually continuous functions of the variable $x$.

The probabilistic approach to the Feynman-Kac formula is to use the characterization of the semigroup by the fact that it gives a solution to the heat equation, and use simple martingale and integration by parts arguments to complete the derivation from a stochastic calculus point of view. We refer the interested reader to [47].

In view of the right hand side of formula (II.34), it will be crucial to know that:

$$\sup_{x\in \mathbb{R}^d} \mathbb{E}_x\{e^{r\int_0^t V^-(X_s)\,ds}\} < \infty, \tag{II.35}$$

for all $r > 0$ and all $t > 0$. Indeed, such an assumption implies that the free semigroup controls the perturbed one for Hölder's inequality implies that:

$$\begin{aligned} |[e^{-tH}f](x)| &\leq \mathbb{E}_x\{|f(X_t)|^p\}^{1/p} \sup_{x\in \mathbb{R}^d} \mathbb{E}_x\{e^{-p'\int_0^t V(X_s)\,ds}\}^{1/p'} \\ &\leq [e^{-tH_0}|f|^p](x)^{1/p} \sup_{x\in \mathbb{R}^d} \mathbb{E}_x\{e^{p'\int_0^t V^-(X_s)\,ds}\} \end{aligned} \tag{II.36}$$

which shows that the operators $e^{-tH}$ inherit the regularity properties of $e^{-tH_0}$ stated in the previous section. Moreover, a classical argument using only the right continuity of the sample paths and assumption (II.35) implies that these semigroups are stongly continuous on all the $L^p$–spaces with $p < \infty$.

Condition (II.35) looks very artificial. Also, it would be difficult to check if it was not for the following sufficient condition due to Khasminski:

**Lemma II.3.6** *Condition (II.35) is equivalent to the following condition:*

$$\lim_{t\to 0} \sup_{x\in \mathbb{R}^d} \mathbb{E}_x\{\int_0^t V^-(X_s)ds\} = 0. \tag{II.37}$$

*Proof:*
Obviously we need only to show that (II.37) implies (II.35). To do this one first expands the exponential and one uses the Markov property to rewrite the expectations:

$$\mathbb{E}_x\left\{\left(\int_0^t V^-(X_s)ds\right)^k\right\}$$

in a form where the assumption (II.37) comes in naturally. The details can be found for example in [294]) or [47]. ■

We now sketch an argument which we borrow from Carmona [43], to show that the function $e^{-tH}f$ is a bounded continuous function whenever $t > 0$, $f$ is in some $L^p$–space, $V^- \in K_d$ and $V^+ \in K_{d,loc}$. For each $\epsilon \in (0,t)$ we can use the Markov property and get:

$$[e^{-tH}f](x) = \mathbb{E}_x\{f(X_t)e^{-\int_\epsilon^t V(X_s)ds}\}$$

$$
\begin{aligned}
&+ \mathbb{E}_x\{f(X_t)e^{-\int_\epsilon^t V(X_s)ds}[e^{-\int_0^\epsilon V(X_s)ds} - 1]\} \\
=\ & [e^{-\epsilon H_0}(e^{-(t-\epsilon)H} f)](x) + \text{remainder}.
\end{aligned}
$$

The first term in the above right hand side is a bounded continuous function because of the smoothing properties of the free semigroup operator $e^{-\epsilon H_0}$ which we reviewed in SectionII.1. The remainder can be estimated by:

$$
\begin{aligned}
&|\text{ remainder }| \\
&\leq \|f\|_\infty \sup_{x\in \mathbb{R}^d} \mathbb{E}_x\{e^{2\int_0^t V^-(X_s)ds}\}^{1/2} \sup_{x\in \mathbb{R}^d} \mathbb{E}_x\{(e^{-\int_0^\epsilon V(X_s)ds} - 1)^2\}^{1/2}
\end{aligned}
$$

and the proof will be complete when we prove that for each compact set $K$ one has:

$$
\lim_{\epsilon \searrow 0} \sup_{x\in K} \mathbb{E}_x\{(e^{-\int_0^\epsilon V(X_s)ds} - 1)^2\} = 0.
$$

The proof is easy but rather technical and the interested reader is refered to [43] for details.

We summarize the above discussion in a proposition for future reference.

**Proposition II.3.7** *If $1 \leq s \leq p \leq \infty$ and $t > 0$, then under the above assumptions on the potential function $V$, i.e. when $V^- \in K_d$ and $V^+ \in K_{d,loc}$, the operators $e^{-tH}$ map the space $L^s(\mathbb{R}^d, dx)$ boundedly into the space $L^p(\mathbb{R}^d, dx)$ and the elements of the range are bounded continuous functions. Moreover, for each $1 \leq p < \infty$, $\{e^{-tH}; t \geq 0\}$ is a strongly continuous semigroup of bounded operators on $L^p(\mathbb{R}^d, dx)$ and these operators are self-adjoint in the particular case $p = 2$.*

The above proposition implies in a strikingly simple way the following important result:

**Corollary II.3.8** *All the eigenfunctions of the operator $H$ are bounded and continuous.*

**Remark II.3.9** *The self-adjointness of the operator $e^{-tH}$ can be understood probabilistically by the following formula:*

$$
\begin{aligned}
&\int_{\mathbb{R}^d} \mathbb{E}_x\left\{f(X_0)\overline{g(X_t)}e^{-\int_0^t V(X_s)ds}\right\} dx \\
&\qquad = \int_{\mathbb{R}^d} \mathbb{E}_x\left\{g(X_0)\overline{f(X_t)}e^{-\int_0^t V(X_s)ds}\right\} dx,
\end{aligned}
$$

*for $f \in L^p(\mathbb{R}^d, dx)$ and $g \in L^{p'}(\mathbb{R}^d, dx)$. It is an immediate consequence of the invariance of the process of Brownian motion (and more generally of any symmetric Lévy process) under time reversal when the initial distribution is taken to be the Lebesgue measure of $\mathbb{R}^d$. This form of the self-adjointness shows that the adjoint (or transpose) of the operator $e^{-tH}$ as an operator on $L^p$ is the same operator $e^{-tH}$ but regarded now as an operator on $L^{p'}$.*

**Remark II.3.10** *It is of crucial importance to keep in mind that all the operators $H = H_0 + V$ defined for $V^-$ in the $K_d$ class are bounded below self-adjoint operators. The infimum of the spectrum, say $\lambda_0$ can be computed from the large time behaviour of the sample paths of the process of Brownian motion (or more generally of the relevant Lévy process) via the formula:*

$$\lambda_0 = -\lim_{t\to\infty} \frac{1}{t} \log \mathbb{E}_x\{e^{-\int_0^t V(X_s)\,ds}\}, \tag{II.38}$$

*which is discussed in [290] and [291] in detail. This boundedness below rules out some interesting models, and it would be nice if one could extend the probabilistic approach to include them.*

The semigroup operators in a bounded domain can also be computed with path integrals. Indeed, a simple perturbation argument gives easily:

$$[e^{-t(H_{0,\Lambda}^{(D)}+V)}f](x) = \mathbb{E}_x\{f(X_t)e^{-\int_0^t V(X_s)ds}; t < T_\Lambda\}$$

whenever $V$ is bounded. In fact one can show that this formula remains true when $V$ is merely in $\mathbf{K}_d$. For details see the work of Aizenman and Simon [3] or [294].

We now consider, simultaneously for a bounded domain $\Lambda$ and for the whole space $\Lambda = \mathbb{R}^d$ (in which case $T_\Lambda = \infty$ identically), the problem of the existence of integral kernels for the semigroup operators. From a probabilistic point of view things are very simple for one can use conditioning to see that:

$$[e^{-t(H_{0,\Lambda}^{(D)}+V)}f](x) = \int_{\mathbb{R}^d} e^{-t(H_{0,\Lambda}^{(D)}+V)}(x,y)f(y)\,dy,$$

where the integral kernel $e^{-t(H_{0,\Lambda}^{(D)}+V)}(x,y)$ is given by the conditional expectation:

$$e^{-t(H_{0,\Lambda}^{(D)}+V)}(x,y) = \frac{1}{\sqrt{4\pi t}^d}\mathbb{E}_x\{e^{-\int_0^t V(X_s)ds}; t < T_\Lambda | X_t = y\}e^{-|x-y|^2/4t}.$$

We use from time to time the notation $\mathbb{E}_{0,x,t,y}\{\cdot\}$ for the conditional expectation $\mathbb{E}_x\{\ \cdot\ |X_t = y\}$ (i.e., the expectation with respect to the probability measure $\mathbb{P}_{0,x,t,y}$ of the Brownian motion starting from $x$ at time 0 and conditioned to be at $y$ at time $t$). Under the usual assumptions, that is for $V^+ \in K_{d,loc}$ and $V^- \in K_d$, the above discussion of the smoothing properties of the semigroups implies that these kernels are bounded and jointly continuous in $(x, y)$ for each fixed $t > 0$.

Note that when $\Lambda$ is bounded, the boundedness of the kernel implies that the operators $\exp[-t(H_{0,\Lambda}^{(D)} + V)]$ are Hilbert-Schmidt on $L^2(\Lambda, dx)$ and consequently their spectra must be of the form $\{e^{-t\lambda_j};\ j \geq 1\}$ for some sequence $\{\lambda_j;\ j \geq 1\}$ increasing to $+\infty$. Notice that the latter is nothing but the spectrum of the operator $H_{0,\Lambda}^{(D)} + V$. We will use these facts in Chapter VI to construct the integrated density of states for random Schrödinger operators.

The following result will play a crucial role in the generalized eigenfunction expansions to be proven below.

**Proposition II.3.11** *If $V^- \in K_d$, $V^+ \in K_{d,loc}$ and $\phi$ is a real valued function in $L^2(\mathbb{R}^d, dx)$, then for any $t > 0$ the operator $e^{-tH}M_\phi$ is a Hilbert-Schmidt operator on $\mathcal{H} = L^2(\mathbb{R}^d, dx)$. The same conclusion holds for $E(A)M_\phi$ if $A$ is a bounded Borel subset of $\mathbb{R}$ where as usual $E$ denotes the spectral resolution of the identity of the operator $H$.*

*Proof:*
Recall that $e^{-tH}$ maps boundedly from $L^2$ into $L^\infty$ so that:

$$||e^{-tH}||_{2,\infty} = \sup_{x\in\mathbb{R}^d} \int |e^{-tH}(x, y)|^2\, dy < \infty.$$

Consequently, for any $\phi \in L^2$ we have:

$$\int |\phi(x)|^2 |e^{-tH}(x, y)|^2\, dx\, dy < \infty.$$

which proves that the operator $M_\phi e^{-tH}$ (and hence its adjoint $e^{-tH}M_\phi$) is a Hilbert-Schmidt operator since it has a square integrable kernel. To conclude the proof we pick, for each fixed bounded Borel set $A$, a number $\alpha > 0$ large enough so that the characteristic function of $A$ is bounded from above by the function $\lambda \hookrightarrow \alpha e^{-\lambda t}$. The spectral theorem gives:

$$< E(A)f, f > \leq \alpha < e^{-tH}f, f >,$$

for all $f \in L^2(\mathbb{R}^d, dx)$ which implies that $E(A)$ is also Hilbert-Schmidt. ∎

We now come back to the case of the lattice $\mathbb{Z}^d$

**Proposition II.3.12** *Let $V$ be a real function on $\mathbb{Z}^d$ and let $H = H_0 + V$ on $\mathcal{H} = \ell^2(\mathbb{Z}^d)$. Then for each $t \geq 0$, $f \in \mathcal{H}$ and $\mathbf{n} \in \mathbb{Z}^d$ we have:*

$$[e^{-tH} f](\mathbf{n}) = e^{2dt} \mathbb{E}_{\mathbf{n}}^{(d)} \{ f(X_t) e^{-\int_0^t V(X_s)ds} \}, \tag{II.39}$$

*and:*

$$[e^{-itH} f](\mathbf{n}) = e^{2dt} \mathbb{E}_{\mathbf{n}}^{(d)} \{ f(X_t) i^{N(t)} \exp[-i \int_0^t V(X_s)ds] \} \tag{II.40}$$

*Proof:*
The details are left as an exercise to the reader for the important ideas have already been detailed above in the case of the free Hamiltonians for which $V \equiv 0$. ∎

**Remark II.3.13** *In the lattice case, $\ell^2(\mathbb{Z}^d) \subset \ell^\infty(\mathbb{Z}^d)$ and consequently the fact that the semigroup operators $e^{-tH}$ map $\ell^2(\mathbb{Z}^d)$ into $\ell^\infty(\mathbb{Z}^d)$ is trivial. In particular, Proposition II.3.11 is still true. Moreover, the boundedness and the continuity of the eigenfunctions are also obvious.*

# II.4 Eigenfunctions

Properties like the smoothness of the semigroup or the unitary group generated by an operator, the asymptotic behavior of the eigenfunctions, ... depend on the actual Hilbert space on which the operator is considered. They are not invariant under unitary transformations and they are consequently not spectral characteristics in the usual sense. They are nevertheless of crucial importance because of their physical interpretation in the study of localization. Moreover, on the top of being objects of interest of their own, they are very useful tools in the mathematical investigation of the operator because they provide very often the necessary estimates and controls on the quantities of interest.

## II.4.1 $L^2$-Eigenfunctions

We first review some of the most typical examples of asymptotic behavior for eigenfunctions corresponding to energies in the point spectrum of the operator. In fact, the examples we discuss deal only with the discrete spectrum of isolated eigenvalues of finite multiplicity.

The first example is the standard generalization of the classical models for harmonic and anharmonic oscillators. We use path integral proofs eventhough the results are well known and usually proved by other methods.

**Theorem II.4.1** *Let $H_0$ be the negative self-adjoint Laplacian in $\mathbb{R}^d$ and let $H = H_0 + V$ on $\mathcal{H} = L^2(\mathbb{R}^d, dx)$ for a potential function $V$ such that $V^- \in K_d$, $V^+ \in K_{d,loc}$ and:*

$$\lim_{|x|\to\infty} V^+(x) = \infty. \tag{II.41}$$

*Then the semigroup operators $e^{-tH}$ are compact, the spectrum of $H$ is discrete and all the eigenfunctions decay exponentially at $\infty$.*

*Proof:*
We only give the main steps of the proof. The sample path properties of the process of Brownian motion imply that for each fixed $t > 0$ we have:

$$\lim_{r\to\infty} \sup_{x\in\mathbb{R}^d} \mathbb{P}_x\{\sup_{0\le s\le t} |X_s - x| > r\} = 0.$$

Using the assumptions one then checks that:

$$\lim_{r\to\infty} \sup_{x\in\mathbb{R}^d} \mathbb{E}_x\{e^{-\int_0^t V(X_s)ds}; |X_t| > r\} = 0.$$

which in turn implies that:

$$\lim_{r\to\infty} \sup_{\|f\|\le 1} \int_{|x|>r} |[e^{-tH} f](x)|^2\, dx = 0.$$

Putting this last estimate together with the fact that the operators $e^{-tH}$ map $L^2$ into $L^\infty$ gives the compactness claim.
We now prove the second claim of the theorem. Let $\psi \in L^2(\mathbb{R}^d, dx)$ be an eigenfunction for an eigenvalue $\lambda$, i.e. $H\psi = \lambda\psi$. The functional calculus for self-adjoint operators implies that $e^{-tH}\psi = e^{-t\lambda}\psi$ for all $t > 0$. The results of the previous section imply that the function $\psi$ is bounded and continuous and that, using the Feynman-Kac formula, we have:

$$e^{-t\lambda}\psi(x) = \mathbb{E}_x\{\psi(X_t)e^{-\int_0^t V(X_s)ds}\}, \tag{II.42}$$

for all $x \in \mathbb{R}^d$. This implies that the process $\{M_t;\, t \ge 0\}$ defined by:

$$M_t = \psi(X_t)e^{t\lambda}e^{-\int_0^t V(X_s)ds}, \tag{II.43}$$

is a $\mathbb{P}_x$-martingale. Using the optional stopping theorem we get:

$$\psi(x) = \mathbb{E}_x\{\psi(X_{t\wedge\tau})e^{-t\wedge\tau\lambda}e^{-\int_0^{t\wedge\tau} V(X_s)ds}\}, \tag{II.44}$$

Note that (II.44) is nothing but (II.42) with the deterministic time $t$ replaced by the random time $t \wedge \tau$. Let us now choose the stopping time $\tau$ to be the first time the path gets at a distance at least $r > 0$ from its starting point, i.e.:

$$\tau = \inf\{t > 0; |X_t - x| > r\}.$$

Then we have:

$$\begin{aligned} |\psi(x)| &\leq \|\psi\|_\infty \mathbb{E}_x\{e^{t\wedge\tau\lambda}e^{-\int_0^{t\wedge\tau} V(X_s)\,ds}\} \\ &\leq \|\psi\|_\infty \mathbb{E}_x\{e^{-(v-\lambda)t\wedge\tau}\} \\ &\leq \|\psi\|_\infty \left[e^{-(v-\lambda)t} + e^{-\alpha r}\mathbb{E}_0\{e^{\alpha \sup_{0\leq s\leq t}|X_s|}\}\right] \end{aligned} \tag{II.45}$$

for any $\alpha > 0$ provided we set $v = \inf_{|y-x|\leq r} V(y)$. Using standard integrability properties of the process of Brownian motion we get, for $\alpha > 0$ small enough the existence of positive constants $c_1$ and $c_2$ for which we have:

$$|\psi(x)| \leq \|\psi\|_\infty \left[e^{-(v-\lambda)t} + e^{-\alpha r}c_1 e^{c_2 t}\right]. \tag{II.46}$$

We can now choose $r = \rho_1|x|$, with $\rho_1 \in (0,1)$, and $t = \rho_2|x|$, for $\rho_2 > 0$ small enough so that $c_2\rho_2 < \rho_1\alpha$ and this completes the proof. ∎

**Remark II.4.2** *The proof of the compactness which we gave is very general. It merely used the fact that the operator $H_0$ generates a Markov process with right continuous sample paths with some uniformity in the starting point. Consequently, it applies to all the generalized Laplacians we discussed in the previous section, i.e. all the generators of Lévy processes. The proof of the exponential decay is also quite general. The derivation of (II.45) only used the fact that the operator $H_0$ generates a Markov process, but the next step requires some estimates on the sample paths of this process. The exponential bound leading to (II.46) is satisfied by Lévy processes having an exponentially decaying Lévy measure. This applies in particular to the case of the relativistic free Hamiltonian $H_0 = \sqrt{-\Delta + m^2} - m$.*

The next example deals with a simple case of the so-called two-body problem.

**Theorem II.4.3** *Let $H_0$ be the self-adjoint negative Laplacian on $\mathcal{H} = L^2(\mathbb{R}^d, dx)$ and let $H = H_0 + V$ be a Schrödinger operator with a potential function $V$ locally in $K_d$ and satisfying:*

$$\lim_{|x|\to\infty} V(x) = 0. \tag{II.47}$$

*Then the essential spectrum of $H$ is $[0,\infty)$ and the eigenfunctions corresponding to the negative eigenvalues, if any, decay exponentially at $\infty$.*

*Proof:*
The first claim is an easy consequence of the Weyl's characterization of the essential spectrum. See also Problem II.5.1. Let us now prove the claim about the eigenfunction decay. Let us assume that $H\psi = \lambda\psi$ with $\lambda < 0$ and $\psi \in L^2(\mathbb{R}^d, dx)$. We use the same martingale argument as before but we choose the stopping time $\tau$ to be the first hitting time $T_r$ of the ball of radius $r$ centered at the origin, i.e.

$$T_r = \inf\{t > 0;\ |X_t| < r\}.$$

We obtain:

$$\begin{aligned} |\psi(x)| &= |\mathbb{E}_x\{\psi(X_{t\wedge T_r})e^{\lambda t\wedge T_r}e^{-\int_0^{t\wedge T_r} V(X_s)\,ds}\}| \\ &\leq \|\psi\|_\infty \mathbb{E}_x\{e^{\lambda t\wedge T_r}e^{\int_0^{t\wedge T_r} V^-(X_s)\,ds}\} \\ &\leq \|\psi\|_\infty \mathbb{E}_x\{e^{-\beta t\wedge T_r}\}, \end{aligned} \tag{II.48}$$

where $r$ is chosen large enough so that $V^-(y) < \epsilon$ for $|y| > r$ and where $\epsilon > 0$ is chosen small enough so that $\beta \equiv -(\lambda + \epsilon) > 0$. By the monotone convergence theorem we have:

$$\lim_{t\to\infty} \mathbb{E}_x\{e^{-\beta t\wedge T_r}\} = \mathbb{E}_x\{\lim_{t\to\infty} e^{-\beta t\wedge T_r}\} = \mathbb{E}_x\{e^{-\beta T_r}\}$$

and hence:

$$|\psi(x)| \leq \|\psi\|_\infty \mathbb{E}_x\{e^{-\beta T_r}\}. \tag{II.49}$$

We know from the classical probabilistic potential theory of Markov processes (see for example [29]) that:

$$\mathbb{E}_x\{e^{-\beta T_r}\} = \int_{\mathbb{R}^d} G_0(-\beta, x, y)\, d\mu^\beta_{B(0,r)}(y),$$

for some nonnegative Radon measure $\mu^\beta_{B(0,r)}$ called the $\beta$-capacitory measure of the ball $B(0,r)$. We used the notation $G_0(z,\cdot,\cdot)$ for the free Green's function at energy $z \in \mathbb{C}$, i.e. for the kernel of the resolvent operator $[H - zI]^{-1}$. Consequently we have:

$$|\psi(x)| \leq c_{d,r} \sup_{|y|\leq r} G_0(-\beta, x, y), \tag{II.50}$$

and the desired result follows. ∎

**Remark II.4.4** *The above proof works in full generality up to the estimate (II.50). In the case of the relativistic free Hamiltonian* $H_0 = \sqrt{-\Delta + m^2} - m$, *the Green's function decays exponentially and the result is (quantitatively) the same as for the usual Laplacian discussed in the above theorem. On the other hand, the results are different for the other generalized Laplacians which we discussed in Section II.1. Indeed, in the case of the free Hamiltonian* $H_0 = \sqrt{-\Delta}^{\alpha}$ *the Green's function decays only polynomially and so do the eigenfunctions.*

## II.4.2 The Periodic Case

Our next example concerns the case of periodic potentials. It is of crucial importance in solid state physics. A potential function $V : \mathbb{R}^d \to \mathbb{R}$ (resp. $V : \mathbb{Z}^d \to \mathbb{R}$ is said to be periodic if there exist $d$ linearly independent vectors $a_1, \cdots, a_d$ in $\mathbb{R}^d$ such that:

$$V(x + \sum_{j=1}^{d} k_j a_j) = V(x), \tag{II.51}$$

for all $x \in \mathbb{R}^d$ (resp. $x \in \mathbb{Z}^d$) and all integers $k_1, \cdots, k_d$ in $\mathbb{Z}$. We will not discuss the lattice case in the present general setting and, for the sake of simplicity, we will assume that the vectors $a_j$ are in fact the vectors of the canonical basis $\{e_1, \cdots, e_d\}$ of $\mathbb{R}^d$.

A periodic potential does not tend to $\infty$ nor does it converge to 0. So the spectral characteristics of the corresponding Schrödinger operators have to be different from those of the examples discussed above. In fact, the spectrum is typically purely absolutely continuous (i.e. $E = E_{ac}$) and consequently there is no $L^2$-eigenfunction. One can try to develop some intuition from the particular case of the free Hamiltonian $H_0$. In this case, all the solutions of the eigenvalue equation $Hf = \lambda f$ can be determined explicitly. The operator can be reconstructed from complex exponentials, the so-called *plane waves* and the spectral properties of the operator can be recovered from there. This is relatively elementary in one dimension and it is more involved in higher dimension. The following discussion is presented for the sake of completeness. It is patterned after the complete analysis given in volume IV of the Reed and Simon series. We will not give any proof here, but we will give a complete analysis of the one-dimensional case in Chapter VII.

For each $\alpha_1, \cdots, \alpha_d$ in $[0,1)$ we define the operator $H_0(\alpha)$ on $\tilde{\mathcal{H}} = L^2(C, dx)$ as the negative Laplacian $-\Delta$ with boundary conditions:

$$f(x+e_j) = e^{2i\pi\alpha_j} f(x) \qquad \text{and} \qquad \frac{\partial f}{\partial x_j}(x+e_j) = e^{2i\pi\alpha_j}\frac{\partial f}{\partial x_j}(x),$$

for $j = 1, \cdots, d$, whenever $x$ and $x + e_j$ both belong to the closure $\overline{C}$ of the unit cell $C = [0,1)^d$. For each $\alpha \in [0,1)^d$, $V$ is $H_0(\alpha)$-bounded with relative bound 0 so that the operator $H(\alpha) = H_0(\alpha) + V$ is self-adjoint on the domain of $H_0(\alpha)$. It is possible to show that the operators $H_0(\alpha)$ (and hence the operators $H(\alpha)$ have compact resolvent. So the operators $H(\alpha)$ have purely discrete spectra and the eigenvalues $\lambda_j(\alpha)$ and the corresponding eigenfunctions $\psi_j^\alpha(x)$ are measurable functions of $\alpha$. If one sets:

$$\tilde{\varphi}_j(\alpha) = \int_{\mathbb{R}^d} \varphi(x)\overline{\psi_j^\alpha(x)}\, dx$$

for $\varphi$ in the Schwartz space $\mathcal{S}(\mathbb{R}^d)$ of smooth rapidly decreasing functions, then one has:

$$\int_{\mathbb{R}^d} |\varphi(x)|^2\, dx = \sum_j \int_{[0,1)^d} |\tilde{\varphi}_j(\alpha)|^2\, d\alpha$$

and the function $\varphi$ can be recovered by the inversion formula:

$$\varphi(x) = \sum_j \int_{[0,1)^d} \tilde{\varphi}_j(\alpha)\overline{\psi_j^\alpha(x)}\, d\alpha.$$

The operation $\tilde{}$ can be extended by continuity into a unitary isomorphism from $L^2(\mathbb{R}^d, dx)$ onto the direct sum $\bigoplus_j L^2([0,1)^d, d\alpha)$ and the periodic Schrödinger operator $H = -\Delta + V$ satisfies:

$$\tilde{H\varphi}_j(\alpha) = \lambda_j(\alpha)\tilde{\varphi}_j(\alpha)$$

for all $\varphi \in \mathcal{D}(H)$. This last relation gives an eigenfunction expansion of the operator $H$ of a form which proved to be very useful in scattering theory. We will need (and prove in the next section) a different kind of eigenfunction expansion result which applies in full generality to most (if not all) Schrödinger operators. In any case a little extra work using the theory of direct integral decomposition of Hilbert spaces is needed to control the spectrum of these periodic Schrödinger operators, and it can be proved that this spectrum is absolutely continuous in great generality. We will give a complete proof of this fact in our study of the one-dimensional case in Chapter VII. The reader interested in the multidimensional case is refered to Section XIII.16 of volume IV of the Reed and Simon series [272].

## II.4.3 Generalized Eigenfunction Expansions

We noticed that the eigenfunctions corresponding to eigenvalues away from the essential spectrum decay exponentially. The $L^2$-eigenfunctions corresponding to eigenvalues embedded in the continuous spectrum are more difficult to estimate in general. It has been shown that they still decay exponentially when they are *threshold eigenvalues* of many body Hamiltonians. But the general case is unclear. We also saw that the generalized eigenfunctions of the periodic Schrödinger operators were typically bounded. We are about to show that these generalized eigenfunctions are polynomially bounded outside a set of spectral measure 0. This result is a by-product of a general *eigenfunctions expansion* theorem for Carleman operators. This result is not in the text books we quoted so far. It can be found in the book of Berezanski [28]. Our presentation follows Simon's survey paper [294].

We already noticed that whenever the self-adjoint operator $H$ has discrete spectrum $\{\lambda_1, \lambda_2, \cdots\}$, the resolution of the identity of $H$, say $E$, can be written in the form:

$$E(A) = \sum_{\lambda_j \in A} < f_j, \cdot > f_j \, \delta_{\lambda_j}$$

where $\{f_j; j \geq 1\}$ is any orthonormal basis of corresponding eigenfunctions and $A$ is any bounded Borel subset of $\mathbb{R}$. Such a spectral projection has the integral kernel:

$$E(A, x, y) = \sum_{\lambda_j \in A} f_j(x) f_j(y)$$

which can be written in the form $E(A, x, y) = \int_A E(\lambda, x, y)\sigma(d\lambda)$ where $\sigma = \sum_{j \geq 1} \delta_{\lambda_j}$. Note that the measure $\sigma$ and the spectral resolution of the identity of $H$ have the same sets of measure 0 and that any other measure having the same sets of measure 0 could be used to desintegrate the kernel of the spectral projections. Consequently we proved that, at least for operators with discrete spectrum:

*for all the $\lambda \in \mathbb{R}$ outside a negligible set for the spectral resolution of the identity, there exists an operator $E(\lambda)$ such that the projection valued measure $E(d\lambda)$ desintegrates in the form $E(\lambda)\sigma(d\lambda)$. These operators have an integral kernel $E(\lambda, x, y)$ and for almost every $\lambda$ with respect to $\sigma$ and almost every $x$, the function $E(\lambda, x, \cdot)$, (which is nothing but $f_j(x)f_j(\cdot)$ if $\lambda = \lambda_j$) is an eigenfunction of $H$.*

We now discuss the possibility, in the general case, to desintegrate the projection valued measure $E(d\lambda)$ with respect to a measure $\sigma(d\lambda)$. Note that this measure cannot be uniquely defined. Indeed, any nonnegative measure having the same null sets as $E$ can do the job. We will try to impose on this measure $\sigma$ to be minimal in some sense, and this will determine uniquely the measure class of $\sigma$. The fiber operators $E(\lambda)$ are now constructed as integral operators from a space smaller than $L^2(\mathbb{R}^d, dx)$ into a space larger than $L^2(\mathbb{R}^d, dx)$. These spaces can actually be taken to be weighted $L^2$-spaces. The $E(\lambda)$'s are still integral operators but now, for each bounded Borel subset $A$ of $\mathbb{R}$, the function

$$\int_A E(\lambda, x, y)\sigma(d\lambda)$$

of $x$ and $y$ is simply a weak kernel for the bounded operator $E(A)$ on $L^2(\mathbb{R}^d, dx)$. Nevertheless, the $E(\lambda, x, \cdot)'s$ are still solutions of the eigenvalue equation $Hf = \lambda f$. They are called *generalized eigenfunctions* since they are no longer in $L^2(\mathbb{R}^d, dx)$ but only in a weighted $L^2$-space. Moreover, the size of the weight provides some information on the asymptotic behavior of these generalized eigenfunctions at infinity.

**Theorem II.4.5** *Let us assume as before that the potential $V$ is in the class $K_d$ in the sense that $V^+ \in K_{d,loc}$ and $V^- \in K_d$. Then for each fixed $\delta > d/2$ there exists a nonnegative Radon measure $\sigma$ on $\mathbb{R}$ which has the same null sets as the spectral resolution of the identity of the operator $H = H_0 + V$, and a desintegration of $E$ with respect to $\sigma$ in the sense that there exists a jointly measurable function $(\lambda, x, y) \hookrightarrow E(\lambda, x, y)$ which is jointly continuous in $(x, y)$ for $\sigma$-almost every $\lambda$ and such that:*

$$< \phi(H)f, g >= \int \phi(\lambda) \left[\int \int E(\lambda, x, y) f(x)\overline{g(y)}\, dx dy\right] d\sigma(\lambda), \quad \text{(II.52)}$$

*for all bounded measurable functions $\phi$ on $\mathbb{R}$ and for all $f$ and $g$ in $C_c^\infty(\mathbb{R}^d)$. Moreover for $\sigma$-almost every fixed $\lambda \in \mathbb{R}$ and for each fixed $y \in \mathbb{R}^d$, the function $x \hookrightarrow E(\lambda, x, y)$ is a generalized eigenfunction in the sense that:*

$$HE(\lambda, \cdot, y) = \lambda E(\lambda, \cdot, y).$$

*in the sense of distributions. Finally, for each bounded Borel subset A of $I\!R$ there exists a positive constant $c_A$ such that:*

$$|E(\lambda, x, y,)| \leq c_A(1+|x|^2)^{\delta/2}(1+|y|^2)^{\delta/2}, \tag{II.53}$$

*for all $\lambda \in A$.*

*Proof:*
Let us fix $\delta > d/2$. By Proposition II.3.11, for each bounded Borel set $A \in \mathcal{B}_{I\!R}$ the operator $E(A)(1+|x|^2)^{-\delta/2}$ is Hilbert-Schmidt and consequently, the operator $\tilde{E}(A) = (1+|x|^2)^{-\delta/2}E(A)(1+|x|^2)^{-\delta/2}$ is trace class. We can then use the properties of the desintegration of trace class operator valued measures proven in Problem I.7.14 and conclude the existence of a nonnegative measure $\sigma$ on $\mathbb{R}$ having the same null sets as $E$, and for $\sigma$-almost all $\lambda \in \mathbb{R}$, of a trace class operator $\tilde{E}(\lambda)$ such that $\tilde{E}(A) = \int_A \tilde{E}(\lambda)\, d\sigma(\lambda)$ for all bounded Borel sets $A$. We get:

$$< \phi(H)f, g >= \int \phi(\lambda) < \tilde{E}(\lambda)[(1+|x|^2)^{\delta/2}f], (1+|x|^2)^{\delta/2}g > d\sigma(\lambda), \tag{II.54}$$

for all bounded measurable functions $\phi$ on $\mathbb{R}$ and for all $f$ and $g$ in $C_c^\infty(\mathbb{R}^d)$ by a simple monotone class argument. Note also that the operators $\tilde{E}(\lambda)$ have a jointly continuous kernel because of the results proved in Problem II.5.3 at the end of this chapter and that we get (II.52) from (II.54) and (II.53) simply by setting $E(\lambda, x, y) = (1+|x|^2)^{\delta/2}\tilde{E}(\lambda, x, y)(1+|x|^2)^{\delta/2}$. It now remains to prove that the elements of the range of the fiber operator $E(\lambda)$ are generalized eigenfunctions (i.e. solutions in the distributiuon sense of the eigenvalue equation for the energy $\lambda$). If $\varphi = E(\lambda)f$ for some $f \in C_c^\infty(\mathbb{R}^d)$ and if $g \in C_c^\infty(\mathbb{R}^d)$ then we have:

$$< \varphi, [H - \lambda I]g >=< (1+|x|^2)^{\delta/2}\tilde{E}(\lambda)(1+|x|^2)^{\delta/2}f, [H-\lambda I]g >$$

which is 0 because $\tilde{E}(\lambda)(1+|x|^2)^{\delta/2}[H-\lambda I] = 0$. Indeed, for all $f, g \in C_c^\infty(\mathbb{R}^d)$ we have:

$$\begin{aligned}
&< \tilde{E}(\lambda)(1+|x|^2)^{\delta/2}[H-\lambda I]f, g > \\
&= \lim_{A \searrow \{\lambda\}} \frac{1}{\sigma(A)} < \tilde{E}(A)(1+|x|^2)^{\delta/2}[H-\lambda I]f, g > \\
&= \lim_{A \searrow \{\lambda\}} \frac{1}{\sigma(A)} < E(A)[H-\lambda I]f, (1+|x|^2)^{\delta/2}g > \\
&= \lim_{A \searrow \{\lambda\}} \frac{1}{\sigma(A)} \int_A (\lambda' - \lambda) < E(d\lambda')f, (1+|x|^2)^{\delta/2}g > \\
&= 0. \blacksquare
\end{aligned}$$

**Remark II.4.6** *1. It is easy to see that formula (II.52) remains true for all the measurable functions $\phi$ which are bounded on the spectrum of $H$ and all functions $f$ and $g$ in the weighted $L^2$-space $L^2(\mathbb{R}^d, (1+|x|^2)^{\delta/2}dx)$.*

*2. It is not always possible to obtain a desintegration of the projection valued measure $E(d\lambda)$ in operators on $L^2(\mathbb{R}^d, dx)$. The above result gives a desintegration when it is regarded as taking values in the space of operators from the weighted Hilbert space $L^2_\delta = L^2(\mathbb{R}^d, (1+|x|^2)^{\delta/2}dx)$ into $L^2_{-\delta} = L^2(\mathbb{R}^d, (1+|x|^2)^{-\delta/2}dx)$. Indeed, for $\sigma$-almost every $\lambda$ the function $(x, y) \hookrightarrow E(\lambda, x, y)$ is a kernel for the operator $E(\lambda)$ when the latter is viewed as an operator from $L^2_\delta$ into $L^2_{-\delta}$ but it is merely a weak kernel for the operator $E(\lambda)$ is not defined on $L^2(\mathbb{R}^d, dx)$ in general.*

*3. The above theorem shows that for almost all $\lambda \in \mathbb{R}$ with respect to $E$ there exist eigenfunctions in the sense of distributions, namely the functions $x \hookrightarrow E(\lambda, x, y)$, and (II.53) shows that these generalized eigenfunctions are polynomially bounded. This fact will play a crucial role in the investigation of the spectral properties of random Schrödinger operators. It is interesting to notice that the existence of polynomially bounded solutions (in the sense of distributions) of the eigenvalue equation is in fact characteristic of the energies in the spectrum. See [292] and [294] for details.*

*4. Note also that this is still true in the lattice case and that, according to the last remark of the previous section, proving it requires only to desintegrate the spectral resolution of the identity for the existence of kernels and the fact that they give generalized eigenfunctions are obvious in this case.*

## II.5 Problems

**Problem II.5.1** *Let $H_0$ be the free Hamiltonian corresponding to the operator of multiplication in Fourier space by the function $F(p)$ for some function $F$ appearing in the Lévy-Kintchine formula, and let us assume that:*

$$\lim_{|p|\to\infty} F(p) = 0.$$

*Show that for each function $V$ of the class $K_d$ the assumption:*

$$\lim_{|x|\to\infty} V(x) = 0$$

*implies that $H$ is a compact perturbation of the operator $H_0$ and consequently that $\Sigma_{ess}(H) = [0, \infty)$.*

*Hint: compute the kernels of the resolvent operators of $H$ and $H_0$ in Fourier space and show that their difference is compact.*

**Problem II.5.2** *Let us denote by $H_0$ the free Hamiltonian on the lattice $\mathbb{Z}^d$. The Fourier transform is a unitary map from $\ell^2(\mathbb{Z}^d)$ onto the Hilbert space $L^2(\mathbb{T}^d, d\vartheta_1 \times \ldots \times d\vartheta_k)$ of square integrable functions on the d-torus and we denote by $\widehat{H}_0$ the self-adjoint operator in $L^2(\mathbb{T}^d, d\vartheta_1 \times \ldots \times d\vartheta_k)$ which corresponds to $H_0$ in this unitary equivalence.*

1. *Prove that $\widehat{H}_0$ is the multiplication operator by the function*

$$f(\vartheta) = 2\cos(\vartheta_1) + \ldots + 2\cos(\vartheta_d).$$

2. *Conclude that the spectrum of $H_0$ is equal to $[-2d, 2d]$ and is absolutely continuous.*

*The strip of width $\ell$, say $S_\ell$, is made out of $\ell$ copies of the set of integers $\mathbb{Z}$, i.e. $S_\ell = [1, \ldots, \ell] \times \mathbb{Z}$. Let $\psi$ be a function in $\ell^2(S_\ell)$. Then one defines $[H_\ell^D \psi](r, n)$ by the formula:*

$$\begin{cases} \psi(r, n+1) + \psi(r, n-1) + \psi(r+1, n) + \psi(r-1, n) & \text{if } 1 < r < \ell \\ \psi(1, n+1) + \psi(1, n-1) + \psi(2, n) & \text{if } r = 1 \\ \psi(\ell, n+1) + \psi(\ell, n-1) + \psi(\ell - 1, n) & \text{if } r = \ell \end{cases}$$

*$H_\ell^D$ is called the free Hamiltonian (or Laplace operator) with Dirichlet boundary conditions.*
*We also consider the Laplace operator with periodic boundary conditions. For each function $\psi \in \ell^2(S_\ell)$ we define $[H_\ell^P \psi](r, n)$ by the formula:*

$$\begin{cases} \psi(r, n+1) + \psi(r, n-1) + \psi(r+1, n) + \psi(r-1, n) & \text{if } 1 < r < \ell \\ \psi(1, n+1) + \psi(1, n-1) + \psi(2, n) + \psi(\ell, n) & \text{if } r = 1 \\ \psi(\ell, n+1) + \psi(\ell, n-1) + \psi(1, n) + \psi(\ell - 1, n) & \text{if } r = \ell \end{cases}$$

*The strip $S_\ell$ can be viewed as the discrete group $\Omega \times \mathbb{Z}$ where $\Omega = [1 \ldots \ell]$ is the set of integer modulo $\ell$. The dual group is $\Omega \times \mathbb{T}$ and one can apply the general theory of Fourier transform in this case.*

1. *Prove that the Fourier transform of $H_\ell^P$ is the multiplication operator by the function:*

$$f(r, \vartheta) = 2\cos(2r\pi/\ell) + 2\cos(\vartheta).$$

   *Conclude that the spectrum of $H_\ell^P$ is equal to $[-4, 4]$ and is absolutely continuous.*

2. *One can write*

$$H_\ell^D = \Pi \; H_{\ell+1}^P \; \Pi$$

*where* $\Pi$ *is the projection operator from* $S_{\ell+1}$ *onto* $S_\ell$.

(a) *Prove that* $\widehat{\Pi} = I - W$ *where* $W$ *is defined by:*

$$[Wf](r,\vartheta) = \frac{1}{\ell+1}\sum_{k=1}^{\ell+1} f(k,\vartheta).$$

(b) *Conclude that the spectrum of* $H_\ell^D$ *is absolutely continuous and is equal to* $[-2-2\cos(\frac{\pi}{\ell+1}), 2+2\cos(\frac{\pi}{\ell+1})]$.

**Problem II.5.3** *Let* $\mu$ *be a nonnegative measure on the locally compact metrizable space* $\mathcal{X}$, *and let* $\Gamma$ *be a jointly continuous function on* $\mathcal{X}\times\mathcal{X}$ *for which the kernel* $\{\Gamma(x,y);\, x,y\in\mathcal{X}\}$ *is nonnegative definite and has the representation:*

$$\Gamma(x,y) = \sum_{n\geq 1} f_n(x) f_n(y),$$

*for some sequence* $\{f_n;\, n\geq 1\}$ *of locally square integrable functions on* $\mathcal{X}$ *and where the convergence is in the sense of* $L^2_{loc}(\mathcal{X}\times\mathcal{X}, d\mu(x)d\mu(y))$. *Show that for each fixed compact subset* $K$ *of* $\mathcal{X}$ *one has:*

**a.** *The operator* $\Gamma_K$ *defined on* $L^2(K, d\mu)$ *by:*

$$[\Gamma_K f](x) = \int_K K(x,y) f(y)\, d\mu(y),$$

*is of Hilbert-Schmidt type and if* $\{\varphi_i;\, i\geq 1\}$ *is a complete orthonormal system of eigenvectors corresponding to the eigenvalues* $\{\lambda_i^K;\, i\geq 1\}$, *then one has:*

$$\Gamma(x,y) = \sum_{i=1}^{\infty} \lambda_i^K \varphi_i(x)\varphi_i(y),$$

*for all* $x,y\in K$, *where the series on the right hand side converges uniformly.*

**b.** *The functions* $\varphi_i$*'s are continuous and the series:*

$$\psi_n(x) = \sum_{i=1}^{\infty}\Big(\int_K f_n(x)\overline{\varphi_i(x)}\, dx\Big)\, \varphi_i(x),$$

*defining the functions* $\psi_n$*'s converge uniformly (over* $K$*) which proves that the functions* $\psi_n$*'s are continuous.*

**c.** *$f_n = \psi_n$ for all $n \geq 1$,*

**d.** *The series in the assumption on $\Gamma$ is uniformly convergent over $K \times K$ by Dini's theorem.*

**Problem II.5.4** *This problem is devoted to the proof of various estimates which are used in the proof of the Birman-Kato's Theorem.*
*Let $\phi$ be an element of the absolutely continuous part of a Hilbert space $\mathcal{H}$ relative to a self-adjoint operator $H$ (i.e. $\phi \in \mathcal{H}^H_{ac}$) and let $P$ be the (orthogonal) projection onto the closed subspace generated by the vectors of the form $e^{itH}\phi$. Let us also recall the notation $\sigma_\phi$ for the spectral measure introduced in Chapter I. According to our proof of the spectral theorem, it is a nonnegative measure of total mass $||\phi||^2$ on $\mathbb{R}$ and the range of $P$ is unitary isomorphic to $L^2(\mathbb{R}, d\sigma_\phi)$ in such a way that $\phi$ goes to the constant function 1 and the operator $e^{itH}$ to the operator of multiplication by the function $\lambda \hookrightarrow e^{iy\lambda}$. By definition of $\mathcal{H}^H_{ac}$ the measure $\sigma_\phi$ is absolutely continuous with respect to the Lebesgue measure. We define the set $\mathcal{AC}_\infty(H)$ as the set of those vectors $\phi$ for which the density $d\sigma_\phi(\lambda)/d\lambda$ is actually (essentially) bounded.*

**a.** *Prove that $\mathcal{AC}_\infty(H)$ is dense in $\mathcal{H}^H_{ac}$.*

**b.** *Prove that:*

$$\lim_{t\to\pm\infty} Ce^{-itH}E_{ac}(\mathbb{R}) = 0$$

*in the sense of the strong convergence of operators whenever $C$ is a compact operator. (Hint. First reduce the problem to the case of finite rank and then of rank one operators $C$, and then use part a. above and the Riemann-Lebesgue lemma).*

**c.** *Prove that, for any $\phi \in \mathcal{AC}_\infty(H)$ and any $\psi \in \mathcal{H}$ one has:*

$$\int_{\mathbb{R}} |< e^{-itH}\phi, \psi >|^2\, dt \leq 2\pi ||\psi||^2\, ||\frac{d\sigma_\phi(\lambda)}{d\lambda}||^2_\infty .$$

**d.** *Prove that:*

$$|< \int_0^\infty e^{itH} Q C e^{-i(t+s)H}\, dt\, \phi, \phi >| \leq (\sum_{n\geq 0} \lambda_n)^{1/2}$$

$$\left(\sum_{n\geq 0} \lambda_n \int_s^\infty |< e^{-itH}\phi, \psi_n >|^2\, dt\right)^{1/2} ||Q||\, ||\frac{d\sigma_\phi(\lambda)}{d\lambda}||_\infty$$

*for any trace class operator $C = \sum_{n\geq 0} \lambda_n \psi_n \otimes \varphi_n$ and for any $\phi \in \mathcal{AC}_\infty(H)$.*

**Problem II.5.5** *Let us consider a generalized free Hamiltonian $H_0$ on the lattice $\mathbb{Z}^d$ defined by its matrix elements $H_0(\mathbf{m},\mathbf{n})$ for $\mathbf{m},\mathbf{n} \in \mathbb{Z}^d$. We assume that all these entries are nonnegative, that $H_0(\mathbf{m},\mathbf{n}) = H_0(\mathbf{n},\mathbf{m})$, that $H_0(\mathbf{m},\mathbf{m}) = \beta$ for some fixed $\beta > 0$ and all $\mathbf{m},\mathbf{n} \in \mathbb{Z}^d$ and that for each fixed $\mathbf{m} \in \mathbb{Z}^d$ we have:*

$$\sum_{\mathbf{n}\neq\mathbf{m}} H_0(\mathbf{m},\mathbf{n}) = 1.$$

*Note that the finiteness of this sum is the important hypothesis for the fact that the sum is 1 can be achieved by renormalization.*

**a.** *Check that $H_0$ defines a bounded self-adjoint operator on $\mathcal{H} = \ell^2(\mathbb{Z}^d)$ and that the free Feynman formula (II.26) and the free Feynman-Kac formula (II.27) are still valid in the following forms:*

$$[e^{itH_0}f](\mathbf{n}) = e^{(\alpha+i\beta)t}\mathbb{E}_{\mathbf{n}}\{f(X_t)i^{N(t)}\}$$

*and:*

$$[e^{tH_0}f](\mathbf{n}) = e^{(\alpha+\beta)t}\mathbb{E}_{\mathbf{n}}\{f(X_t)\}$$

*where the stochastic process is the one staying at each point of the lattice an exponential time with parameter $\alpha = 1$ and jumping after that to the point $\mathbf{n}$ with probability $H_0(\mathbf{m},\mathbf{n})$ if its position before the jump is $\mathbf{m}$.*

**b.** *Let us assume that $V$ is a bounded real valued function on the lattice $\mathbb{Z}^d$. Prove the Feynman formula*

$$[e^{it(H_0+V)}f](\mathbf{n}) = e^{(\alpha+i\beta)t}\mathbb{E}_{\mathbf{n}}\{f(X_t)i^{N(t)}e^{i\int_0^t V(X_s)ds}\}$$

*and the Feynman-Kac formula:*

$$[e^{t(H_0+V)}f](\mathbf{n}) = e^{(\alpha+\beta)t}\mathbb{E}_{\mathbf{n}}\{f(X_t)e^{\int_0^t V(X_s)ds}\}.$$

**c.** *Extend the above formulas to general potential functions $V$.*

## II.6 Notes and Complements

The free Hamiltonians presented in this chapter are familiar objects to probabilists and analysts. A systematic study of their semigroups as convolution operators on the Lebesgue spaces can be found in Bochner's treatise [163]. The Lévy-Kintchine formula can be found in any probability text book, but it is more difficult to find its connection to positivity preserving semigroups. We found it in the second Appendix of Section 12 of Chapter XIII of the Reed and Simon series [272]. Volume II of this series [271] is recommended for the self-adjointness of Schrödinger operators as perturbations of the free Laplacians. The corresponding theory for the relativistic Hamiltonians was put on firm grounds by Herbst and Sloan in [146]. The refinements involving the $K_d$ class of potentials were introduced by Aizenman and Simon in [3]. See also [294] and the extension by Carmona, Masters and Simon in [49] to the relativistic cases. The proof of the Faris Lavine essential self-adjointness result is based on commutator manipulations. Although we did not include it, the interested reader will find a complete account in [271]. The analysis of spherically symmetric Schrödinger operators, and the reduction of the problem to the analysis of a family of Schrödinger operators on the half line $(0, \infty)$ is classical. We followed the lines of the second volume of the Reed and Simon series. Our discussion of the Birman-Kato theory is borrowed from Simon's set of notes on Trace Ideals [288]. This theory is marginal to the subject of the book but we decided to include it because of the recent results of Simon and Spencer which we present in the next chapter. The absolutely continuous spectrum of random Hamiltonians is not very well understood and it is not clear yet what are the relevant tools. We chose one. Subsequent developments will tell if it was a good choice.

Path integral formulas played an important role in the investigations of the spectral properties of Schrödinger operators. They are very powerful in guiding the intuition and in leading to many mathematical proofs free of irrelevant technical assumptions. We refer the reader to Simon's book [289] or his review article [294] or to Carmona's review article [48] and to the forthcoming [47] for the generalizations to relativistic operators and Lévy processes. The Feynman-Kac formula for the discrete Schrödinger operators is well known to probabilists. It was systematically used in the investigation of the integrated density of states by Fukushima and his collaborators. See for example [112], [113] and [114]. But the use of path integrals as expectations over the distribution of a stochastic process in the study of the unitary group rather than the semigroup is not standard for the Feynman integrals do not fit in this framework. We learn the possibility to prove and use these formulas in the discrete case from private discussions with S.A. Molcanov in 1981. See also [243] and [65]. We use these path integral formulas to prove the regularity properties of the semigroup and to estimate the decay (or the growth) of the eigenfunctions. We followed the approach of [41] and [49]. The results we presented could have been proved by purely functional analytic methods. Our bias toward probabilistic techniques is responsible for our choice. But the path integral formulas will be needed later on when we investigate the integrated density of states of random systems and some of its asymptotic properties (such as the Lifschitz tails) in Chapter VI.

As we explained in the introduction, Theorem II.4.5 will play a crucial role in the study of the localization phenomenon. It is a result from the theory of eigenfunction expensions of self-adjoint operators. The latter is presented in detail in Berezanskii monograph [28]. The relevant facts have been reset in a nice form by Faris in [91]. We followed the discussion given by Simon in [294].

# III

# ONE-DIMENSIONAL SCHRÖDINGER OPERATORS

Contents:



This chapter is devoted to the investigation of the spectral properties of one-dimensional Schrödinger operators. They deserve a special treatment for the theory of ordinary differential equations ( O.D.E. for short) and its lattice analog, i.e. the theory of finite difference equations, add to the menagerie of tools that can be used. The main new input is the systematic use of the properties of the propagators of the O.D.E. or the transfer matrices in the lattice case. This will be particularly important in the random case which we will study in the following chapters. We also add a special section on quasi one-dimensional systems to discuss some models which are not formally one-dimensional but which can be treated by the same techniques. They usually go under the name of ***band*** or ***strip***.

# III.1 The Continuous Case

Throughout this section $V$ denotes a real valued measurable locally square integrable function on $\mathbb{R}$.

## III.1.1 Essential Self-adjointness

Whenever we are dealing with the one-dimensional case we will use the letter "$t$" instead of the letter "$x$" for the space variable. This abuse of notation is very convenient for it avoids conflicts with the standard notations and the terminology of the theory of O.D.E. and later on with those of the theory of stochastic processes. Indeed, when we are dealing with random potential functions, the space variable of the Schrödinger operator will be the parameter of a stochastic process and probabilists are used to the letter $t$ in this situation.

We assume that the potential function $V$ is locally square integrable. Consequently, one can define the Schrödinger operator $H$ on $C_c^\infty(\mathbb{R})$ by:

$$Hf = H_0 f + Vf = -\ddot{f} + Vf .$$

Note that we use a dot to denote the derivation with respect to the variable $t$. The operator $H$ is obviously symmetric and its deficiency indices are equal because $H$ commutes with complex conjugation. Moreover, if $g \in \mathcal{D}(H^*)$, we have $< -\ddot{f} + Vf, g > = < f, H^* g >$ for all $f \in C_c^\infty(\mathbb{R})$ from which one can get that $\mathcal{D}(H^*)$ is the set of $f \in \mathcal{H} = L^2(\mathbb{R}, dt)$ such that $-\ddot{f} + Vf$ (the second derivative being computed in the sense of distributions) is a square integrable function. Consequently, the values of the deficiency indices and the existence of self-adjoint extensions of $H$ depend strongly on the existence and the properties of the $L^2$-solutions of the O.D.E.

$$-\ddot{f} + Vf = \pm i f .$$

More generally we shall need to understand the properties of the solutions of the eigenvalue equation:

$$-\ddot{f} + Vf = \lambda f \tag{III.1}$$

with $\lambda \in \mathbb{C}$. The solutions of this equation form a two-dimensional vector space. The notion of the Wronskian of two functions gives a convenient way to check the independence of two such solutions.

**Definition III.1.1** *The Wronskian of two differentiable complex functions $u$ and $v$, say $W(u,v)$, is defined as the function $[W(u,v)](t) = u(t)\dot{v}(t) - \dot{u}(t)v(t)$.*

It is very easy to check that the Wronskian of two solutions of the equation (III.1) is a constant function and that this constant is nonzero if and only if the two solutions in question are linearly independent.

The first important result in this respect is the following:

**Lemma III.1.2** *If $\Im m\,\lambda \neq 0$ then at least one solution of the eigenvalue equation (III.1) is square integrable near $+\infty$ and at least one solution is square integrable near $-\infty$. Moreover, if for some $\lambda \in \mathbb{C}$ two linearly independent solutions of (III.1) are square integrable near $+\infty$ (resp. $-\infty$), then for all $\lambda \in \mathbb{C}$ all the solutions of (III.1) are square integrable near $+\infty$ (resp. $-\infty$).*

*Proof:*
The first claim will be proved below in our discussion of the $m$-functions. In order to prove the second claim, let $\lambda_0 \in \mathbb{C}$ and let $\varphi_1$ and $\varphi_2$ be two linearly independent solutions of the eigenvalue equation (III.1) with $\lambda = \lambda_0$ which are square integrable near $+\infty$. We can assume without any loss of generality that their Wronskian is 1. If $\varphi$ is a solution of (III.1) with $\lambda = \lambda_1$, then we have:

$$[-\frac{d^2}{dt^2} + V - \lambda_0]\varphi = (\lambda_1 - \lambda_0)\varphi\,,$$

and the variation of the constants formula gives:

$$\varphi(t) = c_1\varphi_1(t) + c_2\varphi_2(t) + (\lambda_1 - \lambda_0)\int_c^t [\varphi_1(t)\varphi_2(s) - \varphi_1(s)\varphi_2(t)]\varphi(s)\,ds$$

for some constants $c_1$ and $c_2$. Using the notation $||f||_{[a,b]} = [\int_a^b |f(s)|^2 ds]^{1/2}$ and Schwarz inequality one gets:

$$|\varphi(t)| \leq |c_1|\,|\varphi_1(t)| + |c_2|\,|\varphi_2(t)| + M|\lambda_1 - \lambda_0|\,(|\varphi_1(t)| + |\varphi_2(t)|)\,||\varphi||_{[c,t]}$$

provided that $M$ is chosen so that $||\varphi_1||_{[c,\infty)} < M$ and $||\varphi_2||_{[c,\infty)} < M$. Hence we have:

$$||\varphi||_{[c,t]} \leq (|c_1| + |c_2|)M + 2M^2|\lambda_1 - \lambda_0|\,||\varphi||_{[c,t]},$$

from which we get:

$$||\varphi||_{[c,t]} \leq 2(|c_1| + |c_2|)M,$$

provided $c$ is chosen large enough so that $4\,|\lambda_1 - \lambda_0|M^2 < 1$. This gives the desired result since the right hand side is independent of $t$. ■

The above result is at the origin of the following important definition:

**Definition III.1.3** *The potential function $V$ is said to be in the limit circle case at $+\infty$ (resp. $-\infty$) if for some, and hence all $\lambda \in \mathbb{C}$, all the solutions of the eigenvalue equation (III.1) are square integrable near $+\infty$ (resp. $-\infty$). Otherwise it is said to be in the limit point case.*

The terminology *limit-point case, limit-circle case* will be justified below. We shall make the following typical assumption:

$$V(t) \geq -a(t^2 + 1) \tag{III.2}$$

for some $a > 0$ and $|t|$ large enough. This assumption (III.2) gives the essential self-adjointness of the operator $H$ because of the Faris-Lavine sufficient condition for essential self-adjointness given in Proposition II.2.4. We shall denote by the same symbol $H$ the unique self adjoint extension. Since we did not prove PropositionII.2.4 and since we present the O.D.E. approach to the spectral theory of the operators in the continuous case of $\mathbb{R}$, we give a proof of this self-adjointness result using O.D.E. techniques. In fact, the main consequence of our assumption (III.2) is given by the following elementary result.

**Lemma III.1.4** *Assumption (III.2) implies that for any $\lambda \in \mathbb{C}$, any solution $\psi$ of the eigenvalue equation (III.1) which is square integrable near $+\infty$ (resp. $-\infty$) necessarily satisfies:*

$$\int^{+\infty} \frac{|\dot{\psi}(t)|^2}{t^2}\, dt < \infty \quad (\text{resp.} \int_{-\infty} \frac{|\dot{\psi}(t)|^2}{t^2}\, dt < \infty)\,. \tag{III.3}$$

*Proof:*
Our assumptions imply that:

$$\int_{t_1}^{t} \frac{\ddot{\psi}(s)\psi(s)}{s^2} ds \geq -c \int_{t_1}^{\infty} |\psi(s)|^2 \, ds$$

for some positive constants $c$ and $t_1 < t$, provided they are large enough. Integrating the left hand side by parts we get:

$$-\frac{\dot{\psi}(t)\psi(t)}{t^2} + \int_{t_1}^{t} \frac{\dot{\psi}(s)}{s^2} ds - 2\int_{t_1}^{t} \frac{\dot{\psi}(s)\psi(s)}{s^3} ds \leq c' \qquad \text{(III.4)}$$

for some constant $c' > 0$. Using Schwarz inequality to control the right most integral of (III.4) we get:

$$-\frac{\dot{\psi}(t)\psi(t)}{t^2} + h(t) - c_2\sqrt{h(t)} \leq c_1$$

if we set:

$$h(t) = \int_{t_1}^{t} \frac{|\dot{\psi}(s)|^2}{s^2} \, ds\,.$$

Here $c_1$ and $c_2$ are two positive constants. We now conclude that $h(t)$ cannot converge to $+\infty$ when $t \to \infty$ for this would imply that $\dot{\psi}(t)\psi(t) \geq t^2 h(t)/2$ for $t$ large enough and consequently that $\dot{\psi}(t)$ and $\psi(t)$ have the same sign for $t$ large enough which would contradict the square integrability of $\psi$. ∎

**Corollary III.1.5** *Assumption (III.2) implies that for any $\lambda \in \mathbb{C}$ one cannot have two linearly independent solutions of (III.1) which are both square integrable near $+\infty$ (resp. $-\infty$).*

*Proof:*
Let $\varphi$ and $\psi$ be linearly independent solutions of (III.1). We assume that they are normalized so that their Wronskian $W(\varphi, \psi)$ is the constant 1. Since the left hand side of:

$$\frac{1}{t} = \varphi(t)\frac{\dot{\psi}(t)}{t} - \psi(t)\frac{\dot{\varphi}(t)}{t}$$

is not integrable, Schwarz inequality and (III.3) imply that $\varphi$ and $\psi$ cannot both be square integrable near $+\infty$. ∎

The above discussion immediately implies:

**Corollary III.1.6** *Assumption (III.2) implies that the operator $H$ is essentially self-adjoint on $C_c^\infty(\mathbb{R})$.*

A different sufficient condition is given in Problem III.6.1. The following lemma is a convenient tool to determine the spectrum of the operator $H$.

**Lemma III.1.7** *If we assume (III.2) and if $\lambda \in \mathbb{R}$ is such that all the solutions of the eigenvalue equation (III.1) and their derivatives are bounded in a neighborhood of $+\infty$ (resp. $-\infty$), then $\lambda \in \Sigma(H)$.*

*Proof:*
Corollary III.1.5 gives the existence of a solution $\varphi$ of (III.1) which is not square integrable near $+\infty$. The proof in the case of $-\infty$ would be similar. Now let $\varphi_n = \varphi g_n / ||\varphi g_n||$ where $g_n \in C_c^\infty(\mathbb{R})$ is supported in $[0, n+1]$, is identically 1 on $[1, n]$ and is such that

$$\sup_{t \in \mathbb{R}, n \geq 1} \max(|g_n(t)|, |\dot{g}_n(t)|, |\ddot{g}_n(t)|) < \infty.$$

Then we have:

$$-\ddot{\varphi}_n(t) + [V(t) - \lambda]\varphi_n(t) = \frac{1}{||\varphi g_n||}[-\ddot{g}_n(t)\varphi(t) - 2\dot{g}_n(t)\dot{\varphi}(t)]$$

so that the $L^2$-norm of $-\ddot{\varphi}_n(t) + [V(t) - \lambda]\varphi_n(t)$ is bounded above by a constant divided by the $L^2$-norm of $\varphi g_n$. The latter goes to $\infty$ as $n \to \infty$ by our choice of $\varphi$ and our construction of the $g_n$'s. Hence we can use Weyl's condition for a point to be in the spectrum (recall Problem I.7.4 of Chapter I) to conclude that $\lambda \in \Sigma(H)$. ∎

It will happen fairly often that one solution of the eigenvalue equation decays exponentially at $+\infty$ or $-\infty$. This is typically the case when the energy $\lambda$ is not in the spectrum of the operator. In any case one can use the Wronskian to show that every other linearly independent solution has to explode exponentially. This situation is usually refered to as *exponential dichotomy* and we shall come back to it in Chapter VII when we study periodic potentials. The notion of *subordinate solution* which we introduce now is more general and it will be shown that it is the right tool to characterize the various spectra in terms of the asymptotic behavior of the solutions of the eigenvalue equation.

**Definition III.1.8** *A nonzero solution $\varphi$ of the eigenvalue equation (III.1) is said to be subordinate at $+\infty$ if any solution $\psi$ which is linearly independent of $\varphi$ satisfies:*

$$\lim_{t \to +\infty} \frac{||\varphi||_{[0,t]}}{||\psi||_{[0,t]}} = 0.$$

*The notion of subordinacy of a solution at $-\infty$ is defined similarly.*

As a general rule, the spectral characteristics of the Schrödinger operators on the whole space will be approximated by the corresponding quantities in bounded domains. For this reason we introduce the operators in bounded intervals.

### III.1.2 The Operator in an Interval

The following notation will help us deal with the boundary conditions of operators in bounded intervals. If $f$ is a continuously differentiable function we set $\tilde{f}$ for the vector valued function defined by $\tilde{f}(t) = (\dot{f}(t), f(t))$ or:

$$\tilde{f}(t) = \begin{bmatrix} f(t) \\ \dot{f}(t) \end{bmatrix}$$

in matrix notation. If $x$ and $y$ are vectors in $\mathbb{R}^2$ we will denote by $x \sim y$ the fact that $x$ and $y$ correspond to the same element in the projective space $P(\mathbb{R}^2)$. We will also use the notation $x \sim \alpha$ when $x \in \mathbb{R}^2$ is a representative of the equivalence class $\alpha \in P(\mathbb{R}^2)$.

For all real numbers $a$ and $b$ such that $-\infty < a < b < +\infty$ we denote by $\Lambda$ the interval $\Lambda = [a, b]$. We will call it a *box* by analogy with the multidimensional case. Eventhough it will play no essential role in what follows, it will be convenient to assume from time to time that the box $\Lambda$ contains the origin. Now, for any elements $\alpha$ and $\beta$ of the projective space $P(\mathbb{R}^2)$ we define the operator $H^\Lambda_{\alpha,\beta}$ on the domain $\mathcal{D}(H^\Lambda_{\alpha,\beta}) = \{f \in L^2(\Lambda, dt);\ f$ is differentiable, $\dot{f}$ is absolutely continuous, $-\ddot{f} + Vf \in L^2(\Lambda, dt)$ and $\tilde{f}(a) \sim \alpha$ and $\tilde{f}(b) \sim \beta\}$ by the formula:

$$H^\Lambda_{\alpha,\beta} f = -\ddot{f} + Vf\,.$$

Notice that the boundary conditions $\tilde{f}(a) \sim \alpha$ and $\tilde{f}(b) \sim \beta$ take the more familiar form $f(a)\cos\alpha - \dot{f}(a)\sin\alpha = f(b)\cos\beta - \dot{f}(b)\sin\beta = 0$ when we identify the projective space with the interval $[0, \pi)$ and we use the parametrization giving $(\sin\alpha, \cos\alpha) \sim \alpha$. In any case, the operator $H^{\Lambda}_{\alpha,\beta}$ is a self-adjoint operator and its resolvent operator is compact. There are many ways to prove this claim. The simplest one is certainly to check that the resolvent operator is of Hilbert-Schmidt type because it has a bounded kernel. Indeed, its kernel is given by formula (III.9) below. Another possibility is to use the Sturm-Liouville oscillation theory given in problems III.6.2 and III.6.3 at the end of the chapter. Still another one is to use path integral methods; recall for example the discussion of the operators in bounded domains in Section II.3. In any case, the spectrum is a bounded below discrete set. Let us denote by:

$$\lambda_0 < \lambda_1 \leq \lambda_2 \leq \cdots$$

the eigenvalues and by:

$$f_0, f_1, f_2, \cdots$$

a corresponding complete orthonormal system of eigenfunctions. Note that the multiplicity of each of the eigenvalues is 1 for the space of solutions of the eigenvalue equation (III.1) is two dimensional and $a \neq b$. We then set:

$$\begin{aligned} \xi^{\Lambda}_{\alpha,\beta} &= \sum_{j\geq 0} |f_j(0)|^2 \, \delta_{\lambda_j} \\ \zeta^{\Lambda}_{\alpha,\beta} &= \sum_{j\geq 0} |\dot{f}_j(0)|^2 \, \delta_{\lambda_j} \\ \eta^{\Lambda}_{\alpha,\beta} &= \sum_{j\geq 0} f_j(0)\overline{\dot{f}_j}(0) \, \delta_{\lambda_j} \end{aligned} \tag{III.5}$$

where we used the notation $\delta_\lambda$ for the Dirac measure at $\lambda$. Notice that one can always write:

$$f_j(\cdot) = f_j(0)\, y_1(\cdot\,, \lambda_j) + \dot{f}_j(0)\, y_2(\cdot\,, \lambda_j)$$

where for each $\lambda \in \mathbf{C}$ the function $t \hookrightarrow y_1(t,\lambda)$ (resp. $t \hookrightarrow y_2(t,\lambda)$) is the unique solution of the equation:

$$-\ddot{y}(t) + [V(t) - \lambda]y(t) = 0 \tag{III.6}$$

which satisfies:

$$y_1(0,\lambda) = 1 \qquad \text{and} \qquad \dot{y}_1(0,\lambda) = 0 \tag{III.7}$$

(resp.

$$y_2(0,\lambda) = 0 \qquad \text{and} \qquad \dot{y}_2(0,\lambda) = 1). \tag{III.8}$$

For each $\lambda \in \mathbb{C}$ the set $\{y_1(\cdot,\lambda), y_2(\cdot,\lambda)\}$ is a natural basis of the vector space of solutions of the eigenvalue equation (III.1). Now, for each $\lambda \in \mathbb{C}$, we set:

$$G^{\Lambda}_{\alpha,\beta}(\lambda, s, t) = \frac{1}{W(u,v)} \begin{cases} u(s)v(t) & \text{whenever} \quad a \leq s \leq t \leq b \\ u(t)v(s) & \text{whenever} \quad a \leq t \leq s \leq b \end{cases} \tag{III.9}$$

where $u$ denotes any solution of (III.1) which satisfies the boundary condition at $a$, i.e. which satisfies $\tilde{u}(a) \sim \alpha$, and $v$ denotes any solution of (III.1) which satisfies the boundary condition at $b$, i.e. which satisfies $\tilde{v}(b) \sim \beta$. Note that the above definition makes sense for all the $\lambda \in \rho(H^{\Lambda}_{\alpha,\beta})$. Indeed, it is easy to check that, if $\tilde{u}(a) \sim \alpha$ and $\tilde{v}(b) \sim \beta$, then $W(u,v) = 0$ if and only if $\lambda$ is an eigenvalue of $H^{\Lambda}_{\alpha,\beta}$. Consequently, $G^{\Lambda}_{\alpha,\beta}(\lambda, s, t)$ is a jointly continuous function of $(s,t)$ in $\Lambda \times \Lambda$ as long as $\lambda$ is in the resolvent set of $H^{\Lambda}_{\alpha,\beta}$. Also, the definition of $G^{\Lambda}_{\alpha,\beta}(\lambda, s, t)$ implies the existence and the continuity of $\frac{\partial^2}{\partial s \partial t} G^{\Lambda}_{\alpha,\beta}(\lambda, s, t)$ on $\Lambda \backslash \{(s,t) \in \Lambda \times \Lambda;\ s = t\}$. In particular, for each fixed $s \in \Lambda$ the function $t \hookrightarrow G^{\Lambda}_{\alpha,\beta}(\lambda, s, t)$ is continuously differentiable on $\Lambda \backslash \{s\}$ and its derivative has a discontinuity of the first kind at the point $t = s$ with a jump of size 1. Similarly for $G^{\Lambda}_{\alpha,\beta}(\lambda, s, t)$ when regarded as a function of $s \in \Lambda \backslash \{t\}$ for $t$ fixed. It is remarkable that the second derivative $(s,t) \hookrightarrow \frac{\partial^2}{\partial s \partial t} G^{\Lambda}_{\alpha,\beta}(\lambda, s, t)$ is continuous through the diagonal and we will use the notation $\frac{\partial^2}{\partial s \partial t} G^{\Lambda}_{\alpha,\beta}(\lambda, s, s)$ eventhough we know that this second order derivative does not exist on the diagonal.

Moreover, if $\lambda$ is not an eigenvalue, then it is easy to check that, for each $f \in L^2(\Lambda, dt)$ the function $u$ defined by

$$u(s) = \int_{\Lambda} G^{\Lambda}_{\alpha,\beta}(\lambda, s, t) f(t)\, dt$$

satisfies the boundary conditions appearing in the definition of the domain of the operator $H^{\Lambda}_{\alpha,\beta}$ and that it is a solution of the equation:

$$[H^{\Lambda}_{\alpha,\beta} - \lambda I]u = f.$$

This shows that $G^{\Lambda}_{\alpha,\beta}(\lambda, s, t)$ is the kernel of the resolvent operator of $H^{\Lambda}_{\alpha,\beta}$ for the energy $\lambda$. The function $G^{\Lambda}_{\alpha,\beta}(\lambda, s, t)$ is called the *Green's function* of the problem (III.1) with boundary conditions given at the endpoints $a$ and $b$ by the angles $\alpha$ and $\beta$. The functional calculus tells us that the kernel of the resolvent operator should also be given by:

$$[H^{\Lambda}_{\alpha,\beta} - zI]^{-1}(s,t) = \sum_{j \geq 0} \frac{f_j(s)\, f_j(t)}{\lambda_j - z}$$

where the series on the right hand side converges in the space $L^2(\Lambda \times \Lambda, ds\,dt)$. Notice that the summands are jointly continuous and that the sum is equal to $G^{\Lambda}_{\alpha,\beta}(\lambda, s, t)$ which is jointly continuous. Consequently, this series expansion is uniformly convergent. See Problem II.5.3.

If $f \in L^2(\Lambda, dt)$ we also set:

$$[U^{\Lambda}_{\alpha,\beta}]f(\lambda) = \left( \int_a^b f(t)y_1(t,\lambda)dt, \int_a^b f(t)y_2(t,\lambda)dt \right).$$

The operator $U^{\Lambda}_{\alpha,\beta}$ defined in this way is a unitary isomorphism of $L^2(\Lambda, dt)$ and $L^2(\mathbb{R}, dM^{\Lambda}_{\alpha,\beta})$ where $M^{\Lambda}_{\alpha,\beta}$ is the matrix measure:

$$M^{\Lambda}_{\alpha,\beta} = \begin{bmatrix} \xi^{\Lambda}_{\alpha,\beta} & \eta^{\Lambda}_{\alpha,\beta} \\ \eta^{\Lambda}_{\alpha,\beta} & \zeta^{\Lambda}_{\alpha,\beta} \end{bmatrix}$$

The operator $U^{\Lambda}_{\alpha,\beta}$ transforms the operator $H^{\Lambda}_{\alpha,\beta}$ into the operator of multiplication by the variable $\lambda$. Consequently, if one uses the terminology of SectionI.4 of Chapter I, $U^{\Lambda}_{\alpha,\beta}$ realizes the canonical form of $H^{\Lambda}_{\alpha,\beta}$.

It may seems very strange to use such an unnecessarilyy complicated spectral representation for an operator with a simple spectrum. Indeed, we did not have to construct a canonical representation over an $L^2$-space with respect to a matrix measure since the multiplicity of the operator $H^{\Lambda}_{\alpha,\beta}$ is one. The justification comes from our desire to study the operator on the whole real line. Indeed, such an operator has multiplicity two and we have to use a representation on an $L^2$-space with respect to a matrix measure. We show below that this representation can be obtained via a limiting procedure from the representation of $H^{\Lambda}_{\alpha,\beta}$ which we constructed above.

It is also interesting to notice that the spectral representation which we discussed above is not exactly of the type of those which we introduced in Chapter I. Indeed, we now consider a matrix measure $M = [M_{i,j}]_{i,j=1,2}$ of the form $M_{i,j}(d\lambda) =< E(d\lambda)\varphi_i, \varphi_j >$ with $\varphi_1 = \delta_0$ and $\varphi_2 = \dot{\delta}_0$ while we previously required that the vectors $\varphi_1$ and $\varphi_2$ be in the Hilbert space $\mathcal{H}$ on which the operator is defined. This can be fully justified by first taking $\varphi_1 = \varphi_\epsilon$ and $\varphi_2 = \dot{\varphi}_\epsilon$ where $\{\varphi_\epsilon; \epsilon > 0\}$ is an approximate identity with compact support and then letting $\epsilon \to 0$. Note also that our using the generalized functions $\varphi_1 = \delta_0$ and $\varphi_2 = \dot{\delta}_0$ instead of elements of the Hilbert space $\mathcal{H} = L^2(\Lambda, dt)$ accounts for the fact that the spectral measures entering in the spectral matrix measure $M$ are not finite but merely locally finite (i.e. Radon).

As we already pointed out, most of the information contained in the spectral matrix measure $M^{\Lambda}_{\alpha,\beta}$ can also be found in the trace-measure $\sigma^{\Lambda}_{\alpha,\beta}$ defined by:

$$\sigma^{\Lambda}_{\alpha,\beta} = \xi^{\Lambda}_{\alpha,\beta} + \zeta^{\Lambda}_{\alpha,\beta}\,.$$

This measure can be recovered from its Stieltjes transform as the next result shows:

**Proposition III.1.9** *For any $z \in \mathbb{C}$ with $\Im m z > 0$ one has:*

$$\int_{\mathbb{R}} \frac{d\sigma^{\Lambda}_{\alpha,\beta}(\lambda)}{\lambda - z} = G^{\Lambda}_{\alpha,\beta}(z,0,0) + \frac{\partial^2 G^{\Lambda}_{\alpha,\beta}}{\partial s \partial t}(z,0,0)\,. \tag{III.10}$$

*Proof:*
Notice that the definition of the spectral measures $\sigma^{\Lambda}_{\alpha,\beta}$, $\xi^{\Lambda}_{\alpha,\beta}$ and $\zeta^{\Lambda}_{\alpha,\beta}$ and the properties of the Green's functions which we discussed above imply that:

$$\int_{\mathbb{R}} \frac{d\xi^{\Lambda}_{\alpha,\beta}(\lambda)}{\lambda - z} = \sum_{j\geq 0} \frac{|f_j(0)|^2}{\lambda_j - z} = G^{\Lambda}_{\alpha,\beta}(z,0,0)$$

and :

$$\int_{\mathbb{R}} \frac{d\zeta^{\Lambda}_{\alpha,\beta}(\lambda)}{\lambda - z} = \sum_{j\geq 0} \frac{|\dot{f}_j(0)|^2}{\lambda_j - z} = \frac{\partial^2 G^{\Lambda}_{\alpha,\beta}}{\partial s \partial t}(z,0,0)\,. \blacksquare$$

### III.1.3 Green's and Weyl-Titchmarsh's Functions

This subsection is devoted to the introduction of the so-called Titchmarsch-Weyl functions $m$ and to the investigation of their connections with the Green's function of the Schrödinger operator $H$ on a half axis or on the whole real line $\mathbb{R}$.

Let $\alpha \in [0,\pi)$ be fixed and let $\varphi_\lambda$ be the solution of the eigenvalue equation (III.1) satisfying:

$$\varphi_\lambda(a) = \cos\alpha \qquad \text{and} \qquad \dot{\varphi}_\lambda(a) = -\sin\alpha,$$

and $\psi_\lambda$ the solution of (III.1) satisfying:

$$\psi_\lambda(a) = \sin\alpha \qquad \text{and} \qquad \dot{\psi}_\lambda = \cos\alpha.$$

The functions $\varphi_\lambda(t)$ and $\psi_\lambda(t)$ are real valued, jointly continuous in $(\lambda, t)$, and analytic in $\lambda$ for $t$ fixed. They are chosen so that they satisfy linearly independent boundary conditions at the endpoint $a$ and so that their Wronskian is constant and equal to 1. Also, the solution $\psi_\lambda$ satisfies the left boundary condition which we use for the Schrödinger operator in an interval with left hand point $a$. Now, every solution $\chi$ of the eigenvalue equation (III.1), except $\psi_\lambda$, is of the form $\chi = \varphi_\lambda + m\psi_\lambda$ for some complex number $m$ dependent on $\chi$ and $\lambda$. For each $b \in (a, \infty)$ and $\beta \in (0, \pi)$ we define the number $m = m(\lambda, b)$ as the unique $m$ for which the solution $\chi$ satisfies the boundary condition $\tilde{\chi(b)} \sim -\beta$, i.e.:

$$\chi(b) \cos\beta + \dot{\chi}(b) \sin\beta = 0 .$$

One sees immediately that:

$$m(\lambda, b) = -\frac{\dot{\varphi}_\lambda(b) + \varphi_\lambda(b)\cot\beta}{\dot{\psi}_\lambda(b) + \psi_\lambda(b)\cot\beta} . \tag{III.11}$$

Notice that, for fixed $b \in (a, \infty)$, $m(\lambda, b)$ is a meromorphic function of $\lambda$ which is real for real $\lambda$. It is convenient to set $z = \cot\beta$ and to write:

$$m(\lambda, b) = -\frac{Az + B}{Cz + D} . \tag{III.12}$$

One can now consider $\lambda \in \mathbb{C}$ and $b \in (a, \infty)$ fixed, in which case the complex numbers $A = \varphi_\lambda(b)$, $B = \dot{\varphi}_\lambda(b)$, $C = \psi_\lambda(b)$ and $D = \dot{\psi}_\lambda(b)$ are also fixed, and let $\beta$ vary from 0 to $\pi$ in which case $z$ varies through the whole real line. Formula (III.12) says that $m$ has to be on the circle given by (III.12) in order for the solution $\chi$ to satisfy the boundary condition at $b$. Inverting formula (III.12) gives the equation of the circle:

$$z = -\frac{B + Dm}{A + Cm} \tag{III.13}$$

and simple algebra gives the center:

$$m_0(\lambda, b) = \frac{A\overline{D} - B\overline{C}}{\overline{C}D - C\overline{D}} = -\frac{W(\varphi_\lambda, \overline{\psi_\lambda})(b)}{W(\psi_\lambda, \overline{\psi_\lambda})(b)} \tag{III.14}$$

and the radius:

$$r(\lambda, b) = \frac{|AD - BC|}{|\overline{C}D - C\overline{D}|} = \frac{1}{|W(\psi_\lambda, \overline{\psi_\lambda})(b)|} . \tag{III.15}$$

The interior of the circle $C(\lambda, b)$ is given by:

$$\frac{W(\chi,\chi)}{W(\psi_\lambda,\psi_\lambda)} < 0\,,$$

and if we recall Green's formula:

$$\int_{t_1}^{t_2} [(Hf)\overline{g} - f\overline{Hg}]\,dt = [W(f,g)]_{t_1}^{t_2}$$

we obtain:

$$W(\psi_\lambda,\psi_\lambda)(b) = 2i\Im m\,\lambda \int_a^b |\psi_\lambda(t)|^2\,dt$$

and:

$$W(\chi,\chi)(b) = -2i\Im m\,m(\lambda,b) + 2i\Im m\,\lambda \int_a^b |\chi(t)|^2\,dt\,.$$

Consequently, for $\Im m\,\lambda \neq 0$ we have:

$$m \in C(\lambda,b) \qquad \Longleftrightarrow \qquad \int_a^b |\chi(t)|^2\,dt = \frac{\Im m\,m}{\Im m\,\lambda} \tag{III.16}$$

and:

$$m \text{ in the interior of } C(\lambda,b) \qquad \Longleftrightarrow \qquad \int_a^b |\chi(t)|^2\,dt < \frac{\Im m\,m}{\Im m\,\lambda}\,. \tag{III.17}$$

Notice also that:

$$r(\lambda,b) = \frac{1}{2\,\Im m\,\lambda \int_a^b |\psi_\lambda(t)|^2\,dt}\,.$$

We now draw a few important consequences of the above considerations.

- The characterization (III.17) of the interior of the circle $C(\lambda, b)$ shows that the circle $C(\lambda, b_2)$ is contained in the circle $C(\lambda, b_1)$ whenever one has $a < b_1 < b_2 < \infty$. Thus, for any nonreal $\lambda$, say $\Im m\,\lambda > 0$ for example, the circle $C(\lambda, b)$ converges as $b \to \infty$, either to a circle $C(\lambda)$ or to a point $m(\lambda)$.
- In any case, if $\chi = \varphi_\lambda + m\psi_\lambda$ with $m = m(\lambda)$ or $m \in C(\lambda)$ then for any $b > a$ one has:

$$\int_a^b |\chi(t)|^2\,dt < \frac{\Im m\,m}{\Im m\,\lambda}$$

because $m$ is interior to the circle $C(\lambda, b)$ and this implies that:

$$\int_a^\infty |\chi(t)|^2 \, dt < \frac{\Im m \, m}{\Im m \, \lambda}$$

which in turn shows the existence of a solution of the eigenvalue equation (III.1) which is square integrable near $+\infty$.

- When $C(\lambda, b)$ converges to a circle, the above argument can be used to construct square integrable solutions and it shows that all the solutions of the eigenvalue equation (III.1) are square integrable near $\infty$. On the other hand, when $C(\lambda, b)$ converges to a point, its radius converges to 0 and consequently:

$$\int_a^\infty |\psi_\lambda(t)|^2 \, dt = \infty$$

which shows that the space of solutions of (III.1) is one-dimensional. This is the justification of the terminology limit circle and limit point.

- One can take the limit $b \to \infty$ in the characterization of the circle $C(\lambda, b)$ and one gets in this way a simple characterization of the limit circle when the latter occurs:

$$m \in C(\lambda) \Longleftrightarrow \int_a^\infty |\varphi_\lambda(t) + m\psi_\lambda(t)|^2 \, dt = \frac{\Im m \, m}{\Im m \, \lambda}$$

or equivalently:

$$m \in C(\lambda) \Longleftrightarrow W(\varphi_\lambda + m\psi_\lambda)(+\infty) = 0\,,$$

and of the limit point $m(\lambda)$ when the latter occurs:

$$m(\lambda) = -\lim_{b \to +\infty} \frac{\varphi_\lambda(b)}{\psi_\lambda(b)}\,.$$

The following result gives the analyticity of the $m$-function.

**Proposition III.1.10** *In the limit point case, the limit point $m(\lambda)$ is an analytic function of $\lambda$ in the upper half plane $\Pi_+$ and in the lower half plane $\Pi_-$. $\Im m \, m(\lambda) > 0$ for $\Im m \, \lambda > 0$ and if $m(\lambda)$ has zeros or poles on the real axis, the latter are simple.*

*Proof:*

Let us consider for example that $\lambda$ is restricted to a closed bounded subset $K$ of $\Pi_+$. By restricting the values of $b$ to $b > 1$ we make sure that all the meromorphic functions $m(\,\cdot\,, b, \beta)$ are bounded since they take values in the closed disk whose boundary is $C(\lambda, 1)$. Consequently they are analytic and form by Cauchy's theorem an equicontinuous family. Hence they converge uniformly in $K$. The limit $m$ is thus analytic and satisfies $\Im m\, m > 0$ on $K$. This also implies that the poles of $\lambda \hookrightarrow m(\,\cdot\,)$ on the real axis, if any, are simple and that the corresponding residues are negative. ∎

In the above discussion of the $m$-function, the dependence of $m$ in the end point $a$ and the boundary condition $\alpha$ was not considered. From now on we will restrict the notation $m_\alpha(\lambda)$ to the case of $a = 0$ and the notation $m(\lambda)$ to the case $a = 0$ and $\alpha = \pi/2$. In this case we have $\varphi_\lambda = y_1(\cdot, \lambda)$ and $\psi_\lambda = y_2(\cdot, \lambda)$ and the $L^2$-solution at $+\infty$ becomes $\chi = y_1(\cdot, \lambda) + m y_2(\cdot, \lambda)$, from which one sees that:

$$m(\lambda) = \frac{\dot{\chi}(0)}{\chi(0)}\,. \tag{III.18}$$

Note also that whenever $\alpha = \pi/2$ one always has:

$$m_{a,\alpha=\pi/2}(\lambda) = \frac{\dot{\chi}(a)}{\chi(a)} \tag{III.19}$$

which makes it possible to recover one $m$-function from the other by following the solution $\chi$ between $a$ and 0. We shall explain below why, instead of changing the reading point $a$, we shall rather shift the potential function. Note also that the *old* $m$-function can be recovered from the new one for we have:

$$m_\alpha(\lambda) = \frac{\sin\alpha + m(\lambda)\cos\alpha}{\cos\alpha - m(\lambda)\sin\alpha}.$$

if we denote by $m_{a=0,\alpha}$ the $m$-function constructed above in the case $a = 0$. The $m$-functions are Herglotz functions and one can expect to be able to control their boundary behavior. In fact there is a interesting connection between this boundary behavior and the asymptotic behavior of the solutions of the eigenvalue equation as captured by the notion of subordinate solution. More precisely, we have:

**Proposition III.1.11** *If $\lambda \in I\!R$, the eigenvalue equation (III.1) has a subordinate solution at $+\infty$ if and only if, either the boundary limit $m(\lambda + i0)$ exists as a finite real number (in which case $\chi$ is subordinate at $+\infty$) or if* $\lim_{\epsilon \searrow 0} |m(\lambda + i\epsilon)| = +\infty$ *(in which case $\psi_\lambda$ is subordinate at $+\infty$).*

This result was proven by Gilbert and Pearson. See [121].

For each $t \in \mathbb{R}$ we define the *shift operator* $\mathcal{T}_t$ by:

$$\mathcal{T}_t V : s \hookrightarrow V(t+s) \tag{III.20}$$

and instead of changing the reading point $a$ in the definition of the $m$-function, we now shift the potential function by $a$ and we set $m(\lambda, a)$ for the $m(\lambda)$ computed for the potential function $\mathcal{T}_a V$ instead of $V$. This is the same thing as $m_{a,\alpha=\pi/2}(\lambda)$ which we previously used but the shift notation will be convenient if not necessary when we deal with stationary random potentials in the sequel. Consequently formula (III.19) can be rewritten in the form:

$$f(t, \lambda) = e^{\int_0^t m(s,\lambda)\, ds} \tag{III.21}$$

if we denote by $f$ the unique solution of the eigenvalue equation (III.1) which is square integrable near $+\infty$ and which is equal to 1 at the origin. Note also that this last relation implies that the $m$-function, as a function of $t$, is a solution of the Ricatti equation:

$$\dot{m}(t, \lambda) = V(t) - m(t, \lambda) - m(t, \lambda)^2 . \tag{III.22}$$

The next application which we give is the definition of the operator $H$ on the half-line $[a, \infty)$.

**The operator on a half-line**

Let us recall that, with the notation of the present section we have:

$$G_{\alpha,\beta}^{[a,b]}(\lambda, t, s) = \begin{cases} \psi_\lambda(t)[\varphi_\lambda(s) + m(\lambda, b, \beta)\psi_\lambda(s)] & \text{if } t \le s \\ \psi_\lambda(s)[\varphi_\lambda(t) + m(\lambda, b, \beta)\psi_\lambda(t)] & \text{if } t > s \end{cases} \tag{III.23}$$

and if we are in the limit point case, in the limit $b \to \infty$ the Green's function of the problem in the interval $[a, b]$ converges, irrespectively of the possible values of the angle $\beta$ to:

$$G_\alpha^{[a,\infty)}(\lambda, t, s) = \begin{cases} \psi_\lambda(t)[\varphi_\lambda(s) + m(\lambda)\psi_\lambda(s)] & \text{if } t \le s \\ \psi_\lambda(s)[\varphi_\lambda(t) + m(\lambda)\psi_\lambda(t)] & \text{if } t > s \end{cases} \tag{III.24}$$

A similar convergence holds for the second derivative of the Green's function. If moreover one recalls that the Stieltjes transform of the spectral measure $\sigma_{\alpha,\beta}^{[a,b]}$ is equal to the sum of the Green's function and its second derivative at $(0,0)$ we just showed that these Stieltjes transforms converge and so, the corresponding measures converge vaguely to a nonnegative measure $\sigma_\alpha^{[a,\infty)}$ whose Stieltjes transform is given by the following formula:

$$\int_{I\!R} \frac{\sigma_\alpha^{[a,\infty)}(d\lambda)}{\lambda - z} = G_\alpha^{[a,\infty)}(z,0,0) + \frac{\partial^2 G_\alpha^{[a,\infty)}}{\partial s \partial t}(z,0,0).$$

Notice that this implies the tightness (and hence the convergence) of the spectral measures $\xi_{\alpha,\beta}^{[a,b]}$ and $\zeta_{\alpha,\beta}^{[a,b]}$, and also of the spectral measure $\eta_{\alpha,\beta}^{[a,b]}$ by Schwarz inequality. In fact, the kernel (III.24) is the Green's function of the problem on the half-line $[a,\infty)$ with boundary condition $\alpha$ at the end point $a$. In other words, the operator $H_\alpha^{[a,\infty)}$ defined by $Hf = -\ddot{f} + Vf$ on the space of functions $f$ on $[a,\infty)$ which satisfy the boundary condition given by $\alpha$ at $a$ and which are identically 0 on an interval of the form $[t_0,\infty)$, is essentially self-adjoint and the resolvent operator of its unique self-adjoint extension which we will still denote by $H_\alpha^{[a,\infty)}$, say $[H_\alpha^{[a,\infty)} - zI]^{-1}$, has an integral kernel given by (III.24).

Note that in both cases, the value of the Green's function at $(s,t)$ with $s < t$ can be computed by multiplying the value of any solution of the eigenvalue equation (III.1) satisfying the boundary condition at the left hand point of the box by the value of any solution satisfying the boundary condition at the right hand point and dividing this product by the Wronskian of the two solutions. In this interpretation, the boundary condition at infinity is understood as the square integrability of the function.

This last fact is general and this special form of the Green's function holds in other cases: for example, it is still true in the case of the whole axis as we shall see later on.

We will use the notation $H_\alpha^+$, $G_\alpha^+$ and $\sigma_\alpha^+$ when $a = 0$. In this case the formulas are simpler and we state them for future reference.

$$G_\alpha^+(z,s,t) = \int_{I\!R} \frac{\psi_\lambda(s)\psi_\lambda(t)}{\lambda - z}\sigma_\alpha^+(d\lambda)$$

which becomes:

$$G_\alpha^+(z,s,t) \int_{I\!R} \frac{[y_1(s,\lambda)\sin\alpha + y_2(s,\lambda)\cos\alpha)][y_1(t,\lambda)\sin\alpha + y_2(t,\lambda)\cos\alpha)]}{\lambda - z}\sigma_\alpha^+(d\lambda)$$

in a different notation. From this it follows that:

$$m(z) = -\cot\alpha + \int_{I\!R} \frac{\sigma_\alpha^+(d\lambda)}{\lambda - z}$$

at least for $\alpha \neq 0$. Using the theorems of Fatou and de la Vallée Poussin recalled in the first chapter we can characterize the various spectra of the operator $H_\alpha^+$ up to sets of Lebesgue measure zero. Indeed one can describe, using the boundary behavior of the function $m$, the essential supports (with respect to Lebesgue measure) of the corresponding parts of the spectral measure $\sigma_\alpha^+$. We get:

**Proposition III.1.12** *The essential supports of the various parts of the spectral measure $\sigma_\alpha^+$ are given by:*

$$\begin{aligned}
&(i) && \text{ess-supp } \sigma_\alpha^+ = \{\lambda \in \mathbb{R};\ \Im m\, m(\lambda + i0) \text{ exists and } \\
& && \qquad 0 < \Im m\, m(\lambda + i0) \leq \infty\} \\
&(ii) && \text{ess-supp } \sigma_{\alpha,ac}^+ = \{\lambda \in \mathbb{R};\ \Im m\, m(\lambda + i0) \text{ exists and } \\
& && \qquad 0 < \Im m\, m(\lambda + i0) < \infty\} \\
&(iii) && \text{ess-supp } \sigma_{\alpha,s}^+ = \{\lambda \in \mathbb{R};\ \Im m\, m(\lambda + i0) = \infty\} \\
&(iv) && \text{ess-supp } \sigma_{\alpha,sc}^+ = \{\lambda \in \mathbb{R};\ \Im m\, m(\lambda + i0) = \infty \text{ and } \sigma_\alpha^+(\{\lambda\}) = 0\} \\
&(v) && \text{ess-supp } \sigma_{\alpha,pp}^+ = \{\lambda \in \mathbb{R};\ \Im m\, m(\lambda + i0) = \infty \text{ and } \sigma_\alpha^+(\{\lambda\}) > 0\}.
\end{aligned}$$

It is now possible to use the characterization of the existence of subordinate solutions in terms of the boundary behavior of the $m$-function given in Proposition III.1.11. We get the following equivalent characterizations of the essential supports of the various parts of the spectral measure $\sigma_\alpha^+$.

**Proposition III.1.13** *With the above notation we also have:*

$$\begin{aligned}
&(i) && \text{ess-supp } \sigma_\alpha^+ = \mathbb{R} \setminus \{\lambda \in \mathbb{R};\ \text{a subordinate solution exists which} \\
& && \qquad \text{does not satisfy the b.c. at } 0\} \\
&(ii) && \text{ess-supp } \sigma_{\alpha,ac}^+ = \{\lambda \in \mathbb{R};\ \text{no subordinate solution exists}\} \\
&(iii) && \text{ess-supp } \sigma_{\alpha,s}^+ = \{\lambda \in \mathbb{R};\ \text{a subordinate solution exists which} \\
& && \qquad \text{satisfies the b.c. at } 0\} \\
&(iv) && \text{ess-supp } \sigma_{\alpha,sc}^+ = \{\lambda \in \mathbb{R};\ \text{a subordinate solution exists which} \\
& && \qquad \text{satisfies the b.c. at } 0 \text{ but which is not square integrable at } +\infty\} \\
&(v) && \text{ess-supp } \sigma_{\alpha,pp}^+ = \{\lambda \in \mathbb{R};\ \text{a subordinate solution exists which} \\
& && \qquad \text{satisfies the b.c. at } 0 \text{ and which is square integrable at } +\infty\}.
\end{aligned}$$

*where we wrote b.c. for boundary condition.*

We saw that it was always possible to find an essential support whose closure is equal to the spectrum but the essential supports which we gave in the above propositions will not have this property in general. In fact, it can happen that some of them are not even contained in the corresponding spectra. The dependence of some of these essential supports as functions of the boundary condition $\alpha$ at the origin is studied in Problem III.6.4.

We now consider the problem of the whole real line $(-\infty, +\infty)$.

### The operator on the whole line

As we already explained, we are interested in the operator on the whole real line rather than on a bounded interval $\Lambda$. In other words we are interested in the limit $a \to -\infty$ and $b \to +\infty$.

It is relatively easy to show (see for example the Chapter 2 of Levitan and Sargsjan's book [220]) that, even in the most general situation, the measures $\xi^{\Lambda}_{\alpha,\beta}, \zeta^{\Lambda}_{\alpha,\beta}$ and $\eta^{\Lambda}_{\alpha,\beta}$ form tight families of measures (by which we mean a relatively compact family in the sense of the vague topology for measures). This shows that these families have limit points. On the other hand, assumption (III.2) implies the essential self-adjointness, hence the uniqueness of the self-adjoint extension, and in the present framework, the uniqueness of the possible limit points for these sequences of measures. Consequently, this assumption implies that the measures $\xi^{\Lambda}_{\alpha,\beta}, \zeta^{\Lambda}_{\alpha,\beta}$ and $\eta^{\Lambda}_{\alpha,\beta}$ converge vaguely to measures $\xi$, $\zeta$, and $\eta$ independent of the boundary conditions $\alpha$ and $\beta$. In other words, the essential self-adjointness guarantees the convergence of these sequences. Let us define the matrix measure $M$ by:

$$M = \begin{bmatrix} \xi & \eta \\ \eta & \zeta \end{bmatrix}$$

and let us set:

$$[Uf](\lambda) = (\text{“}\int_{-\infty}^{+\infty}\text{”} f(t)y_1(t,\lambda)dt, \text{“}\int_{-\infty}^{+\infty}\text{”} f(t)y_2(t,\lambda)dt). \qquad \text{(III.25)}$$

The operator $U$ so defined is a unitary isomorphism of $L^2(\mathbb{R}, dt)$ and $L^2(\mathbb{R}, dM(\lambda))$. It transforms the operator $H$ into the operator of multiplication by the independent variable. The above integrals are in quotation marks because they are not integrals in the usual sense. They are meaningful in the $L^2$-sense only. Except for the fact that we used the generalized functions $\delta_0$ and $\dot{\delta}_0$, the unitary operator $U$ gives an explicit construction of the spectral representation discussed in an abstract setting in Chapter I. As we already pointed out, the spectral information on the operator is contained in the nonnegative measure:

$$\sigma(d\lambda) = \text{ trace } M(d\lambda) = \xi(d\lambda) + \zeta(d\lambda).$$

Indeed, for each Borel subset $A$ of $\mathbb{R}$ we have:

$$\big(< E(A)f, f >= 0 \text{ for all } f \in L^2(\mathbb{R}, dt)\,\big) \quad \Longleftrightarrow \quad \sigma(A) = 0.$$

We do not need to argue the convergence of the spectral measures further for we are going to proceed as we did in the case of the half line and prove directly the convergence of the Green's functions. Indeed, our assumption (III.2) applies to $-\infty$ as well as to $+\infty$ and it implies that for each complex number $\lambda \notin \mathbb{R}$ there exists a solution of the eigenvalue equation (III.1) which is square integrable near infinity and a solution which is square integrable near $-\infty$. These solutions are unique up to a multiplicative factor, and consequently, those which are equal to 1 at the origin can be written in the form:

$$t \hookrightarrow f_\pm(t, \lambda) = y_1(t, \lambda) \pm m_\pm(\lambda) y_2(t, \lambda)$$

for some complex number $m_\pm(\lambda)$. This is because of our choice $a = 0$ and $\alpha = \pi/2$ for the $m$-functions. Then, according to the discussion we had in the case of the half line, we expect that the resolvent $[H - \lambda I]^{-1}$ of the operator on the whole line has an integral kernel of the form:

$$G(\lambda, s, t) = \frac{1}{W(f_+(\cdot, \lambda), f_-(\cdot, \lambda))} f_+(t, \lambda) f_-(s, \lambda) \tag{III.26}$$

for $s \le t$. Note that the above Wronskian can be computed in terms of the $m$-functions and one gets:

$$W(f_+(\cdot, \lambda), f_-(\cdot, \lambda)) = -(m_+(\lambda) + m_-(\lambda)).$$

This is indeed the case for the Green's function $G^{[a,b]}_{\alpha,\beta}(\lambda(s,t)$ converges to $G(\lambda, s, t)$ when $a \to -\infty$ and $b \to \infty$ irrespectively of the boundary conditions $\alpha$ and $\beta$ by construction of the $m$-functions $m_+$ and $m_-$.

Let us summarize the procedure we used to get to the operator on the whole real line. We first start from a bounded interval $[a, b]$ on which the potential $V$ is square summable. It was easy in such a situation, to define a self-adjoint operator for each choice of boundary conditions. In the classical literature on the ODE approach to spectral theory one says that the boundary points $a$ and $b$ are regular or/and that the eigenvalue problem is regular on $[a, b]$. We then took the limits $b \to \infty$ and $a \to -\infty$. $+\infty$ and $-\infty$ are regarded as singular boundary points because the potential function $V$ may not be square integrable in a neighborhood of these points. One could consider the problem of a finite boundary point, say $a_0 < a$, in the neighborhood of which $V$ is not square integrable. The limiting procedure $a \searrow a_0$ should be similar to the case $a \to -\infty$ when $V$ is locally square integrable over the whole line. In particular one can define the notion of limit point and limit circle near the singular boundary point $a_0$. The following classical result gives a good handle on the effect of a local singularity of the potential on the essential self-adjointness problem.

**Proposition III.1.14** *Let us assume that the potential function $V$ is continuous and nonnegative near* $0$.
*1. If $V(t) \geq 3/(4t^2)$ near $0$ then $H$ is in the limit point case at* $0$.
*2. If $V(t) \leq (3/4 - \epsilon)t^{-2}$ near $0$ for some $\epsilon > 0$, then $H$ is in the limit circle case at* $0$

The proof of this result can be found in the Appendix to Section X.1 of the second volume of the Reed and Simon series. We shall apply it to the problem of the essential self-adjointness of spherically symmetric Schrödinger operators.

**Remark III.1.15 Essential Self-adjointness for Spherically Symmetric Potentials.** Let us consider a spherically symmetric continuous function $V(x) = v(|x|)$ on $\mathbb{R}^d \backslash \{0\}$. According to the the above proposition and the discussion of Remark II.2.5, all the operators $\tilde{H}_k$ are essentially self-adjoint on $C_c^\infty((0, \infty))$ if:

$$v(r) + \frac{(d-1)(d-3)}{4}\frac{1}{r^2} \geq \frac{3}{4r^2}. \qquad \text{(III.27)}$$

because all the $\kappa_k$'s are nonnegative. On the other hand, if:

$$0 \leq v(r) + \frac{(d-1)(d-3)}{4}\frac{1}{r^2} \leq (\frac{3}{4} - \epsilon)\frac{1}{r^2}. \qquad \text{(III.28)}$$

for some $\epsilon > 0$, then at least one of the $\tilde{H}_k$ is not essentially self-adjoint. Consequently, $-\Delta + V$ is essentially self-adjoint on $C_c^\infty(\mathbb{R}^d \backslash \{0\})$ if $V$ satisfies (III.27) and $-\Delta + V$ is not essentially self-adjoint on $C_c^\infty(\mathbb{R}^d \backslash \{0\})$ if $V$ satisfies (III.28).

The $m$-functions contain all the information regarding the spectral characteristics of the operator. In particular, the spectral measures $\xi$ and $\zeta$ can be recovered very easily because they are the spectral measures appearing in the canonical decompositions of the Herglotz functions:

$$z \hookrightarrow h_1(z) = -\frac{1}{m_+(z) + m_-(z)}$$

and:

$$z \hookrightarrow h_2(z) = \frac{m_+(z)m_-(z)}{m_+(z) + m_-(z)} = \frac{1}{\frac{1}{m_+(z)} + \frac{1}{m_-(z)}}$$

respectively. See Problem III.6.6. The definition of $\sigma$ implies that an essential support for $\sigma_{ac}$ is equal to the set of $\lambda \in \mathbb{R}$ for which $0 < \Im m\, h_1(\lambda + i0) < \infty$ or $0 < \Im m\, h_2(\lambda + i0) < \infty$. Notice that:

$$\Im m\, h_1 = \frac{\Im m\, m_+ + \Im m\, m_-}{|m_+ + m_-|^2} \text{ and } \Im m\, h_2 = \frac{|m_-|^2 \Im m\, m_+ + |m_+|^2 \Im m\, m_-}{|m_+ + m_-|^2}$$

and that $|m_\pm(\lambda + i0)|^2$ and $|m_+(\lambda + i0) + m_-(\lambda + i0)|$ are finite and nonzero for almost every $\lambda \in \mathbb{R}$. Consequently, an essential support of $\sigma_{ac}$ is equivalently given by the set of $\lambda \in \mathbb{R}$ for which $0 < \Im m\, m_+(\lambda + i0) < \infty$ or $0 < \Im m\, m_-(\lambda + i0) < \infty$. Putting this fact together with the Gilbert-Pearson characterization we have:

**Proposition III.1.16** *An essential support of $\sigma_{ac}$ is given by $S_+ \cup S_-$ where:*

$S_\pm = \{\lambda \in \mathbb{R};$ *the eigenvalue equation has no subordinate solution at* $\pm\infty\}$.

*and an essential support for the singular component of $\sigma$ is given by:*

$$S_s = \{\lambda \in \mathbb{R}; \text{ the eigenvalue equation has a solution subordinate at both } +\infty \text{ and } -\infty\}.$$

We conclude with two technical properties of the $m$-functions which we shall need in the sequel.

**Lemma III.1.17**

$$\left| m_\pm(\lambda) - i\sqrt{\lambda} + \int_{\mathbb{R}_\pm} V(t) e^{2i\sqrt{\lambda}|t|}\, dt \right| \leq \frac{1}{dist(\Sigma(H), \lambda)} \int_{\mathbb{R}_\pm} V(t)^2 e^{-2\Im m\sqrt{\lambda}|t|}\, dt$$

*Moreover, if we assume that the potential functions $V_k$ converge in $L^2_{loc}$ to $V^{(0)}$ then the corresponding m-functions $m^{(k)}(\lambda)$ and $m^{(0)}$ satisfy:*

$$\lim_{k\to\infty} m^{(k)}(\lambda) = m^{(0)}(\lambda) \tag{III.29}$$

*the convergence being uniform in $\lambda$ restricted to compact subsets of $\Pi_+$.*

*Proof:*
We prove only the statements concerning the functions $m_+$. Recall that:

$$m_+(\lambda) = \lim_{0<s<t\, t\searrow 0} \frac{\partial^2}{\partial s \partial t} G^+(\lambda, s, t).$$

Using the resolvent identity twice one gets:

$$\begin{aligned}[H^+ - \lambda I]^{-1} = \quad & [H_0^+ - \lambda I]^{-1} - [H_0^+ - \lambda I]^{-1} V [H_0^+ - \lambda I]^{-1} \\ + \quad & [H_0^+ - \lambda I]^{-1} V [H^+ - \lambda I]^{-1} V [H_0^+ - \lambda I]^{-1}\end{aligned}$$

where we used the notation $H_0^+$ for the negative of the second derivative operator with Dirichlet boundary condition at 0, i.e. the operator $H^+$ for $V = 0$. Computing the kernels of both sides and computing the derivatives one gets:

$$m_+(\lambda) = i\sqrt{\lambda} - \int_{I\!R_+} V(t) g_\lambda(t)^2\, dt + < [H^+ - \lambda I]^{-1} V g_\lambda, V g_\lambda >$$

where:

$$g_\lambda(t) = \frac{\partial}{\partial t} G_0^+(\lambda, t, 0) = e^{i\sqrt{\lambda} t}.$$

The desired estimate now follows from:

$$\|[H^+ - \lambda I]^{-1}\| \le \frac{1}{\mathrm{dist}(\Sigma(H), \lambda)}$$

To prove the second claim we use the notation introduced earlier and we add a superscript $^{(k)}$ to emphasize the dependence of the corresponding quantity upon the potential function $V_k$. We want to use the fact that $m_+(\lambda) = \dot\chi_\lambda(0)/\chi_\lambda(0)$ for any solution of the eigenvalue equation which is square integrable near $+\infty$. Let $f \in C_c^\infty(I\!R)$ be supported in the interval $(-\infty, -1)$ and be such that:

$$\int_{I\!R} f_-(s, \lambda) f(s)\, ds \neq 0.$$

Under these conditions:

$$\begin{aligned}[H^{(0)} - \lambda I]^{-1} f(t) &= \int_{\mathbb{R}} G_\lambda(t,s) f(s)\, ds \\ &= \frac{1}{W(f_-(\cdot,\lambda) f_+(\cdot,\lambda))} f_+(t,\lambda) \int_{\mathbb{R}} f_-(s,\lambda) f(s)\, ds\end{aligned}$$

is not identically zero near $+\infty$ and we can choose $\chi_\lambda^{(0)} = [H^{(0)} - \lambda I]^{-1} f$. Since $\lim_{k\to\infty} V_k \varphi = V\varphi$ for every $\varphi \in C_c^\infty(\mathbb{R})$ which is a core for $H^{(0)}$, we can conclude that $[H^{(k)} - \lambda I]^{-1} f$ converges to $[H^{(0)} - \lambda I]^{-1} f$ and one can also choose $\chi_\lambda^{(k)} = [H^{(k)} - \lambda I]^{-1} f$. This implies the convergence of $\chi_\lambda^{(k)}$ to $\chi_\lambda$ in $L^2((-1,\infty), dt)$ and the convergence of $\ddot{\chi}_\lambda^{(k)}$ to $\ddot{\chi}_\lambda$ in $L^1_{loc}$ because of the eigenvalue equation. We now have the pointwise convergence of $\chi_\lambda^{(k)}$and $\dot{\chi}_\lambda^{(k)}$ to $\chi_\lambda^{(0)}$ and $\dot{\chi}_\lambda^{(0)}$ respectively and consequently the pointwise convergence of $m_+^{(k)}$ to $m_+^{(0)}$. The uniform convergence on compact sets is now obvious because of the fact that the $m$-functions are analytic and because of their construction. Recall the inclusion of the circles converging to a point! ∎

### III.1.4 The Propagator

It is clear from the formulas we discussed in the previous section that the knowledge of the functions $m_\pm$, and hence the knowledge of the asymptotic behavior of the solutions of the eigenvalue equation (III.6) plays a crucial role in the study of the spectral measure $\sigma(d\lambda)$. It is convenient to rewrite this eigenvalue equation (III.6) as a first order differential system:

$$\dot{Y}(t) = A_\lambda(t) Y(t) \tag{III.30}$$

with:

$$Y(t) = \begin{bmatrix} y(t) \\ \dot{y}(t) \end{bmatrix} \quad \text{and} \quad A_\lambda(t) = \begin{bmatrix} 0 & 1 \\ V(t) - \lambda & 0 \end{bmatrix}. \tag{III.31}$$

The *propagator* of this system (also called the *time evolution operator*, or the *transfer matrix*) is the unique flow $\{U_\lambda(t,s);\, t,s \in \mathbb{R}\}$ of $2 \times 2$ unimodular matrices satisfying for each $s \geq 0$:

$$\begin{cases} \frac{d}{dt} U_\lambda(t,s) = A_\lambda(t) U_\lambda(t,s), & t \geq s, \\ U_\lambda(s,s) = I. \end{cases} \tag{III.32}$$

where I denotes the $2 \times 2$ identity matrix. The time evolution of all the solutions of the system can then be given in the form:

$$U_\lambda(t,s) Y(s) = Y(t).$$

Note that one has the easy bound:

$$||U_\lambda(t,s)|| \le e^{\int_s^t ||A_\lambda(u,s)||du}. \tag{III.33}$$

We now recall some classical definitions from the classical theory of O.D.E. We give them in the general case of an equation:

$$\dot{Y}(t) = A(t)Y(t), \tag{III.34}$$

for some locally integrable function $t \hookrightarrow A(t)$ with values in the space of $n \times n$ matrices. The existence of a propagator $\{U(t,s);\, t,s \in \mathbb{R}\}$ is given by the general theory of O.D.E.. Recall that $U(t,s) = U(t)U(s)^{-1}$ if one uses the notation $U(t) = U(t,0)$ for the matrix which takes the value of the solution at time $t = 0$ to the value of the solution at time $t$. This matrix $U(t)$ is sometimes called a *fundamental matrix* for the equation (III.34).

The equation (III.34) is said to be:
*stable* near $+\infty$ (resp. $-\infty$) if:

$$\sup_{t\ge 0} ||U(t)|| < \infty \qquad (\text{resp. } \sup_{t\le 0} ||U(t)|| < \infty) \tag{III.35}$$

*uniformly stable* near $+\infty$ (resp. $-\infty$) if:

$$\sup_{t\ge s\ge 0} ||U(t,s)|| < \infty \qquad (\text{resp. } \sup_{t\le s\le 0} ||U(t,s)|| < \infty) \tag{III.36}$$

*strongly stable* near $+\infty$ (resp. $-\infty$) if:

$$\sup_{t\ge 0} ||U(t)|| < \infty \quad \text{and} \quad \sup_{t\ge 0} ||U(t)^{-1}|| < \infty, \tag{III.37}$$

$$(\text{resp.} \sup_{t\le 0} ||U(t)|| < \infty \quad \text{and} \quad \sup_{t\le 0} ||U(t)^{-1}|| < \infty), \tag{III.38}$$

*asymptotically stable* near $+\infty$ (resp. $-\infty$) if:

$$\lim_{t\to+\infty} ||U(t)|| = 0 \qquad (\text{resp. } \lim_{t\to-\infty} ||U(t)|| = 0) \tag{III.39}$$

*uniformly asymptotically stable* near $+\infty$ (resp. $-\infty$) if there exist positive constants $c > 0$ and $\gamma > 0$ such that:

$$||U(t,s)|| \le ce^{-\gamma(t-s)} \quad \text{for all } 0 \le s \le t \tag{III.40}$$

$$(\text{resp. } ||U(t,s)|| \le ce^{\gamma(t-s)} \quad \text{for all } 0 \ge s \ge t). \tag{III.41}$$

We chose to take the propagator between 0 and $t$ in the above definitions but the bound (III.33) shows that one could have used any real number $s \in \mathbb{R}$ instead.

**Example:**
If $A(t) = A$ is constant, then $U(t, s) = e^{(t-s)A}$ and stability is equivalent to uniform stability and asymptotic stability is equivalent to uniform asymptotic stability. Moreover, the equation is stable near $+\infty$ (resp. near $-\infty$) if and only if every characteristic root of the matrix $A$ has real part not greater than 0 (resp. not less than 0) and those with zero real part are semi-simple (recall that a characteristic root $\gamma$ of a matrix $A$ is said to be semi-simple if every vector $a \in \mathbb{R}^n$ which satisfies the equation $(A - \gamma I)^m a = 0$ for some integer $m \geq 1$ necessarily satisfies it for $m = 1$).
It is strongly stable near $+\infty$ (resp. near $-\infty$) if and only if every characteristic root of the matrix $A$ is purely imaginary and semi-simple. Finally, it is asymptotically stable near $+\infty$ (resp. near $-\infty$) if and only if every characteristic root of the matrix $A$ has negative real part (resp. positive real part).

**The free eigenvalue equation:**
The simplest eigenvalue problem which we can solve explicitly is of course the case $V(t) \equiv 0$ of zero potential.
In this case, if $\lambda > 0$ the fundamental matrix $U_\lambda$ is given by:

$$U_\lambda(t) = \begin{bmatrix} \cos t\sqrt{\lambda} & \frac{1}{\sqrt{\lambda}} \sin t\sqrt{\lambda} \\ -\sqrt{\lambda} \sin t\sqrt{\lambda} & \cos t\sqrt{\lambda} \end{bmatrix}, \tag{III.42}$$

from which one easily sees that:

$$\sup_{s,t \in \mathbb{R}} ||U_\lambda(t, s)|| = 2 \max(\lambda, \lambda^{-1}).$$

Now, if $\lambda < 0$ one has:

$$U_\lambda(t) = \begin{bmatrix} \cosh t\sqrt{-\lambda} & \frac{1}{\sqrt{-\lambda}} \sinh t\sqrt{-\lambda} \\ \sqrt{-\lambda} \sinh t\sqrt{-\lambda} & \cosh t\sqrt{-\lambda} \end{bmatrix}, \tag{III.43}$$

from which one easily sees that there exist two linearly independent solutions $Y_+(t)$ and $Y_-(t)$ of the vector form of the eigenvalue problem (III.30) satisfying:

$$\lim_{t \to \pm\infty} \frac{1}{|t|} \log ||Y_\pm(t)|| = -\sqrt{-\lambda},$$

while:

$$\lim_{t\to+\infty} \frac{1}{|t|} \log \|Y(t)\| = \sqrt{-\lambda},$$

if the solution $Y(t)$ is linearly independent of $Y_+(t)$ and:

$$\lim_{t\to-\infty} \frac{1}{|t|} \log \|Y(t)\| = \sqrt{-\lambda},$$

if the solution $Y(t)$ is linearly independent of $Y_-(t)$. This is a particular case of the exponential dichotomy result which we will prove below in the case of periodic Schrödinger operator.

For the sake of completeness we close this section with the discussion of a some quantities often used in the study of one-dimensional Schrödinger operators. They are standard in the literature on Sturm-Liouville problems and they are used in many papers on one dimensional continuous disordered systems. Setting:

$$\begin{cases} y(t) = r(t)\sin\theta(t) \\ \dot{y}(t) = r(t)\cos\theta(t) \end{cases} \tag{III.44}$$

with $r(t) \geq 0$ (in which case $r(t) = \|Y(t)\| = \sqrt{|y(t)|^2 + |y'(t)|^2}$), we can switch to polar coordinates. This operation is called the *Prüffer transformation,* and the functions $r(t)$ and $\theta(t)$ are called the *amplitude* and the *phase* of the solution $y(t)$. With these notations, $y(t)$ is a solution of the eigenvalue equation (III.6) if and only if $r(t)$ and $\theta(t)$ are solutions of the system:

$$\begin{cases} \dot{\theta}(t) = \cos^2\theta(t) + [\lambda - V(t)]\sin^2\theta(t) \\ \dot{r}(t) = -\frac{1}{2}r(t)[\lambda - V(t) - 1]\sin 2\theta(t). \end{cases} \tag{III.45}$$

This system is nonlinear but the second equation can be integrated directly if one assumes the knowledge of the solution of the first one. Indeed:

$$r(t) = r(s)\exp[-\frac{1}{2}\int_s^t [\lambda - V(u) - 1]\sin 2\theta(u)du] \tag{III.46}$$

The choice of the "initial condition $r(s)$" is arbitrary. In fact choosing it amounts to fixing the amplitude of the solution at the point $s$. Remember that we are only interested in the asymptotic behavior of the solutions and the latter will not be qualitatively affected by this choice, as long as $r(s) \neq 0$. We will also consider that the angle $\theta(t)$ is defined modulo $\pi$ instead of modulo $2\pi$ in order to be consistent with the way in which we label the boundary conditions. Also, we will use the following notation from time to time:

$\theta_\lambda(t,a,\alpha)$ will stand for the value of the phase computed at $t$, knowing that it equals $\alpha$ for $t = a$, and

$r_\lambda(t,a,\alpha)$ will stand for the value of the amplitude computed at $t$, knowing that for $t = a$, the amplitude equals 1 and the phase equals $\alpha$. With this convention, formula (III.46) can be rewritten as:

$$r_\lambda(t,a,\alpha) = \exp[-\frac{1}{2}\int_0^t [\lambda - V(u) - 1]\sin 2\theta(u,a,\alpha,\lambda)du]. \quad \text{(III.47)}$$

### III.1.5 Examples

For illustration purposes we give a list of examples of potential functions with the spectral characteristics of the corresponding Schrödinger operators. They have been chosen to illustrate the typical spectral properties one should expect from such systems. Most of the properties we discuss will be proven later in the text. For those which will not be given a proof, the reader is asked to find in the section on Notes and Complements precise references to texts and articles providing the missing links.

**Anharmonic Oscillators:**
Let us assume that $\lim_{|t|\to\infty} V(t) = +\infty$. We already proved in Section II.3 that, even in the general multidimensional setting, the resolvent of $H$ was a compact operator from which it followed that $\Sigma(H)$ is a countable set without finite accumulation points, that $\Sigma(H) = \Sigma_{pp}(H)$ and even $\Sigma(H) = \Sigma_{disc}(H)$, and that $E = E_{pp}$.

**Singular Continuous Spectrum:**
Examples of potential functions $V$ giving rise to purely singular continuous spectrum (i.e., $\Sigma(H) = \Sigma_{sc}(H)$ and $E = E_{sc}$) seems to have been known for quite some time, but the first systematic account was given by Pearson in [266]. In order to gain some understanding of this unusual spectral type we quote a result of his. We do not give a proof for it would require more time and space than what we want to allocate to this *curiosity*.

Let $\chi$ be a real valued integrable function which vanishes outside a bounded interval $[0,a]$. Let $\{t_n;\ n \geq 1\}$ be a strictly increasing sequence of positive numbers and let us define the potential function $V$ by:

$$V(t) = \sum_{n\geq 1} \chi(t - t_n).$$

Pearson shows that the operator $H_0+V$ with Dirichlet boundary condition at the origin has purely singular continuous spectrum. He also shows how the ***sliding bumps*** formed by the shifts of $\chi$ can be scaled down so that the potential $V$ converges to 0 at $\infty$ while still keeping the singularity of the spectrum. He also extends his results to cases of operators on the whole real line $\mathbb{R}$ and on the multidimensional space $\mathbb{R}^d$ by considering spherically symmetric potentials.

**Two Body Case:**
If $V(t)$ is integrable in a neighborhood of $+\infty$ or $-\infty$, we will show in SectionIII.4 below that $[0,\infty) \subset \Sigma(H)$ and that $E = E_{ac}$ on $[0,\infty)$.

**Constant Electric Field:**
In the context of the diffusion theory in a constant electric field it is natural to consider potential functions of the form:

$$V(t) = \kappa t + v(t) \tag{III.48}$$

where $\kappa$ is a real constant and $v$ is a bounded function (usually assumed to have two bounded derivatives). In this case, the spectrum is the whole real line and it is purely absolutely continuous, i.e., $\Sigma(H) = \Sigma_{ac}(H) = \mathbb{R}$ and $E = E_{ac}$. It is interesting to notice that the corresponding problem on the lattice has a completely different solution. Indeed, the multiplication operator by the function $\kappa t$, which has obviously discrete spectrum, has a compact resolvent operator. Moreover, the free Laplacian $H_0$ and the multiplication operator by the potential $v$ are bounded operators. Consequently, the resolvent operator of the full Hamiltonian $H$ is still compact and the spectrum is a fortiori discrete!

**Embedded Eigenvalues:**
There are many examples, among which the famous Wigner-Von Neumann one, of potential functions giving rise to eigenvalues embedded in the continuous spectrum. We simply refer to the Section 13 of Chapter XIII of the Reed and Simon series [272] for a thorough discussion of this technical problem.

**Periodic Potentials:**
Periodic potentials play an important role in the mathematics of solid state physics. For example, in the theory of crystals, the nuclei sit at the sites of a regular (i.e. periodic) sublattice of $\mathbb{R}^3$ and electrons traveling in the crystal feel a periodic potential. We shall give a complete analysis of the spectral

properties of one dimensional periodic Schrödinger operators in Chapter VII. For the time being let us simply repeat what we already mentioned in the previous chapter, namely that these operators have essentially always absolutely continuous spectrum and that finding the spectrum as a set is important for the physical interpretation of the energies in the bands and in the gap in terms of conductance properties of the material. The interested reader is refered to volume IV of the Reed and Simon series [272] for a nice exposé of this mathematical theory and of some of the physics behind it.

**Sturm-Liouville Operators:**
One dimensional Schrödinger operators can be viewed as particular cases of the so-called Sturm-Liouville operators. The latter can be described in the following way. Let us consider three real valued measurable functions $r$, $q$ and $p$ on $[0,\infty)$ and let us assume (for the sake of simplicity) that:

$$p(t) \geq p_-, \qquad t \geq 0$$

and:

$$r(t) \geq r_-, \qquad t \geq 0$$

for some positive constants $p_-$ and $r_-$. The method of quadratic forms described in Chapter I can be used to define:

$$H = -\frac{1}{r(t)}\frac{d}{dt}\left(\frac{1}{p(t)}\frac{d}{dt}\right) + \frac{q(t)}{r(t)} \tag{III.49}$$

as a self-adjoint operator on the Hilbert space $\mathcal{H} = L^2([0,\infty), r(t)dt)$ with a boundary condition at the origin of the form:

$$y(0)\cos\alpha - \dot{y}(0)\frac{\sin\alpha}{p(0)} = 0. \tag{III.50}$$

As before, the eigenvalue equation is advantageously rewritten as a first order differential system:

$$\dot{Y}(t) = A_\lambda(t)Y(t) \tag{III.51}$$

with:

$$Y(t) = (t) = \begin{bmatrix} y(t) \\ \dot{y}(t)/p(t) \end{bmatrix} \quad \text{and} \quad A_\lambda(t) = \begin{bmatrix} 0 & p(t) \\ q(t) - \lambda r(t) & 0 \end{bmatrix}. \tag{III.52}$$

The propagator of this system, still denoted by $\{U_\lambda(t,s);\, t,s \in \mathbb{R}\}$, the amplitude of the solution and the phase, now defined as:

$$\vartheta_\lambda(t) = \tan^{-1} \frac{p(t)\, y(t)}{\dot{y}(t)}$$

can be used in the same way to control the oscillation properties and the asymptotic behavior of the solutions of the eigenvalue problem. We shall not dwell on this theory, we refer the interested reader to the text books of Dunford and Schwarz [92] and Coddington and Levinson [56]for further reading.

**Acoustic Waves in Inhomogeneous Media:**
The study of the propagation of acoustic waves in inhomogeneous media leads to the investigation of the following wave equation:

$$\frac{\partial^2 u}{\partial t^2} - c(\mathbf{x})^2 \rho(\mathbf{x}) \nabla \cdot \left( \frac{1}{\rho(\mathbf{x})} \nabla u \right) = F(t, \mathbf{x}), \tag{III.53}$$

where $\rho(\mathbf{x})$ denotes the density of the fluid at the point $\mathbf{x} = (x, y, z)$, $c(\mathbf{x})$ denotes the velocity of sound in the fluid and $F(t, \mathbf{x})$ a source. The unknown function $u(t, \mathbf{x})$ represents the acoustic pressure at time $t$ at the point $\mathbf{x}$ of the fluid. In fact $u(t, \mathbf{x})$ represents the difference between the actual pressure and the equilibrium pressure because the equation (III.53) is actually derived by linearization of the fluid equations around the equilibrium pressure. Equation (III.53) can be transformed into (or studied via) an initial value problem:

$$\begin{cases} \frac{\partial^2 u}{\partial t^2} - c(\mathbf{x})^2 \rho(\mathbf{x}) \nabla \cdot \left( \frac{1}{\rho(\mathbf{x})} \nabla u \right) = 0 \\ u(0, \mathbf{x}) = f(x), \qquad \frac{\partial u}{\partial t}(0, \mathbf{x}) = g(\mathbf{x}). \end{cases} \tag{III.54}$$

Under the usual assumptions:

$$0 < \rho_- \leq \rho(\mathbf{x}) \leq \rho_+ < +\infty$$

and:

$$0 < c_- \leq c(\mathbf{x}) \leq c_+ < +\infty$$

for some constants $\rho_-$, $\rho_+$, $c_-$ and $c_+$, the method of quadratic forms described in Chapter I can be used to define:

$$H = -c(\mathbf{x})^2 \rho(\mathbf{x}) \nabla \cdot \left( \frac{1}{\rho(\mathbf{x})} \nabla \right)$$

as a self-adjoint operator on the Hilbert space:

$$\mathcal{H} = L^2(\Lambda, \frac{d\mathbf{x}}{c(\mathbf{x})^2 \rho(\mathbf{x})}),$$

for some domain $\Lambda \subset \mathbb{R}^3$. For the purpose of the present discussion we shall consider the case $\Lambda = \mathbb{R}^2 \times [0, \infty)$ and we shall impose Neumann boundary conditions on the boundary hyperplane $z = 0$. In other words, we are considering the effect of a flat ocean bottom, but we assume that the ocean is unlimited above and in the horizontal directions. It is possible to use the functional calculus for self-adjoint operators developed in Chapter one to see that the initial value problem is solved by:

$$u(t, \cdot) = \cos(t\sqrt{H})f + \frac{1}{\sqrt{H}} \sin(t\sqrt{H})g,$$

whenever the initial data $f$ and $g$ satisfy $f \in \mathcal{H}$ and $g \in \mathcal{D}(H^{-1/2})$. The spectral properties of the operator $H$ dictate the asymptotic properties of the solutions and the solution of many scattering problems which are beyond the scope of this book. We simply remark that (generalized) eigenfunction expansions are a standard tool in the solution of the scattering problems. They go under the name of *normal mode expansions.* They are not always easy to control, even when the domain $\Lambda$ is a product space as in the present (academic) model. Nevertheless, there are many problems for which a simplifying assumption can be made to ease the burden of the computations. In particular, this is the case for the *stratified* or *layered media.* For such media, it is natural to assume that the acoustic velocity and the fluid pressure are only functions of the third coordinate $z$, namely that:

$$c(x, y, z) = c(z) \qquad \text{and} \qquad \rho(x, y, z) = \rho(z).$$

Under this extra hypothesis, the operator $H$ can be rewritten in the form:

$$H = -c(z)^2 \left( \frac{\partial^2}{\partial x^2} + \frac{\partial^2}{\partial y^2} + \rho(z) \frac{\partial}{\partial z} \left( \frac{1}{\rho(z)} \frac{\partial}{\partial z} \right) \right)$$

and the eigenvalue problem $H\psi = \lambda\psi$ can be tackled by the method of separation of variables. One looks for solutions of the form:

$$\psi(x, y, z) = e^{i(p_1 x + p_2 y)} \psi(z)$$

for $p = (p_1, p_2) \in \mathbb{R}^2$ and the eigenvalue problem for the wave propagator $H$ reduces to a family of one dimensional Sturm-Liouville eigenvalue problems:

$$-c(z)^2 \rho(z) \frac{d}{dz} \left( \frac{1}{\rho(z)} \frac{d\psi}{dz} \right) - |p|^2 c(z)^2 \psi = \lambda \psi$$

parametrized by $|p|^2 = p_1^2 + p_2^2$. This problem can be explicitly (and completely) solved by the method of this chapter and a (complete) expansion in generalized eigenfunctions can be obtained for the operator $H$ and the spectral resolution of the identity can then be constructed from the generalized eigenfunctions and the spectral properties can then be spelled out. The interested reader is refered to Wilcox's monograph [331] for the details of this study.

## III.2 The Lattice Case

We now consider the one-dimensional lattice analogue of the theory developed in the previous section. In the present situation $\mathcal{H} = \ell^2(\mathbb{Z})$ is a space of double sequences and operators on $\mathcal{H}$ can be regarded as doubly infinite matrices. In particular, operators restricted to boxes will actually be matrices. We adapt somehow our point of view to take full advantage of this fact. All the results we proved in the previous section hold in the lattice case once appropriately reformulated. We review most of them and we give only the proofs which are significantly different from the ones we gave in the continuous case.

A potential function $V$ is now a doubly infinite real sequence $\{V(n); n \in \mathbb{Z}\}$. Recall that the discrete Schrödinger operator $H$ is always self-adjoint on the domain of the multiplication operator by $V$ (it is even essentially self-adjoint on the space of sequences $\psi$ which vanish everywhere except for finitely many integers $n$) and that it acts on the complex sequences $\psi \in \ell^2(\mathbb{Z})$ via the formula:

$$[H\psi](n) = -\psi(n+1) - \psi(n-1) + V(n)\psi(n).$$

We denote by $E$ the spectral resolution of the identity of the operator $H$ and by $\{\delta_n; n \in \mathbb{Z}\}$ the canonical basis of $\mathcal{H}$.

We want to present the lattice analogue of the $m$-functions. They will be of crucial importance in the investigation of the absolutely continuous spectrum and the inverse spectral theory of random Schrödinger operators. In the lattice case, this inverse theory cannot be developed in the context of Schrödinger matrices. Indeed, the reconstruction procedures cannot recover the ones on the secondary diagonal of the matrix and the framework of Jacobi matrices is more appropriate for this kind of problem. For this reason, we generalize slightly the class of operators we consider.

Let us assume that $\{a_n;\ n \in \mathbb{Z}\}$ and $\{b_n;\ n \in \mathbb{Z}\}$ are given sequences of real numbers and let us also assume that $a_n \neq 0$ for all $n \in \mathbb{Z}$. On the Hilbert space $\mathcal{H} = \ell^2(\mathbb{Z})$ we consider the operator $H$ defined formally by:

$$(H\psi)_n = a_n\psi_{n+1} + a_{n-1}\psi_{n-1} + b_n\psi_n. \tag{III.55}$$

To obtain a natural self-adjoint realization of $H$, we proceed as follows. We set

$$\begin{aligned}\mathcal{D}_0 &= \{\psi = (\psi_n)_{n\in\mathbb{Z}} \in \mathcal{H};\ \psi_n = 0 \text{ for all but finitely many } n \in \mathbb{Z}\}\\ \mathcal{D}_1 &= \{\psi = (\psi_n)_{n\in\mathbb{Z}} \in \mathcal{H}; (a_n\psi_{n+1} + a_{n-1}\psi_{n-1} + b_n\psi_n)_n \in \mathcal{H}\}\end{aligned}$$

and we define the operators $H_{min}$ and $H_{max}$ by the formula (III.55) on the domains $\mathcal{D}_0$ and $\mathcal{D}_1$ respectively. One easily checks that $H_{min}$ and $H_{max}$ are symmetric and that $(H_{min})^* = H_{max}$. Moreover, the condition:

$$\sum_{n=1}^{\infty} \frac{1}{a_n^2} = \sum_{n=-\infty}^{-1} \frac{1}{a_n^2} = \infty$$

implies that $H_{min}$ is essentially self-adjoint and that its unique self-adjoint extension is $H_{max}$. The latter will be denoted by $H$ from now on. The proofs of these elementary facts are left as exercises. See Problem III.6.5 As before, the investigation of the eigenvalue equation:

$$a_n\psi_{n+1} + a_{n-1}\psi_{n-1} + b_n\psi_n = \lambda\psi_n \tag{III.56}$$

will be of crucial importance in the study of the spectral properties of the operator $H$. We use the formalism of transfer matrices. If $\psi = (\psi_n)_n$ is any solution of the eigenvalue equation (III.56) we have:

$$\begin{pmatrix}\psi_n\\ \psi_{n-1}\end{pmatrix} = U_\lambda(n)\begin{pmatrix}\psi_0\\ \psi_{-1}\end{pmatrix}$$

with

$$U_\lambda(n) = g_\lambda(n-1)g_\lambda(n-2)\cdots g_\lambda(0) \tag{III.57}$$

and:

$$g_\lambda(k) = \begin{pmatrix}\frac{\lambda - b_k}{a_k} & -\frac{a_{k-1}}{a_k}\\ 1 & 0\end{pmatrix} \tag{III.58}$$

for $k \geq 1$. If $\varphi = (\varphi_n)_n$ is another solution, we can write

$$\begin{pmatrix}\psi_n & \varphi_n\\ \psi_{n-1} & \varphi_{n-1}\end{pmatrix} = U_\lambda(n)\begin{pmatrix}\psi_0 & \varphi_0\\ \psi_{-1} & \varphi_{-1}\end{pmatrix} \tag{III.59}$$

and this gives:

$$a_{n-1}[\psi_n\varphi_{n-1} - \psi_{n-1}\varphi_n] = a_{-1}[\psi_0\varphi_{-1} - \psi_{-1}\varphi_0]$$

by using (III.58), (III.57) and (III.59) to compute the determinant of $U_\lambda(n)$. So, if $\psi$ and $\varphi$ are complex sequences we define their Wronskian $W(\psi, \varphi)$ as the sequence defined by:

$$W(\psi, \varphi)_n = -a_{n-1}[\psi_n\varphi_{n-1} - \psi_{n-1}\varphi_n],$$

for $n \in \mathbb{Z}$. Consequently, we just proved that the Wronskian of two solutions is a constant. One easily sees that this constant is different from zero if and only if the two solutions are linearly independent.

We now proceed to the definition of the Weyl-Titchmarsh $m$-functions.

We define the operator $H^+$ on $\mathcal{H}^+ = \ell^2[1,\infty)$ as the restriction of the operator $H$ to $[1,\infty)$. One can think of $H^+$ as being obtained by adding a Dirichlet boundary condition at 0. In fact, we set

$$\begin{aligned} \mathcal{D}_0^+ &= \{\psi = (\psi_n)_{n\geq 1} \in \mathcal{H}^+;\ \psi_n = 0 \text{ for } n \textit{ large}\} \\ \mathcal{D}_1^+ &= \{\psi = (\psi_n)_{n\geq 1} \in \mathcal{H}^+;\ \sum_{n\geq 1} |a_n\psi_{n+1} + a_{n-1}\psi_{n-1} + b_n\psi_n|^2 < \infty\} \end{aligned}$$

and

$$(H^+\psi)_n = \begin{cases} a_1\psi_2 + b_1\psi_1 & \text{if } n = 1 \\ a_n\psi_{n+1} + a_{n-1}\psi_{n-1} + b_n\psi_n & \text{if } n \geq 2 \end{cases} \tag{III.60}$$

for $\psi \in \mathcal{D}_1^+$, and we denote by $H^+_{min}$ the operator $H^+$ with domain $\mathcal{D}_0^+$ and by $H^+_{max}$ the operator $H^+$ with domain $\mathcal{D}_1^+$. It is easy to check that $H^+_{min}$ and $H^+_{max}$ are symmetric and that $(H^+_{min})^* = H^+_{max}$. Moreover, the condition:

$$\sum_{n\geq 1} \frac{1}{a_n^2} = \infty \tag{III.61}$$

implies that $H^+_{min}$ is essentially self-adjoint and its unique self-adjoint extension is $H^+_{max}$. From now on we will denote this unique self-adjoint extension by $H^+$. For $\lambda \in \Pi_+$, we let $f_{\lambda,0+}$ be the solution of the eigenvalue equation (III.56) which satisfies:

$$f_{\lambda,0+}(0) = 0, \qquad f_{\lambda,0+}(1) = 1$$

and we denote by $f_{\lambda,\infty}$ any solution which is square summable at $+\infty$. Note that the space of such solutions is one-dimensional for all $\lambda$ for which

$\Im m\,\lambda > 0$. Indeed, its dimension is at most one because of the usual Wronskian argument, and it is exactly 1 because one can easily construct a solution of (III.56) which is $\ell^2$ near $+\infty$ by simply extending the sequence $[H^+ - \lambda]^{-1}\delta_1$, which is only defined on $\{1, 2, \cdots\}$, into a solution over the whole axis $\mathbb{Z}$. Now let $G^+(\lambda, m, n)$ be the symmetric kernel defined by:

$$G^+(\lambda, m, n) = \frac{1}{W(f_{\lambda,0+}, f_{\lambda,\infty})} f_{\lambda,0+}(m) f_{\lambda,\infty}(n)$$

when $m \leq n$. Note also that this definition is independent of the particular choice of $f_{\lambda,\infty}$. Now, $G^+(\lambda, \cdot, \cdot)$ is easily seen to be the kernel of the resolvent operator $[H^+ - \lambda I]^{-1}$. Consequently, if one defines the Titchmarsh-Weyl function $m$ by

$$m_+(\lambda) = < [H^+ - \lambda I]^{-1}\delta_1, \delta_1 > \tag{III.62}$$

one has

$$m_+(\lambda) = G^+(\lambda, 1, 1) = -\frac{1}{a_0}\frac{f_{\lambda,\infty}(1)}{f_{\lambda,\infty}(0)}\,. \tag{III.63}$$

For $\Im m\,\lambda > 0$ we have

$$|m_+(\lambda)| \leq \|[H^+ - \lambda I]^{-1}\| \leq \frac{1}{\Im m\,\lambda} \tag{III.64}$$

and, using the first resolvent identity one gets:

$$\Im m\, m_+(\lambda) = \Im m\,\lambda\, \|[H^+ - \lambda I]^{-1}\delta_1\|^2 \geq \frac{\Im m\,\lambda}{|b_1 - \lambda|^2 + a_1^2}. \tag{III.65}$$

Moreover, if for each $k = 0, 1, 2, \cdots$ $\{a_n^{(k)}; n \geq 1\}$ is a sequence of real numbers satisfying condition (III.61) and $\{b_n^{(k)}; n \geq 1\}$ is a sequence of real numbers, and if we denote by $H(k)_+$ the corresponding operator, then, if we also assume that:

$$\lim_{k\to\infty} a_n^{(k)} = a_n^{(0)}$$

and

$$\lim_{k\to\infty} b_n^{(k)} = b_n^{(0)}$$

for each fixed $n = 1, 2, \cdots$, one can use Proposition I.1.8 of Chapter I and conclude that $H_+^{(k)}$ converges to $H_+^{(0)}$ in the strong resolvent sense. Indeed, $\mathcal{D}_0^+$ is a common core and $H_+^{(k)} f$ converges to $H_+^{(0)} f$ for each $f \in \mathcal{D}_0^+$. Then, the definition of the $m$-function, implies that:

$$\lim_{k\to\infty} m_+^{(k)}(\lambda) = m_+^{(0)}(\lambda) \tag{III.66}$$

uniformly in $\lambda$ restricted to compact subsets of $\Pi_+$.

Note that the function $\lambda \hookrightarrow m_+(\lambda)$ belongs to the Herglotz class $\mathcal{H}erg$ defined in Chapter I.

We will need the following technical result.

**Lemma III.2.1** *Let us assume that the sequence $\{(a_n, b_n); n \geq 1\}$ is bounded. Then the function $\lambda \hookrightarrow \log m_+(\lambda) - \log(-1/\lambda)$ is holomorphic near $\lambda = \infty$, it has there the following Taylor expansion*

$$\begin{aligned}\log m_+(\lambda) - \log\frac{1}{-\lambda} &= \frac{1}{\lambda}s_1 + \frac{1}{\lambda^2}(s_2 - \tfrac{1}{2}s_1^2) + \frac{1}{\lambda^3}(s_3 - s_1 s_2 + \tfrac{1}{3}s_1^3)\\ &\quad + \frac{1}{\lambda^4}(s_4 - \tfrac{1}{2}s_2^2 - s_1 s_3 + s_1 s_2^2 s_3 - \tfrac{1}{4}s_1^4) + \cdots\end{aligned}$$

*where we used the notation $s_n = < H_+^n \delta_1, \delta_1 >$, $n = 1, 2, \cdots$ and the above series converges for $|\lambda| \geq ||H^+||$.*

*Proof:*
We first notice that:

$$\begin{aligned}\log m_+(\lambda) &= \log < [H^+ - \lambda I]^{-1}\delta_1, \delta_1 >\\ &= \log\frac{1}{-\lambda} + \log < [I - \frac{1}{\lambda}H^+]^{-1}\delta_1, \delta_1 >\end{aligned}$$

The operator $H^+$ is bounded because of the boundedness assumption on the coefficients $a_n$ and $b_n$. Moreover, the function $f$ defined by:

$$f(z) = \log < [I - zH^+]^{-1}\delta_1, \delta_1 >$$

is holomorphic on the disk $|z| < \frac{1}{2}||H^+||^{-1}$ and:

$$\begin{aligned}f(z) &= \log(1 + \sum_{n=1}^{\infty} s_n z^n)\\ &= \sum_{n=1}^{\infty} s_n z^n - \frac{1}{2}\left(\sum_{n=1}^{\infty} s_n z^n\right)^2 + \frac{1}{3}\left(\sum_{n=1}^{\infty} s_n z^n\right)^3 + \cdots\\ &= s_1 z + (s_2 - \frac{1}{2}s_1^2)z^2 + (s_3 - s_1 s_2 + \frac{1}{3}s_1^3)z^3 + \cdots\end{aligned}$$

which gives the desired result if one chooses $z = 1/\lambda$. ∎

One defines similarly the function $m_-(\lambda)$ by formula (III.62) with the operator $H^+$ replaced by the operator $H^-$ obtained by restricting the operator $H$ to $(-\infty,-1]$ and with $\delta_{-1}$ intead of $\delta_1$. One can define the solutions $f_{\lambda,0-}$ and $f_{\lambda,-\infty}$ and one gets a formula similar to (III.63). Obviously the asymptotic and the regularity properties of the function $m_-$ are the same as those of the function $m_+$.

Now, let us define the symmetric kernel $G(\lambda,m,n)$ by:

$$G(\lambda,m,n)=\frac{1}{W(f_{\lambda,\infty},f_{\lambda,-\infty})}f_{\lambda,\infty}(n)f_{\lambda,-\infty}(m) \qquad \text{(III.67)}$$

when $m\leq n$. One checks as in the continuous case that this is the kernel of the resolvent of the operator $H$. One easily deduces the formula:

$$-\frac{1}{G(\lambda,0,0)}=\lambda-b_0+a_0m_+(\lambda)+a_0a_{-1}m_-(\lambda) \qquad \text{(III.68)}$$

which we shall need later on.

It is proven in Problem III.6.8 that the Green's function decays exponentially. when the energy $\lambda$ is outside the spectrum of the operator $H$. This result implies that the norm of the propagator explodes at least exponentially as the following proposition shows.

**Proposition III.2.2** *Let $\lambda$ be a number (real or complex) lying outside the spectrum of $H$. Then one has:*

$$\liminf_{|t|\to\pm\infty}\frac{1}{|t|}\log\|U_\lambda(t)\|>0$$

*Proof:*
Let us set $\varphi(t)=G(\lambda,0,t)$. It follows from the results of Problem III.6.8 that there exist two constants $C<\infty$ and $\gamma>0$ such that $\|\tilde{\phi}(t)\|\leq C\,e^{-\gamma|t|}$. Let $\psi$ be a solution of the equation $H\psi=\lambda\psi$ such that $\tilde{\psi}(1)$ is not colinear to $\tilde{\varphi}(1)$. Schwarz inequality and the constancy of the Wronskian implies immediately that:

$$\|\tilde{\psi}(t)\|\geq C'\,e^{\gamma|t|}$$

for some positive constant $C'$ and any positive $n$. The same property holds for negative $n$ and this yields the desired result. ∎

**Remark III.2.3** We shall define later the Lyapunov exponents $^{\pm}\gamma(\lambda)$ for the sequence of transfer matrices as the limit as $t\to\pm\infty$ of the logarithm of $||U_\lambda(t)||^{1/|t|}$ whenever this limit exists. The above proposition says that, whenever these Lyapunov exponents exist, one must have:

$$\{\lambda\ ;\ ^{+}\gamma(\lambda) = 0\} \cup \{\lambda\ ;\ ^{-}\gamma(\lambda) = 0\} \subset \Sigma(H)$$

The spectral properties of the operator $H$ will be investigated by means of the solutions of the eigenvalue equation:

$$[H\psi](n) = \lambda\psi(n) \tag{III.69}$$

Let us denote by $p_\lambda(n)$ and $q_\lambda(n)$ the solutions of the equation $H\psi = \lambda\psi$ satisfying the initial conditions $p(0) = q(-1) = 1$ and $p(-1) = q(0) = 0$. It is readily seen that the propagator matrix $U_\lambda(n)$ can be written in the form:

$$U_\lambda(n) = \begin{pmatrix} p_\lambda(n) & q_\lambda(n) \\ p_\lambda(n-1) & q_\lambda(n-1) \end{pmatrix}.$$

**Proposition III.2.4** *With the previous notation we have:*

$$p_n(H)\delta_0 + q_n(H)\delta_{-1} = \delta_n \quad n \in \mathbb{Z}$$

*It follows that $(\delta_0, \delta_{-1})$ is a basis of a generating subspace.*

*Proof:*
This equality is obvious for $n = 0$ and $n = -1$. It is then proved by induction using the relation $H\delta_n = a_n\delta_{n-1} + a_{n-1}\delta_{n+1} + b_n\delta_n$. The last conclusion is an easy consequence of this equality since it implies that the subspace generated by the vectors obtained by computing the powers of $H$ applied to $\delta_0$ and $\delta_{-1}$ is dense in $\mathcal{H}$. ∎

Let $\sigma_{m,n}$ be the complex measure $< E(.)\delta_m, \delta_n >$. The spectral matrix measure $M$ is then defined as the $2 \times 2$ matrix

$$\begin{pmatrix} \sigma_{0,0} & \sigma_{0,-1} \\ \sigma_{-1,0} & \sigma_{-1,-1} \end{pmatrix}$$

and we denote by $\sigma$ the trace of $M$. The matrix measure $M$ is nonnegative in the sense of Definition I.4.3. The spectral type of the operator $H$ is given by $\sigma$ in the sense that for any $\psi \in \mathcal{H}$ the spectral measure $\sigma_\psi$ is absolutely continuous with respect to $\sigma$. More precisely one has:

**Proposition III.2.5** *With the previous notation one has:*

$$\sigma_{m,n} = \begin{pmatrix} p(m) & q(m) \end{pmatrix} \begin{pmatrix} \sigma_0 & \sigma_{-1,0} \\ \sigma_{-1,0} & \sigma_{-1} \end{pmatrix} \begin{pmatrix} p(n) \\ q(n) \end{pmatrix}$$

*Proof:*
The proof is a simple application of Proposition III.2.4. ∎

The Jacobi operator introduced in formula (III.55) has a unitarily equivalent analogue called the wave operator or the Helmotz operator. Let $\{c_n, d_n \,; n \in \mathbb{Z}\}$ be two sequences of real numbers such that $c_n > 0$ for all $n \in \mathbb{Z}$.
Then one defines the Helmotz operator $\tilde{H}$ by the formula:

$$(\tilde{H}\psi)(n) = \frac{1}{c_n}(-\psi(n-1) - \psi(n+1) + d_n\psi(n))$$

The operator $\tilde{H}$ is not selfadjoint for the usual inner product of $\ell^2(\mathbb{Z})$ but it is readily seen that if one introduces the "weighted" space

$$\tilde{\mathcal{H}} = \{\psi \,; \sum_{n=-\infty}^{+\infty} c_n|\psi(n)|^2 < \infty\}$$

which is a Hilbert space for the inner product:

$$< \varphi, \psi >_\sim = \sum_{n=-\infty}^{+\infty} c_n\varphi(n)\overline{\psi(n)},$$

then one has:

**Proposition III.2.6** *The mapping $U$ defined by*

$$(U\psi)(n) = \sqrt{c_n}\psi(n)$$

*is unitary from $\tilde{\mathcal{H}}$ onto $\ell^2(\mathbb{Z})$ and the Helmotz operator is mapped by $U$ on the Jacobi operator defined in (III.55) if we set:*

$$a_n = \frac{1}{\sqrt{c_n c_{n+1}}} \quad , \quad b_n = \frac{d_n}{c_n}$$

It follows from this proposition that the Jacobi and the Helmotz operators are unitary equivalent. The results proved above for the Jacobi matrices should apply to the Helmotz operator. Unfortunately, we will not be able to make much use of this unitary equivalence. Indeed, we are mostly interested in the case of random operators and the assumptions usually made on the coefficients of the Schrödinger and Jacobi matrices, i.e. independence or Markov property, will not be preserved by the unitary mapping $U$ !

We now return to the case of the classical Schrödinger operator defined by:

$$[H\psi](n) = -\psi(n+1) - \psi(n-1) + V(n)\psi(n).$$

Let $a$ and $b$ be two integers with $a < 0 < b$ and let $\Lambda$ be the box $\Lambda = [a, b-1]$. For two real numbers say $\alpha$ and $\beta$, the operator $H^\Lambda_{\alpha\beta}$ is defined in the box $\Lambda$ by the equations:

$$\begin{aligned}(H^\Lambda_{\alpha\beta}\psi)(a) &= \psi(a+1) + (V(a) - \alpha)\psi(a) \\ (H^\Lambda_{\alpha\beta}\psi)(n) &= (H\psi)(n) \qquad \text{for } a < n < b-1 \\ (H^\Lambda_{\alpha\beta}\psi)(b-1) &= -\psi(b-2) + (V(b-1) - \beta)\psi(b-1).\end{aligned}$$

It is easy to check that $H^\Lambda_{\alpha,\beta}$ is defined by a real symmetric matrix on the orthonormal basis $\{\delta_n; n = a, \ldots, b-1\}$ and thus is self-adjoint. If one defines the spectral matrix $M^\Lambda_{\alpha,\beta}$ in the same way as for the whole line. Note that the general theory developed at the beginning of this chapter yields the weak convergence of $M^\Lambda_{\alpha,\beta}$ to $M$. The spectral type of the operator $H^\Lambda_{\alpha,\beta}$ is then given by the trace of the matrix $M^\Lambda_{\alpha,\beta}$ that we denote by $\sigma^\Lambda_{\alpha,\beta}$.

We have already pointed out that the set of real numbers is isomorphic to the projective line up to a point at infinity. To obtain this identification one can write the one dimensional subspace associated to a real number $\gamma$ as the subspace generated by the vector $\binom{1}{\gamma}$ or $\binom{\gamma}{1}$. We also denote by $\gamma$ the associated subspace. If we choose the first mapping to represent the number $\alpha$ and the second to represent the number $\beta$ then any solution $\psi$ of the equation $H^\Lambda_{\alpha\beta}\psi = \lambda\psi$ is the restriction to the box $\Lambda$ of a solution of $H\psi = \lambda\psi$ with boundary conditions $\tilde{\psi}(a) \sim \alpha$ and $\tilde{\psi}(b) \sim \beta$. It follows that $\lambda$ is an eigenvalue of $H^\Lambda_{\alpha\beta}$ iff the vector $v = \binom{1}{\alpha}$ satisfies $U(b,a)v \sim \beta$. If $\alpha$ is a one dimensional subspace of $\mathbb{R}^2$ and $v \in \mathbb{R}^2$ we will write $v \sim \alpha$ if the direction of $\alpha$ is given by $v$.
As in the continuous case, for each complex number $\lambda$ with $\Im m\,\lambda > 0$ we denote by $G^\Lambda_{\alpha,\beta}(\lambda, x, y)$ the kernel of the resolvent operator $(H^\Lambda_{\alpha,\beta} - \lambda I)^{-1}$. For future reference, we state without proof some of the results which we proved in the continuous case and which we shall use in the lattice setting.

**Proposition III.2.7** *The Stieltjes transform of the spectral measure* $\sigma^{\Lambda}_{\alpha\beta}$ *has the following expression:*

$$\int \frac{d\sigma^{\Lambda}_{\alpha\beta}(t)}{t-\lambda} = G^{\Lambda}_{\alpha\beta}(\lambda,0,0) + G^{\Lambda}_{\alpha\beta}(\lambda,-1,-1)$$

**Definition III.2.8** *For two complex sequences* $u$ *and* $v$ *the Wronskian* $W(u,v)$ *is the sequence defined by:*

$$[W(u,v)](n) = u(n)v(n-1) - u(n-1)v(n)$$

**Proposition III.2.9** *If* $u$ *and* $v$ *are two solutions of the equation* $H\psi = \lambda\psi$ *then the Wronskian of* $u$ *and* $v$ *is constant. Moreover if* $\tilde{u}(a) \sim \alpha$ *and* $\tilde{v}(b) \sim \beta$ *then this constant is zero iff* $\lambda$ *is an eigenvalue of* $H^{\Lambda}_{\alpha\beta}$.

**Proposition III.2.10** *Let* $u$ *and* $v$ *be two solutions of the equation* $H\psi = \lambda\psi$ *satisfying the boundary conditions* $\tilde{u}(a) \sim \alpha$ *and* $\tilde{v}(b) \sim \beta$. *Then we have :*

$$G^{\Lambda}_{\alpha\beta}(\lambda,x,y) = \begin{cases} \dfrac{u(x)v(y)}{W(u,v)} & if\, x \leq y \\ \dfrac{u(y)v(x)}{W(u,v)} & if\, x \geq y \end{cases}$$

**Proposition III.2.11** *Let us denote the inner product and the norm restricted to the box* $\Lambda = [a,b]$ *by the subscript* $_\Lambda$. *Then we have:*

*(i) If* $\psi$ *and* $\varphi$ *are two functions defined on* $\Lambda$ *then*

$$< H\psi,\varphi >_\Lambda - < \psi, H\varphi >_\Lambda = W(\psi,\overline{\varphi})_a - W(\psi,\overline{\varphi})_b$$

*(ii) If* $\psi$ *is a solution of the equation* $H\psi = \lambda\psi$ *with* $\tilde{\psi}(a)$ *real then*

$$\Im m\,(\lambda)||\psi||^2_\Lambda = \Im m\,(\psi(b-1)\overline{\psi}(b))$$

## III.3 Approximations of the Spectral Measures

The spectral measure $\sigma$ is the central object of investigation and it is very convenient to approximate it by measures which can be computed in *finite boxes*. As a rule, we shall try to choose approximations which are easier to handle than the measures $\sigma^{\Lambda}_{\alpha,\beta}$ used in the definition of $\sigma$.

We already used the parametrization of the projective space $P(\mathbb{R}^2)$ given by the identification of a one dimensional subspace $\alpha$ with an angle in the interval $[0,\pi)$. One can also identify a one dimensional subspace with the unique real number $\beta$ such that the subspace contains the vector $(\beta, 1)$. This last parametrization does not work for the subspace of direction $(1,0)$ but this is not a problem as long as we integrate functions on the projective space with respect to a continuous measure. Notice that the Cauchy measure on the projective space is mapped onto the normalized Lebesgue measure of $[0,\pi)$ by the first parametrization and onto the usual Cauchy distribution on $\mathbb{R}$ by the second parametrization.

**Lemma III.3.1** *Let $u$ be a non-zero solution of the eigenvalue equation (III.1) with $\Im m\,\lambda > 0$ and let us assume that $\tilde{u}(a)$ is real. Then the function $z \hookrightarrow u(b) - z\dot{u}(b)$ does not vanish for $\Im m\, z \geq 0$. The same conclusion holds in the discrete case for the function $z \hookrightarrow u(b) - zu(b-1)$.*

*Proof:*
The Green's formula which we already recalled during our discussion of the $m$-functions implies that:

$$\Im m\,\lambda \int_{\Lambda} |u(t)|^2\, dt = \Im m\ (u(b)\overline{\dot{u}}(b))$$

which makes it impossible to have a complex number $z$ with $\Im m\, z > 0$ and such that $u(b) = z\dot{u}(b)$. The same proof works for the discrete case using the Green's formula in Proposition III.2.10. ∎

**Theorem III.3.2** *For each $\lambda \in \mathbb{R}$, let $u_\lambda$ be any solution of the eigenvalue equation (III.1) which satisfies the boundary condition $\tilde{u}(a) = \alpha$ and let $m$ be the Cauchy measure on the projective space. Then the averaged spectral measure $\sigma^{\Lambda}_{\alpha} = \int \sigma^{\Lambda}_{\alpha\beta}\, dm(\beta)$ is absolutely continuous with respect to the Lebesgue measure of $\mathbb{R}$ and the density is given by the formula:*

$$\frac{d\sigma^{\Lambda}_{\alpha}(\lambda)}{d\lambda} = \frac{1}{\pi}\frac{||\tilde{u_\lambda}(0)||^2}{||\tilde{u_\lambda}(b)||^2} \tag{III.70}$$

*Proof:*
The strategy of the proof is simple. We first integrate the Stieltjes transform of $\sigma^{\Lambda}_{\alpha\beta}$ with respect to $\beta$ using the properties of the Poisson kernel

of the upper half plane. Then, we use the de la Vallée Poussin theorem to control the limit $\Im m\,\lambda \rightarrow 0$ in order to recover the density. The proof below is written for the continuous case but one has only to replace $\dot{u}(0)$ by $u(-1)$ to obtain the discrete analog. The Wronskian appearing in the definition of the Green kernel can be computed at any point. The computation at the point $b$ yields $W(u,v) = u(b) - \beta\dot{u}(b)$. It follows that this Wronskian viewed as a function of $\beta$ has a non-vanishing extension to the closed upper half plane. Let us set:

$$U_\lambda(b) = \begin{pmatrix} \gamma & \delta \\ \zeta & \eta \end{pmatrix}$$

Since the propagator is unimodular one has:

$$U_\lambda(b)^{-1} = \begin{pmatrix} \eta & -\delta \\ -\zeta & \gamma \end{pmatrix}$$

It follows from Proposition III.2.7 that the Stieltjes transform of $\sigma^{\Lambda}_{\alpha,\beta}$ is the restriction to the real axis of a bounded holomorphic function in the closed upper half-plane defined by:

$$F(z) = \frac{(\eta z - \delta)u(0) + (\gamma - \zeta z)\dot{u}(0)}{(\gamma - \zeta z)u(0) - (\eta z - \delta)\dot{u}(0)} = \frac{N(z)}{D(z)}$$

Notice that, in this formula, the Wronskian has been evaluated at the origin. The integration of $F(\beta)$ with respect to the Cauchy measure $m$ is then immediate since we have the Poisson formula:

$$\int F(\beta)dm(\beta) = F(i).$$

This yields

$$\int \frac{d\sigma^{\Lambda}_{\alpha}(t)}{t - \lambda} = F(i).$$

Taking once more into account the unimodality of the matrix $U_\lambda(b)$, a straightforward computation shows that :

- $\lim_{\Im m\,\lambda \rightarrow 0} N(i)\overline{D(i)} = ||\tilde{u}(0)||^2$
- $\lim_{\Im m\,\lambda \rightarrow 0} |D(i)|^2 = (\gamma u(0) + \delta\dot{u}(0))^2 + (\zeta u(0) + \eta\dot{u}(0))^2$

This proves that the imaginary part of the Stieltjes transform of the measure $\sigma^{\Lambda}_{\alpha}$ has a finite limit at any point of the real axis. Then the theorems of Fatou and de la Vallée Poussin recalled in Chapter I implies that the measure $\sigma^{\Lambda}_{\alpha}$ is absolutely continuous with respect to the Lebesgue measure and the density is given by this limit. ∎

**Remark III.3.3** *The density can be rewritten using the special notation for the amplitude of a solution introduced in the section on the propagator.*

$$\frac{d\sigma^{\Lambda}_{\cdot,\alpha}(\lambda)}{d\lambda} = \frac{1}{\pi}\frac{1}{r_\lambda(a,0,\theta_\lambda(0,a,\alpha))^2}. \tag{III.71}$$

*If $a < 0 < b$ this formula is very useful because it separates the contributions of $\{V(t); t \geq 0\}$ and $\{V(t); t < 0\}$. Indeed, the values of $V(t)$ for $t < 0$ (and the boundary condition $\alpha$ at $a$) enter only in the computation of the angle $\theta = \theta_\lambda(0,a,\alpha)$ while the values of $V(t)$ for $t > 0$ enter only in the computation of $r_\lambda(b,0,\theta)$. Hence, if one can show that as $b$ tends to $\infty$, the amplitude $r_\lambda(b,0,\theta)$ remains bounded away from zero independently of $\theta$ in $[-\pi,+\pi)$ and $\lambda$ restricted to some set, then the limiting spectral measure is absolutely continuous on this set of energies $\lambda$, and the Radon-Nykodym density is bounded on this set of energies. As we already pointed out, the above estimate is particularly convenient to handle potential functions $V(t)$ which have different properties for $t > 0$ and for $t < 0$.*

Obviously one can integrate with respect to $\alpha$ rather than to $\beta$ and obtain the same kind of formula with a "time reversal". One can also integrate with respect to the two boundary conditions to obtain a second kind of averaged spectral measure.

**Proposition III.3.4** *Let us set $\sigma^\Lambda = \int \sigma^\Lambda_\alpha dm(\alpha)$. Then the measure $\sigma^\Lambda$ is absolutely continuous and its density with respect to the Lebesgue measure on $\mathbb{R}$ is given by:*

$$\frac{d\sigma^\Lambda(\lambda)}{d\lambda} = \frac{1}{\pi^2}\int_0^\pi \frac{d\alpha}{||U_\lambda(b)\alpha||^2||U_\lambda(a)\alpha||^2}$$

*Proof:*
Let us define the function $f$ on $P(\mathbb{R}^2)$ by $f(\overline{v}) = ||v||^2||U_\lambda(b)v||^{-2}$ where $v$ is any vector in $\mathbb{R}^2$ such that $v \sim \overline{v}$. Then one has:

$$\begin{aligned}
\int \frac{||U_\lambda^{-1}(a)v||^2}{||U_\lambda(b)U_\lambda^{-1}(a)v||^2}dm(\overline{v}) &= \int f(U_\lambda^{-1}(a)\overline{v})dm(\overline{v}) \\
&= \int f(\overline{v})d(U_\lambda^{-1}(a)m)(\overline{v}) \\
&= \int f(\overline{v})\frac{||v||^2}{||U_\lambda(a)v||^2}dm(\overline{v}) \\
&= \int \frac{||v||^4}{||U_\lambda(b)v||^2||U_\lambda(a)v||^2}dm(\overline{v}),
\end{aligned}$$

which gives the desired result provided we choose the parametrization $\overline{v} \sim (\sin\alpha, \cos\alpha)$. ∎

**Remark III.3.5** *We obtain once more a formula which separates the contributions of $[a,0]$ and $[0,b]$ to the spectral measure. As a consequence one can let $a \to -\infty$ keeping $b$ fixed. This differs from the previous one in the symmetric role played by the contributions of the potential function $V(t)$ for $t > 0$ and for $t < 0$.*

The essential approximation results which we will need are contained in the following:

**Theorem III.3.6** *Let $\phi$ be a continuous function with compact support on $\mathbb{R}$. Then,*

1. *the integrals*

$$\int \phi(\lambda)\, d\sigma_\alpha^\Lambda(\lambda) = \frac{1}{\pi}\int \phi(\lambda)\|U_\lambda^{-1}(a)\alpha\|^2\ \|U_\lambda(b,a)\alpha\|^{-2}\, d\lambda$$

*and*

$$\int \phi(\lambda)\, d\sigma^\Lambda(\lambda) = \frac{1}{\pi^2}\int_{\lambda\in\mathbb{R}}\int_{\alpha\in[0\ \pi]} \phi(\lambda)(\|U_\lambda(a)\alpha\|\ \|U_\lambda(b)\alpha\|)^{-2}\, d\alpha\, d\lambda$$

*converge to $\int\phi(\lambda)\, d\sigma(\lambda)$ when $a \to -\infty$ and $b \to +\infty$.*

2. *for any angle $\alpha \in [0,\pi)$ we have:*

$$\begin{aligned}\lim_{b\to\infty}\int \phi(\lambda)\, d\sigma_\alpha^{[0,b]}(\lambda) &= \lim_{b\to\infty}\frac{1}{\pi}\int \phi(\lambda)\|U_\lambda(b)\alpha\|^{-2}\, d\lambda \\ &= \int \phi(\lambda)\, d\sigma_\alpha^+(\lambda) \qquad \text{(III.72)}\end{aligned}$$

3. *the integral*

$$\frac{1}{\pi}\int_0^\pi \left(\int \phi(\lambda)\|U_\lambda(a)\alpha\|^{-2} d\sigma_\alpha^+(\lambda)\right) d\alpha$$

*converges to $\int\phi(\lambda)\, d\sigma(\lambda)$ when $a \to -\infty$.*

*Proof:*
The property 1. is a direct consequence of the formula obtained for $\sigma^\Lambda$ in Theorem III.3.4 and the weak convergence of the spectral measures in

an interval to the spectral measures of the operator on the half line (which was obtained by the convergence of the corresponding Stieltjes transforms). Property 2. follows easily our discussion of the $m$-function. Finally, Property 3. is a direct consequence of 2. and the formula proved in Proposition III.3.4. ∎

# III.4 Spectral Types

Throughout this section we try to treat simultaneously the continuous and the lattice cases. Recall that the operator $H$ is always self-adjoint in the lattice case while we have to assume that the potential function $V$ is locally square integrable and satisfies a condition of the form (III.2) to make sure that the operator $H$ is essentially self-adjoint on $C_c^\infty(\mathbb{R})$ in the continuous case.

## III.4.1 Absolutely Continuous Spectrum

The first important application of the above integration results on the spectral measures gives a very efficient way to investigate the absolutely continuous spectrum of one-dimensional operators.

**Theorem III.4.1** *Let $\mathcal{I}_0$ be an open subset of $\mathbb{R}$ such that for each compact set $K$ contained in $\mathcal{I}_0$ one has:*

$$\sup_{\lambda\in K,\, t\geq 0} \|U_\lambda(t)\| < \infty, \qquad \text{or} \qquad \sup_{\lambda\in K,\, t\leq 0} \|U_\lambda(t)\| < \infty.$$

*Then, $\mathcal{I}_0$ is contained in the spectrum of $H$ and the spectrum is purely absolutely continuous on $\mathcal{I}_0$ in the sense that $E_{|\mathcal{I}_0} = E_{ac|\mathcal{I}_0}$.*

*Proof:*
The above assumption, together with Lemma III.1.7 imply that $\mathcal{I}_0 \subset \Sigma(H)$. Let us prove the desired result under the assumption:

$$k = \sup_{\lambda\in K,\, t\leq 0} \|U_\lambda(t)\| < \infty,$$

the proof being similar in the other case. For any open subset $\mathcal{I}$ of $\mathcal{I}_0$ one has:

$$\begin{aligned}
\sigma(\mathcal{I}) &\leq \liminf_{b\to\infty} \sigma_\alpha^{[a,b]}(\mathcal{I}) && \text{(III.73)}\\
&\leq \frac{1}{\pi}\liminf_{b\to\infty} \int_{\mathcal{I}} \frac{d\lambda}{\inf_{\theta\in P(\mathbb{R}^2)} \|U_\lambda(t)\theta\|^2} && \text{(III.74)}\\
&\leq \frac{k^2}{\pi}|\mathcal{I}|, && \text{(III.75)}
\end{aligned}$$

which implies the desired conclusion. ■

**Remark III.4.2** *The above argument also shows that the density of the spectral measure $\sigma$ constructed with reading point 0 is locally bounded below away from 0 because for each compact set $K$ contained in $\mathcal{I}$ one has:*

$$\sigma(K) \geq \limsup \int_K \frac{d\lambda}{\sup_{\theta\in P(\mathbb{R}^2)} \|U_\lambda(t)\theta\|^2} \geq \frac{1}{\pi k^2}|K|.$$

**Remark III.4.3** *We notice that the free one-dimensional Hamiltonian is the obvious first application of this result, and consequently we recover the fact that the free Hamiltonian has purely absolutely continuous spectrum. Indeed, $\lambda = 0$ cannot be an eigenvalue for none of the solutions of the eigenvalue equation can be square integrable.*

**Corollary III.4.4** *Let $V$ be a real valued locally integrable function which is bounded below near $-\infty$ and $+\infty$ by a function of the form $-c(t^2+1)$ for some $c > 0$ and which is integrable near $-\infty$ or $+\infty$. Then the spectrum of $H = H_0 + V$ contains $(0,\infty)$ where it is purely absolutely continuous.*

*Proof:*
Let us assume that the potential is integrable near $+\infty$ for example. Using (III.33) and the bound on the free propagator computed earlier we have:

$$\sup_{s,t\geq 0} \|U_\lambda(t,s)\| \leq 2\max(\lambda,\lambda^{-1})e^{2\max(\lambda,\lambda^{-1})\int_0^\infty |V(t)|\,dt}$$

and the conclusion follows from the above theorem. ■

The application of the above theorem to the case of periodic potentials will be given in Chapter VII.

We close this subsection with some negative results proving the absence of the absolutely continuous spectrum. We give them as applications of the Kato-Birman theory by comparing a given operator to an operator without absolutely continuous spectrum. We discuss an example in the lattice case and we give the corresponding continuous result without proof. But first we prove a simple estimate:

**Lemma III.4.5** *If $\{H[i,j];\ i,j \in \mathbb{Z}\}$ is any real symmetric matrix satisfying:*

$$H[i,j] = 0 \text{ if } |i-j| \neq 1 \quad \text{and} \quad |H[i,i\pm 1]| \leq 1,$$

*then for any real sequence $\{V(n);\ n \in \mathbb{Z}\}$ we have:*

$$\|[H+V+iI]^{-1}\delta_n\| \leq \frac{3}{|V(n)+i|}$$

*Proof:*
Using the identity:

$$[H+V+iI]^{-1} = [1+[H+V+iI]^{-1}H][V+i]^{-1}$$

the result follows from the fact that $\|[H+V+iI]^{-1}\| \leq 1$ and $\|H\| \leq 2$. ∎

Here is the application of the Kato-Birman theory which we announced earlier.

**Proposition III.4.6** *If $\{V(n);\ n \in \mathbb{Z}\}$ is any real sequence satisfying:*

$$\limsup_{n\to\infty} V(n) = \limsup_{n\to-\infty} V(n) = \infty,$$

*then the Hamiltonian $H = H_0 + V$ on $\ell^2(\mathbb{Z})$ has no absolutely continuous spectrum, i.e. $\Sigma_{ac}(H) = \emptyset$.*

*Proof:*
Let $\{n_j;\ j \in \mathbb{Z}\}$ be a doubly infinite sequence of integers such that one has $\lim_{j\to\infty} n_j = \infty$, $\lim_{j\to-\infty} n_j = -\infty$ and such that:

$$\sum_{j\in\mathbb{Z}} \frac{1}{|V(n_j)|} < \infty.$$

Notice that the free Hamiltonian $H_0$ can be written in the form $H_0 = \sum_{n\in\mathbb{Z}} B_n$ where all but two of the matrix elements of the operators $B_n$'s are zero and $B_n[n, n+1] = B_n[n+1, n] = 1$. We then define inductively the operators $H_j$'s for $j \geq 1$ by:

$$H_1 = H_0 + V \qquad \text{and} \qquad H_{j+1} = H_j - B_{n_j} - B_{-n_j}$$

Lemma III.4.5 and the resolvent identity imply that:

$$\|[H_j + iI]^{-1} - [H_{j+1} + iI]^{-1}\| \leq 6(|V(n_j)|^{-1} + |V(n_{-j})|^{-1}).$$

Since the rank of this operator is at most 4, its trace class norm satisfies:

$$\|[H_j + iI]^{-1} - [H_{j+1} + iI]^{-1}\|_1 \leq 24(|V(n_j)|^{-1} + |V(n_{-j})|^{-1}).$$

This implies that $\{[H_j + iI]^{-1}; j \geq 1\}$ is a Cauchy sequence in the trace class norm and consequently that the operator $[H + iI]^{-1} - [H_\infty + iI]^{-1}$ is trace class where $H_\infty$ denotes the strong limit of the operators $H_j$'s. Since $H_\infty$ is the direct sum of finite matrices we have $\Sigma_{ac}(H_\infty) = \emptyset$ and the Birman-Kato theorem gives the desired result. ∎

The statement and the proof of the corresponding result in the continuous case are more involved. The corresponding technical estimate is the following:

**Lemma III.4.7** *If $H$ denotes the self-adjoint operator $-d^2/dt^2$ with Dirichlet boundary conditions on some set of points and if $\tilde{H}$ is the same operator with an extra Dirichlet boundary condition at the origin, then for any nonnegative potential function $V$ satisfying $V(t) \geq \lambda^2$ for $|t| \leq T$ with $\lambda > 10$ and $\lambda T > 10$, then the operator $[H+V+I]^{-1} - [\tilde{H}+V+I]^{-1}$ is nonnegative and trace class and its trace is at most of the order of $O(\lambda^{-2}) + O(e^{-2\lambda T})$.*

Using this result it can be proven that:

**Proposition III.4.8** *Let $V$ be a bounded below measurable real valued function for which there exist doubly infinite sequences, $\{t_n; n \in \mathbb{Z}\}$ such that $\lim_{n\to\infty} t_n = \infty$ and $\lim_{n\to-\infty} t_n = -\infty$, and $\{T_n; n \in \mathbb{Z}\}$ and $\{h_n; n \in \mathbb{Z}\}$ such that $\lim_{|n|\to\infty} h_n = \infty$ and $\lim_{|n|\to\infty} h_n T_n^2 = \infty$, and such that:*

1. *$V(t) \geq h_n$ if $|t - t_n| \leq T_n$*

*then the self-adjoint operator $-(d^2/dt^2) + V$ on $L^2(\mathbb{R}, dt)$ has no absolutely continuous spectrum.*

The assumptions mean that there exists a sequence of potential barriers of height $h_n$ and width $2T_n$ centered at the points $t_n$ going up to both infinities, the condition $h_n T_n^2 \to \infty$ meaning that these barriers are effective in the sense that the tunnelling probabilities at energies small compared to $h_n$ go to zero. See [305] for the details.

## III.4.2 Singular Spectrum

We now introduce the essential concept of hyperbolic behavior of the propagator $U_\lambda(t)$ associated to a Schrödinger operator on $\mathbb{R}$ or $\mathbb{Z}$. We assume that:

- $\lim_{n\to\pm\infty} \frac{1}{|n|} \log(1+|V(n)|) = 0$ in the discrete case
- $\lim_{n\to\pm\infty} \frac{1}{|n|} \int_n^{n+1} |V(t)|\, dt = 0$ in the continuous case.

These assumptions are not really restrictive since, at least in the random case, they will be satisfied for almost all trajectories for a wide class of potential processes.

In the discrete case the transfer matrices have already been defined. In the continuous case if we only consider integer values of the time parameter then $U_\lambda(n)$ can be written as a product of transfer matrices $g(n)$ defined by:

$$g_\lambda(n) = \begin{cases} U_\lambda(n+1)U_\lambda(n)^{-1} & \text{if } n \geq 0 \\ U_\lambda(n-1)U_\lambda(n)^{-1} & \text{if } n < 0. \end{cases}$$

It follows from the above assumptions that for any real or complex number $\lambda$ we have:

- $\lim_{n\to\pm\infty} \frac{1}{|n|} \log \|g_\lambda(n)\| = 0$
- if we set $\breve{U}_\lambda(n) = \sup_{t\in[0\ 1]}(\log \|\breve{U}_\lambda(n+t,n)\|)$ then:

$$\lim_{n\to\pm\infty} \frac{1}{|n|} \breve{U}_\lambda(n) = 0.$$

**Definition III.4.9** *Let $\{U_\lambda(t);\, t \in T, \lambda \in \mathbb{R}\}$ be the propagator of the Schrödinger operator $H$ on $T = \mathbb{R}$ or $T = \mathbb{Z}$. The real number $\lambda$ is said to be hyperbolic if there exist real numbers $\gamma^\pm(\lambda) > 0$ such that :*

$$\lim_{n\to\pm\infty} \frac{1}{|n|} \log \|U_\lambda(n)\| = \gamma^\pm(\lambda)$$

*The set of hyperbolic values of* $\lambda$ *will be denoted by* $\mathrm{hyp}(H)$.

The measurability of $\mathrm{hyp}(H)$ is obvious since $U_\lambda(t)$ is a continuous function of $\lambda$.

**Proposition III.4.10** *If* $\lambda \in \mathrm{hyp}(H)$ *then there exist two one dimensional subspaces of* $\mathbb{R}^2$ *say* ${}^+\mathcal{V}(\lambda)$ *and* ${}^-\mathcal{V}(\lambda)$ *such that for a non-zero vector* $v$ *in* $\mathbb{R}^2$ *we have:*

$$v \in^{\pm} \mathcal{V}(\lambda) \iff \lim_{t\to\pm\infty} \frac{1}{|t|} \log \|U_\lambda(t)v\| = -\gamma^{\pm}(\lambda)$$

$$v \notin^{\pm} \mathcal{V}(\lambda) \iff \lim_{t\to\pm\infty} \frac{1}{|t|} \log \|U_\lambda(t)v\| = \gamma^{\pm}(\lambda).$$

*Proof:*
For integer values of $t$ this is just a consequence of the deterministic Oseledec theorem stated in Proposition IV.2.4. For continuous values of the parameter we write for $n \leq t \leq n+1$

$$|\log \|U_\lambda(t)v\| - \log \|U_\lambda(n)v\|| = |\log \frac{\|U_\lambda(t)v\|}{\|U_\lambda(n)v\|}| \leq \check{U_\lambda}(n)$$

and this yields the result for a continuous parameter $t$. ∎

**Proposition III.4.11** *Let* $\lambda \in \mathrm{hyp}(H)$, *then* $\lambda$ *is an eigenvalue of* $H$ *iff* ${}^+\mathcal{V}(\lambda) \cap^- \mathcal{V}(\lambda) \neq \{0\}$.

*Proof:*
Let $\lambda$ be in $\mathrm{hyp}(H)$ and $\psi$ be a non-zero solution of $H\psi = \lambda\psi$. Then there exist positive numbers $\alpha$ and $t_0$ such that :

$$\tilde{\psi}(0) \in^+ \mathcal{V}(\lambda) \cap \mathcal{V}^-(\lambda) \quad \Rightarrow \quad \|\tilde{\psi}(t)\| \leq e^{|t|\alpha} \quad \text{for } |t| \geq t_0$$

$$\tilde{\psi}(0) \notin^+ \mathcal{V}(\lambda) \cap^- \mathcal{V}(\lambda) \quad \Rightarrow \quad \begin{cases} \|\tilde{\psi}(t)\| \geq e^{t\alpha} & \text{for } t \geq t_0 \\ \text{or} & \\ \|\tilde{\psi}(t)\| \geq e^{-t\alpha} & \text{for } t \leq -t_0 \end{cases}$$

It is obvious that $\lambda$ is an eigenvalue in the first alternative and it remains to prove that it is not an eigenvalue in the second. The conclusion is immediate in the discrete case since:

$$\sum_{n=-\infty}^{+\infty} \|\tilde{\psi}(n)\|^2 = 2 \sum_{n=-\infty}^{+\infty} \|\psi(n)\|^2$$

In the continuous time case it is enough to remark that

$$\dot{\psi}(t) = \dot{\psi}(0) + \int_0^t \ddot{\psi}(s)\,ds = \dot{\psi}(0) + \int_0^t (V(s) - \lambda)\psi(s)\,ds$$

Thus if we suppose that $\psi$ is a square integrable function, the Cauchy-Schwarz inequality added to the polynomial bound on the potential imply a polynomial bound on $\dot{\psi}$. It follows that $\psi$ is exponentially increasing in at least a direction of $\mathbb{R}$ and we get a contradiction. ∎

**Theorem III.4.12** *Let $m$ be a nonnegative continuous measure on $\mathbb{R}$ which is carried by* hyp$(H)$. *Then $m$ is orthogonal to the spectral measure $\sigma$ of $H$.*

*Proof:*
The measure $m$ is continuous and the set of eigenvalues of $H$ is at most countable. Thus we can conclude that for $m$-almost all $\lambda$ we have ${}^{+}\mathcal{V}(\lambda) \cap {}^{-}\mathcal{V}(\lambda) = \{0\}$. It follows that for $m$-almost all $\lambda$ and for any non-zero solution of $H\psi = \lambda\psi$, $||\tilde{\psi}(t)||$ is growing exponentially fast in at least one direction of $\mathbb{R}$. But we know from Theorem II.4.5 that for $\sigma$-almost all $\lambda$ there exists a non-zero solution of $H\psi = \lambda\psi$ which is bounded by a polynomial in $t$. The formula:

$$\dot{\psi}(t) = \dot{\psi}(0) + \int_0^t \ddot{\psi}(s)\,ds = \dot{\psi}(0) + \int_0^t (V(s) - \lambda)\psi(s)\,ds$$

gives a polynomial bound on $|\tilde{\psi}(t)|$ because of our assumption on the potential. It follows that the compliment of the set of eigenvalues of $H$ in hyp$(H)$ is of zero $\sigma$-measure and one can conclude that $\sigma$ is orthogonal to $m$. ∎

### III.4.3 Pure Point Spectrum

**Theorem III.4.13** *If the spectral measure $\sigma$ of $H$ is carried by* hyp$(H)$, *then $\sigma$ is pure point. Moreover any eigenvector $\psi_\lambda(t)$ corresponding to an eigenvalue $\lambda$ satisfies:*

$$\lim_{t\to\pm\infty} \frac{1}{|t|} \log ||\tilde{\psi}_\lambda(t)|| = -\gamma^{\pm}(\lambda).$$

*Proof:*

Let us write $\sigma = \sigma_c + \sigma_{pp}$ where $\sigma_{pp}$ is the pure point part of $\sigma$ and $\sigma_c$ the continuous part. Then Theorem III.4.12 implies that $\sigma$ is orthogonal to $\sigma_c$ thus $\sigma = \sigma_{pp}$. Moreover each eigenvalue $\lambda$ is hyperbolic and the conclusion follows from Proposition III.4.11. ■

It seems difficult to believe that the hypothesis of Theorem III.4.13 can be satisfied for a "generic" Schrödinger operator. In fact, for random Schrödinger operators $H(\omega)$ constructed from an ergodic process, for each energy $\lambda$ for which the Lyapunov exponent is positive, i.e. $\gamma(\lambda) > 0$, then $\lambda \in \mathrm{hyp}(H_\omega)$ and ${}^+\mathcal{V}_\omega \cap {}^- \mathcal{V}_\omega(\lambda) = \{0\}$ for almost all $\omega$. It follows from Proposition III.4.11 that such a fixed $\lambda$ is not an eigenvalue of $H(\omega)$ for almost all $\omega$! Nevertheless it happens quite often that for almost all $\omega$, the spectral measures of $H(\omega)$ are carried by $\mathrm{hyp}(H_\omega)$ and thus these measures have to be pure point !!

We now turn our attention to a quite different kind of criterion insuring the pure point spectrum property that does not refer directly to the Oseledec theorem or the slow-growth of the generalized eigenfunctions. Roughly speaking the basic argument is the following: if the integrals of a sequence of integrable functions are exponentially decreasing then almost surely this sequence of functions is pointwise exponentially decreasing.

In the following the unit ball of $\mathbb{R}^2$ is denoted by $B_2$.

**Proposition III.4.14** *Let $\mathcal{I}$ be an open bounded interval of the real line such that there exists $0 < \delta < 1$ with*

$$\int_{\mathcal{I}} \inf_{v \in B_2} \sum_{n=-\infty}^{+\infty} ||U_\lambda(n)v|| \delta^{-|n|} \, d\sigma(\lambda) < \infty$$

*Then $\sigma$ is pure point on $\mathcal{I}$ and for an eigenvalue $\lambda \in \mathcal{I}$ the eigenfunction $\psi$ decays exponentially fast. More precisely for any eigenvalue $\lambda \in \mathcal{I}$ and $\delta < \rho < 1$ there exists $C(\lambda) < \infty$ such that we have :*

$$|\tilde{\psi}(t)| \leq C(\lambda) \rho^{|t|}$$

*Proof:*
First we remark that if we set

$$D_N(\lambda) = \inf_{v \in B_2} \sum_{n=-N}^{N} ||U_\lambda(n)v|| \delta^{-|n|}$$

then $D_N(\lambda)$ is an increasing sequence of continuous functions and the limit $D(\lambda)$ is measurable. An integrable function is almost everywhere finite thus one can deduce from the hypothesis that for $\sigma$-almost all $\lambda \in \mathcal{I}$ there exists a unit vector $v_\lambda \in \mathbb{R}^2$ such that :

$$\sum_{n=-\infty}^{+\infty} ||U_\lambda(n)v_\lambda||\delta^{-|n|} = C(\lambda) < +\infty$$

The proof is complete in the discrete case. In the continuous time case we write for $n \leq t \leq n+1$ :

$$||U(t)v|| = ||U(t,n)U(n)v|| \leq \check{U}(n)||U(n)v||$$

The hypothesis on the sequence $\check{U}_n$ implies that for any positive $\epsilon$ there exists an integer $N$ such that $\check{U}(n) \leq \exp(|n|\epsilon)$ for $|n| \geq N$ and this yields the conclusion. ∎

Using the approximations of the spectral measure obtained by restriction of the operator $H$ to boxes it is possible to give a more tractable criterion. Let $\Lambda$ be a box and $\sigma^\Lambda$ be a positive measure on the real line converging weakly to the spectral measure $\sigma$ when the box $\Lambda$ converges to the whole space.

**Corollary III.4.15** *Let $\mathcal{I}$ be an open bounded interval of the real line such that there exist $0 < \rho < 1$, $C < \infty$ such that for each box $\Lambda$ one can find a measurable mapping $\lambda \hookrightarrow v^\Lambda(\lambda)$ satisfying :*

$$\int_{\mathcal{I}} ||U_\lambda(n)v^\Lambda(\lambda)||\, d\sigma^\Lambda(\lambda) \leq C\rho^{|n|} \qquad \textit{for } n \in \Lambda$$

*Then the hypothesis of Proposition III.4.14 is fulfilled for any $\rho < \delta < 1$.*

*Proof:*
Let $\rho < \delta < 1$, using the notations of the proof of Proposition III.4.14 one obtains that there exists a finite constant $D$ such that for $[-N, N] \subset \Lambda$ one has:

$$\int_{\mathcal{I}} D_N(\lambda)\, d\sigma^\Lambda(\lambda) \leq D$$

The function $\lambda \hookrightarrow \mathbf{1}_{\mathcal{I}}(\lambda)D_N(\lambda)$ is lower semi-continuous. Hence, Fatou's lemma implies that

$$\int_{\mathcal{I}} D_N(\lambda)\, d\sigma(\lambda) \leq D \quad \text{and} \quad \int_{\mathcal{I}} D(\lambda)\, d\sigma(\lambda) = \lim_{N\to\infty} \uparrow \int_{\mathcal{I}} D_N(\lambda)\, d\sigma(\lambda) \leq D$$

and this yields the desired result. ■

In fact, the choice of the vector $v^\Lambda(\lambda)$ is not arbitrary and is imposed by the construction of the Oseledec contractive vector. Namely for a given boundary condition $\alpha$ at the left end point $a$ of the box $\Lambda$ we will choose :

$$v^\Lambda(\lambda) = \frac{U^{-1}(a)\alpha}{||U^{-1}(a)\alpha||}$$

Then if we use the approximations of the spectral measure proved in Theorem III.3.6 we finally obtain the following criterion:

**Theorem III.4.16** *Let $\mathcal{I}$ be an open bounded interval of the real line and $\alpha \in [0, \pi]$ be a fixed angle. Assume that there exist $0 < \rho < 1$ and $C < \infty$ such that for each box $\Lambda$ if we denote by $u$ the solution of $Hu = \lambda u$ with $\tilde{u}(a) = \alpha$ one has :*

$$\int_{\mathcal{I}} ||\tilde{u}(n)|| \; ||\tilde{u}(0)|| \; ||\tilde{u}(b)||^{-2} \, d\lambda \leq C\rho^{|n|} \qquad \textit{for } n \in \Lambda$$

*Then $\sigma$ is pure point on $\mathcal{I}$ and for any $\rho < \delta < 1$ and any eigenvalue $\lambda \in \mathcal{I}$ there exists $C(\lambda) < \infty$ such that the associated eigenfunction $\psi$ satisfies :*

$$|\tilde{\psi}(t)| \leq C(\lambda)\delta^{|t|}$$

It turns out that this criterion is well adapted to the proof of localization in the Markovian random case since it reduces the problem to the spectral properties of operators defining the Laplace Transform.

# III.5 Quasi-one Dimensional Schrödinger Operators

## III.5.1 The Schrödinger Operator in a Strip

Let $\ell$ be a positive integer and $d = 2\ell$. As usual we denote by $T$ the parameter set which will be $\mathbb{R}$ or $\mathbb{Z}$. We define the strip of width $\ell$ as the set $\{(i,t); i = 1 \ldots \ell, t \in T\}$. It is the union of $\ell$ copies of the one

dimensional set $T$. Then the Schrödinger operator on the strip can be viewed as $\ell$ coupled one dimensional operators. We also define the "slice" of index $t$ as the subset $\{(i,t); i = 1 \dots \ell\}$.
Let $V_1(t), \cdots, V_\ell(t)$ be $\ell$ real valued "potential" functions defined on the set $T$. Then one defines the $\ell \times \ell$ potential matrix $V(t)$ by:

$$V_{ij}(t) = \begin{cases} V_i(t) & \text{if } \ i = j \\ -1 & \text{if } \ |i-j| = 1 \\ 0 & \text{if } \ |i-j| > 1 \end{cases}$$

In order to prove spectral properties of the Schrödinger in a strip we will often look at its restriction to "boxes". Then we will use a limiting procedure called "thermodynamic limit", to let these boxes grow to the whole space. Let $a$ and $b$ be two elements of $T$. The box $\Lambda$ is defined as the set $\Lambda = \{(i,t); i = 1 \dots \ell, t \in [a,b]\}$ in the continuous case, and $\Lambda = \{(i,t); i = 1 \dots \ell, t \in [a, b-1]\}$ in the discrete case.

**The Discrete Case**

The Schrödinger operator $H$ on the strip of width $\ell$ acts on the vector sequences $\psi(n)$ with $\ell$ complex components $(\psi_1(n), \cdots, \psi_\ell(n))$ by the formula:

$$[H\psi](n) = -\psi(n+1) - \psi(n-1) + V(n)\psi(n).$$

We remark that this operator restricted to a "slice" with Dirichlet boundary conditions is nothing but the one dimensional Schrödinger operator on the box $[1, \dots, \ell]$.
For each complex number $\lambda$ we define the transfer matrix $g_\lambda(n) \in SP(\ell, \mathbb{R})$ by:

$$g_\lambda(n) = \begin{pmatrix} V(n) - \lambda I_\ell & -I_\ell \\ I_\ell & 0 \end{pmatrix}$$

where $I_\ell$ is the identity matrix of order $\ell$. The propagator matrix $U_\lambda(n)$ is then given by the products:

$$U_\lambda(n) = \begin{cases} g_\lambda(n-1) \ \cdots \ g_\lambda(0) & \text{if } \ n \geq 0 \\ g_\lambda(n)^{-1} \ \cdots \ g_\lambda(-1)^{-1} & \text{if } \ n < 0. \end{cases}$$

We will also consider the propagator matrix with two time parameters defined by $U_\lambda(m,n) = U_\lambda(m)U_\lambda(n)^{-1}$. If we denote by $\tilde{\psi}(n)$ the $d$-dimensional vector $\tilde{\psi}(n) = \begin{pmatrix} \psi(n) \\ \psi(n-1) \end{pmatrix}$ it is easy to check that any solution of the "eigenvalue" equation $H\psi = \lambda\psi$ can be expressed by the formula:

$$\tilde{\psi}(n) = U_\lambda(n)\tilde{\psi}(0)$$

One can also consider the operator $H$ acting on sequences of $\ell \times \ell$ matrices by the formula

$$(HM)_n = -M_{n-1} - M_{n+1} + V(n)M_n$$

Let us denote by $P_n(\lambda)$ and $Q_n(\lambda)$ the solutions of the equation $HM = \lambda M$ satisfying the initial conditions $P_0 = Q_{-1} = I_\ell$ and $P_{-1} = Q_0 = 0$. It is readily seen that the propagator matrix $U_n$ can be written in the form :

$$U_\lambda(n) = \begin{pmatrix} P_n & Q_n \\ P_{n-1} & Q_{n-1} \end{pmatrix}$$

Let $\mathcal{H}$ be the Hilbert space of square summable sequences $\psi$. It follows from the above general discussion that $H$ is a self adjoint operator on $\mathcal{H}$ for $H = H_0 + V$ with $H_0$ bounded since $[H_0\psi](n) = -\psi(n-1) - \psi(n+1)$. Let us denote by $E$ the spectral resolution of the identity of $H$ and by $\{\delta_{in}; i = 1 \cdots \ell, n \in \mathbb{Z}\}$ the canonical basis of $\mathcal{H}$. One defines an $\ell$-system $u$ as a sequence $(u_1, \cdots, u_\ell)$ of vectors in $\mathbb{C}^\ell$. An $\ell \times \ell$ matrix $M$ with operator entries acts on the set of $\ell$-systems by the formula

$$(Mu)_i = \sum_{j=1}^{\ell} M_{ij} u_j .$$

The entries of the matrices $P_\lambda(n)$ and $Q_\lambda(n)$ are polynomial in $\lambda$ thus the operator matrices $P_n(H)$ and $Q_n(H)$ are well defined. Let $\delta_n$ be the $\ell$-system $(\delta_{1n}, \cdots, \delta_{\ell n})$.

**Proposition III.5.1** *With the previous notations we have:*

$$P_n(H)\delta_0 + Q_n(H)\delta_{-1} = \delta_n \quad n \in \mathbb{Z}$$

*It follows that the two $\ell$-systems $(\delta_0, \delta_{-1})$ form a basis of a generating subspace.*

*Proof:*
This equality is obvious for $n = 0$ and $n = -1$. It is then proved by induction using the relation $H\delta_n = -\delta_{n-1} - \delta_{n+1} + V(n)\delta_n$. The last conclusion is an easy consequence of this equality since it implies that the vector subspace generated by the powers of $H$ applied to the $\ell$-systems $\delta_0$ and $\delta_{-1}$ is dense in $\mathcal{H}$. ∎

Let $\sigma_{m,n}$ be the $\ell \times \ell$ matrix valued measure whose $(i,j)$-th entry is the complex measure $< E(.)\delta_{i,m}, \delta_{j,n} >$. The spectral matrix measure $M$ is then defined as the $d \times d$ matrix

$$\begin{pmatrix} \sigma_{0,0} & \sigma_{0,-1} \\ \sigma_{-1,0} & \sigma_{-1,-1} \end{pmatrix}$$

and we denote by $\sigma$ the trace of $M$. The matrix measure $M$ is nonnegative in the sense of Definition I.4.3. Moreover the discussion of Section I.4 of Chapter II implies that the spectral type of the operator $H$ is given by $\sigma$ in the sense that for any $\psi \in \mathcal{H}$ the spectral measure $\sigma_\psi$ is absolutely continuous with respect to $\sigma$.

Let now $a$ and $b$ be two integers with $a < 0 < b$ and let $\Lambda$ be the associated box. For two real symmetric matrices of order $\ell$, say $\alpha$ and $\beta$, the operator $H^\Lambda_{\alpha\beta}$ is defined in the box $\Lambda$ by the equations:

$$\begin{aligned} (H^\Lambda_{\alpha\beta}\psi)(a) &= \psi(a+1) + (V(a) - \alpha)\psi(a) \\ (H^\Lambda_{\alpha\beta}\psi)(n) &= (H\psi)(n) \qquad\qquad \text{for } a < n < b-1 \\ (H^\Lambda_{\alpha\beta}\psi)(b-1) &= -\psi(b-2) + (V(b-1) - \beta)\psi(b-1). \end{aligned}$$

It is easy to check that $H^\Lambda_{\alpha,\beta}$ is defined by a real symmetric matrix on the orthonormal basis $\{\delta_{in}; i = 1 \cdots \ell, n = a \cdots b-1\}$ and thus is self-adjoint. If one defines the spectral matrix $M^\Lambda_{\alpha,\beta}$ in the same way as for the whole strip, the general theory developed at the beginning of this chapter yields the weak convergence of $M^\Lambda_{\alpha,\beta}$ to $M$. The spectral type of the operator $H^\Lambda_{\alpha,\beta}$ is then given by the trace of the matrix $M^\Lambda_{\alpha,\beta}$ that we denote by $\sigma^\Lambda_{\alpha,\beta}$.

We shall see in Chapter IV Section 6 that the set $\mathcal{M}(\ell)$ of real symmetric matrices of order $\ell$ is isomorphic to the boundary $L_\ell$ up to a projective submanifold of lower dimension. To obtain this identification one can write the Lagrangian $\ell$-dimensional subspace associated to a symmetric matrix $\gamma$ as the subspace generated by the columns of the $d \times \ell$ matrix $\binom{I}{\gamma}$ or $\binom{\gamma}{I}$. We also denote by $\gamma$ the associated Lagangian subspace. If we choose the first mapping to represent the matrix $\alpha$ and the second to represent the matrix $\beta$ then any solution $\psi$ of the equation $H^\Lambda_{\alpha\beta}\psi = \lambda\psi$ is the restriction to the box $\Lambda$ of a solution of $H\psi = \lambda\psi$ with boundary conditions $\tilde{\psi}(a) \in \alpha$ and $\tilde{\psi}(b) \in \beta$. If follows that $\lambda$ is an eigenvalue of $H^\Lambda_{\alpha\beta}$ iff there exists a non-zero vector $v \in \alpha$ such that $U_\lambda(b,a)v \in \beta$.

**The Continuous Case**

We define formally the operator $H$ acting on a locally integrable vector function $\psi(t)$ with $\ell$ complex coordinates $(\psi_1(t), \cdots, \psi_\ell(t))$ by the formula:

$$[H\psi](t) = -\ddot{\psi}(t) + V(t)\psi(t)$$

The derivatives are to be understood in the sense of distributions and $V(t)$ is the $\ell \times \ell$ potential matrix defined earlier. If we assume that for $i = 1, \cdots, \ell$ the functions $V_i(t)$ are locally square integrable, then one can define the propagator $U_\lambda(t)$ (which is a matrix in $SP(\ell, \mathbb{R})$) as the solution of the equation:

$$\dot{U}_\lambda(t) = \begin{pmatrix} 0 & I_\ell \\ V(t) - \lambda I_\ell & 0 \end{pmatrix} U_\lambda(t)$$

with the initial value $U_\lambda(0) = I_\ell$. We will also consider the propagator with two time parameters defined by $U_\lambda(s,t) = U_\lambda(s)U_\lambda(t)^{-1}$.
Let $P_t(\lambda)$ and $Q_t(\lambda)$ be the $\ell \times \ell$ matrix solutions of the equation:

$$-\ddot{M}_t + V(t)M_t = \lambda M_t$$

with the boundary conditions $P_0 = \dot{Q}_0 = I_\ell$ and $\dot{P}_0 = Q_0 = 0$.
It is readily seen that the propagator $U_\lambda(t)$ can be written as:

$$U_\lambda(t) = \begin{pmatrix} P_t & Q_t \\ \dot{P}_t & \dot{Q}_t \end{pmatrix}$$

Let us use the notation $\tilde{\psi}$ for the $d$ dimensional vector $\tilde{\psi}(t) = \begin{pmatrix} \psi(t) \\ \dot{\psi}(t) \end{pmatrix}$. Then any solution of the equation $H\psi = \lambda\psi$ can be expressed by means of the propagator $U_\lambda$ as $\tilde{\psi}(t) = U_\lambda(t)\tilde{\psi}(0)$. If we assume that the potential function $V(t)$ is such that the operator $H$ be essentially self-adjoint on the Hilbert space $\mathcal{H} = L^2(\mathbb{R}, dt; \mathbb{C}^\ell)$ of square integrable $\mathbb{C}^\ell$-valued functions on $\mathbb{R}$ (for instance if there exists a positive constant $c$ such that $V_i(t) \geq -c(1+t^2)$ for $i = 1, \ldots, \ell$), then one can construct a spectral matrix for $H$ exactly in the same way as we did in the one dimensional case.
For two real numbers $a < 0 \leq b$ let $\Lambda$ be the associated box and let $\alpha$ and $\beta$ be two Lagrangian subspaces of dimension $\ell$ in $\mathbb{R}^d$. See Chapter IV below for the definition of Lagrangian subspaces. Then we define the domain

$\mathcal{D}^\Lambda_{\alpha,\beta} = \{\psi \in L^2(\Lambda); \psi$ is differentiable, $\dot{\psi}$ is absolutely continuous and $-\ddot{\psi} + V\psi \in L^2(\Lambda), \tilde{\psi}(a) \in \alpha, \tilde{\psi}(b) \in \beta\}$.

The operator $H^{\Lambda}_{\alpha,\beta}$ defined on the domain $\mathcal{D}^{\Lambda}_{\alpha,\beta}$ is self-adjoint and has a compact resolvent. Let $\lambda_1 \leq \lambda_2 \leq \ldots$ be the eigenvalues of $H^{\Lambda}_{\alpha\beta}$ repeated with their multiplicity (which is at most $\ell$) and let ${}^1\psi, {}^2\psi, \cdots$ be a complete orthonormal set of corresponding eigenfunctions. Then each eigenfunction satisfies:

$$^k\tilde{\psi}(t) = U_{\lambda_k}(t)\,^k\tilde{\psi}(0)$$

The spectral matrix $M^{\Lambda}_{\alpha\beta}$ is defined as the $d \times d$ matrix valued measure whose entry $(i,j)$ is given by :

$$\sum_{k=1}^{\infty} {}^k\tilde{\psi}_i(0)\,^k\tilde{\psi}_j(0)\delta_{\lambda_k}$$

The spectral type of the operator $H^{\Lambda}_{\alpha,\beta}$ is then given by the trace of the matrix $M^{\Lambda}_{\alpha,\beta}$ which we denote by $\sigma^{\Lambda}_{\alpha,\beta}$.
The proof of the one dimensional case yields that each entry of the matrix $M^{\Lambda}_{\alpha,\beta}$ converges weakly when $a \to -\infty$ and $b \to +\infty$ to a limit independent of the choice of the boundary conditions $\alpha$ and $\beta$. If we call $M$ the limit matrix then the spectral type of $H$ is given by the measure $\sigma =$trace$(M)$.

**The Green Kernel**

We handle both the continuous and the discrete case simultaneously. The formulas are the same, even if the proofs can be a bit different.
If $\alpha$ is an $\ell$-dimensional subspace of $\mathbb{R}^d$ and $u$ a $d \times \ell$ matrix we will write $u \sim \alpha$ if the range of $u$ is equal to $\alpha$. It is well known that the range of a $d \times \ell$ matrix $h = \binom{A}{B}$ is a Lagrangian subspace of $\mathbb{R}^d$ iff the $\ell \times \ell$ matrix $A'B$ is symmetric and in this case the matrix $gh$ has the same property for any $g \in SP(\ell, \mathbb{R})$.

**Proposition III.5.2** *Assume that the ranges of the $d \times \ell$ matrices $h = \binom{A}{B}$ and $p = \binom{C}{D}$ are two $\ell$-dimensional Lagrangian subspaces of $\mathbb{R}^d$ and let $W$ be the $\ell \times \ell$ matrix $W = D'A - C'B$. Then we have:*

- *The ranges of $h$ and $p$ have a common non-zero vector iff $W$ is singular.*
- *If we assume $W$ non singular then :*

$$AW^{-1}C' = CW'^{-1}A' \quad \text{and} \quad DW'^{-1}A' - BW^{-1}C' = I_\ell$$

*Proof:*

One can write $W = p'Jh$ and a non-zero vector $u$ in the range of $h$ can be written $u = hv$ where $v$ is an $\ell$-dimensional non-zero vector . If $u$ is also in the range of $p$ we obtain immediately that $Wv = 0$. In the reverse way if $Wv = 0$ for a non-zero $\ell$-dimensional vector $v$ then the subspace generated by the range of $p$ and the vector $hv$ is Lagrangian. Since there does not exist Lagrangian subspaces of dimension greater than $\ell$ the vector $hv$ is in the range of $p$. Furthermore the vector $hv$ is non-zero since the columns of $h$ are linearly independent. The equalities given in this proposition are the result of an easy computation. ■

For a complex number $\lambda$ with $\Im m(\lambda) > 0$ we denote by $G^{\Lambda}_{\alpha,\beta}(\lambda, x, y)$ the $\ell \times \ell$ symmetric matrix kernel of the Green function of $H^{\Lambda}_{\alpha,\beta}$. This means that the entry $(i, j)$ of this matrix is the classical Green kernel computed at the points $(i, x)$ and $(j, y)$. One can state and prove exactly in the same way as in the one dimensional case :

**Proposition III.5.3** *The Stieltjes transform of the spectral measure $\sigma^{\Lambda}_{\alpha\beta}$ has the following expression:*

$$\int \frac{d\sigma^{\Lambda}_{\alpha\beta}(t)}{t-\lambda} = \begin{cases} \operatorname{tr}\left(G^{\Lambda}_{\alpha\beta}(\lambda,0,0) + G^{\Lambda}_{\alpha\beta}(\lambda,-1,-1)\right) & \textit{discrete case} \\ \operatorname{tr}\left(G^{\Lambda}_{\alpha\beta}(\lambda,0,0) + \frac{\partial^2}{\partial x \partial y} G^{\Lambda}_{\alpha\beta}(\lambda,0,0)\right) & \textit{continuous case} \end{cases}$$

The definition of the Wronskian is very close to the one given in the one-dimensional case. The main difficulty is that one has to take care of the order of the factors and the transposes in products of matrices. For a function $\Psi(t)$ from $T$ to the set of $\ell \times \ell$ complex matrices we denote by $\tilde{\Psi}(t)$ the $d \times \ell$ matrix $\binom{\Psi(t)}{\Psi(t-1)}$ in the discrete case and $\binom{\Psi(t)}{\dot{\Psi}(t)}$ in the continuous case.

**Definition III.5.4**
- *The Wronskian $W(u, v)$ of two sequences $u$ and $v$ with values in the set of square matrices of order $\ell$ with complex entries is defined as the matrix valued sequence:*

$$[W(u,v)](n) = v'(n-1)u(n) - v'(n)u(n-1) = \tilde{v}'(n)J\tilde{u}(n)$$

- *For two differentiable complex functions $u$ and $v$ with values in the set of square complex matrices of order $\ell$ the Wronskian $W(u, v)$ is defined as the matrix valued function:*

$$[W(u,v)](t) = \dot{v}'(t)u(t) - v'(t)\dot{u}(t) = \tilde{v}'(t)J\tilde{u}(t).$$

**Proposition III.5.5** *If $u$ and $v$ are two $\ell \times \ell$ matrix solutions of the equation $H\Psi = \lambda\Psi$ then the Wronskian of $u$ and $v$ is constant. Moreover if $\tilde{u}(a) \sim \alpha$ and $\tilde{v}(b) \sim \beta$ then this constant matrix is singular iff $\lambda$ is an eigenvalue of $H^{\Lambda}_{\alpha\beta}$.*

*Proof:*
For the constancy of the Wronskian we remark that in the set of $d \times d$ matrices we have the relation:

$$\tilde{u}(t) = U_{\lambda}(t)\tilde{u}(0) \quad \text{and} \quad \tilde{v}(t) = U_{\lambda}(t)\tilde{v}(0)$$

The result now follows from the relation $U'(t)JU_{\lambda}(t) = J$ which expresses that $U_{\lambda}(t) \in SP(\ell, \mathbb{R})$.
We know from Proposition III.5.2 that the Wronskian $W(u,v)$ is singular at some point $t$ iff the range of the matrices $\tilde{u}(t)$ and $\tilde{v}(t)$ have a common non-zero vector in $\mathbb{R}^d$. If we evaluate this Jacobian at the point $b$ we obtain that this Jacobian is singular iff there exists a non-zero vector $v \in \alpha$ such that $U_{\lambda}(b,a)v \in \beta$ and this last condition is equivalent to the fact that $\lambda$ is an eigenvalue of $H^{\Lambda}_{\alpha\beta}$. ∎

**Proposition III.5.6** *Let $u$ and $v$ be two solutions of the matrix equation $H\Psi = \lambda\Psi$ satisfying the boundary conditions $\tilde{u}(a) \sim \alpha$ and $\tilde{v}(b) \sim \beta$. Then we have :*

$$G^{\Lambda}_{\alpha\beta}(\lambda, x, y) = \begin{cases} u(x)\, W(u,v)^{-1}\, v'(y) & \text{if } x \leq y \\ v(x)\, W(v,u)'^{-1}\, u'(y) & \text{if } x \geq y \end{cases}$$

*Proof:*
We first remark that the definition is consistent for $x = y$ and that the exchange of $x$ and $y$ corresponds to a transpose of the Green matrix. In the discrete case for a fixed $y \in \Lambda$ one checks that :

$$\left((H^{\Lambda}_{\alpha,\beta} - \lambda I)G^{\Lambda}_{\alpha,\beta}(\lambda, ., y)\right)(x) = \begin{cases} 0 & \text{if } x \neq y \\ I_{\ell} & \text{if } x = y \end{cases}$$

This equality is obvious for $x < y$ or $x > y$. In the case $x = y$ the right side of the above equality is then equal to:

$$\begin{aligned} -u(x-1)W^{-1}v'(x) + (V(x) - \lambda I)v(x)W'^{-1}u'(x) - v(x+1)W'^{-1}u'(x) \\ = -u(x-1)W^{-1}v'(x) + v(x-1)W'^{-1}u'(x) = I_{\ell} \end{aligned}$$

In the continuous case, if $\psi(t)$ is a continuous function we write:

$$G^{\Lambda}_{\alpha,\beta}\psi(x) = v(x)W'(u,v)^{-1}\int_a^x u'(t)\psi(t)dt + u(x)W(u,v)^{-1}\int_x^b v'(t)\psi(t)dt$$

and a computation of derivatives yields the desired result. ■

Let $\phi$ and $\psi$ be two functions defined on $\Lambda$ with values in $\mathbb{C}^\ell$. We define the restriction of the inner product to the box $\Lambda$ by the formula :

$$< \psi, \phi >_\Lambda = \begin{cases} \sum_{n=a}^{b-1} \psi'(n)\overline{\phi}(n) & \text{in the discrete case} \\ \int_a^b \psi'(t)\overline{\phi}(t)\, dt & \text{in the continuous case} \end{cases}$$

**Proposition III.5.7** *Let $\phi$ and $\psi$ be two functions defined on $\Lambda$ with values in $\mathbb{C}^\ell$. Assuming the square integrability of $\phi$ and $\psi$ in the continuous case we have:*

$$< H\psi, \phi >_\Lambda - < \psi, H\phi >_\Lambda = \tilde{\psi}'(a)J\overline{\tilde{\phi}}(a) - \tilde{\psi}'(b)J\overline{\tilde{\phi}}(b)$$

*If $\psi$ is a solution of the equation $H\psi = \lambda\psi$ then*

$$2i\Im m\,(\lambda)||\psi||_\Lambda^2 = \tilde{\psi}'(a)J\overline{\tilde{\psi}}(a) - \tilde{\psi}'(b)J\overline{\tilde{\psi}}(b)$$

*Proof:*
In the discrete case this is a result of an easy computation. In the continuous case it is enough to apply the formula of integration by parts. We remark that the first relation proves again that $H^\Lambda_{\alpha,\beta}$ is a symmetric operator. ■

## III.5.2 Approximation of the Spectral Measures

Let $H$ be the Schrödinger operator in a strip of width $\ell$ and $a$ and $b$ be two elements of $T$ with $0 \in (a, b)$ (we assume that $b \geq 1$ in the discrete case). Recall that the box $\Lambda$ is defined as the set $\Lambda = \{(i,t); i = 1 \ldots \ell, t \in [a,b]\}$ in the continuous case, and $\Lambda = \{(i,t); i = 1 \ldots \ell, t \in [a, b-1]\}$ in the discrete case. For two $\ell$ dimensional Lagangian subspaces $\alpha$ and $\beta$ we denote by $H^\Lambda_{\alpha\beta}$ the operator $H$ restricted to the box $\Lambda$ with boundary conditions $\alpha$ and $\beta$. We already pointed out that the spectral measure $\sigma$ of the Schrödinger operator can be approximated by the spectral measures $\sigma^\Lambda_{\alpha\beta}$ computed in boxes with arbitrary boundary conditions $\alpha$ and $\beta$. It turns out that the measures $\sigma^\Lambda_{\alpha\beta}$ are not very tractable and we will instead use their averaged value with respect to one or two boundary conditions. At first sight this procedure seems surprising if we recall that our goal is to obtain that $\sigma$ is pure point: We replace the approximating measures $\sigma^\Lambda_{\alpha\beta}$ which are pure point, by averaged measures which are absolutely continuous! The choice of the Cauchy distribution as averaging measure is arbitrary but the essential reason for this choice is that integration with respect to the Cauchy measure reduces to the mean value property of harmonic functions with respect to the Poisson kernel. It is not excluded that other averaging distributions can yield interesting results but it seems that they will certainly give rise to much more complicated formulas.

The results and proofs are essentially the same as in the one dimensional case if we adopt the appropriate system of definitions. The notations and definitions related to the symplectic group and its boundaries can be found in Section 6 of the following chapter.

We identify the $\ell$-dimensional Lagrangian subspace $\beta$ of $\mathbb{R}^d$ with a symmetric matrix of order $\ell$ which we also denote by $\beta$, such that the decomposable $\ell$-vector given by the columns of the $d \times \ell$ matrix $\binom{\beta}{I}$ has the same direction as this subspace. This last parametrization does not work for the subset of Lagrangian $\ell$-vectors with a zero coordinate on the basis vector $e_{\ell+1} \wedge \ldots \wedge e_d$ but this will not be a problem if we want to integrate a function of $\beta$ with respect to a proper measure on $L_\ell$ that is a measure for which any projective proper subspace is negligible. The Cauchy measure on $L_\ell$ is mapped onto the usual Cauchy law on $\mathcal{M}(\ell)$ by this parametrization. For a square matrix $M$ we denote by $M^*$ its adjoint that is the conjugate of the transpose.

The following result about the Poisson kernel of the Siegel upper-half plane can be found for example in the book of Hua[153]. For $k = 0, \ldots, \ell$ we set:

$$S_k = \{Z = X + iY \,;\, X, Y \in \mathcal{M}(\ell)\,,\, Y \geq 0\,,\, \text{rank}(Y) = k\}$$

The Siegel half-plane is defined as the set $S_\ell$ and the union for $k = 0 \ldots \ell$ of the sets $S_k$ is called the closed Siegel half-plane. Let $F(Z)$ be a function which is harmonic in each $S_k$ for $k = 1 \ldots \ell$ and which can be continuously extended to $S_0 \cup \{\infty\}$. Then if $m$ is the Cauchy measure on $\mathcal{M}(\ell)$ we have the Poisson formula:

$$\int F(X)\, dm(X) = F(iI_\ell)$$

For a complex number $\lambda$ with $\Im m\,(\lambda) > 0$ we denote by $G^\Lambda_{\alpha,\beta}(\lambda, x, y)$ the $\ell \times \ell$ symmetric matrix kernel of the Green function of $H^\Lambda_{\alpha,\beta}$. This means that the entry $(i, j)$ of this matrix is the classical Green kernel computed at the points $(i, x)$ and $(j, y)$.

**Proposition III.5.8** *Let $u$ be a non-zero square matrix of order $\ell$ solution of the equation $Hu = \lambda u$ with $\Im m(\lambda) > 0$ and $\tilde{u}(a)$ real. Then in the discrete case the matrix $u(b) - Zu(b-1)$ is non-singular for $Z$ in the closed Siegel half-plane and the same conclusion holds for the matrix $u(b) - Z\dot{u}(b)$ in the continuous case.*

*Proof:*
We give the proof in the discrete case, the proof being exactly the same in the continuous case. If the conclusion was false there would exist a non-zero vector $v \in \mathbf{C}^\ell$ such that if we define $\psi(n) = u(n)v$ then $\psi$ is a solution of $H\psi = \lambda\psi$ and $\psi(b) = Z\psi(b-1)$. It follows from Proposition III.5.7 that

$$2i\Im m(\lambda)||\psi||_\Lambda^2 = \tilde{\psi}'(a)J\overline{\tilde{\psi}}(a) - \tilde{\psi}'(b)J\overline{\tilde{\psi}}(b)$$

Since $\tilde{u}(a)$ is real then $\tilde{\psi}'(a)J\overline{\tilde{\psi}}(a) = 0$ and if we set $Z = X + iY$ then we have:

$$\Im m(\lambda)||u||_\Lambda^2 = -u(b-1)'Y\overline{u}(b-1)$$

The left part of this equatity is strictly positive and the right part nonpositive thus we get a contradiction. ∎

**Theorem III.5.9** *Let $\sigma_{\alpha\beta}^\Lambda$ be the spectral measure of $H_{\alpha\beta}^\Lambda$, $u$ be the square matrix of order $\ell$ solution of the equation $Hu = \lambda u$ with the boundary condition $\tilde{u}(a) \sim \alpha$ and $m$ be the Cauchy measure on the Lagrangian boundary $L_\ell$. Then the averaged spectral measure $\sigma_\alpha^\Lambda = \int \sigma_{\alpha\beta}^\Lambda \, dm(\beta)$ has a density with respect to the Lebesgue measure on $I\!R$ given by the formula:*

$$\frac{1}{\pi}\text{tr}\left([\tilde{u}'(0)\tilde{u}(0)][\tilde{u}'(b)\tilde{u}(b)]^{-1}\right)$$

*Proof:*
The proof follows very closely the one given in Theorem III.3.2 but we need to use here more specific notation.
The value of the Wronskian appearing in Proposition III.5.6 in the Green kernel can be computed at any point. The computation at the point $b$ yields $W(u,v) = u(b) - \beta u(b-1)$ in the discrete case and $W(u,v) = u(b) - \beta\dot{u}(b)$ in the continuous case. It follows from Corollary III.5.8 that this Wronskian viewed as a function of $\beta$ has a non-vanishing extension to the closed Siegel half plane. Let $v$ be a square matrix valued function

solution of $Hv = \lambda v$ with $\tilde{v}(b) \sim \beta$. It follows from Proposition III.5.3 that the Stieltjes transform of $\sigma^{\Lambda}_{\alpha,\beta}$ can be written :

$$\begin{array}{rl} \mathrm{tr}\left(u(0)W^{-1}v'(0) + u(-1)W^{-1}v'(-1)\right) & \text{in the discrete case} \\ \mathrm{tr}\left(u(0)W^{-1}v'(0) + \dot{u}(0)W^{-1}\dot{v}'(0)\right) & \text{in the continuous case} \\ \mathrm{tr}\left(W^{-1}\tilde{v}'(0)\tilde{u}(0)\right) & \text{in both cases} \end{array}$$

This function of $\beta$ is the restriction to the set $S_0 = \mathcal{M}(\ell)$ of a bounded holomorphic function in the closed Siegel half-plane defined by :

$$F(Z) = \mathrm{tr}[W(Z)^{-1}N(Z)]$$

where $N(Z) = \tilde{Z}'U_b'^{-1}\tilde{u}(0)$, $W(Z) = \tilde{Z}'U_b'^{-1}J\tilde{u}(0)$, $\tilde{Z} = \begin{pmatrix} Z \\ I_\ell \end{pmatrix}$
(In this formula the Wronskian has been evaluated at the origin). The integration of $F(\beta)$ with respect to the Cauchy measure $m$ is then immediate from the Poisson formula and this yields:

$$\int \frac{d\sigma^{\Lambda}_{\alpha}(t)}{t-\lambda} = F(iI_\ell).$$

If we denote by $I_d$ the identity matrix of order $d$ we obtain :

$$\begin{array}{rcl} W^*(iI_\ell)N(iI_\ell) & = & \tilde{u}^*(0)J'\overline{U}_\lambda(b)^{-1}(I_d + J)U'(b)^{-1}\tilde{u}(0) \\ W^*(iI_\ell)W(iI_\ell) & = & \tilde{u}^*(0)J'\overline{U}_\lambda(b)^{-1}(I_d + J)U'(b)^{-1}J\tilde{u}(0) \end{array}$$

Taking into account that the matrix $U_\lambda(b)$ is symplectic one easily checks that:

$$\begin{array}{rcl} \lim\limits_{\Im m\,(\lambda)\to 0} J'\overline{U}_\lambda(b)^{-1}U'(b)^{-1}J & = & U'(b)U_\lambda(b) \\ \lim\limits_{\Im m\,(\lambda)\to 0} J'\overline{U}_\lambda(b)^{-1}JU'(b)^{-1}J & = & J \end{array}$$

It follows that:

$$\begin{array}{rcl} \lim_{\Im m\,(\lambda)\to 0} \Im m\,(W^*(iI_\ell)N(iI_\ell)) & = & \tilde{u}'(0)\tilde{u}(0) \\ \lim_{\Im m\,(\lambda)\to 0} W^*(iI_\ell)W(iI_\ell) & = & \tilde{u}'(0)U'(b)U_\lambda(b)\tilde{u}(0) \end{array}$$

The de la Vallée Poussin theorem now yields the result. ∎

One can repeat the procedure used in the one dimensional case and integrate with respect to $\alpha$ rather than to $\beta$ to obtain the same kind of formula with a "time reversal". One can also integrate with respect to the two boundary conditions to obtain an other type of averaged spectral measure.

**Proposition III.5.10** *Let us set* $\sigma^{\Lambda} = \int \sigma^{\Lambda}_{\alpha,\beta} dm(\alpha)\, dm(\beta)$. *Then the measure* $\sigma^{\Lambda}$ *has a density with respect to the Lebesgue measure on* $\mathbb{R}$ *given by:*

$$\frac{1}{\pi}\int_{L_\ell}\left(\frac{||v||}{||U_\lambda(b)v||}\right)^{\ell+1} \operatorname{tr}\left(v'v(v'U'(a)U_\lambda(a)v)^{-1}\right)\, dm(\bar{v})$$

*In this formula the norm is computed in* $\Lambda^{\ell}(\mathbb{R}^d)$.

*Proof:*
Let $v$ be a Lagrangian decomposable $\ell$-vector with direction $\bar{v} = \beta$ and let $f(\overline{v})$ be the function defined on $L_\ell$ by $f(\overline{v}) = \operatorname{tr}\left(v'v(v'U'(a)U_\lambda(a)v)^{-1}\right)$. Then if we first perform an integration with respect to the boundary condition at the left end of the box, one has:

$$\begin{aligned}
\int \sigma^{\Lambda}_{\beta}\, dm(\beta) &= \int \operatorname{tr}\left(v'U'(b)^{-1}U_\lambda(b)^{-1}v(v'U'(a,b)U_\lambda(b,a)v)^{-1}\right) dm(\overline{v}) \\
&= \int f(U_\lambda(b)^{-1}\overline{v})dm(\overline{v}) \\
&= \int f(\overline{v})d(U_\lambda(b)^{-1}m)(\overline{v}) \\
&= \int f(\overline{v})\left(\frac{||v||}{||U_\lambda(b)v||}\right)^{\ell+1} dm(\overline{v}) \\
&= \int \left(\frac{||v||}{||U_\lambda(b)v||}\right)^{\ell+1} \operatorname{tr}\left(v'v(v'U'(a)U_\lambda(a)v)^{-1}\right)\, dm(\bar{v})
\end{aligned}$$

The third equality has been obtained using the result of the computation of the Poisson kernel of the boundary $L_\ell$ given in the chapter "Products of Random Matrices" section 6. ∎

This formula has the advantage of separating the contribution of $[a, 0]$ and $[0, b]$ in the construction of the spectral measure.

**Theorem III.5.11** *Let* $\phi$ *be a continuous function with compact support on* $\mathbb{R}$. *Then :*

1. *the integral* $\int \phi(\lambda)\, d\sigma^{\Lambda}(\lambda)$ *is equal to*

$$\frac{1}{\pi}\int \phi(\lambda)\left(\frac{||v||}{||U_{\lambda}(b)v||}\right)^{\ell+1} \operatorname{tr}\left(v'v(v'U'(a)U_{\lambda}(a)v)^{-1}\right)\, dm(\bar{v})\, d\lambda$$

*and converges to* $\int \phi(\lambda)\, d\sigma(\lambda)$ *when* $a \to -\infty$ *and* $b \to +\infty$.

2. *For any* $a < 0$ *one has the formula :*

$$\int \operatorname{tr}\left(v'v(v'U'(a)U_{\lambda}(a)v)^{-1}\right)\, dm(\bar{v}) = \ell$$

*Proof:*
The property 1. is a direct consequence of the formula obtained for $\sigma^{\Lambda}$ in Theorem III.5.9 and the uniform weak convergence with respect to the boundary conditions of $\int \phi\, d\sigma^{\Lambda}_{\alpha\beta}$ to $\int \phi\, d\sigma$.
To prove 2. we have only to remark that the formula obtained in 1. has to be symmetric with respect to $a$ and $b$ since we can exchange the order of integration with respect to the boundary conditions $\alpha$ and $\beta$. If we write this equality for $b = 0$ then one has $U_{\lambda}(b) = I$ and we obtain :

$$\begin{aligned}\int \operatorname{tr}\left(v'v(v'U'(a)U_{\lambda}(a)v)^{-1}\right)\, dm(\bar{v}) &= \ell \int \left(\frac{||v||}{||U_{\lambda}(a)v||}\right)^{\ell+1} dm(\bar{v}) \\ &= \ell \int d(U_{\lambda}(a)^{-1}m)(\bar{v}) \;=\; \ell\end{aligned}$$

One can remark that we have used the spectral measure to prove this formula but actually this result is more general and the formula remains true if one replaces the matrix $U_{\lambda}(a)$ by any matrix of $SP(\ell, \mathbb{R})$. ∎

In the discrete case let us denote by $E^{\Lambda}_{\alpha,\beta}$ the spectral resolution of the operator $H^{\Lambda}_{\alpha,\beta}$ and for $x, y \in \Lambda$ by $\sigma^{\Lambda}_{(\alpha,\beta),(x,y)}$ the square matrix valued measure of order $\ell$ whose $(i,j)$-th entry is the complex measure $< E^{\Lambda}_{\alpha,\beta}(\cdot)\delta_{i,x}, \delta_{j,y} >$.

**Theorem III.5.12** *Let $u$ be the square matrix of order $\ell$ solution of the equation $Hu = \lambda u$ with the boundary condition $\tilde{u}(a) \sim \alpha$ and $m$ be the Cauchy measure on the Lagrangian boundary $L_{\ell}$. Then for $x, y \in \Lambda$ the averaged spectral measure $\sigma^{\Lambda}_{\alpha,(x,y)} = \int \sigma^{\Lambda}_{(\alpha\beta),(x,y)}\, dm(\beta)$ has a density with respect to the Lebesgue measure on $\mathbb{R}$ given by the formula:*

$$\frac{1}{\pi}\left(u'(x)[\tilde{u}'(b)\tilde{u}(b)]^{-1}u(y)\right)$$

*Proof:*
Using the relation :

$$\int \frac{d\sigma^{\Lambda}_{(\alpha\beta),(x,y)}(t)}{t-\lambda} = G^{\Lambda}_{\alpha\beta}(\lambda, x, y) = \begin{cases} u(x)\, W(u,v)^{-1}\, v'(y) & \text{if } x \leq y \\ v(x)\, W(v,u)'^{-1}\, u'(y) & \text{if } x \geq y \end{cases}$$

one has only to repeat the proof of Theorem III.5.9 but without taking traces. ∎

One can prove a theorem of the same kind in the continuous case if one defines properly the spectral kernels $\sigma^{\Lambda}_{(\alpha,\beta),(x,y)}$ and their relationship with the Green kernel.

### III.5.3 Nature of the Spectrum

**Hyperbolic Behavior**

In the same way as in the one dimensional case we now introduce the concept of hyperbolic behavior of the propagator $U_\lambda(t)$ associated to a Schrödinger operator in a strip of width $\ell$. Recall that in the discrete case the operator $H$ is always self-adjoint. In the continuous case we assume that there exists a constant $c$ such that $|V_i(t)| \leq c(1+t^2)$ for $i = 1, \ldots, \ell$. Then the operator $H$ is essentially self-adjoint.
Moreover we also assume that for $i = 1, \ldots, \ell$ we have :

- $\lim_{n\to\pm\infty} \frac{1}{|n|} \log(1 + |V_i(n)|) = 0$ in the discrete case
- $\lim_{n\to\pm\infty} \frac{1}{|n|} \int_n^{n+1} |V_i(t)|\, dt = 0$ in the continuous case.

In the continuous case the transfer matrices are defined by the same formulas as for the one dimensional operator. It follows from the above assumptions that for any real or complex number $\lambda$ we have :

- $\lim_{n\to\pm\infty} \dfrac{1}{|n|} \log ||g_\lambda(n)|| = 0$
- if we set $\breve{U}_\lambda(n) = \sup_{t\in[0\ 1]}(\log ||U_\lambda(n+t, n)||)$ then one has $\lim_{n\to\pm\infty} \dfrac{1}{|n|} \breve{U}_\lambda(n) = 0$

**Definition III.5.13** *Let us denote by $\{U_\lambda(t); t \in T, \lambda \in I\!R\}$ the propagator of the Schrödinger operator $H$ in a strip of width $\ell$. The real number $\lambda$*

*is said to be hyperbolic if there exist real numbers* $\gamma_1^{\pm}(\lambda) \geq \gamma_2^{\pm}(\lambda) \cdots \geq \gamma_\ell^{\pm}(\lambda) > 0$ *such that for* $p = 1 \cdots \ell$ *:*

$$\lim_{n \to \pm\infty} \frac{1}{|n|} \log ||\Lambda^p U_\lambda(n)|| = \gamma_1^{\pm}(\lambda) + \cdot + \gamma_p^{\pm}(\gamma)$$

*The set of hyperbolic values of* $\lambda$ *will be denoted by* hyp($H$).

In the same way as in the one dimensional case one checks that the Green's function decays exponentially when the energy is outside the spectrum. More precisely:

**Proposition III.5.14** *Let* $H$ *be a Schrödinger operator on the discrete strip of width* $\ell$ *and* $\lambda$ *be a number (real or complex) outside the spectrum of* $H$*. Then one has:*

$$\liminf_{|n| \to \pm\infty} \frac{1}{|n|} \log ||\Lambda^\ell U_\lambda(n)|| > 0$$

*Proof:*
Let us define the $\ell \times \ell$ matrix $\Phi(n) = G_\lambda(0, n)$. It follows from Problem III.6.8 that there exist two constants $C < \infty$ and $\alpha > 0$ such that $||\tilde{\Phi}(n)|| \leq C e^{-\alpha|n|}$ where the norm is computed in $\Lambda^\ell(\mathbb{R})$. Moreover one has for $1 \leq m \leq n$ the relation:

$$\tilde{\Phi}'(m)\, J\, \tilde{\Phi}(m) = \tilde{\Phi}'(n)\, J\, \tilde{\Phi}(n)$$

The right hand side of this equality goes to zero when $n \to +\infty$ hence the $\ell$ dimensional subspace generated by the columns of the $d \times \ell$ matrix $\tilde{\Phi}(n)$ is Lagangian for $n \geq 1$.
Let $\Psi$ be an $\ell \times \ell$ matrix solution of the equation $H\Psi = \lambda\Psi$ with a Lagrangian initial data $\tilde{\psi}(1)$ such that the two Lagrangian subspaces generated by the columns of the matrices $\tilde{\Psi}(1)$ and $\tilde{\Phi}(1)$ have no common nonzero vector. Proposition III.5.2 and the constancy of the Wronskian implies immediately that there exists a nonzero constant $D$ such that:

$$D = |<\tilde{\Psi}(n)\,,\, J\,\tilde{\Phi}(n)>| \leq ||\tilde{\Psi}(n)||\; ||\tilde{J}\,\Phi(n)|| = ||\tilde{\Psi}(n)||\; ||\tilde{\Phi}(n)||$$

for any positive $n$. The same property holds for negative $n$ and this yields the result taking into account that the norm of the matrix $\Lambda^\ell U_\lambda(n)$ can be computed on its restriction to the Lagrangian linear space Lag($\ell$) (see Proposition IV.4.21). ∎

**Remark III.5.15** *Let us anticipate for the sake of the present remark and let us assume the existence for $p = 1 \ldots \ell$ of the Lyapunov exponents $^{\pm}\gamma_p(\lambda)$ which will be defined in the next chapter for the sequence of transfer matrices. Then, the above result implies immediately that:*

$$\{\lambda \,;\, {}^{+}\gamma_1(\lambda) = 0\} \cup \{\lambda \,;\, {}^{-}\gamma_1(\lambda) = 0\} \subset \Sigma(H)$$

*One would like to prove the same statement with $\gamma_1$ replaced by $\gamma_\ell$ but this result cannot be obtained by the method explained above which actually works for any sequence of symmetric $\ell \times \ell$ potential matrices. Using for example constant diagonal matrices it is easily seen that one can have $\Sigma(H) = \{\lambda \;\; \gamma_1(\lambda = 0\}$ but $\gamma_\ell(\lambda) = 0$ for any $\lambda$. It is likely that for Schrödinger matrices the improved statement is true but the proof is missing.*

The following propositions are proved exactly in the same way as in the one dimensional case.

**Proposition III.5.16** *If $\lambda \in \mathrm{hyp}(H)$ then there exist two $\ell$ dimensional subspaces of $\mathbb{R}^d$ say $^{+}\mathcal{V}_\lambda$ and $^{-}\mathcal{V}_\lambda$ such that for a non-zero vector $v$ in $R^d$ we have:*

$$v \in {}^{\pm}\mathcal{V}_\lambda \iff \lim_{t\to\pm\infty} \frac{1}{|t|} \log ||U_\lambda(t)v|| \leq -\gamma_\ell^{\pm}(\lambda)$$

$$v \notin {}^{\pm}\mathcal{V}_\lambda \iff \lim_{t\to\pm\infty} \frac{1}{|t|} \log ||U_\lambda(t)v|| \geq \gamma_\ell^{\pm}(\lambda).$$

**Proposition III.5.17** *Let $\lambda \in \mathrm{hyp}(H)$, then $\lambda$ is an eigenvalue of $H$ iff $^{+}\mathcal{V}_\lambda \cap^{-} \mathcal{V}_\lambda \neq \{0\}$.*

**Theorem III.5.18** *Let $m$ be a nonnegative continuous measure on $\mathbb{R}$ which is carried by $\mathrm{hyp}(H)$. Then $m$ is orthogonal to the spectral measure $\sigma$ of the operator $H$.*

**Theorem III.5.19** *Let $A$ be a Borel subset of the real line. If the restriction to $A$ of the spectral measure $\sigma$ of $H$ is carried by $\mathrm{hyp}(H)$, then $\sigma$ is pure point on $A$. Moreover any eigenvector $\psi_\lambda(t)$ corresponding to an eigenvalue $\lambda \in A$ satisfies:*

$$\lim_{n\to\pm\infty} \frac{1}{|t|} \log ||\tilde{\psi}_\lambda(t)|| \leq -\gamma_\ell^{\pm}(\lambda).$$

It is readily seen that the dimension of an eigenspace is less than or equal to $d$. Actually, using the properties of symplectic matrices one can prove that this dimension is less than or equal to $\ell$. Let ${}^1\psi_\lambda, \ldots, {}^r\psi_\lambda$ be $r$ linearly independent eigenfunctions and let us denote by $\Psi_\lambda(t) = \binom{A_\lambda(t)}{B_\lambda(t)}$ the $d \times r$ matrix with columns vectors ${}^1\tilde{\psi}_\lambda(t), \ldots, {}^r\tilde{\psi}_\lambda(t)$. The relations $\Psi_\lambda(t) = U_\lambda(t)\Psi_\lambda(0)$ and $U_\lambda(t)'JU_\lambda(t) = J$ imply that $A_\lambda(t)'B_\lambda(t) - B_\lambda(t)'A_\lambda(t)$ does not depend on $t$. In the discrete case $A_\lambda(t)$ goes to zero at infinity and one can conclude that $A_0'B_0 - B_0'A_0 = 0$. To obtain the same conclusion in the continuous case we use the fact that if $\psi$ is an eigenfunction then $\psi(t)$ and $t^{-1}\dot{\psi}(t)$ are square integrable at infinity. The Cauchy-Swartz inequality then implies the integrability at infinity of each entry of the matrix $t^{-1}(A_\lambda'(t)B_\lambda(t) - B_\lambda(t)'A_\lambda(t))$ and this yields the desired conclusion. It follows that the decomposable r-vector ${}^1\tilde{\psi} \wedge \ldots \wedge {}^r\tilde{\psi}$ is Lagrangian and thus $r \leq \ell$.
If $\lambda$ is a hyperbolic eigenvalue then the dimension of its eigenspace is equal to the dimension of ${}^+\mathcal{V}_\lambda \cap {}^-\mathcal{V}_\lambda$ and we will prove in the sequel that this dimension is "often" equal to 1.

### The Discrete Strip Case

The one dimensional criterion given in Proposition III.4.14 can hardly be extended to the strip in an easy way. In this situation (at least in the discrete case) we will use another approach to obtain the pure point property of the spectral measure.

**Lemma III.5.20** *The matrix valued measures $\sigma_{m,n}$ are related to the spectral matrix $M$ by the formula :*

$$\sigma_{n,m} = (P_n, Q_n)M \begin{pmatrix} P_m' \\ Q_m' \end{pmatrix}$$

*Proof:*
This formula is a direct consequence of Proposition III.5.1 and the definition of the spectral matrix. ∎

Let us denote by $M(\lambda)$ a matrix of densities of $M$ with respect to its trace $\sigma$. In general $M(\lambda)$ is defined only $\sigma$ almost surely but if $\lambda$ is an eigenvalue then $M(\lambda)$ is uniquely defined.

**Lemma III.5.21** *Let $\lambda$ be an eigenvalue of the operator $H$ of multiplicity $r$. Then the rank of $M(\lambda)$ is equal to $r$ and the eigenspace associated to*

*$\lambda$ is generated by the sequences $\{U_\lambda(n)v; n \in \mathbb{Z}\}$ where $v$ is any column vector of the matrix $M(\lambda)$. Furthermore the dimension of the eigenspace is at most $\ell$.*

*Proof:*
Let ${}^1\psi \ldots {}^r\psi$ be an orthonormal basis of the eigenspace associated to $\lambda$ and denote by $T_n$ the $\ell \times r$ matrix whose columns are ${}^1\psi(n) \ldots {}^r\psi(n)$. It is readily seen that

$$M(\lambda) = \frac{1}{\sigma(\lambda)} \begin{pmatrix} T_0 \\ T_{-1} \end{pmatrix} (T_0', T_{-1}')$$

The matrix $\begin{pmatrix} T_0 \\ T_{-1} \end{pmatrix}$ is of rank $r$ thus $M(\lambda)$ is also of rank $r$. We have already seen that from the Wronskian property we have $r \leq \ell$ and the proof is complete. ∎

**Proposition III.5.22** *Let $\mathcal{I}$ be an open bounded interval of the real line such that there exist $0 < \delta < 1$ and $C < \infty$ satisfying the inequality*

$$\sum_{i=1,j=1}^{\ell} |\sigma_{(i,n),(j,m)}|(\mathcal{I}) \leq C\delta^{|n|}$$

*for $m = -1$ and $m = 0$. Then $\sigma$ is pure point on $\mathcal{I}$ and for an eigenvalue $\lambda \in \mathcal{I}$ any eigenfunction $\psi$ decays exponentially fast. More precisely for any eigenvalue $\lambda \in \mathcal{I}$ and $\delta < \rho < 1$ there exists $C(\lambda) < \infty$ such that we have :*

$$|\tilde{\psi}(n)| \leq C(\lambda)\rho^{|n|}$$

*Proof:*
Let $M(\lambda) = \begin{pmatrix} A(\lambda) & B(\lambda) \\ C(\lambda) & D(\lambda) \end{pmatrix}$. Using Lemma III.5.20 one sees immediately that the hypothesis implies that $\sigma$-almost surely each column of the matrix $M(\lambda)$ can be taken as the initial vector $\tilde{\psi}(0)$ of an exponentially decreasing solution of $H\psi = \lambda\psi$ for one can write :

$$\sigma_{n,0} = (P_n A + Q_n C)\sigma \quad \text{and} \quad \sigma_{n,-1} = (P_n B + Q_n D)\sigma$$

There exists certainly a column of the matrix $M(\lambda)$ which is not zero since the trace of $M(\lambda)$ is equal to $d$. Then one can conclude that $\sigma$ is carried by the eigenvalues of $H$ and is thus pure point. The claim about the exponential decay of any eigenfunctions is a direct consequence of Lemma III.5.21
∎

As usual we will use approximations of the measure $\sigma_{m,n}$ computed in boxes to check the hypothesis of Proposition III.5.22.

**Theorem III.5.23** *Let $\mathcal{I}$ be an open bounded interval of the real line and $\alpha$ be a fixed Lagrangian subspace of $\mathbb{R}^d$. Assume that there exist $0 < \rho < 1$, $C < \infty$ such that for each box $\Lambda$ if we denote by $u$ the square matrix of order $\ell$ solution of $Hu = \lambda u$ with $\tilde{u}(a) \sim \alpha$ and by $\Gamma$ the symmetric positive matrix $\tilde{u}'(b)\tilde{u}(b)$ then one has:*

$$\int_{\mathcal{I}} \sqrt{\operatorname{tr}(\tilde{u}'(0)\tilde{u}(0)\Gamma^{-1}) \operatorname{tr}(\tilde{u}'(n)\tilde{u}(n)\Gamma^{-1})}\, d\lambda \leq C\rho^{|n|}$$

*Then $\sigma$ is pure point on $\mathcal{I}$. Moreover for any eigenvalue $\lambda \in \mathcal{I}$ and $\rho < \delta < 1$ there exists $C(\lambda) < \infty$ such that for any associated eigenfunction $\psi$ one has :*

$$|\tilde{\psi}(n)| \leq C(\lambda)\delta^{|n|}$$

*Proof:*
If we use the approximations constructed in Proposition III.5.9 then Fatou's lemma implies that it is enough to have for $m = -1$ and $m = 0$:

$$\int_{\mathcal{I}} \sum_{i=1,j=1}^{\ell} |\left(u'(m)\Gamma^{-1}u(n)\right)_{i,j}| d\lambda \leq C\rho^{|n|} \qquad \textit{for } n \in \Lambda$$

If $A, B, C$ are square matrices of order $\ell$ with $C = AB$ then we have:

$$\sum_{i=1,j=1}^{\ell} |C_{i,j}| \leq \ell\sqrt{\operatorname{tr}(A'A) + \operatorname{tr}(B'B)}$$

One can choose $A = u'(m)\sqrt{\Gamma^{-1}}$ and $B = \sqrt{\Gamma^{-1}}u(n)$. Applying the above inequality for $m = -1$ and $m = 0$ the Cauchy-Swartz inequality yields the result. ∎

It is certainly possible to extend this theorem to the continuous strip case exactly in the same form. Moreover one remarks that restricting the above formula to the one dimensional case one obtains exactly the same criterion as the one given in Theorem III.4.15 !

## III.6 Problems

**Problem III.6.1** *Let $t \hookrightarrow V(t)$ be a real valued measurable function satisfying:*

$$\int_{I\!R} \frac{V(t)^2}{1+|t|^3}\, dt < \infty.$$

*Use the argument of the proof of Lemma III.1.4 to prove the essential self-adjointness of the corresponding Schrödinger operator on the line.*

**Problem III.6.2** *We assume that the potential $V$ is a real valued locally integrable function on $I\!R$.*
*1. Show that there exists a unique function $I\!R^4 \ni (\lambda, s, t, \alpha) \longrightarrow \vartheta_\lambda(t, s, \alpha) \in I\!R$ which is analytic in $\lambda$ for fixed $s$, $t$ and $\alpha$, which is infinitely differentiable in $\alpha$ for fixed $\lambda$, $s$ and $t$, which is absolutely continuous in $t$ for fixed $\lambda$, $s$ and $\alpha$ and which satisfies:*

$$U_\lambda(t,s) = \begin{bmatrix} \sin\alpha \\ \cos\alpha \end{bmatrix} = \begin{bmatrix} r_\lambda(t,s,\alpha)\sin\vartheta_\lambda(t,s,\alpha) \\ r_\lambda(t,s,\alpha)\cos\vartheta_\lambda(t,s,\alpha) \end{bmatrix}$$

*provided we set:*

$$r_\lambda(t,s,\alpha) = \exp[-\frac{1}{2}\int_s^t [\lambda - V(u) - 1]\sin 2\vartheta_\lambda(u,s,\alpha)\, du].$$

*2. Use the fact that this function $\vartheta$ is given locally by either:*

$$\vartheta_\lambda(t,s,\alpha) = \cot^{-1}\frac{\dot y_\lambda(t,s,\alpha)}{y_\lambda(t,s,\alpha)}$$

*or:*

$$\vartheta_\lambda(t,s,\alpha) = \tan^{-1}\frac{y_\lambda(t,s,\alpha)}{\dot y_\lambda(t,s,\alpha)}$$

*where $y_\lambda(t,s,\alpha)$ denotes the value at $t$ of the solution of the eigenvalue equation $-\ddot y + Vy = \lambda y$ which satisfies $y(s) = \sin\alpha$ and $\dot y(s) = \cos\alpha$, to prove that:*

*i).* $\dot\vartheta_\lambda(t,s,\alpha) = 1 + [\lambda - V(t) - 1]\sin^2\vartheta_\lambda(t,s,\alpha)$.
*ii).*

$$\frac{\partial\vartheta_\lambda}{\partial\lambda}(t,s,\alpha) = \int_s^t \sin^2\vartheta_\lambda(u,s,\alpha)$$
$$\exp[\int_u^t [\lambda - V(u') - 1]\sin 2\vartheta_\lambda(u',s,\alpha)\, du']\, du.$$

*iii).*

$$\frac{\partial \vartheta_\lambda}{\partial \alpha}(t,s,\alpha) = \int_s^t [\lambda - V(u) - 1] \sin 2\vartheta_\lambda(u,s,\alpha)\, du].$$

*3. Show that for $\lambda$, $s$ and $\alpha$ fixed, the function $t \hookrightarrow \vartheta_\lambda(t,s,\alpha)$ can neither tend to a number of the form $k\pi$from above when $t$ increases nor from below when $t$ decreases.*

**Problem III.6.3** *1. Let $V_1$ and $V_2$ be locally integrable real valued functions and let $y_j$ be a solution of the equation $-\ddot{y}+V_j y = 0$ for $j = 1,2$. Let us also assume that $t_1$ and $t_2$ are consecutive zeros of the function $t \hookrightarrow y_1(t)$ and that $V_2(t) \leq V_1(t)$ for almost every $t \in [t_1,t_2]$. Prove that:*

- *either $V_1(t) = V_2(t)$ for almost every $t \in [t_1,t_2]$ and $y_2(t) = \kappa y_1(t)$ for some constant $\kappa$ and all $t \in [t_1,t_2]$,*
- *or $y_2(t_3) = 0$ for some $t_3 \in (t_1,t_2)$.*

*2. Let $t \hookrightarrow V(t)$ be a locally integrable real valued function and let $y_1$ and $y_2$ be linearly independent solutions of the equation $-\ddot{y} + Vy = 0$. Prove that one and only one zero of $y_2$ lies between two consecutive zeros of $y_1$ and vice versa. Conclude that if $V$ is nonnegative on an interval $\mathcal{I}$, then there is no solution with more than one zero in $\mathcal{I}$.*
*3. Let $t_1 < t_2 < \cdots$ be consecutive zeros of a solution $y$ and let $t_0 < t_1$ be such that $y(t_0) \geq 0$ and such that $y$ has no zero in $(t_0,t_1)$. Prove that if the phase $\vartheta(t)$ of the solution $y(t)$ is such that $\vartheta(t_0) \in [0,\pi)$ then we have necessarily for every integer $k \geq 0$, $\vartheta(t_{k+1}) = (k+1)\pi$ and $\vartheta(t) \in (k\pi,(k+1)\pi)$ whenever $t \in (t_k,t_{k+1})$.*
*4. Let $V_1$ and $V_2$ be locally integrable real valued functions and let $\mathcal{I} = [a,b]$ be a bounded interval on which $V_1 \geq V_2$ almost everywhere. Let $y_j$ be a solution of the equation $-\ddot{y}+V_j y = 0$ such that $y_j(a) \geq 0$ and $\vartheta_j(a) \in [0,\pi)$ for $j = 1,2$.*

- *Prove that, if $\vartheta_2(a) \geq \vartheta_1(a)$ then $\vartheta_2(t) \geq \vartheta_1(t)$ for all $t \in [a,b]$ and that one has equality $\vartheta_2(t_0) = \vartheta_1(t_0)$ for some $t_0 \in [a,b]$ if and only if $\vartheta_2(a) = \vartheta_1(a)$ and $V_2(t) = V_1(t)$ for almost every $t \in [a,t_0]$.*
- *Show that if $t_0 < t_1 < \cdots < t_n$ are $n+1$ consecutive zeros of $y_1$ in $\mathcal{I}$ then $y_2$ has at least $n$ zeros in $[t_0,t_n]$ and in fact in $(t_0,t_n)$ unless $y_2(t_0) = 0$ and $V_2(t) = V_1(t)$ for almost every $t \in [t_0,t_n]$.*

**Problem III.6.4** *Show that, up to sets of Lebesgue measure 0, the essential supports of $E_{ac}(H^+_{\alpha_1})$ and $E_{ac}(H^+_{\alpha_2})$ are equal. On the other hand, show that, if $\alpha_1 \neq \alpha_2 \pmod \pi$ then it is possible to choose an essential support $A_1$ for $E_s(H^+_{\alpha_1})$ and an essential support $A_2$ for $E_s(H^+_{\alpha_2})$ such that $A_1 \cap A_2 = \emptyset$.*

**Problem III.6.5** *Let $\{a_n;\, n \in \mathbb{Z}\}$ and $\{b_n;\, n \in \mathbb{Z}\}$ be sequences of real numbers. We assume that $a_n \neq 0$ for all $n \in \mathbb{Z}$.*

1. *Let us set $\mathcal{H} = \ell^2(\mathbb{Z})$ and let us consider the operator $H$ defined formally by:*

$$[H\psi]_n = a_n\psi_{n+1} + a_{n-1}\psi_{n-1} + b_n\psi_n$$

*on various domains in $\mathcal{H}$. For example, we let $H_{min}$ be the operator $H$ with the domain $\mathcal{D}_0$ and $H_{max}$ be the operator $H$ with the domain $\mathcal{D}_1$ where:*

$$\begin{aligned}\mathcal{D}_0 &= \{\psi = (\psi_n)_{n\in\mathbb{Z}} \in \mathcal{H};\ \psi_n = 0 \text{ for all but finitely many } n \in \mathbb{Z}\}\\ \mathcal{D}_1 &= \{\psi = (\psi_n)_{n\in\mathbb{Z}} \in \mathcal{H};\ ([Hf]_n)_{n\in\mathbb{Z}} \in \mathcal{H}\}.\end{aligned}$$

*Show that $H_{min}$ and $H_{max}$ are symmetric, that $(H_{min})^* = H_{max}$ and that the conditions:*

$$\sum_{-\infty}^{-1} \frac{1}{a_n^2} = \sum_{+1}^{+\infty} \frac{1}{a_n^2} = \infty$$

*imply that the operator $H_{min}$ is essentially self-adjoint and that its unique self-adjoint extension is $H_{max}$.*

2. *We now consider the restriction of the operator $H$ to the integer interval $[1,\infty)$. So we set:*

$$[H^+\psi]_n = \begin{cases} a_1\psi_2 + b_1\psi_1 & \text{if } n = 1 \\ a_n\psi_{n+1} + a_{n-1}\psi_{n-1} + b_n\psi_n & \text{if } n \geq 2 \end{cases} \tag{III.76}$$

*we define the operator $H^+_{max}$ as the operator $H^+$ with domain:*

$$\mathcal{D}^+_0 = \{\psi = (\psi_n) \in \ell^2([1,\infty));\ \psi_n = 0 \text{ for all but finitely many } n \geq 1\}$$

*and the operator $H^+_{min}$ as the operator $H^+$ with domain:*

$$\mathcal{D}^+_1 = \{\psi = (\psi_n) \in \ell^2([1,\infty));\ \sum_{n\geq 2} |a_n\psi_{n+1} + a_{n-1}\psi_{n-1} + b_n\psi_n|^2 < \infty\}.$$

*Show as before that the operators* $H^+_{min}$ *and* $H^+_{max}$ *are symmetric, that* $(H^+_{min})^* = H^+_{max}$ *and that the condition:*

$$\sum_{n\geq 1} \frac{1}{a_n^2} = \infty$$

*implies that the operator* $H^+_{min}$ *is essentially self-adjoint and that its unique self-adjoint extension* $H^+$ *is the operator* $H^+_{max}$.

**3.** *Check that an operator* $H_-$ *can be constructed similarly as a self-adjoint realization of the operator* $H$ *on* $(-\infty, -1]$.

**Problem III.6.6** *Let us recall that every Herglotz function* $w$ *has a unique representation of the form:*

$$w(\lambda) = \alpha + \beta\lambda + \int_{\mathbb{R}} \left[\frac{1}{\xi - \lambda} - \frac{\xi}{1+\xi^2}\right] d\sigma(\xi) \tag{III.77}$$

*for some* $\beta > 0$, *some* $\alpha \in \mathbb{R}$ *and some nonnegative measure* $\sigma$ *on* $\mathbb{R}$ *which satisfies:*

$$\int_{\mathbb{R}} \frac{d\sigma(\xi)}{1+\xi^2} < \infty.$$

*The measure* $\sigma$ *is often called the spectral measure of the function* $w$. *Recall Section I.3 of Chapter I.*
*We now assume that we have two Herglotz functions* $w_+$ *and* $w_-$ *and we denote by* $\sigma_+$ *and* $\sigma_-$ *the corresponding spectral measures.*

**1.** *Let us define for* $i, j = 1, 2$, *the functions* $w_{i,j}$ *on the upper half-plane* $\Pi_+$ *by:*

$$\begin{aligned} w_{1,1}(\lambda) &= \frac{w_-(\lambda)w_+(\lambda)}{w_-(\lambda)+w_+(\lambda)} \\ w_{2,2}(\lambda) &= -\frac{1}{w_-(\lambda)+w_+(\lambda)} \\ w_{1,2}(\lambda) = w_{2,1}(\lambda) &= \frac{w_-(\lambda)}{w_-(\lambda)+w_+(\lambda)} \end{aligned}$$

*Show that the function* $\lambda \hookrightarrow w_\xi = \sum_{i,j=1}^2 \xi_i\xi_j w_{i,j}(\lambda)$ *is a Herglotz function for all* $\xi_1$ *and* $\xi_2$ *in* $\mathbb{R}$.

**2.** *Construct a nonnegative matrix measure* $M = [\sigma_{i,j}]_{i,j=1,2}$ *such that the measure* $\sum_{i,j=1}^2 \xi_i\xi_j\sigma_{i,j}$ *is the spectral measure of the Herglotz function* $w_\xi$.

**Problem III.6.7** *In the one dimensional discrete case it is possible to prove directly the existence of polynomially bounded solutions of the eigenvalue equation. Let* $p_\lambda(n)$ *and* $q_\lambda(n)$ *be the two solutions of the equation* $H\psi = \lambda\psi$ *with* $p_\lambda(0) = q_\lambda(-1) = 1$ *and* $p_\lambda(-1) = q_\lambda(0) = 0$*. Moreover we denote by* $\sigma^+_{m,n}$ *and* $\sigma_{m,n}$ *the spectral measures of the operators* $H^+$ *and* $H$ *for the delta functions at the sites* $m$ *and* $n$*. As usual we denote by* $\sigma_n$ *the probability measure* $\sigma_{n,n}$*, by* $\sigma^+ = \sigma^+_0$ *the spectral measure of* $H^+$ *and the spectral measure of* $H$ *is denoted by* $\sigma = \sigma_0 + \sigma_{-1}$*.*

1. *Prove that*

$$\begin{aligned} \sigma^+_{m,n} &= p_\lambda(n)p_\lambda(m)\sigma^+ \\ \sigma_{m,n} &= \begin{pmatrix} p_\lambda(m) & q_\lambda(m) \end{pmatrix} \begin{pmatrix} \sigma_0 & \sigma_{-1,0} \\ \sigma_{-1,0} & \sigma_{-1} \end{pmatrix} \begin{pmatrix} p_\lambda(n) \\ q_\lambda(n) \end{pmatrix} \end{aligned}$$

2. *Deduce from the relation (i) that* $\int p^2(n)\, d\sigma^+ = 1$ *and conclude that for any positive number* $\epsilon$ *there exists a measurable function* $C(\lambda)$ *which is* $\sigma^+$ *almost surely finite and such that :*

$$|p_\lambda(n)| \leq C(\lambda) n^{1/2+\epsilon} \text{ for } n \geq 1$$

3. *Prove that* $\sigma$ *almost surely there exists a positive symmetric matrix* $S$ *of trace* 1 *such that*

$$\sigma_{m,n} = \begin{pmatrix} p_\lambda(m) & q_\lambda(m) \end{pmatrix} S \begin{pmatrix} p_\lambda(n) \\ q_\lambda(n) \end{pmatrix} \sigma$$

   *Let* $S = K \begin{pmatrix} \alpha & 0 \\ 0 & \beta \end{pmatrix} K^*$ *be a diagonalization of* $S$ *with* $\alpha \geq \beta$ *and denote by* $\begin{pmatrix} \psi_0 \\ \psi_{-1} \end{pmatrix}$ *the first column vector of the orthogonal matrix* $K$*. If* $\psi$ *is the solution of the eigenvalue equation constructed with this initial data, prove that* $\sigma_n \geq \frac{1}{2}\|\tilde{\psi}_n\|^2\sigma$ *and conclude in the same way as in (i).*

4. *Extend this proof to the discrete strip using the formula proved in Lemma III.5.20*

**Problem III.6.8** *Let* $H$ *be the Schrödinger operator on the lattice* $\mathbb{Z}^d$ *and for each* $a \in \mathbb{C}^d$ *let us denote by* $P_a$ *the diagonal operator defined by*

$$(P_a\psi)(n) = e^{i<a,n>}\psi(n)$$

1. *Prove that one can write*

$$P_{-a}HP_a = H + Q_a$$
$$P_{-a}[H - \lambda I]^{-1}P_a = (H - \lambda I + Q_a)^{-1}$$

*where* $Q_a$ *is a bounded operator, independent of the potential and satisfying:*

$$\|Q_a\| \leq 2d(e^{|a|} - 1)$$

2. *Let us denote by* $\delta$ *the distance of the (real or complex) number* $\lambda$ *to the spectrum of* $H$. *Prove that for* $|a| < m(\delta) = \log(\frac{\delta}{4d} + 1)$ *one has:*

$$|P_{-a}[H - \lambda I]^{-1}P_a(x,y)| = |e^{e<a,y-x>}G(\lambda,x,y)| \leq \frac{2}{\delta}.$$

3. *Going to imaginary* $a$, *and always denoting by* $\delta$ *the distance of the number* $\lambda$ *from the spectrum of* $H$, *prove that one has:*

$$|G(\lambda,x,y)| \leq \frac{2}{\delta}e^{-m(\delta)|x-y|}$$

**Problem III.6.9** *Let* $H$ *be the Schrödinger operator on the one dimensional lattice and* $p_\lambda(n)$ *be the solution of the equation* $H\psi = \lambda\psi$ *with boundary conditions* $\psi(0) = 1$ *and* $\psi(1) = 0$. *For* $n \geq 1$ *one sets*

$$z_\lambda(n) = \frac{p_\lambda(n-1) + ip_\lambda(n)}{p_\lambda(n-1) - ip_\lambda(n)}$$

*Prove that the complex number* $z_\lambda(n)$ *is of modulus* 1 *and that*

$$\frac{dz_\lambda(n)}{d\lambda} = z_\lambda(n)\omega_\lambda(n)$$

*with* $\omega_\lambda(n) > 0$.

**Problem III.6.10** *Let* $H$ *be the Schrödinger operator on the discrete strip of width* $\ell$ *and let* $P_\lambda(n)$ *be the* $\ell \times \ell$ *matrix solution of the equation* $H\Psi = \lambda\Psi$ *with boundary conditions* $\Psi(0) = I_\ell$ *and* $\Psi(1) = 0$. *For* $n \geq 1$ *one sets*

$$Z_\lambda(n) = (P_\lambda(n-1) + iP_\lambda(n))(P_\lambda(n-1) - iP_\lambda(n))^{-1}$$

*Prove that the matrix* $Z_\lambda(n)$ *is unitary and symmetric and that*

$$\frac{dZ_\lambda(n)}{d\lambda} = Z_\lambda(n)\Omega_\lambda(n)$$

*where* $\Omega_\lambda(n)$ *is a real positive definite symmetric matrix.*

The following problems consider the spectral characteristics of a natural generalization of the free Hamiltonian $H_0$ first to symmetric graphs and second to the homogeneous trees and in particular the Caley trees or Bethe lattices.

**Problem III.6.11** *Let $X$ be a countable set and for each $x \in X$ we assume that a subset of $X$, say $U(x)$ is singled out to represent the set of neighbors of $x$. We assume that this set is finite, that $x \notin U(x)$ and the symmetry relation $y \in U(x)$ iff $x \in U(y)$. For the sake of simplicity we shall also assume that all the set $U(x)$ have the same number of elements. All these assumptions imply that the Laplace operator $\Delta$ defined by:*

$$\Delta f(x) = \sum_{y \in U(x)} f(y)$$

*is a bounded symmetric operator on the Hilbert space $\mathcal{H} = \ell^2(X)$. For a nonreal complex number $\lambda$ one denotes by $G_\lambda$ the Green's function, i.e. the kernel of the resolvent operator $[\Delta - \lambda I]^{-1}$ Let $o$ be a fixed point of $X$ and let $P_o$ be the projection operator on this point i.e. the operator defined by $P_o f = f(o)\delta_o$. We consider the operator $\tilde{\Delta}$ defined by:*

$$\tilde{\Delta} = (I - P_o)\Delta(I - P_o)$$

*and the associated Green's kernel $\tilde{G_\lambda}$.*
*Prove the following relations:*

1. $\tilde{G_\lambda} - G_\lambda = \tilde{G_\lambda}(\Delta - \tilde{\Delta})G_\lambda = G_\lambda(\Delta - \tilde{\Delta})\tilde{G_\lambda}$
2. $\Delta G_\lambda = G_\lambda \Delta = I + \lambda G_\lambda$
3. $\tilde{\Delta}\tilde{G_\lambda} = \tilde{G_\lambda}\tilde{\Delta} = I + \lambda \tilde{G_\lambda}$

*and conclude that if we denote by $g(\lambda)$ the complex number $G_\lambda(o,o)$ then one has:*

1. $\lambda \tilde{G_\lambda}\delta_o + \delta_o = 0$
2. $(G_\lambda - \tilde{G_\lambda})f = g(\lambda)^{-1} G_\lambda f(o) G_\lambda \delta_o + \lambda^{-1} f(o)\delta_o$
3. $f(o) + \lambda G_\lambda f(o) - \sum_{y \in U(o)} \tilde{G_\lambda} f(y) = \dfrac{1 + \lambda g(\lambda)}{g(\lambda)} G_\lambda f(o)$

**Problem III.6.12** *We now consider the particular case of a homogeneous tree $X$ with coordination number (also called valence) $p$. This means that each vertex has exactly $p+1$ neighbors. For technical reasons we first consider the case of the "half-tree" $X^*$ with root $o$. This is a tree with a particular vertex $o$ which has $p$ neighbors while all the other vertices have exactly $p+1$ neighbors.*

1. *Prove that in the case of an half-tree one has:*
   (a) $G_\lambda(o,x) = -g^2(\lambda)$ *for any* $x \in U(o)$
   (b) $pg^2(\lambda) + \lambda g(\lambda) + 1 = 0$
   (c) *The spectral measure of* $\tilde{\Delta}$ *computed at the point* $o$, *say* $\sigma_o^*(d\lambda)$, *is supported by the interval* $[-2\sqrt{p}, 2\sqrt{p}]$, *it is absolutely continuous and its density on the spectrum is given by:*

$$\frac{d\sigma_o^*(d\lambda)}{d\lambda} = \frac{\sqrt{4p-\lambda^2}}{2\pi p}$$

2. *Prove that in the case of the whole tree one has:*
   (a) $G_\lambda(o,x) = \frac{1+\lambda g(\lambda)}{p}$ *for any* $x \in U(o)$
   (b) $(p^2-\lambda^2)g^2(\lambda) + \lambda g(\lambda)(p-2) + p - 1 = 0$
   (c) *The spectral measure of* $\Delta$ *computed at the point* $o$, *say* $\sigma_o(d\lambda)$, *is supported by the interval* $[-2\sqrt{p}, 2\sqrt{p}]$, *it is absolutely continuous and its density on the spectrum is given by:*

$$\frac{d\sigma_o(d\lambda)}{d\lambda} = \frac{p\sqrt{4(p-1)-\lambda^2}}{2\pi((p+1)^2-\lambda^2)}$$

# III.7 Notes and Complements

The existence of square integrable solutions at infinity and the theory of the $m$-functions are cornerstones of the ODE approach to the spectral theory of the Sturm-Liouville operators, of which the one-dimensional Schrödinger operators are a particular case. See for example the classical texts of Coddington and Levinson [56] and of Levitan and Sargsjan [220]. The condition for essential self-adjointness given in Problem III.6.1 was introduced and extensively used by Kotani. See for example [195]. The notion of subordinate solution is very recent. It was introduced by Gilbert and Pearson in [121]. They proved the characterizations of the various spectra in terms of the existence of subordinate solution which we give in the case of the operator on the half line. See also Gilbert [120] and Minami [240] for the case of the whole line. Another fine analysis of the asymptotic properties of the $m$-functions and their consequences for the spectral characteristics of the operator was recently given by Bennewitz in [26].

Aronszajn already noticed in [10] that the absolutely continuous parts of the spectral measures for the problem on a half line were always equivalent, independently of the choice of the boundary condition at the origin. On the contrary, the singular parts are orthogonal. The form of this result which we give in Problem III.6.4 is proven in [121] as Lemma 6. The result and the proof of Problem III.6.8 are adapted from arguments first used by Combes and Thomas in the continuous case. See for example [272].

Problems III.6.2 and III.6.3 present the basic elements of the so-called Sturm oscillation theory. We shall see in Chapter VI that it has nice applications to the integrated density of states of random one dimensional systems. With this idea in mind, this oscillation theory has been extended in several ways. First to the lattice case, and then to the strip and more general systems. See the papers of Delyon and Souillard [77] and of Goldsheid [125] for example.

Proofs of the absolute continuity of the spectrum in the case of a constant electric field can be found in [27], [23] or [43]. The proof in [27] is presumably the best in terms of extensions to multidimensional models. The proof in [43] is restricted to one dimensional problems for it relies on an extension of the result of PropositionIII.4.1 and intricate estimates on the asymptotic behavior of the solutions of the eigenvalue equation (III.1). It has nevertheless a few interesting twists. It allows for example to replace the term $\kappa t$ in (III.48) by something of the form $\kappa t^\alpha$ as long as $\alpha \in [1,2]$ and also, the assumption on the existence of this generalized electric field is needed either near $+\infty$ or near $-\infty$ but not necessarily near both. The result and the proof of Proposition III.4.6 are borrowed from [305]. In [305] Simon and Spencer use the Kato-Birman theory of trace class perturbation to show that some operators do not have absolutely continuous spectrum. The idea is to compare the given potential with a sequence of potential barriers. We gave the result involving a sequence of barriers with heights going to infinity. It is also shown in this paper that it is possible to force the tunnelling probability to go to zero with barriers with bounded heights. The comparison argument involves periodic potentials which we shall discuss in Chapter VII. The motivation of Simon and Spencer was to shed some light on deterministic reasons why some cases of random Schrödinger operators have no absolutely continuous spectrum.

Pearson's discussion of Schrödinger operators with singular continuous spectra can be found in [266].

The main interest of Problem III.6.6 is to relate the spectral characteristics of Schrödinger operators on the half axes $(-\infty,0]$ and $[0,+\infty)$ the those of the operator on the whole real line $\mathbb{R}$. Indeed, if one chooses for $w_+$

and $w_-$ the Weyl-Titchmarsh functions $m_+$ and $m_-$ respectively, then the corresponding spectral measures $\sigma_+$ and $\sigma_-$ are the spectral measures we constructed for the operator on the half lines $[0,\infty)$ and $(-\infty,0]$ with Neuman boundary condition at 0. I.S. Kac proves in [162] a rather precise result on the multiplicity of the spectrum:

For any nonnegative measure $\sigma$ on $\mathbb{R}$ satisfying $\int_{-\infty}^{+\infty} d\sigma(\lambda)/(1+\lambda^2) < \infty$, we denote by $D_\sigma$ the set of $\lambda \in \mathbb{R}$ at which the function $\lambda \hookrightarrow \sigma((-\infty,\lambda])$ has a nonzero derivative. We also denote by $D'_\sigma$ the set of $\lambda \in \mathbb{R}$ at which the function $\epsilon \hookrightarrow w(\lambda + i\epsilon)$ has a finite limit when $\epsilon \searrow 0$. Here, $w$ is any Herglotz function $w$, with spectral measure $\sigma$. There is no ambiguity here for any two such functions can only differ by a linear term. Let us now assume that $\sigma_+$, $\sigma_-$, $M$ and $\sigma$ are as in Problem III.6.6, and let us denote by $H$ the operator of multiplication by the independent variable $\lambda$ on the Hilbert space $L^2(\mathbb{R}, dM)$. Then, Kac's result states that the spectrum of $H$ has multiplicity one if the set $D_{\sigma_+} \cap D_{\sigma_-}$ has Lebesgue measure 0, and multiplicity 2 otherwise. Moreover, in this case, the multiplicity is exactly 2 on the set $D'_{\sigma_+} \cap D'_{\sigma_-}$ and exactly one on its complement.

In particular, the multiplicity is one if either $\sigma_+$ or $\sigma_-$ is singular. We will come back to this multiplicity result in our study of the absolutely continuous spectrum of one-dimensional Schrödinger operators.

The idea to integrate the spectral measure with respect to a boundary condition seems to appear for the first time in the book of Atkinson [11]. Not much use was made of this idea. But workers in random operators found it very useful and used it frequently, in a more or less explicit for. See for example the papers of Goldsheid and Molcanov [126], Royer [278], Lacroix [206], [208] and Carmona [43], [44]. This line culminates with the paper of Kotani [193] where he used the fact that the integral of the spectral measure $\sigma_\alpha^+$ of the half axis $[0,\infty)$ with respect to $d\alpha$ was equal to a multiple of the Lebesgue measure. Here $\alpha$ is the angle defining the boundary condition at 0. This is at the origin of what we called Kotani's trick in Chapter VIII. The systematic use of the idea of averaged spectral measure to the spectral analysis of deterministic (one-dimensional) Schrödinger operators is due to Carmona and all the applications in this area, described in this chapter, are taken from [43]. Some of these results have been re proved since then by a direct analysis of the $m$-function. See for example [67].

The quasi-one dimensional operators can be regarded as a technical generalization of their one dimensional counterparts. They have been considered, especially in the random setting, as a means to approach the two dimensional case via a limiting procedure. Even though this approach did not prove to be successful yet, it had the merit of bringing to light a large variety of interesting mathematical problems. In particular the notion of Lyapunov exponent is more involved and the integration of spectral measures with respect to boundary conditions requires some knowledge about the Lagrangian boundaries of the symplectic group together with the basic properties of the Poisson kernels of these boundaries. Unfortunately the lack of uniformity in the various estimates (or at least the absence of explicit bounds) with respect to the width $\ell$ of the strip retains from any limiting procedure to approach the two dimensional case. Most of the results on the strip are borrowed from Lacroix [207], [208], [209].

# IV

# PRODUCTS OF RANDOM MATRICES

Contents:



Part of the one dimensional or quasi-one dimensional theory of localization can be reduced to the study of products of random matrices. One of the most important result in this direction is the extension to matrix valued random variables of the strong law of large numbers. Unfortunately the identification of the limit (called the Lyapunov exponent) is more complicated than in the classical case of real valued random variables. In particular this limit can no longer be written as a single expectation. Moreover its determination involves the computation of some invariant measure on the projective space. We only assume that the reader has a minimal background in classical probability theory. Most of the material presented in this chapter is self contained.

## IV.1 General Ergodic Theorems

Let $(\Omega, \mathcal{F}, \mathbb{P})$ be a probability space and $T$ be a set which we will choose to be either the set $\mathbb{N}$ of nonnegative integers ot the set $\mathbb{Z}$ of all the integers in the discrete case, or the set of nonnegative real numbers or the whole real line $\mathbb{R}$ in the continuous case. Endowed with a semigroup $\{\theta_t\,;\, t \in T\}$ of measure preserving transformations the system $(\Omega, \mathcal{F}, \theta_t, \mathbb{P})$ will henceforth be called a dynamical system. (In the continuous case a dynamical system is often called a "dynamical flow"). An invariant random variable is a measurable function $Z$ such that $Z \circ \theta_t = Z$ for all $t \in T$. A dynamical system $(\Omega, \mathcal{F}, \theta_t, \mathbb{P})$ is said to be ergodic if every invariant random variable is constant $\mathbb{P}$-a.e. The discussions of the additive and the subadditive theorems which follow are taylored for the applications to one-dimensional random Schrödinger operators. In oreder to construct the integrated density of states of $d$-dimensional operators, we shall have to review in Chapter VI more general sub- and super additive theorems.

**Definition IV.1.1** *A subadditive process is a measurable real valued process $\{X(t)\,;\, t \in T\}$ such that:*

$$X(0) = 0\ , \quad X(t+s) \leq X(t) + X(s) \circ \theta_t \quad s,t \in T$$

*If the inequality is replaced by an equality in the above definition then $X(t)$ is called an additive process.*

The Kingman subadditive ergodic theorem stated below is one of the basic tools in this chapter .

**Theorem IV.1.2** *Let $\{X(n)\,;\, n \geq 0\}$ be a subadditive process with discrete time such that:*

**(i)** *$X(n)$ is integrable for all $n$*

**(ii)** *The sequence $\{\frac{1}{n}\mathbb{E}\{X(n)\}\ ;\ n \geq 1\}$ is bounded below*

*Then there exists an integrable invariant random variable $Z$ such that the sequence $\frac{1}{n}X(n)$ converges $\mathbb{P}$-a.e. and in expectation to $Z$. Moreover $\mathbb{E}\{Z\} = \inf_{n\geq 1} \frac{1}{n}\mathbb{E}\{X(n)\}$.*

*Proof:*
There are a lot of proofs available for the subadditive ergodic theorem. The original one can be found in Kingman [171], and certainly the simplest one has been given by J.M. Steele[312] ∎

In the additive case we remark that $X(n) = \sum_{k=0}^{n-1} X(1)\circ\theta_k$ thus if $X(1)$ is integrable then conditions (i) and (ii) are satisfied and the Kingman Theorem reduces to the Birkhoff Theorem. It is well known that Theorem IV.1.2 is false for continous time (even in the additive case) but it becomes true with some extra regularity assumptions on the process $\{X(t)\,;\, t > 0\}$

**Corollary IV.1.3** *Let $\{X(t)\,;\, t \geq 0\}$ be a subadditive process with continuous time such that*

**(i)** *$X(t)$ is integrable for all $t \in T$,*
**(ii)** *The set $\left\{\frac{1}{t}\mathbb{E}\{X(t)\}\ ;\ t > 0\right\}$ is bounded below,*
**(iii)** *The process $X(t)$ is separable and $M = \sup_{0\leq s<t\leq 1} |X(t-s)\circ\theta_s|$ is integrable.*

*Then $\frac{1}{t}X(t)$ converges $\mathbb{P}$-a.e. and in expectation to to an integrable invariant random variable $Z$ and $\mathbb{E}\{Z\} = \inf_{t>0} \frac{1}{t}\mathbb{E}\{X(t)\}$.*

*Proof:*
For $n = 0, 1, 2\ldots$ we can apply Theorem IV.1.2 to the process $X(n)$. If $n$ is the integral part of $t$ then we can write:

$$X(n+1) - X(n+1-t)\circ\theta_t \leq X(t) \leq X(n) + X(t-n)\circ\theta_n$$

We remark that $|X(n+1-t)\circ\theta_t| \leq M\circ\theta_n$ and $|X(t-n)\circ\theta_n| \leq M\circ\theta_n$. The Borel-Cantelli lemma implies that $\frac{1}{n}M\circ\theta_n$ converges $\mathbb{P}$-a.e. to zero and this yields the result. ∎

The Birkhoff Theorem admits an important improvement which essentially says that when the limit is not obviously zero then it has to be positive and this property will be very useful in order to prove that the limit is positive in the ergodic case.

**Theorem IV.1.4** *Let $\{X(n)\,;\,n \geq 1\}$ be an integrable additive process with discrete time such that $X(n)$ converges $\mathbb{P}$-a.e. to $+\infty$. Then the limit random variable given by the Birkhoff ergodic theorem has a positive expectation.*

*Proof:*
Suppose that the conclusion is false and let us denote by $Z$ the limit. We know that $Z$ is a nonnegative random variable thus if we suppose that it as a nul expectation then $Z$ is zero $\mathbb{P}$-a.e. For a positive real number $\epsilon$ we define the sequence of random variables

$$A_n = m\Big(\bigcup_{k=1}^{n} [X(k) - \epsilon\ ,\ X(k) + \epsilon]\ \Big)$$

where $m(I)$ is the Lebesgue measure of the set $I$. Using the convergence of $\frac{1}{n}X(n)$ to zero one can conclude that $\frac{1}{n}\mathbb{E}\{A_n\}$ also converges to zero. We now write:

$$\begin{aligned} A_n - A_{n-1} \circ \theta &\geq 2\epsilon\, \mathbf{1}_{\{\cap_{k=2}^{n}\{|X(k) - X(1)| > 2\epsilon\}\}} \\ \mathbb{E}\{A_n\} - \mathbb{E}\{A_{n-1}\} &\geq 2\epsilon\, \mathbb{P}\big\{\cap_{k=2}^{n}\{|X(k) - X(1)| > 2\epsilon\}\big\} \\ &\geq 2\epsilon\, \mathbb{P}\big\{\cap_{k=2}^{\infty}\{|X(k) - X(1)| > 2\epsilon\}\big\} \end{aligned}$$

From $\lim \frac{1}{n}\mathbb{E}\{A_n\} = 0$ one obtains that $\mathbb{P}\big\{\cap_{k=2}^{\infty}\{|X(k) - X(1)| > 2\epsilon\}\big\} = 0$ and by stationarity that

$$\mathbb{P}\big\{\cup_{n=1}^{\infty} \cap_{k=1}^{\infty} \{|X(k+n) - X(n)| > 2\epsilon\}\big\} = 0$$

This last relation contradicts the convergence of $X(n)$ to $\infty$. ∎

# IV.2 Matrix Valued Systems

Let $GL(d, \mathbb{R})$ be the group of linear automorphisms of $\mathbb{R}^d$. For a matrix $g$ in $GL(d, \mathbb{R})$, $g'$ denotes the transpose, $g^{-1}$ the inverse, $\mathrm{tr}(g)$ the trace and $\det(g)$ the determinant. Every matrix $g \in GL(d, \mathbb{R})$ has a polar decomposition $g = k(g)a(g)u(g)$ where $u(g)$ and $k(g)$ are orthogonal matrices and $a(g)$ is a diagonal matrix with diagonal entries $a_1(g) \geq a_2(g) \geq \ldots \geq a_d(g) > 0$. The matrix $a(g)$ is well defined since the scalars $a_i(g)$ are the square roots of the eigenvalues of $g'g$ (but in general the matrices $k(g)$ and $u(g)$ are not uniquely defined). We denote by $\Lambda^p(\mathbb{R}^d)$ the vector space of alternating $p$ linear forms on the dual of $\mathbb{R}^d$. For $u_1, u_2, \ldots, u_p$ in $\mathbb{R}^d$ and $f_1, f_2, \ldots, f_p$

in the dual of $\mathbb{R}^d$ let $m$ be the matrix defined by $m_{i,j} = f_i(u_j)$. The formula $u_1 \wedge u_2 \wedge \ldots \wedge u_p \; (f_1, f_2, \ldots, f_p) = \det(m)$ defines an element $u = u_1 \wedge u_2 \wedge \ldots \wedge u_p$ of $\Lambda^p(\mathbb{R}^d)$ called a decomposable $p$-vector. For a $p$-vector $u$ we also denote by $u$ the $d \times p$ matrix of the coordinates of $u_1, u_2, \ldots, u_p$ in the usual basis of $\mathbb{R}^d$. It is well known that:

**Lemma IV.2.1** **(i)** *The decomposable p vectors constructed with the vectors of a basis of $\mathbb{R}^d$ span the linear space $\Lambda^p(\mathbb{R}^d)$*

**(ii)** *The p-vector $u_1 \wedge u_2 \wedge \ldots \wedge u_p$ is nonzero iff the system $(u_1, u_2, \ldots, u_p)$ is linearly independent*

**(iii)** *The linear extension to the whole of the space $\Lambda^p(\mathbb{R}^d)$ of the formula $< u_1 \wedge u_2 \wedge \ldots \wedge u_p \, , \, v_1 \wedge v_2 \wedge \ldots \wedge v_p >= \det(u'v)$ defines an inner product on $\Lambda^p(\mathbb{R}^d)$.*

In the sequel we will always use the norm on $\Lambda^p(\mathbb{R}^d)$ associated to the inner product defined in Lemma IV.2.1(iii). In particular, $||u||^2 = \det(u'u)$.

For $g \in GL(d, \mathbb{R})$ we let $\Lambda^p g$ be the linear automorphism of $\Lambda^p(\mathbb{R}^d)$ defined by $\Lambda^p g(u_1 \wedge u_2 \wedge \ldots \wedge u_p) = gu_1 \wedge gu_2 \wedge \ldots \wedge gu_p$. We note that $\Lambda^p gh = \Lambda^p g \Lambda^p h$ thus if $k$ is an orthogonal matrix so is $\Lambda^p k$. We can deduce from these properties that $||\Lambda^p g|| = a_1(g) a_2(g) \ldots a_p(g)$ and this implies $||\Lambda^p g|| \leq ||g||^p$.

When $d = 2\ell$ there is a subgroup of $GL(d, \mathbb{R})$ of particular interest called the symplectic group of order $\ell$ and denoted by $SP(\ell, \mathbb{R})$.

**Definition IV.2.2** *The symplectic group $SP(\ell, \mathbb{R})$ is the set of matrices $g$ in $GL(d, \mathbb{R})$ satisfying:*

$$g' J g = J \text{ with } J = \begin{pmatrix} 0 & I_\ell \\ -I_\ell & 0 \end{pmatrix} \text{ where } I_\ell \text{ is the identity matrix of order } \ell.$$

We remark that for $\ell = 1$ the symplectic group is the unimodular group $SL(2, \mathbb{R})$ but that $SP(\ell, \mathbb{R})$ is a proper subgroup of $SL(2\ell, \mathbb{R})$ for greater values of $\ell$. One can find more information about the symplectic group in Section 6 of this chapter. The proof of the following lemma is an easy exercise:

**Lemma IV.2.3** *Let $g$ be a matrix in $SP(\ell, \mathbb{R})$. Then one has:*

**(i)** $\det(g) = 1$ *and* $g' \in SP(\ell, \mathbb{R})$

**(ii)** *If $\lambda$ is an eigenvalue of $g$ so is $1/\lambda$*
**(iii)** *In the polar decomposition of $g$ we have $a(g)_i = a(g)_{d-i+1}^{-1}$ for any $i = 1 \dots d$*

For a sequence $\gamma_1 \geq \gamma_2 \geq \dots \geq \gamma_d$ of real numbers we let $\lambda_1 > \lambda_2 > \dots > \lambda_r$ be the set of its distinct values . We call multiplicity of $\lambda_i$ the number of occurence of $\lambda_i$ in the sequence $(\gamma_1, \gamma_2, \dots, \gamma_d)$. The next Theorem is called the "deterministic" Oseledec theorem.

**Theorem IV.2.4** *Let $\{g(n)\,;\, n \geq 0\}$ be a sequence in $GL(d, I\!R)$ and let us assume that there exist numbers $\gamma_1 \geq \gamma_2 \geq \dots \geq \gamma_d$ called Lyapunov exponents such that the product $U(n) = g_{n-1} \cdots g_0$ satisfies for $p = 1, 2, \dots, d$:*

**(i)** $\lim_{n\to\infty} \frac{1}{n} \log ||\Lambda^p g(n)|| = 0$,
**(ii)** $\lim_{n\to\infty} \frac{1}{n} \log ||\Lambda^p U(n)|| = \gamma_1 + \dots + \gamma_p$.

*Let us denote by $r$ the number of distinct Lyapunov exponents, then there exists a strictly decreasing family of subspaces of $I\!R^d$ such that*

**(i)** $I\!R^d = \mathcal{V}^1 \supset \mathcal{V}^2 \supset \dots \supset \mathcal{V}^r \supset \mathcal{V}^{r+1} = \{0\}$,
**(ii)** $v \in \mathcal{V}^i \setminus \mathcal{V}^{i+1} \iff \lim_{n\to\infty} \frac{1}{n} \log ||U(n)v|| = \lambda_i \quad i = 1, 2, \dots, r$,
**(iii)** $\dim \mathcal{V}^i - \dim \mathcal{V}^{i+1} =$ multiplicity of $\lambda_i \quad i = 1, 2, \dots, r$.

*Proof:*
See Exercise IV.7.2 for the case $d = 2$. The original proof is in Raghunatan [270] and a simpler one can be found in Ledrappier [212]. ∎

In the particular case of products of symplectic matrices Lemma IV.2.3(iii) implies that $\gamma_i = -\gamma_{d-i+1}$ for $i = 1, 2, \dots, d$ and since a symplectic matrix has a determinant equal to one we obtain $\gamma_1 \geq \gamma_2 \geq \dots \geq \gamma_\ell \geq 0$

We now introduce the important concept of multiplicative process. Such processes are defined and studied in full generality for example in Bougerol [32],[31] but we will use a simpler definition adapted to our needs.

**Definition IV.2.5** *Let $\{U(t)\,;\, t \in T\}$ be a process taking values in the group $GL(d, I\!R)$ and defined on the dynamical system $(\Omega, \mathcal{F}, \theta_t, I\!P)$. We say that it is a mutiplicative process if:*

- $U(0) = I\ , \qquad U(t+s) = (U(s) \circ \theta_t) U(t) \quad s, t \in T$

*Such a process is said to be regular if:*

- *the random variable* $\left(\log^+ \|U(1)\| + \log^+ \|U^{-1}(1)\|\right)$ *is integrable in the discrete case,*
- *it is a separable stochastic process and the random variable* $M = \left(\sup_{0\le t\le 1} \log^+ \|U(t)\| + \sup_{0\le t\le 1} \log^+ \|U^{-1}(t)\|\right)$ *is integrable in the continuous case.*

We note that $|\log\|g\|\,| \le \log^+\|g\| + \log^+\|g^{-1}\|$ whenever $g \in GL(d, \mathbb{R})$ and in the particular case of matrices in $SP(\ell, \mathbb{R})$ we have $\|g\| = \|g^{-1}\| \ge 1$. Note also that the integrability conditions in (ii) are also satisfied by the exterior powers of the matrices $U(t)$ since $\|\Lambda^p g\| \le \|g\|^p$. In the discrete case a mutiplicative process can be written as a product of matrices of $GL(d, \mathbb{R})$ since if we denote by $g(n)$ the matrix $U(1) \circ \theta_n$ then $U(n) = g(n-1)\dots g(0)$.

The application of the deterministic Oseledec's theorem to multiplicative processes is often refered as the "Random" Oseledec theorem:

**Theorem IV.2.6** *Let* $\{U(t)\,;\, t \in T\}$ *be a regular multiplicative process defined on an ergodic dynamical system. Then there exists an invariant set* $\Omega_0$ *of full* $\mathbb{P}$ *measure such that for* $\omega \in \Omega_0$ *one has:*

**(1)** $\lim_{t\to\infty} \frac{1}{t} \log\|\Lambda^p U(t)\| = \lim_{t\to\infty} \frac{1}{t} \mathbb{E}\{\log\|\Lambda^p U(t)\|\} = \gamma_1 + \dots + \gamma_p$ *for* $p = 1, 2, \dots, d$. *The (possibly random) numbers* $\gamma_1 \ge \gamma_2 \ge \dots \ge \gamma_d$ *are called the Lyapunov exponents of the process* $U(t)$.

**(2)** *Let* $r$ *be the number of distinct Lyapunov exponents, then there exists a strictly decreasing sequence of measurable subspaces of* $\mathbb{R}^d$ *denoted by* $\mathcal{V}_\omega^i$ *such that*

**(i)** $\mathbb{R}^d = \mathcal{V}_\omega^1 \supset \mathcal{V}_\omega^2 \supset \dots \supset \mathcal{V}_\omega^r \supset \mathcal{V}_\omega^{r+1} = \{0\}$,

**(ii)** $v \in \mathcal{V}_\omega^i \setminus \mathcal{V}_\omega^{i+1} \Longleftrightarrow \lim_{t\to\infty} \frac{1}{t} \log\|U(t)v\| = \lambda_i \quad i = 1, 2, \dots, r$,

**(iii)** $\dim \mathcal{V}_\omega^i - \dim \mathcal{V}_\omega^{i+1} =$ multiplicity of $\lambda_i \quad i = 1, 2, \dots, r$,

**(iv)** $\mathcal{V}_{\theta_t\omega}^i = U(t)\mathcal{V}_\omega^i$.

*Proof:*
For $p = 1, \dots, d$ the process $X(t)$ defined by $\log\|\Lambda^p U(t)\|$ is subadditive and the integrability conditions in the Definition IV.2.5 make it possible to apply the Kingman subadditive ergodic Theorem to the process $X(t)$. This yields conclusion (1) and if we define $g(n) = U(1) \circ \theta_n$ then $U(n) =$

$g(n-1)\dots g(0)$ and we obtain the conclusions 2 (i) and (ii) and (iii) from Theorem IV.2.4 at least for discrete time. The same result follows for continuous time if we remark that for $n \le t \le n+1$ one has:

$$\begin{aligned}
& |\log \|U(t)v\| - \log \|U(n)v\|\,| \\
= \ & |\log \frac{\|U(t)v\|}{\|U(n)v\|}| \\
= \ & |\log \frac{\|(U(t-n)\circ\theta_n)U(n)v\|}{\|U(n)v\|}| \\
\le \ & \max(|\log \|U(t-n)\circ\theta_n\|\,|\,, |\log \|U^{-1}(t-n)\circ\theta_n\|\,|) \le M\circ\theta_n.
\end{aligned}$$

It remains only to check the measurability properties of the spaces $\mathcal{V}_\omega^i$ and conclusion 2(iv). They are easy consequences of the fact that $\mathcal{V}_\omega^i = \{v \in \mathbb{R}^d; \lim_{n\to\infty} \frac{1}{n}\log\|U(n)v\| \le \lambda_i\}$. ∎

Note that the Lyapunov exponents are not random when the dynamical system is ergodic. In any case, dynamical systems for which $\{\theta_t\,;\, t\in T\}$ form a group of invertible mappings of $\Omega$ will be called invertible dynamical systems. For such systems one can apply the preceding theorems in the "two directions" of $T$ and it turns out that the limits at $+\infty$ and $-\infty$ are closely related. We will add the superscript $\pm$ to the Lyapunov exponents and subspaces defined in Theorem IV.2.6. We still consider limits for $t \to \pm\infty$ but we now use the normalization factor $\frac{1}{|t|}$ rather than $\frac{1}{t}$.

**Proposition IV.2.7** *Let us denote by $(\Omega, \mathcal{F}, \theta_t, \mathbb{P})$ an invertible ergodic dynamical system and by $U(t)$ a regular multiplicative process. Then:*

**(i)** $^-\gamma_p = -^+\gamma_{d-p+1} \qquad p = 1\dots d,$
**(ii)** $^-\lambda_i = -^+\lambda_{r-i+1} \qquad i = 1\dots r,$
**(iii)** $\mathbb{R}^d = {}^+\mathcal{V}_\omega^i \oplus {}^-\mathcal{V}_\omega^{r-i} \qquad i = 1\dots r-1.$

*The spaces $^\pm\mathcal{V}_\omega^i$ are defined for $\omega$ in an invariant set $\Omega_0$ of full $\mathbb{P}$-measure*

*Proof:*
Because $U(-t) = U^{-1}(t)\circ\theta_{-t}$ the process $U(-t)$ is multiplicative. This also implies $a_p(U(-t)) = a_{d-p+1}^{-1}(U(t)\circ\theta_{-t})$. The invariance of the probability measure $\mathbb{P}$ yields $\mathbb{E}\{\log a_p(U(-t))\} = -\mathbb{E}\{\log a_p(U(t))\}$ proving (i) and (ii). To prove (iii) we remark that we have $\dim {}^\pm\mathcal{V}_\omega^i = (\#$ of exponents $\le^\pm \lambda_i)$ and (ii) implies that the sum of the dimensions of the subspaces appearing in (iii) is equal to $d$. To prove that the intersection of these

subspaces reduces to the zero vector, we pick a unit vector $v$ in ${}^{+}\mathcal{V}_{\omega}^{i} \cap {}^{-}\mathcal{V}_{\omega}^{r-i}$ and we consider the random variables $Z_{i,n}^{\pm}(v) = \frac{1}{n}\log(||U(n)v||/||v||) -^{\pm} \lambda_i$. We can write $v = (U(-n) \circ \theta_n)U(n)v$ and obtain:

$$0 \leq n({}^{-}\lambda_{r-i} +^{+} \lambda_i + Z_{r-i,-n}^{-} \circ \theta_n(U(n)v) + Z_{i,n}^{+}(v))$$

We know from (ii) that ${}^{-}\lambda_{r-i} = -^{+}\lambda_{i+1}$ and from Theorem IV.2.6 that $U(n)v \in^{-} \mathcal{V}_{\theta_n\omega}^{r-i}$. Since $Z_{i,n}^{\pm}(v)$ converges in probability to zero for $v \in^{\pm} \mathcal{V}_{\omega}^{i}$ we get a contradiction. ∎

The decomposition (iii) is often called a "splitting of $\mathbb{R}^d$"

This proposition has an important consequence for multiplicative processes with values in $SP(\ell, \mathbb{R})$. In this case there is no need to put a superscript $\pm$ to the Lyapunov exponents since ${}^{-}\gamma_p = -^{+}\gamma_{d-p+1} =^{+} \gamma_p$ and the "splitting of $\mathbb{R}^d$" has a simpler interpretation.

**Corollary IV.2.8** *Let $(\Omega, \mathcal{F}, \theta_t, \mathbb{P})$ be an invertible ergodic dynamical system and $U(t)$ be a regular multiplicative process with values in $SP(\ell, \mathbb{R})$. Then with probability one for any nonzero vector $v \in \mathbb{R}^d$ we have:*

$$\max\left(\lim_{t\to-\infty} \frac{1}{|t|}\log||U(t)v|| \, , \lim_{t\to+\infty} \frac{1}{t}\log||U(t)v||\right) \geq \gamma_\ell$$

*Proof:*
If $s$ is the number of non negative distinct Lyapunov exponents then $r = 2s$ and we apply (ii) and (iii) of Proposition IV.2.7 with $i = s$. We then remark that if $v \notin^{\pm} \mathcal{V}_{\omega}^{s}$ then the above limits are greater than $\gamma_\ell$. ∎

When the lowest non negative exponent $\gamma_\ell$ is positive this corollary says that for any nonzero vecror $v$ the process $U(t)v$ has an "hyperbolic behavior", i.e. that it is exponentially growing in at least one direction of $T$. One of the essential problems in the sequel will be to prove that $\gamma_\ell$ is positive. We will see in the following sections that this can be done for independent and Markovian models. Obviously the problem is much simpler in the case $d = 2$ where $\gamma_\ell$ is also the upper exponent !

We end this section with the classical construction of a product dynamical system associated with $(\Omega, \mathcal{F}, \theta_t, \mathbb{P})$ and a regular multiplicative process

$U(t)$. This will turn out to be very useful when we try to obtain an explicit formula for the upper Lyapunov exponent $\gamma_1$. Let $P(\mathbb{R}^d)$ be the projective space of $\mathbb{R}^d$, that is the set of directions in $\mathbb{R}^d$ and let $\mathcal{B}$ be its Borel sigma-algebra . For a non-zero vector $v$ in $\mathbb{R}^d$ we denote by $\bar{v}$ the image of $v$ in $P(\mathbb{R}^d)$. $GL(d,\mathbb{R})$ acts on $P(\mathbb{R}^d)$ by the formula $g\bar{v} = \overline{gv}$ and we set:

$$\tilde{\Omega} = \Omega \times P(\mathbb{R}^d), \quad \tilde{\mathcal{F}} = \mathcal{F} \times \mathcal{B}, \quad \tilde{\theta}_t(\omega, \bar{v}) = (\theta_t\omega, U(t,\omega)\bar{v})$$

It is easy to check that we have the semigroup property $\tilde{\theta}_{t+s} = \tilde{\theta}_t \circ \tilde{\theta}_s$ and that the function $v \hookrightarrow \log(||U(t)v||/||v||)$ is actually a function of $\bar{v}$.

**Proposition IV.2.9** *Let us assume that $(\Omega, \mathcal{F}, \theta_t, \mathbb{P})$ is an ergodic dynamical system and that $\tilde{\mathbb{P}}$ is a $\tilde{\theta}_t$-invariant probability measure with projection $\mathbb{P}$ on $\Omega$. Then:*

**(i)** *The function $t \longrightarrow \int \log(||U(t,\omega)v||/||v||)\, d\tilde{\mathbb{P}}(\omega, \bar{v})$ is additive.*

**(ii)** *If $\tilde{\mathbb{P}}\{(\omega, \bar{v})\, ;\, \omega \in \Omega_0, v \in \mathcal{V}_\omega^2\} = 0$ (notations of Theorem IV.2.6) then:*

$$\gamma_1 = \int \log \frac{||U(1,\omega)v||}{||v||}\, d\tilde{\mathbb{P}}(\omega, \bar{v})$$

**(iii)** *If $(\tilde{\Omega}, \tilde{\mathcal{F}}, \tilde{\theta}_t, \tilde{\mathbb{P}})$ is ergodic and $\tilde{\mathbb{P}}\{(\omega, \bar{v})\, ;\, \omega \in \Omega_0, v \notin \mathcal{V}_\omega^2\} > 0$ then formula (ii) holds.*

*Proof:*
Let $X(t,\omega,\bar{v}) = \log(||U(t,\omega)v||/||v||)$ be defined on the dynamical system $(\tilde{\Omega}, \tilde{\mathcal{F}}, \tilde{\theta}_t, \tilde{\mathbb{P}})$. It is readily seen that $X(t)$ is an additive process and the invariance of $\tilde{\mathbb{P}}$ gives conclusion (i). To prove (ii) we remark that $X(1)$ is integrable and the Birkhoff ergodic Theorem implies that the sequence $\frac{1}{n}X(n)$ converges $\tilde{\mathbb{P}}$-a.e. and in expectation to an invariant random variable $Z$. The Oseledec Theorem IV.2.6 implies that for $\mathbb{P}$-almost all $\omega$ and for all $v$ in $\mathbb{R}^d - \mathcal{V}_\omega^2$ this sequence converges to $\gamma_1$ thus this sequence converges to the same limit $\tilde{\mathbb{P}}$-a.e. It follows that $Z = \gamma_1 = \int Z(\omega, \bar{v})\, d\tilde{\mathbb{P}}(\omega, \bar{v})$ and we obtain (ii). The same proof works for (iii) since we already know that $Z$ is constant thus it is enough to have the above convergence on a set of positive $\tilde{\mathbb{P}}$-measure. ∎

We will see in the next sections how to construct measures $\tilde{\mathbb{P}}$ satisfying (ii) or (iii) in the independent and Markovian cases.

# IV.3 Group Action on Compact Spaces

## IV.3.1 Definitions and Notations

Let $G$ be a locally compact $\sigma$-compact metric group with unit $e$ and let $B$ be a metric topological space.

**Definition IV.3.1** *We say that $G$ acts on $B$ if one can associate continuously to each $(g, b)$ in $G \times B$ an element $gb$ of $B$ such that:*

**(i)** $g_1(g_2 b) = (g_1 g_2)b \qquad \forall g_1 , g_2 \in G \, , \, b \in B$
**(ii)** $eb = b \qquad \forall b \in B$

We define the "pseudo-convolution" of a probability measure $\mu$ on $G$ and a measure $\nu$ on $B$ as the unique measure $\mu * \nu$ on $B$ determined by:

$$\mu * \nu(f) = \int f(gb)\, d\mu(b)\, d\nu(g)$$

where $f$ is any bounded measurable function on $B$. We use the same symbol than for the ordinary convolution on $G$ since when $B = G$ the two definitions coincide. We also remark that $(\mu_1 * \mu_2) * \nu = \mu_1 * (\mu_2 * \nu)$.

From now on we fix a probability measure $\mu$ on $G$. The n$^{th}$ power of convolution of $\mu$ on $G$ is denoted by $\mu^n$. We also denote by $T_\mu$ the smallest closed sub-semi-group of $G$ containing the support of $\mu$ and by $G_\mu$ the smallest closed sub-group of $G$ containing the support of $\mu$. A measure $\nu$ on $B$ is said to be $\mu$ invariant if $\mu * \nu = \nu$. If $B$ is compact then it is readily seen that there always exist invariant probability measures. Take for example any weak limit point of the sequence $\frac{1}{n}\sum_{j=1}^{n} \mu^n * m$ where $m$ is any probability measure on $B$. Uniqueness of the invariant measure will be one of the problems discussed in the next section.
On Abelian groups the harmonic analysis of $\mu$ is carried out using the Fourier-Laplace transform associated to the set of exponentials on $G$. Unfortunately general groups (Simple Lie Groups for instance) have very few exponentials. This is why we need to introduce cocycles generalizing the notion of exponentials.

**Definition IV.3.2** *A continuous map* $\sigma$ *from* $G \times B$ *with values into* $(0\ ,\ \infty)$ *is called a cocycle if*

$$\sigma(g_1 g_2, b) = \sigma(g_1, g_2 b)\sigma(g_2, b) \qquad g_1\ , g_2 \in G, \quad b \in B$$

We remark that if $B$ is reduced to a single point then a cocycle is nothing but an exponential. Products and real powers of cocycles are again cocycles and $\sigma(e, b) = 1$ for any $b \in B$. A cocycle $\sigma$ is said $\mu$-integrable if the function $\bar{\sigma}(g) = \sup_{b \in B} \sigma(g, b)$ is $\mu$-integrable. One remarks that if we set $\tilde{\sigma}(g) = \sup_{b \in B} |\log \sigma(g, b)|$ then $\exp(\tilde{\sigma}(g))$ is $\mu$ integrable iff $\sigma$ and $\sigma^{-1}$ are $\mu$ integrable. The following lemma is an easy consequence of the definition of a cocycle.

**Lemma IV.3.3** *Let* $\sigma$ *be a* $\mu$*-integrable cocycle, then the sequences* $a_n$ *and* $b_n$ *defined by:*

$$a_n = \int \bar{\sigma}(g)\, d\mu^n(g) \qquad \textit{and} \qquad b_n = \sup_{b \in B} \int \sigma(g, b)\, d\mu^n(g)$$

*satisfy* $a_{n+m} \le a_n\, a_m$ *and* $b_{n+m} \le b_n\, b_m$.

We will often use the result of the following proposition:

**Proposition IV.3.4** *Let* $\nu$ *be an invariant probability measure and let* $\sigma$ *be a cocycle such that* $\tilde{\sigma}(g)$ *is* $\mu$ *integrable. Then the sequence* $\frac{1}{n} \log \sigma(U(n), b)$ *converges* $\mathbb{P} \otimes \nu$*-almost surely to some random variable* $Z$*. If we assume furthermore that the sequence* $\sigma(U(n), b)$ *converges to* $+\infty$ *for* $\mathbb{P} \otimes \nu$*-almost all* $(\omega\ ,\ b)$*, then the expectation* $\int Z(\omega, b)\, d\mathbb{P}(\omega) d\nu(b)$ *is strictly positive.*

*Proof:*
Let $(\tilde{\Omega}, \tilde{\mathcal{F}}, \tilde{\theta}_t, \tilde{\mathbb{P}})$ be the product dynamical system defined by:

$$\tilde{\Omega} = \Omega \times B, \qquad \tilde{\mathbb{P}} = \mathbb{P} \otimes \nu, \quad \tilde{\theta}(\omega, b) = (\theta(\omega), U_0(\omega) b).$$

The process $X(n) = \log \sigma(U(n), b)$ is additive on $(\tilde{\Omega}, \tilde{\mathcal{F}}, \tilde{\theta}_t, \tilde{\mathbb{P}})$ and the Birkhoff ergodic theorem implies that $n^{-1} X(n)$ converges $\tilde{\mathbb{P}}$-almost surely. The last assertion is an easy consequence of Proposition IV.1.4 ■

**Proposition IV.3.5** *Let $\sigma$ be a cocycle such that*

**(i)** *there exists a positive number $\tau$ such that $\sigma^t$ is integrable for $|t| \leq \tau$,*
**(ii)** *there exists an integer $N$ such that* $\sup_{b\in B}\int \log\sigma(g,b)\, d\mu^N(g) < 0$.

*Then there exists a positive real number $\alpha$ such that for $0 < t \leq \alpha$ there exist positive constants $C_t < +\infty$ and $\rho_t < 1$ such that:*

$$\sup_{b\in B}\int \sigma^t(g,b)\, d\mu^n(g) \leq C_t\, \rho_t^n \qquad n = 1, 2, \ldots$$

*Proof:*
The inequality $\exp(t|\log\sigma|) \leq \sigma^t + \sigma^{-t}$ implies that for any integer $p$ then $\sup_{b\in B}\int \exp(t|\log\sigma(g,b)|)d\mu^p(g)$ and $\sup_{b\in B}\int |\log\sigma(g,b)|^p\; d\mu^p(g)$ are finite. From the inequality:

$$\sigma^t \leq 1 + t\log\sigma + \frac{t^2}{2}|\log\sigma|\exp(t|\log\sigma|) \qquad 0 \leq t \leq \tau$$

one can deduce that, if we set $b_n(t) = \sup_{b\in B}\int \sigma(g,b)^t\, d\mu^n(g)$, then for $0 \leq t \leq \tau$, $a_N(t)$ is not greater than

$$1 + t\sup_{b\in B}\int \log\sigma(g,b)\; d\mu^N(g) + \frac{t^2}{2}\sup_{b\in B}\int |\log\sigma(g,b)|\; e^{t|\log\sigma(g,b)|}\; d\mu^N(g)$$

The hypothesis (ii) implies that for $t$ sufficiently small one has $b_N(t) < 1$ and the conclusion follows from the subadditivity of the sequence $\log b_n(t)$ which implies that this sequence converges to its infimum. ∎

We now define the operators extending the notion of Fourier-Laplace transform of a measure to the setting of a commutative group. For a cocycle $\sigma$ and a complex number $z$ we define formally the operator $T_{\sigma,z}$ on the set $\mathcal{B}(B)$ of bounded measurable function on $B$ by the formula:

$$T_{\sigma,z}f(b) = \int \sigma^z(g,b)f(gb)\, d\mu(g).$$

$T_{\sigma,z}$ is a bounded operator on $\mathcal{B}(B)$ when $|\sigma^z|$ is $\mu$-integrable. It is readily seen that $T^n_{\sigma,z}$ is defined by the same formula but with $\mu$ replaced by $\mu^n$. When $z = 1$ we write $T_\sigma$ instead of $T_{\sigma,z}$ and we only write $T$ when the cocycle $\sigma$ is equal to the constant one. The operator $T_\sigma$ is non negative thus $||T_\sigma|| = \sup_{b\in B}\int \sigma(g,b)\, d\mu$ and $Tf(b) = \int f(gb)\, d\mu(g)$ defines a Markov kernel on $B$.

**Proposition IV.3.6** *Let $\tau$ be a positive real number such that the cocycle $\sigma^t$ is $\mu$ integrable for $|t| \leq \tau$ and let us denote by $\rho(t)$ the logarithm of the spectral radius of $T_{\sigma,t}$ acting on $\mathcal{B}(B)$. Then for $|t| \leq \tau$ we have:*

**(i)** *The function $t \hookrightarrow \rho(t)$ is convex and $\rho(0) = 0$,*
**(ii)** *$\rho(t) \geq t \int \log \sigma(g,b)\, d\mu(g)\, d\nu(b)$ for any invariant probability measure $\nu$ on $B$.*
**(iii)** *If the function $\rho$ admits a derivative at the origin then its value is given by:*

$$\frac{d\rho(t)}{dt}\Big|_{t=0} = \int \log \sigma(g,b)\, d\mu(g)\, d\nu(b)$$

*(Hence this integral is independent of the particular choice of $\nu$)*

*Proof:*
The spectral radius can be computed from the formula:

$$\rho(t) = \lim \frac{1}{n} \log \sup_{b \in B} \int \sigma^t(g,b)\, d\mu^n(g)$$

Hence (i) is a consequence of the Hölder inequality. We now remark that for an invariant measure $\nu$ the quantity $\int \log \sigma(g,b)\, d\mu^n(g)\, d\nu(b)$ is additive in $n$ and thus (ii) is a simple consequence of Jensen's inequality for the logarithm function. The function $\rho(t) - t \int \log \sigma(g,b)\, d\mu(g)\, d\nu(b)$ is non negative so that its derivative at the origine, if any has to be zero proving conclusion (iii). ∎

The next proposition extends a well known property of the Fourier-Laplace transform.

**Proposition IV.3.7** *If there exists $\tau > 0$ such that the cocycle $\sigma^t$ is $\mu$-integrable for $|t| \leq \tau$, then $z \hookrightarrow T_{\sigma,z}$ is an analytic function defined in a neighbourhood of the origin and taking values in the Banach space of bounded operators on $\mathcal{B}(B)$.*

*Proof:*
Note that if we set $D_n f(b) = \int \left(\log \sigma(g,b)\right)^n f(gb)\, d\mu(g)$ then we have the upper bound $\|D_n\| \leq \sup_{b \in B} \int |\log \sigma(g,b)|^n\, d\mu(g) \leq \int \tilde{\sigma}^n(g)\, d\mu(g)$ and thus the series $\sum \frac{1}{n!} |z|^n\, \|D_n\|$ is convergent for $|z| \leq \tau$. ∎

When $\mu$ is a "smooth" measure on $G$ we will see in subsection 2 that these operators have nice properties on the space $C(B)$ of continuous functions but this is no longer the case for general $\mu$ and this is why we will introduce the spaces $\mathcal{L}_\alpha$ of Hölder continuous functions in subsection 3.

## IV.3.2 Laplace Operators on the Space of Continuous Functions

Throughout this subsection we assume that $B$ is a compact metric space and we denote by $C(B)$ the Banach space of continuous functions on $B$.

If $\sigma$ is a $\mu$-integrable cocycle then $T_\sigma$ is a bounded operator on $C(B)$. We denote by $T_\sigma^*$ its transpose. It acts on the measures on $B$ and $\check{T}_\sigma$ denotes the operator associated to the measure $\check{\mu}$ which is the image of the measure $\mu$ under the mapping $g \hookrightarrow g^{-1}$.
If $m$ is a measure on $B$ we denote by

- $gm$ the image of $m$ by the map $b \hookrightarrow gb$,
- $fm$ the measure of density $f$ with respect to $m$ for $f \in C(B)$,
- $< f, h >_m$ the continuous bilinear form $\int fh\, dm$ defined on the Banach space $C(B)$.

Let us assume that the support of $m$ is equal to $B$ and that for any $g \in G$ the measures $gm$ and $m$ have the same null sets. Then if the density $r(g,b) = \frac{dg^{-1}m}{dm}(b)$ is a continuous function of the couple $(g,b)$ it is a cocycle called a $m$-Radon-Nikodym cocycle.

**Proposition IV.3.8** *Let $\sigma$ be a $\mu$-integrable cocycle and let $r$ be a $m$-Radon-Nikodym cocycle. Let us assume that $r\sigma^{-1}$ is $\mu$-integrable. Then, for any positive integer $k$ and any complex number $\lambda$ we have:*

**(i)** $< \check{T}_\sigma f\, , h >_m = < T_{r\sigma^{-1}} h\, , f >_m \qquad f, h \in C(B)$
**(ii)** $\left[\ker(\check{T}_\sigma - \lambda I)^k\right] m \subset \ker(T^*_{r\sigma^{-1}} - \lambda I)^k$

*Proof:*
If $f\, , h \in C(B)$, then:

$$[T^*_{r\sigma^{-1}}(fm)](h) \quad = \quad \int f(b) r(g,b) \sigma^{-1}(g,b) h(gb)\, d\mu(g)\, dm(b)$$

$$\begin{aligned} &= \int f(b)h(g^{-1}b)\sigma^{-1}(g^{-1},b)\,d\check{\mu}(g)\,dgm(b) \\ &= \int f(gb)\sigma(g,b)h(b)\,d\check{\mu}(g)\;dm(b) \\ &= [(\check{T}_\sigma f)m](h) \end{aligned}$$

The assertion (ii) is a simple consequence of (i). Notice that (i) implies $T_r^* m = m$. ∎

We now assume that the action of $G$ on $B$ admits a continuous lifting i.e. that there exists a continuous map $s : B \hookrightarrow G$ and a fixed element $b_0 \in B$ such that $s(b)b_0 = b$ for any $b \in B$. This implies that a sequence $b_n$ in $B$ converges to $b$ iff there exists a sequence $g_n$ in $G$ converging to $e$ such that $b_n = g_n b$. In others words, the uniform structure on $B$ induced by the action of $G$ is identical to its initial uniform structure of compact space.

**Proposition IV.3.9** *Suppose that $\mu$ has a density with respect to a Haar measure on $G$ and that $\sigma$ is an integrable cocycle. Then $T_\sigma$ is a compact operator on $C(B)$.*

*Proof:*
Let $dg$ be a right Haar measure on $G$ and $d\mu = \phi\, dg$ . One can write for $u \in G$

$$T_\sigma f(ub) = \int f(gb)\sigma(g,b)\sigma^{-1}(u,b)\phi(gu^{-1})\,dg$$

and:

$$|T_\sigma f(ub) - T_\sigma f(b)| \le \|f\| \int \sigma(g,b)|\phi(g) - \sigma^{-1}(u,b)\phi(gu^{-1})|\,dg$$

If $\phi$ is a continuous function with compact support then we can conclude that $T_\sigma$ maps a bounded set of $C(B)$ in an equicontinuous set and the compactness follows from the Arzela-Ascoli theorem. In the general case there exists a sequence $\phi_n(g)$ of continuous functions with compact support converging to $\phi(g)\bar{\sigma}(g)$ in $L_1(dg)$. Taking $\psi_n(g) = \bar{\sigma}^{-1}(g)\phi_n(g)$ and denoting by ${}^nT_\sigma$ the operator associated to the measure $\mu_n = \psi_n\, dg$ we obtain:

$$\|{}^nT_\sigma - T_\sigma\| \le \int |\phi(g)\bar{\sigma}(g) - \psi_n(g)|\,dg$$

We can conclude that $T_\sigma$ is compact as limit in norm of compact operators.
∎

One remarks that the above proof is simpler in some cases. Let us assume for example that $B$ is a homogeneous space under the action of a subgroup $K \subset G$ and that $\sigma$ is a $K$ invariant cocycle (that is $\sigma(k,b) = 1$ for $k \in K$ and $b \in B$). Then in the above proof one can choose $u \in K$ and the term $\sigma(u,b)$ disappears. Actually this will be the general situation for $G = SP(\ell, \mathbb{R})$, $K$ the orthogonal subgroup of $G$ and $B$ a boundary of the symplectic group.

If $T$ is a compact operator on a Banach space, $T$ and $T^*$ have the same spectrum and in fact the dimensions of the kernels of $[(T - \lambda I)^k$ and $(T^* - \lambda I)^k$ are the same for any positive integer $k$ and non-zero complex number $\lambda$. This yields:

**Proposition IV.3.10** *Let $\sigma$ be a cocycle and let $r$ be a $m$-Radon-Nikodym cocycle such that $\sigma$ and $r\sigma^{-1}$ are $\mu$ and $\check{\mu}$ integrable. If $\mu$ has a density with respect to a Haar measure on $G$, then for any positive integer $k$ and non zero complex number $\lambda$ we have:*

**(i)** $\left(\ker(\check{T}_\sigma - \lambda I)^k\right)m = \ker(T^*_{r\sigma^{-1}} - \lambda I)^k$
**(ii)** $\dim\left(\ker(\check{T}_\sigma - \lambda I)^k\right) = \dim \ker(T_{r\sigma^{-1}} - \lambda I)^k$

*Proof:*
It is enough to apply Proposition IV.3.8 to the cocycles $\sigma$ and $r\sigma^{-1}$ replacing $\mu$ by $\check{\mu}$ for the latter and to take into account the above remark.
∎

An important consequence of Proposition IV.3.10 is that, when $\mu$ has a density on $G$ then for a $\mu$ and $\check{\mu}$ integrable $m$-Radon-Nikodym cocycle $r$ the operators $T$ and $T_r$ have the same spectral properties . Thus we can "shift" the spectral properties of $T$ which are well known for a Markov operator to the operator $T_r$ which is much more difficult to study.

**Definition IV.3.11** *A Markov operator $T$ on the Banach space $C(B)$ is said aperiodic if 1 is the only eigenvalue of modulus one of $T$.*

The following result can be understood as a generalization of the Perron-Frobenius theorem for positive operators:

**Proposition IV.3.12** *Let $r$ be a $\mu$ and $\check{\mu}$ integrable $m$-Radon-Nikodym cocycle. Let us assume that $\mu$ has a density, that $\mu$ and $\check{\mu}$ admit unique invariant measures $\nu$ and $\check{\nu}$ and that $T$ and $\check{T}$ are aperiodic. Then the invariant measures $\mu$ and $\check{\mu}$ admit continuous densities $\psi$ and $\check{\psi}$ with respect to $m$ and we have the decompositions:*

**(i)** $T^n f = \nu(f) + Q^n f$
**(ii)** $T_r^n f = m(f)\psi + Q_r^n f$

*where the operators $Q$ and $Q_r$ have spectral radius strictly less than one. It follows that the spectral radius of $T_r$ is equal to one and that the sequence $||T_r^n||$ is bounded.*

*Proof:*
The uniqueness of the invariant measure added to the compactness of the operators $T$ given by Proposition IV.3.9 imply that the eigenvalue 1 is of geometric multiplicity one. The algebraic multiplicity is also one because the equation $T^*\tau = \tau + \nu$ has obviously no solution $\tau$ in the space of measures on $B$ (otherwise the total mass of $\nu$ would be zero !). This implies immediately the decomposition of $T$.
From Proposition IV.3.10 we deduce that $T$ and $T_r$ have the same spectral properties. The relation $\check{\nu} = \psi m$ is then a consequence of the equality $\ker(\check{T}^* - I) = \big(\ker(T_r - I)\big)m$. ∎

**Proposition IV.3.13** *Under the hyphothesis of Proposition IV.3.12 if we denote by $\rho(t)$ the logarithm of the spectral radius of $T_{r,t}$ then we have:*

**(i)** *$\rho(t)$ is a convex function on $[-1,1]$ with $\rho(0) = \rho(1) = 0$.*
**(ii)** *If there exists an integer $N$ with $\sup_{b\in B} \int \log r(g,b)\, d\mu^N(g) < 0$, then $\rho(t) < 0$ for $t \in (0\ ,\ 1)$*
**(iii)** $\frac{d\rho(t)}{dt}\big|_{t=0} = \int \log r(g,b)\, d\mu(g)\, d\nu(b)$

*Thus if this integral is negative then $\rho(t) < 0$ for $t \in (0\ ,\ 1)$*

*Proof:*
As already noticed in the proof of Proposition IV.3.6, the convexity of $\rho(t)$ is a consequence of Hölder's inequality and (i) follows from Proposition IV.3.8. Conclusion (ii) is a consequence of Proposition IV.3.5 which implies that we have $\rho(t) < 0$ for small positive $t$. We know that the spectrum of $T$ splits to the simple eigenvalue 1 and a part contained in a disk of radius strictly less than one. Furthermore Proposition IV.3.9 (stated on the space

$C(B)$ rather than on $\mathcal{B}(B)$ ) implies that $z \hookrightarrow T_{r,z}$ is analytic and the analytic perturbation theory (see Kato [167]) yields that $\rho(t)$ is smooth at the origin and (iii) follows from Proposition IV.3.6(iii). ∎

One can apply the preceeding results in the particular case of the action of $GL(d,\mathbb{R})$ on the projective space $B = P(\mathbb{R}^d)$. In this situation there exists a probability measure $m$ invariant under all the rotations of $SO(d,\mathbb{R})$. It is called the Cauchy measure and it satisfies:

$$r(g,\bar{v}) = \frac{dg^{-1}m}{dm}(\bar{v}) = \left(\frac{||v||}{||g\,v||}\right)^d$$

This cocycle is $\mu$ and $\check{\mu}$ integrable whenever the random variables $||g||^d$ and $||g^{-1}||^d$ are $\mu$-integrable.

Using the notations and definitions that will be introduced in section 6 we now state a proposition which will allow us to check the hypotheses of Proposition IV.3.12.

**Proposition IV.3.14** *Assume that the probability measure $\mu$ has a density on the symplectic group $SP(\ell,\mathbb{R})$, then:*

- *There exists a unique $\mu$-invariant probability measure for the operator $T$ acting on each boundary $L_I$.*
- *The operator $T$ acting on $C(L_I)$ is aperiodic.*

*Proof:*
The unicity of the invariant measure will follow from Proposition IV.4.22 and IV.6.2, moreover Proposition IV.3.9 yields the compactness of $T$. To obtain the aperiodicity it is enough to prove this property for the "maximal boundary" associated to $I = \{1\ldots\ell\}$ that we will call $B$. Let $f \in C(B)$ with $Tf = \lambda f$ , $||f|| = 1$ , $|\lambda| = 1$ and let us pick $b \in B$ with $|f(b)| = 1$. The equation $T^n f = \lambda^n f$ first implies that $|f(gb)| = |f(b)|$ for $g \in T_\mu$. Secondly the same eigenvalue equation yields that $f(gb) = \lambda^n f(b)$ when $g$ is in the support of $\mu^n$ say $S_n$. The absolute continuity of $\mu$ implies that it is impossible that all the sets $S_n$ be disjoint since their Cauchy measure is bounded below (See Tutubalin[323] for a computation of the Cauchy measure on $B$ using the Haar measure of $G$) and it follows that $\lambda$ is a root of the unity. The unicity of the invariant measure then implies that $\lambda = 1$.
∎

The decomposition in Proposition IV.3.12 part (i) does not hold in general if we do not assume the existence of a density for $\mu$. If we only assume that $\mu$ is "spread out",i.e. there exists an integer $n$ such that $\mu^n$ is not orthogonal to the Haar measure of $G$, then the operator $T$ is quasi-compact (See Brunel and Revuz [38]). Such an operator has a finite number of eigenvalues of modulus one and their multiplicities are finite. Moreover the rest of the spectrum is contained in a disk of radius strictly less than one. Unfortunately this property is not true for the operators $T_\sigma$.

### IV.3.3 The Laplace Operators on the Space of Hölder Continuous Functions

We still assume that $B$ is a compact metric space.
We denote by $\delta$ the distance of $B$. For a positive real number $\alpha$ we define the space of Hölder continuous functions of order $\alpha$ say $\mathcal{L}_\alpha$ as the set of continuous functions $f$ on $B$ such that:

$$m_\alpha(f) = \sup_{(a,b)\in B\times B\,;\,a\neq b} \frac{|f(a)-f(b)|}{\delta^\alpha(a,b)} < +\infty$$

Endowed with the norm $\|f\|_\alpha = \|f\|_\infty + m_\alpha(f)$ the linear space $\mathcal{L}_\alpha$ is a Banach space. Moreover the map $\alpha \hookrightarrow \mathcal{L}_\alpha$ is nonincreasing and the identity map from $\mathcal{L}_\alpha$ to $\mathcal{L}_\beta$ is continuous for $\beta \leq \alpha$.

Let $\sigma$ be a cocycle defined on the set $\{(a,b)\in B\times B\ a\neq b\}$ by:

$$\sigma(g,(a,b)) = \frac{\delta(ga\,,\,gb)}{\delta(a\,,\,b)}$$

The following proposition is essentially due to Le Page [214]:

**Proposition IV.3.15** *Assume that the cocyle $\sigma$ defined above satisfies:*

**(i)** *There exists a positive number $\tau$ such that $\sigma^t$ is integrable for $|t|\leq\tau$*
**(ii)** $\sup_{(a,b)\in B\ a\neq b}\int \log\sigma(g,(a,b))\,d\mu^N(g) < 0$ *for some integer $N$.*

*Then there exists a positive real number $\alpha_0$ such that for $0<\alpha\leq\alpha_0$ we have:*

**(i)** *$T$ is a bounded operator on $\mathcal{L}_\alpha$,*

**(ii)** *There exist constants $C_\alpha < \infty$ and $\rho_\alpha < 1$ such that:*

$$\|T^n f - \nu(f)\|_\alpha \leq \|f\|_\alpha C_\alpha \rho_\alpha^n \quad for\ n = 1, 2 \ldots \quad and\ f \in \mathcal{L}_\alpha$$

*for any invariant probability measure $\nu$ (In particular this implies the uniqueness of the invariant probability measure).*

*Proof:*
We can write

$$\frac{|Tf(a) - Tf(b)|}{\delta^\alpha(a,b)} \leq m_\alpha(f) \int \bar{\sigma}^\alpha(g)\, d\mu(g)$$

The obvious bound $\|Tf\|_\infty \leq \|f\|_\infty$ and hypothesis (i) yield conclusion (i). We know from Proposition IV.3.5 that for $\alpha$ sufficiently small then one has $\int \sigma^\alpha(g,(a,b))\, d\mu^n(g) \leq C_\alpha \rho_\alpha^n$. Hence for such an $\alpha$ one obtains:

$$\begin{aligned} \frac{|T^n f(a) - T^n f(b)|}{\delta^\alpha(a,b)} &\leq m_\alpha(f) \int \sigma^\alpha(g,(a,b))\, d\mu^n(g) \\ &\leq m_\alpha(f) C_\alpha \rho_\alpha^n \end{aligned}$$

Let now $\nu$ be an invariant probability measure then:

$$\begin{aligned} |T^n f(a) - \nu(f)| &= \left| \int f(ga)\, d\mu^n - \int f(gb)\, d\mu^n(g)\, d\nu(b) \right| \\ &\leq \int |f(ga) - f(gb)|\, d\mu^n(g)\, d\nu(b) \\ &\leq m_\alpha(f) \int \sigma^\alpha(g,(a,b))\, d\mu^n(g)\, d\nu(b) \\ &\leq m_\alpha(f) C_\alpha \rho_\alpha^n \sup_{(a,b)\in B\times B} \delta^\alpha(a,b) \end{aligned}$$

This proves conclusion (ii). ∎

This result gives the exponential convergence of $T^n$ to the rank one operator $N(f) = \nu(f)$. In some sense it is an operator form of the classical Dœblin Theorem for Markov chains.

**Corollary IV.3.16** *Under the hypothesis of Proposition IV.3.15 for $\alpha$ sufficiently small the operator $T$ on $\mathcal{L}_\alpha$ has eigenvalue 1 and the rest of the* *spectrum is contained in a disk of radius strictly less than one. Moreover $T$ admits the decomposition:*

$$T^n f = \nu(f) + Q^n f \qquad f \in \mathcal{L}_\alpha$$

*Where $\nu$ is the invariant measure and $Q$ is an operator on $\mathcal{L}_\alpha$ of spectral radius strictly less than* 1.

We will see in the next section that one can check the hypotheses of Proposition IV.3.15 without any smoothness assumption on $\mu$. Using the theory of analytic perturbations it is even possible to extend this property to the operators $T_{\sigma,z}$ for small values of $z$ and use it to prove central limit and large deviation theorems (See Bougerol-Lacroix [33]). We will not dwell on this kind of result there.

# IV.4 Products of Independent Random Matrices

This section contains the essential results of the theory of products of independent random matrices. Such products yield one of the most interesting example of multiplicative process. Following Ledrappier [212] we obtain in the first subsection some immediate consequences of Proposition IV.2.9. We also give the classical result of Furstenberg [115] establishing the positivity of $\gamma_1$. Under stronger hypotheses we obtain in the second subsection the simplicity of the Lyapunov spectrum. This last part is essentially borrowed from Guivarc'h-Raugi [137].

If $\mu$ is a probability measure on the linear group $GL(d, \mathbb{R})$ we define the following ergodic dynamical system:

- $\Omega$ is the set of sequences $\omega = (\omega_0, \omega_1, \ldots)$ with $\omega_i \in G$
- $\mathbb{P}$ is the infinite product measure $\mathbb{P} = \mu \otimes \mu \ldots$
- $\theta$ is the shift operator $\theta(\omega) = (\omega_1, \omega_2, \ldots)$

The coordinates functions $g(n, \omega) = \omega_n$ form an independent sequence of identically distributed matrix valued random variables and the ergodicity of the dynamical system follows from the classical zero-one law. One sets $U(0) = I$ and $U(n) = g(n-1)\dots g(0)$ for $n \geq 1$. We will always assume the integrability condition:

$$\int \{\log^+ \|g\| + \log^+ \|g^{-1}\|\} \, d\mu(g) < \infty$$

Under these assumptions $\{U(n)\,;\, n \geq 0\}$ is a regular multiplicative process. At times we will also consider the invertible dynamical system constructed in the same way but with the set of all integers as time set. In this case the product $U(n)$ is defined for $n \leq -1$ by $U(n) = g^{-1}(n)\dots g^{-1}(-1)$. It is easy to check that $U(n+m) = (U(m) \circ \theta_n)U(n)$ for any integers $m, n$. We also remark that for a fixed non-zero vector $v \in \mathbb{R}^d$ the sequence $U(n)\bar{v}$ is a Markov chain on $B$ starting at $\bar{v}$. Its transition kernel $T$ is given by the formula $Tf(b) = \int f(gb)\, d\mu(g)$.

## IV.4.1 The Upper Lyapunov Exponent

**Definition IV.4.1** *A subset $S$ of $GL(d, \mathbb{R})$ is said to be strongly irreducible if there is no finite union $W = \bigcup_{i=1}^{r} V_i$ of proper subspaces $V_i$ of $\mathbb{R}^d$ such that $gW = W$ for all $g \in S$. We also say that the probability measure $\mu$ is strongly irreducible if the support of $\mu$ is strongly irreducible.*

One remarks that $\mu$ is strongly irreducible iff $G_\mu$ is strongly irreducible.

**Definition IV.4.2** *A probability measure $\nu$ on $P(\mathbb{R}^d)$ is said to be a proper measure if for any proper subspace $V$ of $\mathbb{R}^d$ we have $\nu(\overline{V}) = 0$.*

**Lemma IV.4.3** *If $\mu$ is strongly irreducible then any invariant probability measure $\nu$ is proper.*

*Proof:*
Let us assume that the conclusion of the lemma is false and let $d_0$ be the smallest dimension of the subspaces $V$ for which $\nu(\overline{V}) \neq 0$ and let $\alpha$ be the supremum of the $\nu$-measures of the subspaces of dimension $d_0$. The set of subspaces $V$ of dimension $d_0$ for which $\nu(\overline{V}) = \alpha$ is finite. Let $W$ be the union of these subspaces. The invariance equation implies that for $\mu$-almost all $g$ we have $gW = W$. This proves the lemma. ∎

**Proposition IV.4.4** *If $\mu$ is strongly irreducible then for any invariant probability measure $\nu$ we have the formula:*

$$\gamma_1 = \int \log \frac{||gv||}{||v||}\, d\mu(g)\, d\nu(\bar{v})$$

*Proof:*
Let $\nu$ be an invariant measure. The probability measure $\tilde{\mathbb{P}} = \mathbb{P} \otimes \nu$ is invariant for the shift operator $\tilde{\theta}$ introduced in Proposition IV.2.9. By the previous lemma $\nu$ is proper and the result now follows from Proposition IV.2.9(ii). ∎

**Lemma IV.4.5** *Let $\Phi$ be the function defined by $\Phi(\bar{v}) = \int \log \frac{||gv||}{||v||}\, d\mu(g)$. Then $\Phi$ is a continuous function on $P(\mathbb{R}^d)$ and we have:*

$$\int \Phi(g\bar{v})\, d\mu^k(g) = \mathbb{E}\{\log \frac{||U(k+1)v||}{||v||}\}$$

*Proof:*
The first conclusion is a consequence of the Lebesgue theorem and the second is the result of an easy computation. ∎

**Proposition IV.4.6** *If $\mu$ is strongly irreducible then for any nonzero vector $v \in \mathbb{R}^d$ we have:*

$$\lim_{n\to\infty} \frac{1}{n} \log \frac{||U(n)v||}{||v||} = \gamma_1 \qquad \mathbb{P}\text{-a.e}$$

*Proof:*
For a nonzero $v \in \mathbb{R}^d$ any limit point $\nu$ of the sequence of probability measures $\nu_{n,v} = n^{-1} \sum_{k=0}^{n-1} \mu^k * \delta_{\bar{v}}$ is invariant and we can write with the notations of the previous Lemma:

$$\begin{aligned}
\gamma_1 &= \int \Phi(\bar{u}) d\nu(\bar{u}) \\
&= \lim \int \Phi(\bar{u})\, d\nu_{n,v}(\bar{u}) \\
&= \lim_{n\to\infty} \frac{1}{n} \mathbb{E}\left\{ \log \frac{||U(n)v||}{||v||} \right\} \\
&\leq \mathbb{E}\left\{ \limsup_{n\to\infty} \frac{1}{n} \log \frac{||U(n)v||}{||v||} \right\}
\end{aligned}$$

The last inequality follows from the Fatou's lemma and the bound

$$|\log \frac{||U(n)v||}{||v||}| \leq \log^+ ||U(n)|| + \log^+ ||U^{-1}(n)||$$

Moreover, we know from the Oseledec theorem that there exists an invariant set $\Omega_0$ of full $\mathbb{P}$-measure for wich the above lim sup is actually a limit and is equal to one of the Lyapunov exponents. This yields the result. Note nevertheless that the set of convergence in this proposition depends upon the choice of $v$ and that there is no contradiction with the Oseledec Theorem ! ∎

Actually the last two propositions can be proved under the weaker hypothesis of mere irreducibility defined below.

**Definition IV.4.7** *A subset $S$ of $GL(d, \mathbb{R})$ is said to be irreducible if there exists no proper subspace $V$ of $\mathbb{R}^d$ such that $gV = V$ for all $g \in S$. We also say that the probability measure $\mu$ is irreducible if its support is irreducible.*

One easily sees that the irreducibility of $\mu$ means that $G_\mu$ is irreducible and that if $G_\mu$ is a connected subgroup of $GL(d, \mathbb{R})$ then the irreducibility implies the strong irreducibility.

**Lemma IV.4.8** *If $\mu$ is irreducible then for any invariant probability measure $\nu$ and any proper subspace $V$ of $\mathbb{R}^d$ we have $\nu(\overline{V}) < 1$.*

*Proof:*
Let us assume that the conclusion is false and let $d_0$ be the smallest dimension of the subspaces $V$ for which $\nu(\overline{V}) = 1$. For a subspace $V$ of dimension $d_0$ and such that $\nu(\overline{V}) = 1$, the invariance equation:

$$\nu(\overline{V}) = \int \nu(g\overline{V})\, d\mu(g)$$

implies that for $\mu$ almost all $g$ we have $gV = V$ and this proves the Lemma. ∎

As a consequence if we only assume that $\mu$ is irreducible one can prove Proposition IV.4.4 and hence Proposition IV.4.6 using a theorem of Breiman [36] which asserts that when $\nu$ is an extremal invariant probability measure then the dynamical system $(\tilde{\Omega}, \tilde{\mathcal{F}}, \tilde{\theta}_t, \tilde{\mathbb{P}})$ is ergodic. Then one can apply Proposition IV.2.9(iii) to an extremal invariant measure to obtain the formula for $\gamma_1$ and then integrate on the set of invariant extremal measures.

**Theorem IV.4.9** *Assume that the probability measure $\mu$ is strongly irreducible and that $G_\mu$ is a non compact subgroup of $SL(d,\mathbb{R})$, then $\gamma_1 > 0$.*

*Proof:*
We only give a sketch of the original proof of Furstenberg [115] since we shall give another proof of this result in the next subsection. First we remark that the strong irreducibility of $\mu$ implies that there is no probability measure $\pi$ on $P(\mathbb{R}^d)$ such that $g\pi = \pi$ for $\mu$-almost all $g$. Actually we can replace the assumptions of the theorem by this property and the irreducibility of $\mu$. If we assume that the invariant measure $\nu$ appearing in the formula of Proposition IV.4.4 has a continuous strictly positive density with respect to the Cauchy measure $m$ on $P(\mathbb{R}^d)$ then we have the cocycle relation:

$$\left(\frac{||v||}{||gv||}\right)^d = \frac{dg^{-1}m}{dm} = \frac{dg^{-1}m}{dg^{-1}\nu}\,\frac{dg^{-1}\nu}{d\nu}\,\frac{d\nu}{dm}$$

and this yields

$$\gamma_1 = -\frac{1}{d}\int \log\left(\frac{dg^{-1}\nu}{d\nu}(b)\right) d\nu(b)d\mu(g).$$

If the conclusion of the Theorem was false Jensen's inequality would imply that $\nu$ is left invariant by $\mu$-almost all $g$ and this gives a contradiction. It still remains to extend this proof to a general $\nu$ and this is the hard part of the work ! (See also Ledrappier [212] for a complete proof ) ∎

When $\mu$ is supported by the symplectic group, one of the crucial questions is the positivity of the exponent $\gamma_\ell$. Unfortunately when $d > 2$ Theorem IV.4.9 says nothing about this problem and this justify the search for stronger results in the next sub-section. We also remark that we cannot replace the strong irreducibility by the irreducibility in this Theorem. (See Problem IV.7.4)

## IV.4.2 The Lyapunov Spectrum

**Definition IV.4.10** *A subset $T$ of $GL(d,\mathbb{R})$ is said to be contractive if there exists a sequence $h_n$ in $T$ such that $h_n||h_n||^{-1}$ converges to a rank* *one matrix. Accordingly, the measure $\mu$ is said to be contractive if $T_\mu$ is contractive.*

Equivalently, $\mu$ is contractive if $\inf_{g \in T_\mu} (a_2(g)a_1^{-1}(g)) = 0$ (recall that $a(g)$ is the diagonal part in the polar decomposition of $g$). One remarks that contractivity cannot be related directly to the group $G_\mu$. This explains why it will be difficult to check this property (since semi-groups are in general less tractable than groups). Nevertheless in the particular case of $SL(2, \mathbb{R})$ one has a very simple criterion:

**Proposition IV.4.11** *If the probability measure $\mu$ is supported by $SL(2, \mathbb{R})$ then $\mu$ is contractive if and only if $G_\mu$ is not compact.*

*Proof:*
It is known that a compact semi-group is actually a group (See [148]). Hence one has only to check that the sequence $h_n||h_n||^{-1}$ has a rank one limit point iff the sequence $||h_n||$ is unbounded. ■

We also remark that if $\mu$ is strongly irreducible and contractive then the image measure $\mu'$ of $\mu$ under the map $g \hookrightarrow g'$ has the same properties. We will denote by $R_n$ the right product $R_n = g(1)g(2)\dots g(n)$. We are now in position to state and prove the fundamental result of Guivarc'h-Raugi:

**Theorem IV.4.12** *Let us assume that $\mu$ is strongly irreducible and contractive. Then there exists a random variable $Z$ with values in the projective space $P(\mathbb{R}^d)$ such that for $\mathbb{P}$-almost all $\omega$ every limit point of the sequence $R_n(\omega)||R_n(\omega)||^{-1}$ is a rank one matrix and the direction of its range is precisely given by $Z(\omega)$.*

*Proof:*
Let $\nu$ be an invariant probability measure on $B = P(\mathbb{R}^d)$ and let $f$ be a continuous function on $B$. The random variables $Z_n = \int f(R_n b)\, d\nu(b)$ form a bounded martingale with respect to the family $\mathcal{F}_n$ of $\sigma$-algebras generated by the random variables $g_1, \dots, g(n)$. Taking a countable dense subset of the set of continuous functions on $B$ we can conclude that the sequence $R_n\nu$ is $\mathbb{P}$-almost surely converging to some probability measure $\nu_\omega$. Let $\lambda$ be the probability measure $\sum_{n=0}^{\infty} 2^{-n-1}\mu^n$. An easy computation

shows that if $Z_n^g = \int f(R_n g b)\, d\nu(b)$ then

$$\sum_{n=1}^{n=\infty} \mathbb{E}\{(Z_{n+r} - Z_n)^2\} \leq 2r||f||^2 \quad \text{and} \quad \sum_{n=0}^{\infty} \int (Z_n - Z_n^g)^2\, d\lambda(g) \ < \infty$$

This implies that the sequence $R_n g\nu$ is $\mathbb{P} \otimes \lambda$-almost surely converging to $\nu_\omega$. Let now $A(\omega)$ be a limit point of the sequence $R_n(\omega)||R_n(\omega)||^{-1}$. Since $\nu$ is proper $h\nu$ is defined for any non zero matrix $h$ and one can conclude that $A(\omega)g\nu = \nu_\omega$ $\mathbb{P} \otimes \lambda$-almost surely. The set of $g$ for which this equality holds being closed we obtain the same relation for $\mathbb{P}$-almost all $\omega$ and any $g$ in $T_\mu$. Taking again in account the fact that $\nu$ is proper and the contractivity property of $\mu$ we obtain that there exists a rank one matrix $h$ such that $A(\omega)h$ is a non zero matrix and $A(\omega)h\nu = \nu_\omega$. Hence $\nu_\omega$ is a point measure $\mathbb{P}$-a.e. and we can conclude from $A(\omega)\nu = \nu_\omega$ that $A(\omega)$ is a rank one matrix whose range has the direction of the support of $\nu_\omega$. Note that this construction implies that the direction of the support of $\nu_\omega$ is independent from the choice of the invariant measure $\nu$. ∎

This theorem is false in general for the left product $U(n)$. Note neverthe-less that the sequence $U(n)||U(n)||^{-1}$ always have rank one limit points (applying the above theorem to the right product of transposes ) but the direction of the range depends of the choice of the limit point !

**Lemma IV.4.13** *Let us assume that $\mu$ is strongly irreducible and contractive. Then for any non zero vector $v \in \mathbb{R}^d$ we have $\mathbb{P}$-a.e.:*

**(i)** $||U(n)v|| \geq C_v||U(n)||$ *where $C_v(\omega)$ is a positive random variable*

**(ii)** $\lim_{n\to\infty} \dfrac{||\Lambda^2(U(n))||}{||U(n)v||^2} = 0.$

*Proof:*
Let $S_n$ be the right product $S_n = g'(1)g'(2)\ldots g'(n)$ and $Z'$ the corresponding random variable from Theorem IV.4.12. We remark that $S_n' = U(n)$ and $||U(n)|| = ||S_n|| = a_1(U(n))$, and writing a polar decomposition $U(n) = k(U(n))a(U(n))u(U(n))$ Theorem IV.4.12 implies immediately that $\mathbb{P}$-a.e. we have the following properties:

(*) The bounded sequence $\dfrac{a_2(U(n))}{a_1(U(n))}$ converges to 0,

(**) The sequence $u'(U(n))\bar{e_1}$ converges to $Z'$.

We can write:

$$\|U(n)v\|^2 = \sum_{k=1}^{n} a_k^2(U(n)) < v\,,\, u'(U(n))e_k >^2$$

$$\frac{\|U(n)v\|^2}{\|U(n)\|^2} = < v\,,\, u'(U(n))e_1 >^2 + \sum_{k=2}^{k=n} \frac{a_k^2(U(n))}{a_1^2(U(n))} < v\,,\, u'(U(n))e_k >^2$$

It follows from (*) and (**) that

$$\lim_{n\to\infty} \frac{\|U(n)v\|}{\|U(n)\|} = |< v, z' >|$$

where $z$ is a unit vector of direction $Z'$. The law of $Z'$, which is the invariant measure for $\mu'$, is proper hence for a fixed non zero $v$ the set $\{\omega\,;\, v$ is orthogonal to the direction of $Z'\}$ is of zero $\mathbb{P}$-measure and this yields conclusion (i). Conclusion (ii) is a consequence of (i) and (*) once we remark that $\|\Lambda^2(U(n))\| = a_1(U(n))a_2(U(n))$. ∎

**Theorem IV.4.14** *Let assume that $\mu$ is strongly irreducible and contractive. Then one has:*

**(i)** *There exists a unique invariant probability measure $\nu$ and $\nu$ is proper.*

**(ii)** $\gamma_1 = \int \log \frac{\|gv\|}{\|v\|}\, d\mu(g)\, d\nu(\bar v)$

**(iii)** $\frac{1}{n}\mathbb{E}\left\{\log \frac{\|U(n)v\|}{\|v\|}\right\}$ *converges uniformly to the exponent $\gamma_1$ with respect to $\bar v \in P(\mathbb{R}^d)$.*

*Proof:*
Recall that a probability measure $m$ on $P(\mathbb{R}^d)$ is said to be proper if $m(\overline{V}) = 0$ for any proper subspace $V$ of $\mathbb{R}^d$. Theorem IV.4.12 implies that for any proper probability measure $m$ on $P(\mathbb{R}^d)$ the sequence $R_n(\omega)m$ converges weakly to the Dirac measure at the point $Z(\omega)$. Let $\nu$ be an invariant probability measure and $f$ be a continuous function on $P(\mathbb{R}^d)$. The sequence $\int f(R_n(\omega)\bar v)\, d\nu(\bar v)$ is a bounded martingale with expectation $\nu(f)$. We know from Lemma IV.4.3 that $\nu$ is proper, thus $\nu$ has to be equal to the law of the random variable $Z$ proving (i). The inequalities $C_v\|U(n)\| \le \|U(n)v\| \le \|U(n)\|$ yield immediately (ii). To prove (iii) it is enough to show that for any sequence $\bar v_n$ converging to a point $\bar v_0$ then the sequence $n^{-1}\mathbb{E}\{\log \|U(n)v_n\|\, \|v_n\|^{-1}\}$ converges to $\gamma_1$. Any weak limit point of the sequence $\nu_n = n^{-1}\sum_{k=0}^{n-1} \mu^n * \delta_{\bar v_n}$ is $\mu$ invariant. Hence

this sequence converges weakly to the unique invariant measure $\nu$. If we apply this sequence of probability measures to the continuous function $\bar{v} \hookrightarrow \int \log \frac{||gv||}{||v||} \, d\mu(g)$ we obtain the result. ■

**Theorem IV.4.15** *Let us assume that $\mu$ is strongly irreducible and contractive. Then we have the strict inequality $\gamma_1 > \gamma_2$.*

*Proof:*
Let $B_1$ be the set of norm one matrices of $\mathbb{R}^d$ and $B_2$ the set of norm one matrices of $\Lambda^2(\mathbb{R}^d)$. Following the lines of Problem IV.7.3 one can prove that there exist probability measures $\nu_1$ and $\nu_2$ on $B_1$ and $B_2$ such that:

$$\lim_{n\to\infty} \frac{1}{n} \log ||U(n)M|| = \gamma_1 \qquad \mathbb{P} \otimes \nu_1 - a.e. \qquad (1)$$

$$\lim_{n\to\infty} \frac{1}{n} \log ||\Lambda^2 U(n)N|| = \gamma_1 + \gamma_2 \quad \mathbb{P} \otimes \nu_2 - a.e. \qquad (2)$$

Notice that the group $G$ acts on $B = B_1 \times B_2$ by $g.(M,N) = (g.M, g.N)$ and so if we define the cocycle $\sigma$ by $\sigma(g,(M,N)) = \frac{||gM||^2}{||\Lambda^2(g)N||}$, it is easy to check that any limit point $\nu$ of the sequence $n^{-1} \sum_{k=1}^{n} \mu^n * (\nu_1 \otimes \nu_2)$ is invariant and admits $\nu_1$ and $\nu_2$ as projections on $B_1$ and $B_2$. Lemma IV.4.13(ii) then implies that

$$\lim \sigma(U(n),(M,N)) \geq \lim \frac{||U(n)M||^2}{||\Lambda^2(U(n))||} = +\infty \qquad \mathbb{P} \otimes \nu - a.e.$$

From Proposition IV.3.4, the sequence $n^{-1} \log \sigma(U(n),(M,N))$ converges $\mathbb{P} \otimes \nu$-a.e.to a random variable having a positive expectation. We now remark that (1) and (2) imply that this limit is $\mathbb{P} \otimes \nu$-almost surely equal to $2\gamma_1 - (\gamma_1 + \gamma_2) = \gamma_1 - \gamma_2$ and this completes the proof. ■

It is worth to mention that Theorem IV.4.15 implies that $\gamma_1$ is strictly positive when $\mu$ is carried by $SL(d,\mathbb{R})$ since in this case the sum of the Lyapunov exponent is nul. In the particular case of $SL(2,\mathbb{R})$, Proposition IV.4.11 shows that this result is actually equivalent to the Furstenberg Theorem IV.4.9. But the most important consequence of Theorem IV.4.15 is the contractive action of the sequence $U(n)$ on the projective space $P(\mathbb{R}^d)$. One defines the projective distance $\delta$ on $P(\mathbb{R}^d)$ by the formula

$$\delta(\bar{x},\bar{y}) = \frac{||x \wedge y||}{||x|| \, ||y||} \qquad x\,,\, y \in \mathbb{R}^d$$

The following proposition shows that the distance between two columns of the matrix $U(n)$ goes to zero exponentially fast.

**Proposition IV.4.16** *If $\mu$ is strongly irreducible and contractive then for non-zero vectors $x, y \in \mathbb{R}^d$ $x \neq y$ one has:*

(i) $\lim_{n\to\infty} \frac{1}{n} \log \frac{\delta(U(n)\bar{x}, U(n)\bar{y})}{\delta(\bar{x}, \bar{y})} < 0 \quad \mathbb{P}-a.e.$

(ii) *there exists an integer $N$ such that* $\sup_{x, y \in \mathbb{R}^d} \int \log \frac{\delta(g\bar{x}, g\bar{y})}{\delta(\bar{x}, \bar{y})}\, d\mu^N < 0$

*Proof:*
From the definition of $\delta$ we obtain

$$\begin{aligned}
\frac{\delta(U(n)\bar{x}, U(n)\bar{y})}{\delta(\bar{x}, \bar{y})} &= \frac{\|U(n)x \wedge U(n)y\|}{\|x \wedge y\|} \frac{\|x\|}{\|U(n)x\|} \frac{\|y\|}{\|U(n)y\|} \\
&\leq \|\Lambda^2(U(n))\| \frac{\|x\|}{\|U(n)x\|} \frac{\|y\|}{\|U(n)y\|}
\end{aligned}$$

The convergence of the sequence $\frac{1}{n} \log \frac{\|U(n)x \wedge U(n)y\|}{\|x \wedge y\|}$ can be deduced from the Oseledec Theorem applied on $\Lambda^2(\mathbb{R}^d)$ and (i) follows from Theorem IV.4.15. To prove (ii) we write:

$$\begin{aligned}
&\frac{1}{n} \int \log \frac{\delta(g\bar{x}, g\bar{y})}{\delta(\bar{x}, \bar{y})}\, d\mu^n \\
\leq\ & \frac{1}{n}\mathbb{E}\{\log \|\Lambda^2(U(n))\|\} - \frac{1}{n}\mathbb{E}\{\log \frac{\|U(n)x\|}{\|x\|}\} - \frac{1}{n}\mathbb{E}\{\log \frac{\|U(n)y\|}{\|y\|}\}
\end{aligned}$$

It only remains to apply again Theorem IV.4.15 together with the uniform convergence proved in Proposition IV.4.14(iii). ■

The uniform contraction in mean of the projective space by the sequence $U(n)$ expressed by this proposition can be viewed as a substitute for the Dœblin condition for Markov chains (which is in general not satisfied under the hypothesis of this section). A consequence of this property is the exponential speed of convergence of the iterates $T^n$ of the kernel of the Markov chain $U(n)\bar{v}$ to the invariant measure $\nu$.

**Proposition IV.4.17** *Let us assume that $\mu$ is strongly irreducible and contractive and that there exists a positive $\tau$ such that $\|g\|^t$ is $\mu$ integrable* *for* $-\tau \leq t \leq \tau$. *Then there exists a positive real number* $\alpha_0$ *such that for* $0 < \alpha \leq \alpha_0$ *we have:*

**(i)** $T$ *is a bounded operator on* $\mathcal{L}_\alpha$ *and*
**(ii)** *there exist* $C_\alpha < \infty$ *and* $\rho_\alpha < 1$ *such that for* $n = 1, 2 \ldots$ $f \in \mathcal{L}_\alpha$ *we have:*

$$\begin{aligned} \|T^n f - \nu(f)\|_\alpha &\leq \|f\|_\alpha C_\alpha \rho_\alpha^n \\ \mathbb{E}\{(\frac{\|v\|}{\|U(n)v\|})^\alpha\} &\leq C_\alpha \rho_\alpha^n \end{aligned}$$

*Proof:*
This result is an immediate consequence of Proposition IV.3.15 and Proposition IV.4.16. ∎

Up to now we have seen that the behavior of the norm of the columns of the matrix $U(n)$ is the same as the norm of $U(n)$ and that it is given by the upper Lyapunov exponent. Actually under the hypothesis of Proposition IV.4.17 one can show that all the coefficients of the matrix $U(n)$ also share this property. This has been proved by Guivarc'h and Raugi in [137] where they first obtain some regularity property of the invariant measure $\nu$. Since the proof is somewhat involved we only state their result. The interested reader will find a proof in Bougerol-Lacroix [33].

**Proposition IV.4.18** *Under the hypothesis of Proposition IV.4.17 we have*

**(i)** $\sup_{y \in \mathbb{R}^d} \int \left(\frac{\|x\|\,\|y\|}{|<x,y>|}\right)^\alpha d\nu(\bar{x}) < +\infty$ *for some positive* $\alpha$
**(ii)** $\lim_{n\to\infty} \frac{1}{n} \log |<U(n)x, y>| = \gamma_1$ $\mathbb{P}$*–a.e.* *for any non zero vectors* $x, y \in \mathbb{R}^d$

There is an easy way to extend the foregoing results to the Lyapunov exponents of higher order and thus to obtain that all the Lyapunov exponents are distinct, in which case we say that the Lyapunov spectrum is simple. For $p = 1 \ldots d-1$, let $\Lambda^p\mu$ be the image of $\mu$ under the mapping $g \hookrightarrow \Lambda^p g$. Using a polar decomposition it is readily seen that the two upper Lyapunov exponents associated with the action of $\Lambda^p GL(d, \mathbb{R})$ on $\Lambda^p(\mathbb{R}^d)$ are equal to $\gamma_1 + \ldots + \gamma_p$ and $\gamma_1 + \ldots + \gamma_{p-1} + \gamma_{p+1}$. It follows from Theorem IV.4.15 that if the action of $\Lambda^p\mu$ on $\Lambda^p(\mathbb{R}^d)$ is strongly irreducible and contractive then $\gamma_p > \gamma_{p+1}$. We also remark that the unique invariant probability measure $\nu_p$ on $P(\Lambda^p(\mathbb{R}^d))$ is actually supported by the set of directions of p-vectors since this set is invariant. Tutubalin was the first one to show that when $\mu$ has a density on $SL(d,\mathbb{R})$ then all the Lyapunov exponents are distinct. See Tutubalin [324] and also Virtser [328]. In this situation $T_\mu$ is an open set of $SL(d,\mathbb{R})$ and the following proposition generalizes the result of Tutubalin.

**Proposition IV.4.19** *If $T_\mu$ contains an open set of $SL(d,\mathbb{R})$ then for $p = 1 \dots d-1$ the probability measure $\Lambda^p\mu$ is strongly irreducible and contractive,from which it follows that the Lyapunov spectrum is simple.*

*Proof:*
The group $SL(d,\mathbb{R})$ is connected and hence one can conclude that $G_\mu$ is equal to $SL(d,\mathbb{R})$ and the strong irreducibility is obvious.
A complex eigenvalue $\lambda$ of a matrix $g$ is called simple if the subspace $\ker(g-\lambda I)$ is one-dimensional and equal to $\ker(g-\lambda I)^2$. This eingenvalue is called dominating if $|\lambda|$ is strictly greater than the modulus of any other eigenvalue of $g$. If $T_\mu$ contains a matrix $g$ with a simple dominating eigenvalue, then using a Jordan decomposition of $g$ one sees that $\mu$ is contractive. The hypothesis implies that $T_\mu$ contains a matrix $g$ with $d$ eigenvalues of distinct moduli ( See Problem IV.7.10) thus $\Lambda^p(T_\mu)$ has a simple dominating eigenvalue for $p = 1 \dots d-1$. ∎

If we keep in mind that we want to handle the case of the symplectic group it is worth to mention that Proposition IV.4.19 is not sufficient! Fortunately there is a way to extend the result to this situation.

**Definition IV.4.20** *Let $p$ be an integer in $(1,\dots,\ell)$*

- *A p-vector $u = u_1 \wedge u_2 \wedge \dots \wedge u_p$ is said to be Lagrangian if for any couple $(i,j)$ of integers in $(1,\dots,p)$ one has $u_i' J u_j = 0$ where $J = \begin{pmatrix} 0 & I \\ -I & 0 \end{pmatrix}$.*
- *Let* Lag$(p)$ *be the subspace of $\Lambda^p(\mathbb{R}^d)$ spanned by the Lagrangian p-vectors. A subset $S$ of $\Lambda^p SP(\ell,\mathbb{R})$ is said to be* Lag$(p)$*-strongly irreducible if its action on the vector space* Lag$(p)$ *is strongly irreducible.*
- *The measure $\Lambda^p\mu$ is said* Lag$(p)$ *strongly irreducible if its support is* Lag$(p)$ *strongly irreducible.*

It is readily seen that Lag($p$) is a stable subspace of $\Lambda^p(\mathbb{R}^d)$ under the action of $SP(\ell, \mathbb{R})$, thus one cannot expect an irreducible action of $\Lambda^p\mu$ on $\Lambda^p(\mathbb{R}^d)$ for $p > 1$. Nevertheless the following proposition allows to overcome this lack of irreducibility.

**Proposition IV.4.21** *Let $p$ be an integer in $\{1 \ldots \ell\}$, $g$ be a symplectic matrix, and $r$ be the dimension of the linear space* Lag($p$). *Let us denote by $\bar{g}$ the element of $GL(r, \mathbb{R})$ associated to the restriction of $\Lambda^p g$ to the invariant subspace* Lag($p$) *of $\Lambda^p(\mathbb{R}^d)$ and by $\bar{\mu}_r$ the image of $\Lambda^p\mu$ by the mapping $\Lambda^p g \hookrightarrow \bar{g}$. Then one has:*

1. *The two first Lyapunov exponents of $\Lambda^p\mu$ and $\bar{\mu}_r$ are equal.*
2. *If $\Lambda^p\mu$ is contractive and* Lag($p$) *strongly irreducible then the probability measure $\bar{\mu}_r$ is contractive and strongly irreducible when we consider the action of $GL(r, \mathbb{R})$ on $\mathbb{R}^r$).*

*Proof:*
A symplectic matrix $g$ has a polar decomposition $g = kau$ in $SP(\ell, \mathbb{R})$ and one has $\Lambda^p g = \Lambda^p k \Lambda^p a \Lambda^p u$. The matrices $\Lambda^p k$ and $\Lambda^p u$ are orthogonal and leave the subspace Lag($p$) invariant. It follows that the norms computed in $\Lambda^p(\mathbb{R}^d)$ and Lag($p$) are the same and one has:

$$\|\bar{g}\| = \|\Lambda^p g\| \quad \text{and} \quad \|\Lambda^2(\bar{g})\| = \|\Lambda^2(\Lambda^p g)\|$$

This yields the first conclusion.
Secondly we remark that the claim about strong irreducibility is obvious. Since $\Lambda^p\mu$ is contractive there exists a sequence $h_n$ in the closed semigroup generated by its support such that the sequence $\frac{h_n}{\|h_n\|}$ converges to a rank one matrix $h$. Obviously Lag($p$) is left invariant by $h$ and the above remark about the norms imply that the sequence $\frac{\bar{h}_n}{\|\bar{h}_n\|}$ converges to $\bar{h}$. This last matrix is of rank a most 1 and is certainly not zero since its norm is 1 hence it is a rank one matrix. ∎

Hence if $\Lambda^p\mu$ is contractive and Lag($p$)-strongly irreducible then $\Lambda^p\mu$ viewed as a probability measure on the linear group of Lag($p$) is strongly irreducible and contractive. Taking in account the result of Problem IV.7.11 one can use the proof of Proposition IV.4.19 to obtain:

**Proposition IV.4.22** *If $T_\mu$ contains an open set of $SP(\ell, \mathbb{R})$ then for $p = 1 \ldots \ell$ the probability measure $\Lambda^p\mu$ is contractive and* Lag($p$)*-strongly irreducible, from which it follows that the Lyapunov spectrum is simple.*

This last proposition does not answer our problem when $\mu$ is too singular. We now state without proof the strongest result in this direction. It is originally due to Goldsheid and Margulis and extended by Guivarc'h and Raugi. See [128] and [138]. The Zariski's closure of a subset $A$ of an algebraic manifold is defined as the smallest of the set of zeroes of the polynomial vanishing on $A$. In other words, it is the smallest algebraic manifold containing $A$. One can check that the Zariski closure of a sub semi-group of $GL(d,\mathbb{R})$ is a group.

**Proposition IV.4.23** *If the Zariski closure of $G_\mu$ is equal to $SL(d,\mathbb{R})$, then for $p = 1 \ldots d-1$ the probability measure $\Lambda^p\mu$ is contractive and strongly irreducible. If the Zariski closure of $G_\mu$ is equal to $SP(\ell,\mathbb{R})$ then for $p = 1 \ldots \ell$ the probability measure $\Lambda^p\mu$ is contractive and* Lag$(p)$-*strongly irreducible. The Lyapunov spectrum is simple in both cases.*

The negation of the strong irreducibility of $\Lambda^p\mu$ can be expressed by a finite number of polynomial equations satisfied on $G_\mu$, thus the strong irreducibility is actually a property of the Zariski closure of $G_\mu$. It is much more difficult to prove that the contractivity can also be checked on the Zariski closure of $T_\mu$. Fortunately it is proven in [148] that the contractivity of the Zariski closuire of $G_\mu$ already implies the contactivity of $T_\mu$. We remark that $G_\mu$ is already an algebraic group under the hypothesis of Proposition IV.4.19 or IV.4.22.

### IV.4.3 Schrödinger Matrices

There is a particular type of symplectic matrix which will be used in the following chapters and which we call "Schrödinger matrix".

**Definition IV.4.24** *Let $x$ be a point in $\mathbb{R}^\ell$, one defines the Schrödinger matrix $H(x)$ by the formula:*

$$H(x) = \begin{pmatrix} Q(x) & -I_\ell \\ I_\ell & 0 \end{pmatrix} \quad \textit{with} \quad [Q(x)]_{i,j} = \begin{cases} x_i & \textit{if } i = j \\ -1 & \textit{if } |i-j| = 1 \\ 0 & \textit{if } |i-j| > 1 \end{cases}$$

*where $I_\ell$ is the identity matrix of order $\ell$.*

In the case $\ell = 1$ the strong irreducibility and contractivity of a measure supported by Schrödinger matrices is easily checked:

**Proposition IV.4.25** *Let $\mu$ be a probability measure on $SL(2, \mathbb{R})$ supported by the set of Schrödinger matrices. If $\mu$ is not concentrated on one point then $\mu$ is strongly irreducible and contractive.*

*Proof:*
The support of $\mu$ contains at least two matrices $H(x)$ and $H(y)$ with $x \neq y$. Looking at the sequences $u^n$ and $v^n$ defined from $u = H^{-1}(x)H(y)$ and $v = H(x)H^{-1}(y)$ it is readily seen that $G_\mu$ is not compact and the contractivity follows from Proposition IV.4.11. If there exists a finite subset $\Sigma$ of the projective line which is stable under the action of $G_\mu$ then applying the sequences $u^n$ and $v^n$ to the points in $\Sigma$ we see that $\Sigma$ has to be equal at the same time to $\bar{e}_1$ and $\bar{e}_2$ which is absurd. ∎

Let $v = \binom{a}{b}$ be a vector in $\mathbb{R}^2$. We identify the projective line $P(\mathbb{R}^2)$ with the set $\{\mathbb{R} \cup \infty\}$ by the mapping $\Psi$:

$$\Psi(\bar{v}) = \begin{cases} \frac{a}{b} & \text{if } b \neq 0 \\ \infty & \text{if } b = 0 \end{cases}$$

The image of a measure $\tau$ on $P(\mathbb{R}^2)$ by the mapping $\Psi$ will be denoted by $\tilde{\tau}$.

**Corollary IV.4.26** *Let $\mu$ be a probability measure on $SL(2, \mathbb{R})$ supported by the set of Schrödinger matrices. Let us assume that the probability measure $\mu$ is not concentrated on one point and that $\mathbb{E}\{\log \|g\| \, d\mu(g)\} < \infty$. Then one has:*

1. *There exists a unique invariant measure $\nu$ on $P(\mathbb{R}^2)$ and its image by the mapping $\Psi$ is carried by $\mathbb{R}$.*
2. *The upper Lyapunov exponent $\gamma$ is strictly positive and is given by the formula:*
$$\gamma = \int \log \frac{\|gv\|}{\|v\|} \, d\nu(\bar{v}) \, d\mu(g)$$
3. *If there exists a positive real number $\eta$ with $\mathbb{E}\{\|g\|^\eta\} < \infty$ then one has:*
$$\gamma = \int_{\mathbb{R}} \log |t| \, d\tilde{\nu}(t)$$

*Proof:*
The first conclusion is a direct consequence of Proposition IV.4.25 and Theorem IV.4.14 since on the projective line, a measure is proper iff it is continuous, hence $\tilde{\nu}\{\infty\} = 0$.
The positivity of the upper lyapunov exponent $\gamma$ is also a direct consequence of Proposition IV.4.25 and Theorem IV.4.15. In order to prove the formula 3. we remark that Proposition IV.4.18(i) implies that for some positive $\alpha$ one has $\int (1+t^2)^\alpha \, d\tilde{\nu}(t) < \infty$ hence $\int \log(1+t^2) \, d\tilde{\nu}(t) < \infty$. Let us denote by $m$ the Cauchy measure on $P(\mathbb{R}^2)$ and let $r$ be the measure on the projective space such that $\tilde{r}$ is the Lebesgue measure on $\mathbb{R}$. Denoting by $f$ the density of $m$ with respect to the measure $r$ one has:

$$\begin{aligned} \frac{dg^{-1}m}{dm}(\bar{v}) = \frac{||v||^2}{||gv||^2} &= \frac{dg^{-1}m}{dg^{-1}r}(\bar{v})\frac{dg^{-1}r}{dr}(\bar{v})\frac{dr}{dm}(\bar{v}) \\ &= \frac{f(g\bar{v})}{f(\bar{v})}\frac{dg^{-1}r}{dr}(\bar{v}) \end{aligned}$$

The invariance of $\nu$ and the integrability of the function $\log(f)$ stated above implies immediately that:

$$\int \log \frac{f(g\bar{v})}{f(\bar{v})} \, d\nu(\bar{v}) \, d\mu(g) = 0$$

Then taking in account the formula (ii) in Theorem IV.4.14 one obtains:

$$\begin{aligned} \gamma = \int \log \frac{||gv||}{||v||} \, d\mu(g) \, d\nu(\bar{v}) &= -\frac{1}{2}\int \log \frac{dg^{-1}r}{dr}(\bar{v}) \, d\mu(g) \, d\nu(\bar{v}) \\ &= \int \log |t| \, d\tilde{\nu}(t) \end{aligned}$$

One obtains the last relation by computing the Jacobian of the mapping $t \hookrightarrow x - 1/t$ wich is equal to $t^{-2}$. This term is independent from the value of the real number $x$ appearing in the Schrödinger matrix hence the double integral reduces to a single integral. ■

This computation will be extented to the strip in Proposition IV.6.6 and a very useful formula relating the lyapunov exponent and the Fourier transform of the distribution of $\mu$ due to Pastur can be found in Problem IV.7.12.

For $\ell > 1$ the situation is much more complicated. If one wants to use the criterion given in Proposition IV.4.22 one sees immediately that the problem is not trivial since the matrices $H(x)$ are very degenerated. Working in the group $SL(d,\mathbb{R})$ gives rise to tremendous computations of Jacobians (even in the case $\ell = 2$). This is why we will use some properties of Lie Algebras.

**Definition IV.4.27** *The Lie Algebra $\mathcal{G}$ of the group $SP(\ell,\mathbb{R})$ is the set of $d \times d$ matrices $g$ of the form $g = \begin{pmatrix} M_1 & M_2 \\ M_3 & -M_1' \end{pmatrix}$ where $M_2$ and $M_3$ are symmetic matrices of order $\ell$.*

We will use the following notations:

- $E_{ij}$ is the $\ell \times \ell$ matrix which has a 1 in the $i$th row and $j$th column and 0 elsewhere.
- $$X_{ij} = \frac{1}{2}\begin{pmatrix} 0 & E_{ij} + E'_{ij} \\ 0 & 0 \end{pmatrix} \qquad Y_{ij} = X'_{ij} \qquad Z_{ij} = \begin{pmatrix} E_{ij} & 0 \\ 0 & -E_{ij} \end{pmatrix}$$

The next identities are easily obtained.

**Lemma IV.4.28** *One has the relations:*

$$[Z_{ij}\ ,\ X_{kr}] = \delta_j^k X_{ir} + \delta_j^r X_{ik} \qquad [Y_{kr}\ ,\ Z_{ij}] = \delta_i^k Y_{rj} + \delta_i^r Y_{kj}$$
$$[X_{ij}\ ,\ Y_{kr}] = \frac{1}{4}(\delta_j^k Z_{ir} + \delta_j^r Z_{ik} + \delta_k^i Z_{jr} + \delta_r^i Z_{jk})$$

It follows that $\mathcal{G}$ is generated by the matrices $X_{ij}$ and $Y_{ij}$ for $|i-j| \leq 1$.

**Lemma IV.4.29** *Let $O$ be an open subset of $\mathbb{R}^\ell$. Then the subgroup of $SL(d,\mathbb{R})$ generated by the set $\{H(x)\,;\, x \in O\}$ is equal to $SP(\ell,\mathbb{R})$.*

*Proof:*
Looking at products of the form $H(x)H^{-1}(y)$ and $H^{-1}(x)H(y)$ it is readily seen that the Lie algebra of the group contains all the matrices $X_{ii}$ and $Y_{ii}$ and this is enough in the case $\ell = 2$. In the case $\ell > 2$ we write:

$$H(x)(\exp X_{ii})H^{-1}(x) = \exp\begin{pmatrix} -Q(x)E_{ii} & Q(x)E_{ii}Q(x) \\ -E_{ii} & E_{ii}Q(x) \end{pmatrix} = \exp A_i(x)$$

Applying this formula to two points $x$ and $y$ we obtain:

$$A_i(x) - A_i(y) = (y_i - x_i)Z_{ii} + X_{i,i+1} + X_{i,i-1} + (x_i^2 - y_i^2)X_{ii}$$

where $X_{ij}$ is nul if the index $j$ is not in $(1, \ldots, \ell)$. A direct application of Lemma IV.4.28 yields that the Lie Algebra of the generated group is equal to $\mathcal{G}$. ■

C.Glaffig [122] has proved that this result holds if we merely assume that one coordinate section of $O$ contains an open set in $\mathbb{R}$ using the same technique.

**Lemma IV.4.30** *Let $O$ be an open subset in $\mathbb{R}^\ell$ and let $X_1 \ldots X_r$ be non-zero vectors of $\mathcal{G}$. Then there exists a finite subset $\{x_1, x_2 \ldots x_p\}$ of $O$ such that for $g_k = H(x_k)$ the rank of the system:*

$$Ad(g_1^{-1})X_k \ , \ Ad(g_1^{-1}g_2^{-1})X_k \ , \ \ldots Ad(g_1^{-1}g_2^{-1} \ldots g_p^{-1})X_k \qquad k = 1 \ldots r$$

*is equal to the dimension of $\mathcal{G}$.*

*Proof:*
For a finite word $\omega = \{g_1, \ldots, g_n\}$ we denote by $V_\omega$ the linear span of the above system of vectors in $\mathcal{G}$. We have $V_{\omega\omega'} = V_\omega + Ad(g_1^{-1}g_2^{-1} \ldots g_n^{-1})V_{\omega'}$ where $\omega\omega'$ is the concatenation of the two words $\omega$ and $\omega'$. Let $V_{\omega_0}$ be such a subspace of maximal dimension associated to a word $h_1 \ldots h_p$. Then $V_{\omega_0\omega} = V_{\omega_0}$ and $Ad(h_1^{-1} \ldots h_p^{-1})V_\omega \subset V_{\omega_0}$ thus $V_\omega \subset V_{\omega_0}$ for any $\omega$. This yields that $Ad(g^{-1})V_{\omega_0} \subset V_{\omega_0}$ for $g \in O$ and hence for any $g \in SP(\ell, \mathbb{R})$ from Lemma IV.4.29. The group $SP(\ell, \mathbb{R})$ is a simple Lie group hence one can concludes that $V_{\omega_0} = \mathcal{G}$. ■

**Proposition IV.4.31** *Let $O$ be an open subset of $\mathbb{R}^\ell$. Then there exist $p = (2\ell + 1)$ points $a_1, a_2 \ldots a_p$ in $O$ such that the mapping $\Phi$ defined by:*

$$\Phi(x_1, x_2, \ldots, x_p) = H(x_p)H(x_{p-1}) \ldots H(x_1)$$

*is a submersion from $\mathbb{R}^{p\ell}$ to $SP(\ell, \mathbb{R})$ near the point $(a_1, a_2 \ldots a_p)$.*

*Proof:*
In the Lemma IV.4.30 we choose the elements $X_i \in \mathcal{G}$ for $i = 1 \ldots \ell$ to be the matrix $X_{ii}$ defined above. Then we obtain the existence of $p$ points $a_i$ with the rank property stated in this lemma. If for $t \in \mathbb{R}^\ell$ we define the matrix

$T(t) = \sum_{k=1}^{\ell} t_i X_{ii}$ then we have the relation $H(x+t) = (\exp T(t))H(x)$ and hence:

$$\begin{aligned} & \Phi(a_1+t_1, a_2+t_2, \ldots, a_p+t_p) \\ = \ & (\exp T(t_p))H(a_p)(\exp T(t_{p-1}))H(x_{p-1})\ldots(\exp T(t_1))H(x_1) \end{aligned}$$

The result now follows from the computation of the differential of the function $\Phi$ at the point $(a_1, a_2, \ldots, a_p)$. The dimension of $\mathcal{G}$ is equal to $\ell \times (2\ell+1)$ thus $p$ is certainly no less than $2\ell+1$. Using the notations of Lemma IV.4.30 and the linear independence of the vectors $X_{ii}$ one sees that the maximal number of generators of $V_{\omega_0}$ is $\ell \times (2\ell+1)$ and this completes the proof. ■

If $\tau$ is a probability measure on $\mathbb{R}^\ell$ let us denote by $\mu$ the image of $\tau$ by the mapping $H$.

**Proposition IV.4.32** *If the support of $\tau$ contains an open set of $\mathbb{R}^\ell$ then $T_\mu$ contains an open set of $SP(\ell, \mathbb{R})$.*

*Proof:*
Direct application of Proposition IV.4.31. ■

**Proposition IV.4.33** *If the law $\tau$ is absolutely continuous with respect to the Lebesgue measure of $\mathbb{R}^\ell$ then $\mu^{2\ell+1}$ is absolutely continuous with respect to the Haar measure of $SP(\ell, \mathbb{R})$.*

*Proof:*
We apply Proposition IV.4.31 choosing $O = \mathbb{R}^\ell$. The function $\Phi$ is an analytical mapping from $(\mathbb{R}^\ell)^{2\ell+1}$ to $SP(\ell, \mathbb{R})$. Its Jacobian is also an analytic function which is not identically zero. If follows that the vanishing set of this Jacobian function is a countable union of submanifolds of lower dimension in $(\mathbb{R}^\ell)^{2\ell+1}$. This completes the proof. ■

Notice that under the hypothesis of Proposition IV.4.33 then $T_\mu$ contains an open set of $SP(\ell, \mathbb{R})$. We end this section by a result stated by Goldsheid and Margulis in [128] (the proof has to appear in an announced paper).

*Let $\Sigma$ be a product set in $\mathbb{R}^\ell$ of the form $A_1 \times \ldots \times A_\ell$ such that each $A_i$ contains at least two points. Then the Zariski closure of the group generated by the set $\{H(x)\,;\, x \in \Sigma\}$ is equal to the whole of $SP(\ell, \mathbb{R})$.*

# IV.5 Markovian Multiplicative Systems

Sequences of independent random matrices have no counterpart in the continuous case and moreover in various models products of random matrices are actually governed by a Markov chain or a Markov process. Hence we are lead to the study of Markovian multiplicative processes both for continuous and discrete time. We will not study these models in full generality but we will try to focus on some of the typical examples arising in the theory of random linear difference or differential operators.

Let $M$ be a compact metric space and $\{X(t)\ ;\ t \in T\}$ be a $M$-valued Markov process. As usual we suppose that $X(t)$ is the $t$th coordinate on the product space $\Omega = M^T$ and we denote by $\{\theta_t\ ;\ t \in T\}$ the semi-group of shift operators such that $X(t+h) = X(t) \circ \theta_h$ and by $\mathcal{F}_t$ the sigma-algebra generated by the the random variables $\{X(s)\ ;\ s \leq t\}$.
We assume some strong regularity properties for the process $X(t)$ namely

- The transition semigroup $Q_t$ is a Feller semi-group, i.e. $Q_t$ maps continous functions into continuous functions. This implies that in the continuous time case the process $X(t)$ has a version with right-continuous sample paths and left-hand limits.
- The kernel $Q_1$ has an unique invariant probability measure $\pi$ whose support is equal to the whole space $M$. Furthermore $Q_1(x, dy) = q(x, y)d\pi(y)$ where $q$ is a $\pi \otimes \pi$ positive function.

In the continuous time case a typical example of such a process $X(t)$ is the following. Let $M$ be a $C^\infty$ compact Riemannian manifold and $A_1(x) \dots A_m(x)$ be $C^\infty$ vector fields on $M$, such that at each point $x \in M$ the Lie algebra generated by these vector fields is equal to the tangent space. If $B^1(t) \dots B^m(t)$ are real independent Brownian motions then the Stratonovitch stochastic differential equation:

$$dX(t) = \sum_{k=1}^{m} A_k(X(t)) \circ d\, B^k(t)$$

has a unique solution which satisfies the above requirements. Moreover, the invariant measure $\pi$ has a density with respect to the Riemannian volume element $dx$ and this density and the kernel $q_t(x, y)$ are smooth positive functions. (See [154] [191] for example).
For any $x \in M$ there is a unique probability measure $\mathbb{P}_x$ on $\Omega$ such that the process $X(t)$ is Markovian and $\mathbb{P}_x\{X(0) = x\} = 1$. In the sequel

the expectation with respect to $\mathbb{P}_x$ will be denoted by $\mathbb{E}_x$. The system $(\Omega, \mathcal{F}, \theta_t, \mathbb{P})$ is an ergodic dynamical system if we define the probability measure $\mathbb{P}$ by the formula $\mathbb{P} = \int \mathbb{P}_x \, d\pi(x)$. We then define the process $U(t)$ in the following way:

- In the discrete case we let $F$ be a continuous function from $M$ to $GL(d, \mathbb{R})$ and we set $g(n) = F(X(n))$. The sequence $U(n)$ is then defined for $n \geq 1$ as the left product $U(n) = g(n-1) \dots g(0)$ and $U(0) = I$.
- In the continuous case we let $F$ be a continuous function from $M$ to the set of $d \times d$ real matrices. The process $U(t)$ is then defined as the (unique) solution of the matrix valued differential equation $dU(t) = F(X(t))U(t)dt$ with the initial condition $U(0) = I$.

In the continuous case the matrix $U(t)$ is actually in $GL(d, \mathbb{R})$ since the solution of the equation $dV(t) = -V(t)F(X(t))dt$ with initial condition $V(0) = I$ satisfies $d(U(t)V(t)) = 0$ hence one has $V(t) = U^{-1}(t)$. Moreover it is readily seen that $U(t)$ is a multiplicative process defined on the ergodic dynamical system $(\Omega, \mathcal{F}, \theta_t, \mathbb{P})$. In the continuous case the integrability properties follow from Gronwald's theorem and the separability follows from the Feller property of $X(t)$. In general $U(t)$ is not a Markov process but we will see below that for each fixed $b \in B = P(\mathbb{R}^d)$ the process $Z(t) = (X(t), U(t)b)$ is Markovian.

**Lemma IV.5.1** *Let $f$ be a measurable bounded function on $\Omega \times B$ and $b \in B$. For $t \in T$ define $f_t(\omega) = f(\theta_t(\omega), U(t, \omega)b)$ then*

$$\mathbb{E}_x \{f_t | \mathcal{F}_t\} = \int f(\omega', U(t)b) \, d\mathbb{P}_{X(t)}(\omega')$$

*Proof:*
Let us first assume that $f(\omega, b) = \phi(\omega)\psi(b)$ then

$$\mathbb{E}_x \{f_t | \mathcal{F}_t\} = \mathbb{E}_x \{(\phi \circ \theta_t) \; \psi(U(t)b) | \mathcal{F}_t\} = \mathbb{E}_{X(t)}\{\phi\} \; \psi(U(t)b)$$

hence the formula is true for such a function $f$ and by a monotone class argument for any bounded measurable function $f$. ∎

**Proposition IV.5.2** *The process $Z(t)$ is Markovian with respect to the increasing sequence of sigma-algebras $\mathcal{F}_t$ and any probability measure $\mathbb{P}_x$. Moreover the semi-group $\{R_t \; ; \; t \in T\}$ defined by*

$$R_t f(x, b) = \mathbb{E}_x\{f(X(t), U(t)b)$$

*is a Feller semi-group on $M \times B$.*

*Proof:*
Let $f$ be a bounded measurable function on $M \times B$ then one can write:

$$\begin{aligned} \mathbb{E}_x\{f(Z(t+h))|\mathcal{F}_t\} &= \mathbb{E}_x\{f(X(h)\circ\theta_t,(U(h)\circ\theta_t)\,U(t)b)|\mathcal{F}_t\} \\ &= \int f(X(h),\omega'),U(h),\omega')U(t)b)\,d\mathbb{P}_{X(t)}(\omega') \\ &= R_h f(X(t),U(t)b) \end{aligned}$$

The semi-group property of $R_t$ is an easy consequence of the above formula and the Feller property follows from the joint continuity of $U(t)$ with respect to initial conditions $X(0)$ and $U(0)$. ∎

We remark that $\theta_t$ is no longer the classical shift operator of the probabilists for the Markov process $Z(t)$ and that $R_1$ always admits invariant probability measures since one can pick any limit point of the sequence of probability measures $\nu_n$ defined by $\nu_n(f) = \frac{1}{n}\sum_{k=0}^{n-1} R_n f(x,b)$.
Using the Markov process $Z(t)$ one can recover most of the results obtained in the case of independent products. Unfortunately the proofs are more involved. In particular checking measurability properties related to the strong irreducibility defined below is a tedious job and we will refer to the work of Bougerol [31] [32] for the details. One can also remark that even in the continuous case all the limit properties can be stated only using the Markov chain obtained by restricting the time parameter to integer values.

## IV.5.1 The Upper Lyapunov Exponent

**Definition IV.5.3** *The process $U$ is said to be not strongly irreducible if there exist an integer $r$ and measurable functions $V_1(x)\ldots V_r(x)$ with values in the set of proper subspaces of $\mathbb{R}^d$ such that if we note $W(x) = \cup_{i=1}^m V_i(x)$ then we have*

$$U(1)W(X_0) = W(X(1)) \qquad \mathbb{P}\text{-}a.e$$

The strong reccurence hypothesis which we made on the process $X$ allows to give a very simple criterion of strong irreducibility for $U$ only involving the properties of the function $F$. We say that a subset $S$ of the set of $d \times d$ matrices is irreducible if there does not exist a proper subspace $V$

of $\mathbb{R}^d$ such that $sV \subset V$ for all $s \in S$. One check immediately that $S$ is irreducible iff the Lie Algebra generated by $S$ is irreducible.

**Proposition IV.5.4** *Let $S = \{F(x)\ ;\ x \in M\}$ and assume that:*

- *$S$ is strongly irreducible as a subset of $GL(d, \mathbb{R})$ in the discrete time case.*
- *$S$ is irreducible as a subset of the set of $d \times d$ matrices in the continuous time case.*

*Then the process $U$ is strongly irreducible.*

*Proof:*
Let us first consider the discrete time case and let us assume that the process $U$ is not strongly irreducible. Then there exists a measurable family $W(x)$ of finite unions of proper subspaces of $\mathbb{R}^d$ such that $U(1)W(X(0)) = W(X(1))$ $\mathbb{P}$-almost surely. If follows that

$$\pi \otimes \pi\big\{(x, y)|W(x) = F(y)^{-1}W(y)\big\} = 1 \ .$$

This implies that the family $W(x)$ is $\pi$ almost surely constant and equal to some $W$. The same relation now yields $F(y)W = W$ for $\pi$ almost all $y$ and this contradicts the strong irreducibility of $S$.
In the continuous case let $H$ be the connected Lie sub-group of $GL(d, \mathbb{R})$ whose Lie algebra is generated by $S$. $H$ is irreducible and being connected it is also strongly irreducible and the conclusion follows from the above discussion. ∎

**Proposition IV.5.5** *Let us denote by $\nu$ an invariant probability measure for the Markov chain $Z(n)$ and by $\nu = \int \delta_x \otimes \nu_x \, d\pi(x)$ a desintegration of $\nu$. If $U$ is strongly irreducible then the probability measure $\nu_x$ is proper for $\pi$-almost all $x$.*

*Proof:*
The proof of this result is very technical and we only give the essentials steps. (See Bougerol [32] for details). One first proves that there exists a continuous (and hence unique) version of the mapping $x \hookrightarrow \nu_x$ and for $x \in M$ one defines:

- $d(x) =$ the minimum dimension of a subspace $V$ with $\nu_x(\overline{V}) > 0$
- $p(x) =$ the maximum value of $\nu_x(\overline{V})$ for a $d(x)$-dimensional subspace $V$

- $W(x) =$ the finite union of the $r(x)$ subspaces $V$ of dimension $d(x)$ such that $\nu_x(\overline{V}) = p(x)$

The first step is to establish the measurability of the functions $d(x)$, $r(x)$, $p(x)$ and to prove that $r(x)$ is a constant function equal to some integer $r$. This allows to consider $r$ measurable functions $V_x^i$ and the invariance equation implies that for $\pi$ almost all $x$ one has $\nu_x(\overline{V_x^i}) = \mathbb{E}_x\{U^{-1}(1)\overline{V_x^i}\}$. It follows that $U(1)W(X(0)) = W(X(1))$ $\mathbb{P}$-a.e. and this contradicts the strong irreducibility of the process $U$. ∎

Let $\tilde{\Omega}$ be the product space $\tilde{\Omega} = \Omega \times B$ and let $\tilde{\theta}$ be the shift operator $\tilde{\theta}_t(\omega, b) = (\theta_t(\omega), U(t,\omega)b)$ then one has:

**Lemma IV.5.6** *Let $\nu$ be an invariant probability measure for the Markov chain $Z(n)$. Then the probability measure $\tilde{\mathbb{P}}$ defined on $\tilde{\Omega}$ by the formula*

$$\tilde{\mathbb{P}} = \int (\mathbb{P}_x \otimes \delta_b)\, d\nu(x,b) = \int (\delta_x \otimes \nu_x)\, d\pi(x)$$

*is $\tilde{\theta}$ invariant and its projection on $\Omega$ is equal to $\mathbb{P}$*

*Proof:*
Let $\phi$ be a bounded measurable function on $\tilde{\Omega}$ and let us set $\Phi(x,b) = \int \phi(\omega, b)\, d\mathbb{P}_x(\omega)$. With the notations of Lemma IV.5.1 one has:

$$\begin{aligned}
\int \phi \circ \tilde{\theta}\, d\tilde{\mathbb{P}} &= \int \phi(\theta(\omega), U(1,\omega)b)\, d\mathbb{P}_x(\omega)\, d\nu(x,b) \\
&= \int \mathbb{E}_x\{\mathbb{E}_x\{\phi(\theta(\omega), U(1,\omega)b) | \mathcal{F}_1\}\}\, d\nu(x,b) \\
&= \int \Phi(X(1), U(1)b)\, d\mathbb{P}_x\, d\nu(x,b) \\
&= \int \phi\, d\tilde{\mathbb{P}}\,. \quad \blacksquare
\end{aligned}$$

**Proposition IV.5.7** *Suppose that the process $U$ is strongly irreducible and that $\nu$ is an invariant probability measure for the Markov chain $Z(n)$. Then we have the formula:*

$$\gamma_1 = \int \mathbb{E}_x\{\log \frac{||U_1 v||}{||v||}\}\, d\nu(x, \bar{v})$$

*Proof:*
The result is a direct consequence of Lemma IV.5.6 , Proposition IV.5.5 and part (ii) of Proposition IV.2.9. ∎

**Proposition IV.5.8** *Let the process $U$ be strongly irreducible. Then for any non-zero vector $v \in \mathbb{R}^d$ we have:*

$$\lim_{n\to\infty} \frac{1}{n} \log \frac{\|U(n)v\|}{\|v\|} = \gamma_1 \qquad \mathbb{P}\text{-}a.e.$$

*Proof:*
The argument is the same as in Proposition IV.4.6 if we use the probability measures $\nu_n$ defined on $M \times B$ by $\nu_n(f) = \frac{1}{n}\sum_{k=0}^{n-1} \mathbb{E}\{f(X(n), U(n)\bar{v})\}$. The Markov property of the process $Z$ implies that

$$\mathbb{E}_x\{\log \frac{\|U(k+1)v\|}{\|U(k)v\|} \mid \mathcal{F}_k\} = \int \log \frac{\|U(1,\omega')U(k)v\|}{\|U(k)v\|}\, d\mathbb{P}_{X(k)}(\omega')$$

and this yields the desired result. ∎

Using the Markov property stated above (for continuous time) one sees that

$$\int \mathbb{E}_x\{\log \frac{\|U(t)v\|}{\|v\|}\}\, d\nu(x,\bar{v}) = t\,\gamma_1$$

Actually this property is fairly general (See Proposition IV.2.9).

## IV.5.2 The Lyapunov Spectrum

In order to prove that the Lyapunov exponents $\gamma_1$ and $\gamma_2$ are different we need to introduce the notion of contractivity for a Markovian multiplicative process.

**Definition IV.5.9** *The process $U$ is said to be contractive if :*

- *The subset $\{F(x)\ ;\ x \in M\}$ generates a contractive semi-group in $GL(d,\mathbb{R})$ in the discrete time case.*
- *The set $\{\exp(tF(x))\ ;\ x \in M\ ,\ t > 0\}$ generates a contractive semi-group in $GL(d,\mathbb{R})$ in the continuous time case.*

One can remark that strong irreducibility and contractivity are easier to handle in the continous time case since it is possible in this situation to work in the Lie algebra of $SP(\ell, \mathbb{R})$.
Assuming strong irreducibility and contractivity it is possible to extend the results we discussed for independent products. The proof follows the lines of the paper of Guivarc'h (See [135]). A complete proof can be found in Bougerol [32]. The following Theorem summarizes the essential results:

**Theorem IV.5.10** *Suppose that the process $U$ is strongly irreducible and contractive then one has:*

**(i)** *The Markov chain $Z(n)$ admits a unique invariant probability measure $\nu$ on $M \times P(\mathbb{R}^d)$.*
**(ii)** *The sequence $\frac{1}{n}\mathbb{E}_x\{\log \frac{\|U(n)v\|}{\|v\|}\}$ converges to $\gamma_1$ uniformly with respect to the variables $(x\,,\,\bar{v}) \in M \times P(\mathbb{R}^d)$.*
**(iii)** *The Lyapunov exponents $\gamma_1$ and $\gamma_2$ are distinct.*

*Proof:*
We only give a sketch of the proof which follows closely the argument used in the theory of independent products (but with more tears ....) One first proves that all the limit points of the sequence $U(n)\|U(n)\|^{-1}$ are rank one matrices. Then using the fact that the sequence $U'(n)\nu_{X(n)}$ is a martingale (where $\nu_x$ is given by the desintegration in Proposition IV.5.5) one obtains the uniqueness of the invariant measure and (ii) is a direct consequence of this. To obtain $\gamma_1 > \gamma_2$ one uses again the Proposition IV.3.4 but with a more complicated dynamical system than in the independent case. ∎

**Definition IV.5.11** *Schrödinger matrices.*

- *In the discrete case we say that $F(x)$ is a Schrödinger matrix if $F = H \circ h$ where $H$ is the function defined in IV.4.24 and $h$ is a continuous function from $M$ to $\mathbb{R}^\ell$.*
- *In the continuous case we say that $F(x)$ is a Schrödinger matrix if $F = \tilde{H} \circ h$ where $\tilde{H}$ is the mapping from $\mathbb{R}^d$ to the Lie algebra of $SP(\ell, \mathbb{R})$ defined by*

$$\tilde{H}(x) = \begin{pmatrix} 0 & I_\ell \\ Q(x) & 0 \end{pmatrix}$$

*and $Q(x)$ is the matrix defined in IV.4.24.*

**Proposition IV.5.12** *Let $A$ be a subset of $\mathbb{R}^\ell$ of the form $A_1 \times A_2 \times \ldots \times A_\ell$ such that each $A_i$ contains more than one point. Then the Lie algebra generated by $\tilde{H}(A)$ is the whole Lie algebra of $SP(\ell, \mathbb{R})$.*

*Proof:*
We remark that the Lie bracket of two matrices $\tilde{H}(x)$ and $\tilde{H}(y)$ is a diagonal matrix. Hence using the notations introduced in the preceding section on obtains all the matrices $Z_{ii}$ and by difference all the matrices $Y_{ii}$. We now remark that:

$$[\tilde{H}(x)\ ,\ Z_{ii}] = 2(X_{ii} + x_i Y_{ii} - Y_{i,i-1} - Y_{i,i+1})$$

where $Y_{i,j}$ is zero if the index $j$ is not in $(1 \ldots \ell)$. A direct application of Lemma IV.4.28 yields the result. ∎

**Proposition IV.5.13** *In the continuous case let us assume that the range of $h$ contains a subset $A$ of $\mathbb{R}^d$ of the form $A_1 \times A_2 \times \ldots \times A_\ell$ such that each $A_i$ contains more than one point. Then the process $\Lambda^p U$ is contractive and* Lag($p$)*-strongly irreducible for $0 \le p \le \ell$. From which it follows that all the Lyapunov exponents are distinct.*

*Proof:*
The Zariski closure of the sub-group generated by the set $\{\exp(tF(x))\ ;\ x \in M\ ,\ t > 0\}$ is a group. Proposition IV.5.12 implies that this group is equal to $SP(\ell, \mathbb{R})$ and Proposition IV.4.23 yields the result. ∎

As already remarked strong irreducibility and contractivity are more difficult to check in the discrete case.

**Proposition IV.5.14** *In the discrete case if the range of $h$ has a non-empty interior in $\mathbb{R}^\ell$ then all the Lyapunov exponents are distinct.*

*Proof:*
Proposition IV.4.31 implies that the semi-group generated by the range of $F$ contains an open set in $SL(d, \mathbb{R})$ thus the result follows from the Proposition IV.4.19 and Proposition IV.5.4 (i). ∎

Notice that the claim of Goldsheid and Margulis given at the end of section 4 would allow to obtain the same criterion as in the continuous time case. Nevertheless as in the independent case we have:

**Proposition IV.5.15** *In the one dimensional case, let us assume that the function $h$ takes at least two distinct values. Then:*

1. *The Markov chain $Z(n)$ admits a unique invariant probability measure $\nu$ on $M \times P(\mathbb{R}^2)$.*
2. *The sequence $\frac{1}{n}\mathbb{E}_x\{\log \frac{\|U(n)v\|}{\|v\|}\}$ converges to $\gamma_1$ uniformly with respect to the variables $(x, \bar{v}) \in M \times P(\mathbb{R}^2)$.*
3. *The Lyapunov exponent $\gamma_1$ is strictly positive.*

*Proof:*
In the discrete time case we only apply Proposition IV.4.25. In the continuous time case the result follows from Proposition IV.5.13. It is also possible to write down a direct proof since Lemma IV.5.12 implies that the Lie algebra generated by the matrices $H(x)$ is the whole Lie algebra of $SL(2, \mathbb{R})$. Hence the strong irreducibility is obvious and contractivity follows from Proposition IV.4.11. ∎

## IV.5.3 Laplace Transform

One can extend to the Markovian case most of the results proved for the Fourier-Laplace operators associated to a product of independent matrices. Careful readers have noticed that in this latter case one has often used the existence of a dual operator associated to the measure $\check{\mu}$. Time reversal is not granted for Markov processes hence we need to assume some extra properties for the process $X$. We will not dwell on the probabilistic construction of a dual process and we will immediately assume that $X(t)$ is already defined for any real or integer values of the parameter $t$. We have to emphasize that theory of stochastic processes, and especially duality theory for Markov processes, invites us to interpret the parameter $t$ as a time parameter. This is very infortunate for the applications to one dimensional Schrödinger operators for in this case the parameter of the stochastic process is actually the space variable. We will come back to this slight conflict between two well established traditions several times in the sequel. In any case, in the situation we have in mind, the process is naturally defined on the whole line. In this subsection $T$ denotes the set of real numbers or the set of integers and $T^+$ is the subset of non-negative elements of $T$. We will assume that the random variable $X(t)$ is the $t$-th coordinate in the product space $\Omega = M^T$ and that the probability measure $\mathbb{P}$ on $\Omega$ is invariant and ergodic with respect to the action of the group of

shift operators $\{\theta_t\,;\,t\in T\}$. For $(s,t)\in T$ with $s\leq t$ we denote by $\mathcal{F}_{s,t}$ the sigma algebra generated by the random variables $\{X(h)\,;\,s\leq h\leq t\}$.
We also assume that there exist two Feller semi-groups $\{Q_t\,;\,t\in T^+\}$ and $\{\check{Q}_t\,;\,t\in T^+\}$ on $M$ such that for each $s'\leq s\leq t\leq t'$ in $T$ and each bounded measurable function $f$ on $M$ we have:

$$\begin{aligned}\mathbb{E}\{f(X(t')\,|\,\mathcal{F}_{s,t}\} &= Q_{t'-t}f(X(t)),\\ \mathbb{E}\{f(X(s')\,|\,\mathcal{F}_{s,t}\} &= \check{Q}_{s-s'}f(X(s)).\end{aligned}$$

For any $x\in M$ the probability $\mathbb{P}$ conditionned by the event $\{X(0)=x\}$ is assumed to have a regular version denoted by $\mathbb{P}_x$. It is easily seen that the common law $\pi$ of the random variables $X(t)$ is invariant for the semi-groups $Q_t$ and $\check{Q}_t$. We also assume that the transition probabilities of these semi-groups have strictly positive densities with respect to $\pi$.
All these assumptions are not independent but, as we already pointed out, we are not aiming at the greatest generality.

In the continuous case the process $U$ is defined for all $t\in T$ and in the discrete time case one defines $U(n)$ for negative values of $n$ by $U(n)=F(X(n))^{-1}\dots F(X(-1))^{-1}$. Recall that $Z(t)=\{(X(t),U(t)b)\,;\,t\in T^+\}$ and $\check{Z}(t)=\{(X(-t),U(-t)b)\,;\,t\in T^+\}$ are then Markov processes.

The stationnarity of the process $\{X(t)\,;\,t\in T\}$ yields immediately:

**Proposition IV.5.16** *The law of the process $\{(X(t),U(t))\,;\,t\in T\}$ is invariant by the shift on the space $\Omega$.*

Note that one should not confuse this property with the classical notion of stationnarity since $U(t)\circ\theta_s\neq U(t+s)$.

Let us denote by $\sigma$ a cocycle on the projective space $B$ and by $z$ a complex number. For $t\in T^+$ one defines the Fourier-Laplace operators on the space $C(M\times B)$ of continuous functions on $M\times B$ by:

$$\begin{aligned}T^t_{\sigma,z}f(x,b) &= \mathbb{E}_x\{f(X(t),U(t)b)\sigma^z(U(t),b)\}\\ \check{T}^t_{\sigma,z}f(x,b) &= \mathbb{E}_x\{f(X(-t),U(-t)b)\sigma^z(U(-t),b)\}\end{aligned}$$

Since $M$ is compact, no integrability property of the coycle $\sigma$ is needed to obtain bounded operators on $C(M\times B)$. Moreover one can prove exactly as in Proposition IV.3.7 that these operators are holomorphic functions of

$z$. The Markov property added to the cocycle relation yield immediately the semi-group property with respect to the parameter $t$.

Actually, in the continuous time case, we only need to deal with integer values of $t$ thus we will use the notations $T_\sigma = T^1_{\sigma,1}$, $\check{T}_\sigma = \check{T}^1_{\sigma,1}$ and $T$, $\check{T}$ for the Markov operators associated to the trivial cocycle $\sigma = 1$. The superscript $*$ will denote dual operators on the Banach space of measures on $M \times B$.
The next propositions are only stated for the operator $T_\sigma$ but they obviously have a counterpart for $\check{T}_\sigma$.

**Proposition IV.5.17** *Let $\sigma$ be a cocycle such that there exists an integer $p$ with*

$$\sup_{(x,b)\in M\times B} \mathbb{E}_x\{\log \sigma(U(p), b)\} < 0$$

*Then there exists a positive real number $\alpha$ such that for $0 < s \leq \alpha$ there exist positive constants $C_s < +\infty$ and $\rho_s < 1$ such that:*

$$\sup_{(x,b)\in M\times B} \mathbb{E}_x\{\sigma^s(U(n), b)\} \leq C_s\, \rho_s^n \qquad n = 1, 2, \ldots$$

*Proof:*
See Proposition IV.3.5. ∎

A probability measure $\nu$ on $M \times B$ is said to be $T$ invariant whenever $T^*\nu = \nu$.

**Proposition IV.5.18** *Let $\sigma$ be a cocycle and for each real number $s$ let us denote by $\rho(s)$ the logarithm of the spectral radius of $T_{\sigma^s}$ acting on $C(M \times B)$. Then we have:*

**(i)** *The function $s \hookrightarrow \rho(s)$ is convex and $\rho(0) = 0$,*

**(ii)** $\rho(s) \geq s \int \mathbb{E}_x\{\log \sigma(U(1), b)\}\, d\nu(x, b)$ *for any $T$-invariant probability measure $\nu$.*

**(iii)** *If the function $\rho$ admits a derivative at the origin then its value is given by:*

$$\frac{d\rho(s)}{dt}\Big|_{s=0} = \int \mathbb{E}_x\{\log \sigma(U(1), b)\}\, d\nu(x, b)$$

*(Hence this integral is independent from the choice of $\nu$)*

*Proof:*
See Proposition IV.3.6 ∎

If $\tau$ is a measure on $M \times B$ we denote by $< f, h >_\tau$ the continuous bilinear form on $C(M \times B)$ defined by $\int fh \, d\tau$. We also denote by $r$ a Radon-Nikodym cocycle associated to a probability measure $m$ on $B$.

**Proposition IV.5.19** *Let $r$ be a m-Radon-Nikodym cocycle. Then, for any positive integer $k$ and any complex number $\lambda$ we have:*

**(i)** $< \check{T}_\sigma f, h >_{\pi\otimes m} = < T_{r\sigma^{-1}} h, f >_{\pi\otimes m} \qquad f, h \in C(M \times B)$
**(ii)** $\left[\ker(\check{T}_\sigma - \lambda I)^k\right](\pi \otimes m) \subset \ker(T^*_{r\sigma^{-1}} - \lambda I)^k$

*Proof:*
Let $f$ and $h$ be two continuous functions on $M \times B$. Then we have:

$$\begin{aligned}
& < T_{r\sigma^{-1}} h, f >_{\pi\otimes m} \\
= & \int T_{r\sigma^{-1}} h(x,b) f(x,b) \, d\pi(x) \, dm(b) \\
= & \int \mathbb{E}_x\{h(X(1), U(1)b) r(U_1, b) \sigma^{-1}(U(1), b)\} f(x,b) \, d\pi(x) \, dm(b) \\
= & \int \mathbb{E}\{h(X(1), b) \sigma(U^{-1}(1), b) f(X(0), U^{-1}(1)b)\} \, dm(b) \\
= & \int \mathbb{E}\{h(X(0), b) \sigma(U(-1), b) f(X(-1), b)\} \, dm(b) \\
= & \int \check{T}_\sigma f(x,b) h(x,b) \, d\pi(x) \, dm(b)
\end{aligned}$$

This proves conclusion (i) and (ii) is an immediate consequence. ∎

**Proposition IV.5.20** *Let $\sigma$ be a cocycle and $r$ be a m-Radon-Nikodym cocycle. Assume that $T_{r\sigma^{-1}}$, $\check{T}_{r\sigma^{-1}}$, $T_\sigma$ and $\check{T}_\sigma$ are compact operators on $C(M \times B)$. Then for any positive integer $k$ and any nonzero complex number $\lambda$ we have:*

**(i)** $\left[\ker(\check{T}_\sigma - \lambda I)^k\right](\pi \otimes m) = \ker(T^*_{r\sigma^{-1}} - \lambda I)^k$
**(ii)** $\dim\left(\ker(\check{T}_\sigma - \lambda I)^k\right) = \dim \ker(T_{r\sigma^{-1}} - \lambda I)^k$

*Proof:*
These formulas are immediate consequence of the compacity and Proposition IV.5.19. ∎

**Proposition IV.5.21** *Let $r$ be a $m$-Radon-Nikodym cocycle. Assume that $T$, $\check{T}$, $T_r$, $\check{T}_r$ are compact operators on $C(M \times B)$, that $T$ and $\check{T}$ have unique invariant probability measures $\nu$ and $\check{\nu}$ and are aperiodic. Then we have the decompositions:*

**(i)** $T^n f = \nu(f) + Q^n f$
**(ii)** $T_r^n f = \pi \otimes m(f)\psi + Q_r^n f \qquad$ *with* $\check{\nu} = \psi(\pi \otimes m)$

*the operators $Q$ and $Q_r$ having spectral radius strictly less than one. It follows that the spectral radius of $T_r$ is equal to one and that the sequence $||T_r^n||$ is bounded.*

*Proof:*
See Proposition IV.3.12. ∎

In the independent case one has proved the compacity of the Laplace operators as a consequence of the absolute continuity of $\mu$ with respect to a Haar measure on $GL(d, \mathbb{R})$. In the Markovian case we will need some smoothness and non-singularity assumptions on the function $F$. It seems difficult to state a compactness criterion in the general Markovian case but in the particular example of Schrödinger matrices associated to a diffusion on a Riemannian manifold, which we discussed earlier in this section, the situation is somewhat simpler. We will see in the sequel that assuming that the function $h$ from $M$ to $\mathbb{R}^\ell$ is a Morse function, then $T_\sigma$ and $\check{T}_\sigma$ are aperiodic compact operators on $C(M \times B)$ admitting unique invariant measures.

# IV.6 Boundaries of the Symplectic Group

In this section $G$ will denote the symplectic group $SP(\ell, \mathbb{R})$ which acts on $\mathbb{R}^d$ for $d = 2\ell$. The compact subgroup of orthogonal matrices in $G$ is denoted by $K$. For a decomposable $p$-vector $u = u_1 \wedge u_2 \wedge \ldots \wedge u_p$ we also denote by $u$ the $d \times p$ matrix with columns vectors $u_1, u_2, \ldots, u_p$ and we will use the notations and definitions of section 2.

We first recall without proof some useful results from linear algebra.

**Lemma IV.6.1** *One has the following properties:*

**(i)** *Let* $u = \{u_1, u_2, \ldots, u_p\}$ *and* $v = \{v_1, v_2, \ldots, v_p\}$ *be two linearly independent systems. They generate the same* $p$ *dimensional subspace of* $\mathbb{R}^d$ *iff* $u_1 \wedge u_2 \wedge \ldots \wedge u_p$ *and* $v_1 \wedge v_2 \wedge \ldots \wedge v_p$ *are colinear in* $\Lambda^p(\mathbb{R}^d)$. *It is also equivalent to the existence of a* $p \times p$ *invertible matrix* $w$ *such that* $u = vw$.

**(ii)** *Let* $u$ *be a decomposable p-vector with matrix* $u = \binom{A}{B}$ *where* $A$ *and* $B$ *are* $\ell \times p$ *matrices. Then* $u$ *is Lagrangian iff the matrix* $A'B$ *is symmetric.*

**(iii)** *Let* $u = u_1 \wedge u_2 \wedge \ldots \wedge u_p$ *be a decomposable p-vector and let* $v = v_1 \wedge \ldots \wedge v_q$ *be a decomposable q-vector with* $p < q$. *Then the subspace generated by* $u$ *is contained in the subspace generated by* $v$ *iff there exists a decomposable* $(q-p)$*-vector* $w$ *such that* $v = u \wedge w$

This lemma makes possible the identification of the image $\bar{v}$ in the projective space $P(\Lambda^p(\mathbb{R}^d))$ of a non-zero decomposable $p$-vector $v$ with a $p$ dimensional subspace of $\mathbb{R}^d$. For an integer $1 \le p \le \ell$ we will denote by $L_p$ the set of $p$ dimensional subspaces of $\mathbb{R}^d$ corresponding to Lagrangian decomposable $p$-vectors (See IV.4.20). For a subset $I = i_1 < i_2 < \ldots < i_p$ of $\{1, \ldots, \ell\}$ we define the flag-manifold:

$$L_I = \{(b_{i_1} \ldots b_{i_p}) \mid b_{i_k} \in L_k \ , \ b_{i_k} \subset b_{i_{k+1}}\}$$

It is known that $L_I$ is a homogeneous compact space under the action of $G$ or $K$. Let $\{e_1 \ldots e_d\}$ be the canonical basis of $\mathbb{R}^d$ and define $e_I = e_{i_1} \wedge \ldots \wedge e_{i_p}$. Then we have $L_I = G\bar{e_I} = K\bar{e_I}$. The flag manifold $L_{\{1\ldots\ell\}}$ is called the maximal boundary or Furstenberg boundary and one can remark that it is isomorphic to $G/H$ where $H$ is the "parabolic" sub-group of $G$ which fixes the element $\bar{e}_{\{1\ldots\ell\}}$. For any $I$ the space $L_I$ admits a unique probability measure which is invariant by all the rotations of $K$. It is called the Cauchy measure and will be denoted by $m_I$. This measure is quasi-invariant under the action of $G$ (that is $gm_I$ has the same nul sets than $m_I$) and the associated Radon-Nikodym cocycle, say $r_I$, is called the Poisson kernel of $L_I$. This Cauchy measure is well known for $L_1$ which is the projective space $P(\mathbb{R}^d)$. Up to a submanifold of lower dimension this manifold is isomorphic to $\mathbb{R}^{d-1}$ and the Cauchy measure has a density with respect to the Lebesgue measure proportional to $(1 + ||x||^2)^{-(d+1)/2}$. We will see below that $L_\ell$, once a submanifold of lower dimension is removed, is isomorphic to the set $\mathcal{M}(\ell)$ of symmetric matrices of order $\ell$ and the Cauchy measure admits a density proportionnal to $(1 + (\det M)^2)^{-(d+1)/2}$ with respect to the Lebesgue measure on $\mathcal{M}(\ell)$.

**Proposition IV.6.2** *Let $\mu$ be a probability measure on $G$ such that for $p = 1 \ldots \ell$ the measure $\Lambda^p \mu$ is contractive and* Lag($p$)*-strongly irreducible then*

**(i)** *There exists a unique invariant probability measure $\nu_I$ on each $L_I$ and*
**(ii)**

$$\gamma_1 + \ldots + \gamma_p = \int \log \frac{||gv||}{||v||} \, d\mu(g) \, d\nu_p(\bar{v})$$

*Proof:*
For each $p$ the set of Lagrangian $p$-vectors is invariant under the action of $G$, this implies that the unique invariant measure $\nu_p$ is supported by this set and hence that the random variable $Z_p$ defined in Proposition IV.4.12 has its range in this set. Applying the classical martingale argument we obtain that if $\nu$ is an invariant measure on $L_I$ then $\nu$ has to be the law of $(Z_1, \ldots, Z_p)$. Conclusion (ii) follows from the fact that the upper Lyapunov exponent on $\Lambda^p(\mathbb{R}^d)$ is the sum $\gamma_1 + \ldots + \gamma_p$. ■

A cocycle $\rho$ on $L_I$ is said to be $K$-invariant if $\rho(g, kb) = \rho(g, b)$ for all $b \in L_I$ and we denote by $\rho_p$ the cocycle defined on $L_p$ by $\rho_p(g, \bar{v}) = ||gv|| \; ||v||^{-1}$ where the norm is taken in $\Lambda^p(\mathbb{R}^d)$.
To carry out the computation of an explicit formula for the Poisson kernel we will need the following theorem proved by Furstenberg and Tzkoni in [118].

**Proposition IV.6.3** *Any $K$-invariant cocycle on $L_I$ is of the form:*

$$\rho(g, b_I) = \prod_{k \in I} \rho_k(g, b_k)^{\lambda_k}$$

*where $\lambda_k$ , $k \in I$ are real numbers.*

If follows from this proposition that the computation of a $K$-invariant cocycle reduces to the computation of a finite number of exponents $\lambda_k$. Let $\theta$ be a $\sigma$-finite measure on $L_I$ which is equivalent to $m_I$ and define $f(b) = \frac{dm_I}{d\theta}(b)$. Then we have $r_I(g, b) = \frac{dg^{-1}\theta}{d\theta}(b) \frac{f(gb)}{f(b)}$. Choosing for $\theta$ the Lebesgue measure in a system of coordinates near a point $b$ with $b = gb$ one has only to compute a Jacobian to obtain $r_I$. Hence we are lead to exhibit a parametrization of the manifold $L_I$. Actually we will only perform this computation for the boundaries $L_p$ and $L_{\ell-1,\ell}$ which give the simplest results. See [208] for further details.

**Proposition IV.6.4** *For* $1 \leq p \leq \ell$ *the Poisson kernel of* $L_p$ *is* $r_p = \rho_p^{-d+p-1}$

*Proof:*
To each Lagangian $p$-vector $u = u_1 \wedge u_2 \wedge \ldots \wedge u_p$ we associate the $d \times p$ matrix of coordinates of $u_1, u_2, \ldots, u_p$ whose transpose has the form $(a', c', b', d')$ where $a$ and $b$ are $(\ell - p) \times p$ matrices, and $c$ and $d$ are square matrices of order $p$. The mapping $\Pi_p$ defined by:

$$\Pi_p(\bar{u}) = (ad^{-1}, bd^{-1}, cd^{-1} - d'^{-1}a'bd^{-1})$$

is one-to-one from the subset of $L_p$ with $d$ invertible onto the product space $\mathcal{S}$ of $(\ell - p) \times p$ matrices times the space of $(\ell - p) \times p$ matrices times the space of $p \times p$ symmetric matrices ( hence the dimension of $L_p$ is $2p(\ell - p) + p(p+1)/2$). Let $\omega_0$ be the point in $L_p$ with coordinates $(0, 0, I)$. Then the matrix $d$ is invertible near $\omega_0$ thus the parametrization $\Pi_p$ is well defined in a neighborhood of this point. Let $g_0$ be the matrix $g_0 = \begin{pmatrix} \Delta & 0 \\ 0 & \Delta^{-1} \end{pmatrix}$ where $\Delta$ is a diagonal matrix with diagonal entries $(1, \ldots, 1, \alpha)$ with $\alpha > 0$. It is easy to check that $g_0 \omega_0 = \omega_0$ and that the mapping of $L_p$ defined by $\bar{u} \hookrightarrow g_0 \bar{u}$ near $\omega_0$ can be read in the chart $\Pi_p$ as the linear mapping $(r, s, N) \hookrightarrow (r\delta, s\delta, \delta N \delta)$ of $\mathcal{S}$ where $\delta$ is a $p \times p$ diagonal matrix of the same form than $\Delta$. The Jacobian of this mapping is equal to $\alpha^{d-p+1}$ and this yields the result since $\rho_p(g_0, \omega_0) = \alpha^{-1}$. ∎

In the case $p = \ell$, the Poisson kernel is well known since $L_\ell$ is the boundary of the symmetric space $G/K$ which is isomorphic to the Siegel upper half-plane. We also remark that the parametrization $\Pi_\ell$ maps the elements of $L_\ell$ of the form $\binom{c}{d}$ with $d$ invertible onto the set of symmetric matrices of order $\ell$ by $\binom{c}{d} \hookrightarrow cd^{-1}$.

**Proposition IV.6.5** *The Poisson kernel of* $L_{\ell-1,\ell}$ *is* $r_{\ell-1,\ell} = \rho_{\ell-1}^{-\ell} \rho_\ell^{-2}$

*Proof:*
We associate to each pair of Lagrangian vectors $x_1 = u_1 \wedge \ldots \wedge u_{l-1}$ and $x_2 = x_1 \wedge u_\ell$ the $d \times \ell$ matrix of coordinates of $(u_1, \ldots, u_\ell)$ of the form $\binom{a}{b}$ where $a$ and $b$ are square matrices of order $\ell$.When $b$ is invertible we denote by $t$ the upper left entry of $ab^{-1}$. The mapping $\tilde{\Pi}$ defined by $\tilde{\Pi}(\bar{x}, \bar{y}) = (\Pi_\ell(\bar{x}), t)$ from the subset of $L_{\ell-1,\ell}$ with $b$ invertible is one to one and onto the manifold $\mathbb{R}^{\ell-1} \times \mathbb{R}^{\ell-1} \times \mathbb{R}^{\ell(\ell-1)/2} \times \mathbb{R}$. Hence the dimension of $L_{\ell-1,\ell}$ is equal to $2(\ell-1)+\ell(\ell-1)/2$. If we choose $(\omega_1, \omega_2) \in L_{\ell-1,\ell}$ with coordinates $(0, 0, I, 1)$ then the parametrization $\tilde{\Pi}$ is well defined near this point. Let now $g_0$ be a diagonal matrix with diagonal entries $(\beta, 1, \ldots, 1, \alpha)$. Then $(\omega_1\omega_2)$ is preserved by $g_0$ and using the chart $\tilde{\Pi}$, one can see that the mapping $(\bar{x}1, \bar{x}_2) \hookrightarrow g(\bar{x}_1, \bar{x}_2)$ is locally the linear mapping $(r, s, N, t) \hookrightarrow (\beta r\delta, \beta^{-1}s\delta, \delta N\delta, \beta^2 t)$ where $\delta$ is the diagonal matrix of order $\ell - 1$ with entries $(1, \ldots, 1, \alpha)$. The Jacobian of this mapping is equal to $\alpha^{\ell+2}\beta^2$ and this yields the result since $\rho_\ell(g_0, \omega_2) = \alpha^{-1}\beta^{-1}$ and $\rho_{\ell-1}(g_0, \omega_1) = \alpha^{-1}$. ∎

In the case of Schrödinger matrices one can apply these relations to compute the Lyapunov exponents. We now assume that the measure $\mu$ satisfies the hypothesis of Proposition IV.6.2 and moreover, that there exists a positive real $\tau$ such that $\int \|g\|^\tau \, d\mu(g)$ is finite. Using Proposition IV.6.4 the boundary $L_\ell$ up to a projective submanifold is isomorphic to the set $\mathcal{M}(\ell)$ of symmetric matrices of order $\ell$. Hence the proper invariant measure $\nu_\ell$ can be viewed as a measure $\tilde{\nu}_\ell$ on $\mathcal{M}(\ell)$.

**Proposition IV.6.6** *The sum of the nonnegative Lyapunov exponents is given by:*

$$\gamma_1 + \ldots + \gamma_\ell = \int_{\mathcal{M}(\ell)} \log|\det M| \, d\tilde{\nu}_\ell(M)$$

*Proof:*
We know from Proposition IV.6.2 that the sum of nonnegative Lyapunov exponents is given by the integral of the cocycle $\rho_\ell$ with respect to $\mu \otimes \nu_\ell$. Because of Proposition IV.6.4 we can replace $\log \rho_\ell$ by $-1/(\ell + 1) \log r_\ell$. If $f(b)$ is the density of the Cauchy measure with respect to the Lebesgue measure $\theta$ associated to the parametrization $\Pi_\ell$, then Proposition IV.4.18 implies that this function is $\nu_\ell$-integrable. The cocycle relation yields that the sum of the nonnegative Lyapunov exponents is the integral of $-\frac{1}{\ell+1} \log \frac{dg^{-1}\theta}{d\theta}$. For a Schrödinger matrix $g = \begin{pmatrix} Q & -I \\ I & 0 \end{pmatrix}$ the action on $\mathcal{M}$ can be written $M \hookrightarrow Q - M^{-1}$ and the Jacobian of this mapping is

equal to $|\det M|^{-(\ell+1)}$. This completes the proof. ∎

Let us also notice that using the parametrization $\Pi_{\ell-1}$ the proper invariant measure $\nu_{\ell-1}$ can be viewed as a measure $\tilde{\nu}_{\ell-1}$ on the manifold $\mathcal{N}(\ell) = \mathbb{R}^{\ell-1} \times \mathbb{R}^{\ell-1} \times \mathbb{R}^{\ell(\ell-1)/2}$.

**Proposition IV.6.7** *With the notations of Proposition IV.6.5 the sum of the $\ell - 1$ first Lyapunov exponents is given by:*

$$\gamma_1 + \ldots + \gamma_{\ell-1} = \int_{\mathcal{N}(\ell)} \log|\det(N + rs')|\, d\tilde{\nu}_{\ell-1}(r, s, N)$$

*Proof:*
The Poisson kernel $r_{\ell-1}$ is not very tractable. In the present situation it is more convenient to introduce the cocycle $\sigma = r_{\ell-1,\ell} r_{\ell-1}^{-1} = \rho_{\ell-1}^2 \rho_\ell^{-2}$. Then, using again the cocycle property of the Poissons kernels one obtains $\gamma_\ell = -1/2 \int g'(t)\, d\nu_{\ell-1,\ell} \otimes d\mu$ where $g(t)$ is the action on the coordinate $t$ of a matrix $g \in SP(\ell, \mathbb{R})$. For a point of coordinates $(r, s, N, t)$ we write

$$M = ab^{-1} = \begin{pmatrix} t & w' \\ w & S \end{pmatrix}, \quad g = \begin{pmatrix} Q & -I \\ I & 0 \end{pmatrix}, \quad Q = \begin{pmatrix} x & e' \\ e & R \end{pmatrix}$$

Then one has $g(t) = x - \frac{\det S}{\det M}$ with $w = r - ts$ and $S = N + tss'$. With some elementary algebra one checks $g'(t) = [\frac{\det(N + rs')}{\det M}]^2$ and the formula follows from the above considerations. ∎

We remark that the classical double integral for a Lyapunov exponent reduces to a single integral with respect to the invariant measure. If $\mu$ is associated to a product of Cauchy laws on $\mathbb{R}$ it is possible to obtain an explicit formula for the sum of the nonnegative exponents using Proposition IV.6.6 (See [208]). Unfortunately the computation using Proposition IV.6.7 seems to be more complicated.

## IV.7 Problems

One can find a complete solution of most of the following problems in [33].

**Problem IV.7.1** *Let $U(n) = g(n)\dots g(1)$ for a sequence of products of matrices $g(n) \in GL(2, \mathbb{R})$. We assume that there exist constants $C$ and $\delta$ with*

$$\lim_{n\to\infty} \frac{1}{n} \log|\det(U(n))| = \delta \qquad \max(\|U(n)^{-1}\| \ , \ \|U(n)\|) \le C^n$$

*For a non zero vector $v \in \mathbb{R}^2$ we define the Lyapunov function*

$$L(v) = \limsup_{n\to\infty} \frac{1}{n} \log \|U(n)v\|$$

*Prove that*

1. *$L(v)$ is a finite number and $L(v_1 + v_2) \le \max\big(L(v_1)\ ,\ L(v_2)\big)$.*
2. *The function $L$ has a two points range $\{\lambda_1, \lambda_2\}$ (with possibly $\lambda_1 = \lambda_2$) and $\lambda_1 + \lambda_2 \ge \delta$*
3. *If $\lambda_1 + \lambda_2 = \delta$ then the* lim sup *in the definition of the function $L$ is actually a limit.*

**Problem IV.7.2** *Let $U(n) = g(n)\dots g(1)$ for a sequence of products of matrices $g(n) \in GL(2, \mathbb{R})$. We assume that:*

$$\begin{aligned} \lim_{n\to\infty} \frac{1}{n} \log\|g_n\| &= 0 \\ \lim_{n\to\infty} \frac{1}{n} \log|a_1(U(n))| &= \gamma_1 \\ \lim_{n\to\infty} \frac{1}{n} \log|a_2(U(n))| &= \gamma_2 \end{aligned}$$

1. *Check that the hypothesis of Problem IV.7.1 are satisfied and that we have the relations $\lambda_1 = \gamma_1$ and $\gamma_2 \le \lambda_2 \le \gamma_1$.*
2. *Conclude that if $\gamma_1 = \gamma_2$ then $\lim n^{-1} \log \|U(n)v\| = \gamma_1 = \gamma_2$ for any vector $v \ne 0$.*
3. *We now suppose that $\gamma_1 > \gamma_2$. Let us denote by $a_{1,n} = a_1(U(n))$ and $a_{2,n} = a_2(U(n))$ be the squareroots of the eigenvalues of the matrix $U'(n)U(n)$ and let $v_{1,n}$ , $v_{2,n}$ be corresponding unit eigenvectors (with polar angles in $[0\ ,\pi[$). We will denote by $w_n$ the projection of $v_{2,n}$ on $v_{1,n+1}$.*
   *Prove the following properties:*
   (a) $\|U(n+1)v_{2,n}\| \ge a_{1,n+1}\|w_n\|$

(b) $\|w_n\| \leq \|g(n+1)\| a_{2,n} a_{1,n+1}^{-1}$

(c) $\|v_{2,n+1} - v_{2,n}\| \leq 2\|w_n\|$

(d) $v_{2,n}$ *converges as* $n \to \infty$ *to a vector* $v_2$ *such that* $L(v_2) = \gamma_2$

(e) *Conclude that* $n^{-1} \log \|U(n)v\|$ *converges to* $\gamma_2$ *if the vector* $v$ *is proportional to* $v_2$ *and to* $\gamma_1$ *if not.*

**Problem IV.7.3** *The goal of this problem is to show that without any irreducibility assumption on* $\mu$ *it is possible to recover the upper Lyapunov exponent as an ergodic limit. This trick has been used by Furstenberg-Kesten in [116].*

*Let* $B$ *be the compact space of* $d \times d$ *matrices of norm one. The group* $G = GL(d, \mathbb{R})$ *acts on* $B$ *by* $g.M = \frac{gM}{\|gM\|}$ *(Here we put a dot to distinguish the action of* $G$ *on* $B$ *from the ordinary product of matrices).*

1. *Let* $\nu$ *be an invariant probability measure on* $B$ *and* $\tilde{\mathbb{P}} = \mathbb{P} \otimes \nu$. *Using Proposition IV.3.4 prove that* $n^{-1} \log \|U(n)M\|$ *converges* $\tilde{\mathbb{P}}$*-a.e to an* $\tilde{\mathbb{P}}$*-integrable random variable* $Z$.
2. *Let* $I$ *be the identity matrix of order* $d$ *and* $\nu$ *be any weak limit point of the sequence* $n^{-1} \sum_{k=1}^{n} \mu^n * \delta_I$. *Check that* $\nu$ *is invariant and that* $M \hookrightarrow \int \log(g.M)\, d\mu(g)$ *is a continuous function and prove that* $\mathbb{E}\{Z\} = \gamma_1$.
3. *Deduce from the inequality* $Z \leq \lim n^{-1} \log \|U(n)\|$ *that* $Z = \gamma_1$ $\tilde{\mathbb{P}}$*-a.e.*

**Problem IV.7.4** *For* $\lambda > 1$ *one defines the matrices* $A$ *and* $B$ *of* $SL(2, \mathbb{R})$ *by:*

$$A = \begin{pmatrix} \lambda & 0 \\ 0 & 1/\lambda \end{pmatrix} \qquad B = \begin{pmatrix} 0 & -1 \\ 1 & 0 \end{pmatrix}$$

*Let* $\mu$ *be the probability measure on* $SL(2, \mathbb{R})$ *defined by* $\mu(A) = p$, $\mu(B) = 1 - p$, *for some* $p \in (0\ ,\ 1)$.

1. *Prove that* $\mu$ *is irreducible.*
2. *Prove that* $\nu = \frac{1}{2}(\delta_{\bar{e}_1} + \delta_{\bar{e}_2})$ *is the unique invariant probability measure (where* $\{e_1, e_2\}$ *is the canonical basis of* $\mathbb{R}^2$*).*
3. *Prove that* $\gamma_1 = 0$.
4. *Prove that in the limit case* $p = 1$ *we have* $\gamma_1 = \log \lambda \neq 0$.

**Problem IV.7.5** *The following example is taken from Cohen-Newman [57]. It gives one of the very few examples for which it is possible to compute $\gamma_1$. It assumes that $\mu$ is a probability measure on $GL(d, \mathbb{R})$ such that the law of $\|gv\|$ is the same for any unit vector $v \in \mathbb{R}^d$.*

1. *Prove that for a fixed $v$ the $Y(n) = \log \|U(n)v\| - \log \|U(n-1)v\|$ are independent and identically distributed random variables.*
2. *Use the classical law of large numbers and Proposition IV.4.6 to obtain $\gamma_1 = \mathbb{E}\{Y(1)\}$.*

**Problem IV.7.6** *Let $z = x+iy$ be a complex number with a positive imaginary part and let $C_z$ be the Cauchy measure of density $t \hookrightarrow \frac{1}{\pi}\frac{y}{(t-x)^2+y^2}$. We define $\mu$ to be the image of $C_z$ under the mapping $t \hookrightarrow \begin{pmatrix} t & -1 \\ 1 & 0 \end{pmatrix}$*

1. *Let $a$ and $b$ be two complex numbers such that the function $z \hookrightarrow az+b$ never vanishes for $\Im m z < 0$. Prove that:*

$$\int \log |at+b| \, dC_z(t) = \log |a\bar{z}+b|$$

2. *Prove that $C_a * C_b = C_{a+b}$ and that $gC_z = C_{gz}$ whenever $g \in SL(2, \mathbb{R})$. Here $z \hookrightarrow gz$ is the homographic action of $g$ on the complex plane.*
3. *Let $u$ be the solution of the equation $u + u^{-1} = z$ with positive imaginary part. Prove that $C_u$ is invariant (Actually the invariant measure is unique)*
4. *Prove that $\gamma_1 = \log |u|$*

**Problem IV.7.7** *Let us assume that the probability distribution $\mu$ on the group $GL(d, \mathbb{R})$ is strongly irreducible and contractive. Then $\mu$ admits a unique invariant probability distribution $\nu$ on the projective space $P(\mathbb{R}^d)$ which is a continuous measure. Prove that $\nu$ is of pure type i.e. $\nu$ is either absolutely continuous or singular continuous.*
**Hint:** *if $\nu = \nu_{a.c.} + \nu_{s.}$ is the Lebesgue decomposition of $\nu$ prove that $\mu * \nu_{a.c.}$ is absolutely continuous. Remarking that the inequality $\mu * \nu_{a.c.} \leq \nu_{a.c.}$ is actually an equality conclude that if $\nu_{a.c.}$ is nonzero then $\nu = \nu_{a.c.}$.*

**Problem IV.7.8** *Prove that if $G_\mu$ is a non compact subgroup of $SL(2,\mathbb{R})$ then $\mu$ is strongly irreducible if and only if for any non zero vector $v$ in $\mathbb{R}^2$ the set $\{g\bar{v}\,;\, g \in G_\mu\}$ has more than two elements.*

**Problem IV.7.9** *The purpose of this problem is to show that the positivity of the exponent $\gamma_1$ can be proved using the spectral properties of the Fourier-Laplace transform introduced in section 3. Let*

$$r(g,\bar{v}) = \frac{dg^{-1}m}{dm}(\bar{v}) = \left(\frac{||v||}{||gv||}\right)^d$$

*be the Radon-Nikodym cocycle associated to the Cauchy measure $m$ on $P(\mathbb{R}^d)$ and let $Q = T_{r,1/2}$ be the associated Laplace operator.*

1. *Prove that $Q$ is a bounded operator on the Hilbert space $L^2(m)$ and that $||Q|| \leq 1$.*
2. *Prove that $\log ||Q^n|| \geq -(d/2)\int ||g||\, d\mu^n$ and conclude that the logarithm of the spectral radius of $Q$ acting on $L^2(m)$ is greater than $-(d/2)\gamma_1$.*
3. *It is known that this spectral radius is equal to 1 if and only if the group $G_\mu$ is amenable (See Guivarc'h-Derrienic [82]). Conclude that if $G_\mu$ is non amenable then $\gamma_1$ is strictly positive.*

**Problem IV.7.10** *If $U$ is an open subset of $SL(d,\mathbb{R})$ then there exists an integer $m$ such that $U^m$ contains a matrix with $d$ eigenvalues of distinct moduli.*

1. *Prove this property for $d = 2$. Hint: The only case of interest deals with $U$ contained in the subset of elliptic matrices. In this case one remarks that one can find a rational rotation in $U$.*
2. *Using a Jordan decomposition extend this property to all $d$.*

**Problem IV.7.11** *One wants to prove that if $U$ is an open subset of the group $SP(\ell,\mathbb{R})$ then there exists an integer $m$ such that $U^m$ contains a matrix with $d$ eigenvalues of distinct moduli.*

1. *Prove that a symplectic matrix has a Jordan decomposition in the symplectic group $SP(\ell,\mathbb{R})$*
2. *Use the same proof than in Problem IV.7.10 to conclude.*

**Problem IV.7.12** *Let* $g(n) = \begin{pmatrix} V(n) & 1 \\ -1 & 0 \end{pmatrix}$ *be a sequence of Schrödinger matrices associated to a real i.i.d. sequence* $\{V(n)\,;\, n \in \mathbb{Z}\}$ *such that* $\log(1 + |V(0)|$ *is integrable, and let us denote by* $\gamma$ *the upper Lyapunov exponent. Assuming that the potentials are not constant we know from Proposition IV.4.26 that one has:*

$$\gamma = \int_{\mathbb{R}} \log(|t|)\, d\tilde{\nu}(t)$$

*where* $\tilde{\nu}$ *is the projection of the invariant measure* $\nu$ *on the real line.*
*Let* $\nu_{\boldsymbol{T}}$ *be the image measure of* $\tilde{\nu}$ *on the torus* $[-\pi/2, \pi/2]$, *by the mapping* $t = \tan(\varphi)$ *and define the function*

$$g(s, \vartheta) = \int_{\boldsymbol{T}} \exp\big(is \tan(\varphi + \vartheta)\big)\, d\nu_{\boldsymbol{T}}(\varphi)$$

1. *Prove that one can write:*

$$\gamma = \int_{\boldsymbol{T}} \log|\tan(\varphi)|\, d\nu_{\boldsymbol{T}}(\varphi) = \int_{-\pi/2}^{0} \left( \int_{\boldsymbol{T}} \tan(\varphi + \vartheta))\, d\nu_{\boldsymbol{T}}(\varphi) \right) d\vartheta$$

2. *Using the invariance equation, prove that* $g(s, 0) = h(s) g(s, -\pi/2)$ *where* $h$ *is the Fourier transform of the distribution of potentials.*
3. *Assuming the existence of a second order moment for the probability distribution* $\tilde{\nu}$ *prove that* $g$ *satisfies the partial differential equation:*

$$\frac{\partial^2 g}{\partial s^2} = g + \frac{i}{s} \frac{\partial g}{\partial \vartheta}$$

4. *Assume moreover that the Fourier transform of* $\tilde{\nu}$ *vanishes at infinity. Multiplying the above indentity by the complex conjugate of* $g$ *say* $g^*$, *substracting the complex conjugated expression and integrating over* $s \in [0, \infty)$ *prove that one has:*

$$\begin{aligned} \frac{d}{d\vartheta} \int_0^{\infty} |g(s, \vartheta)|^2\, \frac{ds}{s} &= 2ig(0, \vartheta) \frac{\partial g^*}{\partial s}(0, \vartheta) \\ &= -2 \int_{\boldsymbol{T}} \tan(\varphi + \vartheta)\, d\nu_{\boldsymbol{T}} \end{aligned}$$

5. *Conclude that under the above hypothesis one has:*

$$\begin{aligned} 2\gamma &= \int_0^{\infty} \left( |g(s, -\pi/2)|^2 - |g(s, 0)|^2 \right) \frac{ds}{s} \\ &= \int_0^{\infty} |H(s)|^2 (1 - |h(s)|^2)\, \frac{ds}{s} \end{aligned}$$

*where* $H(s) = g(s, -\pi/2) = \int_{\mathbb{R}} \exp(-is/u)\, d\tilde{\nu}(u)$. *Under this last form, the positivity of the Lyapunov exponent is easily checked.*

## IV.8 Notes and Comments

The theory of products of random matrices initiated by Bellman, has been fully developed by P.Bougerol, H. Furstenberg, Y. Guivarc'h, H. Kesten, E. Le Page and A. Raugi. We have tried to find the shortest way to obtain the essential results needed in the study of the spectral theory of one and quasi-one dimensional Schrödinger operators. Most of the results and proofs for the i.i.d. case can also be found in the book of P. Bougerol and J. Lacroix [33].

The subadditive ergodic theorem is used in conjunction with the deterministic Oseledec's theorem to prove the existence of Lyapunov exponents. Some other approaches to the random Oseledec's theorem are available, in particular M. Raghunatan [270] gave a self-contained proof of this fundamental theorem. We followed Ledrappier [212] to show that this theorem yields immediately the almost sure growth of the norm of the column vectors of the product matrix with an exponential speed given by the upper Lyapunov exponent. The same method also yields a representation formula involving some invariant measure on the projective space. An earlier approach due to Furstenberg and Kesten [116] and Furstenberg [115], based on a refinement of the classical ergodic theorem for Markov chains give the same results. These properties also follow from the basic work of Guivarc'h and Raugi [137] which we will present below. We chose the first method since it can be applied to the general case of Markovian products with minor changes.

The theory of Fourier-Laplace operators associated to a Markov chain goes back to Doeblin and Fortet and has been developed in the framework of random walks by the Russian school, see Nagaev, Virtser [327] and Tutubalin [323], [324] in example. These authors essentially assume some Doeblin type condition insuring an exponential speed of convergence of iterates of the Markov kernel to the invariant distribution of probability. The spectral theory of the Laplace operators on the space of continuous functions, assuming that the probability distribution has a density, is classical and more information can be found for example in Lacroix [208]. Brunel and Revuz [38] have studied the same question in the case of a "spread-out" distribution and have established the close connexion with the notion of quasi-compact operators. Derrienic and Guivarc'h have obtained in [82] a very powerful result, relating the spectral radius of the operator $T_1$ acting on $L^2(m)$ ($m$ is the Cauchy measure on the projective space) to the amenability of the group generated by the support of the distribution of probability. The theory of the Laplace transform on spaces of Hölder continuous functions is essentially borrowed from Le Page [214]. The fundamental result is the exponential convergence on this space of the iterates of the Markov kernel to the invariant distribution without any assumption of regularity on the distribution of the matrices.

The presentation of the theory of i.i.d. products in Section 3 is entirely due to Guivarc'h and Raugi [137]. We only have removed the algebraic framework of their general theory of random products with values in semi-simple Lie groups since we will only use subgroups of $SL(d, \mathbb{R})$ and essentially the symplectic group $SP(\ell, \mathbb{R})$. The simplicity of the Lyapunov spectrum was proved earlier by Virtser in [328], assuming the existence of a density for the distribution of the matrices. The proof of the celebrated theorem of Furstenberg [115] insuring the positivity of the upper Lyapunov exponent (or equivalently the non-triviality of the spectrum) is only sketched since we prove a much stronger result in this direction (but also with stronger assumptions...). We have not tried to study in details the properties of the invariant measure. In particular the proof of the integrability of the function $\vartheta \hookrightarrow |\sin(\vartheta - \vartheta_0)|^\alpha$ for some negative $\alpha$ with respect to the invariant measure $\nu(d\vartheta)$ on the torus $\mathbb{T}$ is out of the scope of this book. This property is very commonly used to obtain a simple representation formula for the Lyapunov exponent in the Schrödinger case, and also to apply the ergodic theorem to the $\nu$ integrable function $\log(|x|)$ (considering now the invariant measure on the projective line). The interested reader may found this proof (and much more) in the work of Guivarc'h and Raugi cited above. The criterion of strong irreducibility and contractivity using the notion of Zariski's closure of the group generated by the distribution of the matrices, given by Goldsheid and Margulis in [128], was considerably extended by Guivarc'h and Raugi in [138]. In particular they give a complete description of the multiplicity of the Lyapunov spectrum. The proof of the absolute continuity of some power of convolution of the probability distribution on the symplectic group $SP(\ell, \mathbb{R})$ induced by a Schrödinger matrix is borrowed from Lacroix [207]. Some similar computations were performed by Le Page in [215] for a group equal to the product of a finite number of copies of $SL(2, \mathbb{R})$. There is an announced proof of the criterion given by Goldsheid and Margulis in [128] for the particular case of Schrödinger matrices on a strip. Indeed, this criterion asserts that the Zariski closure of the set of Schrödinger matrices in a strip generated in example by independent Bernouilli distributions in the diagonal of the $\ell \times \ell$ dimensional matrix $Q$ is equal to $SP(\ell, \mathbb{R})$.

The section on Markovian multiplicative systems was strongly influenced by the work of Bougerol [31], [32]. The positivity of the upper Lyapunov exponent in this situation was already investigated by Royer [277] and a general method to prove the simplicity of the spectrum has been given by Guivarc'h in [135]. A lot of interesting extensions of the theory of the semigroups generated by the range of a Markov process have been studied by Arnold, Wishtutz, Crauel, Klieman, Kunita etc...in connexion with control theory.

The last section deals with the boundaries of the symplectic group which is a particular case of the general theory of boundaries of semi-simple Lie groups initiated by Furstenberg. A lot of details can be found in Chapter IV of Bougerol & Lacroix [33]. Our construction of the Poisson kernels of the Lagragian boundaries is borrowed from Lacroix [209] and the key ideas can be found in the basic paper of Furstenberg and Tzkoni [118].

It is worth mentioning that the material presented here is only a small part of the theory of random walks on matrix groups. In particular we said nothing about the reducible case studied by Hennion [144], Kifer [169], Furstenberg and Kifer [117] and the case of nonnegative matrices (see Furstenberg and Kesten [116]). The theory of Fourier-Laplace transform is a basic tool in the study of central limit theorems, large deviations etc..The interested reader can find a lot of information in Le Page [214], Bougerol and Lacroix [33] Chapter V, Bougerol [31], Guivarc'h and Raugi [136]. In this last paper the theory of random products is studied under the fairly general assumptions of proximality (generalizing the notion of contractivity) and strong irreducibility. Moreover the classical limit theorems of probability theory can be obtained in the general framework of irreducible Markov kernels satisfying some kind of Doeblin condition.

In some particular cases, explicit computations of the Lyapunov exponents have been carried out. For example by Cohen and Newman in [57] for matrices with normal entries. Also, the case of i.i.d. Schrödinger matrices associated to a Bernouilli distribution supported by the two points 0 and $\beta$ with $\mathbb{P}\{0\} = p$, $0 < p < 1$ have been studied in [230] by F. Martinelli and L. Micheli. See Chapter VI for details.

# V

# ERGODIC FAMILIES OF SELF-ADJOINT OPERATORS

Contents:



Throughout this chapter $(\Omega, \mathcal{F})$ is a fixed measurable space and $\mathcal{H}$ is a separable complex Hilbert space on which all the operators to be considered are defined. We want to investigate the measurability properties of operator valued functions of the form:

$$\omega \hookrightarrow H(\omega)$$

from $\Omega$ into the space of operators on $\mathcal{H}$. We are mainly interested in self-adjoint operators and, as we explained in Chapter I, mostly in are unbounded ones. With this in mind we choose a notion of measurability of operator valued functions which relies on the functional calculus of self-adjoint operators and on the properties of bounded functions of these possibly unbounded operators. We believe that all the technical difficulties relative to these measurability problems have been carefully swept under the rug in most of the research litterature. This is one of the reasons why we decided to study these problems thoroughly.

The second part of the chapter deals with the various spectra of the operators $H(\omega)$ when an ergodic flow acts on $(\Omega, \mathcal{F})$ and when the function $H(\omega)$ is stationary in a natural sense. We also pay a special attention to the particular case of the random Schrödinger operators which are the main object of investigation of this book.

Finally, we discuss some regularity properties of the Lyapunov exponents of one dimensional and quasi-one dimensional Schrödinger operators. The theory of products of random matrices developed in the previous chapter is applied to the investigation of the dependence of these exponents upon the energy parameter.

## V.1 Measurability Concepts

The Hilbert space $\mathcal{H}$ being separable, its Borel $\sigma$-field $\mathcal{B}_{\mathcal{H}}$ is equal to the smallest $\sigma$-field for which all the functions $\mathcal{H} \ni f \to < f, g > \in \mathbb{C}$ for $g \in \mathcal{H}$ are measurable. Hence, a function $\Omega \ni \omega \to f(\omega) \in \mathcal{H}$ is measurable (for the Borel $\sigma$-field $\mathcal{B}_{\mathcal{H}}$) if and only if all the functions

$$\Omega \ni \omega \to < f(\omega), g > \in \mathbb{C}$$

are measurable. Consequently, for a function $\omega \hookrightarrow T(\omega)$ in the space $L(\mathcal{H})$ of bounded linear operators on $\mathcal{H}$, the following properties are equivalent:

**(i)** $\forall f \in \mathcal{H}$, $\forall g \in \mathcal{H}$, $\Omega \in \omega \to < f, T(\omega)g > \in \mathbb{C}$ is measurable,

**(ii)** $\forall f \in \mathcal{H}$, $\Omega \ni \omega \to T(\omega)f \in \mathcal{H}$ is measurable,

**(iii)** $\Omega \ni \omega \to T(\omega) \in L(\mathcal{H})$ is measurable for the Borel $\sigma$-field $\mathcal{B}_{L(\mathcal{H})}$ of $L(\mathcal{H})$ endowed with the uniform norm.

See Problem V.5.1 for details of the proof.

**Definition V.1.1** *We shall say that a function $\Omega \ni \omega \to T(\omega) \in L(\mathcal{H})$ is measurable if it satisfies any one of the above three properties.*

The functional calculus for self-adjoint operators and especially the following proposition makes the extension of this definition to unbounded operators possible.

**Proposition V.1.2** *If for each $\omega \in \Omega$, $H(\omega)$ is a self-adjoint operator on $\mathcal{H}$ and $\{E(A,\omega);\ A \in \mathcal{B}_{\mathbb{R}}\}$ denotes its resolution of the identity, the following three properties are equivalent:*

**(i)** $\Omega \ni \omega \to E(A,\omega) \in L(\mathcal{H})$ *is measurable for all* $A \in \mathcal{B}_{\mathbb{R}}$,
**(ii)** $\Omega \ni \omega \to e^{itH(\omega)} \in L(\mathcal{H})$ *is measurable for all* $t \in \mathbb{R}$,
**(iii)** $\Omega \ni \omega \to [H(\omega) - zI]^{-1} \in L(\mathcal{H})$ *is measurable for all* $z \in \mathbb{C} \setminus \mathbb{R}$.

Notice that a simple monotone class argument shows that i) is in fact equivalent to the measurability of the function $\Omega \ni \omega \to E(A,\omega) \in L(\mathcal{H})$ for all Borel sets $A$ of the form $(-\infty, a]$ for $a \in \mathbb{R}$, or even for $a$ in any dense subset of $\mathbb{R}$

*Proof:*
For each $\omega \in \Omega$, $f$ and $g$ in $\mathcal{H}$ and $t \in \mathbb{R}$ the functional calculus gives:

$$< f, e^{itH(\omega)} g >= \int_{\mathbb{R}} e^{it\lambda} < E(d\lambda,\omega) f, g > \tag{V.1}$$

and i) $\Rightarrow$ ii) follows by approximating the right hand side by integrals of step functions. If $\lambda$ is real and if $\Im m\, z < 0$, we have:

$$\frac{1}{\lambda - z} = i^{-1} \int_0^\infty e^{i\lambda t} e^{-izt} dt$$

(and the corresponding formula with $-i$ instead of $i$ if $\Im m\, z > 0$). Then, the functional calculus gives, for each $\omega \in \Omega$:

$$< f, [H(\omega) - zI]^{-1}, g >= i^{-1} \int_0^\infty e^{izt} < f, e^{-itH(\omega)} g > dt \tag{V.2}$$

(and the corresponding formula in the case $\Im m\, z < 0$). This proves ii) $\Rightarrow$ iii). Let us now assume that $a$ and $b$ are fixed real numbers such that $-\infty < a < b < +\infty$ and for each $a > 0$ let us define the function $F_{a,b,\alpha}$ by:

$$F_{a,b,\alpha}(\lambda) = \frac{1}{2i\pi} \int_a^b [\frac{1}{t - i\alpha - \lambda} - \frac{1}{t + i\alpha - \lambda}] dt.$$

If we let $\alpha$ tend to 0, then, $F_{a,b,\alpha}(\lambda)$ converges to 0 if $\lambda \notin (a,b)$, to 1/2 if $\lambda = a$ or $\lambda = b$ and to 1 if $\lambda \in (a,b)$. Consequently, $\lim_{\alpha \to 0} F_{a-\beta,b+\beta,\alpha}(\lambda)$ converges monotonically to the characteristic function of the interval $(a,b)$

when $\beta$ tends to 0. Since $F_{a,b,\alpha}(\lambda)$ is uniformly bounded as $\alpha \searrow 0$ we conclude that (using again the functional calculus for self-adjoint operators):

$$< E((a,b),\omega)f,g >= \lim_{\beta\searrow 0}\lim_{\alpha\searrow 0}\frac{1}{2i\pi}$$
$$\int_{a-\beta}^{b+\beta} < f,([(t-i\alpha)I - H(\omega)]^{-1}$$
$$- [(t+i\alpha)I - H(\omega)]^{-1})g > dt \qquad \text{(V.3)}$$

which proves the measurability of $\omega \hookrightarrow E(A,\omega)$ for $A=(a,b)$. The general form of iii) $\Rightarrow$ i) is then obtained via a monotone class argument. ∎

This proposition justifies the following choice for the definition of measurability for (possibly unbounded) operator valued functions.

**Definition V.1.3** *A function $\omega \hookrightarrow H(\omega)$ with values in the set of self adjoint operators on $\mathcal{H}$ is said to be measurable if any of the three equivalent properties of the above proposition is satisfied.*

Eventhough very elementary, the following criteria will be of practical importance in the sequel.

**Proposition V.1.4** *Let $\{H_n; n \geq 1\}$ be a sequence of functions from $\Omega$ into the set of self-adjoint operators on $\mathcal{H}$ and let $H$ be another such function. If the sequence $\{H_n(\omega); n \geq 1\}$ converges to $H(\omega)$ in the strong resolvent sense for each $\omega \in \Omega$ and if each function $H_n(\omega)$ is measurable, then the function $H(\omega)$ is also measurable.*

*Proof:*
For each $z \in \mathbb{C}\setminus\mathbb{R}$, $\omega\in\Omega$ and $f$ and $g$ in $\mathcal{H}$ we have (recall the definition of the strong resolvent convergence):

$$\lim_{n\to\infty} < f,[H_n(\omega)-zI]^{-1}g > = < f,[H(\omega)-zI]^{-1}g >$$

which shows that the right hand side is a measurable function of $\omega$ as limit of measurable functions. ∎

We stressed in Chapter I the importance of the Lebesgue decomposition of a self-adjoint operator and of its resolution of the identity. We now address the following crucial problem:

- Does the Lebesgue decomposition of self-adjoint operators preserve the measurability?

Because of our definition of the measurability this problem can be rewritten in the following simpler say: does the Lebesgue decomposition of a measurable resolution of the identity give measurable components? The answer has to be positive, without the shadow of a doubt, but in order to check the mathematical details we need some definitions and an elementary lemma from classical measure theory.

We will denote by $\mathcal{B}_{\mathcal{M}}$ the smallest $\sigma$-field of subsets of the space $\mathcal{M}(\mathbb{R})$ of complex (finite) measures on $\mathbb{R}$, for which all the functions $\mathcal{M}(\mathbb{R}) \ni \mu \hookrightarrow \mu(A) \in \mathbb{C}$ are measurable. With this definition, for each $f$ and $g$ in $\mathcal{H}$ the function:

$$\omega \hookrightarrow < f, E(\cdot, \omega)g >$$

from $\Omega$ into $\mathcal{M}(\mathbb{R})$ is measurable whenever $E(\cdot, \omega)$ is the resolution of the identity of a measurable self-adjoint operator $H(\omega)$. The nice thing about our choice for this $\sigma$-field is that it gives the measurability of the components of the Lebesgue decomposition of scalar measures. Indeed:

**Lemma V.1.5** *The three functions $\mu \to \mu_{pp}, \mu \to \mu_{ac}$ and $\mu \to \mu_{sc}$ from $\mathcal{M}(\mathbb{R})$ into itself are measurable with respect to the $\sigma$-field $\mathcal{B}_{\mathcal{M}}$.*

*Proof:*
The sum of two measures and the product of a measure by a scalar are obviously measurable operations. Consequently, it is enough to show that $\mu \hookrightarrow \mu_s = \mu_{pp} + \mu_{sc}$ and $\mu \hookrightarrow \mu_c = \mu_{ac} + \mu_{sc}$ are measurable. Moreover we can first decompose $\mu$ into the sum of its real part and its imaginary part and then decompose each of them into positive and negative parts. All these operations are measurable so that we can restrict ourselves to nonnegative measures $\mu$, and show the measurability of the restrictions of the functions defined in the statement of the present lemma to the subspace $\mathcal{M}_+(\mathbb{R})$ of nonnegative measures.
Let $\mathcal{I}$ be the set of all finite unions of real intervals with rational endpoints. $\mathcal{I}$ is countable and generates $\mathcal{B}_{\mathbb{R}}$. Also, it is easy to check that for each $A$ in $\mathcal{B}_{\mathbb{R}}$ we have:

$$\mu_s(A) = \lim_{n\to\infty} \sup_{I\in\mathcal{I}, |I|<1/n} \mu(A \cap I)$$

where we used the notation $|\cdot|$ for the Lebesgue measure. This proves the measurability of the function $\mu \hookrightarrow \mu_s(A)$ for all $A \in \mathcal{B}_{\mathbb{R}}$, and hence of $\mu \hookrightarrow \mu_s$.

To prove the measurability of $\mu \hookrightarrow \mu_c$ it suffices to show the measurability of the function $\mu \longrightarrow \mu_c(A)$ for each fixed Borel set $A$ of the form $A = [a, b)$ with $-\infty < a < b < +\infty$. The latter follows from:

$$\mu_c(A) = \lim_{\delta \searrow 0} f_\delta(\mu)$$

where the limit is monotone nonincreasing and where $f_\delta(\mu)$ is defined as the monotone limit of the nondecreasing family of $f_{\delta,n}$ where:

$$f_{\delta,n}(\mu) = \sum_{0 \leq k < 2^n - 1} \mu(I_{n,k}) \mathbf{1}_{[0,\delta)}(\mu(I_{n,k}))$$

and

$$I_{n,k} = [a + k2^{-n}(b-a), a + (k+1)2^{-n}(b-a)]$$

for $k = 0, 1, \cdots, 2^n - 1$. This completes the proof because the measurability of $f_{\delta,n}(\mu)$ is obvious. ∎

We now set:

**Definition V.1.6** *We say that a subresolution $E$ of the identity of $\mathcal{H}$ is measurable on $(\Omega, \mathcal{F})$ if for each $\omega \in \Omega$, $E(\omega)$ is a subresolution of the identity of $\mathcal{H}$ on $(I\!R, \mathcal{B}_{I\!R})$ such that the function:*

$$\Omega \ni \omega \longrightarrow < E(\omega)f, g > \in \mathcal{M}(I\!R)$$

*is measurable for all $f$ and $g$ in $\mathcal{H}$.*

Hence, Definition V.1.3 simply says that a function $\omega \hookrightarrow H(\omega)$ with values in the set of self-adjoint operators on $\mathcal{H}$ is measurable if and only if the corresponding resolution of the identity is measurable in the sense of the above definition. Moreover, Lemma V.1.5 gives:

**Proposition V.1.7** *If $\omega \hookrightarrow E(\omega)$ is a measurable resolution of the identity of $\mathcal{H}$ and if for each $\omega \in \Omega$ we denote by $E_{pp}(\omega), E_{ac}(\omega)$ and $E_{sc}(\omega)$ the components of its Lebesgue decomposition then $E_{pp}, E_{ac}$ and $E_{sc}$ are measurable subresolutions of the identity of $\mathcal{H}$.*

# V.2 Spectra of Ergodic Families

This section deals with the study of the various spectra of measurable self-adjoint operators $H(\omega)$ defined on an ergodic dynamical system. From now on $(\Omega, \mathcal{F}, \mathbb{P})$ is a fixed complete probability space and $T$ is a subsemigroup of the additive group $\mathbb{R}^d$. The reader can think of $T$ as of $\mathbb{R}^d$ itself, or $\mathbb{Z}^d$, or else, $\mathbb{R}_+$ or $\mathbb{Z}_+$ in the one dimensional case. Moreover we assume that a fixed ergodic family $\{\theta_x; x \in T\})$ of measure preserving transformations has also been chosen. $\mathcal{H}$ will again be a fixed separable Hilbert space and $\{\mathcal{T}_x; x \in T\}$ will be assumed to be a unitary representation of $T$ on $\mathcal{H}$ in the following sense: $\mathcal{T}_0 = I$ and for each $x \in T$, $\mathcal{T}_x$ is a unitary operator on $\mathcal{H}$ such that:

$$\mathcal{T}_{x+y} = \mathcal{T}_x \mathcal{T}_y$$

for all $x$ and $y$ in $T$. We will say that the representation is *total* if the set $F$ of all the $f \in \mathcal{H}$ for which there exists an infinite subset $\{x_k(f); k \geq 1\}$ of $T$ such that

$$< \mathcal{T}_{x_k(f)} f, \mathcal{T}_{x_j(f)} f >= \delta_{j,k} \qquad j,k \geq 1 \tag{V.4}$$

is total in $\mathcal{H}$. The classical example (which is at the origin of this definition) is $\mathcal{T}_x f = f(\cdot - x)$ on $\mathcal{H} = L^2(\mathbb{R}^d, dx)$. In this case $F$ contains the set of normalized functions with compact support. Finally, we assume that:

> $\{\theta_x; x \in T\}$ is a fixed ergodic subsemigroup of the group of all the automorphisms of the measurable space $(\Omega, \mathcal{F})$ which preserve the probability measure $\mathbb{P}$.

This ergodicity assumption gives measure 0 or 1 to all the elements of $\mathcal{F}$ left invariant, up to $\mathbb{P}$-null sets, by all the $\theta_x$'s for $x$ in $T$.

Now that the decor is set, we can start the investigations of the various spectra. The first result we prove is very elementary but it will turn out to be very useful.

**Lemma V.2.1** *If $\omega \hookrightarrow E(\omega)$ is a measurable function in the set of orthogonal projections of $H$ which satisfies:*

$$E(\theta_x \omega) = \mathcal{T}_x^* E(\omega) \mathcal{T}_x \tag{V.5}$$

*for all $\omega \in \Omega$ and $x \in T$ then there exists a (possibly infinite) nonnegative integer $r$ such that for $\mathbb{P}$-almost all $\omega \in \Omega$ the rank of $E(\omega)$ is equal to $r$.*

*Moreover, if the representation $\{\mathcal{T}_x;\, x \in T\}$ is total, we must have $r = 0$ or $r = \infty$, and in fact:*

$$r = 0 \quad \Leftrightarrow \quad \forall f \in F,\ \mathbb{E}\{< Ef, f >\} = 0 \quad \Leftrightarrow \quad E = 0 \quad \mathbb{P} - a.s.$$

*and consequently:*

$$r = \infty \quad \Leftrightarrow \quad \forall f \in F, \quad \mathbb{E}\{< Ef, f >\} > 0.$$

As usual we use the notation $\mathbb{E}\{\cdot\}$ for the expectation with respect to the probability measure $\mathbb{P}$, and a.s. as an abbreviation for almost surely.

*Proof:*
Let $\{f_k; k \geq 1\}$ be an orthonormal basis of $\mathcal{H}$. For each $\omega \in \Omega$ we set:

$$r(\omega) = \text{tr } E(\omega) = \sum_{k\geq 1} < E(\omega) f_k, f_k > .$$

$r$ is a nonnegative random variable and for each $x \in T$ we have:

$$\begin{aligned} r(\theta_x\omega) &= \sum_{k\geq 1} < E(\theta_x\omega) f_k, f_k > \\ &= \sum_{k\geq 1} < E(\omega)\mathcal{T}_x f_k, \mathcal{T}_x f_k > \\ &= \text{tr } E(\omega) \\ &= r(\omega) \end{aligned}$$

where we used the assumption (V.5) and the fact that $\mathcal{T}_x$ is unitary. The ergodicity implies that $r$ is a constant $\mathbb{P}$-a.s. Let $f \in F$ and let $\{x_k(f); k \geq 1\}$ satisfy (V.4). Since $\{\mathcal{T}_{x_k(f)} f; k \geq 1\}$ is an orthonormal system in $\mathcal{H}$, for each $\omega \in \Omega$ we have:

$$r = r(\omega) \geq \sum_{k\geq 1} < E(\omega)\mathcal{T}_{x_k(f)} f_k, \mathcal{T}_{x_k(f)} f_k >= \sum_{k\geq 1} < E(\theta_{x_k(f)}\omega) f, f >,$$

and taking expectations of both sides we obtain:

$$r \geq \sum_{k\geq 1} \mathbb{E}\{< Ef, f >\} = \begin{cases} 0 & \text{if } \mathbb{E}\{< Ef, f >\} = 0 \\ \infty & \text{if } \mathbb{E}\{< Ef, f >\} > 0. \end{cases}$$

If $r < \infty$, for each $f \in F$ we must have $\mathbf{E}\{< Ef, f >\} = 0$, and hence $< Ef, f >= 0$ $\mathbb{P}$-a.s. Since we can assume without loss of generality that

$F$ is countable, we can interchange the"for each $f \in F$" and the "$\mathbb{P}$-a.s.". So, for $\mathbb{P}$-almost all $\omega \in \Omega$ we have:

$$\forall f \in F, \quad < E(\omega)f, f >= 0$$

and hence:

$$\forall f \in \mathcal{H}, \quad < E(\omega)f, f >= 0$$

since $F$ is total. This gives $E(\omega) = 0$ by polarization and hence $r = 0$. This also completes the proof. ∎

The first application of the above lemma is the following.

**Proposition V.2.2** *Let $\omega \hookrightarrow E(\omega)$ be a measurable subresolution of the identity such that (V.5) is satisfied by the function $\omega \hookrightarrow E(A,\omega)$ for each $A$ in $\mathcal{B}_{\mathbb{R}}$. Then there exists a closed subset $\Sigma$ of $\mathbb{R}$ such that the topological support of the projection valued measure $E(\cdot,\omega)$ is $\Sigma$ for $\mathbb{P}$-almost all $\omega \in \Omega$.*

*Proof:*
Recall that the topological support of $E(\omega)$ is the set of $\lambda \in \mathbb{R}$ for which each neighborhood $A$ of $\lambda$ is such that $E(A,\omega) \neq 0$, or equivalently, $\text{tr}E(A,\omega) \neq 0$. But since this trace is independent of $\omega$ we may think that this set of $\lambda$'s is also independent of $\omega$. Let us prove this fact. For each rational numbers $\lambda$' and $\lambda$" such that $\lambda < \lambda''$ we let $\Omega_{\lambda',\lambda''}$ be a set in $\mathcal{F}$ of full $\mathbb{P}$-measure and $r_{\lambda',\lambda''}$ a nonnegative (possibly infinite) integer such that:

$$\forall \omega \in \Omega_{\lambda',\lambda''}, \quad \text{tr } E((\lambda'',\lambda''),\omega) = r_{\lambda',\lambda''}.$$

We then set:

$$\Sigma = \{\lambda \in \mathbb{R}; \forall \lambda',\lambda'' \in \mathbb{Q},\ \lambda' < \lambda < \lambda'' \Rightarrow r_{\lambda',\lambda''} \neq 0\}$$

and:

$$\Omega' = \bigcap_{\substack{\lambda',\lambda'' \in \mathbb{Q} \\ \lambda' < \lambda''}} \Omega_{\lambda',\lambda''}$$

Then, $\Omega' \in \mathcal{F}$, $\mathbb{P}(\Omega') = 1$ and supp $E(\omega) = \Sigma$ for all $\omega \in \Omega'$. ∎

We now discuss an easy consequence of the above result. In order to state and prove it we need an extra assumption:

- $\Omega$ is assumed to be a Polish space, i.e. a separable metrizable complete topological space, $\mathcal{F}$ is its Borel $\sigma$-field and the function $\omega \hookrightarrow E(\omega)$ is continuous in the following sense: the measures $< E(\cdot,\omega_n)f, f >$ converge weakly to the measure $< E(\cdot,\omega)f,f >$ whenever $\omega_n$ converges to $\omega \in \Omega$ and $f \in \mathcal{H}$.

Note that the measurability of $\omega \hookrightarrow E(\omega)$ follows from this stronger assumption. The following proposition is elementary. It shows that the sets $\Sigma$'s are nondecreasing functions of the topological support of the probability measures $\mathbb{P}$. More precisely, let $\mathbb{P}_1$ and $\mathbb{P}_2$ be two probability measures on $(\Omega, \mathcal{F})$ which are invariant and ergodic for the action of $\{\theta_x; x \in T\}$. Proposition V.2.2 gives the existence of two closed subsets $\Sigma_1$ and $\Sigma_2$ of $\mathbb{R}$ such that the topological support of $E(\cdot,\omega)$ (which we will denote $\text{supp}E(\cdot,\omega)$) is equal to $\Sigma_1$ for $\mathbb{P}_1$-almost all $\omega \in \Omega$ and equal to $\Sigma_2$ for $\mathbb{P}_2$-almost all $\omega \in \Omega$.

**Corollary V.2.3** *Under the above assumptions we have:*

$$\text{supp}\ \mathbb{P}_1 \subset \text{supp}\mathbb{P}_2 \quad \Rightarrow \quad \Sigma_1 \subset \Sigma_2$$

*Proof:*
As we just noticed, according to Proposition V.2.2 we have the existence of $\Omega_j \in \mathcal{F}$ such that $\mathbb{P}_j(\Omega_j) = 1$ and such that $\text{supp}E(\cdot,\omega) = \Sigma_j$ for all the $\omega$'s in $\Omega_j$ for $j = 1,2$. Let $\omega$ be chosen in $\Omega_1 \cap \text{supp}\ \mathbb{P}_1$ which is not empty because it has $\mathbb{P}_1$-measure one. We have $\text{supp}\ E(\cdot,\omega) = \Sigma_1$ but also $\omega \in \text{supp}\ \mathbb{P}_2$ and hence $\omega \in \overline{\Omega_2}$ since the topological support of $\mathbb{P}_2$ is the smallest closed set with $\mathbb{P}_2$-measure one. Hence, there exists a sequence $\{\omega_n; n \geq 1\}$ in $\Omega_2$ which converges to $\omega \in \Omega$. Our continuity assumption implies the convergence of the measure $< E(\cdot,\omega_n)f,f >$ to $< E(\cdot,\omega)f,f >$ for each $f \in \mathcal{H}$. If $A$ is an open subset of $\mathbb{R}$ contained in the complement of $\Sigma_2$, we must have $< E(A,\omega_n)f,f >= 0$ for all $n \geq 1$ since $\omega_n \in \Omega_2$, and hence $< E(A,\omega)f,f >= 0$ in the limit. Since $f$ was arbitrary, this proves that $E(\cdot,\omega)$ vanishes on the complement of $\Sigma_2$ and consequently that its topological support is contained in $\Sigma_2$. ∎

We now apply the above results to the particular case of the Lebesgue decomposition of random self-adjoint operators.

**Proposition V.2.4** *Let $\omega \hookrightarrow H(\omega)$ be a measurable function in the set of self-adjoint operators on $\mathcal{H}$ and let us assume that for each $A \in \mathcal{B}_{\mathbb{R}}$ the* *function* $\omega \hookrightarrow E(A,\omega)$ *satisfies (V.5) in the sense that:*

$$E(A,\theta_x\omega) = \mathcal{T}_x^* E(A,\omega)\mathcal{T}_x. \tag{V.6}$$

*Then, there exist closed subsets* $\Sigma$, $\Sigma_{pp}$, $\Sigma_{ac}$ *and* $\Sigma_{sc}$ *of* $\mathbb{R}$ *such that for* $\mathbb{P}$-*almost all* $\omega \in \Omega$ *we have* $\Sigma(H(\omega)) = \Sigma$, $\Sigma_{pp}(H(\omega)) = \Sigma_{pp}$, $\Sigma_{ac}(H(\omega)) = \Sigma_{ac}$ *and* $\Sigma_{sc}(H(\omega)) = \Sigma_{sc}$.

*Proof:*
For each $\omega \in \Omega$ the definitions of the subresolutions of the identity $E_{pp}(\cdot,\omega)$, $E_{ac}(\cdot,\omega)$ and $E_{sc}(\cdot,\omega)$, and the uniqueness of the Lebesgue decomposition of measures on $(\mathbb{R},\mathcal{B}_{\mathbb{R}})$ imply that the functions $\omega \hookrightarrow E(\cdot,\omega)$, $\omega \hookrightarrow E_{pp}(\cdot,\omega)$ and $\omega \hookrightarrow E_{ac}(\cdot,\omega)$ satisfy the assumptions of Proposition V.2.2.
∎

**Remark V.2.5** *The results and the proofs of Propositions V.2.2 and V.2.4 above also give the existence of a closed subset* $\Sigma_c$ *of* $\mathbb{R}$ *and of two subsets* $\Sigma_{ess}$ *and* $\Sigma_{disc}$ *of* $\mathbb{R}$, *the union of which is* $\Sigma$, *and such that we have* $\mathbb{P}$-*almost surely:*

$$\Sigma_c(H(\omega)) = \Sigma_c, \quad \Sigma_{ess}(H(\omega)) = \Sigma_{ess} \quad \text{and} \quad \Sigma_{disc}(H(\omega)) = \Sigma_{disc}$$

**Remark V.2.6** *Using formulas (V.1), (V.1) and (V.3) in the same way as in the proof of Proposition V.1.2 one easily checks that for a function* $\omega \hookrightarrow H(\omega)$ *in the set of self-adjoint operators on* $\mathcal{H}$ *the following properties are equivalent:*

**i)** *for each* $A \in \mathcal{B}_{\mathbb{R}}$ *the function* $\omega \hookrightarrow E(A,\omega)$ *is measurable and satisfies*

$$E(A,\theta_x\omega) = \mathcal{T}_x^* E(A,\omega)\mathcal{T}_x$$

*for all* $x \in T$.

**ii)** *for each* $t \in \mathbb{R}$ *the function* $\omega \hookrightarrow e^{itH(\omega)}$ *is measurable and satisfies*

$$e^{itH(\theta_x\omega)} = \mathcal{T}_x^* e^{itH(\omega)}\mathcal{T}_x \tag{V.7}$$

*for all* $x \in T$

**iii)** *for each* $z \in \mathbb{C}\setminus\mathbb{R}$ *the function* $\omega \hookrightarrow [H(\omega) - zI]^{-1}\mathcal{T}_x$

$$[H(\theta_x\omega) - zI]^{-1} = \mathcal{T}_x^*[H(\omega) - zI]^{-1}\mathcal{T}_x \tag{V.8}$$

*for all* $x \in T$.

When either one of the above properties is satisfied we say that the function $\omega \hookrightarrow H(\omega)$ (which is measurable according to Definition II.2) satisfies:

$$H(\theta_x\omega) = \mathcal{T}_x^* h(\omega)\mathcal{T}_x \tag{V.9}$$

for all $x \in T$. The property (V.9) can also be checked directly. Indeed:

**Proposition V.2.7** *Let $D$ be a dense subspace of $\mathcal{H}$ which is a common core for all the operators $H(\omega)$ and such that:*

$$H(\theta_x\omega)f = \mathcal{T}_x^* H(\omega)\mathcal{T}_x f$$

*for all $f \in D$, $x \in T$ and $\omega \in \Omega$. Under these conditions the function $\omega \to H(\omega)$ satisfies (V.9)*

*Proof:*
Let us fix $z \in \mathbb{C} \setminus \mathbb{R}$, $x \in T$ and $\omega \in \Omega$ and let us check property iii) of Remark V.2.6 above. If $f$ belongs to the domain of $H(\theta_x\omega)$ there exists a sequence $\{f_n; n \geq 1\}$ in $D$ which converges to $f$ and such that $H(\theta_x\omega)f_n$ converges to $H(\theta_x\omega)f$ as $n \to \infty$. Consequently we must have:

$$\begin{aligned}
&\mathcal{T}_x^*[H(\omega) - zI]^{-1}\mathcal{T}_x[H(\theta_x\omega) - zI]f \\
&= \lim_{n\to\infty} \mathcal{T}_x^*[H(\omega) - zI]^{-1}\mathcal{T}_x[zI - H(\theta_x\omega)]f_n \\
&= \lim_{n\to\infty} \mathcal{T}_x^*[H(\omega) - zI]^{-1}[H(\omega) - zI]\mathcal{T}_x f \\
&= f. \blacksquare
\end{aligned}$$

**Proposition V.2.8** *Let us assume that the function $\omega \hookrightarrow H(\omega)$ in the set of self-adjoint operators on $\mathcal{H}$ is measurable and satisfies (V.9) and that the representation $\{\mathcal{T}_x; x \in T\}$ is total. Then:*

**i)** $\Sigma_{disc} = \phi$

**ii)** *If the spectral multiplicity of $H(\omega)$ is $\mathbb{P}$-almost surely finite, then, for each $\lambda \in \mathbb{R}$, $\lambda$ is $\mathbb{P}$-almost surely not an eigenvalue of $H(\omega)$.*

*Proof:*
Both i) and ii) are consequences of the following remark. When $\lambda \in \mathbb{R}$ is fixed, the rank of the projection $E(\{\lambda\}, \omega)$ is finite either because of the definition of $\Sigma_{disc}$ or because of the assumption on the spectral multiplicity of $H(\omega)$. Consequently, this rank has to be zero according to Lemma V.2.1.
$\blacksquare$

**Remark V.2.9** *We will see later that the above proposition applies to one dimensional ergodic Schrödinger operators. Nevertheless, ii) should not be understood as the absence of eigenvalues, since as we will see in Chapter VIII, the spectrum is expected to be (and actually is very often) pure point. ii) simply means that the eigenvalues are very much dependent on $\omega$ and when the latter varies, they "move so fast" that a fixed real (nonrandom) number does not see them almost surely.*

The above result, and especially its part ii) raises the question of the spectral multiplicity of the random self-adjoint operators. It seems that many early investigators in the field of random Schrödinger operators thought that under very mild conditions (for example the usual stationarity and ergodicity assumptions and some assumption insuring some actual randomness in the problem) the spectral multiplicity of the random Schrödinger operators should be almost surely equal to 1. More ambitiously one could conjecture that:

- for each $f$ in $\mathcal{H}$ there exists a set $\Omega_f \in \mathcal{F}$ with probability 1 and such that $f$ is a cyclic vector for the operator $H(\omega)$ for all $\omega$ in $\Omega_f$.

This conjecture is formulated without any reference to Schrödinger operators and as such it cannot be true for general stationary ergodic self-adjoint operators satisfying (V.9).
In fact, it is commonly believed that the pure point spectrum only should be of spectral multiplicity 1 (i.e., the spectral multiplicity of $H_{pp}(\omega)$ should be almost surely equal to 1) while the absolutely continuous spectrum should be almost surely of infinite multiplicity (i.e., the spectral multiplicity of $H_{ac}(\omega)$ should almost surely be infinite) as soon as the dimension as strictly greater than 1.

*None of these conjectures has been settled in general!*

We now investigate the essential support of our stationary (i.e., satisfying (V.9)) random self-adjoint operators. First we prove a technical lemma.

**Lemma V.2.10** *Let $\mathcal{F}'$ be the Borel $\sigma$-field of a Polish space $\Omega'$ and let $\mu'$ be a nonnegative finite measure on $(\Omega', \mathcal{F}')$. Then, the functions:*

$$\mathcal{M}_1(\Omega', \mathcal{F}') \ni \mu \hookrightarrow d\mu_{ac}/d\mu' \in L^1(\Omega', \mathcal{F}', \mu')$$

*and*

$$L^1(\Omega', \mathcal{F}', \mu') \ni f \hookrightarrow \mathbf{1}_{\{f \neq 0\}} \in L^1(\Omega', \mathcal{F}', \mu')$$

*are measurable. $\mathcal{M}_1(\Omega', \mathcal{F}')$ still denotes the set of probability measures and $\mu_{ac}$ the absolute continuous component of $\mu$ in the Lebesgue decomposition of $\mu$ with respect to $\mu'$.*

*Proof:*
The function $L^1(\Omega', \mathcal{F}', \mu') \ni f \rightarrow f \cdot \mu' \in \mathcal{M}(\Omega', \mathcal{F}')$ defined by $[f \cdot \mu'](A) = \int_A f d\mu'$ for $A \in \mathcal{F}'$ is measurable because for each $A \in \mathcal{F}'$ the function $f \rightarrow [f \cdot \mu'](A)$ is continuous, hence measurable. Moreover, it is one-to-one. Since $L^1(\Omega', \mathcal{F}', \mu')$ and $\mathcal{M}(\Omega', \mathcal{F}')$ are Polish spaces, its range is a Borel subset of $\mathcal{M}(\Omega', \mathcal{F}')$ and the inverse function is also measurable. If one composes the latter with the function $\mu \hookrightarrow \mu_{ac}$ (which is easily seen to be measurable by a proof along the same lines as in Lemma V.1.5) the first claim of the lemma follows.
To prove the second claim, for each integer $n \geq 1$, let us first consider the continuous even function $f_n$ on $\mathbb{R}$ defined by $f_n(t) = 1$ if $t > 1/n$, $f_n(0) = 0$, and which is linear between 0 and $1/n$, and second the function $F_n$ defined by:

$$L^1(\Omega', \mathcal{F}', \mu') \ni f \rightarrow F_n(f) = f_n \circ f \in L^1(\Omega', \mathcal{F}', \mu').$$

We then notice that:

$$||F_n(f) - F_n(g)||_{L^1} \leq n||f - g||_{L^1}$$

which proves that $F_n$ is continuous and hence measurable. Furthermore, if $f \in L^1(\Omega', \mathcal{F}', \mu')$ we have:

$$||F_n(f) - \mathbf{1}_{\{f \neq 0\}}||_{L^1} \leq \int_{\{0<|f|<1/n\}} |f| d\mu'$$

which tends to zero as $n$ tends to infinity. This shows that the function $\mathbf{1}_{\{f \neq 0\}}$ is the pointwise limit of the $F_n$. Its measurabilaity follows. ∎

**Proposition V.2.11** *Let $\omega \hookrightarrow E(\omega)$ be a measurable subresolution of the identity of $\mathcal{H}$ which satisfies part i) of Remark V.2.6 and let $\mu$ be a $\sigma$-finite nonegative measure on $\mathbb{R}$. Then there exists a Borel subset $\Sigma_\mu$ of $\mathbb{R}$ such that for each $\omega \in \Omega$ the $\mu$-essential support of the projection valued measure $E(\omega)$ is equal to $\Sigma_\mu$ up to sets dof $\mu$-measure 0.*

The result above and the proof below apply as well to the case of a measurable projection valued measure $E(\omega)$ on a Polish space $(\Omega', \mathcal{F}')$. Note also that since the $\mu$-essential support of $E(\omega)$ depends only on the sets of

$\mu$-measure 0, we can assume without loss of generality that the measure $\mu$ is actually finite.

*Proof:*
Let $\{f_k; k \geq 1\}$ be an orthonormal basis for $\mathcal{H}$ and for each $\omega \in \Omega$ let us define the probability measure $\mu^\omega$ by:

$$\mu^\omega = \sum_{1 \leq k < \infty} 2^{-k} < E(\cdot, \omega) f_k, f_k > .$$

Now, if $x \in T$ we have:

$$\mu^{\theta_x \omega} = \sum_{1 \leq k < \infty} 2^{-k} < E(\cdot, \omega) \mathcal{T}_x f_k, \mathcal{T}_x f_k > .$$

This shows that $\mu^{\theta_x \omega}$ can be obtained from $E(\cdot, \omega)$ and $\{\mathcal{T}_x f_k; k \geq 1\}$ in the same way as $\mu^\omega$ is obtained from $E(\cdot, \omega)$ and $\{f_k; k \geq 1\}$. Consequently, $\mu^{\theta_x \omega}$ and $\mu^\omega$ have the same null sets (namely those of $E(\cdot, \omega)$), $\omega \to \mu^\omega$ is measurable because $\omega \to E(\cdot, \omega)$ is measurable and thus,

$$\Omega \ni \omega \to \mathbf{1}_{\{d(\mu^\omega)_{ac}/d\mu \neq 0\}} \in L^1(\mathbb{R}, d\mu)$$

is measurable because of the preceeding lemma. This function depends only on the null sets of the measure $\mu^\omega$ and, according to what we saw, it must be invariant and hence constant $\mathbb{P}$-almost surely. The proof is complete because $E(\cdot, \omega)$ and $\mu^\omega$ have the same $\mu$-essential support. ∎

The above proposition has been designed for the following application.

> *If $\omega \hookrightarrow H(\omega)$ is a measurable function in the set of self-adjoint operators on $H$ which satisfies (V.9), there exists a set $\Sigma_{Leb}$ defined up to a set of Lebesgue measure 0, which carries for $\mathbb{P}$-almost all $\omega \in \Omega$ the absolutely continuous component $E_{ac}(\cdot, \omega)$ of the resolution of the identity of $H(\omega)$.*

The absolute continuous spectrum of $H(\omega)$ is concentrated on $\Sigma_{Leb}$ in the sense that if $A \in \mathcal{B}_{\mathbb{R}}$ we must have:

$$A \cap \Sigma_{Leb} = \emptyset \quad \Rightarrow \quad E_{ac}(A, \omega) = 0.$$

We will see in Chapter VII a result analog to Corollary V.2.3 for the essential support of the absolutely continuous spectra of random Schrödinger operators in one dimension: these essential supports are nonincreasing functions of the topological support of the probability measures $\mathbb{P}$. It would be nice to know wether or not this result is general or if it actually depends on the special characteristics of the dimension 1.

We end this section with an improvement of Lemma V.2.1 which will apply to the case of the lattice models and to the continuous one dimensional and quasi one dimensional cases. We will assume that $\mathcal{H} = L^2(T, m)$ where $m$ is the Haar measure of $T$, and that $\mathcal{T}_x$ is given by $\mathcal{T}_x f = f(\cdot - x)$.

**Lemma V.2.12** *On the top of the assumptions of Lemma V.2.1 we assume that for each $\omega \in \Omega$ the projection $E(\omega)$ has a weak kernel $E(x, y, \omega)$ which is a jointly continuous function of $(x, y) \in T \times T$. Then:*

$$E(\cdot) = 0 \ \mathbb{P} - a.s \quad \Leftrightarrow \quad \mathbb{E}\{E(0,0,\cdot)\} = 0,$$

*and hence:*

$$\text{rank} E(\cdot) = \infty \ \mathbb{P} - a.s. \quad \Leftrightarrow \quad \mathbb{E}\{E(0,0,\cdot)\} > 0.$$

Recall that a bounded operator $K$ is said to have a *weak kernel* $K(x, y)$ if

$$< Kf, g >= \int \int K(x,y) f(y) \bar{g}(x) dy \, dx$$

for all continuous functions $f$ and $g$ with compact support.

*Proof:*
The kernel has to be nonnegative on the diagonal because of its continuity. Consequently,

$$\mathbb{E}\{E(0,0,\cdot)\} = 0 \quad \Leftrightarrow \quad E(0,0,\omega) = 0 \ \mathbb{P} - a.s. \text{ in } \omega \in \Omega \qquad \text{(V.10)}$$

which implies that:

$$[\forall k \geq 1, E(0,0,\theta_{x(k)}\omega) = 0] \ \mathbb{P} - a.s. \text{ in } \omega \in \Omega \qquad \text{(V.11)}$$

where $\{x(k); k \geq 1\}$ is any countable dense subset of $T$, and:

$$[\forall k \geq 1, \ E(x(k), x(k), \omega) = 0] \ \mathbb{P} - a.s. \text{ in } \omega \in \Omega \qquad \text{(V.12)}$$

which in turn implies that:

$$[\forall x \in T, E(x,x,\omega) = 0] \ \mathbb{P} - a.s. \ \text{ in } \omega \in \Omega, \qquad \text{(V.13)}$$

because of the continuity of the kernel. Using Schwarz inequality we obtain:

$$[\forall x, y \in T, E(x,y,\omega) = 0] \ \mathbb{P} - a.s. \text{ in } \omega \in \Omega$$

which implies that $E(\omega) = 0$ $\mathbb{P}$-a.s. in $\omega \in \Omega$, and the proof is complete because the converse is obvious. ∎

# V.3 The Case of Random Schrödinger Operators

The purpose of this section is to present the Schrödinger operators

$$H(\omega) = H_0 + V(\cdot, \omega) \tag{V.14}$$

with a potential function depending on a parameter $\omega$ running through a probability space. We first consider the continuous case of $H_0 = -\Delta$ on $L^2(\mathbb{R}^d, dx)$. We give sufficient conditions for the essential self-adjointness on $C_c^\infty(\mathbb{R}^d)$ and for the measurability in the sense given in the first part of this Chapter. We then illustrate the results on examples the spectral properties of which will be investigated in details later on. We remark that the corresponding problems are trivially solved in the discrete case of the lattice $\mathbb{Z}^d$. We end this section with a discussion of the determination of the almost sure spectrum as a set.

**Proposition V.3.1** *Let $\{V(x);\, x \in \mathbb{R}^d\}$ be a real valued measurable process on a complete probability space $(\Omega, \mathcal{F}, \mathbb{P})$, let us assume for each $\omega \in \Omega$ that the function $x \hookrightarrow V(x,\omega)$ is locally square integrable and that the operator (V.14) is essentially self-adjoint on $C_c^\infty(\mathbb{R}^d)$. Then the function $\omega \hookrightarrow H(\omega)$ is measurable.*

*Proof:*
Let us first assume that for each $\omega$, the path or the function:

$$\mathbb{R}^d \ni x \hookrightarrow V(x,\omega) \in \mathbb{R}$$

is bounded. For each $t \in \mathbb{R}$, the Trotter product formula gives:

$$< e^{itH(\omega)} f, g > = \lim_{n\to\infty} < [e^{i(t/n)H_0} e^{i(t/n)V(\cdot,\omega)}]^n f, g >$$

for all $f$ and $g$ in $\mathcal{H} = L^2(\mathbb{R}^d, dx)$. Using the measurability of the process $V$ one easily shows the measurability of the right hand side. This concludes the proof in the case of bounded potential functions. In the general case, for each integer $k \geq 1$, we define the potential process $V^{(k)} = \{V^{(k)}(x); x \in \mathbb{R}^d\}$ by;

$$V^{(k)}(x) = [V(x) \vee (-k)] \wedge k.$$

It has bounded paths and this implies that the function:

$$\omega \to H^{(k)}(\omega) = H_0 + V^{(k)}(\cdot, \omega)$$

is measurable. Since $D = C_c^\infty(\mathbb{R}^d)$ is common core for (almost) all the $H^{(k)}(\omega)$ and since, for each $f \in D$ we have:

$$H(\omega)f = H_0 f + V(\cdot, \omega)f = \lim_{k \to \infty} H_0 f + V^{(k)}(\cdot, \omega)f$$

in $\mathcal{H}$, we can conclude that $H^{(k)}(\omega)$ converges to $H(\omega)$ in the strong resolvent sense and the measurability of the function $\omega \hookrightarrow H(\omega)$ follows Proposition V.1.2. ∎

The above proof applies to the lattice case and gives:

- if $\{V(n);\, n \in \mathbb{Z}^d\}$ is any family of real valued random variables on a complete probability space $(\Omega, \mathcal{F}, \mathbb{P})$ then the function $\omega \hookrightarrow H(\omega) = H_0 + V(\cdot, \omega)$ in the space of self-adjoint operators on $\mathcal{H} = \ell^2(\mathbb{Z}^d)$ is measurable.

The following proposition gives a sufficient condition for the assumptions of the above proposition to be satisfied.

**Proposition V.3.2** *Let $\{V(x);\, x \in \mathbb{R}^d\}$ be a real valued measurable process on a complete probability space $(\Omega, \mathcal{F}, \mathbb{P})$ and let us assume that:*

$$\sup_{\mathbf{n} \in \mathbb{Z}^d} \mathbb{E}\{[\int_{C(\mathbf{n})} |V(x)|^p dx]^{r/p}\} < \infty \tag{V.15}$$

*for some $p > p(d)$ and some $r > dp/(2[p - p(d)])$ where $p(d) = 2$ if $d \leq 3$, $p(d) > 2$ if $d = 4$ and $p(d) = d/2$ if $d \geq 5$. Then, the operator $H(\omega) = H_0 + V(\cdot, \omega)$ is $\mathbb{P}$-almost surely essentially self-adjoint on $C_c^\infty(\mathbb{R}^d)$.*

Note that the supremum is not needed whenever the process $\{V(x);\, x \in \mathbb{R}^d\}$ is stationary, or at least $\mathbb{Z}^d$-stationary. This will be particularly useful in the applications.

*Proof:*
We check that the assumptions of Proposition II.2.4 are satisfied $\mathbb{P}$-almost surely. For each $x \in \mathbb{R}^d$ let $\mathbf{n}(x)$ be the unique element of $\mathbb{Z}^d$ for which $x \in C_{\mathbf{n}(x)}$. For each $\omega \in \Omega$ we define the functions $x \hookrightarrow V_1(x,\omega)$ and $x \hookrightarrow V_2(x,\omega)$ by:

$$V_1(x,\omega) = V(x,\omega)\mathbf{1}_{\{V(\cdot,\omega)\geq -|\mathbf{n}(\cdot)|^2\}}(x)$$

and

$$V_2(x,\omega) = V(x,\omega) - V_1(x,\omega).$$

Because of the Faris-Lavine criterion, the proof reduces to showing that $V_2(\cdot,\omega)$ is in $L^{p(d)}_{loc,unif}(\mathbb{R}^d, dx)$ for $\mathbb{P}$-almost all $\omega \in \Omega$.

$$\begin{aligned}
&[\int_{C(\mathbf{n})} |V_2(x)|^{p(d)} dx]^{1/p(d)} \\
&\quad = [\int_{C(\mathbf{n})} |V(x)|^{p(d)} \mathbf{1}_{\{|V(\cdot,\omega)|\geq|\mathbf{n}|^2\}}(x) dx]^{1/p(d)} \\
&\quad \leq [\int_{C(\mathbf{n})} |V(x)^p dx]^{1/p} [\int_{C(\mathbf{n})} \mathbf{1}_{\{|V(\cdot,\omega)|\geq|\mathbf{n}|^2\}}(x) dx]^{1/p(d)-1/p} \\
&\quad \leq [\int_{C(\mathbf{n})} |V(x)|^p dx]^{1/p(d)} |\mathbf{n}|^{-2p(1/p(d)-1/p)}
\end{aligned}$$

where we used $p > p(d)$ to obtain the second inequality. Consequently:

$$\begin{aligned}
&\mathbb{P}\{[\int_{C(\mathbf{n})} |V_2(x)|^{p(d)} dx]^{1/p(d)} > 1\} \\
&\qquad \leq \mathbb{P}\{[\int_{C(\mathbf{n})} |V(x)|^p\, dx]^{1/p(d)} > |\mathbf{n}|^{2p(1/p(d)-1/p)}\} \\
&\qquad \leq |\mathbf{n}|^{-2kp(1/p(d)-1/p)} \mathbb{E}\{[\int_{C(\mathbf{n})} |V(x)|^p\, dx]^{k/p(d)}\}
\end{aligned}$$

for any $k > 0$. We then choose $k = rp(d)/p$ and the exponent of $|\mathbf{n}|$ is smaller than $-d$. The right hand side is consequently the general term of a convergent series because of the assumption (V.15). We conclude by using the first Borel-Cantelli lemma. ∎

The following corollaries illustrate the usefulness of the above proposition. The proof of the first one of them is obvious and we omit it.

**Corollary V.3.3** *Let $\{V(x);\ x \in I\!R^d\}$ be a real valued measurable stochastic process on a complete probability space $(\Omega, \mathcal{F}, I\!P)$ and let us assume that either:*

**i)** *it is stationary and $I\!E\{|V(0)|^k\} < \infty$ for some $k > p(d) + d/2$,*
*or:*

**ii)** *$V(x) = \sum_{\mathbf{n}\in \mathbb{Z}^d} v_{\mathbf{n}} f(x-\mathbf{n})$ for some $f \in \ell^1(L^p(I\!R^d, dx))$ and a stationary sequence $\{v_{\mathbf{n}};\ \mathbf{n} \in \mathbb{Z}^d\}$ such that $I\!E\{|v_0|^r\} < \infty$ with $p$ and $r$ as in the above proposition.*

*Then, the operator $H(\omega) = H_0 + V(\cdot, \omega)$ is $I\!P$-almost surely essentially self-adjoint on $C_c^\infty(I\!R^d)$.*

**Corollary V.3.4** *Let us assume that:*

$$V(x,\omega) = \int_{I\!R^d} f(x-y)\mu(dy,\omega) \tag{V.16}$$

*for a measurable function $\omega \hookrightarrow \mu(\cdot,\omega)$ from $\Omega$ into $\mathcal{M}(I\!R^d)$ and for some $f \in \ell^1(L^p(\mathbf{R}^d, dx))$ and let us assume furthermore that:*

$$\sup_{\mathbf{n}\in \mathbb{Z}^d} I\!E\{|\mu(\cdot,\omega)|(C(\mathbf{n})))^r\} < \infty \tag{V.17}$$

*with $p$ and $r$ as in Proposition V.3.2. Then, the operator $H(\omega) = H_0 + V(\cdot,\omega)$ is $I\!P$-almost surely essentially self-adjoint on $C_c^\infty(I\!R^d)$.*

*Proof:*
We show that the stochastic process $\{V(x);\ x \in \mathbb{R}^d\}$ so defined satisfies the assumptions of Proposition V.3.2. Jensen's inequality gives:

$$\begin{aligned}
[\int_{C(\mathbf{n})} |V(x)|^p dx]^{1/p} &\le \int_{I\!R^d} [\int_{C(\mathbf{n})} |f(x-y)|^p dx]^{1/p} |\mu(\cdot,\omega)|(dy) \\
&= \sum_{\mathbf{m}\in \mathbb{Z}^d} [\int_{C(\mathbf{n})-C(\mathbf{m})} |f(x)|^p dx]^{1/p} |\mu(\cdot,\omega)|(C(\mathbf{m}))
\end{aligned}$$

and hence:

$$\begin{aligned}
&\mathbb{E}\{([\int_{C(\mathbf{n})} |V(x)|^p dx]^{1/p})^r\}^{1/r} \\
&\qquad \le \sum_{\mathbf{m}\in \mathbb{Z}^d} [\int_{C(\mathbf{n})-C(\mathbf{m})} |f(x)|^p dx]^{1/p} \mathbb{E}\{|\mu|(C(m))^r\}^{1/r}
\end{aligned}$$

$$\leq c \sum_{\mathbf{m}\in\mathbb{Z}^d} [\int_{C(\mathbf{n})-C(\mathbf{m})} |f(x)|^p dx]^{1/p}$$
$$\leq c\|f\|_{\ell^1(L^p(\mathbb{R}^d,dx))}$$

which proves the desired result since the above right hand side does not depend on $\mathbf{n}$. ∎

## V.3.1 Examples

We now present the particular random potentials which we will use most frequently in the sequel.

**0.** The first example we give is the original model proposed by Anderson. It deals with the discrete case of the lattice $\mathbb{Z}^d$, and the potential is simply an i.i.d. sequence $\{V(\mathbf{n}); \mathbf{n} \in \mathbb{Z}^d\}$ of real valued random variables. We will make several technical assumptions on the common distribution of the random variables $V(\mathbf{n})$. In particular we will often need the existence of a density. On the contrary, we will give a special attention to the case where the potential takes only finitely many values. The particular case $d = 1$ with a distribution concentrated on two points only will be called the Bernouilli model.

**1.** Let $v$ be an almost periodic function on $T$ where the set $T$ is either $\mathbb{R}$ or $\mathbb{Z}$ and let $\Omega$ be the closure in $C_b(T)$ of the relatively compact set $\{\mathcal{T}_t v; t \in T\}$. The shift extends to $\Omega$ and $\Omega$ becomes a compact metrizable group. Let $\mathcal{F}$ be its Borel $\sigma$-field and let $\mathbb{P}$ be the normalized Haar measure. Then $(\Omega, \mathcal{B}, \mathbb{P}, \{\mathcal{T}_t; t \in T\})$ is an ergodic dynamical system and $\{V(t); t \in T\}$ can be regarded as an ergodic stationary stochastic process on the probability space $(\Omega, \mathcal{F}, \mathbb{P})$ with bounded continuous paths and such that $\mathcal{T}_t V(0) = V(t)$.

**2.** Let $\{X_t; t \in T\}$ be a stationary Gaussian field. Here $T$ can be any of the sets $\mathbb{R}^d$ or $\mathbb{Z}^d$. In the continuous case of $T = \mathbb{R}^d$ we also assume that the process is continuous in quadratic mean. In any case we set:

$$V(t) = F(X_t) \tag{V.18}$$

for any measurable real valued function $F$ on $\mathbb{R}$. Then $\{V(t); t \in T\}$ is a (measurable when $T = \mathbb{R}^d$) stationary real valued process. These potential processes are ergodic whenever the spectral measure of the process $\{X_t; t \in T\}$ is continuous. See for example [131].

**3.** Let $\chi$ be a bounded measurable real valued function on $\mathbb{R}$ which vanishes outside the interval $[0, 1]$ and let $\{\xi_n; n \in \mathbb{Z}\}$ be an independent identically distributed (i.i.d. for short) sequence of real valued random variables on a complete probabality space $(\Omega, \mathcal{F}, \mathbb{P})$ and let us set:

$$V(t,\omega) = \sum_{n\in\mathbb{Z}} \xi_n(\omega)\chi(t-n) \tag{V.19}$$

for $\omega \in \Omega$ and $t \in \mathbb{R}$. $\chi$ can be regarded as a potential well and the potential function $V(\cdot,\omega)$ as a sequence of wells of the same type but with random depth $\xi_n(\omega)$.

4 Let $\{X_t; t \in \mathbb{R}\}$ be the stationary process of Brownian motion on a compact connected Riemannian manifold $M$ (i.e., the stationary Markov process whose infinitesimal generator is the Laplace Beltrami operator of the manifold), let $F : M \to [0,1]$ be a Morse function (we will assume that $\inf_{m\in M} F(m) = 0$ and $\sup_{m\in M} F(m) = 1$ for simplicity), and we define the random potential $\{V(t);\ t \in \mathbb{R}\}$ by formula (V.18) as before. This example may not have a very strong physical significance but it plays an important role since it is the first one for which a rigorous proof of the localization phenomenon was given. We will call it the model of the Russian school.

The above examples give stationary ergodic potential functions, but they are more or less random. In fact we gave them in increasing order of randomness. Indeed, the almost periodic functions (and especially the periodic ones) give deterministic processes while the last two examples give strongly mixing ones. The third example is interesting because its randomness can be varied by adjusting the covariance of the Gaussian process. Let us also remark that except for some of the cases from the third example, all of the potential functions presented so far can be regarded, via formula (V.18), as the images of simple Markov processes by a real valued function on the state space of the process. We will often say that these random potential functions are of the Markovian type. Notice also that the first and the third examples above have natural multidimensional generalizations which are of some physical interest while the two other ones are purely academic.

Finally, we dwell on a particular case which we will investigate in full detail in Chapter VI when we study the asymptotic behavior of the distribution of states.

5. Let $\mu$ be an infinitely divisible random measure with independent increments on $\mathbb{R}^d$. This is a random variable $\mu$ on some complete probability space $(\Omega, \mathcal{F}, \mathbb{P})$ with values in the space $\mathcal{M}_+(\mathbb{R}^d)$ of nonnegative measures on $\mathbb{R}^d$ such that for each integer $n$ there exist

similar random variables $\mu_1, \cdots, \mu_n$ such that $\mu$ and $\mu_1 + \cdots + \mu_n$ have the same distribution on $\mathcal{M}_+(\mathbb{R}^d)$ and such that for each finite collection $A_1, \cdots, A_m$ of mutually exclusive Borel subsets of $\mathbb{R}^d$ the nonnegative random variables $\mu(A_1), \cdots, \mu(A_m)$ are independent.

These random measures are characterized by their Laplace transforms which are given for each function $f$ in the space $C_c(\mathbb{R}^d)$ of continuous functions on $\mathbb{R}^d$ with compact support, by:

$$\mathbb{E}\{e^{-<f,\mu>}\} = \exp[- < f, \nu > + \int_{\mathbb{R}^d} [1 - e^{-tf(x)}] g(dt, dx)] \quad (V.20)$$

for some nonnegative measures $\nu$ on $\mathbb{R}^d$ and $g$ on $\mathbb{R}_+ \times \mathbb{R}^d$, where we used the notation $< \cdot, \cdot >$ for the duality between functions and measures, i.e.,

$$< f, \nu > = \int f(x) \nu(dx).$$

We will concentrate on the Poisson case which can be characterized by $\nu = 0$ and the fact that the measure $g$ is concentrated on $\{1\} \times \mathbb{R}^d$. In this case, $\mu$ is a point measure with probability one and formula (V.20) can be rewritten as:

$$\mathbb{E}\{e^{-<f,\mu>}\} = \exp[- \int_{\mathbb{R}^d} [1 - e^{-tf(x)}] m(dx)] \quad (V.21)$$

where $m$ is a (nonrandom) nonnegative measure on $\mathbb{R}^d$ called the intensity of the random measure $\mu$. Note that the measure $m$ can be obtained via the formula:

$$m(A) = \mathbb{E}\{\mu(A)\}$$

where $A$ is any Borel subset of $\mathbb{R}^d$. Note also that for each $A \in \mathcal{B}_{\mathbb{R}^d}$, $\mu(A)$ is a Poisson random variable with parameter $\mu(A)$. Hence, condition (V.17) can be rewritten as:

$$\sup_{\mathbf{n} \in \mathbb{Z}^d} e^{-m(C(\mathbf{n}))} \sum_{0 \leq k < \infty} \frac{k^r}{k!} m(C(\mathbf{n}))^k < \infty \quad (V.22)$$

Finally, we remark that the random measure $\mu$ is stationary (for the appropriate action of the translations of $\mathbb{R}^d$ if and only if the intensity measure $m$ is translation invariant. Consequently, we will concentrate on the case where $m$ is the usual Lebesgue measure to make sure that the potential process $\{V(x);\, x \in \mathbb{R}^d\}$ is stationary. It will also be ergodic because there is enough independence in Poisson measures. In this case the essential self-adjointness condition (V.22) is necessarily satisfied for it can be rewritten as:

$$\sum_{0 \le k < \infty} \frac{k^r}{k!} < \infty.$$

We now check the almost sure essential self-adjointness on $C_c^\infty(\mathbb{R}^d)$ (and hence the measurability according to Proposition V.1.2) of the continuous random Schrödinger operators constructed from the random potential functions we presented in the above examples.

The examples 1 and 4 are trivial since the sample paths of the potential function are bounded and hence, for each $\omega \in \Omega$, the operator $H(\omega)$ is self-adjoint on the domain of the free Hamiltonian $H_0$.
If the absolute value of the function $F$ in the third example is bounded above by an exponential function, then it is easy to see that for each $k > 0$ we have:

$$\mathbb{E}\{|V(0)|^k\} < \infty$$

and the desired essential self-adjointness follows.
The existence of the $r$-th moment for the common distribution of the $\xi_n$ gives the same estimate as above and one concludes in the same way.
Finally, if $\mu$ is a stationary Poisson random measure like in the fifth example, our essential self-adjointness conditions involve only the $r$-th moment of a Poisson random variable with parameter $|C(\mathbf{0})| = 1$, and consequently, the supremum is finite for all $r$'s. Hence these assumptions are satisfied whenever $f \in \ell^1(L^p(\mathbb{R}^d, dx))$ for some $p > p(d)$.

We close this section with a few results on the determination of the almost-sure spectrum $\Sigma$ as a set.

Let us first consider the case of a self-adjoint operator $H$ which is the sum of two self-adjoint operators $K_1$ and $K_2$ one of them, say $K_1$, being bounded. Adding a constant to $K_1$ and subtracting this constant from $K_2$ if necessary, one can assume without loss of generality that the spectrum of $K_1$ is symmetric about 0, i.e. that $\Sigma(K_1) = [-||K_1||, ||K_1||]$. Using ProblemI.7.3 from Chapter I one sees that every $\lambda \in \Sigma(H)$ satisfies $d(\lambda, \Sigma(K_2)) \le ||K_1||$ which shows that $\lambda \in \Sigma(K_1) + \Sigma(K_2)$.

We want to use the above result in the case of the ergodic Schrödinger operators. Recall that the spectrum of the operator of multiplication $M_V$

by a function $V$ is equal to the essential range of the function $V$. Moreover, if this function $V$ is the sample path of a stationary and ergodic measurable stochastic process $\{V(x);\, x \in T\}$ with $T = \mathbb{Z}^d$ in the discrete case of the lattice $\mathbb{Z}^d$ and $T = \mathbb{R}^d$ in the continuous case, then the almost sure spectrum is the topological support of the distribution at one site of the potential, i.e. the probability measure $\mu$ defined by:

$$\mu(A) = \mathbb{P}\{V(0) \in A\} \qquad A \in \mathcal{B}_{\mathbb{R}}$$

where $\mathcal{B}_{\mathbb{R}}$ denotes the Borel $\sigma$-field ot $\mathbb{R}$. See ProblemV.5.2.

a. in the discrete case of the lattice $\mathbb{Z}^d$ we can choose $K_1 = H_0$ and $K_2 = M_V = V$ and get:

$$\Sigma \subset [-2d, +2d] + \text{supp}(\mu) \tag{V.23}$$

b. in the continuous case of $\mathbb{R}^d$ and of a bounded potential we can choose $K_1 = M_V = V$ and $K_2 = H_0$ and get:

$$\Sigma \subset [\inf \text{supp}(\mu), \infty). \tag{V.24}$$

Note that a simple approximation argument can be used to show that this last inclusion is actually true without the assumption on the boundedness of the potential.

Now the problem is to show that the inclusion signs can in fact be replaced by equality signs in most of the (random) cases. In fact the reader is presumably aware of the fact that this cannot hold in full generality for periodic potentials give an ovbious counterexample to any attempt to prove equality in the general ergodic case. Some sort of independence assumption is needed in order to prove the desired equalities. In fact we shall find convenient to single out an assumption in terms of the topological support of the distribution of the process.

In order to avoid technical measurability problems we assume that, in the continuous case $T = \mathbb{R}^d$, the process $\{V(x);\, x \in T\}$ has continuous sample paths. In other words we assume that the potential is almost surely continuous.

With this in mind, we assume furthermore that the sample space $\Omega$ is the space of real valued continuous functions on $T$ endowed with the topology of the uniform convergence on the compact subsets of $T$ and we assume that $\mathcal{F}$ is the corresponding Borel $\sigma$-field. This terminology is certainly pompous in the discrete case of the lattice $T = \mathbb{Z}^d$ but it is nevertheless convenient to cover both cases at once. In this setting, the process $\{V(x);\ x \in T\}$ is simply chosen to be the coordinate process given by $V(x,\omega) = \omega(x)$, $\theta_x$ is the shift $\theta_x(\omega) = \omega(x + \cdot)$ and $\mathbb{P}$ is an ergodic invariant probability measure. We formulate our assumption in term of the topological support of $\mathbb{P}$.

Recall that we use the notation $\mu$ for the distribution of the potential at one site, i.e. the marginal of $\mathbb{P}$ defined by:

$$\mu(A) = \mathbb{P}\{V(0) \in A\}$$

for any Borel subset $A$ of the real line $\mathbb{R}$. We now define the set $\mathrm{Cst}(\mu)$ as the set of constant functions on $T$ which are equal to an element of the topological support of $\mu$, i.e.

$$\mathrm{Cst}(\mu) = \{f \in \Omega;\ \exists \beta \in \mathrm{supp}(\mu),\ f(T) = \{\beta\}\ \},$$

and our assumption is then formulated as:

$$\mathrm{Cst}(\mu) \subset \mathrm{supp}\mathbb{P}. \tag{V.25}$$

**Proposition V.3.5** *Under the assumption (V.25) one has:*

$$\Sigma = \Sigma(H_0) + \mathrm{supp}(\mu) \tag{V.26}$$

*$\mathbb{P}$-almost surely, or in other words:*

$$\Sigma = [-2d, +2d] + \mathrm{supp}(\mu) \tag{V.27}$$

*in the discrete case of the lattice $T = \mathbb{Z}^d$ and:*

$$\Sigma = [\inf \mathrm{supp}(\mu), +\infty) \tag{V.28}$$

*in the continuous case of $T = \mathbb{R}^d$.*

*Proof:*

Assumption (V.25) means that for every $\beta \in \mathrm{supp}(\mu)$, $\epsilon > 0$ and $L > 0$ one has:

$$\mathbb{P}\{\sup_{|x| \leq L} |V(x) - \beta| \leq \epsilon\} > 0,$$

and Poincaré's reccurrence lemma gives that:

$$\mathbb{P}\{\exists \text{ infinitely many boxes } \Lambda_{\epsilon,L} \text{ of size } L^d \\ \text{so that } \sup_{x \in \Lambda_{\epsilon,L}} |V(x) - \beta| \leq \epsilon\} = 1,$$

and using a countable set of $\epsilon$'s going to 0 and a countable set of $L$'s going to $\infty$ one gets:

$$\mathbb{P}\{\forall\epsilon > 0, \forall L > 0, \exists \text{ infinitely many boxes } \Lambda_{\epsilon,L} \\ \text{of size } L^d \text{ so that } \sup_{x\in\Lambda_{\epsilon,L}} |V(x) - \beta| \leq \epsilon\} = 1. \quad \text{(V.29)}$$

Let us now choose $\lambda = \alpha + \beta$ with $\alpha \in \Sigma(H_0)$ and $\beta \in \text{supp}(\mu)$. Using PropositionI.1.5 one gets the existence of a sequence $\{f_n;\, n \geq 1\}$ of unit vectors in $\mathcal{H}$ such that $||H_0 f_n - \alpha f_n|| \to 0$ as $n \to \infty$. In fact, it is possible to choose these $f_n$'s with compact support. In the case $T = \mathbb{Z}^d$ one uses the fact that the operator $H_0$ is bounded while one uses the fact that $\alpha \in \Sigma_{ess}$ in the continuous case $T = \mathbb{R}^d$. Notice that $||H_0 f_n - \alpha f_n||$ does not change if one replaces $f_n$ by any of its translates. Consequently, one can, with probability one, shift the functions $f_n$'s in such a way that their supports are contained in the boxes $\Lambda_{\epsilon,L}$ appearing in (V.29) in which case one can make sure that $||H_0 f_n + V f_n - (\alpha + \beta) f_n|| \to 0$ as $n \to \infty$. This shows that $\lambda = \alpha + \beta$ is in the spectrum of the operator $H(\omega)$ for all the $\omega$'s in a set of probability one and this completes the proof. ∎

It is important to notice that assumption (V.25) is always satisfied in the i.i.d. case when $T = \mathbb{Z}^d$. i.e. in the Anderson model, and that it is satisfied in some continuous cases such as the Russian school model.

# V.4 Regularity Properties of the Lyapunov Exponents

This section is devoted to the discussion of some general results on the existence and the smoothness of the Lyapunov exponents of one dimensional (and quasi one dimensional) Schrödinger operators with stationary ergodic potentials. The properties we present below could have been proved in Chapter IV had we added a section on product of random matrices depending on a parameter. But since most of the interesting results depend heavily on the fact that the matrices are of a *Schrödinger type* we decided to include them in the present chapter.

Let $\{H(\omega); \omega \in \Omega\}$ be an ergodic family of Schrödinger operators on the strip of width $\ell$. This means that there exist an invertible dynamical ergodic system $(\Omega, \mathcal{F}, \theta_t, \mathbb{P})$ and $\ell$ real valued random variables $\{V_i(\omega); i = 1 \ldots \ell\}$ such that $V_i(t, \omega) = V_i \circ \theta_t(\omega)$ for $t \in T$. We also make the usual assumptions about the existence of Lyapunov exponents that is:

- In the discrete case we assume that $\sum_{i=1}^{\ell} \mathbb{E}\{\log(|V_i| + 1)\} < +\infty$
- In the continuous case we assume the separability of the $\mathbb{R}^\ell$ valued process $\{V_i(t, w); i = 1 \ldots \ell, t \in T\}$ and that

$$\sum_{i=1}^{\ell} \mathbb{E}\{|V_i(0)| < +\infty$$

Then for a fixed value of the energy $\lambda$, the propagator $\{U_\lambda(t, \omega);\ t \in T\}$ is a regular invertible multiplicative process with values in $SP(\ell, \mathbb{R})$ in the sense of DefinitionIV.2.5. The random Oseledec theorem stated in PropositionIV.2.6 gives the existence of $\ell$ nonnegative Lyapunov exponents $\gamma_1(\lambda) \geq \ldots \geq \gamma_\ell(\lambda)$.

We know from Proposition V.2.2 that there exists a closed subset $\Sigma \subset \mathbb{R}$ such that for $\mathbb{P}$-almost all $\omega$ the spectrum of $H(\omega)$ is equal to $\Sigma$. Propositions III.2.2 in the one dimensional case or III.5.14 in the strip case immediately implies the following result :

**Proposition V.4.1** *Assuming the above integrability properties we have:*

$$\{\lambda\,;\ \gamma_1(\lambda) = 0\} \subset \Sigma$$

As we already pointed out, this result is certainly not optimal for the strip.

In the one dimensional independent case and in the model of the Russian scholl the Lyapunov exponent is given by an explicit formula only involving the invariant measure.

**Theorem V.4.2** *Let $H = H_0 + V$ be a one dimensional Schrödinger operator.*

1. *In the i.i.d. lattice case, assuming that the potentials are not constant, then the Lyapunov exponent is strictly positive. Moreover if the distribution of potential has a positive moment this exponent is equal to:*

$$\gamma(\lambda) = \int \log|t|\; d\tilde{\nu}_\lambda(t)$$

*where $\tilde{\nu}_\lambda$ is the restriction to the real line of the unique invariant measure $\nu_\lambda$ on the projective line.*

2. *In the Russian school model, i.e. for $V(t) = F(X(t))$ where $X(t)$ is the stationary Brownian motion on some compact connected Riemannian manifold $M$, the Lyapunov exponent is given by the formula:*

$$\gamma(\lambda) = -\frac{1}{2}\int_{M\times T}(\lambda - F(x) - 1)\sin 2\vartheta \, d\nu_\lambda(x,\vartheta)$$

*where $\nu_\lambda$ is the unique invariant measure on $M \times \mathbb{T}$ and $\gamma(\lambda)$ is strictly positive as soon as the function $F$ is not constant.*

*Proof:*
Part 1. is a direct consequence of Proposition IV.4.26.
To prove part 2. it is enough to use the formula:

$$r(t) = r(0)\exp\{-\frac{1}{2}\int_0^t (\lambda - F(X(s)) - 1)\sin 2\vartheta(s)\, ds\}$$

proved in Chapter III where $r(t)$ and $\vartheta(t)$ are the amplitude and the phase variables of the "Prüffer" transformation. The Birkhoff ergodic theorem added to the uniqueness of the invariant measure yields the result. ∎

In this section we will investigate the regularily properties of the exponents $\gamma_i(\lambda)$ viewed as functions of the parameter $\lambda$.

In the general ergodic case the subharmonicity of the partial sums is an easy consequence of the definition. In the independent or Markovian case we will prove that each function $\lambda \hookrightarrow \gamma_i(\lambda)$ $i = 1\ldots\ell$ is continuous under the classical assumptions of strong irreducibility and contractivity which guarantee the uniqueness of the invariant measure. Finally we prove, still in the independent case, local Hölder continuity and the smoothness of the upper Lyapunov exponent, under some extra conditions on the distribution of the potentials. For this last part we follow Le Page [215], [216].

## V.4.1 Subharmonicity

The notion of Lyapunov exponents can be extended to any complex value of the energy $\lambda$ if one defines them by the subadditive limits:

$$\gamma_1(\lambda) + \cdots + \gamma_p(\lambda) = \lim \frac{1}{n}\mathbb{E}\{\log ||\Lambda^p U_\lambda(n)||\} \qquad \text{for } p = 1\cdots\ell$$

We first state some classical definitions and properties related to subharmonic functions.

**Definition V.4.3** *A function $f : \mathbb{C} \hookrightarrow [-\infty, +\infty)$ is said to be subharmonic if:*

- *$f$ is upper semicontinuous that is : $f(x) \geq \limsup_{y \to x} f(y)$*
- *for any $\lambda \in \mathbb{C}$ and $r > 0$, $f(\lambda)$ is not greater than the mean value of $f$ on the circle of center $\lambda$ and radius $r$.*

We gather some essential properties of subharmonic functions in a lemma for convenience.

**Lemma V.4.4** *One has the following prperties:*

**i)** *If $f$ and $g$ are two subharmonic functions almost everywhere equal with respect to the Lebesgue measure of $\mathbb{R}^2$ then they are equal everywhere.*
**ii)** *If $f_n$ is a locally bounded below sequence of subharmonic functions then the pointwise infimum of this sequence is a subharmonic function.*
**iii)** *If $z \hookrightarrow M(z)$ is an holomorphic mapping with values in the set of $d \times d$ matrices, then $z \hookrightarrow \log ||M(z)||$ is a subharmonic function.*

*Proof:*
Integrating with respect to the radius of the circle appearing in the definition we see that the value of a subharmonic function $f(\lambda)$ is no greater than its mean value on a ball centered at the point $\lambda$. Taking into account the upper semicontinuity of $f$ if follows that $f(\lambda)$ is actually the limit of these mean value when the radius of the ball is going to zero. This proves the conclusion (i). Conclusion (ii) is a simple consequence of the Lebesgue's theorem since an upper semicontinuous function is bounded above on a compact set. The property (iii) is a classical result of analytic function theory. See for example the text by Hoffman [149]. ∎

**Proposition V.4.5** *For any $p = 1 \cdots \ell$ the partial sum $\lambda \hookrightarrow \gamma_1(\lambda) + \cdots + \gamma_p(\lambda)$ is a subharmonic function.*

*Proof:*
From the property iii) of the above lemma it is readily seen that $\lambda \hookrightarrow \log ||U_\lambda(n)||$ is subharmonic. The Fatou's lemma implies that the expectation of this function is also subharmonic. Then it remains only to use property ii) of the above lemma together with the fact that the sequence whose limit gives the partial sums of Lyapunov exponents is nonnegative and subadditive. ∎

This property will be very useful in order to prove the so called "Herbert-Jones-Thouless" formula which implies that the partial sum $\gamma_1(\lambda) + \cdots + \gamma_\ell(\lambda)$ is actually an analytic function of $\lambda$ outside the support of the integrated density of states.

A very nice example of application of subharmonicity in the theory of Lyapunov exponents have been given by M.Herman [147] in the case of the so called "almost Mathieu operator". Let $H(\omega)$ be the random Schrödinger operator on the one dimensional lattice defined by:

$$H(\omega)\psi(n) = -\psi(n-1) - \psi(n+1) + \beta\cos(2\pi n\alpha + \omega)\ \psi(n)$$

where $\alpha$ is an irrational number and $\beta$ a real parameter. This operator can be regarded as a random Schrödinger operator over the ergodic dynamical system $(\Omega, \mathcal{F}, \theta_t, \mathbb{P})$ where $\Omega$ is the torus $[0\ ,\ 2\pi]$, $\mathbb{P}$ is the Lebesgue measure and $\theta$ is the rotation of angle $\alpha$.

**Proposition V.4.6** *In the case of the almost Mathieu equation one has* $\gamma(\lambda) \geq \log(\frac{|\beta|}{2})$ *for any complex number* $\lambda$.

*Proof:*
For any complex number $z$ we set $z_n = ze^{2n\pi\alpha}$ and we define the matrix $Z_n(z)$ by the formula:

$$Z_n(z) = \begin{pmatrix} \frac{\beta}{2}(z_n^2+1) - \lambda z_n & -z_n \\ z_n & 0 \end{pmatrix}$$

Observing that $Z_n(e^{i\omega}) = z_n g_\lambda(n)$ we obtain:

$$||Z_{n-1}(e^{i\omega})\dots Z_0(e^{i\omega})|| = ||U_\lambda(n)||$$

from which it follows that:

$$\begin{aligned} \gamma_n(\lambda) &= \mathbb{E}\{\log(||U_\lambda(n)||)\} \\ &= \frac{1}{2\pi}\int_0^{2\pi} \log(||Z_{n-1}(e^{i\omega})\dots Z_0(e^{i\omega})||)\ d\omega. \end{aligned}$$

The function $\lambda \hookrightarrow \log(||Z_{n-1}(e^{i\omega})\dots Z_0(e^{i\omega})||)$ is subharmonic (see property (iii) of LemmaV.4.4) and this yields:

$$\gamma_n(\lambda) \geq \log(||Z_{n-1}(0)\dots Z_0(0)||) = n\log(\frac{|\beta|}{2})$$

The result now follows from $\gamma(\lambda) = \lim \frac{1}{n}\gamma_n(\lambda)$. ∎

This result has been extended (see [308]) to more general potentials of the form

$$V(n,\omega) = \beta\big(\cos(2\pi n\alpha + \omega) + \epsilon f(2\pi n\alpha + \omega)\big)$$

where $f$ is real analytic on some strip $|\Im m\, z| < \delta$ for $\delta > 0$ and of period 1. Then the same conclusion holds for large $|\beta|$ and small $|\epsilon|$.

An important consequence of this proposition is the positivity of the Lyapunov exponent as soon as $|\beta| > 2$ and we will see that this property implies the singularity of the spectrum of the operator $H(\omega)$ for almost all $\omega$. Such an operator has no eigenvalues for $|\beta| < 2$ for any $\omega$ (see Problem V.5.4 for a proof). Moreover it can be shown that if $\alpha$ is a Liouville number then for any $\omega$ the above operator has no eigenvalues (See [14]). It follows that for such a rotation $\alpha$ and for $|\beta| > 2$ the almost Mathieu operator has almost surely a singular continuous spectrum.

## V.4.2 Continuity

We already proved that the Lyapunov exponents are upper semi-continuous. It is reasonable to expect more, namely that these exponents are in fact continous in most of the classical models. We prove it for the Anderson model on the strip, i.e. in the case of i.i.d. potentials on the discrete strip. Hence we assume that the sequence $\{(V_1(n), \cdots, V_\ell(n));\ , n \in \mathbb{Z}\}$ of $\mathbb{R}^\ell$-valued random variables is independent and identically distributed.

One first proves the continuity of the upper exponent $\gamma_1(\lambda)$. Then one extends this property to any partial sum by looking at the action of the symplectic group on the exterior powers of $\mathbb{R}^d$. This gives the continuity of each individual exponent $\gamma_p(\lambda)$ as a by-product.
In order to apply the results of Chapter IV we need that the distribution $\mu_\lambda$ of the matrix $g_\lambda(n) \in SP(\ell, \mathbb{R})$ be strongly irreducible and contractive on each Lagragian subspace Lag$(p)$ for $p = 1 \cdots \ell$ for each fixed $\lambda$. In the one dimensional case, that is for $\ell - 1$, it is enough to assume that the potential function takes at least two values. For greater values of $\ell$ we have to assume a stronger condition, namely that the support of the distribution of $(V_1, \cdots, V_\ell)$ contains an open set in $\mathbb{R}^\ell$. Then PropositionIV.4.32 implies that the semi-group $T_{\mu_\lambda}$ generated by the support of $\mu_\lambda$ contains an open set of $SP(\ell, \mathbb{R})$ and PropositionIV.4.22 yields the desired irreducibility and

contractivity properties.

In the sequel we denote by $\mathcal{I}$ a compact interval of the real line and by $P(\mathbb{R}^d)$ the projective space. We define the function:

$$\Phi_\lambda(\overline{v}) = \mathbb{E}\{\log \frac{||g_\lambda v||}{||v||}\} \qquad \overline{v} \in P(\mathbb{R}^d)$$

**Lemma V.4.7** *One has:*

**(i)** *The mapping $\overline{v} \hookrightarrow \Phi_\lambda(\overline{v})$ is continuous on $P(\mathbb{R}^d)$*
**(ii)** *There exists a finite constant $C$ such that*

$$\sup_{\overline{v}\in P(\mathbb{R}^d)} |\Phi_\lambda(\overline{v}) - \Phi_{\lambda'}(\overline{v})| \leq C|\lambda - \lambda'| \qquad \lambda, \lambda' \in \mathcal{I}$$

*As a consequence the mapping $(\lambda, \bar{v}) \hookrightarrow \Phi_\lambda(\bar{v})$ is continuous on $\mathbb{R} \times P(\mathbb{R}^d)$.*

*Proof:*
The inequality $|\Phi_\lambda(\overline{v})| \leq \mathbb{E}\{\log ||g_\lambda||\}$ together with the Lebesgue theorem and the integrability of the logarithm of the norm of the matrix $g_\lambda$ yield conclusion (i).
To prove (ii) we remark that:

$$\begin{aligned} |\Phi_\lambda(\overline{v}) - \Phi_{\lambda'}(\overline{v})| &\leq \mathbb{E}\{|\log \frac{||g_{\lambda'} v||}{||g_\lambda v||}|\} \\ &\leq \mathbb{E}\{\log ||g_{\lambda'} g_\lambda^{-1}||\} \end{aligned}$$

If one sets $h = g'_\lambda g_\lambda^{-1} = I_d + (\lambda - \lambda')\begin{pmatrix} 0 & I_\ell \\ 0 & 0 \end{pmatrix}$ it is easy to check that $||h|| \leq 1 + |\lambda - \lambda'|$ ∎

**Corollary V.4.8** *The function $\lambda \hookrightarrow \gamma_p(\lambda)$ is continuous for any $p = 1 \cdots \ell$*

*Proof:*
As we already noticed it is enough to prove this property for $p = 1$. PropositionIV.4.14 implies the uniqueness of the invariant probability measure $\nu_\lambda$ for $\mu_\lambda$ on the projective space $P(\mathbb{R}^d)$. The uniqueness of this invariant measure implies immediately that the mapping $\lambda \hookrightarrow \nu_\lambda$ is weakly

continuous. If $\lambda_n$ is a sequence converging to $\lambda$ the integral representation given in PropositionIV.4.14 together with conclusion (ii) imply that:

$$\begin{aligned}\lim_{\lambda_n\to\lambda}\gamma_1(\lambda_n) &= \lim_{\lambda_n\to\lambda}\nu_{\lambda_n}(\Phi_{\lambda_n})\\ &= \lim_{\lambda_n\to\lambda}\nu_{\lambda_n}(\Phi_\lambda)+\nu_{\lambda_n}(\Phi_{\lambda_n}-\Phi_\lambda)\\ &= \nu_\lambda(\Phi_\lambda)=\gamma_1(\lambda)\end{aligned}$$

The proof is now complete. ∎

We have already noticed the uniform convergence of the expectation of the cocycle $||gv||/||v||$ on the projective space. The following proposition shows that this convergence also holds uniformly with respect to a compact set of energy.

**Proposition V.4.9** *One has :*

$$\lim_{n\to+\infty}\frac{1}{n}I\!E\{\log\frac{||U_\lambda(n)v||}{||v||}\}=\gamma_1(\lambda)$$

*uniformly with respect to $\lambda\in\mathcal{I}$ and $\overline{v}\in P(I\!R^d)$.*

*Proof:*
Let $\nu_{n,\lambda,\overline{v}}$ be the sequence of probability measures on $P(\mathbb{R}^d)$ defined by $\nu_{n,\lambda,\overline{v}}=\frac{1}{n}\sum_{k=0}^{n-1}\mu_\lambda^k*\delta_{\overline{v}}$. If we pick two sequences $\overline{v}_n\to v$ in $P(\mathbb{R}^d)$ and $\lambda_n\to\lambda$ in $\mathcal{I}$ then one checks easily that any limit point of the sequence $\nu_{n,\lambda_n,\overline{v}_n}$ is $\mu_\lambda$ invariant and hence equal to $\nu_\lambda$. It follows that this sequence of probability measures converges weakly to $\nu_\lambda$. If we define the sequence of functions

$$h_n(\lambda,\overline{v})=\frac{1}{n}\mathbb{E}\{\log\frac{||U_\lambda(n)v||}{||v||}\}$$

a straightforward computation yields that

$$h_n(\lambda,\overline{v})=\nu_{n,\lambda,\overline{v}}(\Phi_\lambda)$$

Moreover the argument used in the proof of LemmaV.4.7 gives the continuity of the function $h_n$ on the compact space $\mathcal{I}\times P(\mathbb{R}^d)$. In order to prove that $h_n$ converges uniformly on $\mathcal{I}\times P(\mathbb{R}^d)$ to the continuous function $\gamma_1(\lambda)$ it is then enough to prove that $h_n(\lambda_n,\overline{v}_n)$ converges to this limit but:

$$\begin{aligned}\lim_{n\to\infty}h_n(\lambda_n,\overline{v}_n) &= \lim_{n\to\infty}\nu_{n,\lambda_n,\overline{v}_n}(\Phi_{\lambda_n})\\ &= \lim_{n\to\infty}\left(\nu_{n,\lambda_n,\overline{v}_n}(\Phi_\lambda)+\nu_{n,\lambda_n,\overline{v}_n}(\Phi_{\lambda_n}-\Phi_\lambda)\right)\\ &= \nu_\lambda(\Phi_\lambda)=\gamma_1(\lambda)\end{aligned}$$

The same uniform convergence can be proved for the sum $\gamma_1 + \ldots + \gamma_p$ on the Lagrangian boundary $L_p$ ■

The continuity property of the exponents and the above uniform convergence can be extended to the Markovian case provided one can check contractivity and strong irreducibility of the exterior powers of the propagator process $\{\Lambda^p U_\lambda(t);\, t \in T\}$ for $p = 1 \cdots \ell$. The corresponding criterion have been stated in Chapter IV.

### V.4.3 Local Hölder Continuity

Since the following results will only be used in the independent case, we will restrict ourself to this situation. However they can be extended to the Markovian case. In order to be able to work on spaces of Hölder continuous functions we add to the hypotheses of the preceding subsection the following extra assumptions:

- There exists a strictly positive real number $\eta$ such that $\sum_{i=1}^{\ell} |V_i|^\eta$ is integrable.

This implies that for any $\lambda$ the random variable $\sum_{i=1}^{\ell} ||\Lambda^p g_\lambda||^\eta$ is also integrable. We still denote by $P(\mathbb{R}^d)$ the projective space and by $\nu_\lambda$ the unique $\mu_\lambda$-invariant probability measure on $P(\mathbb{R}^d)$. The transition operator of the Markov chain $U_\lambda(n)\bar{x}$, say $T_\lambda$, and the projector on the space of constant functions, say $N_\lambda$, are defined by :

$$T_\lambda f(\bar{x}) = \int f(g\bar{x}) d\mu_\lambda(g), \qquad N_\lambda f(\bar{x}) = \nu_\lambda(f)$$

for any bounded measurable function $f$ on $P(\mathbb{R}^d)$.

**Definition V.4.10** *A real function $\gamma$ defined on $\mathbb{R}$ is said to be locally Hölder continuous if for any compact interval $\mathcal{I}$ there exist a finite constant $C$ and a strictly positive real number $\alpha$ such that :*

$$|\gamma(\lambda) - \gamma(\lambda')| \leq C|\lambda - \lambda'|^\alpha \qquad \lambda, \lambda' \in \mathcal{I}$$

In order to prove that all Lyapunov exponents are locally Hölder continuous, it is enough to prove this property for the upper Lyapunov exponent. The stategy is as follows: first we prove the local Hölder continuity of the mapping $\lambda \hookrightarrow \nu_\lambda(f)$ for any Hölder continuous function. Secondly we use the Furstenberg formula to derive the local Hölder continuity of the Lyapunov exponent.

In the following $\mathcal{I}$ is a fixed compact interval of $\mathbb{R}$ and most of the constants will depend on this interval. We denote by $\delta(\bar{x}, \bar{y})$ the usual distance on $P(\mathbb{R}^d)$ and by $||f||_\alpha$ the norm of the function $f$ in the Hölder space $\mathcal{L}_\alpha$ defined in Chapter IV.

**Lemma V.4.11** *There exists* $\alpha_0 > 0$ *such that for* $0 < \alpha \leq \alpha_0$ *there exist* $\rho_\alpha < 1$ *and* $C_\alpha < +\infty$ *such that:*

$$\sup_{\lambda \in \mathcal{I}, \bar{x}, \bar{y} \in P(\mathbb{R}^d), \bar{x} \neq \bar{y}} \mathbb{E}\{\frac{\delta(U_\lambda(n)\bar{x}, U_\lambda(n)\bar{y})^\alpha}{\delta(\bar{x}, \bar{y})^\alpha}\} \leq C_\alpha \rho_\alpha^n$$

*Proof:*
From the definition of the distance one obtains :

$$\begin{aligned} & \mathbb{E}\{\log \frac{\delta(U_\lambda(n)\bar{x}, U_\lambda(n)\bar{y})}{\delta(\bar{x}, \bar{y})}\} \\ \leq\ & \mathbb{E}\{\log(||\Lambda^2 U_\lambda(n)||)\} - \mathbb{E}\{\log \frac{||U_\lambda(n)x||}{||x||}\} - \mathbb{E}\{\log \frac{||U_\lambda(n)y||}{||y||}\}. \end{aligned}$$

The uniform convergence proved in PropositionV.4.9 added to the continuity of the Lyapunov exponent and the inequality $\gamma_1(\lambda) > \gamma_2(\lambda)$ imply that there exists an integer $p$ such that

$$\sup_{\lambda \in \mathcal{I}, \bar{x}, \bar{y} \in P(\mathbb{R}^d); \bar{x} \neq \bar{y}} \int \log \frac{\delta(g\bar{x}, g\bar{y})}{\delta(\bar{x}, \bar{y})}\, d\mu_\lambda^p(g) < 0$$

The proof is now a consequence of the general theorem on negativity of cocycles proved in PropositionIV.3.5. ∎

**Corollary V.4.12** *There exists* $\alpha_0 > 0$ *such that for* $0 < \alpha \leq \alpha_0$ *there exist* $\rho_\alpha < 1$ *and* $C_\alpha < +\infty$ *such that:*

$$\sup_{\lambda \in \mathcal{I}} ||T_\lambda^n f - N_\lambda(f)||_\alpha \leq C_\alpha ||f||_\alpha \rho_\alpha^n \qquad n = 1, 2, \ldots$$

*Proof:*
Using the preceding lemma the proof is identical to PropositionIV.3.15 if one takes into account the uniformity of the bounds with respect to the parameter $\lambda$. ∎

In general the mapping $\lambda \hookrightarrow T_\lambda$ is not continuous from $\mathcal{I}$ to the set of bounded operators on the Banach space $\mathcal{L}_\alpha$ but the following weaker property can be viewed as continuity of this mapping if we consider $T_\lambda$ as a bounded operator from $\mathcal{L}_\alpha$ to $\mathcal{L}_{\alpha/2}$.

**Proposition V.4.13** *There exists $\alpha_0 > 0$ such that for $0 < \alpha \leq \alpha_0$ there exists $C_\alpha < +\infty$ with:*

$$||T_\lambda f - T_{\lambda'} f||_{\alpha/2} \leq C_\alpha ||f||_\alpha |\lambda' - \lambda|^{\alpha/2} \qquad \text{for } \lambda, \lambda' \in \mathcal{I} \ , \ f \in \mathcal{L}_\alpha$$

*Proof:*
For $f \in \mathcal{L}_\alpha$ and $\lambda, \lambda' \in \mathcal{I}$ one has:

$$\begin{aligned} |T_\lambda f(\bar{x}) - T_{\lambda'} f(\bar{x})| &\leq m_\alpha(f) \mathbb{E}\{\delta(g_\lambda \bar{x}, g_{\lambda'} \bar{x})^\alpha\} \\ &\leq m_\alpha(f) \mathbb{E}\{||g_\lambda||^\alpha \, ||g_{\lambda'}||^\alpha\} |\lambda - \lambda'|^\alpha \end{aligned}$$

The integrability assumption on the potential implies the existence (for $\alpha$ small enough) of a constant $A_\alpha$ such that

**(i)** $|T_\lambda f(\bar{x}) - T_{\lambda'} f(\bar{x})| \leq A_\alpha \, m_\alpha(f) |\lambda - \lambda'|^\alpha$

This relation gives immediately a bound for the uniform norm $||T_\lambda - T_{\lambda'}||_\infty$. One can also write:

$$\begin{aligned} |T_\lambda f(\bar{x}) - T_\lambda f(\bar{y})| &\leq m_\alpha(f) \, \mathbb{E}\{\delta(g_\lambda \bar{x}, g_\lambda \bar{y})^\alpha\} \\ &\leq m_\alpha(f) \, \mathbb{E}\{||g_\lambda||^{2\alpha}\} \delta(\bar{x}, \bar{y})^\alpha . \end{aligned}$$

Using again the integrability assumption on the potential one obtains the existence (for $\alpha$ small enough) of a constant $B_\alpha$ such that:

**(ii)** $|T_\lambda f(\bar{x}) - T_\lambda f(\bar{y})| \leq B_\alpha \, m_\alpha(f) \delta^\alpha(\bar{x}, \bar{y})$

If we set $C_\alpha = \max(B_\alpha, A_\alpha)$ one can deduce from the inequalities (i) and (ii) that:

$$|(T_\lambda - T_{\lambda'}) f(\bar{x}) - (T_\lambda - T_{\lambda'}) f(\bar{y})| \leq 2 C_\alpha \, m_\alpha(f) \min\big(\delta(\bar{x}, \bar{y})^\alpha, |\lambda - \lambda'|^\alpha\big)$$

To obtain the desired conclusion it suffices to remark that:

$$\min\big(\delta(\bar{x}, \bar{y})^{\alpha/2}, |\lambda - \lambda'|^\alpha \delta(\bar{x}, \bar{y})^{-\alpha/2}\big) \leq |\lambda - \lambda'|^{\alpha/2} \quad \blacksquare$$

**Proposition V.4.14** *There exists $\alpha_0 > 0$ such that for $0 < \alpha \leq \alpha_0$ there exists $C_\alpha < +\infty$ with :*

$$|\nu_\lambda(f) - \nu_{\lambda'}(f)| \leq C_\alpha |\lambda - \lambda'|^{\alpha/2} \|f\|_\alpha \qquad \text{for } \lambda, \lambda' \in \mathcal{I} \ , \ f \in \mathcal{L}_\alpha$$

*Proof:*
Let $\alpha_0$ be the minimum of the two numbers $\alpha_0$ given by CorollaryV.4.12 and PropositionV.4.13. For $\alpha$ satisfying $0 < \alpha \leq \alpha_0$ let us denote by $\Gamma_\alpha$ the circle of center $z = 1$ and radius $r_\alpha = (1 - \rho_\alpha)/2$ where $\rho_\alpha$ is given by CorollaryV.4.12. Writing $T_\lambda = N_\lambda + Q_\lambda$ as before, we obtain for $z \in \Gamma_\alpha$ the expansion:

$$R_{\lambda,z} = (T_\lambda - zI)^{-1} = \frac{N_\lambda}{z(1-z)} + \sum_{n \geq 0} \frac{Q_\lambda^n}{z^{n+1}}$$

This implies $\|R_{\lambda,z}\|_\alpha \leq 4/(1 - \rho_\alpha^2) + 2C_\alpha/(1 - \rho_\alpha)$ where $C_\alpha$ is given by CorollaryV.4.12.Thus we obtain an uniform bound on the $\mathcal{L}_\alpha$ norm of the Green kernel for $z \in \Gamma_\alpha$ and $\lambda \in \mathcal{I}$.
If we take into account the resolvent and Cauchy identities:

$$R_{\lambda,z} - R_{\lambda',z} = R_{\lambda',z}(T_{\lambda'} - T_\lambda) R_{\lambda,z} \quad \text{and} \quad N_\lambda = \frac{1}{2i\pi} \int_{\Gamma_\alpha} R_{\lambda,z} \, dz$$

the result is a direct application of PropositionV.4.13. ∎

**Theorem V.4.15** *There exist constants $\alpha > 0$ and $C < +\infty$ such that :*

$$|\gamma_p(\lambda) - \gamma_p(\lambda')| \leq C |\lambda - \lambda'|^\alpha \quad \text{for} \quad p = 1 \ldots \ell \ , \ \lambda, \lambda' \in \mathcal{I}$$

*Proof:*
With the notations of LemmaV.4.7 we have:

$$\begin{aligned} \gamma_1(\lambda) - \gamma_1(\lambda') &= \nu_\lambda(\Phi_\lambda) - \nu_{\lambda'}(\Phi_{\lambda'}) \\ &= (\nu_\lambda - \nu_{\lambda'})(\Phi_{\lambda'}) + \nu_\lambda(\Phi_\lambda - \Phi_{\lambda'}) \end{aligned}$$

The result will follow from PropositionV.4.14 and LemmaV.4.7 once we prove that $\sup_{\lambda \in \mathcal{I}} \|\Phi_\lambda\|_\alpha < +\infty$ for some $\alpha > 0$. Notice first that the inequality $|\Phi_\lambda(\bar{x})| \leq \mathbb{E}\{\log \|g_\lambda\|\}$ gives the desired bound for the uniform norm of $\Phi_\lambda$.
Secondly for two directions $\bar{x}$ and $\bar{y}$ one can find two unit vectors $x$ and $y$ such that the angle between $x$ and $y$ is not greater than $\pi/2$. It follows

that $\delta(\bar{x},\bar{y}) \geq \frac{\sqrt{2}}{2}||x-y||$. If we denote by $h$ the rotation which maps $x$ onto $y$ we have $||h-I|| \leq \frac{\sqrt{2}}{2}\delta(\bar{x},\bar{y})$ and one can write:

$$\begin{aligned} |\Phi_\lambda(\bar{y}) - \Phi_\lambda(\bar{x})| &\leq \mathbb{E}\{|\log\frac{||g_\lambda hx||}{||g_\lambda x||}|\} \\ &\leq \mathbb{E}\{\log||g_\lambda h g_\lambda^{-1}||\} \\ &= \mathbb{E}\{\log(1+||g_\lambda(h-I)g_\lambda^{-1}||)\} \\ &\leq \mathbb{E}\{\log(1+||h-I||\ ||g_\lambda||^2)\} \\ &\leq \mathbb{E}\{\log(1+\frac{\sqrt{2}}{2}\delta(\bar{x},\bar{y})||g_\lambda||^2)\} \\ &\leq C\delta(\bar{x},\bar{y})^\alpha. \end{aligned}$$

One gets the last inequality by taking into account the fact that for any $0 < \alpha \leq 1$ and $t \geq 0$ one has $\log(1+t) \leq C_\alpha t^\alpha$ and using the integrability condition on the norm of the potentials for $\alpha$ sufficiently small. ∎

This result can be extented in an obvious way to the partial sums of Lyapunov exponents and hence to any individual exponent since one has assumed the strong irreducibility and the contractivity of $\mu_\lambda$ on each Lagrangian space $\mathrm{Lag}(p)$ for $p = 1\ldots\ell$. The only difference is that now the dependence of the matrix $\Lambda^p(g_\lambda)$ upon the parameter $\lambda$ is not as simple as in the case of the matrix $g_\lambda$ itself. Nevertheless it is easily seen that all the preceding arguments go through if one merely assumes that the matrix $g_\lambda$ satisfies the following property:

- for any bounded interval $\mathcal{I}$ there exists a random variable $C(\omega)$ with a positive moment and a positive number $\alpha$ such that :

$$||g_\lambda(\omega) - g_{\lambda'}(\omega)|| \leq C(\omega)\,|\lambda-\lambda'|^\alpha \quad \lambda,\lambda' \in \mathcal{I}$$

In order to check that the matrix $\Lambda^p(g_\lambda)$ has this property one can write:

$$\Lambda^p g_\lambda - \Lambda^p g_{\lambda'} = \Lambda^p g_\lambda(\Lambda^p(g_{\lambda'}g_\lambda^{-1}) - I_r)$$

where $r$ is the dimension of $\Lambda^p(\mathbb{R}^d)$. Moreover one has

$$h = g_{\lambda'}g_\lambda^{-1} = I_d + (\lambda-\lambda')\begin{pmatrix} I_\ell & 0 \\ 0 & 0 \end{pmatrix}$$

It is now easy to check that one has $\Lambda^p h = I_r + (\lambda-\lambda')g_{\lambda,\lambda'}$ where the matrix $g_{\lambda,\lambda'}$ is such that $\sup_{\lambda,\lambda'\in\mathcal{I}}||g_{\lambda,\lambda'}|| < \infty$.

Actually it has been remarked in [303] that this result is optimal in the one dimensional case when the potentials obey a Bernouilli distribution on the two real numbers $a$ and $b$. Then for any positive $\alpha$ there exists a choice of $a$ and $b$ such that $\gamma(\lambda)$ cannot be Hölder continuous of order greater than $\alpha$ on the spectrum of the associated Schrödinger operator.

### V.4.4 Smoothness

It is not difficult to find examples for which the upper Lyapunov exponent is not smooth. A sufficient condition (certainly not optimal) for smoothness requires the absolute continuity of the distribution of the potentials with respect to the Lebesgue measure on $\mathbb{R}$.

**Theorem V.4.16** *Let us assume that the $\mathbb{R}^\ell$-valued sequence of random variables $\{(V_1(n), \cdots, V_\ell(n));\ n \in \mathbb{Z}\}$ is i.i.d. and that the common distribution has an absolutely continuous distribution with a finite first moment. Then the upper Lyapunov exponent $\gamma_1(\lambda)$ is a $C^\infty$ function.*

We only give a proof of this theorem in the one dimensional case in order to avoid defining the differentiable structure of the general Lagrangian boundaries. Moreover we will only use this result in this simple case. A complete proof can be found in Le Page [216].

In this subsection the projective space $P(\mathbb{R}^2)$ will be identified with the torus $\mathbb{T} = [0, \pi]$ assigning to the direction $\bar{v}$ its polar angle modulo $\pi$. The set of $p$ times continuously derivable functions on $\mathbb{T}$ is denoted by $C^p(\mathbb{T})$. It is a Banach space for the norm:

$$||f||_p = \sum_{k=0}^{p} \sup_{\vartheta \in \mathbb{T}} |D^k f(\vartheta)|$$

where $D$ is the operator of (partial) derivation $D = \partial/\partial\vartheta$.
Recall that a cocycle $\chi$ associated to the action of $G = SL(2, \mathbb{R})$ on the projective line $P(\mathbb{R})^2$ is said to be rotation invariant if $\chi(h, \bar{v}) = 1$ for any rotation $h$ and any nonzero vector $v \in \mathbb{R}^2$. Let $\rho(g, \bar{v})$ be the cocycle

$$\rho(g, \overline{v}) = \left(\frac{||gv||}{||v||}\right)^2$$

Any rotation invariant cocycle $\chi$ on $P(\mathbb{R}^2)$ can be written $\chi = \rho^t$ for some real number $t$. Using the identification of the projective space with the torus, the action of a matrix $g \in SL(2, \mathbb{R})$ on $\mathbb{T}$ is given by

$$g\vartheta = \cot^{-1}\left(\frac{a\cos\vartheta + b\sin\vartheta}{c\cos\vartheta + d\sin\vartheta}\right) \quad , \quad g = \begin{pmatrix} a & b \\ c & d \end{pmatrix}$$

and the cocycle $\rho$ can be written:

$$\rho(g, \vartheta) = (a\cos\vartheta + b\sin\vartheta)^2 + (c\cos\vartheta + d\sin\vartheta)^2$$

In the following, we denote by $r$ the classical classical Radon-Nikodym cocycle defined by $r(g, \vartheta) = \rho^{-1}(g, \vartheta) = dg^{-1}\vartheta/d\vartheta$ where $d\vartheta$ is the normalized Lebesgue measure on $\mathbb{T}$.

**Definition V.4.17** *A probability measure $\mu$ on $G$ is said to have moments of all orders if $\int ||g||^\alpha \, d\mu(g) < +\infty$ for any positive $\alpha$.*

The translation group $\{\mathcal{T}_g; g \in G\}$ acts on $C(\mathbb{T})$ by the formula $\mathcal{T}_g f(\vartheta) = f(g\vartheta)$. In the following we denote by $\chi_g$ the mapping $\vartheta \hookrightarrow \chi(g, \vartheta)$. The following lemma summarizes some technical results needed in the study of the Laplace transform on $C^p(\mathbb{T})$.

**Lemma V.4.18** *Let $p$ be a nonnegative integer. Then one has:*

1. *$D\mathcal{T}_g f = r_g \mathcal{T}_g Df, \qquad f \in C^1(\mathbb{T})$.*
2. *For any $g \in G$ then $\mathcal{T}_g$ is a bounded linear operator on $C^p(\mathbb{T})$ and the mapping $g \hookrightarrow \mathcal{T}_g$ is continuous.*
3. *Let $\chi$ be a rotation invariant cocycle. Then the mapping $g \hookrightarrow \chi_g$ is continuous from $G$ into $C^p(\mathbb{T})$.*
4. *Let $\mu$ be a probability measure on $G$ having moments of all orders. Then for any real $\alpha$ one has:*

$$\int ||\chi_g||_p^\alpha \, d\mu(g) < \infty \quad \text{and} \quad \int ||\mathcal{T}_g||_p^\alpha \, d\mu(g) < \infty$$

*where $||\mathcal{T}_g||_p$ denotes the norm of $\mathcal{T}_g$ as an operator on $C^p(\mathbb{T})$.*

*Proof:*

Conclusion 1. follows from an easy computation. One can also use the definition of a Radon-Nikodym cocycle to directly obtain this result. Conclusion 2. is derived from the Liebnitz formula:

$$D^{p+1}(\mathcal{T}_g f) = \sum_{k=0}^{p} \binom{k}{p} D^k(\mathcal{T}_g Df) D^{p-k} r$$

by a simple induction argument. Part 2. and the formula $\chi_{hg} = \chi_g \mathcal{T}_g \chi_h$ can be used to prove 3. by induction. Recall that any rotation invariant cocycle $\chi$ can be written $\chi = \rho^t$ for some real number $t$. Note also that one has the uniform bound $\rho(g, \vartheta) \leq ||g||^2$. Hence, the integrability of any power of $||g||$ and Hölder's inequality can be used to prove that it is enough to prove 3. for the cocycle $\rho$ to obtain the same result for any $\chi$. Using the explicit expression of $\rho$ given above it is easily seen that one has:

$$\sup_{\vartheta \in \boldsymbol{T}} |D^k \rho(g, \vartheta)| \leq 2^k ||g||^2$$

and this proves the first part of 4. Applying this result to the Radon-Nikodym cocycle $r$ the second part of 4. follows from the Liebnitz formula. ∎

For a given rotation invariant cocycle $\chi$ the Laplace Transform $T$ is formally defined by:

$$Tf(\vartheta) = \int \chi(g, \vartheta) f(g\vartheta) d\mu(g).$$

By Proposition IV.3.9, $T$ is a compact operator on the space $C(\mathbb{T}) = C^0(\mathbb{T})$ when $\mu$ is absolutely continuous with respect to the Haar measure of $G$ and has a second order moment. In fact a stronger result holds assuming that $\mu$ has moments of all orders: the operator $T$ is compact on any space $C^p(\mathbb{T})$.

**Proposition V.4.19** *Let $\mu$ be a probability measure on $G$. Let us assume that $\mu$ has moments of all orders and is absolutely continuous with respect to the Haar measure of $G$. Then for any nonnegative integer $p$ the operator $T$ is compact on the Banach space $C^p(\mathbb{T})$.*

*Proof:*
We first remark that the relation $\chi_g = \mathcal{T}_g \chi_{g^{-1}}^{-1}$ implies that the operator $T$ can be rewritten as:

$$Tf = \int \mathcal{T}_g(\chi_{g^{-1}}^{-1} f)\, \varphi(g) dg$$

where $\varphi$ is the density of $\mu$ with respect to a Haar measure $dg$ on $G$.
Let $f \in C^p(\mathbb{T})$. Lemma V.4.18 allows us to differentiate $p$ times with respect to $\vartheta$ under the integral sign defining the operator $T$. Hence one gets for any $0 \leq k \leq p$ :

$$D^k Tf = \int D^k \mathcal{T}_g(\chi_{g^{-1}}^{-1} f)\, \varphi(g) dg$$

and

$$\sup_{\vartheta \in \mathbf{T}} |D^k Tf(\vartheta)| \leq \|f\|_k \int \|\mathcal{T}_g\|_k \, \|\chi_{g^{-1}}^{-1}\| \, \varphi(g) dg$$

In order to check the compactness of $T$ on $C^p(\mathbb{T})$, one has to prove that for any $0 \leq k \leq p$ and any bounded set $K \subset C^p(\mathbb{T})$, the set $\{D^k f ; f \in K\}$ is equicontinuous in $C(\mathbb{T})$. Any element $\vartheta' \in \mathbb{T}$ can be written $\vartheta' = h\vartheta$ for some rotation $h$ thus the differences $D^k f(\vartheta') - D^k f(\vartheta)$ can be rewritten as $\mathcal{T}_h D^k f(\vartheta) - D^k f(\vartheta)$. Taking into account the relation $D^k \mathcal{T}_h = \mathcal{T}_h D^k$ implied by Lemma V.4.18 2. one has:

$$\mathcal{T}_h D^k Tf = \int D^k \mathcal{T}_{gh}(\chi_{g^{-1}}^{-1} f)\, \varphi(g) dg$$

Taking into account the relation $\chi^{-1}(hg^{-1}, \cdot) = \chi(g^{-1}, \cdot)$ the change of variable $g \hookrightarrow gh^{-1}$ yields:

$$\mathcal{T}_h D^k Tf = \int D^k \mathcal{T}_g(\chi_{g^{-1}}^{-1} f)\, \varphi(gh^{-1}) dg$$

It follows that:

$$\sup_{\vartheta \in \mathbf{T}} |D^k Tf(h\vartheta) - D^k Tf(\vartheta)|$$
$$\leq \|f\|_k \int \|Dk\mathcal{T}_g\|_k \|\chi_{g^{-1}}^{-1}\|_k |\varphi(gh^{-1}) - \varphi(g)| dg.$$

The equicontinuity is now obvious at least if the density $\varphi$ is continuous with compact support. One finishes the proof as in Proposition IV.3.9 by approximating the function $\varphi$ by continuous functions with compact support in the space $L^1(G, \|\mathcal{T}_g\|_k \|\chi_{g^{-1}}^{-1}\|_k \, dg)$. ∎

When $\mu$ has a density on $G$, there exists a unique invariant measure $\nu$ on $\mathbb{T}$ and we have already proved that $\nu$ has a continuous density with respect to the Lebesgue measure on $\mathbb{T}$. Using Proposition V.4.19 we will prove that this density is actually $C^\infty$.

**Corollary V.4.20** *Let $\mu$ be a probability measure on $G$ and let us assume that $\mu$ has moments of all orders and is absolutely continuous with respect to a Haar measure of $G$. Then the invariant measure $\nu$ on $\mathbb{T}$ has a $C^\infty$ density with respect to the Lebesgue measure.*

*Proof:*
Let $\check{T}_2$ be the Laplace transform associated to the measure $\check{\mu}$ and the Radon-Nikodym cocycle $r$. By Proposition IV.3.12 the spectral radius of the operator $\check{T}_2$ acting on $C(\mathbb{T})$ is equal to one. Hence the compact operator $\check{T}$ acting on $C^p(\mathbb{T})$ has a spectral radius no greater than one. Moreover the mapping

$$f \hookrightarrow \int f(\vartheta)d\vartheta$$

is a continuous linear functional on $C^p(\mathbb{T})$ and one has $\int \check{T}_2 f(\vartheta)d\vartheta = \int f(\vartheta)d\vartheta$. It follows that 1 is an eigenvalue of $\check{T}_2^*$ on $(C^p(\mathbb{T}))^*$ hence of $\check{T}_2$ acting on $C^p(\mathbb{T})$. We already know that the density $\psi$ of the invariant measure $\nu$ is the solution of $\check{T}_2\psi = \psi$ on $C(\mathbb{T})$ hence $\psi$ is $C^\infty$. ∎

We now suppose that the matrix $g_\lambda$ is the transfer matrix of a one dimensional Schrödinger operator associated to an i.i.d. sequence of potentials. We also assume the following two conditions:

- The common distribution of the potentials has a density $\varphi$ with respect to the Lebesgue measure.
- $\mathbb{E}\{|V(0)|^\alpha\} = \int |t|^\alpha \phi(t)\, dt < +\infty$ for any positive $\alpha$.

Then for each real $\lambda$ the distribution $\mu_\lambda$ of the matrix $g_\lambda$ has moments of all orders. Obviously the same property is true for $\mu_\lambda^3$. Moreover we have seen that $\mu_\lambda^3$ has a density with respect to the Haar measure of $SL(2,\mathbb{R})$. (See Proposition IV.4.33 or the proof of Theorem VIII.3.5 for a simpler argument). It follows from Proposition V.4.19 that for any rotation invariant cocycle the operator $T_\lambda^3$ is compact on $C^p(\mathbb{T})$ for any real real $\lambda$ and any nonnegative integer $p$. Moreover $\mu_\lambda$ admits a unique invariant measure $\nu_\lambda$ which has a $C^\infty$ density.

In order to obtain the smoothness of the mapping $\lambda \hookrightarrow \gamma(\lambda)$ we use again the method of the preceding subsection where we proved the Hölder continuity of the Lyapunov exponent. In other words we deduce the regularity properties of the Lyapunov exponent from the regularity of the invariant measure.

Let us define the operator

$$T_\lambda f(\vartheta) = \int f(g\vartheta) d\mu_\lambda(g).$$

In the following, $\mathcal{I}$ is a fixed bounded open interval. Recall that it has been proved in Corollary V.4.12 that there exists a positive $\alpha$ and $0 < c < 1$ such that for any $\lambda \in \mathcal{I}$ the spectrum of $T_\lambda$ acting on $\mathcal{L}_\alpha$ splits into the eigenvalue $\{1\}$ and a part contained in the disc of radius $c$. Compactness of the operator $T_\lambda^3$ implies the same spectral property for $T_\lambda$ acting on $C^p(\mathbb{T})$ for any $p$. It follows that for any $p$ one can find a circle $\Gamma$ of center 1 in the complex plane, such that any $z \in \Gamma$ lies in the resolvent set of $T_\lambda$ acting on $C^p(\mathbb{T})$ for any $\lambda \in \mathcal{I}$. Denoting by $R_{\lambda,z}$ the resolvent of $T_\lambda$ acting on $C^p(\mathbb{T})$ one has:

$$\nu_\lambda(f) = \frac{1}{2\pi i} \int_\Gamma R_{\lambda,z} f \, dz \qquad f \in C^p(\mathbb{T})$$

One can remark that the same argument works for $T_\lambda$ viewed as an operator from $C^{p+k}(T)$ into $C^p(T)$.

Let $\chi$ be the cocycle $\chi(g, \bar{v}) = \frac{\|gv\|}{\|v\|}$ and $\Phi$ be the function defined by:

$$\Phi(\lambda, \vartheta) = \mathbb{E}\{\log \chi(g_\lambda, \vartheta)\}$$

Then the Lyapunov exponent $\gamma(\lambda)$ is given by:

$$\gamma(\lambda) = \int \Phi(\lambda, \vartheta) \, d\nu_\lambda(\vartheta) = \frac{1}{2\pi i} \int_\Gamma R_{\lambda,z} \Phi(\lambda, \cdot) \, dz$$

We know from Proposition V.4.7 that $\Phi$ is a continuous function of $\vartheta$ but this is not enough to conclude. The essential steps to prove the smoothness of the Lyapunov exponent are the following:

1. The mapping $(\lambda, \vartheta) \hookrightarrow \Phi(\lambda, \vartheta)$ is $C^\infty$ on $\mathcal{I} \times \mathbb{T}$.
2. There exists an appropriate couple of Banach spaces such that the mapping $\lambda \hookrightarrow R_{\lambda,z}$ is $C^\infty$ when the resolvent is viewed as a continuous operator between these spaces.

3. One can indefinitely differentiate under the integral sign in the formula giving $\gamma$.

One begins by a technical lemma the proof of which is similar to the proof of Lemma V.4.18. The only improvement is that we have to take care of the uniformity with respect to the parameter $\lambda \in \mathcal{I}$.

**Lemma V.4.21** *Let $p$ and $k$ be two nonnegative integers.*

1. *For any cocycle $\chi$ and positive number $\alpha$ one has:*

$$\mathbb{E}\left\{\sup_{\lambda\in\mathcal{I}} \|\frac{d^k}{d\lambda^k}\chi(g_\lambda,\cdot)\|_p^\alpha\right\} < +\infty.$$

2. *There exists a constant $c_{p,k}$ such that*

$$\mathbb{E}\left\{\sup_{\lambda\in\mathcal{I}} \|\frac{d^k}{d\lambda^k}f(g_\lambda,\cdot)\|_p\right\} \leq c_{p,k}\|f\|_{p+k}$$

*for any $f \in C^{p+k}(\mathbb{T})$.*

*Proof:*
For any positive $\alpha$, $\sup_{\lambda\in\mathcal{I}}\|g_\lambda\|^\alpha$ is integrable. Using Hölder's inequality one sees that it is enough to prove 1. for the cocycle

$$\rho(g,\vartheta) = \left((V-\lambda)\cos\vartheta - \sin\vartheta\right)^2 + (\sin\vartheta)^2$$

since the latter is quadratic in $\lambda$. One can write:

$$g_\lambda\vartheta = \cot^{-1}(V-\lambda-\tan\vartheta)$$

If follows immediately from this relation that $d^k g_\lambda\vartheta/d\lambda^k$ is bounded on any bounded set $\mathcal{I}$, the bound being uniform with respect to the value of the potential. This implies that there exists a constant $c_k$ such that

$$\sup_{\vartheta\in\boldsymbol{T}} |\frac{d^k}{d\lambda^k}f(g_\lambda\vartheta)| \leq c_k\|f\|_k$$

Applying this relation to the functions $D^s f$ for $s = 1\ldots p$ one obtains 2. ∎

Using Part 1. of this lemma and Hölder's inequality it is easy to see that one can differentiate under the integral sign defining the function $\Phi$. One gets:

**Proposition V.4.22** *The function* $(\lambda, \vartheta) \hookrightarrow \Phi(\lambda, \vartheta)$ *is* $C^\infty$ *on* $\mathcal{I} \times \mathbb{T}$.

For a nonnegative integer $k$ we define formally the operator $T_\lambda^{(k)}$ by the formula

$$T_\lambda^{(k)} f = \frac{d^k}{d\lambda^k} T_\lambda f.$$

Let $f \in C^k(\mathbb{T})$. Using Lemma V.4.21 with $p = 0$ one sees that $T_\lambda^{(k)} f$ can be defined inductively by derivation of the integral and one obtains:

$$T_\lambda^{(k)} f(\vartheta) = \mathbb{E}\left\{\frac{d^k}{d\lambda^k} f(g_\lambda \vartheta)\right\} \quad , \quad f \in C^k(\mathbb{T}).$$

For two nonnegative integers $p$ and $k$ let $C(k,p)$ be the Banach space of continuous linear mappings from $C^k(\mathbb{T})$ to $C^p(\mathbb{T})$.

**Proposition V.4.23** *For any nonnegative integers $p$ and $k$ one has:*

1. $T_\lambda^{(k)} \in C(k+p, p)$.
2. *The mapping* $\lambda \hookrightarrow T_\lambda$ *from* $\mathcal{I}$ *to* $C(p,p)$ *is continuous.*
3. *There exists a constant* $c_{k,p}$ *such that for any* $f \in C^{p+k+2}$ *one has:*

$$\sup_{\lambda, \lambda' \in \mathcal{I}} \|T_\lambda^{(k)} f - T_{\lambda'}^{(k)} f - (\lambda - \lambda') T_\lambda^{(k+1)} f\|_p \leq c_{k,p} |\lambda - \lambda'|^2 \|f\|_{k+p+2}.$$

*Proof:*
Conclusion 1. is a simple consequence of part 2. of Lemma V.4.21.
Let us set $g = \begin{pmatrix} v & -1 \\ 1 & 0 \end{pmatrix}$. Then one can write:

$$T_\lambda f(\vartheta) - T_{\lambda'} f(\vartheta) = \int f(g\vartheta)(\varphi(v - \lambda) - \varphi(v - \lambda'))\, dv$$

where $\varphi$ is the density of the one site distribution of the potentials. Using the previous notation one sees that:

$$\|T_\lambda f - T_{\lambda'} f\|_p \leq \|f\|_p \sum_{j=0}^{p} \int \|\mathcal{T}_g\|_p |\varphi(v - \lambda) - \varphi(v - \lambda')| dv$$

and conclusion 2. follows from Lebesgue's theorem and the integrability of $\|\mathcal{T}_g\|_p$. Writing the Taylor formula with integral remainder for the function

$$\lambda \hookrightarrow \frac{d^k}{d\lambda^k} f(g_\lambda \vartheta) \quad , \quad f \in C^{k+p+2}$$

and taking expectation one obtains that the left hand side of the desired inequality in 3. is not greater that:

$$(\lambda - \lambda')^2 \mathbb{E}\left\{\int_0^1 (1-t)\frac{d^{k+2}}{d\lambda^{k+2}} f(g_\eta \vartheta) dt\right\}$$

where $\eta = \lambda + t(\lambda - \lambda')$. Lemma V.4.21 2. yields the desired result. ∎

We now state a result of operator theory whose proof can be found in Problem V.5.6.

**Lemma V.4.24** *Let $E_1$ and $E_2$ be two Banach spaces with $E_1 \subset E_2$ and such that the inclusion mapping from $E_1$ into $E_2$ is continuous. Let $\mathcal{L}(E_i, E_j)$ be the Banach space of bounded linear operators from $E_i$ to $E_j$ and let $\{T_\lambda \,;\, \lambda \in \mathcal{I}\}$ be a family of operators such that:*

- *The $\mathcal{L}(E_1, E_1)$ and $\mathcal{L}(E_2, E_2)$ valued mappings $\lambda \hookrightarrow T_\lambda$ are continuous.*
- *The $\mathcal{L}(E_1, E_2)$ valued mapping $\lambda \hookrightarrow T_\lambda$ has a derivative $T'_{\lambda_0}$ at the point $\lambda_0$.*

*Let $z_0$ be a common point in the resolvent sets of $T_\lambda$ acting on $E_1$ and $E_2$. Let us denote by $R_{\lambda,z_0}$ the resolvent operator of $T_\lambda$ on $\mathcal{L}(E_1, E_1)$ or $\mathcal{L}(E_2, E_2)$. Then the $\mathcal{L}(E_1, E_2)$ valued mapping $\lambda \hookrightarrow R_{\lambda,z_0}$ is well defined in a neigborhood of $\lambda_0$ and has a derivative at the point $\lambda_0$ equal to $R_{\lambda_0,z_0} T'_{\lambda_0} R_{\lambda_0,z_0}$*

We are now in a position to conclude our program, namely to prove Theorem V.4.16, at least in the one dimensional case.

**Proposition V.4.25** *For $z \in \Gamma$ and $\vartheta \in \mathbb{T}$ the mapping*

$$H_{z,\vartheta}(\lambda) = R_{\lambda,z}\Phi(\lambda, \vartheta)$$

*is $C^\infty$ on $\mathcal{I}$ and each derivative is a continuous function on $\mathcal{I} \times \Gamma \times \mathbb{T}$.*

*Proof:*
Let $p$ be an nonnegative integer, then $\Phi(\lambda, \cdot) \in C^{p+2}(\mathbb{T})$. If one applies Lemma V.4.24 to the couple of Banach spaces $E_1 = C^{p+2}(\mathbb{T})$ and $E_2 = C^p(\mathbb{T})$ then part 3. of Lemma V.4.23 implies that $T_\lambda$ is derivable when

viewed as an element of $C(p+2,p)$. It follows that the $C^p(\mathbb{T})$ valued mapping $\lambda \hookrightarrow R_{\lambda,z}\Phi(\lambda,\cdot)$ has a derivative equal to:

$$R_{\lambda,z}T'_\lambda R_{\lambda,z}\Phi(\lambda,\cdot) + R_{\lambda,z}\frac{\partial}{\partial\lambda}\Phi(\lambda,\cdot)$$

Obviously this function is continuous with respect to all the parameters $(\lambda, z, \vartheta)$ since the $C(k,k)$ valued mapping $(\lambda, z) \to R_{\lambda,z}$ is continuous for any $k$. It now remains to iterate the process of derivation to obtain the desired result. ∎

# V.5 Problems

**Problem V.5.1** *Prove the first claim of the chapter, i.e. for any function $\omega \hookrightarrow T(\omega)$ in the space $L(\mathcal{H})$, the following properties are equivalent:*

**(i)** $\forall f \in \mathcal{H}, \forall g \in \mathcal{H}, \ \Omega \in \omega \hookrightarrow < f, T(\omega)g > \in \mathbb{C}$ *is measurable,*
**(ii)** $\forall f \in \mathcal{H}, \ \Omega \ni \omega \hookrightarrow T(\omega)f \in \mathcal{H}$ *is measurable,*
**(iii)** $\Omega \ni \omega \hookrightarrow T(\omega) \in L(\mathcal{H})$ *is measurable for the Borel $\sigma$-field $\mathcal{B}_{L(\mathcal{H})}$ of $L(\mathcal{H})$ endowed with the uniform norm.*

**Problem V.5.2** *Let $V$ be a stationary and ergodic measurable stochastic process $\{V(x);\ x \in T\}$ with $T = \mathbb{Z}^d$ in the discrete case of the lattice $\mathbb{Z}^d$ and $T = \mathbb{R}^d$ in the continuous case, and let us consider the random self-adjoint operator $M_V$ of multiplication by the function $V$ on $\mathcal{H} = \ell^2(\mathbb{Z}^d)$ or $\mathcal{H} = L^2(\mathbb{R}^d, dx)$ respectively. Prove that the almost sure spectrum is the topological support of the distribution at one site of the potential ,i.e. the probability measure $\mu$ defined by $\mu(A) = \mathbb{P}\{V(0) \in A\}$ for $A$ in the Borel $\sigma$-field of $\mathbb{R}$.*

**Problem V.5.3** *Let us denote by $H_0$ the Laplace operator on the strip of width $\ell$. On can consider this operator as associated to an invertible ergodic dynamical system with space $\Omega$ reduced to a single point. The transfer matrices $g_\lambda(n)$, $n = 0, 1, \ldots$ do not depend on $n$ and are equal to :*

$$g_\lambda = \begin{pmatrix} S - \lambda I_\ell & -I_\ell \\ I_\ell & 0 \end{pmatrix}$$

*where $S$ is the matrix of the one dimensional Laplace operator restricted to the box $[1, \ldots, \ell]$ with Dirichlet boundary conditions.*

1. *Prove that the complex number $z$ is an eigenvalue of $g_\lambda$ if and only if*
$$z + 1/z = -\lambda + e$$
*for $e$ an eigenvalue of $S$.*
2. *Prove that the eigenvalues of $S$ are*
$$e_k = -2\cos(\frac{k\pi}{\ell+1}) \qquad k = 1 \ldots \ell$$
3. *Conclude that the Lyapunov exponent $\gamma_k(\lambda)$ of $H$ is equal to $\log(|z_k|)$ where $z_k$ is the solution of modulus greater or equal than one of the equation $z + 1/z = e_k - \lambda$.*
4. *Using the result of Problem II.5.2 prove that :*
$$\Sigma(H) = \{\lambda \; ; \; \gamma_\ell(\lambda) = 0\}$$

**Problem V.5.4** *A complete solution of this problem can be found in [69]. Let us denote by $H_\beta(\omega)$ the Almost-Mathieu operator associated to the sequence of potentials $V(n, \omega) = \beta\cos(2n\pi\alpha + \omega)$. Let $\lambda$ be an eigenvalue (if any) of $H_\beta(\omega)$ associated to the normalized eigenvector $\psi$. Let us denote by $\mathbb{T}$ the torus $[0, 2\pi]$ and by $\hat{\psi}(t)$ the function in $L^2(\mathbb{T}, dt)$ whose Fourier coefficients are given by the sequence $\psi(n)$ namely:*
$$\hat{\psi}(t) = \sum_{n=-\infty}^{+\infty} \psi(n)\, e^{2int}$$
*It follows from the theory of Andre and Aubry (see [7]) that the sequence*
$$\phi_t(n) = \hat{\psi}(2n\pi\alpha + t)e^{in\omega}$$
*is a solution of the eigenvalue equation :*
$$H_{4/\beta}(t)\psi = \frac{2\lambda}{\beta}\,\psi$$
*for almost all $t$ with respect to the Lebesgue measure.*

1. *Using the fact that $\phi$ is a unit vector in $L^2(\mathbb{T}, dt)$ prove that there exists a measurable function $C(t)$, almost everywhere finite with respect to the Lebesgue measure, such that:*
$$|\phi_t(n)| \leq |n|\, C(t)$$

2. *Using the Oseledec's theorem, conclude that the Lyapunov exponent associated to the operator $H_{4/\beta}$ vanishes at the point $\frac{2\lambda}{\beta}$.*
3. *Taking in account Proposition V.4.6 prove that the almost Mathieu operator has no eigenvalues for $|\beta| < 2$.*
4. *In the case $\beta = 2$ conclude that the eigenvalues of $H_2(\omega)$ lie in the set where the Lyapunov exponent vanishes.*

**Problem V.5.5** *Let $H_\epsilon = H_0 + \epsilon F(X(t))$ be the Schrödinger operator introduced by the Russian school with a coupling constant $\epsilon > 0$. Recall that $X(t)$ is a stationary process of Brownian motion on some compact connected Riemannian manifold $M$ and $F$ is a smooth function on $M$. It has been proved in [9] that if $F$ satisfies the additional condition $\mathbb{E}\{F(X(t))\} = 0$ then if $\lambda > 0$ one has:*

$$\gamma_\epsilon(\lambda) = \frac{\pi\phi(2\sqrt{\lambda})}{4\lambda}\epsilon^2 + O(\epsilon^3)$$

*where $\phi$ is the density of the spectral measure of the stationary second order process $F(X(t))$.*

1. *Let $\sigma_F(\cdot)$ be the spectral measure $< E^L(\cdot)F, F >$ of the Laplace Beltrami operator $L$ of $M$ acting on $L^2(M, dx)$. Using the Fourier transform of the Cauchy law and the formula expressing the semigroup of transition probabilities of the Brownian motion as an exponential of the operator $L$, prove that:*

$$\phi(t) = \frac{1}{\pi}\int_0^\infty \frac{x}{x^2 + t^2}\, d\sigma_F(x)$$

2. *Conclude that for $\lambda > 0$ one has :*

$$\gamma_\epsilon(\lambda) = \frac{\epsilon^2}{4\lambda}\int_0^\infty \frac{x}{x^2 + 4\lambda}\, d\sigma_F(x) + O(\epsilon^3)$$

   *Hence the Lyapunov exponent defined in Chapter VIII, Section 5 for a non stationary potential of the form $a(t)F(X(t))$ where the function $a(t)$ satisfies $\lim_{t\to\pm\infty} a(t) = 0$, coincides with the coefficient of $\epsilon^2$ in the asymptotic expansion of $\gamma_\epsilon(\lambda)$.*

**Problem V.5.6** *We adopt in this problem the notation and the hypotheses of Lemma V.4.24. Let $\lambda_0$ be fixed in $\mathcal{I}$ and $z_0$ be a complex number such that* *the resolvent operator $R_{\lambda_0,z_0}$ exists as a bounded operator on $E_1$ and $E_2$. In the following we omit the index $z_0$ in the notation of the resolvent operator. It follows from the hypothesis of continuity of the mapping $\lambda \hookrightarrow T_\lambda$ on $E_1$ and $E_2$ that there exists a neighborhood of $\lambda_0$ in which $R_\lambda$ is well defined and such that the mapping $\lambda \hookrightarrow R_\lambda$ is continuous on $E_1$ and $E_2$. Let us define the operators:*

$$\begin{aligned}
\Delta(\lambda) &= \frac{R_\lambda - R_{\lambda_0}}{\lambda - \lambda_0} - R_{\lambda_0} T'_{\lambda_0} R_{\lambda_0} \\
\Delta_1(\lambda) &= (R_\lambda - R_{\lambda_0}) \frac{T_\lambda - T_{\lambda_0}}{\lambda - \lambda_0} R_{\lambda_0} \\
\Delta_2(\lambda) &= R_{\lambda_0} (\frac{T_\lambda - T_{\lambda_0}}{\lambda - \lambda_0} - T'_{\lambda_0}) R_{\lambda_0}
\end{aligned}$$

1. *Using the resolvent identity prove that:*

$$\Delta(\lambda) = \Delta_1(\lambda) + \Delta_2(\lambda)$$

2. *Let us denote by $||A||_1$, $||A||_2$, $||A||_{1,2}$ the norms of an operator $A$ on $E_1$, $E_2$ and from $E_1$ into $E_2$ respectively. Using the derivability of the operator $T_\lambda$ which reads:*

$$||T_\lambda - T_{\lambda_0} - (\lambda - \lambda_0) T'_{\lambda_0}||_{1,2} = |\lambda - \lambda_0| \, \Upsilon(\lambda - \lambda_0)$$

*where the function $\Upsilon$ satisfies $\lim_{\lambda \to 0} \Upsilon(\lambda) = 0$ prove that:*

$$\begin{aligned}
||\Delta_1(\lambda)||_{1,2} &\leq \Upsilon(\lambda - \lambda_0) \, ||R_{\lambda_0}||_2 \, ||R_{\lambda_0}||_1 \\
||\Delta_2(\lambda)||1,2 &\leq ||R_\lambda - R_{\lambda_0}||_2 \, (||T'_{\lambda_0}||_{1,2} + \Upsilon(\lambda - \lambda_0))
\end{aligned}$$

3. *Deduce from the above inequalities the result of Lemma V.4.24.*

**Problem V.5.7** *Let us call "plane waves" two sequence $u(n)$ and $v(n)$ of the form $u(n) = e^{in\vartheta}$ and $v(n) = e^{-in\vartheta}$ where $\vartheta$ is a real number different from a multiple of $\pi$. Then any sequence $\psi(n)$ can be "projected" on the waves $u$ and $v$ according to:*

$$\begin{pmatrix} \psi(n+1) \\ \psi(n) \end{pmatrix} = \alpha(n) \begin{pmatrix} u(n+1) \\ u(n) \end{pmatrix} + \beta(n) \begin{pmatrix} v(n+1) \\ v(n) \end{pmatrix} = T(n) \begin{pmatrix} \alpha(n) \\ \beta(n) \end{pmatrix}$$

*where $T(n) = \begin{pmatrix} u(n+1) & v(n+1) \\ u(n) & v(n) \end{pmatrix}$ is an invertible matrix. We set $Z_\lambda(n) = T^{-1}(n) g_\lambda(n) T(n-1)$ and $\Pi_\lambda(n) = Z_\lambda(n-1) \cdots Z_\lambda(0)$. In these* *formulas the matrix $g_\lambda(n)$ is the transfer matrix of the one dimensional Schrödinger operator.*

1. *Prove that $\Pi_\lambda(n) = T^{-1}(n)U_\lambda(n)T(-1)$. Conclude that if $\psi$ satisfies the difference equation $(H\psi)(n) = \lambda\psi(n)$ for $n \in [0, \cdots, N-1]$ then one has:*
$$\begin{pmatrix} \alpha(N) \\ \beta(N) \end{pmatrix} = \Pi_\lambda(N) \begin{pmatrix} \alpha(-1) \\ \beta(-1) \end{pmatrix}$$
*This means that if we "input" the wave of coordinates $\begin{pmatrix} \alpha(-1) \\ \beta(-1) \end{pmatrix}$ into the box $\Lambda = [0, \cdots, N-1]$, then the "output" wave at site $N$ has coordinates $\begin{pmatrix} \alpha(N) \\ \beta(N) \end{pmatrix}$.*
2. *Let $SU(1,1)$ be the subgroup of matrices in $SL(2, \mathbb{C})$ of the form*
$$\begin{pmatrix} \alpha & \beta \\ \bar{\beta} & \bar{\alpha} \end{pmatrix}$$
*Prove that if $g \in SL(2, \mathbb{R})$ then $T^{-1}(n)gT(n-1)$ is in $SU(1,1)$. Conclude that the matrices $Z_\lambda(n)$ and hence $\Pi_\lambda(n)$ are in $SU(1,1)$.*
3. *The reflection and transmission coefficients $t_N$ and $r_N$ of the box $\Lambda = [0, \cdots, N-1]$ are defined by the equation:*
$$\begin{pmatrix} t_N \\ 0 \end{pmatrix} = \Pi_\lambda(N) \begin{pmatrix} 1 \\ r_N \end{pmatrix}$$
*and the conductance of $\Lambda$ is defined as the ratio $C_N = \left|\frac{t_N}{r_N}\right|^2$.*
*Prove that $C_N = \frac{1}{2}(||\Pi(N)e_1||^2 - 1)^{-1}$ where $e_1$ is the first basis vector of $\mathbb{R}^2$.*

In the sequel we will set $y = T(-1)e_1$.

4. *We now suppose that $H(\omega)$ is an ergodic family of Schrödinger operators.*
   (a) *From the inequalities*
$$||T(n)||^{-1}||U_\lambda(n)y|| \leq ||T^{-1}(n)U_\lambda(n)y|| \leq ||T^{-1}(n)||\ ||U_\lambda(n)y||,$$
$$||T(n)|| \leq 2 \quad \text{and} \quad ||T^{-1}(n)|| \leq |\sin\vartheta|^{-1}$$
   *prove that we can write*
$$\log||T^{-1}(n)U_\lambda(n)y|| = \log||U_\lambda(n)y|| + K(n)$$
   *where the sequence of random variables $K(n)$ is bounded.*

(b) *Remark that one can write:*

$$\frac{1}{n}\log C_n = \frac{Y(n)}{n} - \frac{2}{n}\log ||U_\lambda(n)y||$$

*where*

$$Y(n) = \log 2 - 2\log K(n) - \log\big(1 - (||U_\lambda(n)y|| + K(n))^{-1}\big)$$

5. *Recall that the Lyapunov exponent $\gamma(\lambda)$ is well defined when the integrability condition $\mathbb{E}\{\log(1 + |V(0)|)\} < \infty$ is satisfied. Under this assumption, prove that:*

$$\lim_{n\to+\infty} \frac{1}{n}\log C_n = -2\gamma(\lambda) \quad \mathbb{P} - a.e.$$

*Hint:*

$$||\Pi_\lambda(N)e_1|| = ||T^{-1}(N)U_\lambda(N)T(-1)e_1|| = ||T^{-1}(N)U_\lambda(N)y||$$

6. *We now assume that the sequence of potentials is i.i.d., that their common distribution is not carried by a single point and that there exists $\alpha > 0$ with $\mathbb{E}\{|V(0)|^\alpha\} < \infty$.*
*Prove that there exists a strictly positive number $\sigma(\lambda)$ such that the the sequence $\frac{1}{\sqrt{n}}\big(\log C_n + 2n\gamma(\lambda)\big)$ converges in distribution to a centered Gaussian law with variance $\sigma^2(\lambda)$.*
*Hint: write*

$$\frac{1}{\sqrt{n}}(\log C_n + 2n\gamma(\lambda)) = -\frac{2}{\sqrt{n}}(\log ||U_\lambda(n)y|| - n\gamma(\lambda)) + \frac{Y(n)}{\sqrt{n}}$$

*and use the Central Limit Theorem A.II.3.6 in [33] and the Borel-Cantelli's lemma to conclude.*

## V.6 Notes and Complements

The very first part of the chapter deals with a rigorous discussion of the various measurability concepts necessary to the analysis of random Schrödinger operators. These technical problems were ignored in most of the early papers on the subject. The measurability problems are usually easy to solve in the discrete case of the lattice $\mathbb{Z}^d$, but they are more involved in the continuous case. The first serious attempts to settle these questions are dure to Kunz and Souillard, Kirsch and Martinelli and Carmona. They apperared in [204], [176], [177] and [45]. We followed the analysis of this last reference.

The second topic of the chapter is the study of the various spectra of ergodic families of self-adjoint operators. We prove that, at least in this case, all these spectra are nonrandom. The main ideas are due to Pastur. They can be traced back to his fundamental paper [261]. They appeared in English only 6 years later in[263]. They have been used and generalized in many directions. See for example [204] and [176], [177] and [45]. Our discussion is patterned after this last reference. Note that essentially nothing but the definition of the ergodicity has been used in the first part of the chapter. Note also that the nonrandomness of the spectrum is not true in general. In the one-dimensional case of Sturm-Liouville operators, Molcanov and Seidel proved in [242] that a random negative isolated eigenvalue could be present in some cases. English, Kirsch, Schroder and Simon investigated in [96] the properties of a multidimensional Schrödinger operator with a random potential $\{V(x,\omega);\ x \in \mathbb{R}^d, \omega \in \Omega\}$ such that $V(x,\omega) = V^+(x,\omega)$ for $x$ in a one half space and $V(x,\omega) = V^-(x,\omega)$ in the complimentary subspace, where $\{V^+(x,\omega);\ x \in \mathbb{R}^d, \omega \in \Omega\}$ and $\{V^-(x,\omega);\ x \in \mathbb{R}^d, \omega \in \Omega\}$ are two (possibly uncorrelated) ergodic random potential. The stationarity in the directions orthogonal to the separating hyperplane is enough to show that the spectrum is still nonrandom. In fact the set $\Sigma$ can be shown to contain the union of the nonrandom spectra $\Sigma^+$ and $\Sigma^-$ of the stationary ergodic Schrödinger operators corresponding to the potentials $V^+$ and $V^-$. This union is called the bulk spectrum by the authors. The inclusion is strict in some cases and the other energies are called the surface energies. We shall come back to this model in the Notes and Complements of Chapter VI when we discuss the integrated density of states.

The systematic use of the Lebesgue decomposition of the resolution of the identity and PropositionV.1.7 are borrowed from Carmona[45]. Finally, the reader will notice that Corollary V.2.3 is nothing but Theorem 1 of Kotani [192]. Also, PropositionV.3.2 is due to Kirsch and Martinelli [180].

Let us now comment on the closed set $\Sigma$ which is the spectrum of almost all the operators $H(\omega)$. According to Corollary V.2.3, this set is increasing with respect to the topological support of the distribution of the potential function $\{V(x); x \in \mathbf{R}^d\}$. This result is due to Kotani [192]. In the same paper the essential support of the absolute continuous spectrum is shown to be decreasing in this topological support of the distribution of the potential function in the one dimensional case and as we already pointed out it would be nice to have a proof of this result in the general multidimensional case. The characterization of $\Sigma$ as a set gave rise to various publications. The part of PropositionV.3.5 dealing with the lattice case of $\mathbb{Z}^d$ it due to Kunz and Souillard, see [204]. In some continuous cases it can be shown that $\Sigma$ is the union of the spectra of the deterministic Schrödinger operators with periodic potential chosen from the topological support of the distribution of the random potential function $\{V(x); x \in \mathbb{R}^d\}$ in path space. See [177]. Finally, we note that the case of almost periodic potentials has been investigated by many functional analysts without appealing to any probability theory. Roughly speaking one can say that the further it is from a periodic function, the more frequencies one needs to approximate it by trigonometric polynomials and the more gaps will appear in the spectrum $\Sigma$. In fact $\Sigma$ has a strong tendency to become a Cantor set as we shall explain in Chapter VII.

The second part of this chapter is devoted to the regularity properties of Lyapunov exponents. The main ideas (and proofs) are due to Le Page. The proofs of the Hölder continuity and of the smoothness can be found in [216]. One can also find a proof of smoothness of the invariant measure in the one dimensional Anderson model by B. Simon and M. Taylor in [303]. In this work the authors use heavily the theory of Sobolev spaces associated to the (left and right) Laplace operator on $SL(2,\mathbb{R})$. They assume that the distribution of potential has a compact support and has a density $h$ with the following properties:
*There exist $g \in L^1(\mathbb{R})$ and $\alpha > 0$ such that the Fourier transforms $\hat{h}$ and $\hat{g}$ of $h$ and $g$ satisfy $\hat{g}(t) = (1+t^2)^{\alpha/2}\hat{h}(t)$.*
Under these assumptions, the probability measure $\mu_\lambda$ on $SL(2,\mathbb{R})$ lies in some Sobolev space and the smoothing action of the convolution on $SL(2,\mathbb{R})$ implies that the invariant measure is $C^\infty$. Using the same technics on the group $SP(\ell,\mathbb{R})$, C. Glaffig has extended this proof to the strip in [122]. One can remark that the proof given in this chapter contains all the previous works on the subject.
The subharmonicity of the Lyapunov exponents pointed out by M. Herman [147] have been studied by W. Craig and B. Simon in [62]. They make use of this property to obtain some lower bounds on the decay of eigenfunctions. The beautiful proof of the positivity of the exponent in the "almost Mathieu model" has been borrowed from the paper of M. Herman cited above and this method of minoration can be used in much more general situations. The first general results about the behavior of the Lyapunov exponent in this model are due to G. Andre and S. Aubry [7]. Using some

Fourier transform thecnics they establish an important relation between the Lyapunov exponent and the parameters of this operator namely:

$$\gamma(\beta,\alpha,\lambda)=\gamma(\frac{4}{\beta},\alpha,\frac{\beta\lambda}{2})+\log(\frac{|\beta|}{2})$$

One can recover the positivity of the exponent as a byproduct of this relation . This model will be studied in more details in Chapter VII.

After the paper of B. Derrida and E. Gardner [80] a lot of attention have been given to the so called "weak disorder expansion" of the Lyapunov exponent that is for small $\beta$. Let us consider the following model :

$$[H_\beta\psi](n)=-\psi(n+1)-\psi(n-1)+\beta V(n)\psi(n)$$

where $V(n)$ is an i.i.d. sequence with an absolutely continuous distribution and moments of all orders. For any nonzero coupling constant $\beta$, the Lyapunov exponent $\gamma_\beta(\lambda)$ is positive and the unique invariant probability measure $\nu_{\beta,\lambda}$ has a $C^\infty$ density $\phi_{\beta,\lambda}$ with respect to the Cauchy measure $m$ on the projective space.
For $\lambda\in[-2,+2]$ one sets $\lambda=-2\cos\pi\alpha$ and we remark that for $\beta=0$ the matrix $\begin{pmatrix}\lambda & -1\\ 1 & 0\end{pmatrix}$ is conjugate to a rotation matrix of angle $\pi\alpha$ hence one has two distinct limit cases:

- $\alpha$ is not rational in which case $m$ is the unique invariant measure.
- $\alpha$ is rational in which case there exists an infinite number of invariant measures.

In any case one has $\gamma_{0,\lambda}=0$. It is not difficult to check that the proof of the subharmonicity of the Lyapunov exponent with respect to $\lambda$ can also works to prove the same property with respect to the parameter $\beta$. Since a nonnegative subharmonic function is continuous on is vanishing set it follows that $\lim_{\beta\to0}\gamma_{\beta,\lambda}=0$ . In the first case the unicity of the invariant measure when $\beta=0$ obviously implies that the invariant measure $\nu_{\beta,\lambda}$ weakly converges to $m$ when $\beta\to0$. The situation is more complicated in the second case. Let us assume that $\mathbb{E}\{V(0)\}=0$ and that the density of the invariant measure admits the "formal" expansion:

$$\phi_{\beta,\lambda}=\sum_{n=0}^{\infty}\frac{\beta^n}{n!}\phi_\lambda^{(n)}$$

Then Klein and Bovier have obtained the following results in [34]:

- If $\alpha$ is irrational then it is possible to compute all the coefficients in the above expansion by an induction procedure.
- If $\alpha = p/q$ with $p$ and $q$ relatively prime integers then:

    1. The invariance equation allows to compute all the coefficients in the above expansion up to the order $q-1$ (but not the coefficient of order $q$).
    2. It is possible to find an extra equation satisfied by the invariant density which allows to compute all the coefficients in the above expansion. These coefficients agree with those computed in 1. up to the order $q-1$

An other kind of "expansion" of the Lyapunov exponent (or of the invariant measure) was proposed by Luck in [225]. Let $V(n)$ be a sequence of potentials such that correlations of any order exist and let $H$ be the associated Schrödinger operator. For any complex number $\lambda$ let us define the complex number $t$ such that

- $\Re t > 0$ , $\Im m\, t \in (-\pi, +\pi)$ if $|\lambda| > 2$
- $t = iy$ with $0 \leq y \leq \pi$ if $|\lambda| \leq 2$

In any case one has the relation $\lambda = -2\cosh(t)$. Then for $\Im m\, \lambda > 0$ one gets the expansion:

$$\gamma(\lambda) = \sum_{n=0}^{\infty} a_n(\lambda)$$

where each $a_n(\lambda)$ involves only correlations of order $n$ for the sequence of potentials. It is possible to compute all the terms in particular one has $a_0 = t$. In the case $a_1 = 0$ ("zero mean") the singularities of $a_2(\lambda)$ when $\Im m\, \lambda \to 0$ seem to be related to the spectral properties of the operator $H$. Such an analysis have been carried out in a mathematical framework by Bellissard[22] in the case of potentials generated by a Morse sequence.

The study of the asymptotic behavior of the conductance has been initated by K. Ishii and Matsuda [155], O'Connor [255] and Verheggen [326]. Using the Central Limit Theorem on the group $SP(\ell, \mathbb{R})$ Jhonston and Kunz have extended these technics to the strip in [158].

# VI

# THE INTEGRATED DENSITY OF STATES

Contents :



This chapter is entirely devoted to the study of the so-called integrated density of states. It is an object of interest for various reasons. It is of physical importance for it can be measured experimentally in some cases. On the top of its physical appeal, the integrated density of states is a very interesting mathematical object which deserves to be investigated for its own sake. Many challenging mathematical problems remain open in this respect. Finally, several proofs of technical results crucial to the study of the spectral properties of the random Hamiltonians, and in particular of the localization, rely very heavily on estimates of the integrated density of states.

We prove its existence for fairly general families of stationary ergodic random potentials. As usual, the discrete case of the lattice $\mathbb{Z}^d$ is easier than the continuous case.

We give several proofs of the existence. We also show that the topological support of the integrated density of states is almost surely equal to the spectrum of the operator. In particular, it has to vanish at the left most point of this spectrum. One may then ask the following natural question: how fast does it vanish, or in other words, how many low energy states are there? Using nonrigorous heuristic arguments, the physicist I.M. Lifschitz argued that in many cases the vanishing was exponential in some power of the energy and since then this power has been called the Lifschitz exponent. We study several models for which the existence and the computation of this exponent can be carried out rigorously. One possible approach is to use Abelian-Tauberian arguments to transform the problem into the evaluation of the large time behavior of the Laplace transform of the integrated density of states. This Laplace transform can be expressed as a path integral. The Laplace method to evaluate integrals and the theory of large deviations can then be used. Another approach relies on a direct estimation of the energy levels of the system confined to finite boxes by means of the Dirichlet Neumann bracketing technique introduced in Chapter I.

In the continuous case, proving that the integrated density of states is absolutely continuous (and hence, has an actual density) is a difficult problem which has no general answer yet. In fact, it is not even known that this measure has no point masses in the multidimensional continuous case! There are nevertheless many smoothness results. They deal mostly with the one dimensional cases and the lattice cases for which satisfactory answers are available. We review the most important ones in the last section.

## VI.1 Existence Problems

### VI.1.1 Setting of the Problem

Let $H = H_0 + V$ be a $d$-dimensional Schrödinger operator and let us try to count the possible energy levels of the system. A problem of a mathematical nature arises immediately. The spectrum of $H$ is very often continuous above some energy or, in some other cases, the set of energy levels is dense, as is most of the time the case for disordered systems. The source of the problem contains the elements of a solution: the mathematical model involves a system in the whole (infinite) space while real systems are actually finite, even though large. A possible way out is the study of the system in bounded domains, say $\Lambda$. Indeed, if the operator $H$ is suitably defined on $L^2(\Lambda, dx)$, note that this will involve the choice of boundary conditions, its spectrum will be discrete according to the results we proved in Chapter II and it will be feasible to count the eigenvalues (which are the energy levels of the system). We look for a volume independent quantity to give an account of very large systems, so we divide this count by the volume of the domain $\Lambda$ and try to see if the result converges when $\Lambda$ is very large to some sort of average number of energy levels per unit volume. Such a quantity has been experimentally measured for crystals. The corresponding mathematical models are based on periodic potential functions and consequently, they fit into the framework of random potentials. Many physicists called this quantity the *density of states.* This terminology is very natural since it gives an average count per unit volume. Unfortunately it is very misleading from the mathematical point of view. Indeed, this quantity happens to be the distribution function of a measure and we will discuss cases where this measure is not absolutely continuous, or in other words, it does not have a density in the usual mathematical sense. Fortunately most of the theoretical physicists and mathematical physicists recognized this problem and in order to avoid any possible confusion they coined the term *integrated density of states* instead. We believe that *distribution function of states* would be more appropriate. Nevertheless, we will try to stick to the most commonly used terminology.

## VI.1.2 Path Integral Approach

As before we assume that a complete probability space $(\Omega, \mathcal{F}, \mathbb{P})$ and an ergodic flow $\{\theta_x; x \in T\}$ of measure preserving transformations are given. We consider a measurable function $v : \Omega \to \mathbb{R}$ and we define the random potential $V$ by:

$$V(x,\omega) = v(\theta_x \omega) \qquad x \in T,\ \omega \in \Omega. \tag{VI.1}$$

We first consider the continuous case $T = \mathbb{R}^d$.

We assume that the function $x \hookrightarrow V(x,\omega)$ is in $K_{d,loc}$ for each $\omega \in \Omega$. The reader can think of this function as being bounded if he wants to postpone the treatment of local singularities. We have already noticed that for each bounded domain $\Lambda$, the operator $H_\Lambda^{(D)}(\omega) = H_{0,\Lambda}^{(D)} + V(\cdot,\omega)$ with Dirichlet boundary conditions has a discrete spectrum, say:

$$-\infty < \lambda_0^{(D)}(\Lambda,\omega) < \lambda_1^{(D)}(\Lambda,\omega) \leq \lambda_2^{(D)}(\Lambda,\omega) \leq \cdots$$

We define the measure $\nu_{\Lambda,\omega}^{(D)}$ by:

$$\nu_{\Lambda,\omega}^{(D)} = \frac{1}{|\Lambda|} \sum_{j\geq 0} \delta_{\lambda_j^{(D)}(\Lambda,\omega)}. \tag{VI.2}$$

Its distribution function is given by $N_{\Lambda,\omega}^{(D)}(\lambda)/|\Lambda|$ where $|\Lambda|$ denotes the volume of the domain $\Lambda$ and where the eigenvalue counting function $N_{\Lambda,\omega}^{(D)}(\lambda)$ is defined by:

$$N_{\Lambda,\omega}^{(D)}(\lambda) = \#\{\text{eigenvalues of } H_{\Lambda}^{(D)}(\omega) \leq \lambda\}. \tag{VI.3}$$

We now let the domain $\Lambda$ tend to $\mathbb{R}^d$ in a certain sense and we try to show that the quantity $N_{\Lambda,\omega}^{(D)}(\lambda)$ converges to a limit $N_{\omega}^{(D)}(\lambda)$ (presumably independent of $\omega$ because of the ergodicity assumption), which should be interpreted as the number of possible states per unit volume with energy at most $\lambda$.
We restrict ourselves to rectangular domains $\Lambda$ of the form $\prod(-a_j, a_j)$ with $a_j > 0$, and we say that $\Lambda \to \mathbb{R}^d$ when all the $a_j$ converge to $+\infty$ simultaneously. We aim to show that $\mathbb{P}$-almost surely in $\omega \in \Omega$, the measures $\nu_{\Lambda,\omega}^{(D)}$ converge vaguely to a limiting measure $\nu^{(D)}$ which we will most of the time identify with its distribution function:

$$N^{(D)}(\lambda) = \nu^{(D)}((-\infty, \lambda])$$

and which we will call the *integrated density of states.* The first existence result which we give uses path integral methods and we urge the reader to refresh his memory with the notations introduced in Chapter II.

**Theorem VI.1.1** *Let us assume that the function*

$$\mathbf{W} \times \Omega \ni (w, \omega) \to \exp[\int_0^t V^-(X_s(w), \omega) ds]$$

*is in* $L^r(\mathbf{W}\times\Omega, \mathbb{P}_0\otimes\mathbb{P})$ *for some* $r > 2$ *and all* $t > 0$. *Then* $\mathbb{P}$*-almost surely in* $\omega \in \Omega$*, the measures* $\nu_{\Lambda,\omega}^{(D)}$ *converge vaguely to a nonnegative measure* $\nu^{(D)}$ *(independent of* $\omega$*) as* $\Lambda \to \mathbb{R}^d$. *Moreover the Laplace transform of* $\nu^{(D)}$ *is given by:*

$$\mathcal{L}(\nu^{(D)}, t) = \frac{1}{(4\pi t)^{d/2}} \mathbb{E}_{0,0,t,0} \otimes \mathbb{E}\{\exp[-\int_0^t V(X_s(w), \omega) ds]\}. \tag{VI.4}$$

*Proof:*
It is sufficient to show that for $\omega$ in some set $\Omega_1$ of $\mathbb{P}$-probability one, the Laplace transform $\mathcal{L}(\nu_{D,\omega}^{(D)}, t)$ converges pointwise to the right hand side of (VI.4). In fact, for each $t > 0$ we have:

$$\mathcal{L}(\nu_{\Lambda,\omega}^{(D)}, t) \quad = \quad \int_{\mathbb{R}} e^{-\lambda t} \nu_{\Lambda,\omega}^{(D)}(d\lambda)$$

$$
\begin{aligned}
&= \frac{1}{|\Lambda|}\sum_{j\geq 0} e^{-\lambda_j^{(D)}(\Lambda,\omega)t} \\
&= \frac{1}{|\Lambda|}\operatorname{tr}[e^{-t H_\Lambda^{(D)}(\omega)}] \\
&= \frac{1}{|\Lambda|}(4\pi t)^{-d/2}\int_\Lambda \mathbb{E}_{0,x,t,x}\{e^{-\int_0^t V(X_s,\omega)ds}; T_\Lambda > t\}dx
\end{aligned}
$$

because the kernel of the operator $e^{-t H_\Lambda^{(D)}(\omega)}$ is continuous. Consequently we have:

$$
\mathcal{L}(\nu_{\Lambda,\omega}^{(D)}, t) \;=\; (A_\Lambda) - (B_\Lambda)
$$

with:

$$
(A_\Lambda) = \frac{1}{|\Lambda|}(4\pi t)^{-d/2}\int_\Lambda \mathbb{E}_{0,x,t,x}\{\exp[-\int_0^t V(X_s,\omega)ds]\}dx
$$

and

$$
(B_\Lambda) = \frac{1}{|\Lambda|}(4\pi t)^{-d/2}\int_\Lambda \mathbb{E}_{0,x,t,x}\{\exp[-\int_0^t V(X_s,\omega)ds]; T_\Lambda \leq t\}dx.
$$

For each fixed $t > 0$ we can apply a multidimensional form of the ergodic theorem to $(A_\Lambda)$ and the limit is exactly the right hand side of (VI.4). Consequently we need only to show that $(B_\Lambda)$ converges to 0. First we remark that:

$$
\begin{aligned}
(B_\Lambda) \;\leq\; & \left(\frac{1}{|\Lambda|}(4\pi t)^{-d/2}\int_\Lambda \mathbb{E}_{0,x,t,x}\{\exp[-(r'/2)\int_0^t V(X_s,\omega)ds]\}dx\right)^{2/r'} \\
& \left(\frac{1}{|\Lambda|}(4\pi t)^{-d/2}\int_\Lambda \mathbb{P}_{0,x,t,x}\{T_\Lambda \leq t\}dx\right)^{(r'-2)/r'}
\end{aligned}
$$

for each $r'$ such that $2 < r' < r$, and second that:

$$
\lim_{\Lambda\to\mathbb{R}^d} \frac{1}{|\Lambda|}\int_\Lambda \mathbb{P}_{0,x,t,x}\{T_\Lambda \leq t\}dx = 0,
$$

and finally that the supremum over $\Lambda$ of the first factor is an integrable (and hence finite almost everywhere) random variable by a multidimensional maximal ergodic lemma. This proves that $(B_\Lambda)$ tends to 0 for $\mathbb{P}$-almost all $\omega \in \Omega$. Hence, for each $t > 0$, there exists a measurable set $\Omega_t \in \mathcal{F}$ such that $\mathbb{P}(\Omega_t) = 1$ and

$$
\lim_{\Lambda\to\mathbb{R}^d} \mathcal{L}(\nu_{\Lambda,\omega}^{(D)}, t) = \frac{1}{(4\pi t)^{d/2}}\mathbb{E}_{0,0,t,0}\otimes\mathbb{E}\{exp[-\int_0^t V^-(X_s(w),\omega)ds]\}
$$

for each $\omega \in \Omega_t$. Making the $\mathbb{P}$-almost sure independent of $t$ requires some extra work. The crucial step is to prove the existence of a random distribution function $\lambda \hookrightarrow F_\omega(\lambda)$ satisfying:

$$\int_{\mathbb{R}} e^{-t\lambda} F_\omega(\lambda)\, d\lambda$$

which controls the approximations in the sense that $N^{(D)}_{\Lambda,\omega}(\lambda) \leq F_\omega(\lambda)$ for all $\lambda \in \mathbb{R}$ and all $\Lambda$. This follows a simple application of the multidimensional version of the maximal ergodic lemma. The reader interested in the details may consult [251]. ∎

**Remark VI.1.2** Part of the above proof applies to the stationary not necessarily ergodic case. Under this weaker assumption the conclusion is that the measures $\nu^{(D)}_\Lambda$ defined by $\nu^{(D)}_\Lambda(A) = \mathbb{E}\{\nu^{(D)}_{\Lambda,\omega}(A)\}$ converge vaguely toward the integrated density of states $\nu^{(D)}$.
The above proof also applies to the case of any random potential which is merely $\mathbb{Z}^d$-stationary instead of $\mathbb{R}^d$-stationary. The only modifications consist in picking domains $\Lambda$ which are unions of elementary cubes $C_{\mathbf{n}}$ and replacing the integrals by sums over the cubes which form $\Lambda$.

In the above proof we made two arbitrary choices which do not have any physical justification. The first one concerns the choice of rectangles for domains $\Lambda$. We could have decided on other geometric shapes and other ways to converge to $\mathbb{R}^d$. The existence of the limit and the actual value of this limit should not depend on these choices. We will not elaborate on that but the reader familiar with the mathematics of statistical mechanics will easily convince himself that this problem can easily be fixed. More importantly, one should notice that the sequence $\{\Lambda_n;\ n \geq 1\}$ is not required to fill in the whole space. In fact, it is enough that $\lim_{n\to\infty} |\Lambda_n| = \infty$ and that $\lim_{n\to\infty} |\Lambda_{n+1}\backslash\Lambda_n|/|\Lambda_n| = 0$ for the above argument to apply. See also the notion of regular sequence defined in the discussion of the super-additive ergodic theorem in the following section.

A more subtle difficulty comes from the choice of the boundary conditions for the operator in the bounded domain $\Lambda$. We picked Dirichlet boundary conditions, but the choice of Neumann boundary conditions is at least as natural. Our proof can be adapted to handle this case by using reflected Brownian motion instead of the killed one, but the estimates are more involved. Also, it is not clear that the limits are the same. In fact, there are examples for which these limits are different. This is the case for hyperbolic spaces. Indeed, the size of the boundaries of the balls is of the same order as the size of the balls themselves and their contributions do not disappear in the limit. These two types of boundary conditions are in some sense extremal, but there are many other possible choices. One could think of various mixed boundary conditions, or periodic ones, or.... The existence of the limit and the independence from all these choices are subtle problems which we shall not address here. We shall nevertheless skip the superscript $(D)$ and ignore the possible dependence on our choice of the Dirichlet boundary conditions.

We now relate the measure $\nu$ which we just constructed to the spectral measures of the operators $H(\omega)$. We show that the topological support of $\nu^{(D)}$ coincides with the almost sure spectrum of the $H(\omega)'s$.

**Proposition VI.1.3** *For each strictly positive continuous function $f$ on $\mathbb{R}^d$ such that $\int f(x)^2 dx = 1$, and for each bounded Borel set $A$ in $\mathbb{R}^d$, the operator $M_f E(A,\omega) M_f$ is trace class $\mathbb{P}$-almost surely in $\omega$. Furthermore, we have:*

$$\nu(A) \geq \mathbb{E}\{\mathrm{tr}[M_f E(A,\omega) M_f]\}. \tag{VI.5}$$

*with equality when the potential is uniformly bounded below. In any case, the topological support of $\nu$ equals the almost sure spectrum, i.e.,* $\operatorname{supp}\nu = \Sigma$.

*Proof:*
Let us first assume that the potential is (uniformly) bounded below in the sense that there exists a constant $c \in \mathbb{R}$ such that $V(x,\omega) \geq c$ for all $x \in \mathbb{R}^d$ and $\omega \in \Omega$.
The first claim is then a trivial consequence of Proposition II.3.11 for the product of two Hilbert-Schmidt operators is trace class. Also, the right hand side of (VI.5) defines a nonnegative Radon measure on the interval $[c, \infty)$ and its Laplace transform is given by:

$$\mathcal{L}(\mathbb{E}\{\mathrm{tr}[M_f E(\cdot,\omega) M_f]\}, t) = \mathbb{E}\{\mathrm{tr}[M_f e^{-tH(\omega)} M_f]\}$$

by the spectral theorem. This is nothing but $\mathcal{L}(\nu, t)$ if one follows the above computations and if one recalls (VI.4). We thus proved the first half of the proposition. Note also that we have:

$$\int_{\mathbb{R}} F(\lambda) d\nu(\lambda) = \mathbb{E}\{\mathrm{tr}[M_f\, F(H(\omega)) M_f]\} \tag{VI.6}$$

for each continuous function $F$ with compact support.
To study the general case we set $V^{(n)}(x,\omega) = V(x,\omega) \vee (-n)$. For each integer $n$, the random self-adjoint operator $H^{(n)}(\omega) = H_0 + V^{(n)}(x,\omega)$

is stationary and ergodic so that its spectrum is almost surely equal to some nonrandom closed subset $\Sigma^{(n)}$ of $\mathbb{R}$. The main assumption of Theorem VI.1.1 is obviously satisfied and consequently the integrated density of states of the operator $H^{(n)}$, say $\nu^{(n)}$, exits. Finally we remark that the first half of the present proposition applies to $H^{(n)}(\omega)$. This shows that $\nu^{(n)}$ converges vaguely to $\nu$ because the corresponding Laplace transforms converge. Hence, if $F$ is a continuous function with compact support:

$$\begin{aligned} \int_{\mathbb{R}} F(\lambda)\,\nu(d\lambda) &= \lim_{n\to\infty} \int_{\mathbb{R}} F(\lambda)\,\nu^{(n)}(d\lambda) \\ &\geq \mathbb{E}\{\mathrm{tr}[M_f F(H(\omega))M_f]\} \end{aligned}$$

by Fatou's lemma and the strong resolvent convergence of $H^{(n)}(\omega)$ to $H(\omega)$. In particular, this shows that the operator $M_f F(H(\omega))M_f$ is $\mathbb{P}$-almost surely trace class. If $A$ is a bounded Borel subset of $\mathbb{R}$ we can bound its characteristic function $\mathbf{1}_A$ from above by such a function $F$ and the above estimate shows that the operator $M_f E(A,\omega)M_f$ is almost surely trace class.

We now prove the last claim concerning the topological support of the integrated density of states. If $U$ is any open set contained in $\Sigma^c$ we have:

$$\nu(U) \leq \lim\inf_{n\to\infty} \nu^{(n)}(U) = 0$$

because $U$ is necessarily contained in $\Sigma^c_{(n)}$ for large enough $n$. This proves that the topological support of $\nu$ is contained in $\Sigma$. To prove the other inclusion we let $\lambda$ be any point in $\Sigma$ and we choose a nonnegative continuous function $F$ with compact support such that $F(\lambda) > 0$. Since $\Sigma = \Sigma(H(\omega))$ $\mathbb{P}$-a.s. we must have $F(H(\omega)) \neq 0$ $\mathbb{P}$-a.s. We have to prove that:

$$\int_{\mathbb{R}} F(\lambda)\,d\nu(\lambda) > 0$$

and consequently that $\lambda$ is in the topological support of $\nu$. If the above integral vanishes, then the bound (VI.5) implies that $M_f F(H(\omega))M_f = 0$ $\mathbb{P}$-a.s. and consequently that $F(H(\omega)) = 0$ $\mathbb{P}$-a.s. because $f$ has been chosen strictly positive everywhere. This is a contradiction. The proof is now complete. ∎

**Remark VI.1.4** For each fixed $\omega \in \Omega$ and each fixed bounded domain $\Lambda \subset \mathbb{R}^d$ the function $F \hookrightarrow (1/|\Lambda|)\mathrm{tr}[M_\Lambda F(H(\omega))M_\Lambda]$ defines a nonnegative measure, say $\mu_{\Lambda,\omega}$ on $\mathbb{R}$. We use the notation $M_\Lambda$ for the multiplication operator by the indicator function $\mathbf{1}_\Lambda$ of the domain $\Lambda$. Notice that:

$$\int_{\mathbb{R}} F(\lambda)\,d\mu_{\Lambda,\omega}(\lambda) = \frac{1}{|\Lambda|}\mathrm{tr}[M_\Lambda F(H(\omega))M_\Lambda]$$

while we have:

$$\int_{\mathbb{R}} F(\lambda)\, d\nu_{\Lambda,\omega}^{(D)}(\lambda) = \frac{1}{|\Lambda|}\mathrm{tr}[F(H_\Lambda^{(D)}(\omega))].$$

In other words the measures $\mu_{\Lambda,\omega}$'s localize to the region $\Lambda$ the Hamiltonian $H(\omega)$ of the system in the whole space while the measures $\nu_{\Lambda,\omega}^{(D)}$'s are related to the Hamiltonian of the system in the box $\Lambda$. But the differences between these two ways to localize the system to a box should wash out in the thermodynamic limit $\Lambda \to \mathbb{R}^d$. The following computation shows that this is actually the case and that the measures $\mu_{\Lambda,\omega}$'s can be used as an alternate way to define the integrated density of states. Indeed, the difference between the Laplace transforms of the measures $\mu_{\Lambda,\omega}$ and $\nu_{\Lambda,\omega}^{(D)}$ can be computed with the Feynman-Kac formula for the operator $H(\omega)$ in the whole space and the Feynman-Kac formula for the operator $H_\Lambda^{(D)}(\omega)$. We get:

$$\begin{aligned}\lim_{\Lambda\nearrow\mathbb{R}^d} \mathcal{L}(\mu_{\Lambda,\omega},t) - \mathcal{L}(\nu_{\Lambda,\omega}^{(D)},t) &= \lim_{\Lambda\nearrow\mathbb{R}^d}(B_\Lambda)\\ &= 0,\end{aligned}$$

for each $t \in \mathbb{R}$, where we used the notation $(B_\Lambda)$ from the proof of Theorem VI.1.1 above, and the result of its estimation.

**Remark VI.1.5** If for each $\omega$ and each bounded Borel set $A$ the spectral projection $E(A,\omega)$ has a jointly continuous kernel $E(A,\omega,x,y)$ then it is easy to show that:

$$\nu(d\lambda) = \mathbb{E}\{E(d\lambda,\omega,0,0)\} \tag{VI.7}$$

by using the above ideas and a simple approximation argument. The reader may also want to take a look at Sections C6 and C7 of [294] for results in the same spirit.

These remarks prompt us to say a few words about the lattice case. Indeed, the existence problem can be avoided by choosing formula (VI.7) as a definition of the integrated density of states without any assumption on the potential. One can then use the functional calculus and the Feynman-Kac formula to compute its Laplace transform and show that it is given by a formula similar to (VI.4). In fact we have:

$$\mathcal{L}(\nu,t) = e^{2dt}\kappa(t)\mathbb{E}_{0,0,t,0}\otimes\mathbb{E}\{\exp[-\int_0^t V(X_s(w),\omega)ds]\}. \tag{VI.8}$$

where $\kappa(t)$ is the probability that a Brownian path starting from the origin is back at the origin at time $t$. Note that this quantity satisfies:

$$\kappa(t) \approx \frac{1}{\sqrt{4\pi t}^d}. \tag{VI.9}$$

The only difference between (VI.4) and (VI.8) is simply due to our choice to remove the diagonal part of the free Hamiltonian from $H_0$. Note that the proof that the topological support of the integrated density of states is equal to the almost sure spectrum can be made even simpler if one starts with formula (VI.7). So the only missing part is the fact that the integrated density of states appears as the limit of energy level counts in finite boxes. This property can be worked out backwards by showing that the appropriate Laplace transforms converge to (VI.8). We do not give the details for we will discuss this point in detail in our functional analytic approach to the problem.

### VI.1.3 Functional Analytic Approach

We now present a second existence proof of the integrated density of states. Not that we need to go through more technical arguments, but the various steps of the control of the new limiting procedure are instructive and very useful in the actual proofs of several results we want to discuss.

As before, we consider first the continuous case $T = \mathbb{R}^d$.

The present proof is drastically different from the previous one. It is based on the properties of the Dirichlet-Neumann bracketing we introduced in Section 5 of Chapter I and in Problems I.7.19 and I.7.20. We will discuss its origin in the Section 4 below and in the Notes and Complements at the end of the chapter.

We assume that the potential process $\{V(x);\ x \in \mathbb{R}^d\}$ satisfies condition (V.15) giving the essential self-adjointness of the corresponding random Schrödinger operator. Moreover, for each (bounded) domain $\Lambda \subset \mathbb{R}^d$ the multiplication operator by the function $\mathbf{1}_\Lambda V$ is in the class $K_d$ and, in particular, is a relatively bounded form perturbation of $H_0$ with relative bound 0 as we saw in Section 2 of Chapter II. In fact we have for each $\epsilon > 0$ the existence of a (finite) constant $b_\epsilon(V)$ such that:

$$\left| \int_\Lambda V(x)|f(x)|^2\,dx \right| \leq \epsilon \int_\Lambda |\nabla f(x)|^2\,dx + b_\epsilon(V)||f||^2_{L^2(\Lambda,dx)} \tag{VI.10}$$

for all $f \in \mathcal{H}^1(\Lambda)$ the Sobolev space of all the functions in $L^2(\Lambda, dx)$ the distributional gradient of which has a square integrable norm. This property is an easy consequence of the classical Sobolev's embedding theorem.

We now recall some notations from Chapter I. For each bounded domain $\Lambda$, we consider the operators $H_\Lambda^{(D)}(\omega) = H_{0,\Lambda}^{(D)} + V(\cdot,\omega)$ and $H_\Lambda^{(N)}(\omega) = H_{0,\Lambda}^{(N)} + V(\cdot,\omega)$ with Dirichlet and Neumann boundary conditions respectively. These operators have discrete spectrum, say:

$$-\infty < \lambda_0^{(\cdot)}(\Lambda,\omega) < \lambda_1^{(\cdot)}(\Lambda,\omega) \leq \lambda_2^{(\cdot)}(\Lambda,\omega) \leq \cdots$$

and we define the normalized counting measure $\nu_{\Lambda,\omega}^{(\cdot)}$ by:

$$\nu_{\Lambda,\omega}^{(\cdot)} = \frac{1}{|\Lambda|}\sum_{j\geq 0} \delta^{(\cdot)}_{\lambda_j(\Lambda,\omega)}. \tag{VI.11}$$

Its distribution function is given by $N_{\Lambda,\omega}^{(\cdot)}(\lambda)/|\Lambda|$ where $|\Lambda|$ denotes the volume of the domain $\Lambda$ and where the eigenvalue counting function $N_{\Lambda,\omega}^{(\cdot)}(\lambda)$ is defined by:

$$N_{\Lambda,\omega}^{(\cdot)}(\lambda) = \#\,\{\text{eigenvalues of } \ H_\Lambda^{(\cdot)}(\omega) \leq \lambda\}. \tag{VI.12}$$

Here and in the following we use a superscript $(\cdot)$ for both Dirichlet $(D)$ and Neumann $(N)$ boundary conditions. Notice that the decoupling properties of the Dirichlet and Neumann boundary conditions (recall Corollary I.5.3) imply that the quantities $N_{\Lambda,\omega}^{(\cdot)}(\lambda)$ viewed, for $\lambda$ fixed, as stochastic processes indexed by the regions $\Lambda$, would be additive if it was not for the effects of the boundaries. Using the order relations given in PropositionI.5.5, one sees that $\{N_\Lambda^{(D)}(\lambda)\}_\Lambda$ is superadditive while $\{N_\Lambda^{(N)}(\lambda)\}_\Lambda$ is subadditive.

We pause to explain why these simple remarks are so important at this stage. The super and subadditivity of these processes is the crucial factor in the existence of the limits which guarantee the existence of the integrated density of states. Indeed, the classical Birkhoff ergodic theorem which guarantees the existence of the limit for normalized additive processes has been extended to some super and subadditive processes and these extensions can be used in the present situation. For the sake of completeness we give the definition of the super and subadditive process and we state without proof a form of the superadditive ergodic theorem which suits our needs.

We denote by $\mathcal{J}$ the family of rectangular boxes of the form:

$$[\mathbf{a},\mathbf{b}) = \prod_{j=1}^{d}[a_j, b_j)$$

with $\mathbf{a}$ and $\mathbf{b}$ in $\mathbb{Z}^d$.

**Definition VI.1.6** *We say that a family $\{X_\Lambda;\ \Lambda \in \mathcal{J}\}$ of random variables indexed by $\mathcal{J}$ is a superadditive process if:*

*i)* $X_\Lambda(\theta_t\omega) = X_{\Lambda-t}(\omega), \quad t \in T, \ \omega \in \Omega,$
*ii)* $X_{\Lambda_1\cup\Lambda_2} \geq X_{\Lambda_1} + X_{\Lambda_2}$
*iii)* $\gamma(X) = \sup_{\Lambda\in\mathcal{J}} \frac{1}{|\Lambda|}\mathbb{E}\{X_\Lambda\} < \infty.$

*The family $\{X_\Lambda;\ \Lambda \in \mathcal{J}\}$ is said to be a subadditive process if $\{-X_\Lambda;\ \Lambda \in \mathcal{J}\}$ is superadditive and in this case the number $\gamma(X)$ is defined as the infimum of the family of normalized expectations appearing in condition iii) above.*

Note that the notion of subadditive process which we just defined is a natural generalization to the multidimensional parameter case of the notion of subadditive process used in Chapter IV to show the existence of the Lyapunov exponents. We can now state the Ackoglu-Krengel superadditive ergodic theorem in the form we want to use it.

**Theorem VI.1.7** *If $\{X_\Lambda;\ \Lambda \in \mathcal{J}\}$ is a superadditive (resp. subadditive) process in the sense of the above definition, then the limit:*

$$\lim_{\Lambda\to\mathbb{R}^d} \frac{1}{|\Lambda|} X_\Lambda$$

*exists and is almost surely equal to the number $\gamma(X)$.*

**Remark VI.1.8** The limit $\Lambda \to \mathbb{R}^d$ has to be understood in the following sense. We restrict ourselves to a nondecreasing sequence of boxes $\Lambda_L$ such that there exist $c > 0$ and another nondecreasing sequence of boxes, say $\{\Lambda'_L;\ L \geq 1\}$, such that $\mathbb{R}^d = \cup_{L\geq 1}\Lambda'_L$, $\Lambda_L \subset \Lambda'_L$ and $0 < |\Lambda'_L| \leq c|\Lambda_L|$ for all $L \geq 1$. Such sequences are called *regular*.

We now check the integrability condition *iii*) appearing in the definition of a superadditive and of a subadditive process in the case of the finite box normalized counting measures defined above with the Dirichlet and the Neumann boundary conditions. Let us use the fact that a bounded domain $\Lambda$ is contained in a finite union of unit cubes centered at points of the integer lattice $\mathbb{Z}^d$, say $\Lambda \subset \cup_{\mathbf{n}\in I} C_{\mathbf{n}}$. Then:

$$N_\Lambda^{(D)} \leq N_{\cup_{\mathbf{n}\in I}C_{\mathbf{n}}}^{(D)} \leq N_{\cup_{\mathbf{n}\in I}C_{\mathbf{n}}}^{(N)} = \sum_{\mathbf{n}\in I} N_{C_{\mathbf{n}}}^{(N)}$$

because of the discussion of Section 5 of Chapter I, and especially of Corollary I.5.3. Consequently, using the stationarity of the potential one gets:

$$\mathbb{E}\{N_\Lambda^{(D)}\} \le |\Lambda| \mathbb{E}\{N_{C_0}^{(N)}\}$$

which shows that we need only to show that $\mathbb{E}\{N_{C_0}^{(N)}\} < \infty$. A simple Sobolev's type argument gives the existence, for each $\epsilon > 0$ of a constant $c_\epsilon$ such that:

$$V \le \epsilon H_{C_0}^{(N)} + c_\epsilon ||V||_{L^p(C_0,dx)}^{p/(p-d/2)}$$

in the sense of quadratic forms on $L^2(C_0, dx)$. For the sake of simplicity we will use the notation $c_\epsilon(V) = c_\epsilon ||V||_{L^p(C_0,dx)}^{p/(p-d/2)}$. This implies:

$$N_{C_0}^{(N)}(\lambda) \le N_{0,C_0}^{(N)}\left(\frac{\lambda + c_\epsilon(V)}{1-\epsilon}\right)$$

as a simple consequence of the min-max principle which we discussed in Section 5 of Chapter I. The classical Weyl's asymptotic formula for the eigenvalues of the Laplacian in bounded domains (see ProblemI.7.24) says that $N_{0,C_0}^{(N)}(\lambda) \le c\lambda^{d/2}$ so that:

$$\mathbb{E}\{N_{C_0}^{(N)}(\lambda)\} \le c\mathbb{E}\{(\lambda + c_\epsilon(V))^{d/2}\}$$

which is finite because $\mathbb{E}\{[\int_{C_0} |V(x)|^p \, dx]^{d/(2p-d)}\} < \infty$ by assumption.

The net result of the above discussion is the existence, for each fixed $\lambda \in \mathbb{R}$ of a set $\Omega_\lambda \in \mathcal{F}$ of probability one on which the limits:

$$N^{(D)}(\lambda) = \lim_{\Lambda \to \mathbb{R}^d} \frac{1}{|\Lambda|} N_{\Lambda,\omega}^{(D)}(\lambda) \tag{VI.13}$$

and:

$$N^{(N)}(\lambda) = \lim_{\Lambda \to \mathbb{R}^d} \frac{1}{|\Lambda|} N_{\Lambda,\omega}^{(N)}(\lambda) \tag{VI.14}$$

exist and are nonrandom. Moreover these limits satisfy:

$$N^{(D)}(\lambda) = \sup_\Lambda \frac{1}{|\Lambda|} \mathbb{E}\{N_{\Lambda,\omega}^{(D)}(\lambda)\} \tag{VI.15}$$

and:

$$N^{(N)}(\lambda) = \inf_\Lambda \frac{1}{|\Lambda|} \mathbb{E}\{N_{\Lambda,\omega}^{(N)}(\lambda)\} \tag{VI.16}$$

Now the problem is to show that the set of full measure $\Omega_\lambda$ can be chosen independent of $\lambda$ as long as $\lambda$ is a point of continuity of the limit. In other words we want to show that the normalized counting measures $\nu^{(D)}_{\Lambda,\omega}$ and $\nu^{(N)}_{\Lambda,\omega}$ converge in the sense of the vague convergence of measures to nonrandom limits $\nu^{(D)}$ and $\nu^{(N)}$ on a set of full measure. The necessary argument is very simple. Indeed, one can choose the set $\tilde{\Omega}$ as the intersection of all the sets $\Omega_\lambda$ of full probability measure when $\lambda$ runs through the rationals and redefine the limiting functions $N^{(\cdot)}$ in order to make them right continuous. This could change their values on the at most countable set of discontinuities of the original limiting functions. It is then very easy to check that for each $\omega \in \tilde{\Omega}$ and each continuity point $\lambda$, the family $|\Lambda|^{-1}N^{(\cdot)}_{\Lambda,\omega}(\lambda)$ converges because the limits exist for all rational numbers $\lambda$ and these limits are monotone nondecreasing functions of $\lambda$.

The next step is to show that the two candidates we have for the integrated density of states are actually the same, namely that $N^{(D)} = N^{(N)}$. If such an equality can be proved, the symmetry between the Dirichlet and the Neumann boundary conditions becomes one of the main attractions of the present existence proof. Unfortunately, the equality $N^{(D)} = N^{(N)}$ does not hold in general. In particular, it is false for hyperbolic spaces. See the Notes and Complements at the end of the chapter. We do not know if it holds in the full generality of the continuous case which we are currently studying. We will see that it trivially holds in full generality in the lattice case of $\mathbb{Z}^d$.

It seems that the limiting procedure for the Neumann normalized counting measures is more difficult to control. This is presumably why the integrated density of states is very often defined as the limit of the Dirichlet normalized counting measures and that, only later, and only in some particular cases, the independence with respect to the boundary conditions is argued. Recall for example our discussion following the path integral existence proof of the previous section. In any case, we shall skip very often the superscript $(D)$ in the notation of the limit of the Dirichlet counting measures and act as if the integrated density of states constructed from the Dirichlet boundary conditions was independent of our particular choice of boundary conditions.

Notice that formulas (VI.15) and (VI.16) give the following sandwich bound. For any bounded domain $\Lambda$ we have:

$$\frac{1}{|\Lambda|}\mathbb{E}\{N^{(D)}_{\Lambda,\omega}(\lambda)\} \leq N(\lambda) \leq \frac{1}{|\Lambda|}\mathbb{E}\{N^{(N)}_{\Lambda,\omega}(\lambda)\} \qquad \text{(VI.17)}$$

The latter will be very useful in computing actual estimates.

The above approach was applied to the lattice case by Simon in [300]. We sketch the main steps of a slightly different form which we borrowed from Fukushima [112] and Mezincescu[238]. In particular we will rely on the version of the Dirichlet-Neumann bracketing given in Problem I.7.20. Using the results proven in this problem for each fixed $\omega \in \Omega$ we obtain that for each fixed energy $\lambda$, the process $\{N_\Lambda^{(\infty)}(\lambda); \Lambda \in \mathcal{J}\}$ is superadditive while the process $\{N_\Lambda^{(0)}(\lambda); \Lambda \in \mathcal{J}\}$ is subadditive. Even though we use the same notations as in the continuous case, the boxes $\Lambda$ are now subsets of the lattice $\mathbb{Z}^d$. The integrability condition is trivially satisfied for the function $\lambda \hookrightarrow N_\Lambda^{(X)}(\lambda)$ is bounded by the constant 1 since it is for each $\omega \in \Omega$ and $\Lambda \in \mathcal{J}$ the distribution function of a probability measure. Consequently, the superadditive ergodic theorem gives, for each fixed energy $\lambda$, the existence of the integrated densities of states $N^{(\infty)}(\lambda)$ and $N^{(0)}(\lambda)$ corresponding to the extreme boundary conditions. It is easy to prove that these two quantities are equal using the estimate $0 \leq N_{\Lambda_k}^{(0)}(\lambda) - N_{\Lambda_k}^{(\infty)}(\lambda) \leq 2dk^{d-1}$ for all $\lambda$ where $\Lambda_k$ denotes the hypercube with sides of length $k$, the origin as a vertex and the coordinate axes as sides, which we proved in part iv) of ProblemI.7.20. Finally, the set of full probability which depends a priori on the energy $\lambda$ is made independent of $\lambda$ in the usual way. Note that we still have the sandwich estimate:

$$\frac{1}{|\Lambda|}\mathbb{E}\{N_{\Lambda,\omega}^{(\infty)}(\lambda)\} \leq N(\lambda) \leq \frac{1}{|\Lambda|}\mathbb{E}\{N_{\Lambda,\omega}^{(0)}(\lambda)\} \qquad \text{(VI.18)}$$

The original proof of Fukushima did not use the superadditive ergodic theorem but merely the multidimensional version of the Birkoff additive ergodic theorem in the way it was originally done by Benderskii and Pastur for the one-dimensional case. See the details below.

This proof also leads to the result given in Remark VI.1.5 pertaining to the equality of the integrated density of states and the expected value of the kernel of the spectral resolution of the identity. We skip the details but the interested reader can check that the result is easily proved for the finite boxes $\Lambda$ and it remains true in the limit because of the uniform bound provided by the fact that we are dealing with distribution functions of probability measures.

## VI.2 Asymptotic Behavior and Lifschitz Tails

We showed that the topological support of the integrated density of states is equal to the almost sure spectrum $\Sigma$. Consequently, the function $N(\lambda)$ is constant outside $\Sigma$. We investigate in this section the behavior of $N(\lambda)$

when $\lambda$ approaches the boundary of $\Sigma$ from the inside. We shall mostly consider the way in which $N(\lambda)$ vanishes when $\lambda$ tends to the infimum of the spectrum.

## VI.2.1 Tauberian Arguments

In the continuous case, the easiest asymptotic result on the behavior of the integrated density of states concerns the high energy limit. Indeed, the behavior of the integrated density of states is trivial in this limit for the kinetic energy dominates the potential energy and the integrated density of states has the same asymptotic behavior as the free one. More precisely we have:

**Theorem VI.2.1** *Under the essential self-adjointness assumption (V.15) we have:*

$$\lim_{\lambda\to\infty} \frac{1}{\lambda^{d/2}} N(\lambda) = \frac{\tau_d}{(2\pi)^d} \tag{VI.19}$$

*where $\tau_d$ denotes the volume of the unit ball in $\mathbb{R}^d$, or in other words:*

$$\lim_{\lambda\to\infty} \frac{N(\lambda)}{N_0(\lambda)} = 1 \tag{VI.20}$$

*if we recall the computation of the free integrated density of states given in Problem VI.5.2.*

*Proof:*
From the construction of the integrated density of states we have:

$$\lim_{t\searrow 0} t^{d/2}\mathcal{L}(\nu, t) = \frac{1}{(4\pi)^{d/2}}$$

and the result follows from a simple application of a Tauberian theorem (see for example [289]) which gives:

$$\lim_{\lambda\to\infty} \lambda^{-d/2} N(\lambda) = \frac{1}{(4\pi)^{d/2}\Gamma(1+d/2)}. \quad \blacksquare$$

We showed that under very general conditions, the topological support of the integrated density of states is equal to the almost sure spectrum $\Sigma$. Consequently, the distribution of states vanishes at the left hand edge of $\Sigma$ which can very well be $-\infty$ (as in the Gaussian case for example), and according to the heuristic arguments of Lifschitz, it should decay with a specific exponential behavior. The aim of this section is to show that this is perfectly sound.

**Proposition VI.2.2** *Let $\{V(x);\ x \in \mathbb{R}^d\}$ be a stationary ergodic Gaussian process whose covariance function, say $r(x-y)$, is continuous. Then, the integrated density of states satisfies:*

$$\lim_{\lambda \to -\infty} \lambda^{-2} \log N(\lambda) = -\frac{1}{2r(0)}. \qquad \text{(VI.21)}$$

*Proof:*
We can assume without any loss of generality that the process is centered. Using Jensen's inequality we obtain:

$$\begin{aligned}
\mathbb{E} \otimes \mathbb{E}_{0,0}\{\exp[r \int_0^t V(X_s,\omega)ds]\} &\le \frac{1}{t}\int_0^t \mathbb{E} \otimes \mathbb{E}_{0,0}\{\exp[rt\,V(X_s,\omega)]\}dx \\
&= \frac{1}{t}\int_0^t \mathbb{E}_{0,0}\{\exp[t^2 r^2 r(0)/2]\}ds \\
&= \exp[t^2 r^2 r(0)/2] \\
&< +\infty
\end{aligned}$$

which gives the existence of the integrated density of states. The almost sure spectrum is the whole real line, i.e. $\Sigma = \mathbb{R}$, for the support of the Gaussian distribution is $\mathbb{R}$. Moreover:

$$\begin{aligned}
&\mathbb{E} \otimes \mathbb{E}_{0,0,t,0}\{\exp[\frac{1}{2}\int_0^t V(X_s,\omega)ds]\} \\
&\quad = \mathbb{E}_{0,0,t,0}\{\exp[\frac{1}{2}\int_0^t \int_0^t r(X_s - X_u)duds]\} \\
&\quad \le \exp[\frac{r(0)t^2}{2}]
\end{aligned}$$

which, together with (VI.4) implies:

$$\limsup_{t \to \infty} \frac{1}{t^2} \log \mathcal{L}(\lambda, t) \le \frac{1}{2} r(0). \qquad \text{(VI.22)}$$

On the other hand we have:

$$\begin{aligned}
&\liminf_{t \to \infty} \frac{1}{t^2} \log \mathcal{L}(\lambda, t) \\
&= \liminf_{t \to \infty} \frac{1}{t^2} \log \mathbb{E} \otimes \mathbb{E}_{0,0,t,0}\{\exp[-\int_0^t V(X_s,\omega)ds]\} \\
&= \liminf_{t \to \infty} \frac{1}{t^2} \log \mathbb{E} \otimes \mathbb{E}_{0,0,t,0}\{\exp[-\int_0^t V(X_s,\omega)ds];\ \sup_{0 \le s \le t} |X_s| < a\}
\end{aligned}$$

$$\leq \liminf_{t\to\infty} \frac{1}{t^2} \log(\exp([\int_{|x|\leq 2a} r(x)dx]\frac{1}{2}t^2)\mathbb{P}_{0,0,t,0}\{\sup_{0\leq s\leq t} |X_s| < a\})$$
$$\leq \inf_{|x|\leq 2a} \frac{1}{2}r(x) + \lim_{t\to\infty} \frac{1}{t^2} \log \mathbb{P}_{0,0,1,0}\{\sup_{0\leq s\leq 1} |X_s| < at^{-1/2}\}$$
$$= \frac{1}{2}r(0)$$

since $a$ is arbitrary, $r$ is continuous and the above probability is of the order $t^{d/2}e^{-c(a)t}$ for each fixed $a$ where $c(a)$ is a positive constant depending only on $a$ and the dimension $d$. This completes the proof of:

$$\lim_{t\to\infty} \frac{1}{t^2} \log \mathcal{L}(\nu, t) = \frac{1}{2r(0)}$$

from which we get (VI.21) by a simple Tauberian argument. ∎

### Interband light absorption coefficient

Kirsch and Pastur recently gave an interesting extension of the ideas and techniques we described so far. See [181]. They propose a rigorous treatment of a possible measure of the intensity of light absorption under a given energy. Their model requires the analysis of a *two band Hamiltonian* with a random potential to model the *doping* of a semiconductor. We refer the interested reader to [174] for details on the physical motivations, we present only the main steps of the mathematical analysis.

Let $\{V(x);\, x \in \mathbb{R}^d\}$ be a real valued stationary ergodic stochastic process and let us consider the random Schrödinger operators:

$$H^{\pm}(\omega) = H_0 \pm V(\cdot, \omega),$$

which we assume to be essentially self-adjoint for the purpose of the discussion. The interband light absorption coefficient at energy $\lambda$ is the number $a(\lambda)$ given by:

$$a(\lambda) = \sum_{\lambda^{\pm}\in\Sigma(H^{\pm})} | < \varphi_{\lambda^+}, \varphi_{\lambda^-} > |^2 \delta_{\lambda-(\lambda^+-\lambda^-)}$$

where $\varphi_{\lambda^{\pm}}$ denote the eigenfunctions of $H^{\pm}$ corresponding to the energies $\lambda^{\pm}$. Obviously, this expression does not make any mathematical sense. There are many reasons for that: the spectra of the operators $H^{\pm}$ need not be pure point in which case the *eigenfunctions* $\varphi_{\lambda^{\pm}}$ are not square integrable, moreover, the summation above need not make sense in general .... The main steps of the approach used to give a meaningful definition of the integrated density of states can be used in the present situation. We first consider the interband light absorption coefficient up to a given energy level rather than at an energy level. The effect is to get rid of the delta functions in the above definition. Next we consider the problem in bounded domains where all the spectra are discrete and there exist complete orthonormal systems of eigenfunctions. The final step is to control the *thermodynamic limit* of the domains increasing to cover the whole space.

More precisely, for each box $\Lambda \in \mathcal{J}$ we consider the operators $H_\Lambda^{\pm,(D)}(\omega) = H_\Lambda^{(D)} \pm V(\cdot,\omega)$. From now on we skip the superscript $(D)$ for we shall only consider the Dirichlet boundary conditions. For each domain $\Lambda \in \mathcal{J}$, the operator $H_\Lambda^{\pm}(\omega)$ has a discrete spectrum, say:

$$-\infty < \lambda_0^{\pm}(\Lambda,\omega) < \lambda_1^{\pm}(\Lambda,\omega) \le \lambda_2^{\pm}(\Lambda,\omega) \le \cdots$$

and we shall denote by $\varphi_{\lambda_j^{\pm}(\Lambda,\omega)}$ the normalized eigenfunction corresponding to the eigenvalue $\lambda_j^{\pm}(\Lambda,\omega)$, counting multiplicity as usual. We can now define the distribution function $\lambda \hookrightarrow A_{\Lambda,\omega}(\lambda)$ by the formula:

$$A_{\Lambda,\omega}(\lambda) = \sum_{\lambda_j^{+}(\Lambda,\omega)+\lambda_j^{-}(\Lambda,\omega)\le\lambda} |< \varphi_{\lambda_j^{+}(\Lambda,\omega)}, \varphi_{\lambda_j^{-}(\Lambda,\omega)} >|^2.$$

This quantity measures the intensity of light absorption below energy $\lambda$ in the system restricted to the box $\Lambda$. Since such a system contains a number of electrons of the order of the volume of the box, the quantity which can make sense in the thermodynamic limit is:

$$A(\lambda) = \lim_{\Lambda \nearrow \mathbb{R}^d} \frac{1}{|\Lambda|} A_{\Lambda,\omega}(\lambda) \tag{VI.23}$$

which we shall call the *interband light absorption coefficient* whenever it exists. As before this limit is nonrandom because of the ergodicity assumption.

We give the main steps of the existence proof and an asymptotic result quite similar to PropositionVI.2.2 above. We shall assume that the random potential satisfies the integrability condition:

$$\mathbb{E}\{e^{-t|V(0)|}\} < \infty$$

for all $t \in \mathbb{R}$, and:

$$\int_{C(0)} \mathbb{E}\{e^{-t|V(x)|}\}\, dx < \infty$$

when the random potential is merely $\mathbb{Z}^d$ ergodic. This assumption is more restrictive than necessary but it will be convenient in the present discussion.

In any case, both operators $H^{\pm}$ are essentially self-adjoint and we can use the Feynman-Kac formula.

The Laplace transform of the quantity $A_{\Lambda,\omega}(\lambda)$ is easily computed and we get:

$$\begin{aligned}\mathcal{L}(A_{\Lambda,\omega},t) &= \operatorname{tr}[e^{-tH_\Lambda^+(\omega)}e^{-tH_\Lambda^-(\omega)}] \\ &= \int_\Lambda dx \int_\Lambda dy\, \mathbb{E}_{0,x,t,y}\{e^{-\int_0^t V(X_s,\omega)ds}; T_\Lambda > t\} \\ &\qquad \mathbb{E}'_{0,y,t,x}\{e^{-\int_0^t V(X'_s,\omega)ds}; T'_\Lambda > t\} \\ &\qquad p_t(x-y)^2 \end{aligned} \qquad \text{(VI.24)}$$

where $\{X_t;\, t \geq 0\}$ and $\{X'_t;\, t \geq 0\}$ are two independent processes of Brownian motion. We used a prime $'$ on the quantities (expectations, hitting times, ...) depending on the second process. Recall that $p_t(x-y)$ denotes the Brownian motion transition probability. It follows very easily from formula (VI.24) that, for each fixed $t \in \mathbb{R}$, the stochastic process $\{\mathcal{L}(A_{\Lambda,\omega},t);\, \Lambda \in \mathcal{J}\}$ is superadditive and the superadditive ergodic theorem gives the convergence of the Laplace transform for each fixed $t$. The usual extra work is necessary to get to the actual almost sure vague convergence of the measures with distribution functions $\frac{1}{|\Lambda|}A_{\Lambda,\omega}(\cdot)$. This proves the existence of the interband light absorption coefficient and gives the following expression for its Laplace transform:

$$\mathcal{L}(A,t) = \frac{1}{(4\pi)^d}\int_{\mathbb{R}^d} dx \mathbb{E}\otimes\mathbb{E}_{0,x,t,x}\otimes\mathbb{E}'_{0,x,t,x}\{e^{-\int_0^t [V(X_s,\omega)-V(X'_s,\omega)]ds}\}$$

The similarity of this Laplace transform with the Laplace transform of the integrated density of states suggests that it might be possible to evaluate some asymptotics of the interband light absorption coefficient by first computing asymptotics of its Laplace transform and then reverting them by Tauberian theorems. This is indeed the case for Gaussian random potentials for one can prove:

$$\lim_{t\to\infty}\frac{1}{t^2}\log\mathcal{L}(A,y) = r(0)$$

by proving upper and lower bounds very much in the same way the corresponding bounds were proved for the Laplace transform of the integrated density of states, and consequently one obtains:

$$\lim_{\lambda\to-\infty}\frac{1}{\lambda^2}\log A(\lambda) = -\frac{1}{4r(0)}.$$

**Poisson type potentials**

We now come back to the example of a random potential given by the integral of a test function (which one can think of as a smooth potential well) with respect to a random measure. We consider the particular case of a random potential $\{V(x); x \in \mathbb{R}^d\}$ given by:

$$V(x) = \int_{\mathbb{R}^d} f(x-y)\mu(dy) \tag{VI.25}$$

where $\mu$ is a Poisson random measure on $\mathbb{R}^d$, the intensity of which is the Lebesgue measure, and where $f$ is a nonnegative continuous function on $\mathbb{R}^d$ which is not identically zero and such that $f(x) = 0(|x|^{-d+\lambda})$ when $|x| \to \infty$ for some $\lambda > 0$. Such a function $f$ is trivially in $\ell^1(L^p(\mathbf{R}^d, dx))$ for all $p > 1$ and consequently for each $\omega \in \Omega$, the operator $H(\omega) = H_0 + V(\cdot, \omega)$ is essentially self adjoint on $C_c^\infty(\mathbb{R}^d)$, the function $\omega \to H(\omega)$ is measurable. The nonnegativity of the potential function gives $\Sigma \subset [0, \infty)$ and it is not difficult to show that the decay of $f$ implies that there is actually equality. Consequently, the topological support of the integrated density of states is also equal to $[0, \infty)$ because of Proposition VI.1.3.
We try to find an equivalent of $N(\lambda)$ as $\lambda \searrow 0$. As in the Gaussian case, we first find the asymptotic of the Laplace transform of $\nu$ as $t$ tends to infinity and we use a Tauberian argument to obtain the desired result for $N(\lambda)$. In the present situation we have for each $w \in \mathbf{W}$:

$$\mathbb{E}\{\exp[-\int_0^t V(X_s(w))ds]\} = \mathbb{E}\{e^{-<\varphi,\mu>}\}$$

if we set $\varphi(y) = \int_0^t f(X_s(w) - y)ds$. Hence:

$$\mathbb{E}\{\exp[-\int_0^t V(X_s(w), \cdot)ds]\} = \exp[-\int_{\mathbb{R}^d} [1 - e^{-\int_0^t f(X_s(w)-x)ds}]dx$$

according to formula V.21. If we recall (VI.4) we obtain:

$$\mathcal{L}(\nu, t) = \frac{1}{(2\pi t)^{d/2}} \mathbb{E}_{0,0,t,0}\{\exp[-\int_{\mathbb{R}^d} [1 - e^{-\int_0^t f(X_s(w)-x)ds}]dx\}. \tag{VI.26}$$

Let us remark that, for each positive $\epsilon$, if we let the function $f$ be equal to $+\infty$ on the ball of radius $\epsilon$ centered at the origin and to 0 outside this ball, then for each fixed Brownian path, the function $x \to 1 - \exp[-\int_0^t f(X_s - x)ds]$ equals the characteristic function of the famous *"Wiener saussage"* of width $\epsilon$. Motivated by the problem of the Lifschitz exponents, Donsker and Varadhan studied its asymptotics in their celebrated article (see [86]).

Among other things they proved that if we assume furthermore that the function $f$ is actually $0(|x|^{-(d+2)})$ then:

$$\lim_{t\to\infty}\frac{1}{t^{d/(d+2)}}\log \mathbb{E}_{0,0}\{\exp[-\int_{\mathbb{R}^d}[1-e^{-\int_0 f(X_s-x)ds}]dx\}$$
$$=-(d+2)(\frac{2\gamma_d}{d})^{d/(d+2)}/2$$

where $\gamma_d$ is the smallest eigenvalue of $H_0^{(D)}$ when $D$ is the unit ball of $\mathbb{R}^d$. It is easy to see that the result remains unchanged if we replace the expectation $\mathbb{E}_{0,0}$ by the expectation $\mathbb{E}_{0,0,t,0}$ and that consequently one also has:

$$\lim_{t\to\infty}\frac{1}{t^{d/(d+2)}}\log \mathcal{L}(\nu,t)=-\frac{d+2}{2}(\frac{2\gamma_d}{d})^{d/(d+2)},$$

and using a Tauberian argument one gets:

**Proposition VI.2.3** *If $\mu$ is a Poisson random measure on $\mathbb{R}^d$ the intensity of which is Lebesgue measure, if $f$ is a continuous nonnegative function on $\mathbb{R}^d$ and if $f(x)=0(|x|^{-(d+2)})$ when $|x|\to\infty$, then the integrated density of states satisfies:*

$$\lim_{\lambda\to 0}\lambda^{d/2}\log N(\lambda)=-\gamma_d^{d/2}. \tag{VI.27}$$

**Remark VI.2.4** PropositionVI.2.3 above deals with the case of Poisson random measures and it would be interesting to know if a similar result holds for more general infinitely divisible random measures. Note also that PropositionVI.2.3 has been extended by Okura in [256] to the case of *stable* free Hamiltonians $H_0=|p|^\alpha$.

Formula (VI.27) above tells us that, at least in the sense of exponential equivalents, we have:

$$N(\lambda)\sim e^{-c\lambda^{-\delta}} \tag{VI.28}$$

for some positive constants $c$ and $\delta$. $\delta$ is often called the Lifschitz exponent of the model. The existence of a formula of the type (VI.28) is usually difficult to prove. Nevertheless we show below that it is relatively easy to prove weakened forms of (VI.28). But first we give some useful preliminary asymptotic results in the lattice case.

**Proposition VI.2.5** *Let $\{V(\mathbf{n});\ \mathbf{n}\in\mathbb{Z}^d\}$ be a real valued stationary process and let us set:*

$$F(\lambda)=\mathbb{P}\{V(0)\leq\lambda\}\quad \lambda\in\mathbb{R},$$

*for the distribution function of the potential at one site. Then the integrated density of states satisfies:*

$$F(\lambda - 2d) \leq N(\lambda) \leq F(\lambda + 2d). \qquad \text{(VI.29)}$$

*for each* $\lambda \in \mathbb{R}$.

*Proof:*
The result is an immediate consequence of the min-max principle applied to the analogue of the obvious estimate:

$$M_{V-2d} \leq H_0 + V \leq M_{V+2d}$$

(in the sense of quadratic forms) in finite boxes. ∎

The above proposition has the following immediate consequences:

**Corollary VI.2.6** *For any real constants a and C we have:*
*i)* $\lim_{\lambda \to -\infty} |\lambda|^a N(\lambda) = C \Longleftrightarrow \lim_{\lambda \to -\infty} |\lambda|^a F(\lambda) = C$
*and:*
*ii)* $\lim_{\lambda \to \infty} |\lambda|^a [1 - N(\lambda)] = C \Longleftrightarrow \lim_{\lambda \to -\infty} |\lambda|^a [1 - F(\lambda)] = C$
*when* $\alpha > 0$ *and:*
*iii)* $\lim_{\lambda \to -\infty} |\lambda|^{-a} \log[1 - N(\lambda)] = C \Longleftrightarrow \lim_{\lambda \to -\infty} |\lambda|^a \log[1 - F(\lambda)] = C$
*when* $\alpha < 0$.

The derivation of the above proposition and of its corollary is very elementary and, as a consequence, one should not expect its content to be far reaching. We now consider a more subtle analysis of the asymptotic behavior of the integrated density of states near the edges of the spectrum.

## VI.2.2 The Anderson Model

Let us consider an i.i.d. sequence $\{V(\mathbf{n}); \mathbf{n} \in \mathbb{Z}^d\}$ of real valued random variables with common distribution $\mu$, and the corresponding discrete Schrödinger operator $H(\omega) = H_0 + V(\cdot, \omega)$ on the Hilbert space $\mathcal{H} = \ell^2(\mathbb{Z}^d)$. Let us assume that the support of $\mu$ is bounded and let us set $a = \inf \operatorname{supp} \mu$ and $b = \sup \operatorname{supp} \mu$. Then $\lambda_0 = \inf \Sigma(H(\omega)) = a - 2d$ and $\lambda_\infty = \sup \Sigma(H(\omega)) = b + 2d$. Since the support ot the integrated density of states $N$ is equal to $\Sigma(H(\omega))$, $N$ vanishes near $\lambda_0$ and converges to 1 near $\lambda_\infty$. These limits are attained in a singular way and the asymptotic equivalents of $N$ in these limiting regimes are known under the names of *Lifschitz singularities* or *Lifschitz tails*.

The physicist Lifschitz claimed that the integrated density of states $N(\lambda)$ should behave near $\lambda = \lambda_0$ like $\exp[-c(\lambda - \lambda_0)^{-d/2}]$ for some positive constant $c$, and $1 - N(\lambda)$ should behave in a similar way near $\lambda = \lambda_\infty$. The results we proved above using the asymptotic results of Donsker and Varadhan on the Wiener saussage give an example of a situation in which one can even determine the constant $c$. We now proceed to the proof of a weaker form for the Anderson model. But first we give Lifschitz' intuitive argument. In order for the Hamiltonian $H(\omega)$ to have an eigenvalue between $\lambda_0$ and a number $\lambda$ slightly larger than $\lambda_0$, call $\varphi$ a corresponding normalized eigenfunction, the kinetic energy $< H_0\varphi, \varphi >$ must be near its minimum $-2d$ and the potential energy $< V(\cdot,\omega)\varphi, \varphi >$ must be near its minimum value $a$. For the kinetic energy to be minimum we need the state $\varphi$ to be spread out at least over a region of size $|\lambda - \lambda_0|^{-1/2}$ (which has $|\lambda - \lambda_0|^{-d/2}$ sites). The probability that the potential is near its minimum over this region is a small number, say $e^{-c}$ to the power the number of sites because of the independence and this probability is thus of the order of $\exp[-c(\lambda - \lambda_0)^{-d/2}]$.

We prove rigorously the following form of Lifschitz' claim.

**Theorem VI.2.7** *Let us assume that $a \neq b$ (i.e. the potential takes at least two values) and that:*
*i)* $\mu([a, a+\epsilon]) \geq c\epsilon^\delta$
*ii)* $\mu([b-\epsilon, b]) \geq c\epsilon^\delta$
*for some constants $c$ and $\delta$ and $\epsilon$ sufficiently small. Then one has:*

$$\lim_{\lambda \searrow \lambda_0} \frac{\log|\log N(\lambda)|}{\log(\lambda - \lambda_0)} = -\frac{d}{2} \tag{VI.30}$$

$$\lim_{\lambda \nearrow \lambda_\infty} \frac{\log|\log(1 - N(\lambda))|}{\log(\lambda_\infty - \lambda)} = -\frac{d}{2} \tag{VI.31}$$

*Proof:*
We prove only the first of the two formulas for the second one can be derived from the first one by unitary transformations. The proof is very much in the same spirit as the proofs of the large deviation results in probability theory. We prove that the right hand side is both an upper bound and a lower bound for the left hand side by using the sandwich estimate (VI.18) to get upper and lower bounds on the integrated density of states of the whole system.

Let us set $\lambda = \lambda_0 + \epsilon$ (recall that $\lambda_0 = -2d + a$) and let $\epsilon \to 0$.We consider boxes of the form $\Lambda = \{-\ell, \cdots, +\ell\} \times \cdots \times \{-\ell, \cdots, +\ell\}$. A computation similar to the one of ProblemI.7.21 shows that the eigenvalues of the operator $H^{(\infty)}_{\Lambda,0}$ are sums of quantities of the form $2\cos(k_i\pi/2\ell)$ and that, consequently, the lowest eigenvalue of $H^{(\infty)}_{\Lambda,0}$ belongs to the interval $[-2d, -2d+\epsilon/2)$ for $\ell$ large enough. Consequently, the operator $H^{(\infty)}_{\Lambda} = H^{(\infty)}_{\Lambda,0} + V(\cdot,\omega)$ will have at least one eigenvalue in the interval $[\lambda_0, \lambda_0+\epsilon]$ whenever the potential satisfies $V(\mathbf{n}) < a+\epsilon/2$ for all the sites $\mathbf{n}$ in $\Lambda$. Using the sandwich estimate we get:

$$\begin{aligned} N(\lambda_0+\epsilon) &\geq \frac{1}{|\Lambda|}\mathbb{E}\{N^{(\infty)}_{\Lambda,\omega}(\lambda_0+\epsilon)\} \\ &\geq \frac{1}{|\Lambda|}\mathbb{P}\{V(\mathbf{n}) < a+\epsilon/2 \text{ for all } \mathbf{n}\in\Lambda\} \\ &\geq \frac{1}{|\Lambda|}\mathbb{P}\{V(0) < a+\epsilon/2\}^{|\Lambda|} \\ &\geq \frac{1}{|\Lambda|}c^{|\Lambda|}(\epsilon/2)^{\delta|\Lambda|} \end{aligned}$$

because of our assumption. Taking logarithms of both sides, choosing $\ell \equiv \epsilon^{-1/2}$ as explained in the heuristic discussion of Lifschitz's argument and taking once more logarithms we get:

$$\liminf_{\epsilon\searrow 0} \frac{\log|\log N(\lambda_0+\epsilon)|}{\log(\epsilon)} \geq -\frac{d}{2}.$$

This proves the first half of the desired formula.

The result of ProblemI.7.21 shows that the lowest eigenvalue of the operator $H^{(0)}_{\Lambda,0}$ is zero and that the next eigenvalue is $d$-fold degenerate and is equal to $-2(d-1)-2\cos(\pi/2\ell)$. Consequently one can find a sequence of integers $\ell$ such that $\ell \approx \epsilon^{-1/2}$ and such that this next eigenvalue of the Neumann free Hamiltonian $H^{(0)}_{\Lambda,0}$ in the box $\Lambda$ remains larger that $-2d+\epsilon$. Since $V \geq a$, this implies that the next to lowest eigenvalue of the Neumann operator $H^{(0)}_{\Lambda}(\omega) = H^{(0)}_{\Lambda,0}+V(\cdot,\omega)$ is larger than $\lambda_0+\epsilon$. Using the sandwich estimate we get:

$$\begin{aligned} N(\lambda_0+\epsilon) &\leq \frac{1}{|\Lambda|}\mathbb{E}\{N^{(0)}_{\Lambda,\omega}(\lambda_0+\epsilon)\} \\ &\geq \frac{1}{|\Lambda|}\mathbb{P}\{\lambda_0(H^{(0)}_{\Lambda}(\omega)) \leq \lambda_0+\epsilon\} \end{aligned} \qquad \text{(VI.32)}$$

where we used the notation $\lambda_0(H^{(0)}_{\Lambda}(\omega))$ for the lowest eigenvalue of the operator $H^{(0)}_{\Lambda}(\omega)$. We now prove that:

$$\{\lambda_0(H^{(0)}_{\Lambda}(\omega)) \leq \lambda_0+\epsilon\} \subset \{\frac{1}{|\Lambda|}\sum_{\mathbf{n}\in\Lambda} \mathbf{1}_{\{V(\mathbf{n})>a+\epsilon'\}} < \delta\} \qquad \text{(VI.33)}$$

Indeed, if $\omega$ is fixed in the compliment of the set on the right hand side, namely if the relative frequency $f_\Lambda(\omega)$ in $\Lambda$ of the sites at which the potential is larger than $a+\epsilon'$ is greater than or equal to $\delta$, then one can define the potential function $W(\mathbf{n},\omega)$ as being equal to $a$ when $V(\mathbf{n},\omega) < a+\epsilon'$ and to $\ell^{-2}$ otherwise, where we choose $\ell > \ell_0 = [\epsilon^{-1/2}]$. For such an $\omega$ we have $W(\mathbf{n},\omega) \leq V(\mathbf{n},\omega)$ for $\mathbf{n} \in \Lambda$ so that $\lambda_0(H^{(0)}_{\Lambda,0} + W(\cdot,\omega)) \leq \lambda_0(H^{(0)}_\Lambda)$. If we let $\varphi$ be the $\ell^2(\Lambda)$-normalized constant function on $\Lambda$, then $H^{(0)}_{\Lambda,0}\varphi = -2d$,

$$< [H^{(0)}_{\Lambda,0} + W(\cdot,\omega)]\varphi,\varphi >= (a-2d) + f_\Lambda(\omega)\ell^{-2}$$

and

$$< [H^{(0)}_{\Lambda,0} + W(\cdot,\omega)]^2\varphi,\varphi >= (a-2d)^2 + 2(a-2d)f_\Lambda(\omega)\ell^{-2} + f_\Lambda(\omega)\ell^{-4}.$$

Moreover, one has:

$$\lambda_1(H^{(0)}_{\Lambda,0} + W(\cdot,\omega)) \geq \lambda_1(H^{(0)}_\Lambda) \geq (a-2d) + 4\ell^{-4}$$

so that we actually have:

$$\lambda_1(H^{(0)}_{\Lambda,0} + W(\cdot,\omega)) - < [H^{(0)}_{\Lambda,0} + W(\cdot,\omega)]\varphi,\varphi > \geq 3\ell^{-2}$$

where we used the notation $\lambda_1(K)$ for the first eigenvalue of the operator $K$ to be greater than the lowest one. Temple's inequality (recall ProblemI.7.1) gives:

$$\lambda_0(H^{(0)}_{\Lambda,0} + W(\cdot,\omega)) \geq (a-2d) + \frac{2}{3}f_\Lambda(\omega)\ell^{-2}.$$

Since:

$$\lambda_0(H^{(0)}_{\Lambda,0} + W(\cdot,\omega)) \geq \lambda_0(H^{(0)}_\Lambda)$$

this implies that $\omega$ cannot be in the left hand side of (VI.33) if $f_\Lambda(\omega) \leq \delta$ and $2\delta\ell^{-2}/3 > \epsilon$. Coming back to (VI.32) we have that:

$$\begin{aligned} N(\lambda_0+\epsilon) &\leq \ell^{-d}\mathbb{P}\{\frac{1}{|\Lambda|}\sum_{\mathbf{n}\in\Lambda}\mathbf{1}_{\{V(\mathbf{n})>a+\epsilon'\}} < \delta\} \\ &\leq \frac{1}{|\Lambda}e^{-\epsilon"|\Lambda|} \end{aligned}$$

for some $\epsilon" > 0$ by the classical Cramer's large deviations result. This gives the desired lower bound because of the choice of $\ell$ as a function of $\epsilon = \lambda - \lambda_0$. ∎

In addition to the assumptions we made above to compute the Lifschitz tails of the integrated density of states of the Anderson model, let us also assume that:

$$\text{supp } \mu = [a,a'] \cup [b',b]$$

where $b' - a' > 4d$ and also that:

$$\mu([a' - \epsilon, a']) \geq c\epsilon^{\delta} \qquad \mu([b', b' + \epsilon]) \geq c\epsilon^{\delta}$$

Note that the almost sure spectrum is now:

$$\Sigma(H(\omega)) = [a - 2d, a' + 2d] \cup [b' - 2d, b + 2d]$$

and that it has two bands separated by a gap. The Lifschitz argument can be adapted to predict the behavior of the integrated density of states near these edges. The corresponding asymptotics go under the name of ***internal Lifschitz tails.*** The results of the above theorem can be extended to include the case of the internal tails.

**Theorem VI.2.8** *Under the above assumptions we have:*

$$\lim_{\lambda \nearrow a'+2d} \frac{\log|\log[N(a'+2d) - N(\lambda)]|}{\log(a'+2d-\lambda)} = -\frac{d}{2} \tag{VI.34}$$

$$\lim_{\lambda \searrow b'-2d} \frac{\log|\log[N(\lambda) - N(b'-2d)]|}{\log(\lambda - b' + 2d)} = -\frac{d}{2} \tag{VI.35}$$

Even though the above result was given only in the case of a single gap, it does extend to cases of several gaps in the support of $\mu$ as long as they are large enough (i.e. larger than $4d$) to create a gap in the spectrum of $H(\omega)$. Hamiltonians with a potential energy which is the sum of a periodic potential and a random one give other examples of models with gaps in the spectrum and for which the investigation of the internal Lifschitz tails is a natural problem. They occur for example in the study of an electron in a crystal with impurities. If the support of the one site distribution of the purely random part of the potential is contained in a compact set with diameter smaller than the size of the gaps of the periodic potential, then the spectrum of the full operator will still have gaps. For such an operator, the integrated density of states exists and its topological support is still equal to the almost sure spectrum. Hence it is constant on these gaps and one could try to determine as before the behavior of the integrated density of states near these gap edges. The mathematics are more technical and more difficult because of the presence of the periodic potential and the only known result concerns the infimum of the spectrum. We describe it now.

Let $U \in L^p_{loc}(\mathbb{R}^d, dx)$ with $p = 2$ for $d \leq 3$, $p > 2$ for $d = 4$ and $p = d/2$ for $d \geq 5$, be a periodic function with period $\mathbf{a} \in \mathbb{Z}^d$. This periodic potential $U$ models the effects of a periodic crystal. Let us now consider the random Schrödinger operator $H(\omega) = H_0 + U + V(\cdot, \omega)$ where the random potential

$V$ which models the impurities in the system is of the form:

$$V(x,\omega) = \sum_{\mathbf{n}\in\mathbb{Z}^d} \xi_{\mathbf{n}}(\omega)\,\chi(x-\mathbf{n}) \qquad \text{(VI.36)}$$

where $\{\xi_{\mathbf{n}};\ \mathbf{n}\in\mathbb{Z}^d\}$ is an i.i.d. sequence with common distribution $\mu$ having compact support and where the well function $\chi \in L^p(\mathbb{R}^d, dx)$ with $p$ as above, is nonnegative for large $|x|$ and satisfies:

$$\chi(x) = O(|x|^{-d-\alpha})$$

for some $\alpha > 0$ as $|x| \to \infty$. The result is the following:

**Theorem VI.2.9** *If we denote by $a$ the infimum of the support of the distribution $\mu$ and if we assume that $\mu(\{a\}) < 1$ and that $\mu([a, a+\epsilon]) = O(\epsilon^\delta)$ for some $\delta \geq 0$ as $\epsilon \searrow 0$, then the integrated density of states $N$ has a Lifschitz singularity at the bottom of the spectrum $\lambda_0 = \inf\Sigma(H)$ given by:*

$$\lim_{\lambda\searrow\lambda_0} \frac{\log|\log N(\lambda)|}{\log(\lambda-\lambda_0)} = -\frac{d}{\min(2,\alpha)}. \qquad \text{(VI.37)}$$

This result is stated in the continuous case but it remains valid in the discrete case of the lattice $\mathbb{Z}^d$.

## VI.3 More on the Lattice Case

We now concentrate on the lattice case $T = \mathbb{Z}^d$ and we give some smoothness results which have no known analogues in the continuous case.

Let $\{V(\mathbf{n});\ \mathbf{n}\in\mathbb{Z}^d\}$ be a real valued stochastic process and let us assume that it is of the form $V(\mathbf{n},\omega) = v(\theta_{\mathbf{n}}\omega)$ for some measurable function $v : \Omega \to \mathbb{R}$. For each $\omega \in \Omega$ we consider the self-adjoint operator $H(\omega) = H_0 + V(\cdot,\omega)$ on the Hilbert space $\mathcal{H} = \ell^2(\mathbb{Z}^d)$ where $H_0$ is the usual discrete free Hamiltonian, and where $V(\cdot,\omega)$ is the operator of multiplication by the real function $\mathbf{n} \hookrightarrow V(\mathbf{n},\omega)$. We already pointed out that the integrated density of states exists in full generality. Moreover, since the "delta functions" are in the Hilbert space $\mathcal{H}$, each bounded operator has a kernel, and this kernel is necessarily continuous. Consequently, a trivial adaptation of Remark VI.1.5 to the lattice case gives:

$$\nu(\cdot) = \mathbb{E}\{E(\cdot,\omega)\delta_0, \delta_0 >\} = \mathbb{E}\{< E(\cdot,0,0,\omega)\}. \qquad \text{(VI.38)}$$

This formula remains true if one substitutes any $\mathbf{n}$ for 0 because of the stationarity of the potential. This can be used to give a simple proof, in the present situation, of the fact that the topological support of the integrated density of states is equal to the almost sure spectrum. Finally, if $\hat{\nu}$ denotes the Fourier transform of $\nu$, we have:

$$\begin{aligned}\hat{\nu}(t) &= \mathbb{E}\{\int_{\mathbb{R}} e^{it\lambda} < E(d\lambda,\omega)\delta_0,\delta_0 >\} \\ &= \mathbb{E}\{< e^{itH(\omega)}\delta_0,\delta_0 >\}\end{aligned}$$

which we can compute using Molcanov's formula (II.40):

$$\begin{aligned}&= e^{2dt}\mathbb{E}\otimes\mathbb{E}^{(d)}_{0,0}\{\delta_0(X_t)i^{N(t)}\exp[i\int_0^t V(X_s,\omega)ds]\} \\ &= e^{2dt}\mathbb{E}^{d}_{0,0}\{\delta_0(X_t)i^{N(t)}\mathbb{E}\{\exp[i\int_0^t V(X_s,\omega)ds]\}\}. \qquad \text{(VI.39)}\end{aligned}$$

It is possible in some cases to compute or estimate the innermost expectation and then to get information on the integrated density of states by inverting its Fourier transform. We now give some examples.

**Proposition VI.3.1** *Let $\{V(\mathbf{n});\mathbf{n}\in\mathbb{Z}^d\}$ be i.i.d., and let us assume that the Fourier transform $\chi(t)$ of the common distribution decays exponentially in the sense that:*

$$\forall t\in\mathbb{R},\qquad |\chi(t)|\le c'e^{-c|t|} \qquad \text{(VI.40)}$$

*for some positive constants $c'$ and $c$ such that $c'+2d<c+1$. Then the integrated density of states has an analytic extension to a strip $\{z\in\mathbb{C};\ |\Im m z|<\alpha\}$ for some $\alpha>0$.*

Let us first give an interpretation of the assumption and the conclusion in the light of the theory of analytic extensions. The assumption means that the distribution of $V(\mathbf{0})$ has a density which possesses an analytic extension to a strip and that proving the desired result amounts to showing that the Fourier transform of $\nu$ decays exponentially as in (VI.40).

*Proof:*
If the Brownian path $w$ is fixed we have:

$$\int_0^t V(X_s(w),\omega)ds = \sum_{y\in\mathbb{Z}^d} V(y,\omega)L(t,y,w) \qquad \text{(VI.41)}$$

where $L(t,y,w)$ denotes the time spent at the site $y$ by the Brownian path $w$ before time $t$:

$$L(t,y,w)=\int_0^t \mathbf{1}_{\{y\}}(X_s(w))ds.$$

The right hand side of (VI.41) is a sum of independent random variables on the probability space $(\Omega, \mathcal{F}, \mathbb{P})$, whence:

$$\mathbb{E}\{\exp[i\int_0^t V(X_s,\omega)ds]\} = \prod_{y\in\mathbb{Z}^d} \chi(L(t,y))$$

and consequently:

$$\begin{aligned} |\hat{\nu}(t)| &\leq e^{2dt}\mathbb{E}^{(d)}_{0,0}\{\prod_{y\in\mathbb{Z}^d} |\chi(L(t,y))|\} \\ &\leq e^{2dt}\mathbb{E}^{(d)}_{0,0,}\{\prod_{y\in\mathbb{Z}^d, L(t,y)>0} c'e^{-cL(t,y)}\} \\ &= e^{(2d-c)t}\mathbb{E}^{(d)}_{0,0}\{c'^{\#(t)}\} \end{aligned}$$

where $\#(t)$ denotes the number of different sites visited by the Brownian path before time $t$. Obviously we have $\#(t) \leq N(t)$, and since $\mathbb{E}^{(d)}_{0,0}\{c'^{N(t)}\} = e^{(c'-1)t}$ we must also have:

$$|\hat{\nu}(t)| \leq e^{(2d-c+c'-1)t}$$

which gives the desired result. ∎

As an application of the above result let us assume that each $V(\mathbf{n})$ is a mean zero Gaussian random variable with density:

$$f_\sigma(x) = \frac{1}{\sqrt{2\pi\sigma^2}}e^{-x^2/2\sigma^2}.$$

In this case, $\chi(t) = \exp[-\sigma^2t^2/2]$ and for each $a > 0$, the inequality (VI.40) holds with $c = a$ and $c' = \exp[a^2/2\sigma^2]$. This gives:

**Corollary VI.3.2** *In the mean zero Gaussian i.i.d. case, the density of states exists and has an analytic extension to a strip for large disorder (i.e. for large enough $\sigma$).*

**Remark VI.3.3** We believe that the independence at each site of the potential should not be needed for the analyticity of the density of states in the Gaussian case. If $\{V(\mathbf{n}); \mathbf{n} \in \mathbb{Z}^d\}$ is a mean zero stationary Gaussian process, simple properties of its spectral measure should force the density of states to exist and to be analytic. Recall that the ergodicity of the process is implied by the continuity of this spectral measure.

There are very few smoothness results in the multidimensional case. We quote an interesting result proven by Bovier, Campanino, Klein and Perez. See [35].

**Proposition VI.3.4** *If the random potential* $\{V(\mathbf{n});\, \mathbf{n} \in \mathbb{Z}^d\}$ *is an i.i.d. family of real valued random variables and if the Fourier transform* $\chi$ *of the distribution of* $V(0)$ *is such that the function* $t \hookrightarrow (1+|t|)^{d+n}\chi(t)$ *is integrable for some integer* $n \geq 0$, *then there exists* $\epsilon_0 > 0$ *such that the integrated density of states* $\lambda \hookrightarrow N_\epsilon(\lambda)$ *of the random operator* $H = \epsilon H_0 + V$ *is* $(n+1)$*-times continuously differentiable provided* $0 < \epsilon < \epsilon_0$.

The proof begins with a rewriting of the problem using a *supersymmetric replica trick* to express the Green's function and then it becomes very technical for the next step is a cluster expansion. The above result can be improved in the case of the uniform distribution.

**Proposition VI.3.5** *If the distribution of* $V(0)$ *is uniform on a bounded interval, then there exists* $\epsilon_0 > 0$ *such that the integrated density of states* $\lambda \hookrightarrow N_\epsilon(\lambda)$ *is* $C^\infty$.

**The Lloyd model:**
The above estimations become exact calculations in the case where the potential $V(\mathbf{n})$ has at each site a *Cauchy distribution*, the density of which is (we restrict ourselves to the one dimensional case $d=1$ for simplicity):

$$f_a(x) = \frac{1}{\pi}\frac{a}{x^2+a^2} \tag{VI.43}$$

for some $a > 0$. In this case we have $\chi(t) = e^{-a|t|}$ and formula (VI.39) gives:

$$\begin{aligned}
\hat{\nu}(t) &= e^{2dt}\mathbb{E}^d_{0,0}\{\delta_0(X_t)i^{N(t)}\mathbb{E}\{\exp[i\int_0^t V(X_s,\omega)ds]\}\} \\
&= e^{2dt}\mathbb{E}^d_{0,0}\{\delta_0(X_t)i^{N(t)}\prod_{y\in\mathbb{Z}^d}\chi(L(t,y))\} \\
&= e^{2dt}\mathbb{E}^{(d)}_{0,0,}\{\delta_0(X_t)i^{N(t)}\}e^{-a|t|} \\
&= < e^{itH_0}\delta_0, \delta_0 > e^{-a|t|} \\
&= J_0^{(d)}(t)e^{-a|t|}.
\end{aligned}$$

where we are using the usual notation $J_0^{(d)}(t)$ for Bessel functions. This proves that $\nu$ is the convolution of the Cauchy distribution and the integrated density of states of the free Hamiltonian. This calculation was originally made by Lloyd and this model is often called the Lloyd model. See Problem VI.5.5 for more on this model.

When $\{V(\mathbf{n}); \mathbf{n} \in \mathbb{Z}^d\}$ is an i.i.d. family and the distribution of $V(0)$ has a bounded density, a simple functional analytic argument shows that the integrated density of states is actually absolutely continuous and its density is bounded. This result is due to Wegner. See [329]. It will play a crucial role in the proof of localization for multidimensional systems for it will be the only way to control the probability that a given energy $\lambda$ is near the spectrum of the Hamiltonian restricted to a given box.

**Proposition VI.3.6** *If the random potential $\{V(\mathbf{n}); \mathbf{n} \in \mathbb{Z}^d\}$ is i.i.d. and if the distribution of $V(0)$ has a bounded density, then the integrated density of states is absolutely continuous and its density is bounded.*

*Proof:*
Let $\mathcal{I}$ be a bounded open interval. Then:

$$\nu(\mathcal{I}) \leq \liminf_{\Lambda \to \mathbb{Z}^d} \mathbb{E}\{\nu_\Lambda^{(D)}(\mathcal{I})\}$$

so that the proof reduces to showing that the right hand side is bounded by the length of the interval $\mathcal{I}$ times a constant independent of the box $\Lambda$. But this is an immediate consequence of PropositionVIII.1.8 which shows that when one takes the expectation, if one integrates first with respect to the potential at one site, then the spectral measures gain a bounded density, the bound depending only on the bound on the density of the potential at one site. ∎

**Remark VI.3.7** In general, the integrated density of states is continuous. In one dimension, this follows immediately from the fact that a given energy $\lambda$ cannot be an eigenvalue with positive probability. In the general case, it can be checked using a simple remark due to Delyon and Souillard. This remark is based on the equivalent form of the construction of the integrated density of states which we gave in RemarkVI.1.4. In fact, the integrated density of states is generally slightly better than continuous. It is locally log-Hölder continuous as defined below. This result was first proven by Craig and Simon and we discuss it now.

**Definition VI.3.8** *A real valued function $\lambda \hookrightarrow N(\lambda)$ is said to be locally log-Hölder continuous if for any positive $r$ there exists a finite constant $C_r$*

*such that:*

$$|N(\lambda) - N(\lambda')| \leq C_r \frac{1}{\log(|\lambda - \lambda'|^{-1})}$$

*as long as* $|\lambda| \leq r$ *and* $|\lambda - \lambda'| < 1$.

**Proposition VI.3.9** *Let* $N$ *be a continuous probability distribution on* $\mathbb{R}$ *such that the function* $t \hookrightarrow \log|t - \lambda|$ *is* $N$*-integrable for any real* $\lambda$ *and we denote by* $F(\lambda)$ *its integral.*

1. *Let us assume that* $N$ *has a bounded support and that* $F(\lambda)$ *is nonnegative for any real number* $\lambda$. *Then the function* $\lambda \hookrightarrow N(\lambda)$ *is locally log-Hölder continuous.*
2. *If the function* $F(\lambda)$ *is Hölder continuous then the function* $N(\lambda)$ *has the same property.*
3. *If the function* $F(\lambda)$ *is* $C^\infty$ *then the function* $N(\lambda)$ *has the same property.*

*Proof:*
Let $\lambda$ and $\lambda'$ be two fixed real numbers with $|\lambda - \lambda'| < 1$. Let $A = \{t\,;\,|t - \lambda| \geq 1\}$ then if the support of $N$ is contained in the set $[-c, c]$ one has:

$$0 \leq \int_A \log|t - \lambda|\, dN(t) \leq \log(1 + c + |\lambda|).$$

It follows that the integral on $A^c$ also exists and is nonpositive. This integral is certainly less that the integral over the set $\{t\,;\,|t - \lambda| \leq |\lambda - \lambda'|\}$ which is less than $\log(|\lambda - \lambda'|)|N(\lambda) - N(\lambda')|$. Hence one obtains the inequality:

$$0 \leq \log(1 + c + |\lambda|) + \log(|\lambda - \lambda'|)|N(\lambda) - N(\lambda')|$$

This implies the first conclusion.
The following conclusions are direct applications of the general theory of the Hilbert transform. We review some of the important facts of this theory (the proofs of which can be found in [252]). The Hilbert transform of a square integrable function $\psi$ is defined by the formula :

$$\mathcal{T}\psi(x) = \frac{1}{\pi} \lim_{\epsilon \downarrow 0} \int_{|x-t|>\epsilon} \frac{\psi(t)}{x - t}\, dt$$

Then one has:

1. $\mathcal{T}\psi$ is a square integrable function and $\mathcal{T}^2\psi = -\psi$ almost everywhere with respect to the Lebesgue measure.

2. If $\psi$ is Hölder continuous on some interval $[-a\,,\,a]$ then $\mathcal{T}\psi$ is Hölder continuous on $[-a/2\,,\,a/2]$.
3. If $\psi$ is $C^\infty$ on some interval $[-a\,,\,a]$ then $\mathcal{T}\psi$ is $C^\infty$ on $[-a/2\,,\,a/2]$.

One can add that if $\psi$ is analytic on some interval $(a\,,\,b)$ then $\mathcal{T}\psi$ has the same property. This result is proved in [216] but we will not use it. Let $a$ be a positive number and let us define the function $\psi(t) = N(t)\mathbf{1}_{\{|t|\geq 4a\}}$. A simple integration by parts yields:

$$\begin{aligned}\pi\mathcal{T}\psi(x) &= -\psi(4a)\log|4a-x| + \psi(-4a)\log|4a+x| \\ &\quad +F(x) - \int_{|t|>4a}\log|x-t|\;dN(t)\end{aligned}$$

If $F(x)$ is locally Hölder continuous (respectively $C^\infty$) then $\mathcal{T}\psi$ has the same property on the interval $[-2a\,,\,2a]$ hence $\mathcal{T}^2\phi$ has the same property on the interval $[-a\,,\,a]$. The continuity of the function $N$ together with the equality $\mathcal{T}^2\psi = -N$ almost everywhere give the desired result. ∎

We now prove the log-Hölder continuity result which we announced earlier.

**Theorem VI.3.10** *Let $\{V(\mathbf{n});\,\mathbf{n}\in\mathbb{Z}^d\}$ be an ergodic stationary real valued stochastic process satisfying the usual integrability condition:*

$$\mathbb{E}\{\log(1+|V(0)|)\} < \infty.$$

*Then the integrated density of states of the random Schrödinger operator $H = H_0 + V$ on the Hilbert space $\mathcal{H} = \ell^2(\mathbb{Z}^d)$ is locally log-Hölder continuous.*

*Proof:*
With an elementary, though tedious, approximation argument we can assume without any loss of generality that the support of the distribution of $V(0)$ is bounded. Then, for each $L > 0$, we consider the restriction $H^{\Lambda_L}$ of $H$ to the box:

$$\Lambda_L = \{\mathbf{n} = (n_1, n_2, \cdots, n_d) \in \mathbb{Z}^d;\; |n_2| \leq L, \ldots, |n_d| \leq L\}.$$

For all practical purposes this restriction operator behaves like a discrete Schrödinger operator in a strip of width $\ell = (2[L]+1)^{d-1}$. In particular, its integrated density of states, say $N^{\Lambda_L}$, satisfies the Thouless formula which will be proven in the next section. Note that this integrated density of states may be random if the shift in the direction of the first coordinate axis is not ergodic. In any case $N^{\Lambda_L}$ converges toward the integrated density of states $N$ of the operator $H$ as $L \to \infty$ and consequently we have:

$$\int \log|\lambda - \lambda'|\, dN(\lambda') \geq \limsup_{L\to\infty} \int \log|\lambda - \lambda'|\, dN^{\Lambda_L}(\lambda') \geq 0$$

because of the Thouless formula for the strip, the fact that the sum $\gamma_1(\lambda) + \cdots + \gamma_\ell(\lambda)$ of Lyapunov exponents is nonnegative and because of part 1. of the above proposition. ∎

# VI.4 The One Dimensional Cases

We first consider the continuous one-dimensional case.

## VI.4.1 The Continuous Case

Let us rapidly show how one can go about proving the existence of the integrated density of states in the one dimensional case. Let $N^{[a,b]}_{\alpha,\beta}(\lambda)$ be the number of eigenvalues of the operator $H^{[a,b]}_{\alpha,\beta}$ which are in $(-\infty, \lambda]$. We use freely the notations of Chapter III and we skip the dependence in $\omega$ when convenient. Note that a real number $\lambda$ is an eigenvalue of $H^{[a,b]}_{\alpha,\beta}$ if and only if:

$$\vartheta_\lambda(b, a, \alpha) = \beta + k\pi$$

for some $k \in \mathbb{Z}$. Using various monotonicity properties of the phase $\vartheta$ with respect to some of its arguments (the so-called Sturm oscillation theory reviewed in Problems III.6.2 and III.6.3) it is easy to check that:

**i)** $N^{[a,b]}_{0,0}(\lambda) \leq N^{[a,b]}_{\alpha,\beta}(\lambda) \leq N^{[a,b]}_{\pi/2,\pi/2}(\lambda)$

for most of the angles $\alpha$ and $\beta$. This is the one-dimensional version of the Dirichlet-Neumann bracketing discussed in Chapter I. Note that the Dirichlet and Neumann boundary conditions correspond to $\alpha = \beta = 0$ and $\alpha = \beta = \pi/2$ respectively. Moreover:

**ii)** $0 \leq N_{a,b,\pi/2,\pi/2}(\lambda) - N_{a,b,0,0}(\lambda) \leq 2$

and if $a < b < c$ we also have:

**iii)** $N^{[a,c]}_{\pi/2,\pi/2}(\lambda) \leq N^{[a,b]}_{\pi/2,\pi/2}(\lambda) + N^{[b,c]}_{\pi/2,\pi/2}(\lambda)$

and

$N^{[a,b]}_{0,0}(\lambda) + N^{[b,c]}_{0,0}(\lambda) \leq N^{[a,c]}_{0,0}(\lambda).$

One can now prove the existence of the integrated density of states.

For those boundary conditions $\alpha$ and $\beta$ satisfying i), one uses iii) to bound each quantity of the form $(2hk)^{-1}N_{\alpha,\beta}^{[-hk,+hk]}(\lambda)$ where $h$ is a fixed positive integer, by quantities of the form:

$$\frac{1}{2hk}\sum_{-k\leq j\leq k-1} N_{\pi/2,\pi/2}^{[jh,(j+1)h]}(\lambda)$$

from above, and from below by quantities of the form:

$$\frac{1}{2hk}\sum_{-k\leq j\leq k-1} N_{0,0}^{[jh,(j+1)h]}(\lambda)$$

which are additive and to which one can apply the individual Birkhoff ergodic theorem. We obtain:

$$\begin{aligned}
\frac{1}{h}\mathbb{E}\{N_{0,0}^{[0,h]}(\lambda)\} &\leq \liminf_{k\to\infty}\frac{1}{2hk}N_{\alpha,\beta}^{[-hk,+hk]}(\lambda) \\
&\leq \limsup_{k\to\infty}\frac{1}{2hk}N_{\alpha,\beta}^{[-hk,+hk]}(\lambda) \\
&\leq \frac{1}{h}\mathbb{E}\{N_{\pi/2,\pi/2}^{[0,h]}(\lambda)\}.
\end{aligned}$$

One then concludes the proof by using the fact that the right most term and the left most one differ by at most $2h^{-1}$ because of property ii) above and the fact that $h$ can be chosen arbitrarily large.

The above proof was given for historical reasons: it is basically the first published existence proof. Note that the modern way to conclude the proof is to apply a subadditive ergodic theorem to $N_{0,0}^{[\cdot,\cdot]}(\lambda)$ and a superadditive ergodic theorem to $N_{\pi/2,\pi/2}^{[\cdot,\cdot]}(\lambda)$ and to check that the limits are the same by using ii). In both cases one uses the argument given in the multidimensional case to show that the set of full $\mathbb{P}$-measure can be chosen independent of $\lambda$. The integrability of $N_{\alpha,\beta}^{[a,b]}(\lambda)$ in $\omega$ is justified by the obvious upper bound:

$$N_{\alpha,\beta}^{[a,b]}(\lambda) \leq \frac{1}{\pi}|\vartheta_\lambda(b,a,\alpha)| \tag{VI.44}$$

and the fact that:

$$\vartheta_\lambda(t) = \vartheta_\lambda(0) + \int_0^t [\cos^2\vartheta_\lambda(s) + (\lambda - V(s))\sin^2\vartheta_\lambda(s)]ds. \tag{VI.45}$$

Integrating (VI.45) and using the stationarity we obtain:

$$\begin{aligned}
&\frac{1}{b-a}\mathbb{E}\{\vartheta_\lambda(b,a,\alpha)\}\\
&\qquad = \frac{1}{b-a}\mathbb{E}\{\vartheta_\lambda(b-a,0,\alpha)\}\\
&\qquad = \frac{\alpha}{b-a} + \frac{1}{b-a}\int_0^{b-a}\mathbb{E}\{\cos^2\vartheta_\lambda(s,0,\alpha)\\
&\qquad\qquad + [\lambda - V(s)]\sin^2\vartheta_\lambda(s,0,\alpha)\}\,ds\\
&\qquad = \frac{\alpha}{b-a} + \int\int_{\mathbb{R}\times P(\mathbb{R}^2)} F_\lambda(q,\vartheta)\mu_{\alpha,\lambda}^{(b-a)}(dq,d\vartheta)
\end{aligned}$$

where $F_\lambda(q,\vartheta) = \cos^2\vartheta + [\lambda - q]\sin 2\vartheta$ and where:

$$\mu_{\alpha,\lambda}^{(L)}(dq,d\vartheta) = \frac{1}{L}\int_0^L \mathbb{P}\{V(s)\in dq, \vartheta_\lambda(s,0,\alpha)\in d\vartheta\}\,ds.$$

Hence, if the joint distribution of the potential and the phase has a limit in the Cesaro sense, say $\pi_\lambda$, for $\lambda\in\mathbb{R}$ and $\alpha\in P(\mathbb{R})$ fixed, the integrated density of states has the representation:

$$N(\lambda) = \frac{1}{\pi}\int\int_{\mathbb{R}\times P(\mathbb{R}^2)} F_\lambda(q,\vartheta)\pi_\lambda(dq,d\vartheta). \tag{VI.46}$$

We did not write the dependence in $\alpha$ of the limiting measure because we expect this limit is independent of the choice of the boundary conditions. This formula is very useful in the Markovian case for we are able to compute or at least to estimate the limiting measure in this case.

The existence of limit points $\pi_{\lambda,\alpha}$ for which formula (V.32) holds is quite general. Indeed the torus $\mathbb{T}$ is compact and consequently, the obvious equality:

$$\begin{aligned}
\mu_{\lambda,\alpha}^{(L)}([-n,+n]\times\mathbb{T}) &= \frac{1}{L}\int_0^L \mathbb{P}\{|V(s)|\le n\}\,ds\\
&= \mathbb{P}\{|V(0)|\le n\}
\end{aligned}$$

shows that the family $\{\mu_{\lambda,\alpha}^{(L)}; L>0\}$ is tight, hence the existence of the limit points for which (V.32) holds. When the family is not convergent, the limit points depend on the choices of the boundary condition $\alpha$ and formula (V.32) is less useful.

The integrated density of states has been investigated for operators slightly different from ours. Either because the second order term is not the second derivative any more, or because the perturbation is a random measure rather than a potential function. These operators have to be defined by means of their quadratic forms, but the results are essentially the same. The interested reader is refered to the works of Kotani [189], Fukushima and Nakao [114], Okura [256] and Thompson [318] for example.

Sturm oscillation theory also tells us that $N_{\alpha,\beta}^{[a,b]}(\lambda)$ is equal to (or differs at most by one from) $\pi$ times the number of zeros of the function $t \hookrightarrow \theta_\lambda(t,a,\alpha)$ on the interval $[a,b]$. Hence, and we already made this remark in the derivation above, the integrated density of states is equal, up to a factor $\pi$, to the almost sure limit of $t^{-1}\vartheta_\lambda(t,a,\alpha)$. This limit does not depend on $a$ or $\alpha$. It is called the *rotation number* for obvious reasons. It was cleverly used in Johnson and Moser [157] to label the gaps of Schrödinger operators with almost periodic potentials. See Chapter VII. This work has been extended to the lattice case by Herman in [147] and Delyon and Souillard in [77].

It is possible to use the ideas presented in this section to precise the asymptotic behavior of $N(\lambda)$ for large $\lambda$. The multidimensional result of Theorem VI.2.1 says that $N(\lambda) \sim \pi^{-1}\sqrt{\lambda}$ in the one dimensional continuous case. In fact one has:

**Theorem VI.4.1** *For any stationary ergodic random potential* $\{V(t);\, t \in \mathbb{R}\}$ *such that* $\mathbb{E}\{|V(0)|\} < \infty$ *one has:*

$$|N(\lambda) - \frac{1}{\pi}\sqrt{\lambda}| \leq \frac{\mathbb{E}\{|V(0)|\}}{\pi}\frac{1}{\sqrt{\lambda}}.$$

*Moreover, if we assume that* $\mathbb{E}\{|V(0)|^2\} < \infty$ *then we also have:*

$$N(\lambda) = \frac{1}{\pi}\sqrt{\lambda}| - \frac{\mathbb{E}\{V(0)\}}{2\pi\sqrt{\lambda}} + o(\frac{1}{\sqrt{\lambda}}).$$

*Proof:*
The first claim is a trivial consequence of formula (VI.45). The proof of the second one is more technical. We do not give it here because we shall prove more general estimates in the next chapter. The interested reader can find the details in the Appendix A of [66]. ∎

### Attraction-Repulsion of the Energy Levels

We close this subsection with some words on the important problem of the attraction (or repulsion) of the energy levels. Many incorrect claims can be

found in the physics litterature on this problem and it is very difficult to understand what is actually happening on a rigorous mathematical level. Consequently we restrict ourselves to a rather loose discussion of some mathematical results which have been proven in one dimension for the *very recurrent* random potentials of the Russian school. We refer the reader to the work of Molcanov [244] and Genkova et al. [133] for details.
They investigate the asymptotic behavior of the normalized spectral interval lengths:

$$I_k(a) = \frac{\lambda_{k+1}(a) - \lambda_k(a)}{\mathbb{E}\{\lambda_{k+1}(a) - \lambda_k(a)\}}$$

between the $k$-th and $(k+1)$-th eigenvalues of the eigenvalue problem on $[-a, +a]$. They choose to work exclusively with Dirichlet boundary conditions and consequently we drop the superscript $(D)$ from the notation. We want to answer the following questions: does $I_k(a)$ converge as $a \to \infty$? If yes, in which sense? What is the limit? Can one estimate the speed of convergence? Can one compute large deviation probabilities?...
Let us assume that $I_k(a)$ converges in distribution to some probability measure $\mu_k$ concentrated on $[0, \infty)$. We say that there is:

**a.** repulsion of the energy levels if $\lim_{\lambda \to 0} \lambda \mu_k([0, \lambda]) = 0$.

**b.** no interaction between the energy levels if $\lim_{\lambda \to 0} \mu_k([0, \lambda]) = c_k \neq 0$.

**c.** attraction of the energy levels if $\lim_{\lambda \to 0} \lambda^{-1} \mu_k([0, \lambda]) = \infty$.

Also we may want to study the joint distribution of several consecutive normalized spectral intervals, and their limit correlation. In fact the mathematical problem we just defined is meaningless because the $k$-th energy level $\lambda_k(a)$ converges to the infimum of the spectrum $\Sigma$ when $a \to \infty$. To avoid this problem we choose $k$ as a function of $a$ in such a way that $\lambda_{k(a)}(a)$ converges to a number $\lambda$ when $a \to \infty$ and we talk about the attraction-repulsion of the energy levels at the energy $\lambda$. It is shown in the papers mentioned above that for each $\lambda_0$ in $\Sigma$, the marginals of the process $\{\tilde{\nu}_\lambda(a); \lambda \geq 0\}$ defined by:

$$\tilde{\nu}_\lambda(a) = \#\{\text{eigenvalues of } H_{0,0}^{[-a,a]}(\omega) \text{ in } (\lambda_0, \lambda_0 + \lambda/2a]\}$$

converge as $a \to \infty$ to those of a Poisson process with parameter $N'(\lambda_0) = [d\nu/d\lambda](\lambda_0)$. More precisely if $0 \leq a_1 < b_1 \leq a_2 < b_n$, $\lambda_0 \in \Sigma$ and $n, k_1, k_2, \cdots, k_n$ are integers, then:

$$\lim_{a \to \infty} \mathbb{P}\{\#\text{eigenvalues of } H_{0,0}^{[-a,a]} \text{ in } I_1 = k_1, \cdots,$$
$$\#\text{eigenvalues of } H_{0,0}^{[-a,a]} \text{ in } I_n = k_n\}$$

$$= e^{-[d\nu/d\lambda](\lambda_0)(b_1-a_1)} \frac{([d\nu/d\lambda](\lambda_0)(b_1-a_1))^{k_1}}{k_1!} \cdots$$
$$e^{-[d\nu/d\lambda](\lambda_0)(b_n-a_n)} \frac{([d\nu/d\lambda](\lambda_0)(b_n-a_n))^{k_n}}{k_n!}$$

where we set $I_j = (\lambda_0 + a_j/(2a), \lambda_0 + b_j/(2a)]$ for $j = 1, \cdots n$. This "Poissonian" asymptotic behavior obviously implies the nonrepulsion of the energy levels. Note that some operators of the form:

$$H(\omega) = -\frac{d}{dt} a(t,\omega) \frac{d}{dt}$$

have been shown to exhibit the phenomenon of repulsion of the energy levels. An intuitive explanation of the above results is the following. The interaction between two energy levels has to be understood in perturbation, when the levels are very close to each other. This interaction should be proportional to the overlap of the corresponding wave functions. Repulsion should occur when the latter is important, and this is the case when the states are extended while, when the states are localized, the interaction is smaller and repulsion should not occur. Indeed, for random potentials giving exponential localization, it is possible to determine numerically the eigenfunctions in bounded intervals and check that they are localized in further and further regions when the corresponding energies are closer and closer. This is also responsible for the insulator behavior of the system exhibited in Kunz and Souillard in [204] where it is shown that the electric conductivity of this system is vanishing. Nevertheless, the localization length depends on the energy and it happens that it is not small enough to be responsible for this phenomenon to occur at certain energies. This is the cause of some of the apparent paradoxes which lead to the incorrect claims which can be found in the literature.

## VI.4.2 The Lattice Case

The following integrability condition will be assumed throughout this subsection:

$$\mathbb{E}\{\log(1 + |V(0)|)\} < \infty \tag{VI.47}$$

in the case of the lattice $\mathbb{Z}$ and:

$$\sum_{i=1}^{\ell} \mathbb{E}\{\log(1 + |V_i(0)|)\} < \infty \tag{VI.48}$$

in the case of the discrete strip of width $\ell$.

**Proposition VI.4.2** *Under the integrability condition (VI.47), the function $h_\lambda(t) = \log|t-\lambda|$ is $N$-integrable for any $\lambda$ with $\Im m\,\lambda \neq 0$. Moreover $h_\lambda^+(t)$ is $N$-integrable for any real $\lambda$ and the function $\lambda \hookrightarrow \int h_\lambda(t)\,dN(t)$ is subharmonic in the complex plane.*

*Proof:*
For $\lambda$ nonreal it is enough to prove that the function $t \hookrightarrow \log(1+|t|)$ is $N$-integrable. Let ${}^{\pm}H(\omega)$ be the diagonal operators $\pm 2I + V$ and ${}^{\pm}N_\omega$ be the associated integrated density of states. The nonreal following relations are easily checked:

$$\begin{aligned} {}^{-}H(\omega) &\leq H(\omega) \leq^{+} H(\omega) \\ {}^{+}N(\lambda) &\leq N(\lambda) \leq^{-} N(\lambda) \\ \int h(t)\, d^{\pm}N(t) &= \mathbb{E}\{h(\pm 2 + V(0))|\} \end{aligned}$$

The last relation is a direct consequence of the Birkhoff ergodic theorem. It holds for any nonnegative Borel measurable function $h$ with $\mathbb{E}\{h(\pm 2 + V(0))|\} < \infty$. It follows from the second inequality that for an increasing function $g$ one has $N(g) \leq^{+} N(g)$ and that for a decreasing function $f$ one has $N(f) \leq^{-} N(f)$. If one chooses $g(t) = \log(1+t)\mathbf{1}_{\{t\geq 0\}}$ and $f(t) = \log(1-t)\mathbf{1}_{\{t\leq 0\}}$ then we have:

$$\int \log(1+|t|)\,dN(t) \leq^{+} N(g) +^{-} N(f) = 2\mathbb{E}\{\log(3+|V(0)|)\}$$

and this yields the desired result.
For a real $\lambda$ one has:

$$h_\lambda^+(t) = \mathbf{1}_{\{|t-\lambda|\geq 1\}} h_\lambda(t) \leq \mathbf{1}_{\{|t-\lambda|\geq 1\}} \log|t-(\lambda+i\epsilon)|$$

This implies immediately the integrability of $h_\lambda^+$. The subharmonicity is a consequence of the Fatou's theorem and the subharmonicity of the function $\lambda \hookrightarrow \log|t-\lambda|$. ∎

Remark that this result is obvious in the case of bounded potentials since in this case the probability measure $N$ is compactly supported. Moreover the extension to the strip is straightforward. We are now ready to prove the so-called Thouless formula in the lattice case.

**Proposition VI.4.3** *Let $H(\omega)$ be an ergodic family of Schrödinger operators on the one dimensional lattice satisfying the integrability condition*

$$\mathbb{E}\{\log(1+|V(0)|)\} < \infty.$$

*Then for each complex number $\lambda$ one has:*

$$\gamma(\lambda) = \int Log|t - \lambda| \, dN(t)$$

*Proof:*
Let $\Lambda$ be a box of length $n$. The argument of Proposition VI.4.2 can be applied to the operator restricted to the box $\Lambda$. Let $N_\omega^\Lambda$ be the normalized eigenvalue counting function of the operator $H^\Lambda(\omega)$. Unbounded potentials require special attention for the measures $\mathbb{E}\{N^\Lambda\}$ are not carried by a fixed compact set and the convergence of the integrals of the unbounded continuous function $t \hookrightarrow \log|t - \lambda|$ for $\Im m\lambda \neq 0$ is not garanteed by the weak convergence toward the distribution $N$. Let ${}^\pm H^\Lambda(\omega)$ be the diagonal operators $\pm 2I + V$ and let us denote by ${}^\pm N_\omega^\Lambda$ the corresponding normalized eigenvalue counting function. The following relations are then obvious:

$$\begin{aligned} {}^-H^\Lambda(\omega) &\leq H^\Lambda(\omega) \leq^+ H^\Lambda(\omega) \\ {}^+N_\omega^\Lambda(\lambda) &\leq N_\omega^\Lambda(\lambda) \leq^- N_\omega^\Lambda(\lambda) \\ \int f(\lambda)\, d^\pm N_\lambda^\Lambda(\lambda) &= \frac{1}{n}\sum_{k=0}^{n-1} f(V(k,\omega) \pm 2) \end{aligned}$$

Note that the last relation holds for any function $f$ and that the last two relations remain valid after taking expectations. Let $m$ be an integer greater than 2. Applying the above inequalities to the decreasing function $f_m(\lambda) = \log(1-\lambda)\mathbf{1}_{\{\lambda \leq -m\}}$ and to the increasing function $g_m(\lambda) = \log(1+\lambda)\mathbf{1}_{\{\lambda \geq m\}}$ one obtains:

$$\int_{|\lambda| \geq m} \log(1+|\lambda|)\, d(\mathbb{E}\{N^\Lambda\})(\lambda) \leq 2 \int_{|V(0,\omega)| \geq m-2} \log(3+|V(0,\omega)|)\, d\mathbb{P}(\omega)$$

The crucial point in this inequality is the uniform integrability of the function $\log(1+|\lambda|)$ with respect to the averaged distributions $\mathbb{E}\{N^\Lambda\}$ granted by the integrability of $\log(1+|V(0)|)$.
Let $\lambda$ be a complex number with $\Im m\,\lambda \neq 0$ and let $n > 1$ be an integer. The functions $p_\lambda(n)$ and $q_\lambda(n)$ are polynomials in $\lambda$ with a leading coefficient of norm 1. Moreover their roots are the eigenvalues of the operator $H$ restricted to the boxes $\Lambda = [0\,,\, n-1]$ and $\Lambda' = [1\,,\, n]$ respectively. It follows that (using the notation introduced in Chapter III for the entries of the propagator matrix):

$$\begin{aligned} \int \log|t-\lambda|\, dN^\Lambda(t) &= \frac{1}{n}\log|p_\lambda(n)|, \\ \int \log|t-\lambda|\, dN^{\Lambda'}(t) &= \frac{1}{n-1}\log|q_\lambda(n)|. \end{aligned}$$

Taking into account the above uniform integrability one obtains:

$$\lim_{n\to\infty} \frac{1}{n}\mathbb{E}\{\log|p_\lambda(n)|\} = \lim_{n\to\infty} \frac{1}{n}\mathbb{E}\{\log|q_\lambda(n)|\} = \int \log|t-\lambda|\, dN(t)$$

Hence we have proved that the expectation of the logarithm of the absolute value of each entry of the propagator matrix:

$$U_\lambda(n) = \begin{pmatrix} p_\lambda(n+1) & q_\lambda(n+1) \\ p_\lambda(n) & q_\lambda(n) \end{pmatrix}$$

has the same behavior. Since we know that $\lim_{n\to\infty} n^{-1}\log\|U_\lambda(n)\| = \gamma(\lambda)$, we can conclude that one has $\gamma(\lambda) = \int \log|t-\lambda|\, dN(t)$. for $\Im m\,\lambda \neq 0$. both sides of this formula are subharmonic on the complex plane. For the Lyapunov exponent $\gamma(\lambda)$ this is just the result of Proposition V.4.5 and the subharmonicity of $\lambda \hookrightarrow \int \log|t-\lambda| dN(t)$ is given in Proposition VI.4.2. Since two subharmonic functions which are equal almost everywhere with respect to the Lebesgue measure of the complex plane are equal (see Proposition V.4.4), we obtain the desired result. ∎

In a similar way one can prove the Thouless formula for the strip.

**Proposition VI.4.4** *Let $H(\omega)$ be an ergodic random Schrödinger operator on the strip of width $\ell$ satisfying the integrability condition (VI.48). Then for each complex number $\lambda$ one has:*

$$\gamma_1(\lambda) + \ldots + \gamma_\ell(\lambda) = \ell \int \log|t-\lambda|\, dN(t)$$

*Proof:*
The only difference with the one dimensional proof is that we have to deal with the matrix $\Lambda^\ell(U_\lambda(n))$. Let $\alpha$ and $\beta$ be two $\ell$ dimensional Lagrangian subspaces of $\mathbb{R}^d$ and, as usual, let us denote by $\alpha$ and $\beta$ two $d\times\ell$ matrices whose column vectors form a basis of these subspaces. We know from Proposition III.5.2 that $\lambda$ is an eigenvalue of $H^\Lambda_{\alpha,\beta}$ if and only if one has:

$$\det\big(\beta'\, U_\lambda(n)\,\alpha\big) = 0.$$

Let $\lambda$ be a complex number with $\Im m\,\lambda \neq 0$. Repeating the argument of Proposition VI.4.3 one obtains that

$$\lim_{n\to\infty} \frac{1}{n\ell}\mathbb{E}\{\log|\det(\beta'\, U_\lambda(n)\,\alpha)|\} = \int \log|t-\lambda|\, dN(t)$$

for any couple of Lagrangian vectors $\alpha$ and $\beta$. Let $e_1, \ldots, e_d$ be the canonical basis of $\mathbb{R}^d$. For each subset $I = \{i_1, \ldots, i_\ell\}$ of $\{1 \ldots d\}$, we denote by $e_I$ the decomposable $\ell$-vector $e_{i_1} \wedge \ldots \wedge e_{i_\ell}$. The above formula implies that

$$\lim_{n\to\infty} \frac{1}{n\ell} \mathbb{E}\{\log |(\Lambda^\ell U_\lambda(n))_{I,J}|\} = \int \log|t-\lambda|\, dN(t)$$

for any couple $(I, J)$ such that $e_I$ and $e_J$ are Lagrangian. Unfortunately not all the entries of the matrix $\Lambda^\ell(U_\lambda(n))$ are of this type. Nevertheless the Lagrangian linear space $\mathrm{Lag}(\ell)$ is generated by the Lagrangian decomposable $\ell$-vectors and this linear space is left invariant by any matrix of $SP(\ell, \mathbb{R})$. Repeating the argument of Proposition IV.4.21 one can conclude that the norm of $\Lambda^\ell(g)$ computed on $\Lambda^\ell(\mathbb{R}^d)$ is equal to the norm computed in $\mathrm{Lag}(\ell)$. Each entry of the matrix of the linear mapping $\Lambda^\ell(U_\lambda(n))$ restricted to $\mathrm{Lag}(\ell)$ is of the form $\det(\beta' U_\lambda(n)\, \alpha)$ for two Lagrangian $\ell$-vectors $\alpha$ and $\beta$. Hence each entry of this matrix has the same behavior and we can conclude that:

$$\begin{aligned}\gamma_1(\lambda) + \ldots + \gamma_\ell(\lambda) &= \lim_{n\to\infty} \frac{1}{n} \mathbb{E}\{\log ||\Lambda^\ell U_\lambda(n)||\} \\ &= \ell \int \log|t-\lambda|\, dN(t)\end{aligned}$$

As before we use the subharmonicity to extend this formula to a real $\lambda$. ∎

**Proposition VI.4.5** *Let us assume that the potential is i.i.d., that the distribution of $V(0)$ is not concentrated on a single point, and that it satisfies the integrability condition:*

$$\mathbb{E}\{\log(1 + |V(0)|) < \infty\}$$

*Then:*

$$N(\lambda) = \nu_\lambda(P(\mathbb{R})^-)$$

*where $\nu_\lambda$ is the unique invariant probability measure on $P(\mathbb{R})$ and where $P(\mathbb{R}^2)^- = \{\bar{v} \in P(\mathbb{R}^2)\,;\, v = \binom{a}{b}\,,\, ab < 0$ or $a = 0\}$.*

*Proof:*
We know from Proposition IV.4.26 that for any real number $\lambda$ there exists a unique invariant probability measure $\nu_\lambda$ on $P(\mathbb{R}^2)$. Let $v$ be the vector $\binom{1}{0}$. For each $n \geq 0$ we define the probability measure $\nu_{\lambda,n}$ by $\nu_{\lambda,n} = \frac{1}{n+1}\sum_{k=0}^{n} \mu_\lambda^k * \delta_{\bar{v}}$ as a measure on the projective space $P(\mathbb{R}^2)$. We know that the uniqueness of the invariant measure implies the weak convergence

of the sequence $\nu_{\lambda,n}$ to $\nu_\lambda$. If we denote by $\mathbf{1}_{P(\mathbb{R}^2)^-}$ the indicator function of the set $P(\mathbb{R}^2)^-$, it is readily seen that:

$$\nu_{\lambda,n}(P(\mathbb{R}^2)^-) = \frac{1}{n+1}\mathbb{E}\{\sum_{k=0}^{n} \mathbf{1}_{P(\mathbb{R}^2)^-}(U_\lambda(k)\bar{v})\}$$

We know from Problem VI.5.3 that the sum $\sum_{k=0}^{n} \mathbf{1}_{P(\mathbb{R}^2)^-}(U_\lambda(k)\bar{v})$ is equal to the number of eigenvalues of the operator $H$ restricted to the box $\Lambda = [0\,,\, n]$ which are no greater than $\lambda$. Hence:

$$\nu_{\lambda,n}(P(\mathbb{R}^2)^-) = \mathbb{E}\{N^\Lambda(\lambda)\}$$

and the desired result follows from the fact that $\nu_{\lambda,n}$ converges weakly to $\nu_\lambda$, $\mathbb{E}\{N^\Lambda\}$ converges weakly to $N$ and that $\nu_\lambda$ and $N$ are continuous measures. ∎

The Thouless formula makes it possible to transfer the regularity properties of the Lyapunov exponents to the integrated density of states. In particular it can be used to prove the local-Hölder continuiuty of the integrated density of states in the case of the strip. Recall that this result was used in the proof of TheoremVI.3.10 which deals with the same result in the general case of the lattice $\mathbb{Z}^d$.

**Corollary VI.4.6** *The integrated density of states of any ergodic random Schrödinger operator in the strip of width $\ell$ satisfying the integrability condition (VI.48) is locally log-Hölder continuous.*

*Proof:*
This result is a direct consequence of the Thouless formula for the strip, the non negativity of the sum $\gamma_1(\lambda) + \ldots + \gamma_\ell(\lambda)$, part 1. of Proposition VI.3.9 above and a standard approximation argument which we skip for the sake of simplicity. ∎

**Corollary VI.4.7** *Let $H(\omega)$ be a random Schrödinger operator in the strip of width $\ell$ with an i.i.d. sequence as a potential and such that there exists a positive $\nu$ with*

$$\sum_{i=1}^{\ell} \mathbb{E}\{|V_i(0)|^\nu\} < \infty.$$

*If we assume furthermore that the support of the distribution in $\mathbb{R}^\ell$ of the vector $(V_1(0), \ldots, V_\ell(0))$ contains an open set, then the integrated density of states is locally Hölder continuous.*

*Proof:*
This is just a consequence of Theorem V.4.15 which gives the local Hölder continuity of the first $p$ Lyapunov exponents whenever the distribution $\mu_\lambda$ is strongly irreducible and contractive on the Lagrangian linear spaces Lag($k$) for $k = 1 \ldots p$. This last condition is certainly fulfilled with $p = \ell$ in view of Propositions IV.4.32 and IV.4.22. ∎

In the one dimensional case the strong irreducibility and the contractivity properties are easier to check and one can weaken the above hypothesis.

**Corollary VI.4.8** *Let $H(\omega)$ be a random Schrödinger operator on the one dimensional lattice $\mathbb{Z}$ constructed from an i.i.d. sequence for which there exists a positive $\nu$ with $\mathbb{E}\{|V(0)|^\nu\} < \infty$ and such that the distribution of $V(0)$ is not concentrated on a single point. Then the integrated density of states is locally Hölder continuous.*

*Proof:*
See Proposition IV.4.25. ∎

**Corollary VI.4.9** *Let $H(\omega)$ be a random Schrödinger operator on the one dimensional lattice $\mathbb{Z}$ constructed from an i.i.d. sequence for which there exists a positive $\nu$ with $\mathbb{E}\{|V(0)|^\nu\} < \infty$ and such that the distribution of $V(0)$ is absolutely continuous. Then the integrated density of states is a $C^\infty$ function.*

*Proof:*
It is a straightforward application of Theorem V.4.16. ∎

We would like to emphasize that the smoothness results we gave are not the best one can get. We proved them using the smoothness of the Lyapunov exponent proved in Chapter V, but as we explain in the Notes and Complements at the end of the present chapter, there are other methods to tackle this problem. We quote two results which were obtained after a reformulation of the problem using the so-called replica trick and with a clever use of functional analytic estimates from harmonic analysis.

**Proposition VI.4.10** *Let $H(\omega)$ be a Schrödinger operator on the one dimensional lattice $\mathbb{Z}$ constructed from an i.i.d. sequence for which the Fourier transform $\hat{\mu}$ of the one-site distribution $\mu$ of the potential is of class $C^{(k)}$ and the derivatives satisfy:*

$$|\hat{\mu}^{(j)}(p)| = O(\frac{1}{1+|p|^\alpha}) \tag{VI.49}$$

*for some $\alpha > 0$, all $p \in \mathbb{R}$ and $j = 1, \ldots, k$. Then the integrated density of states is of class $C^{(k)}$.*

This result is due to Campanino and Klein. These authors also proved that the integrated density of states is analytic in a strip about the real axis whenever the Fourier transform $\hat{\mu}$ is exponentially decaying. It is interesting to notice that assumption (VI.49) essentially says that some convolution power of the measure $\mu$ has a bounded density. It is possible to show that the corresponding distribution on $SL(2, \mathbb{R})$ has also a power which has a bounded density and that it is conceivable that one could extend the machinery developed to prove the smoothness of the Lyapunov exponent (and consequently of the integrated density of states) to apply to this case. This can indeed be done and Campanino-Klein's result can be recovered this way.

The above result has been extended, first by Klein and Speis and then by Klein, Lacroix and Speis to the case of the strip.

**Proposition VI.4.11** *Let $H(\omega)$ be a random Schrödinger operator on the strip of width $\ell$ constructed from an i.i.d. sequence for which the Fourier transform $\hat{\mu}$ of the one-site distribution $\mu$ of the potential is of class $C^{(k)}$ with bounded derivatives which satisfy:*

$$\lim_{|p|\to\infty} \hat{\mu}^{(j)}(p) = 0 \tag{VI.50}$$

*for $j = 1, \ldots, k$. Then the integrated density of states is of class $C^{([(k+1)/2])}$.*

See the notes and complements below for precise references.

We close this chapter with a discussion of a particular model who attracted the attention of many because of its simplicity and its resistance to a complete analysis for a long period of time. It is the case of the so-called Bernouilli potential $\{V(n);\, n \in \mathbb{Z}\}$ made of an i.i.d. sequence of random variables taking only the two values 0 and 1. We set $p = \mathbb{P}\{V(0) = 0\}$ in which case we also have $1 - p = \mathbb{P}\{V(0) = 1\}$. We are interested in the random Schrödinger operator $H_\beta = H_0 + \beta V$ with a coupling constant $\beta > 0$. We use a subscript $\beta$ to emphasize the dependence of the usual quantities (Lyapunov exponent, integrated density of states ...) upon the coupling constant. According to PropositionV.3.5 the spectrum of $H_\beta(\omega)$ is almost surely equal to:

$$\Sigma_\beta = [-2, +2] \cup [\beta - 2, \beta + 2].$$

The result we present now is in the negative direction in the sense that it shows that the integrated density of states can be singular. We prove:

**Theorem VI.4.12** *For any open set $\mathcal{I} \subset \Sigma$ satisfying:*

$$\inf_{\lambda \in \mathcal{I}} \gamma_\beta(\lambda) > \log 4 \tag{VI.51}$$

*the restriction of the integrated density of states $\nu_\beta = dN_\beta$ to $\mathcal{I}$ is singular continuous.*

This result strongly suggests that the integrated density of states is singular continuous when the coupling constant $\beta$ is large enough. Indeed, condition (VI.51) should be satisfied in this regime. Indeed, Thouless' formula gives:

$$\gamma_\beta(\lambda) = \int_{-2}^{+2} \log|\lambda' - \lambda|\, dN_\beta(\lambda') + \int_{\beta-2}^{\beta+2} \log|\lambda' - \lambda|\, dN_\beta(\lambda')$$

when $\lambda \in [-2, +2]$. For the sake of the discussion we restrict ourselves to the large coupling constant regime and we consider only energies in the lowest of the two bands of the spectrum. The second integral is easily seen to be equivalent to $(1-p)\log\beta$ when $\beta \to \infty$. This is in full accordance with the naive intuition for the term $\log\beta$ is the contribution to the Lyapunov exponent of a single potential barrier with height $\beta$ and $1-p$ is the density of these barriers. Unfortunately, the first integral has a negative singularity for $\lambda' = \lambda$ and a subtle argument is required to show that this first integral is actually a lower order term when $\beta \to \infty$. Obviously, the control of this integral requires the knowledge of the local regularity properties of the integrated density of states which we proved. The net result is that the naive conjecture is OK and in fact:

**Theorem VI.4.13** *We have:*

$$\lim_{\beta\to\infty} \inf_{\lambda \in (-2,+2)} \frac{1}{\log\beta}\gamma_\epsilon(\lambda) > k_p \tag{VI.52}$$

*for some finite constant:*

$$k_p > \frac{(1-p)^2}{2(1+p^2(1-p))}$$

*and the limit can explicitly be computed for the dense set of energies of the form:*

$$\lambda = 2\cos(\frac{k}{L}\pi) \tag{VI.53}$$

*with $k$ and $L$ relatively prime integers satisfying $0 < k < L$, in which case:*

$$\lim_{\beta\to\infty} \frac{1}{\log\beta}\gamma_\beta(\lambda) = 1 - p - (1-p)^2 \frac{p^{L-1}}{1-p^L}.$$

We will not prove this result. We refer the reader to the Notes and Complements below for precise references.

*Proof of Theorem VI.4.12* We use the following fact which we shall prove later in our discussion of the exponential localization for the Bernouilli model. On a set of full probability the spectrum is pure point with multiplicity one in $\mathcal{I}$ and there exists an integer $\ell$ such that whenever $\lambda \in \mathcal{I}$ is an eigenvalue and $\varphi_\lambda$ is the corresponding real valued normalized eigenfunction then one has:

$$|\varphi_\lambda(x)| \leq e^{-2\gamma_\beta(\lambda)|n|}$$

whenever $|n| \geq \ell$. Note that our assumption implies that one can replace the Lyapunov exponents $\gamma_\beta(\lambda)$ in the above inequality by a single number $\gamma > \log 4$. For each eigenvalue $\lambda \in \mathcal{I}$ we define the function $\tilde{\varphi}_\lambda$ as being equal to $\varphi_\lambda$ in the box $[-\ell, \ell]$ and to 0 outside this box. Note that:

$$1 = ||\varphi_\lambda||^2 \leq ||\tilde{\varphi}_\lambda||^2 + 2 \sum_{n=\ell+1}^{\infty} e^{-4\gamma|n|}$$

which implies that:

$$||\tilde{\varphi}_\lambda||^2 \geq 1 - \frac{2}{1-e^{-4\gamma}} e^{-4\gamma(\ell+1)}.$$

Since we also have:

$$||[H^{[-\ell.+\ell]} - \lambda I]\tilde{\varphi}_\lambda|| \leq e^{-2\gamma(\ell+1)}$$

we can conclude that:

$$d(\lambda, \Sigma(H(\omega))) \leq e^{-\gamma\ell}$$

for all such eigenvalues. This implies that the set of eigenvalues contained in $\mathcal{I}$ is contained in the set:

$$\mathcal{I}_\gamma = \mathcal{I} \cap \bigcap_{k \geq 1} \bigcap_{\ell > k} \Sigma_\beta^\ell$$

where $\Sigma_\beta^\ell$ denotes the set of real numbers at a distance less than $e^{-\gamma\ell}$ of the set of all the possible eigenvalues of the operator $H^{[-\ell,+\ell]}$ when we consider all the possible values of the potential in the box $[-\ell, +\ell]$. Note that on one hand we have $N(\mathcal{I}_\gamma) = N(\mathcal{I})$ by construction, while on the other hand we have $|\mathcal{I}_\gamma| = 0$ because:

$$|\Sigma_\beta^\ell| \leq (2\ell+1)2^{2\ell+1}2e^{-\gamma\ell} = 4(2\ell+1)e^{-(\gamma - \log 4)\ell}$$

which is the general term of a convergent series because of our assumption. This completes the proof. ∎

## VI.5 Problems

**Problem VI.5.1** *Let us replace, in any of the existence proofs of the integrated density of states, the random potential $V$ by $V+p$ where $p$ is a bounded deterministic function which tends to 0 at infinity. Prove that, in both cases, (by which we mean continuous and discrete) the integrated density of states exists and is still equal to the same quantity.*

**Problem VI.5.2** *Choose boxes $\Lambda$'s of the form $\prod_{1\leq j\leq d}[-a_j, +a_j]$ to compute explicitly the spectra of the free Hamiltonians $H_\Lambda^{(D)}$ and $H_\Lambda^{(N)}$ with Dirichlet and Neumann boundary conditions in $\Lambda$ and show that the integrated density of states $N_0$ of the free Hamiltonian $H_0$ exists and is equal to:*

$$N_0(\lambda) = \begin{cases} \tau_d(2\pi)^{-d}\lambda^{d/2} & \text{if } \lambda \geq 0 \\ 0 & \text{if } \lambda < 0 \end{cases} \tag{VI.54}$$

*where we used the notation $\tau_d$ for the volume of the unit ball in $\mathbb{R}^d$.*
*Hint: Use the asymptotics of the eigenvalues of the Laplacian in bounded domains proved in ProblemI.7.24. The computations are done in Section XIII.15 of volume IV of the Reed and Simon series.*

**Problem VI.5.3** *Let $H^\Lambda$ be the one dimensional discrete Schrödinger operator restricted to a box $\Lambda = [0, n]$. The number of eigenvalues of $H^\Lambda$ no greater than some real $\lambda$ is equal to the number of nonnegative eigenvalues of the operator $H^\Lambda - \lambda I$. We adopt the Gauss decomposition method in order to rewrite the quadratic form associated to the symmetric matrix $H^\lambda - \lambda I$ as a sum (or difference) of squares. Let us denote by $\delta_\lambda(0), \cdots, \delta_\lambda(n)$ the coefficient of the squares in the Gauss decomposition of the quadratic form:*

$$Q(x) = \sum_{k=0}^{n} \gamma_k x^2(k) - \sum_{k=0}^{n-1} x_k x_{k+1}$$

*associated to the matrix $H - \lambda I$ with $\gamma_k = (V(k) - \lambda)$.*
*Let $p_\lambda(n)$ be the solution of $H\psi = \lambda\psi$ with initial conditions $\psi(0) = 1$ and $\psi(-1) = 0$.*

1. *Assuming that $p_\lambda(k) \neq 0$ for $k \in [1, n]$ prove that the sequence $\{\delta_\lambda(k); k \in [0, n]\}$ satisfies the same recursion relation as the se-*

*quence* $r_\lambda(k) = \dfrac{p_\lambda(k+1)}{p_\lambda(k)}$, *i.e.*

$$\delta_\lambda(0) = \gamma(0),\ \delta_\lambda(k+1) = V(k+1) - \lambda - \frac{1}{\delta_\lambda(k)} \quad \text{for } k \geq 0$$

2. *If* $p_\lambda(k) = 0$ *for some* $k \in [1, n]$ *then* $r_\lambda(k-1) = 0$, $r_\lambda(k)$ *is infinite and* $r_\lambda(k+1) = \gamma(k)$. *Taking into account that the coefficient of* $x_{k-1}^2$ *is nul at the* $k^{th}$ *step of the Gauss decomposition process, prove that* $\delta_\lambda(k-1)\delta_\lambda(k) = -1$ *and that* $\delta_\lambda(k+1) = \gamma(k+1)$.
3. *Using the notations of Proposition VI.4.5 conclude from* 1. *and* 2. *that the number of indices* $k \in [0, n]$ *for which the sequence* $U_\lambda(k)\binom{1}{0}$ *belongs to the set* $P(\mathbb{R})^-$ *is equal to the number of eigenvalues of the operator* $H^\Lambda$ *no greater than* $\lambda$.

**Problem VI.5.4** *The following "Maryland model" has been studied in detail in [100] and we reproduce here the computation of the density of states. For a positive* $\beta$ *and an irrational* $\alpha$ *we consider the potential* $V(n) = \beta\, tg(\pi n\alpha + \omega)$. *Mimicking the construction we made in the case of almost periodic sequences, this potential can be realized on the ergodic invertible dynamical system* $\Omega = \{[0,\pi]\setminus\{\frac{\pi}{2} - n\pi\alpha;\ n \in \mathbb{Z}\}$ *endowed with the normalized Lebesgue measure, the operator* $\theta$ *being the rotation of angle* $\alpha$. *This model is not exactly quasi-peridic for the potential is not bounded. Nevertheless, many explicit computations can be performed in this situation. We consider the ergodic family of Schrödinger operators on the one dimensional lattice* $H = H_0 + V$. *Let* $U$ *be the operator* $(U\psi)(n) = e^{2i\pi n\alpha}\psi(n)$ *and* $z$ *be the complex number* $e^{2i\omega}$.

1. *Prove that* $H = H_0 + i\beta(zU - I)(zU + I)^{-1}$
2. *Prove that for* $\Im m\,\lambda > 0$ *one has:*

$$G_\lambda = (zU + I)(I + zCU)^{-1}G_{0,\lambda+i\beta}$$

*where* $C = G_{0,\lambda+i\beta}(H_0 - (\lambda - i\beta)I)$ *and* $G_{0,z}$ *is the Green kernel of the operator* $H_0$.

3. *Observing that* $\|C\| < 1$ *expand the operator* $(I + zCU)^{-1}$ *in a power series in* $z$ *and prove that* $\mathbb{E}\{G_\lambda\} = G_{0,\lambda+i\beta}$.

*Conclude that the integrated density of states is absolutely continuous and is given by* $N = N_0 * C_{i\beta}$ *where* $C_{i\beta}$ *is the Cauchy law defined in Problem IV.7.6.*
*Hint: For a real* $\lambda$ *one has:*

$$\frac{dN}{d\lambda} = \lim_{\epsilon\downarrow 0} \Im m \int \frac{dN(t)}{t - (\lambda + i\epsilon)} = \Im m \int \frac{N_0(t)}{t - \lambda - i\beta}$$

**Problem VI.5.5** *The "Lloyd model" has been studied in [296]. We gave in the text the essential steps of an alternative computation of the density. We consider the ergodic family of Schrödinger operators $H = H_0 + V$ on the one dimensional lattice associated to the potential $V(n) = Y_n$ where $Y_n$ is a sequence of real independent random variables with a common law $C_{i\beta}$.*

1. *Let $I$ be a finite set of integers ( not necessarily all distinct) and $\{s_k\,;\, k \in I\}$ be a set of nonnegative numbers with $\sum_{k\in I} s_k = 1$. Prove that $\mathbb{E}\{\exp(it\sum_{k\in I} s_k Y_k)\} = \exp(\beta|t|)$*
2. *The following result is sometimes refered to as the "Dyson's formula". Let $A$ and $B$ be two square matrices with complex entries and $t$ be a real number. Prove that $e^{t(A+B)} = e^{tB} + \int_0^t e^{s(A+B)} A e^{(t-s)B}\, ds$. Iterating this relation, expand $e^{t(A+B)}$ in a convergent series.*
3. *Let $\Lambda$ be a box. Applying 1. and 2. to the operators $H^\Lambda$ and $H_0^\Lambda$ prove that*

$$\mathbb{E}\{e^{itH^\Lambda}\} = e^{-\beta|t|} e^{itH_0^\Lambda} \quad \text{and} \quad \mathbb{E}\{e^{itH}] = e^{-\beta|t|} e^{H_0}$$

4. *Conclude from 3. that $N = N_0 * C_{i\beta}$*

**Problem VI.5.6** *Let $\tilde{H}$ be a Schrödinger operator on the one dimensional lattice associated to a sequence $\{\tilde{V}(n)\,;\, n \in \mathbb{Z}\}$ of complex potentials satisfying $\Im m\, \tilde{V}(n) \leq 0$ for all $n$.*

*Prove that for any box $\Lambda$ the operator $\tilde{H}^\Lambda$ has no eigenvalue $\lambda$ such that $\Im m\, \lambda > 0$.*

*Prove that for $k = 0 \ldots n$ then $\tilde{p}_\lambda(n+1)$ is a linear function of the potential $\tilde{V}(k)$.*

**Problem VI.5.7** *Let $H = H_0 + V$ where the potentials $V(n)$ form an i.i.d. sequence of random variables distributed according to the Cauchy law $C_{i\delta}$. For a box $\Lambda = [0\,,\, n]$ and any $r \in [0 \ldots n+1]$ we define the new potentials:*

$$\tilde{V}(k) = \begin{cases} -i\delta & \text{if } 0 \leq k \leq r-1 \\ V(k) & \text{if } r \leq k \leq n \end{cases}$$

*Let $\{\tilde{p}_{r,\lambda}(k)\,;\, k = 0 \ldots n+1\}$ be the solution of the eigenvalue equation associated to the potentials $\{\tilde{V}(k)\,;\, k = 0 \ldots n\}$.*

*Using the result* 1. *of Problem IV.7.6 and the above results prove that for* $\Im m\,\lambda > 0$ *one has:*

$$\mathbb{E}\{\tilde{p}_{r,\lambda}(n+1)\} = \mathbb{E}\{\tilde{p}_{r+1,\lambda}(k)\} \quad r = 0 \ldots n$$

*Iterating this relation conclude that:*

$$\mathbb{E}\{p_\lambda(n+1)\} = \mathbb{E}\{\tilde{p}_{n+1,\lambda}(k)\}$$

*If we assign the subscript* 0 *to the Lyapunov exponent and the integrated density of states of the free Laplacian* $H_0$ *prove that for* $\Im m\,\lambda > 0$ *one has:*

$$\gamma(\lambda) = \gamma_0(\lambda + i\delta) \quad \text{and} \quad \int \log|t-\lambda|\, dN(t) = \int \log|t-\lambda-i\delta|\, dN_0(t)$$

*Using the relation :*

$$\int \log|t - \lambda - i\delta|\, dN_0(t) = \int \log|t - \lambda|\, d[N_0 * C_{i\delta}](t)$$

*conclude again that* $N = N_0 * C_{i\delta}$.

*This computation can be extended to the strip in an easy way (See[210])*

**Problem VI.5.8** *Let* $\{(V(n), Z(n))\,;\, n \in \mathbb{Z}\}$ *be an* $\mathbb{R}^2$ *valued sequence of i.i.d. random variables such that:*

- *The distribution of* $Z(0)$ *is carried by the positive half-line and its support is bounded away from zero.*
- *The random variables* $Z(0)$ *and* $\log(1 + |V(0)|)$ *are integrable.*

*Recall that the "Helmotz operator"* $H$ *is then defined by :*

$$(H\psi)(n) = \frac{1}{Z(n)}(-\psi(n-1) - \psi(n+1) + V(n)\psi(n))$$

1. *Remarking that the transfer matrices associated to the eigenvalue equation* $H\psi = \lambda\psi$ *are given by* $g_\lambda(n) = \begin{pmatrix} V(n) - \lambda Z(n) & -1 \\ 1 & 0 \end{pmatrix}$, *extend to this case Proposition V.4.5 to obtain the subharmonicity of the exponent* $\gamma(\lambda)$.
2. *Repeating the argument of Proposition VI.4.3 prove the following "Thouless formula":*

$$\gamma(\lambda) = \mathbb{E}\{Z(0)\} + \int \log|t - \lambda|\, dN(t)$$

*Hint: The coefficient of the leading term $\lambda^n$ in the polynomial function $p_\lambda(n)$ is now $(-1)^n Z(0)Z(1)\dots Z(n-1)$ and apply the Birkhoff ergodic theorem.*

**Problem VI.5.9** *Let $H_\beta = H_0 + \beta V$ be the one dimensional Schrödinger operator associated to a stationary bounded process $\{V(n)\,;\, n \in \mathbb{Z}\}$ and a nonzero real parameter $\beta$. Then Proposition VI.4.3 implies that:*

$$\gamma_\beta(\lambda) = \int \log|t-\lambda|\, dN_\beta(t)$$

1. *Let $\tilde{N}_\beta$ be the image measure of $N_\beta$ by the mapping $t \hookrightarrow \frac{t}{\beta}$. Prove that*

$$\gamma_\beta(\beta\lambda) - \log|\beta| = \int \log|t-\lambda|\, d\tilde{N}_\beta(t)$$

2. *Taking into account that $\tilde{N}_\beta$ is the distribution of states of the operator $\frac{1}{\beta}H_0 + V$, prove that if we denote by $N_V$ the distribution of states of the diagonal operator $V$ one has :*

$$N_V(\lambda - \frac{2}{\beta}) \le \tilde{N}_\beta(\lambda) \le N_V(\lambda + \frac{2}{\beta})$$

   *Conclude that $\tilde{N}_\beta$ converges weakly to $N_V$ when $|\beta| \to \infty$.*

3. *Deduce from 1. and 2. that for $\lambda$ sufficiently large*

$$\lim_{|\beta|\to\infty} \left(\gamma_\beta(\beta\lambda) - \log|\beta|\right) = \int \log|t-\lambda|\, dN_V(t)$$

4. *The same property for any $\lambda$ is more difficult to obtain since the existence of the right hand side of the above formula is not granted. Moreover the convergence of $\tilde{N}_\beta$ applied to the function $t \hookrightarrow \log|\lambda - t|$ also fails. Prove that the conclusion 3. remains valid if there exists $\alpha > 0$ and $C < \infty$ such that:*

$$|\tilde{N}_\beta(\lambda) - \tilde{N}_\beta(\lambda')| \le C|\lambda - \lambda'|^\alpha$$

   *for $|\beta|$ sufficiently large. Prove that this estimate holds with $\alpha = 1$ in the independent case, assuming that the law of the potentials has a bounded density (use Wegner's estimates).*

## VI.6 Notes and Complements

Our path integral existence proof of the integrated density of states is borrowed from Nakao [251]. It has in common with most of the functional analytic proofs (see for example the work of Avron and Simon [13] or Shubin's work on partial differential operators with almost periodic coefficients [286]) to rely heavily on the regularity properties of the semigroup generated by the operator. This existence proof should be viewed as the result of the efforts of L.A. Pastur in U.S.S.R. and of M. Fukushima and his collaborators in Japan. See [260], [110], [113], [114] and [251].
The existence proof we give in the section on the one dimensional case is historically the first one. It is due to Benderskii and Pastur. See [24] and [25]. They relied on the Sturm oscillation theory to compare the number of eigenvalues less than a given level to the same quantity for a different choice of boundary conditions, and they showed that these numbers were additive up to controllable errors. The Birkhoff additive ergodic theorem could then be applied to conclude the proof. This argument has been extended to the lattice $\mathbb{Z}^d$ by Fukushima in [112] from where we took the construction given in the first section. This proof can be adapted to the multidimensional continuous case, the main difficulty being the identification of the limits constructed with the Dirichlet and the Neumann boundary conditions. In fact, the Kirsch-Martinelli proof which we gave in the first section can be viewed as an extension of the Benderskii-Pastur original proof. See also the earlier work of Slivnyak [307]. Kirsch and Martinelli gave the existence proof we presented in our subsection on the functional analytic approach. Their work gives a neat form to the ideas behind the bracketing and it shows that it is very natural if one wants to investigate the problem of the Lifschitz tails along the lines proposed by Lifschitz himself. See [178]. In fact, the seminal work of Lifschitz is presumably at the origin of the long lasting interest of his student L.A. Pastur in this problem.

Simon gave the lattice analogue of the Kirsch-Martinelli method. See [300]. Our discussion of what we call the functional analytic approach follows Kirsch's survey [174].

The existence of the integrated density of states is essentially due to one form or another of the additive ergodic theorem:

$$N(\lambda) = \lim_{\Lambda \to \mathbb{R}^d} \frac{1}{|\Lambda|} \sum_{k \geq 0} \mathbf{1}_{(-\infty,\lambda]}(\lambda_k^{(D)}(\Lambda,\omega))$$

where the $\lambda_k^{(D)}(\Lambda,\omega)$ are the Dirichlet eigenvalues of our random operators on the bounded domain $\Lambda$. The natural question to ask is that of a corresponding Central Limit Theorem, namely, do the random variables:

$$\tilde{N}^{(D)}_{\Lambda,\omega}(\lambda) = |\Lambda|^{1/2}(N^{(D)}_{\Lambda,\omega}(\lambda) - \mathbb{E}\{N^{(D)}_{\Lambda,\omega}(\lambda)\}$$

converge in distribution when $\Lambda \to \mathbb{R}^d$, and can we identify the limit? A positive answer has been given for the case of the example of the Russian school when the manifold $M$ is chosen to be the unit circle. In fact, in this case a functional form of the central limit theorem holds: for each compact interval $\mathcal{I}$ of $\mathbb{R}$, the stochastic process $\{N_{[-k,+k]}(\lambda); \lambda \in \mathcal{I}\}$ converges in law to a mean zero Gaussian process with continuous sample paths and the covariance function of which is computable. See Reznikova [275] for details. This work has been extended by Le Page to the lattice $\mathbb{Z}$ when the potential is an i.i.d. sequence in [214].
The fact that the topological support of the integrated density of states is the almost sure spectrum $\Sigma$ is an interesting result by itself since it tells us that the information in the set $\Sigma$ is contained in the integrated density of states. The idea of the proof of Proposition VI.1.3 was borrowed from [293] where the same results are proved for uniformly bounded below almost periodic potentials. Nevertheless the reader should not be misled. The integrated density of states is a very poor predictor of the spectral properties of the operators $H(\omega)$. We shall see simple examples of random potentials depending on some parameter, for which the integrated density of states is independent of the parameter while the spectra are almost surely pure point for some value of the parameter and almost surely singular continuous for other values of the parameter. This was noticed by Simon in [296]. Also we expect that some models which have non trivial absolutely continuous and pure point components in their spectrum have an analytic density of states which shows that the integrated density of states does not see the transition from one spectral type to another. See for example [205]. The functional analytic construction of the integrated density of states which we gave in the discrete case and CorollaryVI.2.6 have been borrowed from [112]. See also [276].

Our discussion of the results on the Lifschits tails for the Anderson model follows very closely the article [300] of B. Simon. It seems that the use of the Dirichlet-Neumann bracketing in the investigation of the Lifschitz tails was initiated by Harris in [141]. TheoremVI.2.8 is due to Mezincescu [237] and the proof has been simplified by Simon [302]. Theorem VI.2.9 on the Lifschitz tail in the case of periodic plus random potentials was first proven by Kirsch and Simon in [182] under the additional restriction that $d = 1$ or the periodic potential $U$ is reflection invariant. The general case given in TheoremVI.2.9 was obtained by Mezincescu in [238]. The improvement is due in part to the use of a mixed boundary conditions to replace the inappropriate Neumann part of the Dirichlet-Neumann bracketing.

The asymptotic behavior of $N(\lambda)$ at the edge of the spectrum has been studied for more general operators. Indeed, formally symmetric differential operators with singular coefficients can be defined as self adjoint operators by the method of quadratic forms. As an example one can consider Schrödinger operators for which the potential $\{V(t); t \in \mathbb{R}\}$ is white noise. The interested reader is refered to [189], [114], [110] or [318]. As we mentioned in the text, the connection between the Lifschitz tails and the large deviation results on the asymptotic of the Wiener saussage was exploited by Okura who considers the case of the free Hamiltonians of the form $H_0 = |p|^\alpha$ in [256]. It has also been exploited by Sznitman in a series of recent works [317], [315] and [316]. This author considers the operator $H_0 = -\Delta$ in the space $\mathbb{R}^d$ with spherical obstacles about points which form a sample from a Poisson point process. The construction of the integrated density of states can be done in the same way as we did it for random Schrödinger operators and its asymptotics near the bottom of the spectrum can still be investigated by means of the corresponding control of the asymptotics of the Wiener saussage. Sznitman also considers the case of the hyperbolic space for which he shows, among other things, that the integrated density of states depends on the choice of the boundary conditions.

The smoothness of the integrated density of states is a difficult problem which attracted the attention of many. Corollary VI.3.2 is a result of Constantinescu, Fröhlich and Spencer. See [58]. In a "Note added in proof" to their paper, the authors mention this short proof due to Molcanov and a similar one based on a Duhamel expansion of the unitary group which is due to Simon. For small disorder (i.e. for small $\sigma$) the question of the analytic extension has not been settled. In fact, there is a controversy in the physics community regarding this problem and it would be nice to find the mathematical truth.

The smoothness of the integrated density of states has been mainly investigated in the discrete case. The main reason is that the technical difficulties are much more difficult to overcome. In particular, absolutely nothing is known of the smoothness of the integrated density of states in the multidimensional continuous case. Moreover, few models have been completely solved in the one-dimensional case. Those which have can be divided into two categories. The first one is composed of potentials of the lattice type like the sequences of wells with i.i.d. heights and the techniques of the lattice case $\mathbb{Z}^d$ can be used successfully. The second one is essentially limited to the model of the Russian school and the integrated density of states is easily seen to be smooth. See for example [244] and [133].

We now concentrate on reviewing the results known in the discrete case. The most elementary smoothness result is certainly TheoremVI.3.6. It is due to Wegner. See [329]. It is remarkable in its simplicity, in its generality and in the depth of its consequences. This result is one of the very few dimension independent ones. Some extensions to continuous models and to non-i.i.d. potentials and to continuous models exist but they are very limited. See nevertheless [201]. The $C^\infty$ of the integrated density of states was first proved by Simon and Taylor in [303]. Their assumptions were somewhat restrictive and their paper initiated a wave of interest in the problem. The next improvement came from the simultaneous work of Klein and Campanino [39] and March and Sznitman [227]. It seems that the first of these two papers contains results which are slightly stronger. The reason is a clever choice of a scale of interpolating Banach spaces in which the smoothness of the Fourier transform of the invariant measure could be controlled. The Klein-Campanino approach has been extended to the case of the strip in [183]. On the other hand, [227] contains a wealth of information on the connection between excursion theory for continuous time Markov processes on the lattice $\mathbb{Z}^d$ and the integrated density of states.

There are essentially only two routes to the smoothness of the integrated density of states in one-dimension (and for the quasi one dimensional models). The first one is to prove first smoothness of the Lyapunov exponent and then to use Thouless formula to recover smoothness properties of the integrated density of states. The other approach relies on another formula giving the integrated density of states namely the expression in terms of the invariant measure of the Markov chain on the projective space. The bulk of the work is then to prove smoothness of this invariant measure and then to go back to the integrated density of states. We adopted the first approach for it is more natural from the point of view of the theory of products of random matrices which we developed in ChapterIV. Natural means here that the strategy and the functional analytic tools were already developed, they only had to be pushed a little bit further. The second approach may be richer in the sense that the functional analysis can be introduced in many ways. As harmonic analysis on groups as in the work of Simon and Taylor, as an outcome of a path integral method as in the work of March and Sznitman or as a byproduct of the so-called replica trick as in the work of Klein and Campanino and the like. In any case, the functional analytic problems have to be solved and no matter how they came about, they are difficult and their solution is the crux of the smoothness of the integrated density of states.

All the results we reviewed are positive but Simon made several early remarks to the effect that the integrated density of states could be singular, in particular in the case of potentials taking only finitely many values. See for example the appendix of [303]. This appendix gives the strongest case by making rigorous an argument of Halperin which shows that, given an order $\alpha > 0$, one can choose the coupling constant $\beta > 0$ large enough so that the integrated density of states of the Bernouilli random Hamiltonian $H_\beta$ is not log-Hölder continuous of order $\alpha$. The set of reprints [222] edited by Lieb and Mattis is also an excellent source for early work on this subject. Following the idea of [303], Carmona, Klein and Martinelli proved TheoremVI.4.12 in [50]. Precisely, they show that the integrated density of states had a singular continuous component in the case of Bernouilli potentials with large coupling constants. The proof that the integrated density of states was actually singular continuous was completed by Martinelli and Michelli in [230] where a complete proof of TheoremVI.4.13 can be found. An interesting, though nonrigorous analysis of the behavior of the integrated density of states near the special energies can be found in [253]. Note that the behavior of the integrated density of states of the Bernouilli model has also been investigated for small coupling constant. In particular, Derrida and Gardner give a convincing argument for the existence of the density of states. See [80]. It would be nice to have a rigorous mathematical account of this transition. Finally we quote a result of Campanino and Perez (see [40]) concerning a slightly different model. They prove that the integrated density of states $N(\lambda)$ of one dimensional disordered chains (one dimensional finite difference operators with off-diagonal disorder) has a density and that this density has a singularity at $\lambda = 0$ at least of the order of $1/(|\lambda|(\log|\lambda|)^4$. Their proof relies on a detailed analysis of the recurrence properties of the associated Markov chain. The idea of Problem VI.5.1 is taken from [226] where a Thouless formula is also proved for random potentials perturbed by a deterministic function going to zero. Finally, we note that Molcanov and Seidel showed in their study of random Sturm-Liouville operators [242] that the integrated density of states exists and can be proven to be smooth in some *nonstationary* models.

# VII

# ABSOLUTELY CONTINUOUS SPECTRUM AND INVERSE THEORY

Contents :



We assume that a complete dynamical system $(\Omega, \mathcal{F}, \mathbb{P}, \{\theta_t; t \in T\})$ is fixed throughout the chapter. In other words, we assume that $(\Omega, \mathcal{F}, \mathbb{P})$

is a complete probability space and $\{\theta_t; t \in T\}$ is an ergodic group of automorphisms of the measurable space $(\Omega, \mathcal{F})$ which leave invariant the measure $\mathbb{P}$. Except for the short last section, we shall not consider the multidimensional case and we concentrate on the cases $T = \mathbb{R}$ and $T = \mathbb{Z}$. We will add most of the time the assumption of ergodicity of the flow $\{\theta_t; t \in T\}$ of transformations. All the potential functions considered in this chapter are assumed to be defined on such a dynamical system. They are even assumed to be of the form $V(t,\omega) = v(\theta_t\omega)$ for some measurable function $v : \Omega \to \mathbb{R}$ whenever possible.

The aim of the present chapter is to study the absolutely continuous spectrum and more precisely, the essential support of the absolutely continuous component $E_{ac}(\cdot,\omega)$ of the resolution of the identity of $H(\omega) = H_0 + V(\cdot,\omega)$.

We begin with a discussion of the so-called $w$-function. We then concentrate on the one dimensional periodic potentials. They give the simplest possible examples of random ergodic stationary potentials and it is well known that their spectra are absolutely continuous. We give a self-contained and complete analysis using the ordinary differential equation approach developed in Chapter III and the Floquet theory of periodic differential equations which we review. We also discuss almost periodic models.

We proved in Chapter V that the essential support of the absolutely continuous component of the resolution of the identity is independent of $\omega \in \Omega$. We show now that it is actually equal (in the one dimensional case) to the essential closure of the set $\{\lambda \in \mathbb{R};\ \gamma(\lambda) = 0\}$ where the Lyapunov exponent $\gamma(\lambda)$ vanishes. We shall also prove that *nondeterministic* random potentials cannot support absolutely continuous spectrum. We then try to answer the following natural question: which among the *less random* potentials do not have absolutely continuous spectrum.

We devote a section to the basic elements of an inverse spectral theory for one dimensional random Schrödinger operators. We treat separately the continuous and the lattice cases and we do not discuss operators in a strip.

Finally we discuss one dimensional potentials taking finitely many values and we give a simple multidimensional criterion for the absence of absolutely continuous spectrum in the lattice case. This last result is to be compared with the deterministic critera we gave in Chapter III and the estimates of the Green's function which we will prove in Chapter IX.

# VII.1 The w-function

The $w$-function is a useful tool in the investigation of the absolutely continuous spectrum of the stationary Schrödinger operators. It was first introduced by Johnson and Moser in the almost periodic case and extensively used by Kotani in his thorough analysis of the general random case. Although the ideas are exactly the same, we separate once more the study of the lattice case from the continuous case. Indeed, the notation and the technicalities are different. But first, we review some extra properties of Herglotz functions which we shall need in the sequel.

## VII.1.1 More on Herglotz Functions

We need to refine some of the representation theorems presented in Chapter I and we investigate the ranges of the $w$-functions. We introduce the following definition for the sake of convenience.

**Definition VII.1.1** *A function $w$ is said to belong to the class $\mathcal{W}$ if $w \in \mathcal{H}erg$ and $w' \in \mathcal{H}erg$.*

The following representation theorem will play a crucial role in the sequel.

**Proposition VII.1.2** *$w \in \mathcal{W}$ if and only if $w$ has the representation:*

$$w(\lambda) = \alpha + \beta\lambda + \int \left( \frac{1}{\xi - \lambda} - \frac{\xi}{1+\xi^2} \right) N(\xi)\, d\xi \tag{VII.1}$$

*for some $\alpha \in \mathbb{R}$, some $\beta \geq 0$ and some nonnegative nondecreasing function $N$ on $\mathbb{R}$ satisfying:*

$$\int_{\mathbb{R}} \frac{1}{1+\xi^2} N(\xi)\, d\xi < \infty.$$

*If moreover $w$ has the asymptotic:*

$$w(iy) = \log \frac{1}{-iy} + O(1) \tag{VII.2}$$

*as $y \to +\infty$, then $w$ can actually be written in the form:*

$$w(\lambda) = \int \log \frac{1}{\xi - \lambda} d\nu(\xi) \tag{VII.3}$$

*for some probability measure $\nu$ on $\mathbb{R}$ which satisfies:*

$$\int \frac{1}{1+\xi^2} d\nu(\xi) < +\infty. \tag{VII.4}$$

*Proof:*
It is obvious that a function $w$ of the form (VII.1) is in the class $\mathcal{W}$. Conversely, since $w' \in \mathcal{H}erg$, the measure $\pi^{-1}\Im m\, w'(\xi + i\epsilon)d\xi$ converges vaguely as $\epsilon \downarrow 0$ to a nonnegative measure, say $\nu$, on $\mathbb{R}$ which satisfies:

$$\int \frac{1}{1+\xi^2} d\nu(\xi) < +\infty.$$

Similarly, $w \in \mathcal{H}erg$ implies that the measure $\pi^{-1}\Im m\, w(\xi+i\epsilon)d\xi$ converges vaguely to a nonnegative measure $\tilde{\nu}$ on $\mathbb{R}$. Note that $\nu$ has to be the derivative of $\tilde{\nu}$ in the sense of distributions. The usual representation of Herglotz functions gives:

$$w(\lambda) = \alpha + \beta\lambda + \int \left(\frac{1}{\xi - \lambda} - \frac{\xi}{1+\xi^2}\right) d\tilde{\nu}(\xi)$$

for some $\alpha \in \mathbb{R}$ and some $\beta \geq 0$, and consequently:

$$w(\lambda) = \alpha + \beta\lambda + \int \left(\frac{1}{\xi - \lambda} - \frac{\xi}{1+\xi^2}\right) N(\xi)d\xi \qquad \text{(VII.5)}$$

if one writes $N(\xi) = \nu((-\infty, \xi])$ for the distribution function of the measure $\nu$. We now prove the last claim of the proposition. Taking derivatives of both sides and integrating by parts, one gets:

$$w'(\lambda) = \beta + \int \frac{1}{\xi - \lambda} d\nu(\xi).$$

Taking imaginary parts of both sides of (VII.5), one gets:

$$\Im m\, w(\lambda) = \beta \Im m\, \lambda + \Im m\, \lambda \int \frac{N(\xi)}{|\xi - \lambda|^2} d\xi$$

and:

$$\Im m\, w(iy) = \beta y + y \int \frac{N(\xi)}{\xi^2 + y^2} d\xi$$

by setting $\lambda = iy$ with $y > 0$. Our assumption (VII.2) implies $\beta = 0$. Now:

$$\begin{aligned} w(\lambda) - w(i) &= \int \left(\frac{1}{\xi - \lambda} - \frac{1}{\xi - i}\right) N(\xi)d\xi \\ &= \int \left(\log \frac{1}{\xi - \lambda} - \log \frac{1}{\xi - i}\right) d\nu(\xi). \end{aligned}$$

Taking the real part of both sides and setting $\lambda = yi$ with $y \to +\infty$, one gets

$$\log y + O(1) = \int \log\left(1 + \frac{y^2 - 1}{\xi^2 + 1}\right) d\nu(\xi)$$

according to assumption (VII.2). This implies that $\nu(\mathbb{R}) = 1$ and (VII.4) at the same time. Consequently $w(\lambda)$ is necessarily of the form:

$$w(\lambda) = \alpha + \int \log \frac{1}{\xi - \lambda} d\nu(\xi)$$

and assumption (VII.2) gives $\alpha = 0$. ∎

**Remark VII.1.3** Note that any $w \in \mathcal{W}$ satisfying assumption (VII.2) is such that:

$$0 \leq \Im m\, w(\lambda) \leq \pi.$$

The following comparison result will play an important role in the sequel.

**Proposition VII.1.4** *Let $\nu_1$ and $\nu_2$ be probability measures on $\mathbb{R}$ which satisfy:*

**(a)** $\int \xi d\nu_1(\xi) = \int \xi d\nu_2(\xi) = 0$
**(b)** $\int \xi^2 d\nu_1(\xi) < +\infty$
**(c)** *$\nu_2$ has compact support.*

*If the holomorphic functions $w_1$ and $w_2$ defined on $\Pi_+$ by:*

$$w_1(\lambda) = \int \log \frac{1}{\xi - \lambda} d\nu_1(\xi) \quad \text{and} \quad w_2(\lambda) = \int \log \frac{1}{\xi - \lambda} d\nu_2(\xi)$$

*satisfy $w_1(\Pi_+) \subset w_2(\Pi_+)$, then one has:*

$$\int \xi^2 d\nu_1(\xi) \geq \int \xi^2 d\nu_2(\xi).$$

The proof relies on the following technical lemmas.

**Lemma VII.1.5** *Let $\nu$ be a probability measure on $\mathbb{R}$ satisfying $\int \xi d\nu(\xi) = 0$ and $\int \xi^2 d\nu(\xi) < +\infty$ and let $w$ be the holomorphic function defined on $\Pi_+$ by:*

$$w(\lambda) = \int \log \frac{1}{\xi - \lambda} d\nu(\xi).$$

*Then, there exists a probability measure $\sigma$ on $I\!R$ satisfying:*

$$\int \lambda\, d\sigma(\lambda) = 0 \quad \text{and} \quad \int \lambda^2 d\sigma(\lambda) \le \int \lambda^2 d\nu(\lambda) \tag{VII.6}$$

*and such that the function $u(\lambda) = e^{w(\lambda)}$ has a representation of the form:*

$$u(\lambda) = \int \frac{1}{\xi - \lambda} d\sigma(\xi). \tag{VII.7}$$

*Proof:*
We first assume that $\nu$ has compact support. In this case one has:

$$\Im m\, w(\lambda) = \int \arg \frac{1}{\xi - \lambda} d\nu(\lambda)$$

so that we must have $0 \le \Im m\ w(\lambda) \le \pi$ for all $\lambda \in \Pi_+$. This implies that $u$ is in the Herglotz class $\mathcal{H}erg$ and admits the usual representation:

$$u(\lambda) = \alpha + \beta\lambda + \int \left( \frac{1}{\xi - \lambda} - \frac{\xi}{1 + \xi^2} \right) d\sigma(\xi). \tag{VII.8}$$

Now, as $\lambda \to \infty$, we have:

$$w(\lambda) = \log \frac{1}{-\lambda} + \frac{1}{2\lambda^2} \int \xi^2 d\nu(\xi) + \cdots$$

and consequently

$$\begin{aligned} u(\lambda) &= -\frac{1}{\lambda} \exp[\frac{1}{2\lambda^2} \int \xi^2 d\nu(\xi) + \cdots] \\ &= -\frac{1}{\lambda}[1 + \frac{1}{2\lambda^2} \int \xi^2 d\nu(\xi) + \cdots]. \end{aligned} \tag{VII.9}$$

Identifying (VII.8) with (VII.9), we get (VII.7) and:

$$\int d\sigma(\xi) = 1, \quad \int \xi d\sigma(\xi) = 0, \quad \int \xi^2 d\sigma(\xi) = \frac{1}{2} \int \xi^2 d\nu(\xi), \quad \cdots$$

which completes the proof when $\nu$ has compact support. In the general case, one approximates $\nu$ by probability measures $\nu_k$ having compact supports. To each $\nu_k$ corresponds a $\sigma_k$ because of the first part of the proof. We choose for $\sigma$ any limit point of the sequence $\{\sigma_k; k \ge 1\}$. The rest is plain. ∎

**Remark VII.1.6** The above argument shows that we have the equality

$$\int \xi^2 d\sigma(\xi) = \frac{1}{2}\int \xi^2 d\nu(\xi)$$

whenever $\int \xi^4 d\nu(\xi) < +\infty$.

**Lemma VII.1.7** *Let $u_1$ and $u_2$ be conformal maps from $\Pi_+$ into $\Pi_+$ of the form (VII.7) for probability measures $\sigma_1$ and $\sigma_2$ on $\mathbb{R}$ satisfying:*

$$\int \xi d\sigma_1(\xi) = \int \xi d\sigma_2(\xi) = 0,$$

$$\int \xi^2 d\sigma_1(\xi) < +\infty, \quad \text{and} \quad \int \xi^2 d\sigma_2(\xi) < \infty.$$

*Then we have*

$$\int \xi^2 d\sigma_2(\xi) \leq \int \xi^2 d\sigma_1(\xi)$$

*whenever $u_1(\Pi_+) \subset u_2(\Pi_+)$.*

*Proof:*
Set $u(\lambda) = u_2^{-1}(u_1(\lambda))$ for $\lambda \in \Pi_+$. $u$ maps conformally $\Pi_+$ into itself and admits the usual representation:

$$u(\lambda) = \alpha + \beta\lambda + \int (\frac{1}{\xi - 1} - \frac{\xi}{1+\xi^2}) d\sigma(\xi).$$

Now, our assumptions on $\sigma_1$ and $\sigma_2$ imply that:

$$u_1(\lambda) = \frac{1}{-\lambda}[1 + \frac{1}{\lambda^2}\int \xi^2 d\sigma_1(\xi) + O(\frac{1}{|\lambda|^2})]$$

and

$$u_2(\lambda) = \frac{1}{-\lambda}[1 + \frac{1}{\lambda^2}\int \xi^2 d\sigma_2(\xi) + O(\frac{1}{|\lambda|^2})]$$

as $\lambda \to \infty$ while keeping $\Im m\, \lambda \geq c > 0$. This implies that

$$u(\lambda) = \lambda + \frac{1}{\lambda}\int \xi^2 d\sigma_1(\xi) + O(\frac{1}{|\lambda|})$$

in the same regime. Consequently, one has:

$$u(\lambda) = \lambda + \int \frac{1}{\xi - \lambda} d\sigma(\xi)$$

with

$$0 \leq \int d\sigma(\xi) = \int \xi^2 d\sigma_1(\xi) - \int \xi^2 d\sigma_2(\xi). \blacksquare$$

*Proof of Proposition VII.1.4*
The functions $w_1$ and $w_2$ map conformally $\Pi_+$ into $\{z \in \Pi_+; 0 \leq \Im m\, z \leq \pi\}$. Indeed, $\Im m\, w_1'(\lambda) > 0$ and $\Im m\, w_2'(\lambda) > 0$. Consequently, the functions $u_1(\lambda) = e^{w_1(\lambda)}$ and $u_2(\lambda) = e^{w_2(\lambda)}$ map conformally $\Pi_+$ into itself and the above lemmas give:

$$\frac{1}{2}\int \xi^2 d\nu_1(\xi) \geq \int \xi^2 d\sigma_1(\xi) \geq \int \xi^2 d\sigma_2(\xi) = \frac{1}{2}\int \xi^2 d\nu_2(\xi)$$

where we used the compactness of the support of $\nu_2$ to derive the last equality. $\blacksquare$

**Remark VII.1.8** The ideas of the above proof can be applied to the case of two $w$-functions $w_1$ and $w_2$ in $\mathcal{W}$ which satisfy the asymptotic:

$$w_j(\lambda) = i\sqrt{\lambda} + \frac{\sigma_j^2}{2i\sqrt{\lambda}\lambda} + o(|\lambda|^{-3/2})$$

as $\lambda \to \infty$ for $j = 1, 2$. In this case, the conclusion is that $w(\Pi_+) \subset w_2(\Pi_+)$ implies that $\sigma_1^2 \geq \sigma_2^2$. In this case one uses the functions $u_j(\lambda) = -w_j(\lambda)^2$ and $u(\lambda) = u_2^{-1}(u_1(\lambda))$. Then the function $u$ has the asymptotic:

$$u(\lambda) = \lambda - \frac{\sigma_1^2 - \sigma_2^2}{\lambda} + o(|\lambda|^{-1})$$

and the representation:

$$u(\lambda) = \lambda + \int \frac{d\nu(\xi)}{\xi - \lambda}$$

from which one conludes that $\sigma_1^2 - \sigma_2^2 = \int d\nu(\xi) \geq 0$. $\blacksquare$

## VII.1.2 The Continuous Case

We first concentrate on the case $T = \mathbb{R}$. We use freely the terminology and the notation of Chapter III.

Throughout this section (and in the rest of this chapter whenever we discuss the continuous case) we implicitely assume that the random potential $\{V(t);\, t \in \mathbb{R}\}$ satisfies:

$$\int_{\mathbb{R}} \frac{V(t)^2}{1+|t|^3}\, dt < \infty \tag{VII.10}$$

$\mathbb{P}$-almost surely. This is trivially the case when the random potential is stationary and has a finite second moment in the sense that $\mathbb{E}\{V(0)^2\} < \infty$. Under this condition, $\omega \hookrightarrow V(\cdot,\omega)$ defines a measurable map from $\Omega$ into $\Omega^{(c)} = L^2_{real}(\mathbb{R}, dt/(1+|t|^3))$ and the probability measure $\mathbb{P}$ is mapped into a probability measure on this real Hilbert space. The shift $\theta_t$ is mapped into the usual shift $f \hookrightarrow f(\cdot + t)$ and the original process is mapped into some version of the coordinate process. In fact, from the point of view of random Schrödinger operators, no information is lost by this change of probability basis. We shall very often find it convenient to assume that the probability space $(\Omega, \mathcal{F}, \mathbb{P})$ on which the random potential is defined is in fact the one we just described. We shall use the special notation $\Omega^{(c)}$ to emphasize that we are working with this particular realization of the random potential but we shall not introduce new notation for the $\sigma$-field $\mathcal{F}$ (which will be chosen as the Borel $\sigma$-field of $\Omega^{(c)}$) and the shift. This point of view is extremely convenient when we deal with properties of the *topological support* of the probability measure $\mathbb{P}$. In this case we need $\mathcal{F}$ to be the Borel $\sigma$-field of a (Polish) topology on $\Omega^{(c)}$. Recall the theorems proved in Chapter V on the comparison of the spectra corresponding to several random potentials. The right setting for this sort of problem is to realize the various random potentials as the same measurable function $(t,\omega) \hookrightarrow V(t,\omega)$ on $T \times \Omega$ for some fixed probability space $(\Omega, \mathcal{F})$, only the probability measures being different. The comparisons are then done via the toplogical supports of these probability measures. Choosing $\Omega^{(c)}$ as a separable Hilbert space is especially handy because its Borel $\sigma$-field is the Borel $\sigma$-field of the weak topology as well, and this weak topology has many compact sets. This last fact will be crucial when we try to find limit points of sequences of random potentials.

Unfortunately, working with a probability measure on the Hilbert space $\Omega^{(c)}$ has some drawbacks. The main one is the following. If we start with a stationary (real valued) stochastic process $\{V(t);\, t \in \mathbb{R}\}$, the expectations $\mathbb{E}\{|V(0)|^k\}$ make sense whether or not they are finite. On the other hand, if we start with a probability measure $\mathbb{P}$ on the space $\Omega^{(c)}$, then the above moments do not make sense any more since the elements of $\Omega^{(c)}$ are no longer functions but merely equivalence classes of functions and it does not make sense to compute them at a given point $t$. In this new setting, the relevant notion of moment is $\mathbb{E}\{\int_0^1 |V(s)|^k\, ds\}$. We will try to be careful and not to ignore these technical details in the sequel. Nevertheless, since the spectral properties of two Schrödinger operators $H_0 + V_1$ and $H_0 + V_2$ cannot be distinguished if $V_1 = V_2$ almost everywhere, we can assume that, even when one starts with a probability measure $\mathbb{P}$ on $\Omega^{(c)}$, one can extract a version of the coordinate process which is defined for every $t \in \mathbb{R}$ as a random variable.

Note that all the measurability results of Chapter V remain valid even when we start with a probability measure on the path space $\Omega^{(c)}$.

We consider the $m$-functions $m_\pm(z)$ along the orbits of the dynamical system $\{\theta_t; t \in T\}$. We set:

$$z_\pm(t, z, \omega) = m_\pm(z, \theta_t\omega).$$

For each fixed $z$ and $\omega$, the function $t \hookrightarrow z_\pm(t, z, \omega)$ satisfies a Ricatti equation:

$$\dot{z}_\pm(t, z, \omega) = \pm[V(t, \omega) - z_\pm - z_\pm(t, z, \omega)^2]. \qquad \text{(VII.11)}$$

Note also that:

$$z_\pm(t, z, \omega) = \pm\frac{\dot{f}_\pm(t, z, \omega)}{f_\pm(t, z, \omega)}$$

since:

$$f_\pm(t, z, \omega) = \exp[\pm\int_0^t z_\pm(s, z, \omega)ds]. \qquad \text{(VII.12)}$$

Let us now define the function $w$ which will play a crucial role in the investigation of the absolute continuous spectrum and the inverse theory. We set:

$$w(z) = \frac{1}{2}\mathbb{E}\{m_+(z) + m_-(z)\} = -\frac{1}{2}\mathbb{E}\{\frac{1}{G(z, 0, 0)}\} \qquad \text{(VII.13)}$$

Lemma III.1.17 implies that the above expectations make perfectly good sense. Since $m_+(z)$ and $m_-(z)$ are analytic on $\Pi_+$ and have positive imaginary parts, $w$ has the same properties: it is also a Herglotz function. The importance of this function lies in the fact that the real part and the imaginary part of its boundary values on the real axis determine the Lyapunov exponent and the integrated density of states of the system. We prepare for the proof of this fundamental result by presenting first a few technical lemmas.

**Lemma VII.1.9** *$w \in \mathcal{W}$ and we have:*

$$w(z) = \mathbb{E}\{m_+(z)\} = \mathbb{E}\{m_-(z)\} \tag{VII.14}$$

$$w'(z) = \mathbb{E}\{G(z,0,0)\} \tag{VII.15}$$

*Proof:*
Notice that $m_+(z, \theta_t\omega) - m_-(z, \theta_t\omega)$ is the derivative with respect to $t$ of the function $t \hookrightarrow \log G(z,t,t,\omega)$. This implies that:

$$\int_a^b [m_+(z,\theta_t\omega) - m_-(z,\theta_t\omega)]dt = \log G(z,0,0,\theta_b\omega) - \log G(z,0,0,\theta_a\omega)$$

for all $a < b$ if one remarks that $G(z,t,t,\omega) = G(z,0,0,\theta_t\omega)$. One gets (VII.14) by taking the expectation of both sides. In order to show (VII.15) we first notice that the function $h$ defined by:

$$h(\omega) = \frac{1}{2}G(z,0,0,\omega)[\int_0^\infty f_+(t,z,\omega)^2 dt - \int_{-\infty}^0 f_-(t,z,\omega)^2 dt]$$

satisfies:

$$-\frac{\partial}{\partial z}\frac{1}{2G(z,t,t,\omega)} = G(z,t,t,\omega) + \frac{\partial h}{\partial t}(\theta_t\omega)$$

which gives, for all $a < b$:

$$(b-a)\mathbb{E}\{-\frac{\partial}{\partial z}\frac{1}{2G(z,t,t,\omega)}\} = (b-a)\mathbb{E}\{G(z,0,0)\}$$

which concludes the proof since:

$$w'(z) = \frac{\partial \mathbb{E}\{-[2G(z,0,0)]^{-1}\}}{\partial z}.$$

Note that we interchanged freely integration signs and derivations without special care. Everything can be justified by using the estimates given in Lemma III.1.17. ∎

Notice that:

$$\begin{aligned} \lim_{t\to\infty} \frac{1}{t}\int_0^t m_+(z,\theta_s)\,ds &= \mathbb{E}\{m_+(z)\} \\ &= w(z). \end{aligned} \tag{VII.16}$$

Putting this together with the definition (VII.12) of the function $f_\pm$ we get that, in some intuitive sense, the asymptotic behavior of the $L^2$-solution at $+\infty$ is given by:

$$f_+(t,z,\omega) \approx e^{\Re e\, w(z)t} e^{i\Im m\, w(z)t} \tag{VII.17}$$

**Lemma VII.1.10** *We have:*

$$\mathbb{E}\{\frac{1}{\Im m\, m_\pm(z)}\} = -\frac{2}{\Im m\, z}\Re e\, w(z)$$

*which implies in particular that* $\Re e w(z) < 0$. *Also:*

$$\frac{1}{-\Im m\, z}\Re e\, w(z) - \Im m\, w'(z) > 0 \tag{VII.18}$$

*and:*

$$-2\Re e\, w(z)\, \Im m\, w(z) \geq \Im m\, z. \tag{VII.19}$$

*Proof:*
Equating imaginary parts of both sides of the Ricatti equation (III.22) we obtain:

$$\Im m\, \dot{z}_+(t,z,\omega) = 2\Re e\, z_+(t,z,\omega)\Im m\, z_+(t,z,\omega) - \Im m\, z$$

and this gives:

$$\frac{d}{dt}\log \Im m\, z_+(t,z,\omega) + \frac{\Im m\, z}{\Im m\, z_+(t,z,\omega)} = -2\,\Re e\, z_+(t,z,\omega)$$

or in other words:

$$\frac{d}{dt}\log \Im m\, m_+(z,\theta_t\omega) + \frac{\Im m\, z}{\Im m\, m_+(z,\theta_t\omega)} = -2\,\Re e\, m_+(z,\theta_t\omega).$$

Taking expectations of both sides gives the desired result for $m_+(z)$. The corresponding result for $m_-(z)$ is shown in the same way. The two inequalities can be proven directly. Indeed:

$$\begin{aligned}
&-\frac{1}{\Im m z}\Re e\, w(z) - \Im m\, w'(z) \\
&= \frac{1}{4}\mathbb{E}\{\frac{1}{\Im m\, m_+(z)} + \frac{1}{\Im m\, m_-(z)}\} + \mathbb{E}\{\Im m\, \frac{1}{m_+(z)+m_-(z)}\} \\
&= \frac{1}{4}\mathbb{E}\{[\frac{1}{\Im m\, m_+(z)} + \frac{1}{\Im m\, m_-(z)}] \\
&\qquad \frac{[\Re e\, m_+(z) + \Re e\, m_-(z)]^2 + [\Im m\, m_+(z) - \Im m\, m_-(z)]^2}{|m_+(z)+m_-(z)|^2}\} \\
&> 0.
\end{aligned}$$

Moreover,

$$\begin{aligned}
\frac{\Im m\, w(z)}{\Im m\, z} &= \frac{1}{\Im m\, z}\mathbb{E}\{\Im m\, m_+(z)\} \\
&= \int_0^\infty \mathbb{E}\{\exp[2\int_0^t \Re e\, m_+(z,\theta_u\omega)du]\}dt \\
&\geq \int_0^\infty \exp[2\int_0^t \mathbb{E}\{\Re e\, m_+(z)\}du]dt \\
&= \int_0^\infty e^{2t\,\Re e\, w(z)}dt \\
&= \frac{-1}{2\,\Re e\, w(z)}.
\end{aligned}$$

where we successively, took expectations on both sides of the relation:

$$\int_0^\infty |f_+(t,z,\omega)|^2 dt = \frac{\Im m\, m_+(z,\omega)}{\Im m\, z},$$

used Jensen's inequality and finally the definition of the function $w$. ∎

Coming back to formula (VII.15) we see that:

$$\begin{aligned}
w'(z) &= \mathbb{E}\{G(z,0,0)\} \\
&= \mathbb{E}\{\int_\Sigma \frac{1}{\lambda - z}E(d\lambda,0,0,)\} \\
&= \int_{\mathbb{R}} \frac{1}{\lambda - z}\nu(d\lambda) \\
&= \int_{\mathbb{R}} \frac{N(\lambda)}{(\lambda - z)^2}d\lambda \qquad \text{(VII.20)}
\end{aligned}$$

by a simple integration by parts because $\int(|\lambda|+1)^{-1}\nu(d\lambda) < \infty$. Integrating both sides of (VII.20), we get:

$$w(z) = \alpha + \int_{\mathbb{R}} \frac{(1+\lambda z)N(\lambda)}{(\lambda - z)(1+\lambda^2)}d\lambda \qquad \text{(VII.21)}$$

for some $\alpha \in \mathbb{C}$. Since $w(z) = (-1/2)\mathbb{E}\{G(z,0,0)^{-1}\}$ implies that $w(\overline{z}) = \overline{w(z)}$, we must have $\Im m\, \alpha = 0$. Hence, if $\lambda \in \mathbb{R}$, we have for each $\epsilon > 0$:

$$\begin{aligned}
\Im m\, w(\lambda + i\epsilon) &= \epsilon \int_{\mathbb{R}} \frac{N(\lambda')}{(\lambda - \lambda')^2 + \epsilon^2}d\lambda' \\
&= \int_{\mathbb{R}} \frac{N(\lambda + \epsilon u)}{1 + u^2}du.
\end{aligned}$$

We already noticed that the integrated density of states was continuous in the one dimensional case because with probability one, a given real number cannot be an eigenvalue. Hence, passing to the limit $\epsilon \to 0$ in the above equality we obtain the first part of the following proposition.

**Proposition VII.1.11** *The Herglotz function $w(z)$ satisfies:*

$$N(\lambda) = \frac{1}{\pi}\Im m\, w(\lambda + i0) \qquad and \qquad \gamma(\lambda) = -\Re e\, w(\lambda + i0) \qquad \text{(VII.22)}$$

*for all $\lambda \in \mathbb{R}$ where $N(\lambda)$ and $\gamma(\lambda)$ are as usual the integrated density of states and the Lyapunov exponent. Moreover we also have:*

$$\gamma(\lambda) = -\alpha + \int_{\mathbb{R}} \log\left|\frac{\lambda - x}{x - i}\right| dN(x) \qquad \text{(VII.23)}$$

*for some real constant $\alpha$.*

Under the assumptions of this section we have:

$$\begin{aligned} \mathbb{E}\{\sup_{0\le t\le 1} \log^+ \|U_\lambda^\omega(t)\|\} &\le \mathbb{E}\{\int_0^1 \|A_\lambda(s)\| ds\} \\ &\le \mathbb{E}\{\int_0^1 |V(s)|\, ds\} + |\lambda| \\ &< \infty, \end{aligned}$$

and the existence of the Lyapunov exponent:

$$\gamma(\lambda) = \lim_{t\to\pm\infty} \frac{1}{|t|} \log \|U_\lambda^\omega(t)\| \qquad \text{(VII.24)}$$

follows. Notice also that formula (VII.23) is the continuous version of the Thouless formula which we proved in the lattice case.

*Proof:*
Formula (VII.21) gives:

$$\begin{aligned} -\Re e\, w(z) &= -\alpha + \int_{\mathbb{R}} \frac{(1+\lambda z)N(\lambda)}{(z-\lambda)(1+\lambda^2)} d\lambda \\ &= -\alpha + \int_{\mathbb{R}} \log|\frac{x-z}{x-i}| dN(x) \end{aligned}$$

by integration by parts. The first inequality of the preceeding lemma gives:

$$\int_{\mathbb{R}} \log|\frac{x-z}{x-i}| dN(x) < \alpha \qquad \text{(VII.25)}$$

but if we set $z = -i\epsilon$ with $\epsilon \in (0,1)$, we get:

$$-\log|\frac{x-i\epsilon}{x-i}| = \log(1+x^2) - \frac{1}{2}\log[(1+x^2)\epsilon^2 + x^2 + x^4]$$

which is positive and increases to $-\log|x/(x-i)|$ as $\epsilon$ decreases to 0. This shows that $\log|x|$ is locally integrable with respect to $dN(x)$. Choosing now $z = \lambda + i\epsilon$ with $\epsilon > 0$ we get:

$$\begin{aligned} -\Re e\, w(\lambda+i\epsilon) &= -\alpha + \int_{I\!R} \log|\frac{x-\lambda-i\epsilon}{x-\lambda}|dN(x) \\ &\quad + \int_{I\!R} \log|\frac{x-\lambda}{x-i}|dN(x). \end{aligned}$$

Using the fact that (VII.25) implies that the integral

$$-\int_{|\lambda-x|\leq 1} \log|\lambda-x|dN(\lambda)$$

is bounded from above by a constant independent of $\lambda$ whenever the latter is restricted to a compact set, one justifies taking the limit $\epsilon \longrightarrow 0$ to get (VII.23). In order to identify the real part of the $w$-function we notice that the functions $\lambda \hookrightarrow \gamma(\lambda)$ and $\lambda \hookrightarrow -2\Re e\, w(\lambda)$ are subharmonic in the whole complex plane and that they coincide outside the real axis because of formula (VII.12) and because we also have:

$$\gamma(\lambda) = \inf_{t>0} \frac{1}{t}\mathbb{E}\{\log||U_\lambda(t)||\}.$$

Consequently, they coincide everywhere. This concludes the proof. ∎

**Remark VII.1.12** The above calculations also show that for each compact set $K$, there is a positive constant $c(K)$ such that:

$$|N(\lambda_1) - N(\lambda_2)| \leq \frac{c(K)}{\log|\lambda_1-\lambda_2|}$$

for all $\lambda_1$ and $\lambda_2$ in $K$, namely that the integrated density of states is also log-Hölder continuous in the continuous case.

### VII.1.3 The Lattice Case

We first describe the setting of the present section. $(\Omega, \mathcal{F}, \mathbb{P})$ is a complete probability space and $\theta$ is a bimeasurable invertible transformation of $\Omega$

which preserves $\mathbb{P}$. On such a probability space we consider a stationary sequence $\{(a_n, b_n); n \in \mathbb{Z}\}$ of random variables such that:

**(i)** The common distribution of the $a_n$'s, say $\mu_a$, is concentrated on $(0,\infty)$ and satisfies:

$$\int_{(0,\infty)} (x^2 + |\log x|)\mu_a(dx) < \infty. \tag{VII.26}$$

**(ii)** The common distribution of the $b_n$'s, say $\mu_b$, satisfies:

$$\int_{\mathbb{R}} x^2 \mu_b(dx) < +\infty. \tag{VII.27}$$

The above assumptions will be in force throughout this section and any time we consider the lattice case in this chapter. Note that the technical discussion we had in the continuous case about the choice of a canonical path space is irrelevant in the present situation. Indeed, one can choose $\Omega^{(d)} = [(0,\infty) \times \mathbb{R}]^{\mathbb{Z}}$ without any loss of generality and without worrying about the regularity of the sample paths of the process since the parameter set is $\mathbb{Z}$!

We first note that:

$$\mathbb{P}\{\sum_{n=1}^{\infty} \frac{1}{a_n^2} = \infty\} = 1$$

because, if this was not the case, we would have

$$\mathbb{P}\{\sum_{n=1}^{\infty} \frac{1}{a_n^2} < \infty,\ a_1 < M\} > 0$$

for $M > 0$ large enough, and Poincaré's recurrence lemma would imply:

$$\mathbb{P}\{\sum_{n=1}^{\infty} \frac{1}{a_n^2} < +\infty, \quad a_m < M \text{ for infinitely many } m\} > 0$$

which is impossible. Consequently, for $\mathbb{P}$-almost all $\omega \in \Omega$, one can define the self-adjoint operator $H^+(\omega)$ on $\mathcal{D}_1^+$ by formula (III.60) and this operator is essentially self-adjoint on $\mathcal{D}_0^+$. One defines similarly self-adjoint operators $H^-(\omega)$ and $H(\omega)$ on $\mathcal{D}_1^-$ and $\mathcal{D}_1$, respectively, and they are essentially self-adjoint on $\mathcal{D}_0^-$ and $\mathcal{D}_0$, respectively. For each fixed $\omega \in \Omega$ the definitions chosen in Chapter III for the $m$-functions give:

$$m_\pm(z,\omega) = < [H^\pm(\omega) - zI]^{-1}\delta_{\pm 1}, \delta_{\pm 1} >$$

for $\Im m\, z > 0$ and:

$$m_{\pm}(z,\omega) = -\frac{1}{a_0(\omega)} \frac{f_{z,\pm\infty}(\pm 1,\omega)}{f_{z,\pm\infty}(0,\omega)}$$

according to (III.63). They also imply that:

$$m_{\pm}(z,\theta^n\omega) = -\frac{1}{a_0(\theta^n\omega)} \frac{f_{z,\pm\infty}(n\pm 1,\omega)}{f_{z,\pm\infty}(n,\omega)}$$

and:

$$m_-(z,\theta^{-n}\omega) = -\frac{1}{a_{-n}a_{-n-1}}\left[z - b_{-n}(\omega) + \frac{1}{m_-(z,\theta^{-n-1}\omega)}\right]. \qquad \text{(VII.28)}$$

Moreover, one also has:

$$m_+(z,\theta^n\omega) = -\frac{1}{a_n^2}\left[z - b_n(\omega) + \frac{1}{m_+(z,\theta^{n-1}\omega)}\right], \qquad \text{(VII.29)}$$

and:

$$\frac{1}{n+1}\log|f_{z,\infty}(n+1,\omega)| = \frac{1}{n+1}\log|f_{z,\infty}(0,\omega)| + \frac{1}{n+1}\sum_{k=0}^{n}\log|a_0 m_+(z)|\circ\theta^k\omega$$

In any case, the Birkhoff ergodic theorem implies that:

$$\begin{aligned}\lim_{n\to+\infty}\frac{1}{n}\log|f_{z,\pm\infty}(\pm n,\omega)| &= \mathbb{E}\{\log|a_0 m_+(z)|\}\\ &= \mathbb{E}\{\log|a_0 m_-(z)|\} \qquad \text{(VII.30)}\end{aligned}$$

whenever the probability measure $\mathbb{P}$ is ergodic for the shift $\theta$. In this case one can also define the integrated density of states $dN(\lambda)$ as the *nonrandom* probability measure that appears as the almost sure limit as $n \to \infty$ of the *random probability measures* $dN_n(\lambda)$, where $N_n(\lambda)$ is the number of eigenvalues per unit volume of the restriction of $H^+$ to $\{0,\cdots,n\}$ which are not greater than $\lambda$. The existence of the integrated density of states in the ergodic case was discussed in the previous chapter.

But for the same reasons as in the continuous case, we do not want to assume the ergodicity of the sequence $\{(a_n,b_n);\, n \in \mathbb{Z}\}$. In this more general setting one can still define the integrated density of state $N(\lambda)$ as the limit of the expectations of the random nondecreasing funtions $N_n(\lambda)$ described above.

One of the main attractions of the integrated density of states is the so-called Thouless formula, which relates to the Lyapunov exponent $\gamma(\lambda)$ defined by:

$$\gamma(\lambda) = \lim_{n\to\infty} \frac{1}{n} \log \|U_\lambda(n,\omega)\|$$

in the ergodic case. The Lyapunov exponent $\gamma(\lambda)$ is defined by:

$$\gamma(\lambda) = \lim_{n\to\infty} \frac{1}{n} \mathbb{E}\{\log \|U_\lambda(n,\omega)\|\}$$

when the dynamical system is not assumed to be ergodic. The limit in the right-hand side exists because the sequence is subadditive. One easily sees that this limit is nonnegative. Moreover, the proof of the Thouless formula which we gave (in the previous chapter) in the ergodic case can be extended to the present situation by taking expectations at the appropriate step of the proof. This formula reads:

$$\gamma(\lambda) = \int \log |\xi - \lambda| dN(\xi) - \mathbb{E}\{\log a_0\}.$$

Using the bounds (III.64) and (III.65), one immediately sees that the quantity:

$$w(z) = \mathbb{E}\{\log m_+(z)\} \tag{VII.31}$$

makes sense for all $z \in \Pi_+$ because assumptions (i) and (ii) imply that

$$\mathbb{E}\{\log(1 + |b_1| + |a_1|)\} < +\infty.$$

Obviously, the function $w$ we just defined is in the Herglotz class. Moreover, we have the following result:

**Proposition VII.1.13** *Under conditions (VII.26) and (VII.27), there exists a probability measure $dN(\lambda)$ on $\mathbb{R}$ such that:*

$$\begin{aligned} \int \lambda dN(\lambda) &= \mathbb{E}\{b_0\} \\ \int \lambda^2 dN(\lambda) &\leq 2\mathbb{E}\{2a_0^2 + b_0^2\} \end{aligned}$$

*and:*

$$w(z) = \int \log \frac{1}{\lambda - z} dN(\lambda), \qquad z \in \Pi_+.$$

*For $z \in \Pi_+$ we also have:*

$$\Re e\ w(z) \leq \mathbb{E}\{\log \frac{1}{a_0}\} < +\infty \tag{VII.32}$$

*Proof:*
The existence of the probability measure $dN$ has been argued in our discussion of the integrated density of states. The Thouless formula and the fact that:

$$-\gamma(\lambda) - \mathbb{E}\{\log a_0\} = \Re e\ w(\lambda) \tag{VII.33}$$

(recall formula (VII.30)) give the integrability of the function $\log(1 + |\lambda|)$ with respect to $dN(\lambda)$ and:

$$-\Re e\ w(z) = \int \log|\lambda - z| dN(\lambda).$$

Hence we have:

$$w(z) = \alpha + \int \log \frac{1}{\lambda - z} dN(\lambda) \tag{VII.34}$$

for some constant $\alpha$. Note that (VII.32) follows from (VII.33) and the fact that the Lyapunov exponent is nonnegative. Let us first assume that the distribution of $(a_0, b_0)$ has a bounded support. Then, Lemma III.2.1 implies that:

$$\begin{aligned} w(z) &= \log \frac{1}{-z} + \frac{1}{z}\mathbb{E}\{s_1\} + \frac{1}{z^2}\mathbb{E}\{s_2 - \frac{1}{2}s_1^2\} + \frac{1}{z^3}\mathbb{E}\{s_3 - s_1 s_2 + \frac{1}{3}s_1^3\} \\ &\quad + \frac{1}{z^4}\mathbb{E}\{s_4 - \frac{1}{2}s_2^2 - s_1 s_3 + s_1^2 s_2 - \frac{1}{4}s_1^4\} + \cdots \\ &= \log \frac{1}{-z} + \frac{1}{z}\mathbb{E}\{b_0\} + \frac{1}{z^2}\mathbb{E}\{a_0^2 + \frac{1}{2}b_0^2\} \\ &\quad + \frac{1}{z^3}\mathbb{E}\{a_0^2 b_0 + \frac{1}{3}b_0^3 + a_0^2 b_1\} + \cdots . \end{aligned}$$

The probability measure $dN$ has compact support, say $K$, because the operators $H(\omega)$ are uniformly bounded. Consequently, using the representation (VII.34), one gets:

$$\begin{aligned} w(z) &= \alpha + \int_K \log \frac{1}{-z}(1 - \frac{\lambda}{z})^{-1} dN(\lambda) \\ &= \alpha + \log \frac{1}{-z} + \sum_{k=1}^{\infty} \frac{1}{k z^k} \int \lambda^k dN(\lambda). \end{aligned}$$

Identifying with the above expansion one obtains:

$$\begin{aligned}
\alpha &= 0 \\
\int \lambda dN(\lambda) &= \mathbb{E}\{b_0\} \\
\int \lambda^2 dN(\lambda) &= \mathbb{E}\{2a_0^2 + b_0^2\} \\
\int \lambda^3 dN(\lambda) &= \mathbb{E}\{3a_0^2 + b_0^3 + 3a_0^2 b_1\} \\
\cdots\cdots & \quad \cdots\cdots
\end{aligned}$$

In the general case where we do not assume that $\{(a_n, b_n); n \in \mathbb{Z}\}$ is uniformly bounded but simply that $\mathbb{E}\{|\log a_0|\} < +\infty$ and $\mathbb{E}\{a_0^2 + b_0^2\} < +\infty$, we use a truncation procedure to reduce the problem to the previous case. More precisely, for each integer $k \geq 1$ we set

$$a_n^{(k)} = a_n \wedge k$$

and

$$b_n^{(k)} = (b_n \wedge k) \vee (-k).$$

Then, according to (VI.37), we have:

$$\lim_{k \to \infty} w^{(k)}(z) = w(z)$$

but since for each $k \geq 1$ we have

$$\int \lambda^2 dN^{(k)}(\lambda) = \mathbb{E}\{2a_0^{(k)2} + b_0^{(k)2}\} \leq \mathbb{E}\{2a_0^2 + b_0^2\}$$

because of the first half of the proof, we conclude easily that

$$w(z) = \int \log \frac{1}{\lambda - z} dN(\lambda)$$

and that

$$\int \lambda^2 dN(\lambda) \leq \mathbb{E}\{2a_0^2 + b_0^2\}. \ \blacksquare$$

**Remark VII.1.14** 1. If the supports of the probability distributions $\mu_a$ and $\mu_b$ are bounded, the above proof shows that:

$$w(z) = \log \frac{1}{-z} + \frac{\mathbb{E}\{b_0\}}{z} + O(|z|^{-2})$$

which implies that, if one assumes furthermore that $\mathbb{E}\{b_0\} = 0$:

$$- 2\cosh w(z) = z + O(|z|^{-1}) \tag{VII.35}$$

uniformly as $|z| \to \infty$ in the upper half plane.

2. The above proposition implies that:

$$w'(z) = \int \frac{dN(\lambda)}{\lambda - z} = \mathbb{E}\{G(z,0,0)\} \tag{VII.36}$$

because $dN(\lambda) = \mathbb{E}\{< E(d\lambda)\delta_0, \delta_0 >\}$, and consequently that $w' \in \mathcal{H}erg$ since $G(z,0,0) =< [H - zI]^{-1}\delta_0, \delta_0 >$ is a function of the Herglotz class. If one recalls the expression of the Green function given in formula (III.68) of Chapter III, equation (VII.36) above implies that:

$$\mathbb{E}\{\Im m \frac{1}{a_0^2 m_+(\lambda + i\epsilon) + a_0 a_{-1} m_-(\lambda + i\epsilon) + \lambda + i\epsilon - b_o}\} = -\frac{d\Re e\, w(\lambda + i\epsilon)}{d\epsilon}. \tag{VII.37}$$

We close this section with some technical results which we will need later.

**Lemma VII.1.15** *If the supports of the probability distributions $\mu_a$ and $\mu_b$ are bounded and if we also assume that $\mathbb{E}\{b_0\} = 0$, then one has:*

$$- 2\sin \Im m\, w(z) \sinh \Re e\, w(z) \geq \Im m\, z. \tag{VII.38}$$

*Proof:*
Let us set $u(z) = -2\cosh w(z)$ to simplify the notation. Let us fix $\epsilon > 0$ and $r > 0$. Notice that the above remark implies that:

$$\Im m\, u(z) \geq \Im m\, z - \epsilon \tag{VII.39}$$

whenever $|z| = r$ and $r$ is large enough. Moreover, if $\Im m\, z = \epsilon$, the right hand side of (VII.39) is zero while $\Im m\, u(\lambda) = 2\sin \pi N(\lambda) \sinh \gamma(\lambda) \geq 0$ for $N(\lambda) \in [0,1]$ and $\gamma(\lambda) \geq 0$. Since both sides of (VII.39) are harmonic functions and since the inequality (VII.39) holds on the boundary of the domain $\{z; |z| < r, \Im m\, z > \epsilon\}$ we can conclude that it holds inside the domain as well. Letting first $r \to \infty$ and then $\epsilon \to 0$ we get that $\Im m\, u(z) \geq \Im m\, z$ in the upper half plane, and this is an equivalent form of (VII.1.15).
∎

**Lemma VII.1.16** *With the usual notation and for $z = \lambda + i\epsilon$ with $\lambda \in \mathbb{R}$ and $\epsilon > 0$, one has:*

$$1. \quad \mathbb{E}\{\log\left(1+\frac{\Im m\, z}{a_0^2 \Im m\, m_+(z,\omega)}\right)\} = \mathbb{E}\{\log\left(1+\frac{\Im m\, z}{a_0 a_{-1} \Im m\, m_-(z,\omega)}\right)\}$$

$$= -2\Re e\, w(z). \qquad \text{(VII.40)}$$

$$2. \quad \mathbb{E}\{\frac{1}{a_0^2 \Im m\, m_+(\lambda+i\epsilon)+\epsilon/2}\} \le \frac{2\gamma(\lambda+i\epsilon)}{\epsilon} \qquad \text{(VII.41)}$$

$$2'. \quad \mathbb{E}\{\frac{1}{a_0 a_{-1} \Im m\, m_-(\lambda+i\epsilon)+\epsilon/2}\} \le \frac{2\gamma(\lambda+i\epsilon)}{\epsilon} \qquad \text{(VII.42)}$$

$$3. \quad \mathbb{E}\{[\frac{1}{a_0^2 \Im m\, m_+(\lambda+i\epsilon)+\epsilon/2} + \frac{1}{a_0 a_{-1} \Im m\, m_-(\lambda+i\epsilon)+\epsilon/2}]$$

$$[a_0^2 \Im m\, m_+(\lambda+i\epsilon) - a_0 a_{-1} \Im m\, m_-(\lambda+i\epsilon)]^2$$

$$+ \frac{[\lambda - b_0 + a_0^2 \Re e\, m_+(\lambda+i\epsilon) + a_0 a_{-1} \Re e\, m_-(\lambda)^2}{|\lambda + i\epsilon - b_0 + a_0^2 m_+(\lambda+i\epsilon) + a_0 a_{-1} m_-(\lambda+i\epsilon)|^2}\}$$

$$\le 4[\frac{\gamma(\lambda+i\epsilon)}{\epsilon} - \frac{\partial\gamma(\lambda+i\epsilon)}{\partial\epsilon}. \qquad \text{(VII.43)}$$

*Proof:*
In order to prove the part of 1 involving $m_+$, we take imaginary parts on both sides of (VII.28) for $n = 1$. Then, we divide both sides by $a_1^2 \Im m\, m_+(z)$ and we take logarithms. This gives:

$$\log\left(1+\frac{\Im m\, z}{a_1^2 \Im m\, m_+(z,\theta\omega)}\right) = \log\frac{1}{-\Im m\, m_+(z,\omega)} - \log(a_1^2 \Im m\, m_+(z,\theta\omega)).$$

Then we take expectations and, using the invariance under the shift one sees easily that the expectation of the right hand side is equal to $-2\mathbb{E}\{\log a_0\} - \mathbb{E}\{\log |m_+(z)|^2\}$ which is the desired result by definition of the function $w$. The part of 1 involving the function $m_-$ is proven similarly.

The estimates in part 2 and $2'$ follow immediately from part 1 and the fact that:

$$\frac{x}{1+x/2} \le \log(1+x)$$

for $x \ge 0$. Finally, we introduce some special notation to simplify the typesetting of the proof of part 3. We set $A_+ = a_0^2 \Im m\, m_+(\lambda+i\epsilon)$, $A_- = a_0 a_{-1} \Im m\, m_-(\lambda+i\epsilon)$ and $A = \lambda + i\epsilon + a_0^2 \Im m\, m_+(\lambda+i\epsilon) + a_0 a_{-1} \Im m\, m_-(\lambda+$

$i\epsilon$). Then we have:

$$\begin{aligned}\left(\frac{1}{A_+}+\frac{1}{A_-}\right)&\frac{(A_+-A_-)^2+(\Re e\,A)^2}{|A|^2}\\&=\left(\frac{1}{A_+}+\frac{1}{A_-}\right)\left(1-4\frac{A_+A_-}{|A|^2}\right)\\&=\frac{1}{A_+}+\frac{1}{A_-}+\Im m\,\frac{1}{A}\end{aligned}$$

where we used several times the fact that $A_+ + A_- = \Im m\, A$. We conclude the proof by taking expectations on both sides and using parts 2 and 2′ to bound the first expectation of the right hand side and formula (VII.37) to bound the second expectation. ∎

# VII.2 Periodic and Almost Periodic Potentials

Periodic potentials are at one hand of the spectrum of the random potentials: they are in some sense the less random of them. As we already explained, the structure of their randomness is trivial in the sense that a Schrödinger operator $H = H_0 + V$ with a periodic potential with period $\tau$ can be viewed as a random operator $H(\omega)$ by choosing $\Omega = [0, \tau)$, the normalized Lebesgue measure as $\mathbb{P}$ and the choice of an element $\omega$ in $\Omega$ consisting merely in the choice of a shift parameter in the interval of length the period chosen for $\Omega$.

## VII.2.1 Floquet Theory

Let us assume that the potential function $V(t)$ is locally integrable and periodic. In our short discussion of the general case in Section II.2 of Chapter II, we explained that the spectrum $\Sigma(H)$ is of the form $\bigcup_{n\geq 1}[a_n, b_n]$ with $b_n \leq a_{n+1}$ and is purely absolutely continuous in the sense that $\Sigma(H) = \Sigma_{ac}(H)$ and $E = E_{ac}$. The intervals $[a_n, b_n]$ are called the *bands* of the spectrum and $(b_n, a_{n+1})$ the *gaps*. The latter are said to be open if $b_n < a_{n+1}$.

We present a complete treatment of the spectral properties of periodic one-dimensional Schrödinger operators. In this setting, the results can be derived in the framework of the so-called Floquet theory of periodic O.D.E. which we describe now.

We assume in the following discussion that the function $t \hookrightarrow V(t)$ is locally integrable and periodic of period $\tau > 0$ and we use freely the notation and the concepts introduced in Chapter III.

It is enlightening to consider the more general vector valued differential equation:

$$\dot{Y}(t) = A(t)Y(t), \tag{VII.44}$$

with $t \hookrightarrow A(t)$ a locally integrable periodic function with period $\tau$ taking values in the space of $n \times n$ matrices. If $t \hookrightarrow Y(t)$ is a solution of this equation, then $t \hookrightarrow Y(t+\tau)$ is also a solution. From this one easily concludes that the fundamental matrix of the system must satisfy:

$$U(t+\tau) = U(t)U(\tau).$$

The matrix $U(\tau)$ is often called the *monodrony matrix* of the periodic equation. Since the matrix $U(\tau)$ is invertible, one can take its logarithm and we can define the matrix:

$$\Gamma = \frac{1}{\tau} \log U(\tau).$$

The matrix valued function $Q$ defined by $Q(t) = U(t)e^{-t\Gamma}$ is periodic. We just proved the so-called Floquet representation:

**Proposition VII.2.1** *If the matrix function $t \hookrightarrow A(t)$ is locally integrable and periodic with period $\tau$, then the fundamental matrix of the differential equation (III.34) has the representation:*

$$U(t) = Q(t)e^{t\Gamma}, \tag{VII.45}$$

*where $\Gamma$ is a constant matrix and where $t \hookrightarrow Q(t)$ is a continuous $n \times n$ matrix valued function with period $\tau$.*

The characteristic roots of the matrix $\Gamma$ are called the *characteristic exponents* of the equation (III.34). Notice that, if $a \in \mathbb{R}^n$ is an eigenvector of the matrix $\Gamma$ corresponding to an eigenvalue $\gamma$ then $Y(t) = Q(t)e^{t\Gamma}a$ is a solution of the form $Y(t) = e^{\gamma t}\tilde{Y}(t)$ where $\tilde{Y}(t) = Q(t)a$ is a periodic (vector valued) function of the variable $t$. Also, exactly as in the case of the equation with constant matrix coefficient $A$, stability is equivalent to uniform stability and asymptotic stability is equivalent to uniform asymptotic stability. Moreover, the necessary and sufficient conditions for stability which we gave in the case of constant coefficients apply to the periodic case provided one replaces the words "characteristic roots of the matrix $A$" by "characteristic exponents of the equation".

Let us use this general result of the theory of periodic O.D.E in the particular case of the eigenvalue equation $-\ddot{y}(y)+V(t)y(t)=\lambda y(t)$ for a periodic potential function $V$. In the present situation the monodrony matrix is:

$$U_\lambda(\tau)=\begin{bmatrix} y_1(\tau,\lambda) & y_2(\tau,\lambda) \\ \dot{y}_1(\tau,\lambda) & \dot{y}_2(\tau,\lambda) \end{bmatrix} \tag{VII.46}$$

and its characteristic roots are the solutions of the equation:

$$\rho^2-d(\lambda)\rho+1=0 \tag{VII.47}$$

where:

$$d(\lambda)=y_1(\tau,\lambda)+\dot{y}_2(\tau,\lambda)\,. \tag{VII.48}$$

Recall that $U(\tau)$ has determinant 1. The number $d(\lambda)$ is called the *discriminant* for the energy $\lambda$. It is a continuous function of $\lambda$. The product of the roots of the characteristic equation (VII.47) is equal to 1 from which it follows that the characteristic exponents (which are proportional to the logarithms of these roots) differ only by a sign factor. Consequently, the characteristic exponents are:

| | | |
|---|---|---|
| complex with nonzero imaginary parts | if | $d(\lambda)\le -2$ |
| pure imaginary | if | $\lvert d(\lambda)\rvert<2$ |
| real | if | $d(\lambda)\ge 2$ |

Consequently, if $|d(\lambda)|\neq 2$, the monodrony matrix has two different characteristic roots whose product is 1 and $\Gamma(\lambda)$ can be chosen so that its characteristic roots, which can be chosen as $\gamma(\lambda)$ and $-\gamma(\lambda)$ for some strictly positive number $\gamma(\lambda)$, do not differ by an integer multiple of $2i\pi/\tau$, and we can find two linearly independent solutions of the eigenvalue equation which are of the form:

$$\tilde{y}_1(t,\lambda)=e^{\gamma(\lambda)t}p_1(t,\lambda)$$

and:

$$\tilde{y}_2(t,\lambda)=e^{-\gamma(\lambda)t}p_2(t,\lambda)$$

where the functions $p_1(\,\cdot\,)$ and $p_2(\,\cdot\,)$ are periodic functions with period $\tau$. In the case of equal characteristic roots, the general result of Proposition

VII.45 gives the existence of two linearly independent solutions of (III.6) of the form:

$$\tilde{y}_1(t,\lambda) = p_1(t,\lambda)$$

and:

$$\tilde{y}_2(t,\lambda) = tp_1(t,\lambda) + p_2(t,\lambda)$$

where, as before, the functions $p_1(\,\cdot\,)$ and $p_2(\,\cdot\,)$ are periodic functions with period $\tau$.

The set $\mathcal{S} = \{\lambda \in \mathbb{R};\ |d(\lambda)| < 2\}$ is an open subset of the real line. It is the union of a family of disjoint open intervals which are called *stability intervals* of the eigenvalue equation (III.6). We now justify this terminology. For each $\lambda \in \mathcal{S}$, the monodrony matrix can be written in the form:

$$U_\lambda(\tau) = P_\lambda \begin{bmatrix} e^{i\gamma(\lambda)} & 0 \\ 0 & e^{-i\gamma(\lambda)} \end{bmatrix} P_\lambda^{-1} \qquad \text{(VII.49)}$$

for some invertible matrix $P_\lambda$ with possibly complex coefficients. Since the coefficients of the monodrony matrix are continuous functions of $\lambda$ and since its characteristic roots are pure imaginary valued continuous functions of $\lambda$, the entries of the matrix $P_\lambda$ can be chosen as continuous functions of $\lambda$, and the same ends up being true for the determinant of $P_\lambda$. Since the latter never vanishes on $\mathcal{S}$, for every compact subset of $\mathbb{R}$ contained in $\mathcal{S}$, say $\Lambda$, we must have:

$$\sup_{n\geq 0,\ \lambda\in\Lambda} \|U_\lambda(\tau)^n\| < \infty$$

from which one easily gets:

$$\sup_{t\geq 0,\ \lambda\in\Lambda} \|U_\lambda(t)\| < \infty$$

which leads to:

**Proposition VII.2.2** *Each compact set $\Lambda$ contained in the stability set $\mathcal{S}$ is a uniform stability set in the sense that:*

$$\sup_{s,t\in\mathbb{R},\ \lambda\in\Lambda} \|U_\lambda(t,s)\| < \infty. \qquad \text{(VII.50)}$$

We now show how the instability intervals are the set of energies $\lambda$ for which the eigenvalue equation (III.6) has an *exponential dichotomy* of its solutions. This is, as we are about to see, a strong form of the deterministic version of the Oseledec theorem.

Let us fix $\lambda$ in the *instability set* $\mathcal{I} = \{\lambda \in \mathbb{R};\ |d(\lambda)| > 2\}$. Since $d(\lambda)^2 - 4 > 0$, the characteristic equation for $U_\lambda(\tau)$ has two distinct real roots, the product of which is 1. Let us call them $\rho(\lambda)$ and $\rho(\lambda)^{-1}$. Note that $\rho(\lambda) > 0$ if $d(\lambda) > 2$ and that $\rho(\lambda) < 0$ if $d(\lambda) < 2$. In any case, if $P_\lambda$ is any matrix whose columns are eigenvectors of $U_\lambda(\tau)$ for the eigenvalues $\rho(\lambda)$ and $\rho(\lambda)^{-1}$ respectively, then we have:

$$U_\lambda(\tau) = P_\lambda \begin{bmatrix} \rho(\lambda) & 0 \\ 0 & \rho(\lambda)^{-1} \end{bmatrix} P_\lambda^{-1} \tag{VII.51}$$

so that:

$$\Gamma(\lambda) = P_\lambda \begin{bmatrix} \gamma(\lambda) & 0 \\ 0 & -\gamma(\lambda) \end{bmatrix} P_\lambda^{-1} \tag{VII.52}$$

where $\gamma(\lambda) > 0$ if $d(\lambda) > 2$ and $\Re e\gamma(\lambda) > 0$ and $\Im m\gamma(\lambda) \neq 0$ if $d(\lambda) < -2$. The Floquet representation gives:

$$U_\lambda(\tau) = Q_\lambda(t) P_\lambda \begin{bmatrix} e^{\gamma(\lambda)t} & 0 \\ 0 & e^{-\gamma(\lambda)t} \end{bmatrix} P_\lambda^{-1} \tag{VII.53}$$

for some $2 \times 2$ $\tau$-periodic matrix valued continuous function $Q_\lambda(t)$. Let us now denote by $\tilde{\Pi}_1$ and $\tilde{\Pi}_2$ the canonical projections of $\mathbb{R}^2$ (i.e. $\tilde{\Pi}_1(x_1, x_2) = x_1$ and $\tilde{\Pi}_2(x_1, x_2) = x_2$), and let us define the projections $\Pi_1(\lambda)$ and $\Pi_2(\lambda)$ by:

$$\Pi_1(\lambda) = P_\lambda \tilde{\Pi}_1 P_\lambda^{-1} \qquad \Pi_1(\lambda) = P_\lambda \tilde{\Pi}_1 P_\lambda^{-1}.$$

Then it is easy to get, using the present form of the Floquet representation, the following:

**Proposition VII.2.3** *For each $\lambda$ in the instability set $\mathcal{I} = \{\lambda \in \mathbb{R};\ |d(\lambda)| > 2\}$ the eigenvalue problem exhibits an exponential dichotomy in the sense that there exists a projection $\Pi_\lambda$ on a one dimensional subspace of $\mathbb{R}^2$ and positive constants $k(\lambda)$ and $\gamma(\lambda)$ such that:*

$$\|U_\lambda(t)\Pi_\lambda U_\lambda(s)^{-1}\| \leq k(\lambda)e^{-\gamma(\lambda)(t-s)}$$

*for $0 \leq s \leq t$ and:*

$$\|U_\lambda(t)(I - \Pi_\lambda)U_\lambda(s)^{-1}\| \leq k(\lambda)e^{\gamma(\lambda)(t-s)},$$

*for $0 \leq t \leq s$.*

We have enough information to completely characterize the spectral properties of one-dimensional periodic Schrödinger operators. Indeed, according to Lemma III.1.7, the stability set has to be included in the spectrum. On the other hand, the instability set cannot intersect the spectrum for no solution of the eigenvalue problem for a value of $\lambda$ in $\mathcal{I}$ can be polynomially bounded because of the exponential dichotomy. Consequently, the spectrum is the closure of the stability set $\mathcal{S}$. Moreover, the spectrum is purely absolutely continuous in $\mathcal{S}$ because of the uniform stability result, Theorem III.4.1 and the fact that the end points cannot be eigenvalues since we proved that no solution could be square integrable when the characteristic equation has a double root.

We saw that the spectrum of a Schrödinger operator with periodic potential is the union of closed intervals possibly separated by open intervals called gaps. The fact that these gaps are actually open and the determination of their location is a problem of importance. It is possible to show that a periodic potential function is necessarily constant if no gap is open, that it is an elliptic function if exactly one gap is open and that it is real analytic if only finitely many gaps are open. There are various mathematical results to the extend that the set of periodic potentials with all of their gaps open is quite large. Explicit examples are nevertheless difficult to come by. Here is a famous one.

**Example: The Mathieu equation**

It is possible to prove that all the gaps are open in the case of the potential function $V(t) = \beta\cos t$ for any nonzero real constant $\beta$. This is a consequence of the fact that there is no value of the coupling constant $\beta$ and of the energy $\lambda$ for which the eigenvalue equation (III.1) has a periodic or an anti-periodic solution (see below for the definition) of period $2\pi$. See p.298-299 of volume IV of the Reed and Simon series for a proof.

It is possible to study the manifold of potentials having a given discriminant. For example, if $d(\lambda)$ is the discriminant of a $C^\infty$ periodic potential $V$ for which the Schrödinger operator $H = H_0 + V$ has exactly $n$ open gaps (n could be infinite) then the set of $C^\infty$ periodic potentials having the same discriminant (and hence the same spectrum) can be endowed with a natural manifold structure which makes it homeomorphic to an $n$-dimensional torus.

Any discussion of spectral properties of periodic potentials can be complete without mention of the famous isospectral flow generated by the Korteweg-de Vries equation. If $V$ is a periodic function, then, the solution $V(t,x)$ of the nonlinear partial differential equation:

$$\frac{\partial V}{\partial t} = 3V\frac{\partial V}{\partial x} - \frac{1}{2}\frac{\partial^3 V}{\partial x^3}$$

with initial condition $V(0,x) \equiv V(x)$ is such that, for each fixed $t$, the function $x \hookrightarrow V(t,x)$ is periodic with the same period and the operator $H_t = H_0 + V(t,\cdot)$ has the same spectrum as the operator $H = H_0 + V$.

We conclude this subsection with a discussion of a couple of eigenvalue problems in a period interval.

**The Periodic and the Anti-periodic Problems:**
Let us consider a piecewise continuous function $V$ with period $\tau$. The eigenvalue problems which we are considering now involve the determination of the spectra of the operators $H_p$ and $H_{ap}$ which we now define. We define $\mathcal{D}(H_a)$ (resp. $\mathcal{D}(H_{ap})$) as the subset of the Hilbert space $L^2([0,\tau],dt)$ of the continuously differentiable functions $f$ with an absolutely continuous derivative such that the function $t \hookrightarrow -\ddot{f} + Vf$ is still in $L^2([0,\tau],dt)$ and such that $f(0) = f(\tau)$ and $\dot{f}(0) = \dot{f}(\tau)$ (resp. $f(0) = -f(\tau)$ and $\dot{f}(0) = -\dot{f}(\tau)$). We then define the operator $H_p$ (resp. $H_{ap}$) on its domain by the usual formula $-\ddot{f} + Vf$. The corresponding eigenvalue problem is called the periodic (resp. anti-periodic) eigenvalue problem. The boundary conditions which we are currently considering are different from those involved in the definition of the operators $H^{\Lambda}_{\alpha,\beta}$ in the sense that each condition involves both endpoints of the interval. But it is still possible to write down explicitely the Green's function, check that it is jointly continuous and that it gives the kernel of the resolvent operator. Consequently, both operators have a discrete spectrum. We denote by:

$$\lambda_0 \le \lambda_1 \le \lambda_2 \le \cdots$$

(resp.

$$\mu_0 \le \mu_1 \le \mu_2 \le \cdots)$$

the eigenvalues of the operators $H_p$ (resp. $H_{ap}$) with the usual convention of repeating twice a degenerate eigenvalue. This last point is very important for the situation is different from the case of separate boundary conditions for which we showed that the eigenvalues were simple. In the present situation of mixed boundary conditions, the eigenvalues can be (and will be) degenerated. Consider for example the case of the periodic boundary value problem for the free Laplacian (i.e. $V(t) \equiv 0$) with $\tau = 2\pi$. We have $\lambda_j = j$

for $j = 0, 1, \cdots$. The eigenvalue $\lambda_0 = 0$ is simple with eigenspace generated by the constant functions but the other eigenvalues are degenerated for the functions $t \hookrightarrow \cos jt$ and $t \hookrightarrow \sin jt$ form a basis of the eigenspace corresponding to the eigenvalue $\lambda_j = j$.

**Theorem VII.2.4** *The eigenvalues $\lambda_j$ and $\mu_j$ of the periodic and anti-periodic problem which we defined above occur in the order:*

$$\lambda_0 < \mu_0 \leq \mu_1 < \lambda_1 \leq \lambda_2 < \mu_2 \leq \mu_3 < \lambda_3 \leq \lambda_4 < \cdots$$

*Moreover,*

- *i) $d(\lambda)$ decreases from 2 to $-2$ in the intervals $[\lambda_{2j}, \mu_{2j}]$,*
- *ii) $d(\lambda)$ increases from $-2$ to 2 in the intervals $[\mu_{2j+1}, \lambda_{2j+1}]$,*
- *iii) $d(\lambda) > 2$ in the intervals $(-\infty, \lambda_0)$ and $(\lambda_{2j+1}, \lambda_{2j+2})$,*
- *iv) $d(\lambda) < -2$ in the intervals $[\mu_{2j}, \mu_{2j+1}]$,*

The above theorem shows that the stability intervals are the intervals $(\lambda_{2j}, \mu_{2j})$ and $(\mu_{2j+1}, \lambda_{2j+1})$ and that the spectrum $\Sigma(H)$ of the operator on the whole real line is the union of the closures of these intervals.

$$\Sigma(H) = [\lambda_0, \mu_0] \cup [\mu_1, \lambda_1] \cup [\lambda_2, \mu_2] \cup [\mu_3, \lambda_3] \cup \cdots$$

None of them is ever absent. On the other hand, the instability intervals or gaps are the interval $(-\infty, \lambda_0)$ and the intervals $(\mu_{2j}, \mu_{2j+1})$ and $(\lambda_{2j+1}, \lambda_{2j+2})$ and the latter could be absent if the characteristic equation (VII.47) happens to have a double root. The interval $(-\infty, \lambda_0)$ is usually called the zero-th gap while the intervals $(\mu_{2j}, \mu_{2j+1})$ and $(\lambda_{2j+1}, \lambda_{2j+2})$ are called the $(2j+1)$-th and the $(2j+2)$-th gaps, whether they are open or not. Notice that a gap is not open when there exists a value of the energy $\lambda$ for which all the solutions of the eigenvalue equation (III.1) have either period $\tau$ or anti-period $\tau$.

A proof of this theorem can be found for example in the fourth volume of the Reed and Simon series [272] or in Eastham's book. [95]

It is also very instructive to consider the eigenvalue problem for the operator on a period interval with Dirichlet boundary conditions at the endpoints. Let us denote by $\delta_1 < \cdots < \delta_n < \cdots$ the corresponding spectrum. Note that for each $j \geq 1$ the eigenvalue $\delta_j$ is in the $j$-th gap of the spectrum (at the location the $j$-th gap should have open in case the gap is actually closed). The following trace formula is a crucial step in the reconstruction procedure of the potential function from the spectral data:

$$V(0) = \lambda_0 + \sum_{j=0}^{\infty}[\lambda_j + \mu_{j-1} - 2\delta_j]. \tag{VII.54}$$

Its proof can be done in two steps. One first proves that if $V$ is measurable, bounded and continuous at the origin, then:

$$\lim_{t\searrow 0} \frac{d}{dt}\int_{\mathbb{R}} [e^{-tH}(x,x) - e^{-tH^{(0)}}(x,x)]dx = -\frac{V(0)}{2}$$

where $H^{(0)}$ is the operator obtained by adding a Dirichlet boundary condition at the origin. The proof can be given using path integrals. Next, one uses some of techniques developed in Chapter III to express the integral of the difference of the two integral kernels in term of the Green function and one finishes the proof by a contour integration around each spectral band. The potential $V(t)$ can be recovered by letting the shift act on the elements of the right hand side so that:

$$V(t) = \lambda_0 + \sum_{j=0}^{\infty}[\lambda_j + \mu_{j-1} - 2\delta_j(t)], \tag{VII.55}$$

provided one can determine the generator of the dynamical system so defined. The system of equations satisfied by the $\delta_j(t)$'s is known under the name of Dubrovin, and solving such a system actually gives the reconstruction of the potential.

Recall the way in which a periodic potential can be regarded as a random potential. We simply set $\Omega = [0,\tau) = \mathbb{R}/\tau\mathbb{Z}$, we set $\mathcal{F}$ for its Borel $\sigma$-field and we consider its (normalized) Haar measure $\mathbb{P}(d\omega) = (1/\tau)d\omega$ as probability measure. In other words, computing the expectation of the function of the (random) potential amounts to averaging over all the possible shifts of the potential. The periodicity implies that the process so-obtained is stationary and ergodic. Consequently we can define the $w$ function, the Lyapunov exponent and the integrated density of states. In particular, we have:

$$w(\lambda) = \frac{1}{\tau}\int_0^{\tau} \frac{dt}{G(\lambda,t,t)}.$$

Floquet's theory tells us that the Lyapunov exponent is zero on the spectrum (i.e. the stability intervals) and strictly positive in the gaps because of the exponential dichotomy of the solutions of the eigenvalue equation.

In fact a more careful analysis shows that, in each gap, this Lyapunov exponent grows from 0 to a local maximum and then decreases from this maximum back to 0. Since the topological support of the integrated density of states measure is equal to the spectrum, the function $\lambda \hookrightarrow N(\lambda)$ is constant outside the spectrum and increases on the spectrum. The following fact, known as *gap labelling theorem,* is remarkable: if $\lambda$ is in a gap, then the $L^2$-solution at $+\infty$ is real and exponentially decaying and this implies that $\Im m\, w(\lambda + i0)$ is an integer multiple of $\pi/\tau$ because of formulas (VII.16) and (VII.17). Consequently, in each gap, the integrated density of states is equal to a constant of the form $j/\tau$ where $j$ is an integer (and $\tau$ the period).

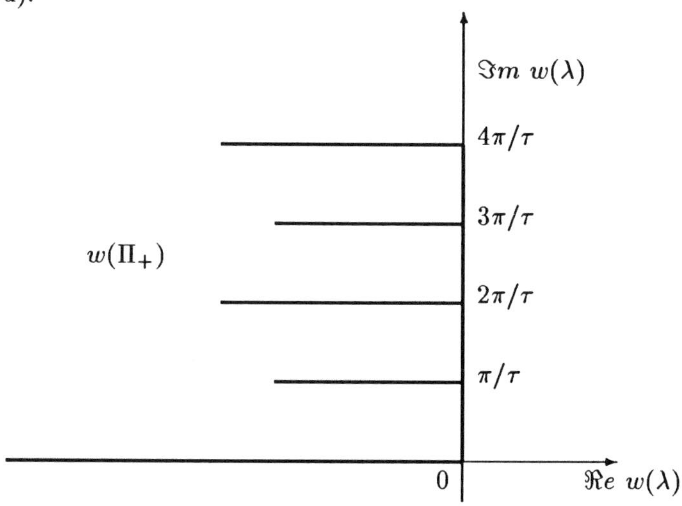


*Range $w(\Pi_+)$ of the function $w$. The Continuous Case.*

The above discussion gives a complete description of the range of the $w$-function of periodic potentials. Indeed, when $\lambda$ varies from $-\infty$ to $+\infty$, we know how the quantities $\gamma(\lambda)$ and $N(\lambda)$ vary, and consequently we know how $w(\lambda + i0)$ varies. Recall Proposition VII.1.11. This gives a parametric description of the boundary of the range $w(\Pi_+)$ of the function $w$. More precisely, when $\lambda$ varies from $-\infty$ to $\lambda_0$, $N(\lambda)$ remains equal to 0 while the Lyapunov exponent decreases from $+\infty$ to 0. Consequently $w(\lambda + i0)$ runs through the real axis from $-\infty$ to 0. Then, when $\lambda$ varies in the first band from $\lambda_0$ to $\mu_0$, the Lyapunov exponent $\gamma(\lambda)$ remains equal to 0 while the integrated density of states $N(\lambda)$ increases from 0 to $1/\tau$, and consequently $w(\lambda + i0)$ runs through the imaginary axis from 0 to $i\pi/\tau$. Then, when $\lambda$

varies in the first gap from $\mu_0$ to $\mu_1$, the Lyapunov exponent $\gamma(\lambda)$ increases from 0 to a local maximum $\gamma_1$ and then decreases to 0 while the integrated density of states $N(\lambda)$ remains constant, and consequently $w(\lambda + i0)$ runs back and forth through the horizontal segment $[-\gamma_1 + i\pi/\tau, i\pi/\tau]$. Then, when $\lambda$ varies in the second band from $\mu_1$ to $\lambda_1$, the Lyapunov exponent $\gamma(\lambda)$ remains equal to 0 while the integrated density of states $N(\lambda)$ increases from $1/\tau$ to $2/\tau$, and consequently $w(\lambda + i0)$ runs through the imaginary axis from $i\pi/\tau$ to $2i\pi/\tau$. We can continue tracing the boundary of the domain $w(\Pi_+)$ in this way. We summarize the results in the above figure.

## VII.2.2 Inverse Spectral Theory

We already quoted several results characterizing the potential function from spectral information such as the fact that there is no open gap, or the existence of only one band or .... The present section is devoted to the discussion of specific results in this spirit. Namely, given spectral data in a form to be specified below, can one recover a periodic function $t \hookrightarrow V(t)$ in such a way that the spectral characteristics of the corresponding periodic Schrödinger operator are the ones we started from. More generally, one asks the question of the existence of such a periodic potential, and if there is no uniqueness, what is the structure of the manifold of solutions. There are many forms of answers. We quote the most appropriate to the use we will make of this theory in the case of general stationary stochastic processes.

To each given real number $\tau > 0$ and to each sequence $\{\gamma_k;\ k \geq 1\}$ of nonnegative numbers one can associate the domain $D$ in the complex plane defined by:

$$D = \{z = x + iy \in \mathbb{C};\ (x < 0 \text{ and } y > 0) \text{ and } (x < -\gamma_k \\ \text{whenever } y = k\pi/\tau \text{ for some integer } k \geq 1)\}.$$

We associate to such a domain $D$ the unique conformal map $w$ of the upper half plane $\Pi_+$ onto $D$ which has the asymptotic $w(y) \approx y$ as $y \to -\infty$.

**Proposition VII.2.5** *Let $D$ be a domain in the complex plane of the form above and let $w$ be the associate conformal map. Then there exists a locally square integrable periodic function $V$ of period $\tau$ for which the $w$-function of the periodic Schrödinger operator $H = H_0 + V$ is the function $w$ we started from if and only:*

$$\sum_{k\geq 1} k^2\gamma_k^2 < \infty. \tag{VII.56}$$

It is easy to see that the condition (VII.56) is equivalent to:

$$\int^{\infty} \sqrt{\xi}\gamma(\xi)\, d\xi < \infty.$$

More generally one can show the equivalences:

$$\sum_{k\geq 1} (k^{n+1}\gamma_k)^2 < \infty \quad \Leftrightarrow \quad \int^{\infty} \xi^{n+1/2}\gamma(\xi)\, d\xi < \infty.$$

The above result is due to Marchenko and Ostrovskii. These authors also prove that:

$$V \in \mathcal{H}^n((0,\tau)) \quad \Leftrightarrow \quad \int^{\infty} \xi^{n+1/2}\gamma(\xi)\, d\xi < \infty.$$

which links the smoothness of the potential to the decay at infinity of the Lyapunov exponent. Here $\mathcal{H}^n((0,\tau))$ denotes the usual Sobolev space of the functions which are square integrable over the period interval $(0,\tau)$ and for which the first $n$ derivatives in the sense of distributions are also square integrable over the period interval.

## VII.2.3 The Lattice Case

Floquet's theory extends without major change to the lattice case. We skip the details and refer the interested reader to the note of Kac and van Moerbeke [163], to [108] or to Problem VII.6.7. We simply present a short discussion of the $w$-function and of its range $w(\Pi_+)$ when the sequence $\{(a_n, b_n); n \in \mathbb{Z}\}$ is periodic. Let us denote by $\tau$ the period. The discriminant:

$$d(\lambda) = \text{ trace } U_\lambda(\tau)$$

is a polynomial of degree $\tau$. Let us denote by $\lambda_0 < \cdots < \lambda_{\tau-1}$ the roots of the equation $d(\lambda) = 2$ and by $\mu_0 < \cdots < \mu_{\tau-1}$ the roots of the equation $d(\lambda) = -2$. We have:

$$\lambda_0 < \mu_0 \leq \mu_1 < \lambda_1 \leq \lambda_2 < \cdots$$

and the spectrum $\Sigma(H)$ of $H$ is the finite union of closed intervals:

$$\Sigma(H) = [\lambda_0, \mu_0] \cup [\mu_1, \lambda_1] \cup \cdots$$

As before expectations are plain averages over a period and the definition of the $w$-function can be rewritten as:

$$w(\lambda) = \frac{1}{\tau} \sum_{n=1}^{\tau} \log m_+(\lambda).$$

The quantity $\pi N(\lambda) = \Im m\, w(\lambda + i0)$ vanishes for $\lambda \in (-\infty, \lambda_0]$, increases from 0 to $\pi/N$ on the first stability interval $[\lambda_0, \mu_0]$, remains constant and equal to $\pi/N$ on the first gap $(\mu_0, \mu_1)$, increases from $\pi/N$ to $2\pi/N$ on the second stability interval $[\mu_1, \lambda_1]$, remains constant and equal to $2\pi/N$ on the second gap $(\lambda_1, \lambda_2)$, increases, .... On the other hand, the Lyapunov exponent $\gamma(\lambda) = \mathbb{E}\{\log(1/a_0)\} - \Re e\, w(\lambda + i0)$ decreases from $+\infty$ to 0 when $\lambda$ increases from $-\infty$ to $\lambda_0$, is identically 0 on the stability intervals of the spectrum, increases from 0 to a strictly positive maximum and decreases from this local maximum to zero on each open gap, and finally increases from 0 to $+\infty$ when $\lambda$ increases from $\sup \Sigma(H)$ to $+\infty$. In particular, we always have:

$$\Re e\, w(\lambda + i0) \leq \mathbb{E}\{\log(1/a_0)\} = -\frac{1}{\tau} \log a_1 \cdots a_\tau$$

with equality when $\lambda$ belongs to the spectrum of $H$. As in the continuous case, the above discussion gives a parametrization of the boundary of the range $w(\Pi_+)$ of the function $w$. The following figure gives a graphical representation of this range.

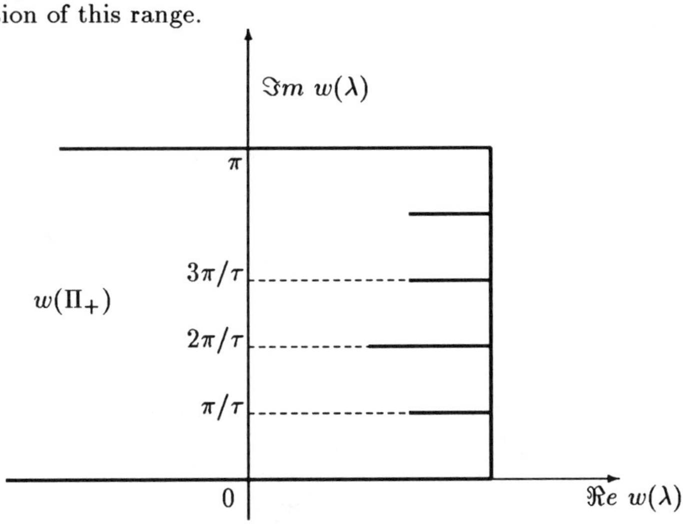


*Range $w(\Pi_+)$ of the function $w$. The Discrete Case.*

Now that we know the exact form of the range of the $w$-function of a periodic Jacobi matrix, we can address the problem of the reconstruction of the periodic coefficients from a function having a range of this type.

More precisely, let us consider a domain $D$ defined by:

$$D = \{x + iy;\, x < c,\, 0 < y < \pi\} \setminus \bigcup_{j=1}^{\tau-1} \{x + ij\pi/\tau; c_j \leq x \leq c\}$$

where $\tau > 0$ is an integer, $c \in \mathbb{R}$ is a constant and the constants $c_j$ are real numbers smaller than $c$ for $j = 1, \ldots, \tau - 1$. We have the following result:

**Lemma VII.2.6** *If the domain $D$ is as above, then there exists a sequence $\{(a_n, b_n);\, n \in \mathbb{Z}\}$ with period $\tau$, such that $a_n > 0$ for all $n \in \mathbb{Z}$, $c = -(1/\tau)\log a_1 \cdots a_\tau$, and for which the corresponding $w$-function satisfies:*

$$w(\Pi_+) = D.$$

*Proof:*
The Riemann mapping theorem gives the existence of a conformal map, say $w$, from $\Pi_+$ onto $D$ such that:

$$w(\lambda) \approx \log \frac{1}{-\lambda}$$

as $\lambda$ goes to infinity in $\Pi_+$. When $\lambda$ varies through $\mathbb{R}$ from $-\infty$ to $+\infty$, then $w(\lambda)$ varies through the boundary of $D$, and setting:

$$d(\lambda) = 2\, \cosh[\tau(w(\lambda) - c)]$$

one gets the following. $\Im m\ w(\lambda) = 0$ (and consequently $d(\lambda) \geq 2$) for $\lambda \in (-\infty, \lambda_0)$ for some $\lambda_0 \in \mathbb{R}$. Next, $\Re e\ w(\lambda) = c$ and $\Im m\ w(\lambda)$ increases from $0$ to $\pi/\tau$, so that $d(\lambda) = 2\cos[\tau\ \Im m\ w(\lambda)]$ decreases from $2$ to $-2$ when $\lambda$ increases from $\lambda_0$ to some $\mu_0 \in \mathbb{R}$. Then, while $\Im m\ w(\lambda)$ remains constant and equal to $\pi/\tau$, $\Re e\ w(\lambda)$ decreases from $c$ to $c_1$ and then increases from $c_1$ to $c$ while $d(\lambda)$ goes from $-2$ to something possibly larger and back to $-2$. This occurs while $\lambda$ increases from $\mu_0$ to some $\mu_1$. Next, $\Re e\ w(\lambda)$ remains constant and equal to $c$ while $\Im m\ w(\lambda)$ increases from $\pi/\tau$ to $2\pi/\tau$ while $d(\lambda)$ increases from $-2$ to $2$. This occurs while $\lambda$ increases from $\mu_1$ to some $\lambda_1 \in \mathbb{R}$. If we keep this process alive we obtain $\lambda_0, \cdots, \lambda_{\tau-1}$ roots of $d(\lambda) = -2$ and $\mu_0, \cdots, \mu_{\tau-1}$ roots of $d(\lambda) = 2$, with the result shown in the figure below. This figure corresponds to the case of an even period $\tau$. When $\tau$ is odd, the last band is $[\lambda_{\tau-1}, \mu_{\tau-1}]$ and $d(\lambda) \leq -2$ for $\lambda \geq \mu_{\tau-1}$. Note that $d(\lambda)$ is real for $\lambda$ real, so that $\overline{d(\overline{\lambda})} = d(\lambda)$ defines an analytic continuation to the lower half-plane. Now $d(\lambda)$ is an entire function. Actually $d(\lambda)$ is a polynomial of degree $\tau$ because $d(\lambda) \approx \lambda^\tau$ for $\lambda \to \infty$, when $\lambda \in \Pi_+$.

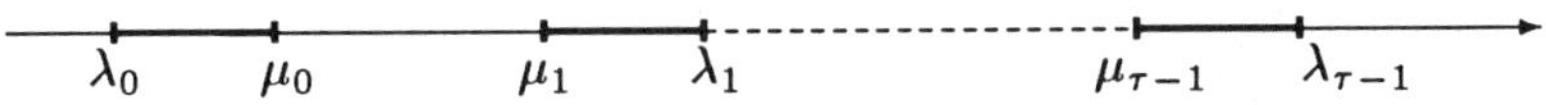


*Reconstruction Procedure. The Discrete Periodic Case.*

Now, we can choose numbers $\tau_1, \cdots, \tau_{\tau-1}$ such that $\tau_j$ belongs to the *jth* gap and impose the constraint $a_1^{(N)} \cdots a_N^{(N)} = e^{-Nc}$. Under these conditions we can use the Kac-van Moerbeke reconstruction procedure given in [164] to finish the proof. See Problem VII.6.8 for details. ∎

### VII.2.4 Almost Periodic Potentials

If $f$ is a function on $\mathbb{R}^d$ (resp. $\mathbb{Z}^d$)) and if $x \in \mathbb{R}^d$ (resp. $\mathbb{Z}^d$)) we use the notation $\mathcal{T}_x f$ for the translated function:

$$[\mathcal{T}_x f](y) = f(y - x) \qquad x \in \mathbb{R}^d, \text{ (resp. } x \in \mathbb{Z}^d).$$

**Definition VII.2.7** *An element $f$ of the real Banach space $C_b(\mathbb{R}^d)$ of bounded continuous functions on $\mathbb{R}^d$ is said to be almost periodic if the set $\{\mathcal{T}_x f;\, x \in \mathbb{R}^d\}$ of all the shifts of $f$, is relatively compact in $C_b(\mathbb{R}^d)$. One defines similarly an almost periodic function on $\mathbb{Z}^d$.*

Let us denote by $\Gamma$ the closure of the shifts of an almost periodic function $f$. There is a unique way to endow $\Gamma$ with the structure of a compact metric group so that $x \hookrightarrow \mathcal{T}_x f$ becomes a continuous homomorphism from $\mathbb{R}^d$ (resp. $\mathbb{Z}^d$) into $\Gamma$. By duality, the dual group $\hat{\Gamma}$ can be viewed as a (countable) subgroup of $\hat{\mathbb{R}}^d = \mathbb{R}^d$. This subgroup is called the *frequency module* of the function $f$. It follows that every almost periodic function $f$ is the uniform limit of linear combinations of complex exponentials $e^{2i\pi\alpha x}$ with $\alpha \in \hat{\Gamma}$. If one sets $\Omega = \Gamma$, if one calls $\mathcal{F}$ the Borel $\sigma$-field of the hull $\Gamma$, if one sets for $\mathbb{P}$ the Harr measure and if one denotes by $\theta_x$ the extension to $\Gamma$ of the shift $\mathcal{T}_x$, then any almost periodic function can be regarded as an ergodic stationary random function. In other words, except for considering the function $f$ itself, one considers all the possible shifts of this function, and one take samples at random according to the Haar measure on the group of all these translates. Unfortunately this random function is not mixing and almost periodic functions cannot have (in general) the behavior one expects from genuinely random functions.

We give two important examples (besides the periodic ones) of almost periodic functions. For the sake of simplicity we restrict ourselves to the case $d = 1$. A function $f$ is said to be *quasi periodic* if the frequency module $\hat{\Gamma}$ is finitely generated. An alternative definition is that $f$ can be written in the form:

$$f(x) = g(\alpha_1 x, \cdots, \alpha_n x)$$

for some continuous function $g$ on $\mathbb{R}^n$ which is periodic with period 1 in each variable. If the $\alpha_i$'s are rationally independent and if $\{\mathbf{m} \in \mathbb{Z}^n;\ \hat{g}(\mathbf{m}) \neq 0\}$ generates $\mathbb{Z}^n$, then $\hat{\Gamma}$ is generated by $\alpha_1,\cdots,\alpha_n$, and $g$ can be viewed as defined on the $n$-torus. Another important class of almost periodic functions is the class of *limit periodic* functions which are uniform limits of periodic functions. A typical example is given by:

$$f(x) = \sum_{n=0}^{\infty} \xi_n \cos(\pi x 2^{-n}), \qquad \text{with} \quad \sum_{n=0}^{\infty} |\xi_n| < \infty.$$

In this case, the frequency module $\hat{\Gamma}$ is the set of dyadic rationals.

In any case, any almost periodic potential $V(t)$ can be viewed as a stationary ergodic random potential and the theory developed so far can be used to obtain results which will be true, *for almost every choice of* $\theta$. This differs slightly from the direct analytical approach to the study of almost periodic potentials in which the results are for *every* $\theta$. Indeed, almost periodic functions are well known for *having mean values,* or put in a probabilistic terminology, there exists for them a version of the ergodic theorem where there is no null set to throw away to guarantee the existence of the limit.

In the same way there are periodic potentials with finitely many open gaps, there are almost periodic potentials with finitely may gaps in their spectra. See for example [91]. But a generic periodic potential has all of its gaps open. Moreover, at least for the class of limit periodic potentials, one can expect that, approximating a generic limit periodic potential by periodic ones, one expects that the periods of these periodic potentials become larger and larger and more and more gaps in the spectrum should open. In this way the spectrum of the original potential should end up being a Cantor set (i.e. a nowhere dense perfect closed set). This intuition was proven to be true in a series of more or less independent works of Chulaevsky [53], Moser [245], Avron and Simon [13], Bellisseard and Testard [19] and Delyon and Souillard [77]. Note that the notion of a *generic property* simply means that there exists a dense $G_\delta$ set of potentials having the property. This is relative to a given class of potentials and this is different (and sometimes weaker) than the notion of *except for a set of measure zero* or *almost surely* for random potentials.

One of the deepest (and most difficult to read) results in the investigation of the spectrum of almost periodic Schrödinger operators is certainly the work of Dinaburg and Sinai [83]. They consider quasi-periodic potentials of the form $V(t) = \beta g(\alpha_1 t, \cdots, \alpha_n t)$ for an analytic function $g$ and for frequencies $\alpha_1, \cdots, \alpha_n$ satisfying:

$$|\sum_{k=1}^{n} m_k \alpha_k| \geq c \exp\left(-\frac{|\mathbf{m}|}{\log^{1+\epsilon}|\mathbf{m}|}\right) \tag{VII.57}$$

for $|\mathbf{m}|$ large enough where $c$ is a positive constant. They prove that for large enough energies $\lambda$'s which are nonresonant in the sense that:

$$|N(\lambda) - \frac{1}{2}\sum_{k=1}^{n} m_k \alpha_k| \geq c \exp\left(-\frac{|\mathbf{m}|}{\log^{1+\epsilon}|\mathbf{m}|}\right), \qquad \mathbf{m} \in \mathbb{Z}^n \tag{VII.58}$$

the eigenvalue equation has two linearly independent solutions of the form $e^{iN(\lambda)t}\chi(t)$ and $e^{-iN(\lambda)t}\overline{\chi(t)}$ for some analytic quasi periodic function $\chi$ with the same frequency module. In particular, conditions (VII.57) and (VII.58) define a Cantor set contained in the upper part of the absolutely continuous spectrum. This result was sharpened by Rüssmann in [283] and extended by Moser and Poschel in [247] where they consider some of the resonant energies $\lambda$'s for which $2N(\lambda) = \sum_{k=1}^{n} m_k \alpha_k$ (recall the gap labelling theorem) as long as it stays away from other resonances. All these proofs are relying in one form or another on the heavy machinery of the Kolmogorov-Arnold-Moser (KAM for short) procedure. These proofs have been recently streamlined by the work of H. Elliasson who proved that the spectrum is actually prely absolutely continuous above a certain energy $\lambda_0(\beta)$, this energy being equal to the infimum of the spectrum when $\beta$ is small.

### Some Inverse Results

The complexity of the spectral structure of Schrödinger operators with almost periodic potentials is such that it is not reasonable to hope for a simple inverse spectral theory. There are nevertheless many results in this direction, some of which will be discussed in the section devoted to the inverse spectral theory of random operators. In the mean time, it is interesting to describe some of the results which will not be covered by this random theory. The work of Pastur and Tkachenko [265] is one of the interesting ones. It is concerned with the *closure* in a super-exponential sense of the space of periodic potentials. Let us define the class of limit periodic potentials they are dealing with. We first fix a sequence $\mathbf{a} = \{a_n;\, n \geq 1\}$ of strictly positive numbers such that, for each integer $n \geq 1$, the ratio $a_{n+1}/a_n$ is an integer which is not less than 2. For such a sequence $\mathbf{a}$ we define the space $\mathcal{V}(\mathbf{a})$ of all the real valued almost periodic functions $V$ for which there exist for each $n \geq 1$ an $a_n$-periodic function $V_n$ such that for any $c > 0$ we have:

$$\lim_{n\to\infty} ||V - V_n||_{\mathcal{B}}\, e^{ca_{n+1}} = 0$$

where the norm $||\cdot||_{\mathcal{B}}$ is defined by:

$$||V||^2_{\mathcal{B}} = \lim_{T\to\infty} \frac{1}{2T} \int_{-T}^{+T} |V(t)|^2\, dt.$$

Let us also define the set $\mathcal{R}(\mathbf{a}) = \cup_{n\geq 1}\mathcal{R}_n$ where $\mathcal{R}_n$ is for each $n \geq 1$ the set defined by:

$$\mathcal{R}_n = \{\frac{k\pi}{a_n};\, k \in \mathbb{Z}, k > 0\}.$$

and $\Gamma(\mathbf{a})$ as the set of positive functions $\gamma$ on $\mathcal{R}(\mathbf{a})$ which satisfy:

$$\lim_{n\to\infty} \left( \sum_{r\in\mathcal{R}(\mathbf{a})\setminus\mathcal{R}_n} r^2\gamma_r^2 \right) e^{ca_{n+1}} = 0$$

for any $c > 0$. In some sense, the set $\Gamma(\mathbf{a})$ is the closure in a super exponential sense of the sequences appearing in the characterization of the ranges of $w$-functions of periodic potentials with period $a_n$. To each $\gamma = \{\gamma_r;\, r \in \mathcal{R}(\mathbf{a})\} \in \Gamma(\mathbf{a})$ we can associate a function $w \in \mathcal{W}$ in the following way. We consider the domain

$$D_\gamma = \{z \in \mathbb{C};\, \Im m\, z > 0,\, \Re e\, z < 0\} \setminus \cup_{r\in\mathcal{R}(\mathbf{a})} [-\gamma_r + ir, ir]$$

Since $D_\gamma$ is simply connected, the Riemann mapping theorem gives the existence of a unique conformal mapping from $\Pi_+$ onto $D_\gamma$ satisfying the usual saymptotic behavior and $w(0) = 0$. On the other hand, given a potential $V \in \mathcal{V}(\mathbf{a})$ one can consider a sequence of $a_n$-periodic potentials as in the definition of the class $\mathcal{V}(\mathbf{a})$ and check that the union of all the sequences $\{\gamma_k^{(n)};\ k \geq 1\}$ given by the Marchenko-Ostrovskii theory form an element $\gamma$ of $\Gamma(\mathbf{a})$. Pastur and Tkachenko proved:

**Proposition VII.2.8** *There exists a unique correspondance between $\mathcal{V}(\mathbf{a})$ and $\Gamma(\mathbf{a})$ as an extension of the correspondence of the periodic case. This correspondence is given by the above procedure associating $\gamma$ to $V$ and the corresponding $w$-function is the function $w$ constructed above.*

It should be pointed out that the rate of uniform approximation by periodic potentials does not need to be super-exponential in the period for the spectrum to be purely absolutely continuous. This was proved by Chulaevsky and Molcanov in [243]. They also proved that the spectrum could be a Cantor set of Lebesgue measure 0 of pure point type with eigenfunctions decaying faster than any polynomial, provided the rate of growth of the periods was much faster than the rate of uniform approximation.

### The Almost Mathieu Equation

One of the most studied examples is without any doubt the generalization of Mathieu's equation obtained by choosing $V(t) = \beta\cos(2\pi\alpha t + \theta)$ for some positive coupling constant $\beta$, some irrational number $\alpha$ and some phase $\theta \in [0, 2\pi)$. As we already explained, if we set $\Omega = [0, 2\pi)$, if we choose $\mathcal{F}$ to be the Borel $\sigma$-field, $\mathbb{P}$ the uniform distribution and the shift operators $\theta_t$ to be addition modulo $2\pi$, then for each fixed $\beta$ and $\alpha$, $V(t, \theta) = V(t)$ can be viewed as a stationary ergodic random potential.

It was proven by Bellissard and Simon in [21] that the spectrum is a Cantor set for all the pairs $(\beta, \lambda)$ in a dense $G_\delta$ subset of $\mathbb{R}^2$. As a consequence of Herman's trick, we proved in Chapter V that the Lyapunov exponent was strictly positive as long as the coupling constant $\beta$ is greater than 2. This implies that, in this case, there is no absolutely continuous spectrum. Moreover, Avron and Simon proved in [14] that, at least for the discrete analogue $H = H_0 + V$ on $\mathcal{H} = \ell^2(\mathbb{Z})$ with $V(n) = \beta\cos(2\pi\alpha n + \theta)$, the spectrum is purely singular continuous for almost every $\theta$ if $\alpha$ is a Liouville number, i.e. if:

$$|\alpha - \frac{p_k}{q_k}| \leq \frac{1}{k^{q_k}}$$

for a sequence $\{q_k;\ k \geq 1\}$ of integers increasing to $\infty$. They simply use an argument of Gordon [129]. Roughly speaking, Gordon's result states that, if a potential function $V$ can be approximated with a super exponential rate of convergence by periodic functions with larger and larger periods then the eigenvalue equation for the corresponding Schrödinger operator $H = H_0 + V$ cannot have square integrable solutions at $\infty$.

Delyon proved in [69] that the almost Mathieu operator does not have any eigenvalue when the coupling constant $\beta$ is strictly smaller than 2 and that, for $\beta = 2$ the eigenvalues (if any) have to belong to the set where the Lyapunov exponent vanishes and that the associated eigenfunctions are not summable. See Problem V.5.4. It follows from the works of Dinaburg and Sinai , Rüssmann and Johnson and Moser and Pöschel quoted above that if $\beta$ is sufficiently small and $\theta$ is *typical* then there exists an absolutely continuous component in the spectrum. Using a duality argument, Bellissard Lima and Testard proved in [19] that there were eigenvalues for large coupling $\beta$. In a paper discussed in the Notes and Complements of the next chapter, Sinai proved that it was possible to overcome the usual limitations of the KAM perturbation technique to handle the problem of the resonances. He proved in [306] that the spectrum is pure point for large coupling $\beta$ and almost every $\alpha$. This result was recently used by Chulaevsky and Delyon to prove (using again a duality argument) in [54] that the spectrum is almost surely purely absolutely continuous for small $\beta$ and almost every $\alpha$.

# VII.3 The Absolutely Continuous Spectrum

The goal of this section is to identify the (almost sure) absolutely continuous spectrum as the set where the Lyapunov exponent vanishes. We try to deal simultaneously with the continuous and the discrete cases.

## VII.3.1 The Essential Support of the Absolutely Continuous Spectrum

The proposition below is the first half of a complete characterization of the absolutely continuous spectrum of one dimensional random Schrödinger operators which we give below in CorollaryVII.3.4. It is known as the Pastur-Ishii result. The simplicity of the proof made it one of the first striking results of the theory. Also, its statement can be formulated in an abstract form in terms of orthogonality of measures on the line and this extra twist shed some interesting light on the result. See the remarks following the proof.

**Proposition VII.3.1** *Let $m$ be a positive measure on the real line such that the Lyapunov exponent $\gamma(\lambda)$ is strictly positive $m$-almost surely. Then for $\mathbb{P}$-almost all $\omega$ the spectral measure $\sigma_\omega$ is orthogonal to the measure $m$.*

*Proof:*
Let $\Omega_1$ be the subset of $\omega \in \Omega$ such that one has:

$$\frac{1}{|n|}\log(1+|V(n,\omega)|) \longrightarrow 0$$

in the discrete case and:

$$\frac{1}{|n|}\int_n^{n+1} |V(t,\omega)|dt \longrightarrow 0$$

in the continuous case. The above assumptions imply that $\Omega_1$ is of full $\mathbb{P}$-measure. Let us define the subset $W \subset \Omega \times \mathbb{R}$ by:

$$W = \{(\omega,\lambda)\,;\,\omega \in \Omega_1,\ \lambda \in \text{hyp}(H(\omega))\}.$$

The set $\text{hyp}(H(\omega))$ of hyperbolic values of $\lambda$ was defined in Chapter III. The function $U_\lambda(n,\omega)$ is continuous with respect to the variable $\lambda$ and measurable with respect to $\omega$. This implies that the set $W$ is measurable for the product $\sigma$-algebra $\mathcal{F} \otimes \mathcal{B}$ where $\mathcal{B}$ is the Borel sigma algebra of $\mathbb{R}$. It follows from the Oseledec theorem and the positivity of $\gamma_\ell(\lambda)$ that for $m$-almost all $\lambda$ one has $\mathbb{P}(W_\lambda) = 1$ where $W_\lambda$ is the section of $W$ for a given value of $\lambda$. The Fubini theorem implies that for $\mathbb{P}$-almost all $\omega$ the measure $m$ is carried by $\text{hyp}(H(\omega))$ and the result follows from Theorem III.4.12 if we first assume that the measure $m$ is continuous. The continuity of the measure $m$ was needed in order to ensure that the countable set of eigenvalues of $H(\omega)$ was $m$-negligible. In the random case we know from Proposition IV.2.8 that a given $\lambda$ is $\mathbb{P}$-almost surely not an eigenvalue of $H(\omega)$ and consequently we can assume without any loss of generality that the measure $m$ is continuous. ∎

**Remark VII.3.2** 1. The above result has the following striking consequence: if $\gamma(\lambda) > 0$ for every $\lambda \in \mathbb{R}$ and if one picks two independent $\omega$ in $\Omega$ according to the distribution $\mathbb{P}$, say $\omega_1$ and $\omega_2$, then the spectral resolutions of the identities $E(\cdot,\omega_1)$ and $E(\cdot,\omega_2)$ are mutually singular in the sense that they are carried by disjoint Borel subsets of $\mathbb{R}$.

2. In the case of a strip of width $\ell$, if the components $V_1, \ldots, V_\ell$ of the random potential satisfy the above integrability properties, then the above

proof applies in extenso provided the assumption is replaced by the strict positivity of the lowest nonnegative Lyapunov exponent $\gamma_\ell$.

We now give the second half of the characterization of the essential support of the absolutely continuous part of the resolution of the identity which we announced earlier. This *converse* is due to S. Kotani.

**Proposition VII.3.3** *Let $\mathcal{I}$ be a Borel subset of the real line such that $\gamma(\lambda) = 0$ for almost every $\lambda \in \mathcal{I}$ with respect to the Lebesgue measure. Then for $\omega$ in a set of probability one and for almost every $\lambda \in \mathcal{I}$ we have:*

$$\frac{d\sigma_{ac}(\cdot,\omega)}{d\lambda}(\lambda) > 0 \qquad \text{(VII.59)}$$

*and:*

$$m_+(\lambda + i0, \omega) = -\overline{m_-(\lambda + i0, \omega)} \qquad \text{(VII.60)}$$

*in the continuous case while in the lattice case we have:*

$$a_0 \Im m\, m_+(\lambda + i0, \omega) = a_{-1} \Im m\, m_-(\lambda + i0, \omega) \qquad \text{(VII.61)}$$

*and:*

$$\lambda - b_0 + a_0^2 \Re e\, m_+(\lambda + i0, \omega) + a_0 a_{-1} \Re e\, m_-(\lambda + i0, \omega) = 0. \qquad \text{(VII.62)}$$

*Proof:*
The function $-d\gamma(\lambda + i\epsilon)/d\epsilon$ is the real part of an Herglotz function (namely $w'(\lambda + i\epsilon)$) and consequently, for almost every $\lambda \in \mathcal{I}$, it has finite limits as $\epsilon \searrow 0$. This implies that:

$$\lim_{\epsilon \searrow 0} \frac{\gamma(\lambda + i\epsilon)}{\epsilon}$$

exists and is finite for almost every $\lambda \in \mathcal{I}$. Using the classical Fatou's lemma of integration theory in the limit $\epsilon \searrow 0$, Lemma VII.1.10 in the continuous case and parts 2. and 2$'$. of Lemma VII.1.16 in the lattice case we get:

$$\begin{aligned} \mathbb{E}\{\frac{1}{a_\pm \Im m\, m_\pm(\lambda + i0)}\} &\leq \lim_{\epsilon \searrow 0} \mathbb{E}\{\frac{1}{a_\pm \Im m\, m_\pm(\lambda + i\epsilon)}\} \\ &= 2 \lim_{\epsilon \searrow 0} \frac{\gamma(\lambda + i\epsilon)}{\epsilon} \\ &< \infty, \end{aligned}$$

where the quantity $a_\pm$ are 1 in the continuous case and $a_+ = a_0^2$ and $a_- = a_0 a_{-1}$ in the lattice case. In any case, this estimate implies that:

$$\Im m\, m_\pm(\lambda + i0) > 0$$

for almost every $\lambda \in \mathcal{I}$ with probability 1. Recall that in the continuous case we have:

$$\Im m\, G(\lambda + i\epsilon, 0, 0) = \frac{\Im m\, m_+(\lambda + i\epsilon) + \Im m\, m_-(\lambda + i\epsilon)}{|\, m_+(\lambda + i\epsilon) + m_-(\lambda + i\epsilon)|^2}$$

and by Fatou's theorem which we recalled in Chapter I, we get:

$$\frac{d\sigma_{ac}(\lambda)}{d\lambda} = \Im m\, G(\lambda + i0, 0, 0) > 0$$

which proves the first claim of the proposition in the continuous case. In the lattice case, we use formula (III.68) from Chapter III to get:

$$\Im m\, G(\lambda + i\epsilon, 0, 0) = \frac{\epsilon + a_0^2 \Im m\, m_+(\lambda + i\epsilon) + a_0 a_{-1} \Im m\, m_-(\lambda + i\epsilon)}{|\, \lambda + i\epsilon - b_0 + a_0^2 m_+(\lambda + i\epsilon) + a_0 a_{-1} m_-(\lambda + i\epsilon)\, |^2}$$

and the same conclusion follows.

We now prove the second claim of the proposition. We already noticed that:

$$\lim_{\epsilon \searrow 0} \frac{\gamma(\lambda + i\epsilon)}{\epsilon} = \lim_{\epsilon \searrow 0} \frac{d\gamma(\lambda + i\epsilon)}{d\epsilon}$$

and that these limits exist and are finite for almost every $\lambda \in \mathcal{I}$. This implies that the right hand side of the estimate in part 3. of Lemma VII.1.16 vanishes in the limit $\epsilon \searrow 0$ and Fatou's lemma implies that, for almost every $\lambda \in \mathcal{I}$, one has:

$$\begin{aligned} \mathbb{E}\{[&\frac{1}{a_0^2 \Im m\, m_+(\lambda + i\epsilon) + \epsilon/2} + \frac{1}{a_0 a_{-1} \Im m\, m_-(\lambda + i\epsilon) + \epsilon/2}] \\ &[a_0^2 \Im m\, m_+(\lambda + i\epsilon) - a_0 a_{-1} \Im m\, m_-(\lambda + i\epsilon)]^2 \\ &\quad + \frac{[\Re e\, (a_0^2 m_+(\lambda + i\epsilon) + a_0 a_{-1} m_-(\lambda) + \lambda - b_0)]^2}{|\lambda + i\epsilon - b_0 + a_0^2 m_+(\lambda + i\epsilon) + a_0 a_{-1} m_-(\lambda + i\epsilon)|^2}\} \\ = 0.& \end{aligned}$$

This is possible only if:

$$a_0(\omega) \Im m\, m_+(\lambda + i0) = a_{-1}(\omega) \Im m\, m_-(\lambda + i0)$$

and

$$\lambda - b_0(\omega) + a_0(\omega)^2 \Re e\, m_+(\lambda + i0, \omega) + a_0(\omega) a_{-1}(\omega) \Re e\, m_-(\lambda + i0, \omega) = 0$$

almost surely in $\omega \in \Omega$. In order to prove the desired result in the continuous case, we recall the identity:

$$\begin{aligned} & -\frac{\Re e\, w(\lambda + i\epsilon)}{\epsilon} - \Im m\, w'(\lambda + i\epsilon) \\ & = \mathbb{E}\{[\frac{1}{\Im m\, m_+(\lambda + i\epsilon)} + \frac{1}{\Im m\, m_-(\lambda + i\epsilon)}] \\ & \qquad ([\Re e\, m_+(\lambda + i\epsilon) + \Re e\, m_-(\lambda + i\epsilon)]^2 \\ & \qquad + \frac{[\Im m\, m_+(\lambda + i\epsilon) - \Im m\, m_-(\lambda + i\epsilon)]^2}{|m_+(\lambda + i\epsilon) - m_-(\lambda + i\epsilon)|^2}\} \end{aligned}$$

which we proved in the proof of Lemma VII.1.10. We can take the limit $\epsilon \searrow 0$ as before and use Fatou's lemma once more. Consequently, for almost every $\lambda \in \mathcal{I}$ we have:

$$\begin{aligned} 0 \;=\; & \mathbb{E}\{[\frac{1}{\Im m\, m_+(\lambda + i0)} + \frac{1}{\Im m\, m_-(\lambda + i0)}] \\ & \qquad ([\Re e\, m_+(\lambda + i0) + \Re e\, m_-(\lambda + i0)]^2 \\ & \qquad + \frac{[\Im m\, m_+(\lambda + i0) - \Im m\, m_-(\lambda + i0)]^2}{|m_+(\lambda + i0) - m_-(\lambda + i0)|^2}\} \end{aligned}$$

and this implies that:

$$\Re e\, m_+(\lambda + i0, \omega) = -\Re e\, m_-(\lambda + i0, \omega) \tag{VII.63}$$
$$\Im m\, m_+(\lambda + i0, \omega) = \Im m\, m_-(\lambda + i0, \omega). \tag{VII.64}$$

$\mathbb{P}$-almost surely in $\omega \in \Omega$. ∎

Putting together the Ishii-Pastur result and Kotani's converse we obtain:

**Corollary VII.3.4** *The essential support of the absolutely continuous part of the resolution of the identity of $H(\omega)$ is almost surely equal to the set $A = \{\lambda \in \mathbb{R}; \gamma(\lambda) = 0\}$ where the Lyapunov exponent vanishes.*

We can precise the above result when the Borel set $\mathcal{I}$ is an open interval.

**Corollary VII.3.5** *If $\gamma(\lambda) = 0$ for almost every $\lambda$ in an open interval $\mathcal{I}$ then $H(\omega)$ has purely absolutely continuous spectrum in $\mathcal{I}$.*

*Proof:*
According to formula (VII.59) the real part of $m_+(\lambda+i0)+m_-(\lambda+i0)$ vanishes almost everywhere on the interval $\mathcal{I}$. Since the function $m_+(\cdot)+m_-(\cdot)$ is analytic and has a positive imaginary part in the upper half plane $\Pi_+$, it can be analytically continued through the interval $\mathcal{I}$ by the Schwarz reflection principle, and the extension does not vanish on $\mathcal{I}$. This shows that $G(z,0,0,\omega)$ is analytic on $\mathcal{I}$. This concludes the proof because $E(\cdot,0,0,\omega)$ is recovered from the boundary values of the imaginary part of the Green's function $G(z,0,0,\omega)$. ∎

The last application of Proposition VII.3.3 shows that the random potential giving absolute continuous spectrum cannot be very random. For the sake of clarity we state a definition to emphasize this lack of randomness.

**Definition VII.3.6** *A stochastic process $\{V(t);t\in T\}$ is said to be deterministic if the $\sigma$-field $\mathcal{F}_{-\infty}$ intersection of all the $\sigma$-fields generated by $\{V(s);s\leq t\}$ when $t\in T$ varies is equal, up to sets of measure 0, to the $\sigma$-field $\mathcal{F}_{\infty}$ generated by $\{V(t);t\in T\}$.*

Since a function on $\Omega$ is measurable with respect to the $\sigma$-field $\{V(s);s\leq t\}$ if and only if it can be written as a measurable function of the $V(s)$ for $s\leq t$, a process is deterministic if and only if the knowledge of an arbitrary end $\{V(s);s\leq t\}$ of a trajectory determines entirely the whole trajectory $\{V(s);s\in T\}$. Let us give a few important examples.

**Examples:**

1. The simplest example of nondeterministic stochastic process is of course given by an i.i.d. sequence $\{V(n);\ n\in\mathbb{Z}\}$ for which the common distribution of the $V(n)$'s is not concentrated on a single point. Indeed, the tail $\sigma$-field $\mathcal{F}_{-\infty}$ is trivial, i.e. $\mathcal{F}_{-\infty}=\{\emptyset,\Omega\}$ and the only possibility for the process to be deterministic is to be constant. This is impossible if the random variables can take at least two values.

2. The next to simplest example is given by a stationary process $\{V(t);\ t\in\mathbb{R}\}$ for which the sample paths $t\hookrightarrow V(t)$'s are analytic functions with probability 1. The proof is obvious and does not depend upon how random the process is. We shall come back to this class of deterministic potentials in the second half of the following section.

3. The last example which we consider concerns almost periodic potentials. If $V(t)$ is an almost periodic function, it is uniformly continuous. If for every $t<0$ the orbit $\{V(s);\ s\leq t\}$ is dense, then $V$ can be reconstructed from its knowledge on $(-\infty,t]\cap T$. This is obviously the case if $V$ is quasi-periodic.

We believe it is also true for general almost periodic.

**Corollary VII.3.7** *If $H(\omega)$ has a nontrivial absolute continuous component for a set of $\omega$'s with positive probability, then the random potential $\{V(t); t \in \mathbb{R}\}$ is a deterministic stochastic process.*

*Proof:*
The knowledge of $\mathcal{F}_{-\infty}$ gives the asymptotic behavior of the potential function $V(t,\omega)$ as $t \to -\infty$. Now, the knowledge of the potential function $V(t,\omega)$ on the negative $t$-axis determines the $m$-function $m_-(z,\omega)$ in the upper half plane and its boundary value on the real axis. If the absolute continuous component of $H(\omega)$ is nontrivial, formula (VII.59) gives us the boundary value of $m_+(z,\omega)$ on a set of strictly positive Lebesgue measure. This is enough to reconstruct the analytic function $m_+(z,\omega)$ in the whole upper half plane from which we can get the potential function $V(t,\omega)$ on the positive $t$-axis. Using the stationarity of the potential function, one can determine the potential function $V(t,\omega)$ on the positive $t$-axis from its knowledge on arbitrary left hand intervals, and thus from the knowledge of $\mathcal{F}_{-\infty}$. Note that we have to use different $m$-functions, namely the ones appropriate for the half intervals in question. It is obvious that we treated very cavalierly the technically difficult measurability problems of the proof. The reader will convince himself that they can be solved with the help of the machinery developed in Chapter V. We consider that the proof is complete. ∎

We conclude this section with a discussion of the actual size of the absolutely continuous spectrum. This search has to be limited to the set $A = \{\lambda \in \mathbb{R}; \gamma(\lambda) = 0\}$ where the Lyapunov exponent vanishes. Since the spectrum is equal to the set of points of increase of the integrated density of states, one tries to estimate the derivative of the integrated density of states on the set $A$. The main result is the following.

**Proposition VII.3.8** *For almost every $\lambda \in A$ we have:*

$$2N(\lambda)\frac{dN(\lambda)}{d\lambda} \geq \frac{1}{\pi^2} \tag{VII.65}$$

*in the continuous case and:*

$$2\pi \sin \pi N(\lambda)\frac{dN(\lambda)}{d\lambda} \geq 1 \tag{VII.66}$$

*in the lattice case.*

*Proof:*
We proved in Lemma VII.1.10 that:

$$-2\Re e\, w(\lambda + i\epsilon)\, \Im m\, w(\lambda + i\epsilon) \geq \epsilon.$$

The first claim follows the argument we used in the proof of Proposition VII.3.3 to take the limit $\epsilon \searrow 0$ for $\lambda$'s such that $\gamma(\lambda) = 0$. The result in the lattice case follows the same argument once we recall the estimate:

$$-2 \sinh \Re e\, w(\lambda + i\epsilon)\, \sin \pi \Im m\, w(\lambda + i\epsilon) \geq \epsilon$$

proved in Lemma VII.1.15 and we notice that:

$$\lim_{\epsilon \searrow 0} \frac{1}{\epsilon} \sinh -\Re e\, w(\lambda + i\epsilon) = \lim_{\epsilon \searrow 0} \frac{1}{\epsilon} - \Re e\, w(\lambda + i\epsilon). \quad \blacksquare$$

**Corollary VII.3.9** *For $a < b$ one has:*

$$N(b)^2 - N(a)^2 \geq \frac{1}{\pi^2} |A \cap (a,b)|$$

*in the continuous case and:*

$$\cos \pi N(a) - \cos \pi N(b) \geq \frac{1}{2} |A \cap (a,b)|$$

*and in particular $|A| \leq 4$ in the lattice case.*

*Proof:*
The above estimates are obvious iterated forms of the estimates (VII.65) and (VII.66). The last estimate follows immediately because $N(a)$ converges to 0 as $a \to -\infty$ and $N(b)$ tends to 1 when $b \to \infty$. $\blacksquare$

**Remark VII.3.10** It is important to keep in mind that the Lebesgue measure of $\Sigma_{ac} \cap (a,b)$ can be much larger than the Lebesgue measure of the set $A \cap (a,b)$. The fact that the set A where the Lyapunov exponent vanishes has measure at most 4 is very interesting for it explains why one expects to have some singular spectrum for almost periodic potentials with a strong coupling constant.

## VII.3.2 Support Theorems and Deterministic Potentials

We learned that there is no absolutely continuous spectrum unless the Lebesgue measure of the set $\{\lambda \in \mathbb{R};\ \gamma(\lambda) = 0\}$ where the Lyapunov exponent vanishes has strictly positive Lebesgue measure. This is only possible when the random potential is deterministic. As defined, determinism seems to indicate a very strong lack of randomness. This is not so. Indeed, as we prove in this subsection, there are many natural examples of deterministic random potentials beside the class of almost periodic potentials which we already encountered. We saw that the almost Mathieu operator does not have absolutely continuous spectrum when $|\beta| > 2$. We show that most of the examples we discuss in this section do not have any absolutely continuous spectrum either. This illustrates the fact that nondeterminism, though sufficient, is very far from necessary for the absence of absolutely continuous spectrum.

Let us consider random potentials of the form:

$$V(x,\omega) = \sum_{n\in\mathbb{Z}} \xi_n(\omega) f(x-n) \tag{VII.67}$$

where $f$ is a nonnegative function decaying fast enough at infinity and $\{\xi_n;\ n \in \mathbb{Z}\}$ is a sequence of i.i.d. real valued random variables. This random potential should be regarded as *very random* when the common distribution of the $\xi_n$'s is not degenerate. Interestingly enough, it is proven in ProblemVII.6.2 that these random potentials are generically deterministic and consequently the above sufficient condition cannot be used to guarantee the absence of absolutely continuous spectrum.

Another very natural class of deterministic potentials is given by the following model: let $\{x_n(\omega);\ n \in \mathbb{Z}\}$ be the points of a stationary Poisson process on the line $\mathbb{R}$ with intensity 1 (we will assume that the points are indexed so that $x_{-1}(\omega) < 0 < x_0(\omega)$ and $x_n(\omega) < x_{n+1}(\omega)$) and let $f(z)$ be an analytic function in a strip $\{z;\ |\Im m z| < \delta\}$ for some $\delta$, which is real on the real axis and which satisfies:

$$|f(z)| \le \frac{c}{(1+|z|)^{2+\epsilon}}$$

for some positive constants $c$ and $\epsilon$. We then consider the random potential:

$$V(x,\omega) = \sum_{n\in\mathbb{Z}} f(x - x_n(\omega)) \tag{VII.68}$$

This random potential should be regarded as very random because of the independence and the randomness of the Poisson process. It is in fact deterministic because the function $x \hookrightarrow V(x,\omega)$ is analytic for $\mathbb{P}$-almost all $\omega \in \Omega$.

The interested reader will find an example of deterministic potential on the lattice $\mathbb{Z}$ in ProblemVII.6.3.

Corollary VII.3.7 does not apply to these models and we need an extra analysis to prove or disprove the existence of a nontrivial absolutely continuous component in the corresponding Schrödinger operator. The results we prove below show that many of them do not have any absolutely continuous spectrum.

The following support theorem for the absolutely continuous spectrum will turn out to be quite appropriate and very fruitful in terms of the variety of models one can handle with its help.

**Lemma VII.3.11** *Let us assume that $\mathbb{P}$ is a probability measure on the special sample space $\Omega^{(\cdot)}$ which satisfies:*

$$\mathbb{E}\{V(0)^2\} < \infty \qquad \textit{and} \qquad \mathbb{E}\{\log(1+|V(0)|)\} < \infty$$

*in the continuous and lattice case respectively. Let us also assume that $v$ is an element of the topological support of $\mathbb{P}$ which satisfies:*

$$\sup_{n\in\mathbb{Z}} \int_n^{n+1} |v(t)|^2\,dt < \infty \qquad \textit{and} \qquad \sup_{n\in\mathbb{Z}} |v(n)| < \infty$$

*in the continuous and lattice case respectively. Then $\Sigma_{ac} \subset \Sigma_{ac}(H_0 + v)$.*

*Proof:*
Since $v$ is in the topological support of $\mathbb{P}$ we can find a sequence $\{v_k;\ k \geq 1\}$ which converges to $v$ in $\Omega^{(\cdot)}$ and such that $\Sigma_{ac} = \Sigma_{ac}(H_0 + v_k)$ and $\Re e\, G(\lambda + i0, 0, 0, v_k) = 0$ almost everywhere on $\Sigma_{ac}$ for all $k \geq 1$. As usual we use the notation $G(z,\cdot,\cdot,v)$ for the Green's function of the operator $H(v) = H_0 + v$. The convergence of $v_k$ toward $v$ implies the local uniform convergence of the Herglotz functions $G(\cdot,0,0,v_k)$ toward the Herglotz function $G(\cdot,0,0,v_k)$ and of the boundary values as well. This implies that $\Re e\, G(\lambda + i0,0,0,v) = 0$ almost everywhere on $\Sigma_{ac}$, from which it follows that $\Im m\, G(\lambda + i0,0,0,v) \neq 0$ almost everywhere on $\Sigma_{ac}$. This implies that $\Sigma_{ac}$ is contained in the absolutely continuous spectrum of $H_0 + v$ by the de la Vallée Poussin theorem. ∎

The above proof actually shows that:

**Corollary VII.3.12** *If $\mathbb{P}_1$ and $\mathbb{P}_2$ are two ergodic probability measures on $\Omega^{(\cdot)}$ which satisfy the conditions of this section we have:*

$$\text{supp } \mathbb{P}_1 \subset \text{supp } \mathbb{P}_2 \quad \Rightarrow \quad \Sigma_{ac}(\mathbb{P}_2) \subset \Sigma_{ac}(\mathbb{P}_1).$$

As we announced, the above result is our main tool in the construction of *deterministic* random potentials with an empty absolutely continuous spectrum.

**Proposition VII.3.13** *Let $H(\omega) = H_0 + V(\cdot, \omega)$ where $\{V(t);\, t \in T\}$ is of the form $V(t,\omega) = F(X_t(\omega))$ for some mean zero stationary ergodic Gaussian process $\{X_t;\, t \in T\}$ which is also assumed to be continuous in quadratic mean in the case $T = \mathbb{R}$, and $F$ is a real valued bounded non-constant continuous function on $\mathbb{R}$. Then $\Sigma_{ac} = \Sigma_{ac}(H(\omega)) = \emptyset$ almost surely.*

*Proof:*
We use the notation of the continuous case $T = \mathbb{R}$ for definiteness but the following argument applies as well to the lattice case $T = \mathbb{Z}$. The covariance $r(t-s) = \mathbb{E}\{X_t X_s\}$ is the Fourier transform of a measure which has no point mass because of the ergodicity assumption. See for example [131] for a proof of this classical result. Moreover our assumptions imply that, for each $t_0 > 0$, the sample paths of the process $\{X_t;\, t \in (-t_0, +t_0)\}$ are almost surely in the space $L^2((-t_0,+t_0), dt)$ and the ergodicity assumption implies that the topological support of its distribution is the whole space $L^2((-t_0,+t_0), dt)$. Since this is true for every $t_0 > 0$ we conclude that the topological support of the distribution of the process $V(t,\omega) = F(X_t(\omega))$ contains the set of functions $t \hookrightarrow v(t)$ satisfying $\inf F \leq v(t) \leq \sup F$ for all $t \in T$. We will see in the next chapter that the topological support of the random potential of the Russian school for which we shall prove exponential localization is exactly this very set of functions $v$. Consequently, the absolutely continuous spectrum is empty for a lot of functions $v$ . The proof can be concluded by appealing to LemmaVII.3.11 above. ∎

We give another example of a random deterministic potential for which the corresponding Schrödinger operator does not have any absolutely continuous spectrum.

**Proposition VII.3.14** *Let us assume that $f$ is a nonnegative nonidenti-*

*cally zero continuous function satisfying:*

$$f(t) \leq \frac{c}{(1+|t|)^{2+\epsilon}}$$

*for some positive constants* $c$ *and* $\epsilon$ *and all* $t \in \mathbb{R}$. *Let the random potential* $\{V(t); t \in \mathbb{R}\}$ *be given by:*

$$V(t,\omega) = \sum_{n\in\mathbb{Z}} \chi(t - x_n(\omega)) \tag{VII.69}$$

*where the* $x_n(\omega)$*'s are the points of a stationary Poisson process on the line. Then* $\Sigma_{ac}(H(\omega)) = \emptyset$ *almost surely.*

*Proof:*
Let $\{x_n;\, n \in \mathbb{Z}\}$ be a doubly infinite nonrandom sequence of real numbers such that $\#\{n \in \mathbb{Z};\, |x_n| \leq k\} \leq ck$ for some constant $c$ and all nonnegative integers $k$. We then define the nonrandom potential $U$ by the formula:

$$U(t) = \sum_{n\in\mathbb{Z}} \chi(t - x_n).$$

The series converges compact uniformly and consequently, for each fixed $\ell > 0$ and $\epsilon > 0$ we can find $k_1$ such that:

$$\sup_{|t|\leq \ell} \left| U(t) - \sum_{|x_n|\leq k_1} \chi(t - x_n) \right| \leq \frac{\epsilon}{3}.$$

The uniform continuity of $\chi$ gives the existence of $\delta > 0$ such that:

$$\sup_{t\in\mathbb{R}} \sum_{|x_n|\leq k_1} |\chi(t - x_n) - \chi(t - x'_n)| \leq \frac{\epsilon}{3}$$

whenever $|x_n - x'_n| < \delta$ for all the $x_n$'s satisfying $|x_n| \leq k_1$. Finally, the assumption on the decay at infinity of the function $\chi$ implies the existence of a constant $k_2 > 0$ such that:

$$\mathbb{P}\{\omega \in \Omega;\, \sup_{|t|\leq \ell} \sum_{|x_n(\omega)|\leq k_1} \chi(t - x_n(\omega)) \leq \frac{\epsilon}{3}\} > 0.$$

Now, if $\omega \in \Omega$ is chosen in the above set of positive probability in such a way that the $x_n(\omega)$'s with $|x_n(\omega)| \leq k_1$ are within $\delta$ of the deterministic $x_n$'s considered above, then one has:

$$\sup_{|t|\leq \ell} |U(t) - \sum_{n\in\mathbb{Z}} \chi(t - x_n(\omega))| \leq \epsilon,$$

which proves that the potential function $U$ belongs to the topological support of the probability $\mathbb{P}$ because this set of $\omega$'s is still of positive measure. Consequently we have:

$$\Sigma_{ac}(H(\omega)) \subset \Sigma_{ac}(H_0 + U)$$

for $\mathbb{P}$-almost all $\omega \in \Omega$. Without any loss of generality we can assume that $\chi(0) > 0$ so that $\chi(t) \geq \epsilon > 0$ for $|t| \leq \delta$. Also, for each $\kappa > 0$ we can choose $x_n = \delta n/\kappa$ which implies that $\chi(t) \geq 2\kappa\epsilon$ and consequently that $\Sigma_{ac}(H_0 + U) \subset [2\kappa\epsilon, \infty)$. This proves that $\Sigma_{ac}(H(\omega)) = \emptyset$ almost surely for $\kappa$ can be chosen arbitrarily large. ∎

Potentials of the form (VII.69) seem to have been first introduced by Herbst and Holland in the study of the Stark effect. Proposition VII.3.14 above shows that these random potentials do not have absolutely continuous spectrum even when the *well* function $\chi$ is analytic. A last remark about the above proposition concerns the Poisson assumption. An inspection of the proof shows that it applies as well to sequences of the form $x_n(\omega) = n + \xi_n(\omega)$ for any i.i.d. sequence $\{\xi_n;\ n \in \mathbb{Z}\}$ of random variables having a common distribution with mean 0 and sufficiently fast decaying tail $\mathbb{P}\{\xi_n \geq t\}$.

## VII.4 Inverse Spectral Theory

The inverse problems are concerned with the recovery of the potential function $V$ from the spectral data. We reformulate this problem to suit the setting of random potentials. The actual sample path $t \hookrightarrow V(t,\omega)$ cannot be recovered unless the spectral information one starts from is given on the precise realization $\omega$ of the randomness (or disorder). If the information we start from involves only moments and expectations of various functions of the random potential, then it is clear that one cannot hope to recover more than the distribution of the stochastic process $\{V(t);\ t \in T\}$. The classical inverse spectral theory, in particular the set of results we recalled in the periodic case, is of the type we just described.

To be specific, the problem we consider is the following:

> given a function $w$, can one find a stochastic process $\{V(t);\ t \in T\}$ such that the $w$-function of the random Schrödinger operator constructed from the random potential $V$ is the function $w$ we started from.

We made the choice to start from the spectral data as it is contained in the $w$-function. We shall argue that this can be regarded as a straightforward generalization of the periodic case. Moreover, solving the problem involves only checking that the $w$-function associated with the random potential to be found is the right one. Notice that this $w$-function is the expectation of some complicated function of the random potential. As such, it depends only on the distribution of the random potential, and certainly not on the particular probability space on which the latter is defined. Consequently our search for a stochastic process can be done in any probability space we wish to work with. We shall take advantage of this extra freedom and choose a convenient measurable space $(\Omega, \mathcal{F})$ and an explicit group of transformations $\{\theta_t;\ t \in T\}$ on which we construct the probability measures we are looking for.

The strategy of the reconstruction proof is the same in the continuous and in the lattice case. It relies on the reconstruction procedures recalled in our discussions of the periodic cases and an approximation argument involving $w$-functions. We present this argument simultaneously in both cases.

**Proposition VII.4.1** *Let $w \in \mathcal{W}$ have the asymptotic:*

$$w(\lambda) = i\sqrt{\lambda} + \frac{\sigma^2}{2i\lambda\sqrt{\lambda}} + o(|\lambda|^{-3/2}) \qquad \text{(VII.70)}$$

$$\textit{(resp.} \qquad w(\lambda) = \log\frac{1}{-\lambda} + \frac{\sigma^2}{\lambda^2} + o(|\lambda|^{-2}) \qquad ). \qquad \text{(VII.71)}$$

*for some positive constant $\sigma$ as $\lambda \to \infty$ in the upper half plane. Then there exists a sequence $\{w_n;\ n \geq 1\}$ in $\mathcal{W}$ such that:*

*i) $w_n$ has the same asymptotic as $w$ with a $\sigma_n > 0$ instead of $\sigma$ but with $\sigma_n \leq \sigma$ for all $n \geq 1$,*
*ii) $w_n$ converges toward $w$ compact uniformly,*
*iii) for each $n \geq 1$ the range $w_n(\Pi_+)$ is a domain $D_n$ of the form (VII.2.1) (resp. of the form (VII.2.3)).*

*Proof:*
Let us first assume that the distribution function $N$ appearing in the representation of the function $w$ is smooth (resp. is smooth and the measure $dN$ has compact support), and let $\{a_n;\ n \geq 1\}$ be a dense subset of $\mathbb{R}$ (resp. $[0, \pi]$). For each $n \geq 1$ we consider the domain $D_n$ contained in $D_0 = \{z \in \mathbb{C};\ \Im m\, z > 0 \text{ and } \Re e\, z < 0\}$ (resp. $D_0 = \{z \in \mathbb{C};\ 0 < \Im m\, z <$

$\pi$ and $\Re e\, z < 0\}$ the boundary of which is the union of the boundary of $D_0$ and of the segments of lines parallel to the real axis connecting the point $ia_j$ to the boundary of $w(\Pi_+)$. The Riemman mapping theorem gives the existence of a conformal map on $\Pi_+$, say $w_n$ with range $D_n$ and the same asymptotic as $w$. Let us denote by $N_n$ the distribution function appearing in the representation of the function $w_n$. Since $w(\Pi_+) \subset w_n(\Pi_+)$, Proposition VII.1.4 and the remark following it imply that:

$$\int dN_n(\xi) = \frac{\sigma_n}{2} \leq \frac{\sigma}{2} \int \xi^2 dN(\xi)$$

(resp.

$$\int \xi^2 dN_n(\xi) \leq \int \xi^2 dN(\xi)\ )$$

for all $n \geq 1$. This implies the existence of a subsequence $\{w_{n_k}; k \geq 1\}$ which converges toward another conformal mapping $\tilde{w}(\lambda)$ from $\Pi_+$ into $\Pi_+$ We show that $\tilde{w} = w$ to conclude the proof.

The smoothness assumption implies the smoothness of the boundary of $w(\Pi_+)$ and the proof of the Caratheodory convergence theorem shows that we actually have $w(\Pi_+) = \tilde{w}(\Pi_+)$, so that $w$ and $\tilde{w}$ are conformal mappings from $\Pi_+$ into $\Pi_+$ with the same range. Hence, the function $u(\lambda)$ defined by $u(\lambda) = \tilde{w}^{-1}(w(\lambda))$ maps $\Pi_+$ onto $\Pi_+$ conformally and is consequently of the form

$$u(\lambda) = \frac{a\lambda + b}{c\lambda + d}$$

for some real numbers $a, b, c$, and $d$ satisfying $ad - bc = 1$. Thus

$$w(\lambda) = \tilde{w}\left(\frac{a\lambda + b}{c\lambda + d}\right)$$

and one can use the various asymptotic regimes to conclude that $a = d = 1$ and $c = b = 0$.

In the general case, we first approximate the probability measure $dN$ by probability measure $dN_\epsilon$ with smooth distribution functions (choose for example:

$$dN_\epsilon(\xi) = \int j_\epsilon(\xi - \eta)\, dN(\eta)$$

for a smooth approximate identity $\{j_\epsilon; \epsilon > 0\}$ on $\mathbb{R}$). The corresponding $w$-functions converge compact uniformly toward the $w$-function we started from and one can use the above argument to conclude the proof. ∎

## VII.4.1 The Continuous Case

According to the previous discussion of the assumptions in force in the sections of this chapter dealing with the continuous case, any random potential we care to work with has sample paths $V(\cdot,\omega) : t \hookrightarrow V(t,\omega)$ in the space $\Omega^{(c)} = L^2_{real}(\mathbb{R}, dt/(1+|t|^3))$ for $\mathbb{P}$-almost all $\omega \in \Omega$ and we can carry the probability measure $\mathbb{P}$ and the group of transformations $\{\theta_t;\, t \in \mathbb{R}\}$ onto $\Omega^{(c)}$. Since all the properties we are interested in depend only on expected values, we can replace $\mathbb{P}$ and by their images under this mapping. We will keep the same notation for these images. In other words, we can assume, without any loss of generality that the random potential is given by a probability measure $\mathbb{P}$ which is stationary (and most of the time ergodic) for the usual shift group of transformations in $\Omega^{(c)} = L^2_{real}(\mathbb{R}, dt/(1+|t|^3))$.

Let us recall that the $w$-function introduced in Section 1 is in the space $\mathcal{W}$ in the sense that both $w$ and $w'$ are Herglotz functions. We also saw that the function $-iw$ was also an Herglotz function and that that the function $w$ had the asymptotic (VII.70) as $\lambda \to \infty$

We introduce a special notation for this space of Herglotz functions. We let $\mathcal{W}^{(c)}$ be the space of functions $w$ such that $w \in \mathcal{H}erg$, $w' \in \mathcal{H}erg$, $-iw \in \mathcal{H}erg$ and which satisfy the asymptotic condition (VII.70). Note that the $w$-function associated with a stationary random potential has been proved to satisfy all these conditions with $\mu = \mathbb{E}\{\int_0^1 V(t)\,dt\}$ and $\tilde{\sigma}$ satisfying $\tilde{\sigma}^2 \le \mathbb{E}\{\int_0^1 |V(t)|^2\,dt\}$.

We prove that to each function $w \in \mathcal{W}^{(c)}$ there exists a (stationary) random Schrödinger operator the corresponding $w$-function of which is the $w$-function we started from. Each such random Schrödinger operator is determined by a probability measure on the path space $\Omega^{(c)}$ which we singled out in this chapter. For this reason it will be convenient to have a special notation for the probability measures on this path space which would be of interest for the purpose of the inverse spectral theory. We shall denote by $\mathcal{P}^{(c)}$ the set of probability measures $\mathbb{P}$ on the Borel $\sigma$-field of the Hilbert space $\Omega^{(c)} = L^2_{real}(\mathbb{R}, dt/(1+|t|^3))$ which are invariant under the shift in the sense that $\mathbb{P}\circ\theta_t^{-1} = \mathbb{P}$ for all $t \in \mathbb{R}$ and which satisfy the integrability condition:

$$\mathbb{E}\{\int_0^1 V(t)^2\,dt \;< \infty.$$

We shall also denote by $\mathcal{P}_e^{(c)}$ the subset of the ergodic elements of $\mathcal{P}^{(c)}$.

**Theorem VII.4.2** *If $w$ is a function in the class $\mathcal{W}^{(c)}$ then there exists a probability space $(\Omega, \mathcal{F}, I\!P)$, a group $\{\theta_t;\, t \in I\!R\}$ of measure preserving transformations and a random variable $v$ such that the $w$-function of the random Schrödinger operator $H(\omega) = -d^2/dt^2 + v(\theta_t\omega)$ is the function $w$ we started from.*

*Proof:*
Let $\{w_n;\, n \geq 1\}$ be the approximating sequence which we constructed in Lemma VII.4.1. Because of the elements of inverse spectral theory which we recalled for periodic potentials we know of the existence for each $n \geq 1$ of a periodic potential whose $w$-function is $w_n$. In the terminology developed for random potentials this means that there exists for each $n \geq 1$ a probability measure $I\!P_n$ on the canonical path space $\Omega^{(c)}$ such that $w_n = w_{I\!P_n}$. The measure $I\!P_n$ is concentrated on smooth and uniformly bounded potentials and consequently we have:

$$\tilde{\sigma}_n^2 = I\!E_{I\!P_n}\{\int_0^1 V(t)^2\, dt\}$$

which implies that:

$$I\!E_{I\!P_n}\{\int_0^1 V(t)^2\, dt\} \leq \tilde{\sigma}^2.$$

This implies that the sequence $\{I\!P_n;\, n \geq 1\}$ of probability measures on $\Omega^{(c)}$ is tight because closed balls are relatively compact for the weak topology. Hence we can choose a subsequence converging toward a probability measure $I\!P$. It is easy to see that $I\!P$ is invariant and satisfies the integrability condition, namely that $I\!P \in \mathcal{P}^{(c)}$. It remains only to check that the corresponding $w$-function $w_{I\!P}$ is the $w$-function $w$ we started from. Note that the function $w_n = w_{I\!P_n}$ converges compact uniformly to the function $w$ and this implies the pointwise convergence of the functions $N_n$ appearing in the decomposition given in PropositionVII.1.2 toward the corresponding function $\tilde{N}$. Using the sandwich estimate and the one dimensional construction of the integrated density of states, this function $\tilde{N}$ can be identified with the integrated density of states of the random Schrödinger operator corresponding to the measure $I\!P$. Using the representation formula given in PropositionVII.1.2 we can conclude that $w(\lambda) - w_{I\!P}(\lambda) = \alpha$ and the estimate (VII.70) gives $\alpha = 0$. This completes the proof. ∎

The above result is rather unsatisfactory in the sense that it does not guarantee the existence of an ergodic random potential giving rise to the

$w$-function one starts from. The only sufficient condition known to this day is the following:

**Theorem VII.4.3** *If on the top of the assumptions of Theorem VII.4.2 we assume that:*

$$\gamma(\lambda) = 0 \qquad \text{for } N\text{-almost all } \lambda \tag{VII.72}$$

*then it is possible to choose an ergodic measure* $I\!P \in \mathcal{P}_e^{(c)}$.

Note that assumption (VII.72) is trivially satisfied in the case of periodic potentials since the Lyapunov exponent vanishes on the spectrum (i.e. the stability intervals) and that the integrated density of states is supported by the spectrum.

*Proof:*
It is a standard fact from ergodic theory (see for example [93]) that there exists a measurable structure on the set $\mathcal{E}$ of extreme points of the convex set $\mathcal{P}$ such that each such probability measure $\mathbb{P}$ can be written as the barycenter of some probability measure $\nu_{I\!P}$ on $\mathcal{E}$:

$$\mathbb{P} = \int_{\mathcal{E}} \mathbb{P}_e \, d\nu_{I\!P}(e).$$

Consequently, if we denote by $\mathbb{P}$ the probability measure whose existence is proved in the above theorem and by $w_e$ the $w$-function of the probability measure $\mathbb{P}_e$, we have:

$$w = \int_{\mathcal{E}} w_e \, d\nu_{I\!P}(e).$$

Now according to a technical result whose proof we shall omit (see for example Lemma 7.11 of [195]) for each fixed $e \in \mathcal{E}$ we also have (using obvious notation):

$$\begin{aligned}
\int_{\Pi_+} |w'(\lambda) - w_e'(\lambda)|^2 \, d\lambda &= -\pi \int_{I\!R} [\gamma(\xi) - \gamma_e(\xi)] \, d[N(\xi) - N_e(\xi)] \\
&\leq \pi \int_{I\!R} \gamma(\xi) \, dN_e(\xi) + \pi \int_{I\!R} \gamma_e(\xi) \, dN(\xi).
\end{aligned}$$

Integrating both sides with respect to $\nu_{I\!P}$ we get:

$$\int_{\mathcal{E}} \int_{\Pi_+} |w'(\lambda) - w_e'(\lambda)|^2 \, d\lambda \, d\nu_{I\!P}(e) \leq 2\pi \int_{I\!R} \gamma(\xi) \, dN(\xi) = 0,$$

because of our assumption (VII.72). This proves that $w = w_e$ $\nu_{I\!P}$-almost surely and this concludes the proof. ∎

So far we did a soft analysis of the inverse problem for random potentials. We set up the decor but we did not touch upon the difficult problems. Let us mention some of them.

In the same way we added an extra assumption (i.e. the *orthogonality* of the Lyapunov exponent and of the integrated density of states) to obtain the ergodicity of the random potential, we may want to add extra properties to the $w$-function we start from, and try to get extra information on the random potential we construct. For example, one can ask which class of $w$-functions can (and actually do) come from a random potential having some absolute continuous spectrum, or when does the reconstruction procedure give an almost periodic potential? One could also ask for properties of the $w$-function which guarantee smoothness of the sample paths of the random potential. We close this section with a discussion without proof of some results in this direction.

**Definition VII.4.4** *A subset $\mathcal{I}$ of $[0,\infty)$ is said to be a Levitan open set if $\mathcal{I} = \cup_{i\geq 1}(\lambda_i, \mu_i)$ in such a way that there exist disjoint bounded intervals $(\alpha_i, \beta_i)$ such that:*

$$(\lambda_i, \mu_i) \subset (\alpha_i, \beta_i) \qquad \textit{and} \qquad \lambda_i - \alpha_i = \beta_i - \mu_i \tag{VII.73}$$

*for all $i \geq 1$ and:*

$$\sup_{i\geq 1} \beta_i - \alpha_i < \infty \qquad \textit{and} \qquad \sum_{i\geq 1} \frac{\mu_i - \lambda_i}{\beta_i - \alpha_i} < \infty. \tag{VII.74}$$

It is possible to prove that, whenever $\mathcal{I}$ is a Levitan open set, then there exists a unique function $w \in \mathcal{W}^{(c)}c$ such that the corresponding functions $\gamma$ and $N$ are absolutely continuous, $\{\xi \in \mathbb{R};\ \gamma(\xi) = 0\} = [0,\infty)\backslash\mathcal{I}$ and $\int \gamma(\xi)\, dN(\xi) = 0$. This result is purely complex analytic. It does not involve Schrödinger operator theory or stochastic processes. See [199] for details. But we can use it to feed Theorem VII.4.3 to get the existence of a probability measure $\mathbb{P}$ on the probability space $\Omega^{(c)}$ in such a way that $w_{\mathbb{P}} = w$. In other words, to each Levitan open set one can associate a random Schrödinger operator. Notice that the $w$-function is uniquely determined by the set $\mathcal{I}$ but that the probability measure $\mathbb{P}$ (and consequently the random potential $V$) are not determined by the set $\mathcal{I}$. In any case, it can be proved that, for any such $\mathbb{P}$, the Schrödinger operator $H(\omega)$ has purely absolutely continuous spectrum for *every* $\omega$ in the topological support of $\mathbb{P}$. Moreover, if one also assumes that:

$$\sum_{i\geq 1} \sqrt{\lambda_i} \frac{\mu_i - \lambda_i}{\beta_i - \alpha_i} < \infty$$

then it is possible to prove that all the potential functions $t \hookrightarrow V(t,\omega)$ are almost periodic differentiable functions as long as $\omega$ remains in the topological support of $\mathbb{P}$. This last result is a mere improvement of a result of Levitan who proved in [217] that, given an open set $\mathcal{I}$ of the Levitan type one could construct an almost periodic potential with spectrum $[0,\infty)\backslash\mathcal{I}$. Note that the inverse theory of the Levitan open sets does not include the class of limit periodic potentials considered by Pastur and Tkachenko in [265] which we discussed earlier in the section on almost periodic potentials. As we stated them, the above results were given by Kotani and Krishna in [199]. The proof of the almost periodicity of the (random) potential is based on an extension of the trace formula to the present situation. This extension says that:

$$V(t) = \sum_{i\geq 1} (\mu_i - \lambda_i - 2\xi(t))$$

where the $t \hookrightarrow \xi_i(t)$ are the solutions of the Dubrovin equation.

As we mentioned earlier in the section on periodic potentials, it follows from a hierarchy of identities and estimates that integrability properties of the Lyapunov exponent are equivalent to smoothness of the potential. Some of these results remain true in the case of general random potentials. For example:

**Theorem VII.4.5** *If $n \geq 3$ and if the sample paths $t \hookrightarrow V(t,\omega)$ are $n$ times differentiable with bounded derivatives up to order $n-3$ and if:*

$$\mathbb{E}\{\int_0^1 |\frac{d^j V}{dt^j}(s)|^2 \, ds\} < \infty, \qquad j = n-2, n-1, n,$$

*then we have:*

$$\int_0^\infty \gamma(\xi)\, \xi^n \sqrt{\xi}\, d\xi < \infty.$$

This result is proven in [199]. Even though one expects that smoothness of the random potential follows from integrability properties of the Lyapunov exponent in general, only very partial results have been proven in this direction. See [199] for such a result for almost periodic potentials of the Levitan type.

## VII.4.2 The Lattice Case

According to the above results, the $w$-function of a random operator $H(\omega)$ constructed from a stationary sequence $\{(a_n, b_n); n \in \mathbb{Z}\}$ satisfying properties (i) and (ii) of Section 1.c is such that:

**(i)** $\Im m\ w \in [0, \pi]$ and $\Re e\ w \in [-\infty, c]$ for some finite constant $c$.

**(ii)** $w(\lambda)$ has an expansion of the form

$$w(\lambda) = \log \frac{1}{-\lambda} + \frac{\alpha_1}{\lambda} + \frac{\alpha_2}{\lambda^2} + 0(\lambda^{-2})$$

in a neighborhood of $\lambda = \infty$ for some real numbers $\alpha_1$ and $\alpha_2$.

**(iii)** $w' \in \mathcal{H}erg$.

Our present goal is, given a function $w$ satisfying these properties, to construct a stationary sequence $\{(a_n, b_n); n \in \mathbb{Z}\}$ satisfying the above properties and such that the corresponding $w$-function is the one we started from.

Our reconstruction procedure relies on the approximation of general $w$-functions satisfying conditions (i), (ii) and (iii) by $w$-functions corresponding to periodic sequences.

The main result of this section is the following:

**Theorem VII.4.6** *Let $w$ be a Herglotz function satisfying properties (i), (ii) and (iii). Then there exists a stationary sequence $\{(a_n, b_n); n \in \mathbb{Z}\}$ such that:*

$$\mathbb{E}\{a_0^2 + |\log a_0| + b_0^2\} < +\infty, \quad c \geq -\mathbb{E}\{\log a_0\} \tag{VII.75}$$

*and:*

$$w(\lambda) = \mathbb{E}\{\log m_+(\lambda)\}. \tag{VII.76}$$

*Proof:*

For each integer $\tau$ we construct the domains $D_\tau$ from $w(\Pi_+)$ as described in the figure VII.2.3 with $c_j = \sup\{x \in \mathbb{R}; x + ij\pi/\tau \in w(\Pi_+)\}$ for $j = 1, \cdots, \tau - 1$. Then we use the form of the reconstruction theorem of Kac and van Moerbeke given in LemmaVII.2.6 to obtain a sequence $\{(a_n^{(\tau)}, b_n^{(\tau)}); n \in \mathbb{Z}\}$ periodic with period $\tau$, and for each integer $\tau$ we redefine this periodic sequence as a stationary process given by the coordinate projections, say $(a_n, b_n)$, on the product space $\Omega^{(d)} = [(0, \infty) \times \mathbb{R}]^{\mathbb{Z}}$ for some probability $\mathbb{P}_\tau$. The corresponding $w$-function:

$$w_\tau(\lambda) = \mathbb{E}_{\mathbb{P}_\tau}\{\log m_+(\lambda)\}$$

satisfies $w_\tau(\Pi_+) = D_\tau$. Notice that:

$$\begin{aligned}
\mathbb{E}_{\mathbb{P}_\tau}\{2a_0^2 + b_0^2\} &= \int \xi^2 dN_\tau(\xi) \\
&\qquad \text{(because of the proof of Proposition VII.1.13)} \\
&\leq \int \xi^2 dn(\xi) \quad \text{(because of Proposition VII.1.4)} \\
&< +\infty.
\end{aligned}$$

This implies the tightness of the family $\{\mathbb{P}_\tau; \tau \geq 1\}$ of probability measures on $\Omega^{(d)}$. let $\mathbb{P}$ be any limit point. For notational convenience we will consider that $\mathbb{P}$ is actually the limit of the whole sequence. $\{(a_n, b_n); n \in \mathbb{Z}\}$ is still stationary for $\mathbb{P}$. Moreover, the above estimate implies that:

$$\begin{aligned}
\mathbb{E}_{\mathbb{P}}\{2a_0^2 + b_0^2\} &\leq \liminf_{\tau \to \infty} \mathbb{E}_{\mathbb{P}_\tau}\{2a_0^2 + b_0^2\} \\
&< +\infty.
\end{aligned}$$

If $c \geq \sup_{\lambda \in \Pi_+} \Re e\ w(\lambda)$, our construction gives:

$$c = \mathbb{E}_{\mathbb{P}_\tau}\{\log \frac{1}{a_0}\} = \mathbb{E}_{\mathbb{P}_\tau}\{\log^+ a_0\}.$$

But we also have:

$$\lim_{\tau \to \infty} \mathbb{E}_{\mathbb{P}_\tau}\{\log^+ a_0\} = \mathbb{E}_{\mathbb{P}}\{\log^+ a_0\} < +\infty.$$

Consequently,

$$\begin{aligned}
\mathbb{E}_{\mathbb{P}}\{\log^+ \frac{1}{a_0}\} &\leq \liminf_{N \to \infty} \mathbb{E}_{\mathbb{P}_\tau}\{\log^+ \frac{1}{a_0}\} \\
&= c + \mathbb{E}_{\mathbb{P}}\{\log^+ a_0\} < +\infty.
\end{aligned}$$

Indeed, Skorokhod's representation theorem gives an interpretation of the weak convergence in terms of almost sure convergence and one can apply Fatou's lemma from integration theory. Hence we have:

$$\mathbb{E}_{\mathbb{P}}\{|\log a_0|\} < +\infty$$

which completes the proof of (VII.75). Using once more Skorokhod's representation theorem, one can assume without any loss of generality that the $(a_n^{(\tau)}, b_n^{(\tau)})$'s, once appropriately redefined on the same probability space, converge almost surely to $\{(a_n, b_n); n \in \mathbb{Z}\}$ as $\tau \to \infty$. Consequently, one also has the almost sure convergence of the corresponding $m$-functions and:

$$\lim_{\tau\to\infty} \mathbb{E}_{\mathbb{P}_\tau}\{\log m_+(\lambda)\} = \mathbb{E}_{\mathbb{P}}\{\log m_+(\lambda)\}$$

by Lebesgue's dominated convergence theorem. This gives (VII.76) and the reconstruction is now complete. ∎

# VII.5 Miscellaneous

## VII.5.1 Potentials Taking Finitely Many Values

We shall see in the next chapter on the study of the pure point spectrum of random one dimensional Schrödinger operators that the potentials taking only finitely many values play a particular role. This is even more so in the lattice case for there are commensurability problems between the period of the lattice and the *periods* created by the fact that the potential takes only finitely many values.

as a consequence of the inverse spectral theory which we developed in this chapter, We present the following *curiosity* as a consequence of the inverse spectral theory which we developed in this chapter.

**Proposition VII.5.1** *Let $\{V(n);\ n \in \mathbb{Z}\}$ be a stationary ergodic sequence of random variables such that the common distribution of the $V(n)$'s is concentrated on a finite subset of $\mathbb{R}$. Then, either the spectrum of the random Schrödinger operator $H = H_0 + V$ has no absolutely continuous component, or the potential $V$ is periodic.*

*Proof:*
We can assume, without any loss of generality that the probability space on which the random potential is given is canonical. We need some notation to describe it advantageously. Let us denote by $S$ the support of the distribution of the potential at one site, by $\mathbb{Z}_-$ the set of negative integers and by $\mathbb{Z}_+$ the set of nonnegative integers. We consider the canonical path space $\Omega^{(d)} = S^{\mathbb{Z}}$. It is a compact separable metric space for the metric:

$$d(V, V') = \sum_{n\in\mathbb{Z}} 2^{-n}|V(n) - V'(n)|.$$

We consider the Borel $\sigma$-field $\mathcal{F}$ and the shift $\mathcal{T}$ defined by $[\mathcal{T}V](n) = V(n+1)$. The probability $\mathbb{P}$ is any stationary ergodic probability measure on the measurable space $(\Omega^{(d)}, \mathcal{F})$. Note that its support is a $\mathcal{T}$-invariant closed subset of $\Omega^{(d)}$.

Now, we assume that the set $A = \{\lambda \in \mathbb{R};\ \gamma(\lambda) = 0\}$ has positive Lebesgue measure and we prove that the support of the measure $\mathbb{P}$ is finite. This will prove the desired result because of the characterization of the essential support of the absolutely continuous spectrum given in Corollary VII.3.4. According to Problem VII.6.10, the maps $\Omega_+^{(d)} \ni V_+ \hookrightarrow V_- \in \Omega_-^{(d)}$ and $\Omega_-^{(d)} \ni V_- \hookrightarrow V_+ \in \Omega_+^{(d)}$ are continuous. Here we denote by $V_+$ (resp. $V_-$) the restriction of $V \in \Omega^{(d)}$ to $\Omega_+^{(d)} = S^{\mathbb{Z}_+}$ (resp.$\Omega_+^{(d)} = S^{\mathbb{Z}_+}$). Consequently, for each $\epsilon > 0$ there exists $\delta > 0$ and an integer $m \geq 1$ such that for $V, V' \in$ supp $\mathbb{P}$ one has:

$$|V(n) - V'(n)| \leq \delta \qquad\qquad n = -1, -2, \cdots, -m$$

implies that $|V(0) - V'(0)| \leq \delta$. Since $S$ is finite, it is possible to choose $\epsilon$ and $\delta$ sufficiently small for the above to be equivalent to:

$$(V(n) = V'(n) \quad n = -1, -2, \cdots, -m) \quad \Rightarrow \quad V(0) = V'(0).$$

Because of the invariance of the support of $\mathbb{P}$, $V_+$ is uniquely determined when $\{V(n);\ n = -1, -2, \cdots, -m\}$ is known. This implies that $\#\Omega_+^{(d)} \leq m^{\#S}$. Moreover, we know from the proof of Corollary VII.3.7 that, because the Lebesgue measure of the set $A$ is positive, the map $V_+ \hookrightarrow V$ is one to one. These two facts imply that the support of $\mathbb{P}$ is finite and the proof is complete. ∎

Note that this result extends to the continuous case if one considers random potentials of the form:

$$V(x) = \sum_{n \in \mathbb{Z}} V(x) \mathbf{1}_{[n,n+1)}(x)$$

with the sequence $\{V(n);\ n \in \mathbb{Z}\}$ as above.

We shall from now on restrict the discussion to the case of potentials taking only two values.

Let us consider first the case of the operator: $H = \epsilon H_0 + V$ on the one dimensional lattice $\mathbb{Z}$ with $\epsilon > 0$ and $V(n) = \mathbf{1}_A(n\alpha + \theta)$ where $A \subset \mathbb{R}$ is an arbitrary interval of the unit circle, $\alpha$ is an irrational number and the addition is understood modulo 1. For such a model one has:

**Proposition VII.5.2** *For almost every $\alpha$ with respect to the Lebesgue measure and for any interval $A$, the spectrum is singular continuous for every $\epsilon$ and almost every $\theta$ with respect to the Lebesgue measure.*

Indeed, using the result of Gordon which we mentioned earlier, Delyon and Petritis proved in [75] that the point component of the spectrum is empty under these conditions. Moreover, since the potential cannot be periodic for $\alpha$ irrational, we can use Proposition VII.5.1 to conclude.

More recently, Zhitomirskaya reproved in [333] the same result if $A$ is equal to half the circle and if $\epsilon$ is small enough. She explicitely constructed a Cantor set of Lebesgue measure 0 for the spectrum of $H$. Her proof is based on a fine analysis of the generalized eigenfunctions for which she constructed a hierarchical scaling structure via approximations of these eigenfunctions corresponding to rational approximations of $\alpha$.

## VII.5.2 A Remark on the Multidimensional Case

We discuss multidimensional localization in Chapter IX. All the results originate in the remarkable work of Frölich and Spencer who proved exponential decay of the Green's function of the Anderson model with Gaussian potential for large disorder or low energy. This proves the absence of diffusion in a sense similar but slightly weaker than the one discussed in this chapter. We now give a simple form due to Simon and Wolff of a remark of Martinelli and Scoppola, who noticed that the main estimate of Spencer and Fröhlich, namely the exponential decay of the Green's function, implies the absence of absolutely continuous spectrum.

**Proposition VII.5.3** *Let us assume that $\{V(\mathbf{n});\, \mathbf{n} \in \mathbb{Z}^d\}$ is an ergodic stationary sequence of real valued random variables and $\mathcal{I}$ is a Borel subset of $\mathbb{R}$ such that for almost every energy $\lambda \in \mathcal{I}$ and $\mathbb{P}$-almost all $\omega \in \Omega$ we have:*

$$\lim_{\epsilon \searrow 0} \sum_{\mathbf{n} \in \mathbb{Z}^d} |G(\lambda + i\epsilon, 0, \mathbf{n})|^2 < \infty.$$

*Then the operator $H(\omega)$ has no absolutely continuous spectrum in $\mathcal{I}$ for $\mathbb{P}$-almost all $\omega \in \Omega$.*

Notice that the exponential decay of the Green's function which is given by the Fröhlich-Spencer estimate which will be proven in the first part of Chapter IX obviously implies the above assumption.

*Proof:*

The proof is a trivial application of the de la Vallée Poussin and Fubini theorems. Indeed, for $\mathbb{P}$-almost all $\omega \in \Omega$ we can use the sufficient condition of ProblemI.7.22 to conclude that the spectral measure $< E(d\lambda)\delta_0, \delta_0 >$ has no absolutely continuous component in $\mathcal{I}$. The same conclusion holds for all the spectral measures $< E(d\lambda)\delta_{\mathbf{n}}, \delta_{\mathbf{n}} >$ because of a simple stationarity argument, and the proof is complete. ∎

# VII.6 Problems

**Problem VII.6.1** *Let $H^+(\omega)$ be an ergodic family of Schrödinger operators on the nonnegative one dimensional lattice. Then one can prove the absence of absolutely continuous spectrum without refering to the existence of polynomially bounded generalized eigenfunctions if one assumes the extra integrability condition $\mathbb{E}\{|V(0)|\} < \infty$.*
*Let us denote by $p_\lambda(n)$ and $q_\lambda(n)$ the two solutions of the equation $H\psi = \lambda\psi$ with initial conditions $p(0) = q(-1) = 1$ and $p(-1) = q(0) = 0$, and by $G^+$ and $\sigma^+$ the Green function and the spectral measure of the operator $H^+$.*

1. *Let $\lambda$ be a complex number with $\Im m\,\lambda > 0$. Prove that:*

$$G^+(\lambda, 0, 0) = \int \frac{d\sigma^+(t)}{t-\lambda} = \lim_{n\to+\infty} -\frac{q_\lambda(n)}{p_\lambda(n)} = \sum_{n=0}^{\infty} \frac{1}{p_\lambda(n)p_\lambda(n+1)}$$

2. *Let $\mathcal{I}$ be a Borel subset of $\mathbb{R}$ such that $\gamma(\lambda)$ is strictly positive almost everywhere on $\mathcal{I}$ with respect to the Lebesgue measure. Let us define the set $W$ by:*

$$W = \{(\omega, \lambda)\,;\, \lambda \in \mathrm{hyp}(H^+(\omega)) \cap \mathcal{I}\ ,\ \lim_{n\to\infty} \frac{1}{n}|V(n,\omega)| = 0\}$$

   *Prove that for $\mathbb{P}$-almost all $\omega$ the $\omega$ section $W_\omega$ of $W$ is of full Lebesgue measure in $\mathcal{I}$.*

3. *Under the hypothesis of 2. prove that for $\lambda \in W_\omega$ then*

$$\lim_{\epsilon\to 0} \Im m \left( \sum_{n=0}^{\infty} \frac{1}{p_{\lambda+i\epsilon}(n)p_{\lambda+i\epsilon}(n+1)} \right) = 0$$

   *Hint : observe that for a real $\lambda$ then $|p_\lambda(n)| \le |p_{\lambda+i\epsilon}(n)|$.*
   *Moreover for $\Im m\,\lambda > 0$ one has:*

$$\frac{1}{p_\lambda(n)p_\lambda(n+1)} + \frac{1}{p_\lambda(n)p_\lambda(n-1)} = (V(n)-\lambda)\frac{1}{p_\lambda(n+1)p_\lambda(n-1)}$$

*Hence for a real $\lambda$ such that $p_\lambda(n) = 0$ one can replace the terms containing $p_\lambda(n)$ by the right hand side of the above inequality and conclude by means of the Lebesgue theorem.*

4. *Using the de la Vallée Poussin's theorem, prove that under the hypothesis of 2., the operator $H^+(\omega)$ has $\mathbb{P}$ almost surely no absolutely continuous spectrum on $\mathcal{I}$.*

**Problem VII.6.2** *For each integer $n \in \mathbb{Z}$ and $m \geq 1$ we define the set:*

$$F_{m,n} = \{(f_j)_{j\in\mathbb{Z}} \in \ell^1(L^2([0,1])); \inf_{(\xi_j)_j \in [-m,+m]^{\mathbb{Z}}} \|f_n - \sum_{j\neq n} \xi_j f_j\| > 0\}$$

1. *Prove that $F_{m,n}$ is a dense open set in $\ell^1(L^2([0,1]))$.*
2. *Check that $F = \bigcap_m \bigcap_n F_{m,n}$ is a dense $G_\delta$ set in $\ell^1(L^2([0,1]))$ and that, if $f = (f_j)_j \in F$ and $\sum_n \xi_n f_n = 0$ for some bounded sequence $(\xi_j)_j$, then the sequence $(\xi_j)_j$ is identically zero.*
3. *Prove that any random potential $V$ of the form:*

$$V(x,\omega) = \sum_{n\in\mathbb{Z}} \xi_n(\omega) f_n(x)$$

*with $f = (f_j)_j \in F$ and $\{\xi_n;\, n \in \mathbb{Z}\}$ is any bounded random sequence, is a deterministic potential.*

**Problem VII.6.3** *Let $\{\xi_n;\, n \in \mathbb{Z}\}$ be an i.i.d. Bernouilli sequence and let $f(n) = 3^{-\alpha(n)}$ where $\alpha$ is the bijection from $\mathbb{Z}$ onto $\mathbb{Z}_+$ defined by $\alpha(0) = 0$, $\alpha(n) = 2n - 1$ if $n \geq 1$ and $\alpha(n) = -2n$ if $n < -1$. Let us then set:*

$$V(n,\omega) = \sum_{m\in\mathbb{Z}} \xi_m(\omega) f(m-n).$$

*Show that the random potential so defined is stationary, ergodic and deterministic for the value of $\xi_m(\omega)$ can be read off the ternary expansion of $V(n,\omega)$ for any $n$.*

**Problem VII.6.4** *Let us consider the free case $H(\omega) = H_0$ on a discrete strip. Taking derivatives in the Thouless formula for the strip prove that:*

$$\Re e G(\lambda + i0, 0, 0) = \frac{d\tilde{\gamma}(\lambda)}{d\lambda}$$

*where the function* $\tilde{\gamma}(\lambda)$ *is defined by:*

$$\tilde{\gamma}(\lambda) = \frac{1}{2\pi}\int_0^{2\pi} \gamma_1(\lambda + 2\cos p)\, dp$$

*where* $\gamma_1(\lambda) = 0$ *for* $|\lambda| \leq 2$ *and* $\gamma_1(\lambda) = \cosh^{-1}(\lambda/2)$ *for* $|\lambda| > 2$ *is the free one dimensional Lyapunov exponent. See Problem V.5.3. Conclude that* $d\tilde{\gamma}/d\lambda$ *does not vanish on* $[-4, +4] = \Sigma_{ac}$ *and that Lemma VII.3.11 cannot be true in higher dimension.*

**Problem VII.6.5** *For each* $\lambda \in \mathbb{R}$ *let us denote by* $\mathcal{V}_{per}(\lambda)$ *the set of continuous periodic functions* $t \hookrightarrow v(t)$ *such that* $\lambda$ *is in a gap of the periodic Schrödinger operator* $H = H_o + v$. *Follow the lines of the proof of Lemma VII.3.11 to show that if* $\lambda \in \mathbb{R}$ *is such that* $\mathbb{P}\{\mathcal{V}_{per}(\lambda)\} > 0$ *then one has* $\lambda \in \Sigma_{ac}$.

**Problem VII.6.6** *Use the proof of Proposition VII.3.14 to prove that* $\Sigma_{ac}(H(\omega)) = \emptyset$ *almost surely if* $H(\omega) = H_0 + V(\cdot, \omega)$ *and* $V(\cdot, \omega) = \sum_{n\in\mathbb{Z}} f(\cdot - x_n(\omega))$ *where* $x_n(\omega) = n + \zeta_n(\omega)$ *for some i.i.d. sequence* $\{\zeta_n;\, n \in \mathbb{Z}\}$ *of normal random variables and where* $f$ *satisfies the same assumption as in Proposition VII.3.14, namely* $f$ *is a nonnegative non-identically zero continuous function satisfying:*

$$f(t) \leq c(1 + |t|)^{-2-\epsilon}$$

*for some* $\epsilon > 0$ *and all* $t \in \mathbb{R}$.

**Problem VII.6.7** *We present in this problem the elements of the Floquet theory of periodic Jacobi matrices. Let* $\{a_n;\, n \in \mathbb{Z}\}$ *and* $\{b_n;\, n \in \mathbb{Z}\}$ *be doubly infinite periodic sequences with period* $\tau$ *of real numbers and let us also assume that* $a_n > 0$ *for all* $n \in \mathbb{Z}$. *For each complex number* $\lambda \in \mathbb{C}$ *let* $y^{(1)}(n, \lambda)$ *and* $y^{(2)}(n, \lambda)$ *be the solutions of the eigenvalue equation:*

$$a_n\, y(n+1) + a_{n-1}\, y(n-1) + b_n\, y(n) = \lambda\, y(n)$$

*satisfying* $y^{(1)}(0, \lambda) = y^{(2)}(1, \lambda) = 1$ *and* $y^{(1)}(1, \lambda) = y^{(2)}(0, \lambda) = 0$.

1. *Show that if a solution of this eigenvalue equation, say* $y$, *is periodic with period* $\tau$ *then* $[y(1), \cdots, y(\tau)]^t$ *is an eigenvector of the* $\tau \times \tau$

*symmetric matrix* $A_\tau^+$ *defined by:*

$$[A_\tau^+]_{i,j} = \begin{cases} b_i & \text{if} \quad i = j \\ a_j & \text{if} \quad i = j-1 \geq 1 \\ a_i & \text{if} \quad j = i+1 \leq \tau \\ a_N & \text{if} \quad (i = \tau, j = 1) \text{ or } (i = 1, j = \tau) \\ 0 & \text{otherwise.} \end{cases}$$

*We denote by* $\lambda_0 < \cdots < \lambda_{\tau-1}$ *the eigenvalues of the matrix* $A_\tau^+$. *Check the trace identities:*

$$\sum_{j=1}^{\tau} b_j = \sum_{i=0}^{\tau-1} \lambda_j$$

*and:*

$$\sum_{j=1}^{\tau} b_j^2 + 2\sum_{j=1}^{\tau} a_j^2 = \sum_{i=0}^{\tau-1} \lambda_j^2.$$

2. *Show that if a solution* $y$ *of the eigenvalue equation is antiperiodic with antiperiod* $\tau$, *that is if* $y(n+\tau) = -y(n)$ *for all integers* $n$, *then* $[y(1), \cdots, y(\tau)]^t$ *is an eigenvector of the* $\tau \times \tau$ *symmetric matrix* $A_\tau^-$ *defined by:*

$$[A_\tau^-]_{i,j} = \begin{cases} b_i & \text{if} \quad i = j \\ a_j & \text{if} \quad i = j-1 \geq 1 \\ a_i & \text{if} \quad j = i+1 \leq \tau \\ -a_N & \text{if} \quad (i = \tau, j = 1) \text{ or } (i = 1, j = \tau) \\ 0 & \text{otherwise.} \end{cases}$$

*We denote by* $\mu_0 < \cdots < \mu_{\tau-1}$ *the eigenvalues of the matrix* $A_\tau^-$.

3. *Show that the formula:*

$$y^{(1)}(k,\lambda)y^{(2)}(k+1,\lambda) - y^{(1)}(k+1,\lambda)y^{(2)}(k,\lambda) = \frac{a_0}{a_k}$$

*holds for every* $k$ *by computing the determinant of the transfer matrix* $U_\lambda(k+1,1)$. *Recall that* $U_\lambda(k+1,1) = U_\lambda(k+1,k)\cdots U_\lambda(2,1)$ *with:*

$$U_\lambda(j+1,j) = \begin{bmatrix} (\lambda - b_j)/a_j & -a_{j-1}/a_j \\ 1 & 0 \end{bmatrix}.$$

4. *Show that* $d(\lambda) = \text{trace } [U_\lambda(\tau,1)]$ *is a polynomial of degree* $\tau$, *that* $d(\lambda) = y^{(1)}(\tau,\lambda) + y^{(2)}(\tau+1,\lambda)$ *and that the* $\lambda_j$*'s are the roots of the equation* $d(\lambda) = 2$ *while the* $\mu_j$*'s are the solutions of the equation* $d(\lambda) = -2$.

5. *Show that* $y^{(1)}(\tau+1,\lambda)$ *is a polynomial of degree* $\tau-1$ *in* $\lambda$ *and that its roots* $\tau_1,\cdots,\tau_{\tau-1}$ *are the eigenvalues of the restriction of the doubly infinite Jacobi matrix to the interval* $[2,\tau]$, *i.e. of the matrix* $A^0_{\tau-1}$ *defined by:*

$$[A^0_\tau]_{i,j} = \begin{cases} b_i & \text{if} \quad i=j \\ a_j & \text{if} \quad i=j-1\geq 1 \\ a_i & \text{if} \quad j=i+1\leq \tau \\ 0 & \text{otherwise} \end{cases}$$

6. *Show that the polynomial* $P(\lambda)=\prod_{j=1}^{\tau-1}(\lambda-\tau_j)$ *satisfies:*

$$P(\lambda) = -\frac{a_\tau\cdots a_1}{a_\tau}\,y^{(1)}(\tau+1,\lambda),$$

*and that:*

$$d(\tau_j) = y^{(1)}(\tau,\lambda) + \frac{1}{y^{(1)}(\tau,\lambda)}$$

*for all* $j<\tau$ *which implies that* $|d(\tau_j)|\geq 2$ *and that the quantities* $\delta_\pm=(d(\tau_j)\pm\sqrt{d(\tau_j)^2-4})/2$ *are real. In other words the* $\tau_j$*'s are in the gaps of the spectrum.*

7. *Define the so-called norming constants* $\ell_j$ *by:*

$$\ell_j = \left(\sum_{k=2}^{\tau}|y^{(1)}(k,\tau_j)|^2\right)^{1/2},$$

*prove the identity:*

$$\ell_j^2 = a_\tau\,y^{(1)}(\tau,\tau_j)\,\frac{d}{d\lambda}y^{(1)}(\tau+1,\tau_j),$$

*and conclude that, because of:*

$$\ell_j^2 = -(a_1\cdots a_\tau)\frac{d(\tau_j)\pm\sqrt{d(\tau_j)^2-4}}{2}\prod_{k\neq j}(\tau_j-\tau_k),$$

*the knowledge of the product* $a_1\cdots a_\tau$, *of the* $\lambda_i$*'s and of the* $\tau_j$*'s determines the norming constants up to the ambiguity remaining in the choice of* $\pm$.

**Problem VII.6.8** *We use the results of the preceding problem to present the various steps of the reconstruction procedure. We set:*

$$\varphi(k,\lambda) = -\frac{a_1}{a_\tau}\, y^{(1)}(k,\lambda) \quad \text{and} \quad r_j = \left(\sum_{k=1}^{\tau} |\varphi(k,\tau_j)|^2\right)^{-1/2}$$

1. *Check that:*

$$r_j^2 = \frac{a_\tau}{a_1^2}\frac{y^{(2)}(\tau+1,\tau_j)}{(d/d\lambda)y^{(1)}(\tau+1,\tau_j)}$$

   *and that* $\{r_j\varphi(\cdot,\tau_j);\ j-1,\cdots,\tau-1\}$ *is a set of normalized eigenvectors of the symmetric matrix* $A^0_{\tau-1}$.
2. *Show that* $\sum_{j=1}^{\tau-1} r_j^2\varphi(k,\tau_j)^2 = 1$ *for* $k=2,3,\cdots,\tau$ *and consequently that* $\sum_{j=1}^{\tau-1} r_j^2 = 1$.
3. *Prove the formulas:*

$$\begin{aligned}
a_1^2 &= -(a_1\cdots a_\tau)\sum_{j=1}^{\tau-1}\frac{\delta(\tau_j)}{P'(\tau_j)} \\
b_1 &= \lambda_1+\cdots+\lambda_\tau-(\tau_1+\cdots+\tau_{\tau-1}) \\
a_\tau^2 &= \frac{1}{2}\left(\sum_{i=1}^{\tau}\lambda_i^2-\sum_{j=1}^{\tau-1}\tau_j^2-b_1^2-2a_1^2\right) \\
a_k &= \sum_{j=1}^{\tau-1}\tau_j r_j^2\varphi(k,\tau_j)\varphi(k+1,\tau_j) \qquad k=2,\cdots,\tau-1 \\
b_k &= \sum_{j=1}^{\tau-1}\tau_j r_j^2\varphi(k,\tau_j)^2 \qquad k=2,\cdots,\tau
\end{aligned}$$

4. *Show that the reconstruction of the sequences* $\{a_n;\ n\in\mathbb{Z}\}$ *and* $\{b_n;\ n\in\mathbb{Z}\}$ *from the spectral data can be done along the following lines. If one knows the product* $(a_1\cdots a_\tau)$*, the* $\lambda_i$*'s and the* $\tau_j$*'s, then one can compute successively, the polynomials* $d(\lambda)$ *and* $P(\lambda)$*, the norming constants* $\ell_j$*'s,* $a_1$*,* $b_1$ *and* $a_\tau$ *and consequently the* $r_j$*'s, and finally by an induction argument the remaining* $a_k$*'s and* $b_k$*'s using the above recursive formulas.*

**Problem VII.6.9** *Let us consider the general Sturm-Liouville operator:*

$$H = -\frac{1}{r(t)}\frac{d}{dt}\left(\frac{1}{p(t)}\frac{d}{dt}\right)+\frac{q(t)}{r(t)}$$

*where $p$, $q$ and $r$ are continuous functions such that $p > 0$ and $r > 0$. We define the function $u$ by $u(t) = \int_0^t p(s)\,ds$ and the function $v$ as the inverse of $u$.*
*1. Prove that the function $w(t) = y(v(t))$ is a solution of the equation:*

$$-\ddot{w}(t) + Q(t)w(t) = 0$$

*where $Q(t) = [q(t) - \lambda r(t)]/p(t)$, whenever $y$ is an eigenfunction of the operator $H$ for the eigenvalue $\lambda$.*
*2. If $\lambda > \inf_t q(t)/r(t)$ and if the function $\lambda \hookrightarrow [q(t) - \lambda r(t)]/p(t)$ is not constant, prove the existence of $t_1, t_2 \in \mathbb{R}$ and $T_1, T_2 > 0$ such that $0$ is not in the essential spectrum of the periodic Schrödinger operator associated to the $(T_1 + T_2)$-periodic potential $V$ defined by:*

$$V(t) = \begin{cases} \frac{q(t_1)-\lambda r(t_1)}{p(t_1)}, & 0 \le t \le T_1 \\ \frac{q(t_2)-\lambda r(t_2)}{p(t_2)}, & T_1 < t \le T_1 + T_2. \end{cases}$$

**Problem VII.6.10** *Let $\{V(t);\, t \in \mathbb{R}\}$ be a stationary ergodic process. We assume that the topological support of the distribution of $V(0)$ is a closed bounded subset of $\mathbb{R}$, say $S$, and we prove the fact (used in the proof of Proposition VII.5.1 in the lattice case) that the maps $\Omega_+^{(d)} \ni V_+ \hookrightarrow V_- \in \Omega_-^{(d)}$ and $\Omega_-^{(d)} \ni V^- \hookrightarrow V^+ \in \Omega_+^{(d)}$ are continuous.*
*1. For any Borel set $B \subset \mathbb{R}$ with positive Lebesgue measure we set: $\Omega(B) = \{V \in \Omega^{(c)};\, m_+(\lambda + i0, V) = \overline{-m_-(\lambda + i0, V)}\ \ a.e. \lambda \in B\}$. Use Lemma III.1.17 to show that $\Omega(B)$ is closed for the weak convergence in $V$.*
*2. For each number $0 < a < \infty$ we set*

$$\Omega_a \text{ ( resp. } \Omega_{\pm,a}) = \{V \in \Omega^{(c)};\, |V(t)| \le a, \text{ a.e.} t \in \mathbb{R} \text{ ( resp. } \mathbb{R}_\pm)\}.$$

*Prove that $\Omega_a$ and $\Omega_{\pm,a}$ are compact metrizable subsets of $\Omega^{(c)}$. Remember that $\Omega^{(c)}$ is endowed with the weak Hilbert space topology.*
*3. Let us set $\Omega_a(B) = \Omega_a \cap \Omega(B)$ and $\Omega_a(B)_\pm = \{V_\pm;\, V \in \Omega_a(B)\}$. Prove that the sets $\Omega_a(B)_\pm$ are also compact.*
*4. Let us assume furthermore that the set $A = \{\lambda \in \mathbb{R};\, \gamma(\lambda) = 0\}$ has positive Lebesgue measure. In this case, we know that $V_-$ determines $V$ and hence $V_+$. Prove that the map $V_- \hookrightarrow V_+$ is continuous.*
*5. Prove the same result in the lattice case used in Proposition VII.5.1.*

**Problem VII.6.11** *Let $H = H_0 + \beta V$ be an ergodic random Schrödinger operator on the lattice $\mathbb{Z}$ (i.e. $V(n,\omega) = v(\theta^n\omega)$ for some real valued measurable function $v$) with a coupling constant $\beta > 0$.*
*1. Prove that:*

$$N(\lambda + \frac{1}{2}) - N(\lambda - \frac{1}{2}) \leq 6 \int_{\lambda-1/2}^{\lambda+1/2} \mathbb{E}\{\frac{1}{\sqrt{(\beta v - \lambda')^2 + 1}}\} \, d\lambda'.$$

*(Hint:use the formula:*

$$\epsilon \int \frac{dN(\lambda')}{(\lambda' - \lambda)^2 + \epsilon^2} = \Im m \, \mathbb{E}\{G(\lambda - i\epsilon, 0, 0)\}$$

*use the resolvent identity to expand the Green function by treating the free part as a perturbation of the potential and use finally the obvious bound:*

$$N(\lambda + \frac{1}{2}) - N(\lambda - \frac{1}{2}) \leq 2 \int \int_{\lambda-1/2}^{\lambda+1/2} \frac{1}{(\lambda" - \lambda')^2 + 1} d\lambda' \, dN(\lambda'').$$

*2. Prove that, if $\beta$ is large enough we have the estimate:*

$$|\{\lambda' \in [\lambda - 1/2, \lambda + 1/2]; \gamma(\lambda') = 0\}| \leq \exp[-\frac{1}{g(\lambda - 1/2) + g(\lambda + 1/2)}]$$

*where $g(\lambda)$ denotes the right hand side of the upper bound proved in the part 1.*

## VII.7 Notes and Complements

The introduction of the $w$-function in the investigation of the absolutely continuous spectrum is due to Johnson and Moser, see [157]. They made explicit in the case of almost periodic potentials the use of some averages which were more or less implicit in the periodic theory. See [246] and [228]. The properties we give in the first section are technical results which belong in a class on complex function theory or classical harmonic analysis. We included them for the sake of easy references and because they are relatively hard to find in a form appropriate to our needs.

The spectral theory of periodic differential (or partial differential) operators is part of the folklore. Eastham's book [95] is an example of a treatise based on the classical Floquet theory of periodic differential equations. Our presentation emphasized the properties of the propagator and the so-called exponential dichotomies of the solutions of the eigenvalue equation. The reader is refered to the Section 16 of Chapter XIII of the fourth volume [272] of the Reed and Simon series for a derivation of these results from the more modern point of view of direct integral decompositions. In the same way we proved the corollary following Theorem III.4.1, one can use the uniform stability result of Proposition VII.2.2 to show the following curiosity: if $V_1$ is a periodic potential and if $\cup_n[a_n, b_n]$ is the spectrum of the corresponding periodic Schrödinger operator and if $\tilde{V} = V_1\mathbf{1}_{[0,\infty)} + V_2\mathbf{1}_{(-\infty,0)}$ for any potential function $V_2$ which is in the limit point case at $-\infty$, then the spectrum of the operator $H = H_0 + \tilde{V}$ contains $\cup_n[a_n, b_n]$and is purely absolutely continuous there. This resul was first proved by Carmona in [43]. It was recently reproven in [67] by means of a fine analysis of the properties of the $m$-function.

The inverse spectral theory for periodic potentials is one of the very richest fields of mathematics. It involves very deep results of an extraordinary beauty. Unfortunately we did not have the time, the space (and the competence) to give a full account. We merely quoted the results we need in our discussion of the random case. The interested reader is invited to look at the fundamental papers of Mc Kean and van Moerbeke [235], Mc Kean and Trubowitz [236]. Moser [246], Levitan [218], [219] and [217], Marchenko and Ostrovsky [228] and the recent book of Pöschel and Trubowitz [269]. The lattice case is treated by Kac and van Moerbeke in [163] and [164].

The spectral theory of almost periodic Schrödinger operator has been the object of intense interest and research in the late seventies and early eighties. In a nice introductory review article [293] Simon compares this wave of interest to the flu! The subject is still very active and another book would be necessary to give a fair account of the field. This was partly done in the book of Cycon et al [64] and should be done in the forthcoming book of Figotin and Pastur [101]. We tried to introduce the reader to the basic facts of almost periodic potentials because they form a very rich class of random potentials. The early works of Dinaburg and Sinai [83], Moser [246], Johnson and Moser [157], Sarnack [284] and Bellissard and Simon [21] discussed the fact that the spectrum should be some sort of a Cantor set. Johnson and Moser introduced in [157] the $w$-function. The latter is sometimes called the Floquet exponent for it plays a crucial role in the (possible) extension of the Floquet theory to non-periodic potentials. The beauty of the matter is that the boundary values of this Herglotz function give the Lyapunov exponent and the integrated density of states of the operator. The integrated density of states (which is often called the rotation number in the literature on almost periodic operators) was shown to play an important role in the paper of Johnson and Moser and was thoroughly studied by Avron and Simon in [13] and by Delyon and Souillard in [77] and [78]. The existence of a Floquet type representation for a basis of solutions of the eigenvalue equation was the main achievement of Dinaburg and Sinai. They had to control a very involved perturbation argument to take care of the small divisor problem. This work was extended by Rüssmann in [283] and Moser and Pöschel in [247].

The discussion of the existence of a Floquet representation for solutions of the eigenvalue equation for energy in the absolutely continuous spectrum was discussed by soft analysis arguments by Deift and Simon in [66]. We shall give more references to works on almost periodic potential in the Notes to Chapter VIII.

In the continuous case, the operator $H = H_0 + \beta V$ was considered by Surace in [313] when $V$ is the sum of two cosine functions. Using a duality argument and an analysis à la Fröhlich, Spencer, Wittwer (which we discuss in the Notes and Complements of the next chapter), he proves that there is no point spectrum for $|\beta|$ small enough. This is a particular case of one of the results of Sarnak who gives in [284] a complete analysis when $V$ is a finite sum of (possibly complex) exponentials.

Proposition VII.3.1 is known as the Ishii-Pastur result because the argument of the proof is due to Pastur, see [261] and [263], but a very similar result was previously proven by K. Ishii in [155] and [156]. Note also that partial results in the same direction were obtained earlier for example by Casher and Lebowitz in [52] and Yoshioka in [332].

The use of the ideas and the results of the theory of products of random matrices in the study of random differential equations goes back to the early seventies, and especially to the work of Borland [78]. One had to wait almost a decade and the far-reaching paper of Kotani [190] for the converse of the Ishii-Pastur result. Kotani's contribution to the understanding of the spectral properties of one-dimensional random Schrödinger operators is remarkable. Among many other things, he gave several very useful mathematical forms of the simple following fact: the more randomnes in the potential, the less absolute continuous spectrum we may expect.

Proposition VII.1.11 was first proved by Kotani but the proof we give is essentially due to Craig and Simon. See [62]. Corollary VII.3.7 says that one cannot expect a nontrivial absolutely continuous component in the spectrum if the random potential is not deterministic. This result has been extended to the strip by Kotani and Simon in [201]. Kotani's theory has been extended to the lattice case by Simon in [298]. See also [213]. Our discussion follows closely Simon's work. More recently it has also been extended by Minami to general random Sturm-Louville operators, the framework of which includes both the lattice and the continuous cases. See [239].

The estimates on the derivative of the integrated density of states on the set where the Lyapunov exponent vanishes can be found in the paper by Deift and Simon [66]. Another estimate of the size of the set where the Lyapunov exponent vanishes is given in Problem VII.6.11. This result is due to Spencer. See [311]. Deift and Simon also discuss in [66] the form of the generalized eigenfunctions corresponding to energies where the Lyapunov exponent vanishes. They prove that the multiplicity of the absolutely continuous spectrum is 2. The subsection on deterministic potentials having no absolutely continuous spectrum is patterned after the work of Kirsch Kotani and Simon [175].

The fact that the absolutely continuous spectrum is equal to the set where the Lyapunov exponent vanishes is only true up to sets of Lebesgue measure zero. This is important to keep in mind for Molchanov and Chulaevsky constructed in [243] a model for which the spectrum is a Cantor set on which the Lyapunov exponent vanishes identically even though there is no absolutely continuous spectrum since the spectrum is pure point (with eigenfunctions decaying faster than any polynomial)!!!

The inverse spectral theory for random Schrödinger operators should be called Kotani's theory for he started the subject (with the small but far reaching note [194]) and participated in most of the subsequent developements, on his own or with collaborators. See for example [195], [51], [196], [197], and [199]. Our discussion of the continuous case follows his original work [194] and [195] and the subsequent work with Krishna [199]. Our presentation of the lattice case follows his work with Carmona [51]. The trace formula (VII.54) has been generalized by Craig in [61] to a large class of almost periodic potentials and some special ergodic random potentials with purely absolutely continuous spectrum. The crucial step is the appropriate definition of the quantities $\lambda_j$, $\mu_j$ and $\delta_j$'s, the determination of their evolution under the shift and the solution of the corresponding system.

Some of the most intriguing results of the inverse spectral theory for periodic potentials come certainly from works on the Korteweg de Vries equation. If the function $(t,x) \hookrightarrow V(t,x)$ is a solution of this nonlinear equation which satisfies $V(0,x) = V(x)$ where $x \hookrightarrow V(x)$ is a given periodic function, then for each fixed $t$, the function $x \hookrightarrow V(t,x)$ is still periodic with the same period and the periodic Schrödinger operator $H(t) = H_0 + V(t, \cdot)$ has the same spectrum as $H = H_0 + V$. The spectral characteristics of a periodic potential are preserved under the Korteweg de Vries flow. See for example [254]. This flow plays a crucial role, whether it is emphasized or not, in the inverse spectral theory of periodic potentials. This theory has an analogue in the lattice case. It is the Toda lattice. See for example the original of Toda [321], the works of Kac and van Moerbeke [163], [164] and [108], and the very recent account by Gesztesy, Holden, Simon and Zhao [119]. Nothing much has been worked out in the random case. Neither in the continuous case, nor in the lattice one. It seems that this should be a very interesting field of (open) investigation!

The discussion of the random potentials taking finitely many values is motivated by the result of Proposition VII.5.1 which is due to Kotani. See [198]. The work of Delyon and Petritis was motivated by an earlier work of Sütö. See [314]. Problems VII.6.2 and VII.6.3 are taken from [175]. Problem VII.6.10 is due to Kotani (private communication).

The last remark which is the object of the last subsection of the chapter was included for the sake of completeness. The results of Chapter IX will imply the absence of absolutely continuous spectrum for large disorder or low energy. One expects the existence of a nontrivial absolute continuous component in other regimes but nothing has been proved rigorously so far. This problem seems to be very hard and we are afraid that we could not shed much light on it. In any case, the result we presented is due to Martinelli and Scoppola. See [231]. The main interest of this paper was to show that there was much more than absence of diffusion in the fundamental paper of Fröhlich and Spencer. Martinelli and Scoppola's argument was streamlined by Simon and Wolff in [304]. The problem of the existence of absolutely continuous spectrum in high dimension is difficult because of the stationarity of the potential and the fact that it does not go to zero. Indeed, if the potential is assumed to go to zero fast enough, then scattering theory techniques can be used to prove the existence of the wave operators and consequently of extended states and of an absolutely continuous component in the spectrum. See for example [203].

There is another model of multidimensional disordered systems for which the absolutely continuous spectrum has been extensively investigated in the physics literature. It is the so-called Bethe lattice (called Caley tree by mathematicians). This model is very popular in statistical mechanics for explicit computations can be performed. It is generally regarded as an infinite dimensional lattice. After the fundamental work of the theoretical physicists Abou-Chacra et al, see [1] and [2], a set of very interesting rigorous results was announced by Kunz and Souillard in [205] but unfortunately, no proof ever appeared in print.

# VIII

# LOCALIZATION IN ONE DIMENSION

Contents :



It was first claimed by P.W.Anderson (1958) in [8] for the multidimensional random Schrödinger operator on a lattice, (associated with i.i.d. potentials) that the spectrum ought to be pure point with exponentially decaying eigenfunctions for a "typical sample" and for "large disorder". It was later conjectured by Mott & Twose in [249] that this property should hold in the one dimensional case at any disorder. This chapter is devoted to the proof of this last conjecture which we will extend to quasi-one dimensional systems.

We first assume some regularity properties on the distribution of the potential process. In the i.i.d case we assume the existence of a nontrivial absolutely continuous component of the one-site potential distribution. In the general ergodic case this assumption will merely be made for the conditional laws. We give three different proofs. First, using the so called "Kotani's trick", one gives a rigorous treatement of an earlier claim of Borland in [30]. Secondly one obtains the same results, using the functional analysis of rank-one perturbations of selfadjoint operators. One of the advantages of this method is that it extends to the multidimensional model with only minor changes. The third proof is an extension of the older approach initiated by the "Russian school". It relies on the spectral theory of the Laplace transform on homogeneous spaces.

In the first case the result is obtained without any reference to limiting procedures along "boxes". It only uses a refinement of the usual Fubini's theorem. This proof is refered as "pointwise theory" (first section). We choose to write this proof directly for the strip because there is absolutely no difference in the treatment of the strip and the one dimensional case. The reader only interested by the line has only to replace "lowest nonnegative Lyapunov exponent" by "Lyapunov exponent" and the Lagrangian boundary $L_\ell$ by the classical projective line. The same remarks hold for the second proof.

On the other hand, the proof by means of Laplace transforms, exposed in the third section, is a bit more involved in the case of the strip. As a consequence we give a separate proof for the one dimensional model. This method requires certainly more soft analysis than the first one, but since it is based on the approximation of the Schrödinger operator by its restrictions to boxes, one obtains as byproducts, some interesting formulas for the density of states. Since the "pointwise proof" yields immediately the decay of the eigenfunctions with a rate given by the Lyapunov exponent, we do not try to re-prove this result in the setting of the second and third sections.

Unfortunately all these previous results do not apply to the simple case of i.i.d. potentials taking only finitely many values. We devote Section IV to this problem. Our solution involves the machinery developed in Chapter IX for the multidimensional systems. This is somewhat embarrassing for it is likely that a simpler approach exists, but up to now, no direct proof has been found.

Let $\{H(\omega)\,;\omega \in \Omega\}$ be an ergodic family of Schrödinger operators in the strip of width $\ell$. This means that there exists an invertible dynamical ergodic system $(\Omega, \mathcal{F}, \theta_t, \mathbb{P})$ and $\ell$ real valued random variables $\{V_i(\omega); i = 1 \ldots \ell\}$ such that $V_i(t,\omega) = V_i(\theta_t(\omega))$ for $t \in T$. When $T = \mathbb{Z}$ this model is to be i.i.d. if $\{V_i(t,\omega)\ ;\ i = 1, \ldots, \ell,\ t \in \mathbb{Z}\}$ is an i.i.d. family of real valued random variables.

We also make the usual integrability assumptions (recall our discussion of the existence of Lyapunov exponents in Chapter IV). More precisely,

- $\sum_{i=1}^{\ell} \mathbb{E}\{\log(|V_i| + 1)\} < +\infty$ in the discrete case.
- the $\mathbb{R}^{\ell}$-valued process $\{V_i(t,w); i = 1 \ldots \ell, t \in T\}$ is separable and $\sum_{i=1}^{\ell} \mathbb{E}\{|V_i(0)|\} < +\infty$ in the continuous case.

Under these assumptions, for a fixed value of $\lambda$, the propagator process $\{U_\lambda(t,\omega);\ t \in T\}$ is a regular invertible multiplicative process with values in $SP(\ell, \mathbb{R})$ in the sense of definition IV.2.5. Moreover the following properties hold for $\mathbb{P}$-almost all $\omega$ and for any $i = 1, \ldots, \ell$ :

$$\lim_{n\to\pm\infty} \frac{1}{|n|} \log(1 + |V_i(n,\omega)|) = 0 \quad \text{in the discrete case}$$

$$\lim_{n\to\pm\infty} \frac{1}{|n|} \int_n^{n+1} |V_i(t,\omega)|dt = 0 \quad \text{in the continuous case.}$$

The random Oseledec theorem stated in Proposition IV.2.6 gives the existence of $\ell$ nonnegative Lyapunov exponents $\gamma_1(\lambda) \geq \ldots \geq \gamma_\ell(\lambda)$ and of the corresponding Oseledec spaces ${}^{\pm}\mathcal{V}_\omega^i$.

In the continuous case we will add the condition $\sum_{i=l}^{\ell} |V_i(t,\omega)| \leq C(\omega)(1+t^2)$, where $C(\omega)$ is a $\mathbb{P}$-a.e. finite random variable, in order to make sure that the operator $H(\omega)$ is essentially self-adjoint for $\mathbb{P}$-almost all $\omega \in \Omega$.

In order to avoid excluding all the above negligible sets in most of the proofs, we will assume throughout this chapter that these $\mathbb{P}$-a.e. properties actually hold for all $\omega \in \Omega$.

# VIII.1 Pointwise Theory

We begin by one of the shortest proofs of localization which, unfortunately, only works for the one dimensional half-axis model. Moreover this result only holds for almost all boundary conditions. One does not assume any regularity of the distribution of potentials. This proof due to Kotani [193] essentially uses the method of integration of the spectral measures with respect to a boundary condition presented in Chapter III. It does not require what we will call below Kotani's trick but only a double application of the classical Fubini theorem.

Let $H(\omega)$ be an ergodic Schrödinger operator on the one dimensional lattice and for a real number $x$ let $H_x^+$ be this operator restricted to the "box" $[0\ ,\ +\infty)$ with the boundary conditions $x$ at the point 0 that is :

$$H_x^+\psi(0) = -\psi(1) + (V(1) - x)\psi(0)$$
$$H_x^+\psi(n) = H\psi(n) \qquad n \geq 1$$

**Proposition VIII.1.1** *Let us assume that the Lyapunov exponent $\gamma(\lambda)$ is strictly positive Lebesgue almost everywhere on a Borel subset $\mathcal{I}$ of the real line. Then for Lebesgue-almost all boundary condition $x$, the operator $H_x^+$ is exponentially localized on $\mathcal{I}$ with decay $\gamma(\lambda)$.*

*Proof:*

It has been remarked in Chapter III that for the half axis the spectral measure of the operator $H_x^+$ say $\sigma_x^+$ is given by the sole spectral measure computed at the point 0. Moreover the Stieltjes transform of this measure can be computed by means of the Weyl function $m_+$ and for $\Im m\lambda > 0$ one obtains:

$$\int \frac{d\sigma_x^+(t)}{t-\lambda} = \frac{m_+(\lambda)}{1 - x m_+(\lambda)}.$$

Let $m$ be the Cauchy measure on $\mathbb{R}$ and let us set $\tilde{\sigma}^+ = \int \sigma_x^+ \, dm(x)$. Since the Weyl function $m_+(\lambda)$ has a positive imaginary part, it is readily seen that the function $z \hookrightarrow \dfrac{m_+(\lambda)}{1 - z\, m_+(\lambda)}$ is a bounded holomorphic function on the closed upper half-plane. Hence, integrating in $x$ with respect to $m$ we obtain:

$$\int \frac{d\tilde{\sigma}^+(t)}{t-\lambda} = \frac{m_+(\lambda)}{1 - i m_+(\lambda)}$$

Using again $\Im m\, m_+(\lambda) > 0$ one sees that the imaginary part of the left side of the above equality is bounded on the upper half-plane and it follows from the de la Vallé Poussin theorem that the measure $\tilde{\sigma}^+$ has a bounded density.

It is obvious that the set of hyperbolic values of $\lambda$ for the operator $H_x^+$ does not depend on $x$ and if we set as usual:

$$W = \{(\omega, \lambda) \in \mathcal{I} \times \Omega\, ;\, \lambda \in \text{hyp}(H_x^+(\omega)) \cap \mathcal{I}\},$$

it follows from the hypothesis that $W$ is of full $\mathbb{P} \otimes \tilde{\sigma}^+$ measure. We conclude with two applications of Fubini's theorem together with the criterion of localization proved in Proposition III.4.13. ∎

This result can be extended without difficulty to the continuous case using the following formula due to Kotani which can be proved exactly in the same way as above, using the theory of $m$-functions developed in Chapter III :

$$\int \sigma_x^+(dy)\, dm(x) = dy$$

One can also extend this result to the strip if one defines properly the Weyl functions in this situation (see [201]) but up to now, it has not been possible to take advantage of this result to prove localization in the bilateral case.

## VIII.1.1 Kotani's Trick

We proved in the general theory of ergodic families of Schrödinger operators that if we denote by $\sigma_\omega$ the spectral measure of the self-adjoint operator $H(\omega)$ then the mapping $\omega \hookrightarrow \sigma_\omega$ is measurable in the sense that $\omega \hookrightarrow \sigma_\omega(\mathcal{I})$ is measurable for any set $\mathcal{I} \in \mathcal{B}$ the Borel $\sigma$-algebra of the real line.

**Lemma VIII.1.2** *Let $W$ be an $\mathcal{F} \otimes \mathcal{B}$ measurable subset of $\Omega \times \mathbb{R}$ and let us denote by $W_\omega$ the $\omega$ section of $W$. Then*

**(i)** *The mapping $\omega \hookrightarrow \sigma_\omega(W_\omega)$ is measurable.*
**(ii)** *The set $W^* = \{\omega\, ; \sigma_\omega$ is carried by $\text{hyp}(H(\omega))\}$ is $\mathcal{F}$ measurable and invariant.*

*Proof:*
If we first consider a product set $W = A \times \mathcal{I}$ then $\sigma_\omega(W_\omega) = \mathbf{1}_A(\omega)\sigma_\omega(\mathcal{I})$ and the measurability of this mapping is obvious. The conclusion (i) is then a simple consequence of the monotone class theorem.
If we choose $W = \{(\omega, \lambda)\, ; \lambda \in \text{hyp}(H(\omega))\}$ then one obtains immediately that $W$ is an $\mathcal{F} \otimes \mathcal{B}$ measurable set, since for each $p \in 1, \ldots, \ell$ the function

$(\omega, \lambda) \hookrightarrow \Lambda^p U_\lambda(n, \omega)$ is measurable in $\omega$ and continuous in $\lambda$ hence jointly measurable. It follows from part (i) that $W^* = \{\omega\,; \sigma_\omega(\mathbb{R} \setminus W_\omega) = 0\}$ is a measurable set.
Obviously $\lambda \in \text{hyp}(H(\omega))$ if and only if $\lambda \in \text{hyp}(H_{\theta(\omega)})$. Moreover the spectral measures $\sigma_\omega$ and $\sigma_{\theta(\omega)}$ have the same null sets since the operators $H(\omega)$ and $H_{\theta(\omega)}$ are unitarily equivalent. This yields the invariance of the set $W^*$. ∎

As a consequence of this lemma, if one can prove that the $\mathbb{P}$-measure of $W^*$ is positive, then this probability is necessarily equal to one because of the ergodicity of the dynamical system. Morever one can replace the measure $\sigma_\omega$ by its restriction $\mathbf{1}_\mathcal{I}(\lambda)\sigma_\omega$ to any $\mathcal{B}$ measurable set $\mathcal{I}$ in the above Lemma.

**Definition VIII.1.3** *Let $\mathcal{I} \in \mathcal{B}$ and $\gamma(\lambda)$ be a nonnegative function defined on $\mathcal{I}$. We say that we have exponential localization on $\mathcal{I}$ with decay at least (resp. equal to) $\gamma(\lambda)$ if for $\mathbb{P}$-almost all $\omega$ we have:*

- *The spectral measure $\sigma_\omega$ is pure point on $\mathcal{I}$ and the function $\gamma(\lambda)$ is $\sigma_\omega$-a.e. strictly positive on $\mathcal{I}$.*
- *For any eigenvalue $\lambda$ of $H(\omega)$ belonging to $\mathcal{I}$ (if any) and any associated eigenfunction $\psi_\lambda(t, \omega)$ one has:*

$$\lim_{t \to \pm\infty} \frac{1}{|t|} \log(|\tilde{\psi}_\lambda(t, \omega)|) \leq -\gamma(\lambda) \text{ resp. } = -\gamma(\lambda)$$

Recall that the notation $\tilde{\psi}(t)$ stands for $\begin{pmatrix} \psi(t) \\ \psi(t-1) \end{pmatrix}$ in the discrete case and $\begin{pmatrix} \psi(t) \\ \dot{\psi}(t) \end{pmatrix}$ in the continuous case.

The following proposition can be viewed as a generalization of the Fubini's theorem to random measures. Let $(\Omega, \mathcal{F}, \mathbb{P})$ be a probability space and $\mathcal{B}$ be the Borel $\sigma$-algebra of $\mathbb{R}$. For each measurable mapping $\omega \hookrightarrow \sigma_\omega$ from $\Omega$ into the set $\mathcal{M}_+(\mathbb{R})$ of Borel positive measures on $\mathbb{R}$ and for each sub $\sigma$-algebra $\mathcal{F}^*$ of $\mathcal{F}$ one defines formally the conditional mapping $\sigma^*_\omega$ by the formula:

$$\sigma^*_\omega(A) = \mathbb{E}\{\sigma.(A)|\mathcal{F}^*\}(\omega)$$

In order to make sure that $\sigma^*_\omega$ is still measurable we need to assume the existence of a regular version of the conditional probability. This is certainly true if $\mathcal{F}^*$ is a separable $\sigma$-algebra and this will be the case in the applications below.

**Proposition VIII.1.4** *Let $m$ be a positive measure on $\mathbb{R}$ such that the measures $\sigma^*_\omega$ are absolutely continuous with respect to $m$ for $\mathbb{P}$-almost all $\omega$. Then if $F(\omega,\lambda)$ is a nonnegative $\mathcal{F}^*\otimes\mathcal{B}$ measurable function, one has:*

$$\left(\mathbb{E}\{\int F(\cdot,\lambda)dm(\lambda)\}=0\right)\Rightarrow\left(\int F(\omega,\lambda)d\sigma_\omega(\lambda)=0\quad \mathbb{P}-a.e.\right)$$

*Proof:*
Let $F$ be the product of an $\mathcal{F}^*$-measurable function of $\omega$ and a $\mathcal{B}$ measurable function of $\lambda$. Using the properties of conditional expectations one sees immediately that

$$\mathbb{E}\{\int F(\omega,\lambda)d\sigma_\omega(\lambda)\}=\mathbb{E}\{\int F(\omega,\lambda)d\sigma^*_\omega(\lambda)\}.$$

By a monotone class argument the same relation holds for a general function $F$. If $\mathbb{E}\{\int F(\omega,\lambda)dm(\lambda)\}=0$ then for $\mathbb{P}$-almost all $\omega$ the function $\lambda\hookrightarrow F(\omega,\lambda)$ is $m$-negligible and our hypothesis implies that this function is also $\sigma^*_\omega$ negligible. The left hand side of the above equality vanishes and the conclusion follows from a new application of Fubini's theorem. ∎

A slightly improved form of this proposition will be very useful in the sequel. Let $\Omega_0$ be an $\mathcal{F}$-measurable subset of $\Omega$ then one can define the "$\Omega_0$-restricted conditional random measure" by the formula:

$${}^0\sigma^*_\omega(A)=\mathbb{E}\{\mathbf{1}_{\Omega_0}(\cdot)\sigma_\cdot(A)|\mathcal{F}^*\}(\omega)$$

**Proposition VIII.1.5** *Let $m$ be a nonnegative measure on $\mathbb{R}$ such that the measures ${}^0\sigma^*_\omega$ are absolutely continuous with respect to $m$ for $\mathbb{P}$-almost all $\omega$. Then if $F(\omega,\lambda)$ is a nonnegative $\mathcal{F}^*\otimes\mathcal{B}$ measurable function one has:*

$$\left(\mathbb{E}\{\int F(\omega,\lambda)dm(\lambda)\}=0\right)\Rightarrow\left(\int F(\omega,\lambda)d\sigma_\omega(\lambda)=0\quad \mathbb{P}-a.e.\ on\ \Omega_0\right).$$

*Proof:*
The argument of Proposition VIII.1.4 applies here if we remark that:

$$\mathbb{E}\{\int F(\cdot,\lambda)d^0\sigma^*_\cdot(\lambda)\}=\mathbb{E}\{1_{\Omega_0}(\cdot)\int F(\cdot,\lambda)d\sigma_\cdot(\lambda)\}\qquad ∎$$

We are now in a position to state and prove our version of the Kotani's criterion. Let us denote by $\mathcal{F}^*$ the sub $\sigma$-algebra of $\mathcal{F}$ generated by the random variables $V_i(t,\omega)$ for $(i,t)$ outside some bounded set.

**Theorem VIII.1.6** *Let us denote by $\Omega_0$ an $\mathcal{F}$-measurable subset of $\Omega$ of positive $\mathbb{P}$ measure, by $\mathcal{I}$ a Borel subset of the real line and by $m$ a nonnegative measure on $\mathcal{I}$ such that:*

**(i)** *The Lyapunov exponent $\gamma_\ell(\lambda)$ is strictly positive $m$-a.e. on $\mathcal{I}$.*
**(ii)** *The measures ${}^0\sigma^*_\omega$ are absolutely continuous with respect to $m$ on $\mathcal{I}$, for $\mathbb{P}$ almost all $\omega \in \Omega$.*

*Then one has exponential localization on $\mathcal{I}$ with decay at least $\gamma_\ell(\lambda)$ ( and equal to $\gamma_1(\lambda) = \gamma(\lambda)$ in the one dimensional case).*

*Proof:*
Let us define the set $W$ in $\Omega \times \mathbb{R}$ by:

$$W = \{(\omega,\lambda)\ : \lambda \in \text{hyp}(H(\omega)) \cap \mathcal{I}\}$$

We have already seen in Lemma VIII.1.2 that $W$ is $\mathcal{F} \otimes \mathcal{B}$ measurable. Fubini's theorem implies that $W$ is of $\mathbb{P} \otimes m$ full measure. Moreover it is $\mathcal{F}^* \otimes \mathcal{B}$ measurable since its definition is independent of the values of the potentials in a bounded set of sites. Proposition VIII.1.4 implies that the measure $\sigma_\omega$ restricted to the set $\mathcal{I}$ is carried by hyp$(H(\omega))$ for $\mathbb{P}$-almost all $\omega$ in $\Omega_0$. It follows from Lemma VIII.1.2 and the ergodicity that this property holds $\mathbb{P}$-a.e. The result now follows from the criterion proved in Proposition III.4.13. ∎

It is clear that the hypothesis (i) is not altered by the "perturbation" of the values of potentials in a bounded set of sites. We will see below that one can force (ii) by such "smooth" perturbations. Such perturbations, which have no effect on the existence and values of the Lyapunov exponents change drastically the behavior of the contractive Oseledec subspaces.

**Remark VIII.1.7** • Let us assume the hypothesis (i) of the above theorem. Using the conclusion (iii) of Proposition IV.2.7 one sees that for a fixed $\lambda$, for $\mathbb{P}$-almost all $\omega$ the two $\ell$-dimensional contractive Oseledec subspaces ${}^{\pm}\mathcal{V}^\ell_\omega(\lambda)$ have only the vector 0 in common. On the other hand the above theorem implies that for $\mathbb{P}$-almost all $\omega$ then for $\sigma_\omega$ all $\lambda$ these subspaces have a nontrivial intersection !!! This inversion with respect to $\lambda$ and $\omega$ of the almost everywhere properties of the Oseledec subspaces actually contains all the "mystery" of the exponential localization for one and quasi-one dimensional systems.

- If we compare this theorem to the Ishii-Pastur result we see that the price one has to pay to go from the singularity of the spectrum to the pure point spectrum property is contained in the hypothesis (ii). In the case of a deterministic process $\mathcal{F}^* = \mathcal{F}$ and ${}^0\sigma^*_\omega$ is the restriction of $\sigma_\omega$ to the set $\Omega_0$. In this situation one cannot expect to check hypothesis (ii) since using the Ishii-Pastur theorem, hypothesis (i) implies that for any fixed measure $m$ then $\sigma_\omega$ is orthogonal to $m$ for $\mathbb{P}$-almost all $\omega$. Thus if one wants to check hypothesis (ii) some randomness is unavoidable.

Let $\mathcal{F}^*_N$ be the $\sigma$-algebra generated by the potentials $V_i(t,\omega)$ fot $|t| \geq N$ and $\mathcal{F}^*_\infty = \cap_{N=0}^{\infty}\mathcal{F}^*_N$. If the $\sigma$-algebra $\mathcal{F}^*_\infty$ is trivial (this happens for example in the independent case) it would be very tempting to choose $\mathcal{F}^* = \mathcal{F}^*_\infty$. With this choice, hypothesis (ii) is automatically satisfied since conditional expectations will reduce to ordinary expectations. Unfortunately although the set $W$ is measurable with respect to the $\sigma$-algebras $\mathcal{F}^*_N \otimes \mathcal{B}$ for any $N$ the measurability of this set with respect to $\mathcal{F}^*_\infty \otimes \mathcal{B}$ is far from obvious. This measurability property (with respect to this $\sigma$-algebra or its completion) is certainly the key property in order to prove localization. Such a result would need a careful study of the dependence upon the energy $\lambda$ in the convergence to the limit in the ergodic theorem....See [330] for a general discussion on this problem of measurability.

It is known that the positivity alone of the Lyapunov exponent $\gamma_\ell(\lambda)$ is not sufficient for localization in the general ergodic case. A counter-example is given by the almost-Mathieu operator. We have seen that such an operator has a positive Lyapunov exponent as soon as the modulus of the coupling constant is greater than two, hence the spectrum is singular. On the other hand it has been proved that such operators have no eigenvalues if the rotation is given by a Liouville number (see [14]) and thus the spectrum is singular continuous.

The careful reader will also notice that the preceding theorem has nothing to do with ergodicity in the case $\Omega_0 = \Omega$ provided we assume the existence of appropriately defined "Lyapunov exponents".

### VIII.1.2 The Discrete Case

In the discrete case the spectral measure $\sigma$ can be approximated by spectral measures of symmetric matrices. This is why we first prove a useful result about the expectation of spectral measures of symmetric matrices.
Let $e_1, \cdots, e_d$ be an orthonormal basis of $\mathbb{R}^d$ and $X$ be a $d \times d$ symmetric real matrix $X$. Then each spectral measure $\sigma_k^X = \sigma_{e_k,e_k}^X$ of $X$ can be viewed as a function of the $d(d+1)/2$ entries of the matrix $X$. Let us denote by $x_1, x_2, \cdots, x_d$ the diagonal entries of the matrix $X$.

**Lemma VIII.1.8** *Let $m$ be a probability measure on $\mathbb{R}$ with a density with respect to the Lebesgue measure, bounded by some constant $C$. Then for $k = 1, \cdots, d$ the averaged probability measure $\int \sigma_k^X \, dm(x_k)$ also admits a density with respect to the Lebesgue measure on $\mathbb{R}$ and this density is bounded by the same constant $C$.*

*Proof:*
There is no loss of generality to suppose $k = 1$. One can write $X = \begin{pmatrix} x_1 & b' \\ b & S \end{pmatrix}$ where $b$ is a $(d-1)$ dimensional vector and $S$ is a $(d-1)\times(d-1)$ symmetric matrix. Let $\lambda$ be a complex number with $\Im m(\lambda) > 0$ and let us set:

$$u = (X - \lambda I)^{-1} e_1 = \begin{pmatrix} u_1 \\ v \end{pmatrix} \quad \text{where } u_1 \in \mathbb{R}$$

Then from the spectral theorem one has:

$$\int \frac{d\sigma_1^X(t)}{t - \lambda} = u_1$$

If we write the set of equations satisfied by the vector $u$ one obtains:

$$x_1 u_1 + b'v = 1 + \lambda u_1 \quad \text{and} \quad u_1 b + Sv = \lambda v.$$

Let us denote by $h$ the $(n-1)$ dimensional vector $h = (S - \lambda I)^{-1} b$, then one has:

$$v = -u_1 h \quad \text{and} \quad u_1 = (x_1 - b'h - \lambda)^{-1}.$$

The spectral theorem applied to the matrix $S$ yields:

$$b'h = \int \frac{d\sigma_{b,b}^S(t)}{t - \lambda}$$

and it follows that the imaginary part of $b'h$ is positive. If we denote by $\varphi$ the bounded density of the law $m$ one can conclude that:

$$\frac{1}{\pi}\Im m \int\int \frac{\varphi(x_1)dx_1\, d\sigma_1^X(t)}{t-\lambda} = \frac{1}{\pi}\Im m \int \frac{\varphi(x_1)dx_1}{(x_1 - b'h - \lambda)} \leq \|\varphi\|_\infty$$

The de la Vallée Poussin theorem yields the desired result. ∎

One can remark that this lemma is very close to the computation of the averaged spectral measure of the Schrödinger operator in a box with respect to boundary conditions but here we integrate the spectral measure at the point $k$ with respect to the law of the diagonal term of the same order $k$ rather than with respect to the farthest right diagonal term.

This lemma together with the general result of Kotani has a direct application to the independent case. Note that the integrability conditions of the beginning of this chapter are certainly satisfied if $\mathbb{E}\{\log(1+|V|)\} < +\infty$.

**Theorem VIII.1.9** *In the independent case, if the common distribution of the potentials is not orthogonal to the Lebesgue measure then:*

**(i)** *the lowest Lyapunov exponent $\gamma_\ell(\lambda)$ is strictly positive for any $\lambda$,*
**(ii)** *one has exponential localization on $\mathbb{R}$ with decay at least $\gamma_\ell(\lambda)$ (equal to $\gamma_1(\lambda)$ for $\ell = 1$).*

*Proof:*
Denoting by $\tau$ the common distribution of the potentials one can write $\tau = \tau_1 + \tau_2$ where $\tau_2$ is a non-zero measure with a bounded density $\varphi$ with respect to the Lebesgue measure. Proposition IV.4.33 implies that the probability measure $\mu_\lambda^{2\ell+1}$ on $SP(\ell, \mathbb{R})$ is greater than a measure with a density and it follows that $T_{\mu_\lambda}$ contains an open set. This last condition together with Proposition IV.4.22 implies that all the Lyapunov exponents are distinct and hypothesis (i) of the Theorem VIII.1.6 is fulfilled.
Let $\Omega_0$ be the subset of $\Omega$ such that $V(0,\omega)$ and $V(-1,\omega)$ are in the support of the measure $\tau_2$. If we denote by $\mathcal{F}^*$ the sub $\sigma$-algebra generated by the column potential vectors $V(t,\omega)$ for $t \neq \{-1, 0\}$, then the conditional expectation,

$$\mathbb{E}\{\mathbf{1}_{\Omega_0}\sigma|\mathcal{F}^*\}$$

is actually the expectation of the spectral measure with respect to the potentials lying in the slices $t = 0$ and $t = -1$ and obeying the law $\tau_2$. Let

$\Lambda$ be a box containing the slices $t = 0$ and $t = -1$ and for $x = (i,t)\,, i = 1\ldots\ell$, $t = -1, 0$ let $\sigma_x^\Lambda$ be the spectral measure of $H^\Lambda$ computed at the point $x$ with arbitrary boundary conditions. Lemma VIII.1.8 implies that if we first integrate with respect to the potential at the site $x$ then this expectation has a density with respect to the Lebesgue measure bounded by $\|\varphi\|$. The weak convergence of $\sigma_x^\Lambda$ to $\sigma_x$ when the box $\Lambda$ goes to infinity and the Fatou's lemma yield the condition (ii) of the theorem VIII.1.6. ∎

**Remark VIII.1.10** In the one dimensional case (that is when $\ell = 1$) one can compute the expectation of the spectral measure without using the above lemma and approximations in boxes. In this case the Green function can be expressed by means of the Weyl $m$-functions and this allows a direct computation (See Problem VIII.6.1). This gives an easy way to prove localization with off-diagonal disorder, in the example for the "Helmotz Operator" introduced in Chapter III. Let $\{c_n, d_n \;:\; n \in \mathbb{Z}\}$ be an $\mathbb{R}^2$ valued i.i.d. sequence of random variables and $H(\omega)$ be theergodic family of Schrödinger operators defined by:

$$H\psi(n) = \frac{1}{c_n}(-\psi(n-1) - \psi(n+1) + d_n\psi(n))$$

The independent sequence of transfer matrices $g_\lambda(n) = \begin{pmatrix} d_n - c_n\lambda & -1 \\ 1 & 0 \end{pmatrix}$ has a positive Lyapunov exponent $\gamma(\lambda)$ as long as the random variable $d_n - \lambda c_n$ is not constant. This occurs only when there exists a real value $\lambda^*$ such that the distribution of $(c, d)$ is supported by the line $d = c\lambda^*$ and $\lambda = \lambda^*$ for at most one value of $\lambda$. The Stieljes transform of the spectral measure $\sigma_{0,0}$ is given by:

$$\int \frac{d\sigma_{0,0}(t)}{t-\lambda} = G(\lambda, 0, 0) = \frac{-1}{m_+(\lambda) + m_(\lambda) + \lambda c_0 - d_0}$$

Let $\mathcal{I}$ be a Borel subset of $\mathbb{R}$ which does not contain $\lambda^*$ (if such a point exists). If one assumes that the random variable $c$ is bounded and that the law of $\lambda c - d$ has a bounded density, this bound being uniform with respect to $\lambda \in \mathcal{I}$ then the de la Vallée Poussin theorem implies that the measure $\sigma_{0,0}$ averaged with respect to the law of $(c_0, d_0)$ is absolutely continuous on $\mathcal{I}$ with a bounded density. The localization property follows from Theorem VIII.1.6. Another example of localization for "off-diagonal disorder" can be found in Problem VIII.6.10.

This last theorem does not give localization for singular potentials but we are not very far from such a result if we allow arbitrarily small "smooth" perturbations of the potentials on a finite number of sites. In the case of independent potentials with the same distribution let $\{\epsilon_i(n); i := 1 \cdots \ell, n = -1, 0\}$ be an independent family of random variables which is also independent of the potential process and such that the common law of the $\epsilon$'s is absolutely continuous with a bounded density. Let us define the new probability measure $\tilde{\mathbb{P}}$ on the space $\Omega$ by replacing the factors of index $-1$ and 0 in $\mathbb{P}$ by the laws of the vectors $\{V_i(n) + \epsilon_i(n)\,; \ell = 1 \cdots \ell, n = -1, 0\}$. These laws are absolutely continuous with a bounded density. If we assume that the original process has a strictly positive Lyapunov exponent $\gamma_\ell(\lambda)$ then the preceding theorem yields the localization property for $\tilde{\mathbb{P}}$ almost all $\omega$. This implies that for almost all values of the perturbations $\epsilon$ (with respect to the product law of the perturbations) the original Schrödinger operator has a pure point spectrum for $\mathbb{P}$-almost all $\omega$! This is a good reason to believe that the localization property occurs in the independent case as long as the lowest Lyapunov exponent is positive. We will prove in the sequel that this is indeed the case. See also the end of this subsection, for a weaker result on the half axis, with the same "flavor".

Let $H(\omega)$ be an ergodic family of random Schrödinger operators in the discrete strip satisfying the integrability conditions stated at the beginning of this chapter. If one wants to prove the localization property following the same lines as in the independent case, the only change is that we have to deal with conditional laws instead of integrated ones. The same proof as in the independent case yields the following theorem:

**Theorem VIII.1.11** *Let $\mathcal{F}^*$ be the sub-$\sigma$-algebra of $\mathcal{F}$ generated by the potentials on sites lying outside some box $\Lambda$ (with at least two slices) and $\mathcal{I}$ be a Borel set of the real line.*
*Assume that:*

**(i)** $\gamma_\ell(\lambda) > 0$ *almost surely on $\mathcal{I}$ with respect to the Lebesgue measure.*
**(ii)** *For $(i,t) \in \Lambda$ the conditional law of $V_i(t,\omega)$ given $\mathcal{F}^*$ has a nonzero absolutely continuous component with a density uniformly bounded with respect to the potentials outside the box $\Lambda$.*

*Then one has exponential localization on $\mathcal{I}$ with decay at least $\gamma_\ell(\lambda)$ (equal to $\gamma_1(\lambda)$ for $\ell = 1$).*

In the one dimensional case (that is when $\ell = 1$) it has been proved in Chapter VII that the Lyapunov exponent is Lebesgue-almost everywhere positive on $\mathbb{R}$ as soon as the potential process is non-deterministic . It is likely that the same property holds for the exponent $\gamma_\ell$ in the strip but up to now no proof is available. See however the work of Simon and Kotani [201] in which it is proved that when the potential process is not deterministic then $\gamma_1(\lambda)$ is positive Lebesgue-almost everywhere). As a consequence, in the one dimensional case, as long as the conditional law of $V(0)$ given a finite number of potentials in a box $\Lambda$ can be computed it is possible to check the hypothesis (ii). This is easy for Gaussian processes. See Problem VIII.6.2.

Following Delyon, Lévy, Souillard [74] we now discuss the multiplicity of the spectrum.

We have seen that the constancy of the Wronskian of two solutions of the eigenvalue equation implies that the dimension of the eigenspaces is at most $\ell$. In the one dimensional case this yields immediately the non degeneracy of the spectrum. This property is much less obvious in the strip since it means that for $\mathbb{P}$-almost all $\omega$, then for $\sigma(\omega)$ all $\lambda$ the two $\ell$ dimensional Oseledec contractive subspaces $^{+}\mathcal{V}$ and $^{-}\mathcal{V}$ have a one dimensional intersection.
Let $x = (i, r)$ be a fixed site and $^{+}\mathcal{V}_{r+1}$ and $^{-}\mathcal{V}_{r-1}$ be the Oseledec contractive subspaces respectively associated to the products $g(n)\dots g(r+1)$ for $n \to +\infty$ and $g^{-1}(n)\dots g^{-1}(r-1)$ for $n \to -\infty$. For a real number $c$ we consider the operator $H + c\delta_x$ (where we also denote by $\delta_x$ the multiplication operator by the function $\delta_x$). It is the perturbed Schrödinger operator at the site $x = (i, r)$ where the potential is now equal to $c + V_i(r)$.
Let us denote by $\mathcal{P}$ the set of polynomially bounded functions on the strip. We define the sets $W$, $W_x^1$ and $W_x^2(c)$ by:

- $W = \{(\omega, \lambda)\,;\, \lambda \in \text{hyp}(H(\omega))\ ,\ \lambda \text{ is not an eigenvalue of } H(\omega)\,\}$
- $W_x^1 = \{(\omega, \lambda)$ ; there exists a unique function $G_x \in \mathcal{P}$ solution of the equation $H(\omega)G_x = \lambda G_x + \delta_x\}$
- $W_x^2(c) = \{(\omega, \lambda)$; there exists at most a function $\psi \in \mathcal{P}$ (up to constant factors) solution of the equation $(H(\omega) + c\delta_x)\psi = \lambda\psi\ \}$

Let us denote by $\mathcal{B}$ the Borel sigma-algebra of $\mathbb{R}$.

**Proposition VIII.1.12** *Assume that $\gamma_\ell(\lambda)$ is almost surely strictly positive with respect to the Lebesgue measure $m$ on $\mathbb{R}$. Then:*

**(i)** *The set $W$ is $\mathcal{F} \otimes \mathcal{B}$ measurable and of full $\mathbb{P} \otimes m$ measure.*

**(ii)** *One has the inclusion $W \subset W_x^1$. Moreover the "pseudo Green's function" $G_x$, defined for $(\omega, \lambda) \in W$ and for any site $x$ satisfies the symmetry relation $G_x(y) = G_y(x)$ and is exponentially decaying with rate at least $\gamma_\ell(\lambda)$.*

**(iii)** *For any real $c$ one has the inclusion $W \subset W_x^2(c)$.*

*Proof:*
The measurability of $W$ is a simple consequence of the joint measurability of the propagator matrix with respect to $(\omega, \lambda)$ and the fact that the set of eigenvalues of $H(\omega)$ is at most countable. The second part of (i) then follows from Fubini's and Oseledec's theorems.
We have already noticed for hyperbolic value of $\lambda$ that "$\lambda$ is not an eigenvalue of $H(\omega)$" is equivalent to the fact that for one (or any ) value of the integer $r$, the two $\ell$ dimensional Oseledec contractive subspaces have a nul intersection and hence satisfy the relation ${}^+\mathcal{V}_{r+1} \oplus {}^-\mathcal{V}_r = \mathbb{R}^d$.
Let $x = (i, r)$. To prove the inclusion in (ii) we remark that for an hyperbolic value of $\lambda$ the polynomially bounded function $G_x = \psi$ has to be exponentially decaying hence it must satisfy the following conditions:

$$\tilde{\psi}(r+1) \in {}^+\mathcal{V}_{r+1}\,, \quad \tilde{\psi}(r) \in {}^-\mathcal{V}_{r-1}\,, \quad \tilde{\psi}(r+1) = g(r)\tilde{\psi}(r) - e_x$$

where the $d$ dimensional vector $e_x$ has all its coordinates nul but the coordinate of index $i$ which is equal to one. Let $v$ be the vector $v = g(r)\tilde{\psi}(r)$. Taking into account the relation $g(r)^-\mathcal{V}_{r-1} = {}^-\mathcal{V}_r$ it is easy to check that there exists a unique vector $v$ satisfying these conditions and $v$ is equal to the component of $e_x$ on the subspace ${}^-\mathcal{V}_r$ in the above splitting of $\mathbb{R}^d$. The computation of the dot products

$$\begin{aligned} < G_x\,, (H - \lambda I)G_y > &= < G_x\,, \delta_y > \\ &= < (H - \lambda I)G_x\,, G_y > \\ &= < \delta_x\,, G_y > \end{aligned}$$

yields immediately the symmetry of the "pseudo Green's function".
To prove (iii) we choose a real number $c$ and $(\omega, \lambda) \in W$. If $\lambda$ is an eigenvalue of $H + c\delta_x$ associated to the eigenvector $\psi$ then $\psi$ is exponentially decaying since $\lambda$ is hyperbolic. For any site $y$ the function $G_y$ exists and is exponentially decaying thus the following dot products are well defined:

$$\begin{aligned} < G_y\,, (H + c\delta_x)\psi > &= \lambda < G_y\,, \psi > \\ &= < (H + c\delta_x)G_y\,, \psi > \\ &= < HG_y + cG_y(x)\delta_x\,, \psi > \\ &= < \lambda G_y + \delta_y + cG_y(x)\delta_x\,, \psi > \end{aligned}$$

This yields the relation $\psi(y) = -cG_y(x)\psi(x)$ and it follows that the function $\psi$ is unique. ∎

Let us denote by $\mathcal{F}^*$ the sigma-algebra generated by the potentials lying outside the site $x$. We say that a nonnegative measure $\mu$ on the Borel sigma-algebra $\mathcal{B}$ is "pseudo-carried" by a set $A$ if there exists a Borel subset of $A$ which carries $\mu$. The next proposition is a slight generalization of the "Kotani's trick" in the independent case.

**Proposition VIII.1.13** *Let $H(\omega)$ be a random Schrödinger operator associated to an i.i.d. family of potentials whose distribution has a bounded density. Let $m$ be the Lebesgue measure and let $W$ be an $\mathcal{F}\otimes\mathcal{B}$ measurable subset of $\Omega\times\mathbb{R}$ of full $\mathbb{P}\otimes m$ measure. Let $\tilde{W}$ be a subset of $\Omega\times\mathbb{R}$ which is independent of the coordinate $x$ and contains $W$. Then for $\mathbb{P}$ almost all $\omega$ the spectral measure $\sigma_x(\omega)$ is peudo-carried by the $\omega$ section of $\tilde{W}$.*

*Proof:*
From Fubini's theorem it is readily seen that there exists $W^* \subset \tilde{W}$ which is of full $\mathbb{P}\otimes m$ measure and $\mathcal{F}^*\otimes\mathcal{B}$ measurable. Then Propositions VIII.1.4 and VIII.1.8 yield the desired result. ∎

**Proposition VIII.1.14** *Let $H(\omega)$ be a random Schrödinger operator in a strip, associated to an i.i.d. family of potentials whose distribution has a bounded density. Then for $\mathbb{P}$-almost all $\omega$ the eigenspaces of $H(\omega)$ are one dimensional.*

*Proof:*
Let $x$ be a fixed site. The set $\tilde{W} = \cap_{c\in\mathbb{R}} W_x^2(c)$ contains $W$, is contained in $W_x^2(0)$ and does not depend upon the potentials at the site $x$. It follows from Propositions VIII.1.12 and VIII.1.13 that for $\mathbb{P}$almost all $\omega$, the spectral measure $\sigma_x(\omega)$ is pseudo-carried by the $\omega$ section of $W_x^2(0)$. We already know that $\sigma_x(\omega)$ is carried by the set of eigenvalues of $H(\omega)$ hence these eigenvalues are simple $\sigma_x(\omega)$ almost surely. If we repeat this argument for any $x$ we obtain the result. Notice that the same idea can be used to prove localization. ∎

Obviously the same result holds under the hypothesis of Theorem VIII.1.11. Up to now, for $\ell > 1$, we have only proved that the decay of the eigenfunctions is given by a positive Lyapunov exponent, that is, an exponential decay with rate *at least* $\gamma_\ell(\lambda)$. In the situation of Proposition VIII.1.14, it is expected that the decay of the eigenfunctions is given by the lowest positive Lyapunov exponent $\gamma_\ell(\lambda)$. Unfortunately there is no proof of this claim.

## VIII.1.3 The General Case

If one tries to repeat the argument used in the discrete case for a real parameter $t$, the essential problem arises from the fact that the computation of the conditional spectral measures cannot be carried out by means of Lemma VIII.1.8. To overcome this difficulty we use the approximations of the spectral measures computed in Chapter III.
Let $X$ be a random variable defined on the probability space $(\Omega, \mathcal{F}, \mathbb{P})$ with values in a measurable space $\mathcal{X}$. Using the notations introduced at the beginning of this section we call "$\Omega_0$-conditional law of $X$ given $\mathcal{F}^*$" the unique random probability measure ${}^0\mathbb{P}_\omega^X$ on $\mathcal{X}$ such that for any real bounded Borel function $f$ defined on $\mathcal{X}$ one has

$$\mathbb{E}\{\mathbf{1}_{\Omega_0} f(X)|\mathcal{F}^*\} = {}^0\,\mathbb{P}_\omega^X(f).$$

In other words ${}^0\mathbb{P}_\omega^X$ is the conditional law of $X$ when the probability measure $\mathbb{P}$ is restricted to $\Omega_0$.

In the following theorem we denote by $\mathcal{F}^*$ the sub-$\sigma$-algebra of $\mathcal{F}$ generated by the potentials outside the box $\{(i,t); i = 1 \cdots \ell, t \in [0,c]\}$ for some positive number $c$.

**Theorem VIII.1.15** *Let $H(\omega)$ be an ergodic family of Schrödinger operators on the strip of width $\ell$, $\mathcal{I}$ be a Borel subset of $\mathbb{R}$ and $\Omega_0$ be a measurable subset of $\Omega$ of positive $\mathbb{P}$-measure. Let us assume that:*

**(i)** *The set of potentials $\{V_i(t,\omega)\}$ is bounded for $\omega \in \Omega_0$, $i \in 1 \ldots \ell$, $t \in [0,c]$.*

**(ii)** *The lowest nonnegative Lyapunov exponent $\gamma_\ell(\lambda)$ is strictly positive almost everywhere on $\mathcal{I}$ with respect to the Lebesgue measure.*

**(iii)** *The $\Omega_0$-conditional law of $U(c)\bar{v}$ given $\mathcal{F}^*$ has a bounded density with respect to the Cauchy measure of $L_\ell$, this bound being uniform with respect to $\lambda \in \mathcal{I}$, $\overline{v} \in L_\ell$ and the potentials outside $[0,c]$.*

*Then one has exponential localization on $\mathcal{I}$ with decay at least $\gamma_\ell(\lambda)$ (equal to $\gamma_1(\lambda)$ if $\ell = 1$).*

*Proof:*
One has only to check the hypothesis (ii) of Theorem VIII.1.6. Let $\phi(\lambda)$ be a continuous function with compact support in $\mathcal{I}$ and $\Lambda$ be the box $\Lambda = [a\ b]$ with $a < 0 < c < b$. In order to simplify the notations we set for $v \in \Lambda^{\ell}(\mathbb{R}^d)$:

$$\psi(\bar{v}) = \operatorname{tr}\left(v'v[v'U'(a)U(a)v]^{-1}\right)$$

Using the notations introduced in Chapter III one can write:

$$\begin{aligned}
&\pi\mathbb{E}\{1_{\Omega_0}\sigma^{\Lambda}(\phi)|\mathcal{F}^*\} \\
&\quad = \int \phi(\lambda)\psi(\bar{v})\mathbb{E}\{1_{\Omega_0}\left(\frac{||v||}{||U(b)v||}\right)^{\ell+1}|\mathcal{F}^*\}\, dm(\bar{v})\, d\lambda \\
&\quad \leq C_1 \int \phi(\lambda)\psi(\bar{v})\mathbb{E}\{1_{\Omega_0}\left(\frac{||U(c)v||}{||U(b,c)U(c)v||}\right)^{\ell+1}|\mathcal{F}^*\}\, dm(\bar{v})\, d\lambda \\
&\quad \leq C_1C_2 \int \phi(\lambda)\psi(\bar{v})\left(\frac{||u||}{||U(b,c)u||}\right)^{\ell+1} dm(\bar{v})\, dm(\bar{u})\, d\lambda \\
&\quad = C_1C_2 \int \phi(\lambda)\psi(\bar{v})\, dm(\bar{v})\, d\lambda \\
&\quad = \ell C_1C_2 \int \phi(\lambda)\, d\lambda
\end{aligned}$$

The first relation is a direct consequence of equality 1. in Theorem III.5.11. To obtain the second relation we write :

$$\frac{||v||}{||U(b)v||} = \frac{||v||}{||U(c)v||}\frac{||U(c)v||}{||U(b,c)U(c)v||}$$

Then it is easy to check from the hypothesis (i) that the ratio $\frac{||v||}{||U(c)v||}$ is bounded for $\omega \in \Omega_0$ and $\lambda$ in the support of $\phi$. The third relation is a simple consequence of the boundedness of the density of the $\Omega_0$ conditional law $g(\bar{u})$ of $\frac{U(c)v}{||U(c)v||}$ given $\mathcal{F}^*$ with respect to the Cauchy measure of $L_\ell$ since one can write:

$$\mathbb{E}\{1_{\Omega_0}\left(\frac{||U(c)v||}{||U(b,c)U(c)v||}\right)^{\ell+1}|\mathcal{F}^*\} = \int \left(\frac{||u||}{||U(b,c)u||}\right)^{\ell+1} g(\bar{u})\, dm(\bar{u})$$

To obtain the fourth relation we remark that for any $g \in SP(\ell, \mathbb{R})$ we have:

$$\int \left(\frac{||u||}{||gu||}\right)^{\ell+1} dm(\bar{u}) = \int d(g^{-1}m)(\bar{u}) = 1$$

The last relation is a direct consequence of the second equality in Theorem III.5.11. The result is now a simple consequence of the Fatou's theorem and the convergence of $\sigma_\omega^\Lambda(\phi)$ to $\sigma_\omega(\phi)$ if we use a nonnegative function $\phi$. ∎

It is possible to use the same trick as in the discrete case to force hypothesis (ii) by "smooth" perturbations of the potentials. That is, given a smooth function $\phi$ with support in $[0\ c]$, one defines the new potentials $V_i(t) + \epsilon_i\phi(t)$ where the real random variables $\epsilon_i$ are independent from the initial potential process with a smooth law. If the law of the original process is absolutely continuous with respect to the the law of the new process then one has exponential localization. (See Problem VIII.6.9).

## VIII.2 Perturbation Theory

The next proof of localization uses the theory of rank-one perturbations of self-adjoint operators and is closely related to the previous one when we used the "Kotani-trick", wriggling the value of the potential at one site. The essential difference is that this approach does not require the $\sigma$ almost sure "slow growth" of the generalized eigenfunction which was fundamental in the preceding section. One can also remark that this proof says nothing about the decay of eigenfunctions. Most of the material presented below is due to B.Simon and T.Wolff [304] and [299]. We begin with some results of functional analysis whose proof can be found in Problems VIII.6.3, VIII.6.4 and VIII.6.5.

Let $H$ be a selfadjoint operator and $Pf = < f, \phi > \phi$ be the projection operator on the normalized function $\phi$. For any real number $c$ we set $H_c = H + cP$ and $\sigma^c$ is the spectral measure of $H_c$ computed on the vector $\phi$. For a real $\lambda$ and a complex $z$ with $\Im m(z) > 0$ one defines the functions $B(\lambda)$ and $F_c(\lambda)$ by:

$$B(\lambda) = \int \frac{d\sigma^0(t)}{(t-\lambda)^2} \qquad F_c(z) = \int \frac{d\sigma^c(t)}{t-z}$$

A direct application of the second resolvent equation immediately yields the fundamental relation:

$$F_c(z) = \frac{F_0(z)}{1 + cF_0(z)} \tag{VIII.1}$$

**Proposition VIII.2.1** *Assume that $\lambda$ is not an eigenvalue of $H$. Then for $c \neq 0$ the following are equivalent:*

1. *$\lambda$ is an eigenvalue of $H_c$.*
2. *$F_0(\lambda + i0) = -c^{-1}$ and $B(\lambda) < \infty$.*

*Proof:*
See Problem VIII.6.3. ∎

The next lemma is very similar to Lemma VIII.1.8 but deals with selfadjoint operators rather than with symmetric matrices.

**Lemma VIII.2.2** *Let $m$ be the Cauchy measure on $\mathbb{R}$, and $\tilde{\sigma} = \int \sigma^c \, dm(c)$ be the averaged spectral measure with respect to $m$. Then $\tilde{\sigma}$ is equivalent to the Lebesgue measure.*

*Proof:*
See Problem VIII.6.4. ∎

The next result is due to Aronszajn and Donoghue [10],[84].

**Proposition VIII.2.3** *Let us define the sets:*

$$X = \{x \; ; \; \Im m \; F_0(x + i0) > 0\}, \quad Y = \{x \; ; \; B(x) < \infty\}, \quad Z = \mathbb{R} \setminus (X \cup Y)$$

*Then for any $c \neq 0$ the absolutely continuous part of $\sigma^c$ say $\sigma^c_{\text{a.c.}}$ is supported on $X$, the pure point part of $\sigma^c$ say $\sigma^c_{\text{p.p.}}$ is supported on $Y$ and the singular continuous part of $\sigma^c$ say $\sigma^c_{\text{s.c.}}$ is supported on $Z$.*

*Proof:*
See Problem VIII.6.5. ∎

One remarks that the sets $X$ and $Y$ have an empty intersection since:

$$\Im m F_0(x + i\epsilon) = \epsilon \int \frac{d\sigma^0(t)}{(t - x)^2 + \epsilon^2} \leq \epsilon B(x)$$

We are now in position to prove the main analytical result, namely:

**Theorem VIII.2.4** *The following are equivalent:*

1. *The spectral measure $\sigma^c$ is pure point for Lebesgue almost all c.*
2. *$B(\lambda) < \infty$ for Lebesgue almost all $\lambda$.*

*Proof:*
From Proposition VIII.2.3 we know that $\sigma^c$ is pure point if and only if $\sigma^c(X \cup Z) = 0$. Hence $\sigma^c$ is pure point for almost all $c$ if and only if $\tilde{\sigma}(X \cup Z) = 0$. From Lemma VIII.2.2 this last property is equivalent to the fact that $(X \cup Z)$ is Lebesgue negligible and this finishes the proof. ∎

We now turn our attention to one dimensional discrete Schrödinger operators. One denotes by $H^+$ and $H^-$ the restriction of the Schrödinger operator to the positive and negative axis with Dirichlet boundary conditions at 0.

**Proposition VIII.2.5** *Let H be a discrete one dimensional Schrödinger operator. Assume that*

- *$\lambda$ is not an eigenvalue of $H$.*
- *$\lambda$ is not an eigenvalue of $H^+$ (or $H^-$).*
- *$\lambda \in \mathrm{hyp}(H)$*

*Then $B(\lambda) < \infty$.*

*Proof:*
We have seen in Proposition VIII.1.12 that if $\lambda$ is hyperbolic then there exists a unique polynomially bounded $G_0$ solution of the equation

$$H\psi = \lambda\psi + \delta_0$$

and that this solution is actually exponentially decreasing at $\pm\infty$. Since $\lambda$ is not an eigenvalue of $H^+$ (or $H^-$) it follows that $G_0(0) \neq 0$. Hence the function $\psi = \frac{G_0}{G_0(0)}$ is an eigenvalue of $H_c$ with $c = -\frac{1}{G_0(0)}$ The result then follows from Proposition VIII.2.1 if one chooses $\phi = \delta_0$. ∎

**Remark VIII.2.6** • *We actually prove that for any hyperbolic $\lambda$, it is possible to wriggle the value of the potential at one site, so that $\lambda$ becomes an eigenvalue of the perturbed operator. It is readily seen, in the example for the free laplacian, that the hyperbolicity of $\lambda$ is essential.*

- *The extension of this proposition to the strip needs a bit more of algebra. For $r \in [1 \dots \ell]$ Proposition VIII.1.12 implies now the existence of a "pseudo Green's function" $G_{r,0}$ solution of the equation*

$$H\psi = \lambda\psi + \delta_{r,0}$$

*and that this solution is actually exponentially decreasing at $\pm\infty$. But the fact that $\lambda$ is not an eigenvalue of $H^+$ only implies that $G_{r,0}$ cannot be simultaneously zero for* all *the points $(i, 0)$, $= 1 \dots \ell$. Hence for $\ell > 1$ it is not granted that $G_{r,0}(r, 0) \neq 0$. This explains why one has to extend the notion of rank one perturbation theory to "$c = \infty$" to handle this situation (or to finite rank perturbations).*

As a consequence we get the localization property when the distribution of potentials is not purely singular. We only prove this property in the simplest case of an i.i.d. sequence of potentials with an absolutely continuous distribution. The extension to nonsingular distributions of potentials and to general stationary ergodic sequences are straightforward if one uses the methods of the preceding section. On the other hand, the extension to the continuous case is more difficult.

**Theorem VIII.2.7** *Let $H(\omega)$ be a one dimensional Schrödinger operator associated to an i.i.d. sequence of potentials with an absolutely continuous distribution. Then for $\mathbb{P}$-almost all $\omega$ the spectrum of $H(\omega)$ is pure point.*

*Proof:*
Let us consider the set:

$$W_1 = \{(\omega, \lambda) \ ; \ \lambda \text{ is not an eigenvalue of } H, \ H^+ \ , \lambda \in \text{hyp}(H)\}$$

For any $\omega$ the set of eigenvalues of $H$ or $H^+$ is countable hence Lebesgue null. We already know that the Lyapunov exponent is positive for any $\lambda$ . It follows from Fubini's and Oseledec's theorems that $W_1$ is of full $\mathbb{P} \otimes m$ measure for $m$ the Lebesgue measure.
Let us define

$$W_2 = \{(\omega, c) \ ; \ \sigma_0^c(\omega) \quad \text{is pure point}\}$$

The standard measurability arguments used in the preceding section show immediately that $W_2$ is a measurable subset of $\Omega \times \mathbb{R}$. From Proposition VIII.2.5 and Fubini's theorem one gets that $W_2$ is of full $\mathbb{P} \otimes m$ measure. Moreover $\sigma_0^c(\omega) = \sigma_0(\omega + c)$ where $\omega + c$ has the same coordinates as

$\omega$ but the coordinate of index 0 which is equal to $\omega_0 + c$. It follows from Fubini's theorem that for $\mathbb{P}$almost all $\omega$ the measure $\sigma_0(\omega)$ is pure point. To check this last claim let $F(c, x, y)$ be a nonnegative measurable function on the measurable space $(\mathbb{R} \times \mathbb{R} \times \mathcal{Y})$ which is $dm \otimes d\mu_x \otimes d\mu_y$ negligible and such that $F(c, x, y) = F(0, x + c, y)$. Then one has:

$$\begin{aligned} 0 &= \int F(c, x, y)\, dm(c) \otimes d\mu_x(x) \otimes d\mu_y(y) \\ &= \int F(0, x + c, y)\, dm(c) \otimes d\mu_x(x) \otimes d\mu_y(y) \\ &= \int F(0, x, y)\, dm(x) \otimes d\mu_y(y) \end{aligned}$$

This implies that $F(0, x, y)$ is $m \otimes d\mu_y$ negligible, hence $\mu_x \otimes \mu_y$ negligible if $\mu_x$ is absolutely continuous. Repeating the argument at the site $-1$ one obtains the desired result. ∎

# VIII.3 Operator Theory

One can consider the Laplace transform $T_\lambda$ associated to the probability measure $\mu_\lambda$ on $SP(\ell, \mathbb{R})$ for any value of the energy $\lambda$. This operator acts on the Banach space of continuous functions on the Lagrangian boundaries of the symplectic group. The spectral properties of this operator have been studied in the chapter "Random matrices" but now one needs some uniformity with respect to the parameter $\lambda$. We first prove the following general result:

**Proposition VIII.3.1** *Let $\lambda \hookrightarrow T_\lambda$ be a continuous mapping from the compact metric set $\mathcal{I}$ in the set of bounded operators on a Banach space and denote by $r(\lambda)$ the spectral radius of the operator $T_\lambda$. Then we have:*

**(i)** *The function $\lambda \hookrightarrow r(\lambda)$ is upper semi-continuous on $\mathcal{I}$.*

**(ii)** *If $r(\lambda) < \delta$ for any $\lambda \in \mathcal{I}$ then there exists $\rho < \delta$ and a constant $C$ such that:*

$$||T_\lambda^n|| \le C\rho^n \quad \textit{for } n = 1, 2, \cdots, \ \lambda \in \mathcal{I}$$

*Proof:*
Let us set $r_n(\lambda) = ||T_\lambda^n||$, then $r_n(\lambda)$ is a continuous function and the sequence $\log r_n(\lambda)$ is subadditive. It follows that $\log r(\lambda) = \lim \frac{1}{n} \log r_n(\lambda) =$

$\inf_n \frac{1}{n} \log r_n(\lambda)$ is upper semi-continuous and this proves (i). The upper semicontinuous function $r$ attains its maximum on the compact set $\mathcal{I}$ hence there exists $\rho' < \delta$ such that $r(\lambda) \leq \rho'$ on $\mathcal{I}$. From $\rho' = \max(\rho', r(\lambda))$ one deduces that $\rho'$ is the pointwise limit of the decreasing sequence $\max(\rho', \inf_{1\leq k\leq n} r_k(\lambda))$. The Dini's theorem implies that this limit is uniform on the compact set $\mathcal{I}$, hence for $\rho' < \rho < \delta$ there exists an integer $N$ such that

$$\inf_{1\leq k\leq N} r_k(\lambda) \leq \rho \quad \text{for all } \lambda \in \mathcal{I}$$

The conclusion now follows from the subadditivity of the sequence $\log r_n(\lambda)$ ∎

**Proposition VIII.3.2** *Let $T$ be a compact operator of spectral radius one on some Banach space. Then the sequence $\|T^n\|$ is bounded if and only if for any eigenvalue $\lambda$ of modulus 1 one has* $\ker(T - \lambda I) = \ker(T - \lambda I)^2$.

*Proof:*
We can write $T^n = Q^n + N^n$ where the spectral radius of $Q$ is strictly less than 1 and $N$ is the restriction of $T$ to the sum of the characteristic spaces corresponding to the eigenvalues of modulus one. The Jordan form of $N$ allows us to say immediately that $N^n$ is bounded iff $N$ is diagonalizable. ∎

For a fixed Schrödinger operator $H$ we have proved in Chapter III that localization occurs as long as one can prove the exponential decay of some real sequence. In the random case we have to handle a sequence of measurable functions of $\omega$. Roughly speaking the following lemma says that if the expectations of a sequence of functions of $\omega$ form a real exponentially decreasing sequence then the sequence of functions is itself almost surely exponentially decreasing.

**Lemma VIII.3.3** *Let $f_n(\omega)$ be a sequence of measurable positive functions such that for some finite constants $C$ and $\rho$ one has:*

$$\mathbb{E}\{f_n\} \leq C\rho^n$$

*Then for any $\delta > \rho$ there exists a $\mathbb{P}$-almost everywhere finite random variable $C(\omega)$ such that*

$$f_n(\omega) \leq C(\omega)\delta^n$$

*Proof:*
One has $\mathbb{E}\{\sum_{n=0}^{\infty} f_n\delta^{-n}\} < +\infty$ and we define the random variable $C(\omega)$ by the formula $C(\omega) = \sum_{n=0}^{\infty} f_n(\omega)\delta^{-n}$. This random variable is integrable hence $\mathbb{P}$-almost everywhere finite and the proof is complete. ■

For the reader's convenience we will give separate proofs of localization in the three following models. First we investigate the simplest case of i.i.d. potentials on the one dimensional lattice. Secondly, using the same techniques, we prove localization in the one dimensional Markov model (discrete and continuous). Finally we extend these proofs to the discrete model in the strip with i.i.d. potentials.

## VIII.3.1 The Discrete I.I.D. Model

We assume in this subsection that the common distribution of the potentials is absolutely continuous with respect to the Lebesgue measure on $\mathbb{R}$ and has a second order moment.

Recall that we define the probability measure $\mu_\lambda$ on the group $SL(2,\mathbb{R})$ as the law of the matrix

$$\begin{pmatrix} V(n)-\lambda & -1 \\ 1 & 0 \end{pmatrix}$$

where $V(n)$ is the potential at the site $n$. If $\chi$ is a $\mu_\lambda$ integrable cocycle on the projective line $B$ then the Laplace transform $T_{\chi,\lambda}$ is defined on the set of continuous functions on $B$ by the formula:

$$T_{\chi,\lambda}f(\bar{x}) = \int f(g\bar{x})\chi(g,\bar{x})d\mu_\lambda(g)$$

Let $m$ be the Cauchy measure on $B$, and denote by $\chi$ the Radon Nikodym cocycle:

$$\chi(g,\bar{v}) = \frac{dg^{-1}m}{dm}(\bar{v}) = (\frac{||v||}{||gv||})^2$$

This cocycle will be fixed in all this subsection, hence for a real $\delta$ we will use the notation $T_{\delta,\lambda}$ for the Laplace transform associated to the cocycle $\chi^{\delta/2}$. Since $\chi(g,\bar{v}) \leq ||g||^2$ the existence of a second order moment for the law of the potentials implies that $T_{\delta,\lambda}$ is a bounded operator on the space $\mathcal{C}(B)$ for any $\lambda \in \mathbb{R}$ and $0 \leq \delta \leq 2$.

**Proposition VIII.3.4** *For any $\delta \in [0, 2]$ the function $\lambda \hookrightarrow T_{\delta,\lambda}$ maps continuously the real line into the Banach space of bounded operators on $C(B)$.*

*Proof:*
If we denote by $\psi(s)$ the density of the law of the potentials one can write:

$$||T_{\delta,\lambda'}f - T_{\delta,\lambda}f|| \leq ||f|| \int_{-\infty}^{+\infty} (1+s^2)^{\delta/2} |\psi(s-\lambda) - \psi(s-\lambda')| ds$$

and the result is a consequence of the Lebesgue theorem since we assume the existence of a second order moment. ∎

The following theorem summarizes the essential results about the Laplace transform. Most of them have been previously proved in Chapter IV.

**Theorem VIII.3.5** *For any real energy $\lambda$ and $0 \leq \delta \leq 2$ let us denote by $r(\delta, \lambda)$ the spectral radius of the operator $T_{\delta,\lambda}$. Then $T^3_{\delta,\lambda}$ is a compact operator on $C(B)$ and we have the decompositions:*

$$\begin{aligned} T^{3n}_{0,\lambda} f &= \nu_\lambda(f) + Q^n_{0,\lambda}(f) \\ T^{3n}_{2,\lambda} f &= m(f)\breve{\phi}_\lambda + Q^n_{2,\lambda}(f) \end{aligned}$$

*where $\nu_\lambda$ is the unique $\mu_\lambda$ invariant probability measure on $B$, $\breve{\phi}_\lambda$ is the density of $\breve{\nu}_\lambda$ with respect to $m$ and the operators $Q_{0,\lambda}$, $Q_{2,\lambda}$ have a spectral radius strictly less than one. Moreover $r(1, \lambda) < 1$*

*Proof:*
We first remark that if the law of the potentials has a density with respect to the Lebesgue measure then $\mu^3_\lambda$ has a density on the group $SL(2, \mathbb{R})$ with respect to the Haar measure. This can be deduced from the general result of Proposition IV.4.33 but in this situation, a simple computation of the Jacobian gives the result since one has:

$$\begin{pmatrix} x & -1 \\ 1 & 0 \end{pmatrix} \begin{pmatrix} y & -1 \\ 1 & 0 \end{pmatrix} \begin{pmatrix} z & -1 \\ 1 & 0 \end{pmatrix} = \begin{pmatrix} xyz - z - x & -yz + 1 \\ xy - 1 & -y \end{pmatrix}$$

If one chooses the chart $\begin{pmatrix} a & b \\ c & d \end{pmatrix} \to (b, c, d)$ (which works for $d \neq 0$) then the Jacobian of the mapping $(x, y, z) \to (b, c, d)$ of $\mathbb{R}^3$ into itself, is equal to $y^2$. The image of an absolutely continuous measure by this mapping is

still absolutely continuous. This yields the answer since the Haar measure on $SL(2,\mathbb{R})$ is equivalent to the Lebesgue measure on the three parameters $(b,c,d)$. (Actually a direct proof shows that $T^2_{\delta,\lambda}$ is a compact operator on $\mathcal{C}(B)$).
It follows from Propositions IV.3.9 and IV.3.14 that $T^3_{\delta,\lambda}$ is a compact aperiodic operator on $\mathcal{C}(B)$. The decompositions have been obtained in Proposition IV.3.12. To prove the last assertion we use the fact that $\delta \hookrightarrow \log r(\delta,\lambda)$ is a convex function on $[0,2]$ which is nul for $\delta = 0$ and $\delta = 2$. Moreover Proposition IV.3.15 implies that for any $\lambda$ there exists a positive number $\alpha$ with $r(\alpha,\lambda) < 1$ thus one can conclude that $r(1,\lambda) < 1$. ■

**Corollary VIII.3.6** *Let $\mathcal{I}$ be a compact subset of the real line. Then there exist constants $C_1, C_2$ and $\rho < 1$ such that for $n \in \mathbb{N}$ and $\lambda \in \mathcal{I}$ we have :*

$$\|T^n_{1,\lambda}\| \le C_1\rho^n \quad , \quad \|T^n_{2,\lambda}\| \le C_2$$

*Proof:*
Using Propositions VIII.3.1 and VIII.3.4 it is enough to prove that for each $\lambda$ the spectral radius of $T_{1,\lambda}$ is strictly less than 1. But this is a direct consequence of Proposition IV.3.13 and the positivity of the Lyapunov exponent.
Let $\lambda_n$ be a sequence converging to $\lambda$, then any weak limit point of the sequence $\nu_{\lambda_n}$ will be $\mu_\lambda$ invariant and thus equal to $\nu_\lambda$ from the unicity of the invariant measure. This implies the weak continuity of the mapping $\lambda \hookrightarrow \nu_\lambda$ from the real line to the set of Radon measures $B$.
This result also implies the continuity of the mapping $\lambda \hookrightarrow \check{\phi}_\lambda$ from the real line to $\mathcal{C}(B)$. From these facts and the continuity of the mapping $\lambda \hookrightarrow T_{\delta,\lambda}$ proved in Proposition VIII.3.4 it follows that the mappings $\lambda \hookrightarrow Q_{0,\lambda}$ and $\lambda \hookrightarrow Q_{2,\lambda}$ are continuous. (In fact these properties can be obtained directly from the theory of continuous perturbations of operators, see Kato[167]). The second uniform bound is then a direct consequence of Proposition VIII.3.5. ■

**Theorem VIII.3.7** *Let $\{V(n)\ ; n \in \mathbb{Z}\}$ be an i.i.d. sequence of real random variables whose common law is absolutely continuous and has a second order moment. Then for $\mathbb{P}$ almost all $\omega$ the operator $H(\omega)$ has a pure point spectrum with exponentially decaying eigenfunctions.*

*Proof:*
Taking into account the criterion III.4.16 and the lemma VIII.3.3 one has only to show that for a fixed angle $\alpha$ and any open bounded set $\mathcal{I}$, there

exists a finite constant $C$ and a positive number $\rho < 1$ such that for any box $\Lambda = [-b, b-1]$ one has for $n \in \Lambda$ :

$$\int_{\mathcal{I}} \mathbb{E}\{||U_\lambda(n,-b)\alpha|| \frac{||U_\lambda(0,-b)\alpha||^2}{||U_\lambda(b,-b)\alpha||^2}\} \, d\lambda \leq C\rho^{|n|}$$

Let us first consider the case $n \geq 0$. If we compute the expectation by successive integration with respect to the sites in $[n, b-1]$, $[0, n-1]$ and $[-b, -1]$, the left hand side of the above inequality can be rewritten as:

$$\int_{\mathcal{I}} \left(T^b_{0,\lambda} T^n_{1,\lambda} T^{b-n}_{2,\lambda} \mathbf{1}\right)(\alpha) \, d\lambda$$

where $\mathbf{1}$ is the function identically equal to one.
For $n \leq 0$ the computation of the expectation by successive integration with respect to the sites $[0, b-1]$, $[n, -1]$ and $[-b, n-1]$ implies that the left side can be rewritten as:

$$\int_{\mathcal{I}} \left(T^{b-|n|}_{0,\lambda} T^{|n|}_{1,\lambda} T^b_{2,\lambda} \mathbf{1}\right)(\alpha) \, d\lambda$$

Taking into account that $T_{0,\lambda}$ is a Markov operator with the uniform bounds proved in Proposition VIII.3.6 one obtains the result. ∎

One can remark that the assumptions on the distribution of potentials in this theorem are very similar to those of Theorem VIII.1.9. It is also possible, using the trick used in the preceding section, to prove this theorem under the weaker assumption that this law is not singular with respect to the Lebesgue measure.

**Remark VIII.3.8** It is possible to prove localization using the operators $T_{\delta,\lambda}$ acting on the space $L^2(m)$ rather than on $C(B)$. Up to now we have fixed the angle $\alpha$ to apply Theorem III.4.16. It is easy to see, using Fatou's lemma, that one can integrate in the variable $\alpha$ with respect to the Cauchy measure $m$ in the formula appearing in this Theorem and then obtain a new form of this criterion:

*Let $\mathcal{I}$ be an open bounded interval of the real line. Assume that there exist two constants $0 < \rho < 1$ and $C < \infty$ such that for each box $\Lambda = [-b, b-1]$ and $n \in \Lambda$ one has :*

$$\int_{\mathcal{I}} \left( \int ||U(n,-b)\alpha|| \; ||U(0,-b)\alpha|| \; ||U(b,-b)\alpha||^{-2} \, dm(\alpha) \right) d\lambda \leq C\rho^{|n|}$$

*Then $\sigma$ is pure point on $\mathcal{I}$ and for any $\rho < \delta < 1$ and any eigenvalue $\lambda \in \mathcal{I}$ there exists $C(\lambda) < \infty$ such that the associated eigenfunction $\psi$ satisfies :*

$$|\tilde{\psi}(t)| \leq C(\lambda)\delta^{|t|}$$

Hence in the proof of Theorem VIII.3.7 it is enough to prove that one has (for example for $n \geq 0$):

$$\int_{\mathcal{I}} \left[ \int_B \left( T^b_{0,\lambda} T^n_{1,\lambda} T^{b-n}_{2,\lambda} \mathbf{1} \right) (\alpha)\, dm(\alpha) \right] d\lambda \leq C\rho^n$$

where $\mathbf{1}$ is the function identically equal to one.
Under this last form one sees immediately that one can use the same estimates as in Corollary VIII.3.6 but with the $L_2(m)$ norm that we will denote by $||\ ||_2$ rather than with the $C(B)$ norm that we will now denote by $||\ ||_\infty$. Using the duality relation in Proposition IV.3.8 (i) that is

$$< \check{T}_{\delta,\lambda} f\ ,\ h >_m = < T_{2-\delta,\lambda} h\ ,\ f >_m$$

we obtain that $||T_{2,\lambda}||_2 = ||\check{T}_{0,\lambda}||_2$ Moreover for $f \in L_2(m)$ one can write:

$$\begin{aligned} |T_{0,\lambda} f(\alpha)|^2 &\leq \int |f|^2(g\alpha)\, dm(\alpha)\, d\mu_\lambda(g) \\ &= \int |f|^2(\alpha) \chi(g,\alpha)\, d\check{\mu}_\lambda(g)\, dm(\alpha) \\ &\leq ||f||_2^2\ ||\check{T}_{2,\lambda}||_\infty \end{aligned}$$

It follows from the estimates of Corollary VIII.3.6 that on a compact set $\mathcal{I}$ of energies the $L_2$ norms of $T^n_{0,\lambda}$ and $T^n_{2,\lambda}$ are uniformly bounded.
On the other hand it is proved in [82] that the spectral radius of $T_{1,\lambda}$ acting on $L_2(m)$ is equal to 1 iff the closed subgroup $G_{\mu_\lambda}$ generated by the support of $\mu_\lambda$ in $SL(2,\mathbb{R})$ is amenable. In our situation this subgroup is the whole of $SL(2,\mathbb{R})$ (See Lemma IV.4.29), which is not amenable. Another way to obtain that the spectral radius is strictly less than one is to check the compactness of the operator $T_{1,\lambda}$ acting on $L_2(m)$ and see that its eigenfunctions are actually continuous. Hence its spectral radius on $L_2(m)$ and $C(B)$ are the same, and the exponential decrease of $T^n_{1,\lambda}$ follows from our estimates on $C(B)$.

**Remark VIII.3.9** In the proof of localization by Kunz & Souillard in [97] and Delyon, Kunz & Souillard in [71] the authors use the same kind of estimates as above but applied to some operators $K_{\delta,\lambda}$ defined as follows:

$$K_{\delta,\lambda} f(\bar{v}) = \int f(g\bar{v}) \zeta(g,\bar{v})^{\delta} \, d\mu_{\lambda}(g) \text{ with } \zeta(g,\bar{v}) = \frac{|< v\,, e_1 >|}{|< gv\,, e_1 >|}$$

where $e_1$ is the first basis vector of $\mathbb{R}^2$. If we denote by $r$ the density of the distribution of the potential and if the projective line minus the point at infinity is identified with the real line by the mapping $v = \binom{a}{b} \hookrightarrow x = \frac{a}{b}$ this operator takes the form:

$$K_{\delta,\lambda} f(x) = \int \frac{1}{|t|^{\delta}} f(t) r(\lambda + t + \frac{1}{x}) \, dt$$

These operators also enjoy a "duality" property of the following form: Let $U$ be the unitary operator of $L^2(dx)$ associated to the mapping $x \hookrightarrow \frac{1}{x}$. If one defines the operators

$$\breve{K}_{\delta,\lambda} f(x) = \int \frac{1}{|t|^{\delta}} f(t) r(\lambda + x + \frac{1}{t}) \, dt$$

a simple change of variables yields

$$U^* K_{\delta,\lambda} U = \breve{K}_{2-\delta,\lambda}$$

The "pseudo-cocycle" $\zeta$ is singular and is no longer rotation-invariant, hence the corresponding "pseudo-Laplace transform" cannot be easily studied on $C(B)$. Nevertheless the same estimates that we obtained in this section can be proved for these operators acting on the space $L^2(\mathbb{R}, dx)$ if we identify the projective line minus infinity with the real line.
See Problem VIII.6.6 for a localization criterion introducing these operators.

We end the study of the one-dimensional i.i.d. case by a formula relating the density of states and the eigenfunctions of the Laplace operator. Recall from Chapter VI that the "distribution of states" $N$ is defined as the averaged spectral measure $N = \mathbb{E}\{\sigma_n\} = \frac{1}{2}\mathbb{E}\{\sigma\}$. (We put the factor $\frac{1}{2}$ in order to obtain a probability measure).

**Proposition VIII.3.10** *Under the hypothesis of Theorem VIII.3.7 the probability measure $N$ has a continuous density with respect to the Lebesgue* *measure given by the formula:*

$$\frac{dN}{d\lambda} = \int_B \phi_\lambda(b)\breve{\phi}_\lambda(b)dm(b)$$

*where* $\phi_\lambda$ *and* $\breve{\phi}_\lambda$ *are the densities of* $\nu_\lambda$ *and* $\breve{\nu}_\lambda$ *with respect to the Cauchy measure* $m$ *on* $B$.

*Proof:*
Let $h$ be a real continuous function with compact support. The Lebesgue theorem and the approximation of the spectral measure computed in Proposition III.3.2 imply immediately that for any nonzero vector $v \in \mathbb{R}^2$ one has:

$$\begin{aligned}\int h(\lambda)dN(\lambda) &= \lim_{b\to+\infty} \frac{1}{2\pi} \int h(\lambda)\mathbb{E}\{\frac{\|U_\lambda(0,-b)v\|^2}{\|U_\lambda(b,-b)v\|^2} d\lambda \\ &= \lim_{b\to+\infty} \frac{1}{2\pi} \int h(\lambda) \left(T^b_{0,\lambda}T^b_{2,\lambda}\mathbf{1}\right)(\bar{v})\, d\lambda\end{aligned}$$

It only remains to use the decompositions of $T^n_{2,\lambda}$ and $T^n_{0,\lambda}$ given in Proposition VIII.3.5 to obtain :

$$\begin{aligned}T^b_{0,\lambda}T^b_{2,\lambda}\mathbf{1} &= T^b_{0,\lambda}\left(\breve{\phi}_\lambda + Q^b_{2,\lambda}\mathbf{1}\right) \\ &= \nu_\lambda(\breve{\phi}_\lambda) + \nu_\lambda Q^b_{2,\lambda}\mathbf{1} + Q^b_{0,\lambda}\breve{\phi}_\lambda + Q^b_{0,\lambda}Q^b_{2,\lambda}\mathbf{1}\end{aligned}$$

It has been proved in Proposition VIII.3.6 that the mappings $\lambda \hookrightarrow Q_{2,\lambda}$ and $\lambda \hookrightarrow Q_{0,\lambda}$ are continuous. Since the spectral radius of these operators is strictly less than one the uniform bound proved in Lemma VIII.3.1 yields the formula.
The continuity of the density of $N$ is a consequence of the continuity of the mapping $\lambda \hookrightarrow \phi_\lambda$ and $\lambda \hookrightarrow \breve{\phi}(\lambda)$ proved in Proposition VIII.3.6. ∎

See Problem VIII.6.8 for another formula of the same kind.

## VIII.3.2 The Markov Model

In the one dimensional Markov case (both discrete and continuous) the proof of localization is essentially the same as for independent potentials since this proof also reduces to checking some spectral properties of the Laplace transform.

For $T = \mathbb{R}$ or $\mathbb{Z}$ let $\{X(t); t \in T\}$ be an ergodic"invertible" Markov process with values in the compact space $M$ satisfying all the assumptions of Section IV.5. The potentials are of the form $V(t) = F(X(t))$ for $F$ a real function defined on $M$ (this is a slight modification of the notations of Section IV.5 in which $F$ was an $SL(2, \mathbb{R})$ valued function). It has been proved in V.3.5 that if $F$ has a lower bound equal to zero then the spectrum of the associated ergodic family of Schrödinger operators is $\mathbb{P}$-almost surely equal to $[0, \infty)$.

Denoting by $B$ the the projective line we have already constructed the Markov process $\{(X(t), U_\lambda(t)b); t \in T\}$ in $(M \times B)$.

For a cocycle $\chi$ on $B$, $t \in T^+$, $f \in C(M \times B)$ the Laplace transform is defined by:

$$\begin{aligned} T^t_{\chi,\lambda} f(x,b) &= \mathbb{E}_x\{f(X(t), U_\lambda(t)b)\chi(U_\lambda(t), b)\} \\ \check{T}^t_{\chi,\lambda} f(x,b) &= \mathbb{E}_x\{f(X(-t), U_\lambda(-t)b)\chi(U_\lambda(-t), b)\} \end{aligned}$$

For $t = 1$ we drop the superscript 1. Let $m$ be the Cauchy measure on $B$, and let us choose for $\chi$ the usual Radon Nikodym cocycle:

$$\chi(g, \bar{v}) = \frac{dg^{-1}m}{dm}(\bar{v}) = \left(\frac{||v||}{||gv||}\right)^2$$

The cocycle $\chi$ will be fixed in all this subsection. For a real $\delta$ we set $T_{\delta,\lambda}$ for the Laplace transform associated with the cocycle $\chi^{\delta/2}$

We will choose at times to represent the projective line as the torus $\mathbb{T} = [0, \pi]$ to be in accordance with earlier papers on the subject. As we said above the proof of localization will work for a large class of Markovian models as long as the operators associated to the Laplace transform satisfy some conditions introduced in the following definition.

**Definition VIII.3.11** *Let us consider the operators $T^t_{\delta,\lambda}$ on the Banach space $C(M \times B)$. We denote by $\mathcal{P}$ the set of following properties :*

1. *For any real $\lambda$ and $\delta \in [0, 2]$, $T_{\delta,\lambda}$ is a compact operator and the mapping $\lambda \hookrightarrow T_{\delta,\lambda}$ is continuous.*
2. *For any positive $\lambda$, the operator $T_{0,\lambda}$ admits a unique invariant probability measure on $M \times B$ say $\nu_\lambda$ and moreover $T_{0,\lambda}$ is aperiodic.*
3. *For any real $\lambda$ there exists $\delta > 0$ such that the spectral radius of $T_{\delta,\lambda}$ is strictly less than 1.*

*We call dual properties the same assumptions on the dual operators $\check{T}_{\delta,\lambda}$.*

These properties are easily obtained in the particular case of a Brownian motion $X(t)$ on the torus $M$ assuming that $F$ is a Morse function, that is:

- $F$ is $C^\infty$ function and $\inf_{x\in M} F(x) = 0$.
- For any point $x$ in $M$ there exists an integer $k$ with $\frac{d^k F}{dx^k}(x) \neq 0$

**Proposition VIII.3.12** *Let $X(t)$ be a Brownian motion on the torus $M$. Then one has:*

1. *The infinitesimal generator of the process $(X(t), U(t)\vartheta)$ is given by:*
$$L = \frac{1}{2}X^2 + Y = -\frac{1}{2}\frac{\partial^2}{\partial x^2} + \left(\cos^2\vartheta + (\lambda - F(x)\sin^2\vartheta\right)\frac{\partial}{\partial\vartheta}$$
2. *If $F$ is a Morse function then the Lie Algebra generated by the vector fields $X$ and $Y$ is two dimensional at any point of $M \times B$.*

*Proof:*
The Prüffer variable $\vartheta(t) = U_\lambda(t)\vartheta$ introduced in Chapter III satisfies the following first order differential equation:

$$\dot\vartheta(t) = \cos^2\vartheta(t) + (\lambda - V(t))\sin^2\vartheta(t)$$

This implies immediately the first formula, since for a $C^2$ function $f$ with compact support on $M \times B$ one can write:

$$f(X(t),\vartheta(t)) - f(x,\vartheta) = f(X(t),\vartheta(t)) - f(x,\vartheta(t)) + f(x,\vartheta(t)) - f(x,\vartheta)$$

Applying the Taylor formula at the second order to the first part of the difference, the relations

$$\lim_{t\to 0}\frac{1}{t}\mathbb{E}\{(X(t)-x)\} = 0 \quad , \quad \lim_{t\to 0}\frac{1}{t}\mathbb{E}\{(X(t)-x)^2\} = 1$$

yield immediately that

$$\lim_{t\to 0}\frac{1}{t}\left(f(X(t),\vartheta(t)) - f(x,\vartheta(t))\right) = \frac{\partial^2}{\partial x^2}f(x,\vartheta)$$

In the same way one can write that

$$\begin{aligned}\lim_{t\to 0}\frac{1}{t}\mathbb{E}\{(\vartheta(t)-\vartheta)\} &= \lim_{t\to 0}\frac{1}{t}\mathbb{E}\{\int_0^t \dot\vartheta(s)ds\} \\ &= \dot\vartheta(0) = \cos^2\vartheta + (\lambda - F(x)\sin^2\vartheta\end{aligned}$$

To prove 2. we remark that if we compute $k$ successive Lie brakets one obtains:

$$[X,[X,[\ldots[X,Y]]]\ldots] = -\sin^2\vartheta\frac{d^kF}{dx^k}(x)\frac{\partial}{\partial\vartheta}$$

Thus the result is clear if $\sin\vartheta \neq 0$ and for $\sin\vartheta = 0$ we remark that $Y = \pm\frac{\partial}{\partial\vartheta}$ and the conclusion follows. ∎

**Proposition VIII.3.13** *Let us assume that $X(t)$ is a Brownian motion on the torus $M$ and that $F$ is a Morse function. Then the property $\mathcal{P}$ and its dual, in Definition VIII.3.11, hold.*

*Proof:*
Proposition VIII.3.12 and the Hörmander theorem [151] imply that the infinitesimal generator $L$ is hypoelliptic. It follows from [154] that the semigroup $\{T^t_{0,\lambda}\,;\, t \geq 0\}$ has a density $p^t_\lambda((x,\vartheta),(x',\vartheta'))$ which is jointly continuous with respect to the variables $(t,\lambda,x,\vartheta,x',\vartheta')$ (and $C^\infty$ with respect to $(x,\vartheta,x',\vartheta')$). This property implies immediately the compactness of the operators $T^t_{\delta,\lambda}$ since these operators have continuous kernels and also the continuity of the mapping $\lambda \hookrightarrow T^t_{\delta,\lambda}$.
It is also proved in [154] that the Markov chain associated with the operator $T_{0,\lambda}$ is Harris recurrent on $M \times B$ with respect to the Riemannian measure. This yields the uniqueness of the invariant measure and the aperiodicity of the transition operator.
The uniqueness of the invariant measure is also a consequence of Proposition IV.5.15 which also implies that the Lyapunov exponent is positive, hence condition 3. in property $\mathcal{P}$ by the standard argument of Proposition IV.5.17. ∎

Actually these properties hold in the general case of a Brownian motion on a connected compact Riemannian manifold $M$.

**Proposition VIII.3.14** *Assuming the property $\mathcal{P}$ of Definition VIII.3.11 and its dual one has the decompositions:*

$$\begin{aligned} T^n_{0,\lambda}f &= \nu_\lambda(f) + Q^n_{\lambda,0}f \\ T^n_{2,\lambda}f &= (\pi\otimes m)(f)\psi_\lambda + Q^n_{\lambda,2}f \end{aligned}$$

*In these formulas $m$ is the normalized Lebesgue measure on $[0,\pi]$, $\psi_\lambda$ is the density with respect to $\pi\otimes m$ of the invariant probability measure $\check{\nu}_\lambda$ of the operator $\check{T}_{0,\lambda}$ and and the operators $Q_{0,\lambda}, Q_{2,\lambda}$ have a spectral radius strictly less than 1.*

*Proof:*
Direct consequences of Proposition IV.5.21. ∎

**Proposition VIII.3.15** *Let us assume that the property $\mathcal{P}$ of Definition VIII.3.11 and its dual hold, then for a bounded interval $\mathcal{I}$ there exist finite constants $C_1$, $C_2$ and $\rho < 1$ such that for $n \in \mathbf{N}$ , $\lambda \in \mathcal{I}$ one has:*

$$(\mathcal{P}_1): \quad ||T^n_{1,\lambda}|| \leq C_1 \rho^n$$
$$(\mathcal{P}_2): \quad ||T^n_{2,\lambda}|| \leq C_2$$

*Proof:*
The proof is exactly the same as in Proposition VIII.3.6. ∎

**Proposition VIII.3.16** *Let us assume that the properties $(\mathcal{P}_1)$ and $(\mathcal{P}_2)$ of Proposition VIII.3.15 hold and let $\mathcal{I}$ be an open bounded interval. Then for $\mathbb{P}$-almost all $\omega$ the spectrum of $H(\omega)$ is pure point on $\mathcal{I}$ with exponentially decaying eigenfunctions.*

*Proof:*
The proof is quite the same as in the independent case if one replaces the successive integrations by conditional expectations. If we consider the Markov process $Z^+_t = Z_{t-b}$ then the Markov property can be written:

$$\mathbb{E}_x\{||U_\lambda(t+s,-b)\alpha||^\delta f(Z^+_{t+s})|\mathcal{F}_{-b,t}\} = ||U_\lambda(t,-b)\alpha||^\delta T^s_{\delta,\lambda} f(Z^+_t)$$

Recall (Proposition III.4.16) that a sufficient condition for exponential localization on the open interval $\mathcal{I}$ is the following :
*For a fixed angle $\alpha$ there exists a finite constant $C$ and $\rho < 1$ such that for any box $\Lambda = [-b\ ,\ b]$ and $n \in \Lambda$ one has:*

$$\int_{\mathcal{I}} \mathbb{E}\{||U_\lambda(n,-b)\alpha|| \frac{||U_\lambda(0,-b)\alpha||^2}{||U_\lambda(b,-b)\alpha||^2}\}\, d\lambda \leq C\rho^n$$

Computing successive conditional expectations rather than expectations the same procedure as in the independent case allows us to rewrite the right handside of the above inequality as:

$$\int_{\mathcal{I}} \left(T^{b-|n|}_{0,\lambda} T^{|n|}_{1,\lambda} T^b_{2,\lambda} \mathbf{1}\right)(x,\alpha)\, d\pi(x) d\lambda \qquad \textit{for } -b \leq n \leq 0$$
$$\int_{\mathcal{I}} \left(T^b_{0,\lambda} T^n_{1,\lambda} T^{b-n}_{2,\lambda} \mathbf{1}\right)(x,\alpha)\, d\pi(x) d\lambda \qquad \textit{for } 0 \leq n \leq b$$

The uniform bounds of $||T^n_{\lambda,2}||$ and $||T^n_{\lambda,1}||$ on the interval $\mathcal{I}$ yield the result. ∎

It is again possible to build a proof using the Laplace transform acting on $L^2(M \times B, dx \otimes dm)$ rather than on $C(M \times B)$. This is what is done by Royer in [278]

Assuming the same hypotheses, one can obtain an explicit formula for the averaged spectral measure.

**Proposition VIII.3.17** *Let us assume that the property $\mathcal{P}$ and its dual hold, then the averaged spectral measure $N = \mathbb{E}\{\sigma\}$ has a continuous density with respect to the Lebesgue measure given by:*

$$\frac{dN}{d\lambda} = \frac{1}{\pi}\int_0^\pi \psi_\lambda(x,\alpha)\breve{\psi}(x,\alpha)d\pi(x)d\alpha$$

*Proof:*
Let $\phi$ be a continuous function with compact support, and $\Lambda$ be the box $[-b, b]$. One has:

$$\begin{aligned} N(\phi) &= \lim_{b\to+\infty} \mathbb{E}\{\sigma^\Lambda_\alpha(\phi)\} \\ &= \lim_{b\to+\infty} \mathbb{E}\int \phi(\lambda)||U_\lambda(0,-b)\alpha||^2||U_\lambda(b,-b)\alpha^{-2}||\, d\lambda \\ &= \lim_{b\to+\infty} \int \phi(\lambda)\left(T^b_{0,\lambda}T^b_{2,\lambda}\mathbf{1}\right)(x,\alpha)d\pi(x)d\lambda \end{aligned}$$

The proof is now the same as in the independent case. ∎

It is interesting to discuss the assumptions made in this subsection and their relationship with those we made in the section devoted to Kotani's trick.
Localization occurs in the Markovian model as long as properties $\mathcal{P}_1$, $\mathcal{P}_2$ appearing in Proposition VIII.3.15 hold. In fact the argument given below shows that $\mathcal{P}_2$ is the most important requirement. First we remark that one can replace in Proposition III.4.14 the term $||U_\lambda(n)||$ by $||U_\lambda(n)||^s$ for a parameter $0 < s < 2$ without changing the proof. It follows that we can replace in the above proof of localization the operator $T^n_{1,\lambda}$ by the

operator $T^n_{s,\lambda}$. If we look to Propositions IV.5.17 and Theorem IV.5.10 which require only strong irreducibility and contractivity (and this is true in the one dimensional case as long as the potentials are not constant !) then we see that for a compact interval $\mathcal{I}$ there exists a real $s > 0$, a finite constant $C$ and $\rho < 1$ such that for $n \in \mathbf{N}$ and $\lambda \in \mathcal{I}$ one has:

$$||T^n_{s,\lambda}|| \leq C\rho^n$$

The uniformity of the constants with respect to $\lambda$ in a compact set can be obtained by improving the uniform convergence stated in Theorem IV.5.10 (ii) in the same way as in Proposition V.4.9.

It follows that the real strong assumption in this subsection is the uniform bound of $||T^n_{2,\lambda}||$.

If we return to the preceding section we see that apart from the positivity of the exponent (which is granted as long as the potential process is non-deterministic) the essential hypothesis is the following : *There exists $c > 0$ such that the conditional law of the angle $U(c)b$ given the $\sigma$-algebra $\mathcal{F}^*$ generated by the potentials outside $[0\ ,\ c]$ has a density uniformly bounded with respect to $\lambda \in \mathcal{I}$, $b \in B$ and the potentials outside $[0, c]$.*

If one denotes by $C$ the bound on this density, one has for $t > c$ :

$$\begin{aligned}
||T^t_{2,\lambda}|| &= \sup_{\bar{v}\in B, x\in M} \mathbb{E}_x\{\frac{||v||^2}{||U_\lambda(t)v||^2}\} \\
&= \sup_{\bar{v}\in B, x\in M} \mathbb{E}_x\{\frac{||v||^2}{||U_\lambda(c)v||^2}\mathbb{E}_x\{\frac{||U_\lambda(c)v||^2}{||U_\lambda(t,c)U_\lambda(c)v||^2}|\mathcal{F}^*\}\} \\
&\leq C \sup_{\bar{v}\in B, x\in M} \mathbb{E}_x\{\frac{||v||^2}{||U_\lambda(c)v||^2}\int \frac{||u||^2}{||U_\lambda(t,c)u||^2}\, dm(\bar{u})\} \\
&= C \sup_{\bar{v}\in B, x\in M} \mathbb{E}_x\{\frac{||v||^2}{||U_\lambda(c)v||^2}\}
\end{aligned}$$

In the case $\Omega_0 = \Omega$ the hypothesis on the boundedness of the potentials $\{V(t,\omega)\ ; t \in [0\ ,\ c]\}$ implies the boundedness of the term $||v||^2||U_\lambda(c)v||^{-2}$ and this shows that the hypotheses of Theorem VIII.1.11 imply the desired bound for the operator $T_{2,\lambda}$. Thus, in the Markovian case, the two methods are more or less equivalent. Unfortunately the above proof of localization seems difficult to extend to the general ergodic case.

## VIII.3.3 The Discrete I.I.D. Model on the Strip

The proof of localization on the strip is a bit more involved than in the one dimensional case. This is due to the fact that matrices are not so easily tractable as real numbers. Nevertheless, the strategy of the proof is the same, namely localization occurs as long as the the spectral radius of operators defining the Laplace transforms satisfy some requirements.

In this subsection we assume that the potentials $\{V_i(n); i = 1 \cdots \ell, n \in \mathbb{Z}\}$ form an i.i.d. family of random real variables whose law is absolutely continuous on $\mathbb{R}$ and has a moment of order $\ell(\ell+1)$. Recall that in this situation we define the probability measure $\mu_\lambda$ on the symplectic group $SP(\ell, \mathbb{R})$ as the law of the $2\ell \times 2\ell$ matrix

$$\begin{pmatrix} V(n) - \lambda I_\ell & -I_\ell \\ I_\ell & 0 \end{pmatrix}$$

where $V(n)$ is the $\ell \times \ell$ potential matrix associated with the Schrödinger operator on a strip of width $\ell$. If $\chi$ is a $\mu_\lambda$ integrable cocycle on a boundary $B$ then the Laplace transform $T_{\chi,\lambda}$ is then defined on the set of continuous functions on $B$ by the formula:

$$T_{\chi,\lambda} f(b) = \int f(gb)\chi(g,b)\, d\mu_\lambda(g)$$

We have already computed the Poisson kernel of the boundary $L_{\ell-1}$ say $r_{\ell-1}$ and it is given by:

$$r_{\ell-1}(g,\bar{y}) = \frac{dg^{-1}m_{\ell-1}}{dm_{\ell-1}}(\bar{y}) = (\frac{||y||}{||gy||})^{\ell+2}$$

The Poisson kernel of the boundary $L_{\ell-1,\ell}$ say $r_{\ell-1,\ell}$ is given by:

$$r_{\ell-1,\ell}(g,(\bar{y},\bar{x})) = (\frac{||y||}{||gy||})^{\ell}(\frac{||x||}{||gx||})^2$$

We will consider the cocycle $\chi(g,(\bar{x},\bar{y}))$ on $L_{\ell-1,\ell}$ defined by:

$$\chi(g,(\bar{y},\bar{x})) = r_{\ell-1,\ell}(g,(\bar{y},\bar{x}))r_{\ell-1}^{-1}(g,\bar{x}) = (\frac{||y||}{||gy||})^{-2}(\frac{||x||}{||gx||})^2$$

One has $\bar{r}_{\ell-1}(g) \le ||g||^{\ell^2+\ell-2}, r_{\ell-1,\ell}(g) \le ||g||^{\ell^2+\ell}$ and $\bar{\chi}(g) \le ||g||^{4\ell-2}$ thus if the distribution of the potentials has a moment or order $\ell(\ell+1)$ all these cocycles are integrable. We define the operator $T_{\delta,\lambda}$ as the Laplace operator on $\mathcal{C}(L_{\ell-1,\ell})$ associated with the cocycle $\chi^{\delta/2}$

In the same way as in the preceding subsection, we introduce the following definition:

**Definition VIII.3.18** *For a fixed interval $\mathcal{I}$ we consider the following properties:*

$(\mathcal{P}_1)$ $\quad \|T^n_{1,\lambda}\| \leq C_1 \rho^n \quad \lambda \in \mathcal{I},\ n \in \mathbf{N}$
$(\mathcal{P}_2)$ $\quad \|T^n_{2,\lambda}\| \leq C_2 \quad \lambda \in \mathcal{I},\ n \in \mathbf{N}$

In order to work with the cocycle $\chi$ we need to introduce the following notations. Let $(e_1, \cdots, e_{2\ell})$ be the canonical basis of $\mathbb{R}^d$ then we set:

$$u_i = e_1 \wedge \cdots \wedge e_{i-1} \wedge e_{i+1} \cdots e_\ell \ , \ u = e_1 \wedge \cdots \wedge e_\ell$$

We will also denote by $u$ the $d \times \ell$ matrix $\binom{I_\ell}{0}$. Any matrix $g \in SP(\ell, \mathbb{R})$ can be written $g = \tilde{g}s$ where $\tilde{g}$ is orthogonal and $s$ is of the form $\begin{pmatrix} A & B \\ 0 & A'^{-1} \end{pmatrix}$

**Lemma VIII.3.19** *Let $g$ and $h$ be two matrices in the group $SP(\ell, \mathbb{R})$. Then one has:*

**(i)** $\operatorname{tr}((u'g'gu)^{-1}) = \sum_{i=1}^{\ell} \chi(g, (\bar{u}_i, \bar{u}))$
**(ii)** $\operatorname{tr}((u'h'hu)(u'h'g'ghu)^{-1}) = \sum_{i=1}^{\ell} \chi(g, \tilde{h}(\bar{u}_i, \bar{u}))$

*Proof:*
The formula (i) is a simple consequence of the definition of the cocycle $\chi$ and of the computation of the inverse of a matrix by means of cofactors. To prove (ii) we remark that $hu = \tilde{h}uA$, then

$$u'h'hu = A'A \ , \ u'h'g'ghu = A'u'\tilde{h}'g'g\tilde{h}uA$$

The invariance of the trace by exchange of the factors implies that the left part of (ii) is equal to $\operatorname{tr}((u'\tilde{h}'g'g\tilde{h}u)^{-1})$ and it remains to apply the formula (i) to the matrix $g\tilde{h}$ taking into account the cocycle relation $\chi(g_1 g_2, b) = \chi(g_1, g_2 b)\chi(g_2, b)$ and the identity $\chi(k, (\bar{u}_i, \bar{u})) = 1$ for any orthogonal matrix $k$. ∎

**Theorem VIII.3.20** *Let $\mathcal{I}$ be a bounded open interval and let us assume that the properties $\mathcal{P}_1$ and $\mathcal{P}_2$ hold on $I$. Then for $\mathbb{P}$-almost all $\omega$ the spectrum of $H(\omega)$ is pure point on $\mathcal{I}$ with exponentially decreasing eigenfunctions.*

*Proof:*
For a positive integer $b$ and for $n \in [-b, b]$ let us set:

$$\Gamma_n = u'U'_\lambda(n,-b)U_\lambda(n,-b)u$$

Applying the criterion of localization given in Proposition III.5.23 we have only to to prove that there exists a finite constant $C$ and $\rho < 1$ such that for any box $\Lambda = [-b\ ,\ b]$ and $n \in \Lambda$ one has:

$$(*) \quad \int_{\mathcal{I}} \mathbb{E}\{\sqrt{tr(\Gamma_0\Gamma_b^{-1})tr(\Gamma_n\Gamma_b^{-1})}\}\, d\lambda \leq C\rho^{|n|}$$

For notational convenience we will set:

$$\begin{array}{ll} X = U_\lambda(n,-b)\ ,\ Y = U_\lambda(0,n)\ ,\ Z = U_\lambda(b,0) & \text{if } -b \leq n \leq 0 \\ X = U_\lambda(0,-b)\ ,\ Y = U_\lambda(n,0)\ ,\ Z = U_\lambda(b,n) & \text{if } 0 \leq n \leq b \end{array}$$

From the Proposition VIII.3.19 we obtain that:

$$\mathrm{tr}(\Gamma_0\Gamma_b^{-1}) = \begin{cases} \sum_{i=1}^{\ell} \chi(ZY, \tilde{X}(\bar{u}_i, \bar{u})) & \text{if } n \geq 0 \\ \sum_{i=1}^{\ell} \chi(Z, \tilde{YX}(\bar{u}_i, \bar{u})) & \text{if } n \leq 0 \end{cases}$$

$$\mathrm{tr}(\Gamma_n\Gamma_b^{-1}) = \begin{cases} \sum_{i=1}^{\ell} \chi(Z, \tilde{YX}(\bar{u}_i, \bar{u})) & \text{if } n \geq 0 \\ \sum_{i=1}^{\ell} \chi(ZY, \tilde{X}(\bar{u}_i, \bar{u})) & \text{if } n \leq 0 \end{cases}$$

It follows, using the Cauchy-Schwarz inequality that for any $n$ :

$$\begin{aligned} &\sqrt{tr(\Gamma_0\Gamma_b^{-1})tr(\Gamma_n\Gamma_b^{-1})} \\ &\quad \leq \sum_{i,j=1}^{\ell} \chi^{1/2}(Z, \tilde{YX}(\bar{u}_i, \bar{u}))\chi^{1/2}(ZY, \tilde{X}(\bar{u}_j, \bar{u})) \\ &\quad = \sum_{i,j=1}^{\ell} \chi^{1/2}(Z, \tilde{YX}(\bar{u}_i, \bar{u}))\chi^{1/2}(Z, Y\tilde{X}(\bar{u}_j, \bar{u}))\chi^{1/2}(Y, \tilde{X}(\bar{u}_j, \bar{u})) \end{aligned}$$

Integrating only with respect to $Z$ and using again the Cauchy-Schwarz inequality one obtains:

$$\mathbb{E}\{\chi^{1/2}(Z, \tilde{YX}(\bar{u}_i, \bar{u}))\chi^{1/2}(Z, Y\tilde{X}(\bar{u}_j, \bar{u}))\}$$

$$\leq \mathbb{E}\{\chi(Z, Y\tilde{X}(\bar{u}_i, \bar{u}))\}\mathbb{E}\{\chi(Z, Y\tilde{X}(\bar{u}_j, \bar{u}))\}$$
$$\leq \begin{cases} ||T_{2,\lambda}^{a-n}|| & \text{if } n \geq 0 \\ ||T_{2,\lambda}^{b}|| & \text{if } n \leq 0 \end{cases}$$

A new integration with respect to $Y$ yields :

$$\mathbb{E}\{\chi^{1/2}(Y, \tilde{X}(\bar{u}_j, \bar{u}))\} \leq ||T_{1,\lambda}^{|n|}||$$

The result is now a direct consequence of the hypothesis of the theorem. ∎

It now remains to prove that under the hypothesis of this subsection the properties $\mathcal{P}_1$ and $\mathcal{P}_2$ hold for any compact set $\mathcal{I}$. Since we have assumed that the law of the potentials has a density it follows from Proposition IV.4.33 that for any $\lambda$ the measure $\mu_\lambda^{2\ell+1}$ has a density on $SP(\ell, \mathbb{R})$. Moreover using the integrability of $||g||^{\ell(\ell+1)}$ it is easy to prove in the same way as in Proposition VIII.3.4 that the function $\lambda \hookrightarrow T_{\delta,\lambda}$ is a continuous maping from the real line to the Banach space of bounded operators on the space of continuous functions on $L_{\ell-1,\ell}$. Hence one can suppose that $\mu_\lambda$ has a density since we can replace $T_{\delta,\lambda}$ by $T_{\delta,\lambda}^{2\ell+1}$ in the proof of $\mathcal{P}_1$ and $\mathcal{P}_2$. From the theory developed in Chapter IV we know that the operator $T_{0,\lambda}$ has a unique invariant measure say $\nu_\lambda$ on $L_{\ell-1,\ell}$. Moreover one has:

$$\begin{aligned}
&\int \log \chi^{1/2}(g, (\bar{y}, \bar{x}))\, d\mu_\lambda(g) d\nu_\lambda(\bar{x}, \bar{y}) \\
&= \int (\log \frac{||gx||}{||x||} - \log \frac{||gy||}{||y||})\, d\mu_\lambda(g) d\nu_\lambda(\bar{x}, \bar{y}) \\
&= (\gamma_1(\lambda) + \ldots + \gamma_{\ell-1}(\lambda)) - (\gamma_1(\lambda) + \ldots + \gamma_\ell(\lambda)) \\
&= -\gamma_\ell(\lambda) < 0
\end{aligned}$$

The second equality follows from the fact that the invariant measure $\nu_\lambda$ projects on the boundaries $L_{\ell-1}$ and $L_\ell$ according to the respective invariant measures and the formula giving the partial sums of Lyapunov exponents in Proposition IV.6.2. Proposition IV.3.5 then implies that for each $\lambda$ there exists a strictly positive real $\alpha$ such that the spectral radius of $T_{\alpha,\lambda}$ is strictly less than one.
On the boundary $L_{\ell-1}$ let us denote by $X_\lambda$ the transition operator of the Markov chain $\bar{y} \hookrightarrow g\bar{y}$ and by $Y_\lambda$ the Laplace operator associated with the Radon-Nikodym cocycle $r_{\ell-1}$. We denote by $Z_\lambda$ the Laplace operator on $L_{\ell-1,\ell}$ associated with the cocycle $r_{\ell-1}$. It follows from the general theory of compact Laplace operators that $X_\lambda$ and $\check{Y}_\lambda$ have the same

spectral properties, and the same conclusion holds for $\check{Z}_\lambda$ and $T_{2,\lambda}$ since $\chi = r_{\ell-1}^{-1} r_{\ell-1,\ell}$. Moreover it is readily seen that for any $n$ $\|\check{Y}^n\| = \|\check{Z}^n\|$ and we can conclude that $X_\lambda, \check{Y}_\lambda, \check{Z}_\lambda, T_{2\lambda}$ have the same spectral radius which is obviously equal to one for the operator $X_\lambda$. If we denote by $r_\lambda(\delta)$ the spectral radius of $T_{\delta,\lambda}$ ,we know that this function is convex, that $r_\lambda(0) = r_\lambda(0) = 1$ and that there exists a positive $\alpha$ with $r_\lambda(\alpha) < 1$. One can conclude that $r_\lambda(1) < 1$ and the Proposition VIII.3.1 yields the property $\mathcal{P}_1$.
The property $\mathcal{P}_2$ is more difficult to check since one does not know, as in the one dimensional case, that 1 is the only eigenvalue of modulus one of $T_{2,\lambda}$ and that this eigenvalue is simple (altough it is likely that these properties are true).To overcome this difficulty we observe that these properties are true for $X_\lambda$ hence for $\check{Y}_\lambda$. It follows that $\mathcal{P}_2$ is true for these operators hence for the operator $\check{Z}_\lambda$. Then Proposition VIII.3.2 and the fact that $\check{Z}_\lambda$ and $T_{2,\lambda}$ have the same characteristic spaces implies that for each $\lambda$ $\|T_{2,\lambda}^n\|$ is a bounded sequence but this does not imply any uniform bound in $\lambda$ (see Problem VIII.6.7). Nevertheless under some restrictions on the density $\phi$ of the common law of the potentials one can obtain this uniformity. For example if we assume that $\sup_{\lambda \in \mathcal{I}} \phi(s-\lambda) \leq \psi(s)$ for an integrable function $\psi$ then it is readily seen that $\mathcal{P}_2$ holds on $\mathcal{I}$ by writing

$$\begin{aligned}
\|T_{2,\lambda}^n\| &= \sup_{b \in B} \int \chi(U_0(n), b) \prod_{i=1}^{n\ell} \phi(s_i - \lambda)\, ds_i \\
&\leq \sup_{b \in B} \int \chi(U_0(n), b) \prod_{i=1}^{n\ell} \psi(s_i)\, ds_i \\
&= \|\Psi_2^n\|
\end{aligned}$$

where $\Psi_2$ is the Laplace operator similar to $T_2$ associated with potentials distributed according to a law of density $\psi$. As we already said it is likely that such a condition is useless...

## VIII.4 Localization for Singular Potentials

In the independent case, the essential hypothesis that we made to prove localization was the absolute continuity of the distribution of the potentials (or at least the existence of a nonzero absolutely continuous part of this distribution). In this section we will consider the case of i.i.d. potentials on the one dimensional lattice and we only assume that :

- The potentials are not constant
- There exists a real number $\eta > 0$ with $\mathbb{E}\{|V(0)|^\eta\} < \infty$

One can remark that there are no regularity assumptions on the law of the potentials, in particular one can consider potentials taking finetely many values, but the price we have to pay is to refer to a highly nontrivial multiscale analysis developed in the next chapter.
An extension of this result to the strip can be found in [18].

For a box $\Lambda = [a\ b]$ we denote by $H^\Lambda$ the operator $H$ restricted to this box with Dirichlet boundary conditions, by $G_\lambda^\Lambda$ its Green kernel and by $\Sigma(H^\Lambda)$ the (finite) spectrum of the operator $H^\Lambda$. We will also define the projection operator $\Gamma_n$ on the site $n$ defined by:

$$(\Gamma_n\psi)(k) = \begin{cases} \psi(n) & \text{if } k = n \\ 0 & \text{if } k \neq n \end{cases}$$

Let ${}^+H^\Lambda = H^\Lambda - i\Gamma_b$ be the operator with boundary condition $i$ at the right endpoint $b$ of $\Lambda$ (this operator is nothing but $H_{0\,i}^\Lambda$ with our previous notations). Similarly one defines the operator ${}^-H^\Lambda$ with boundary condition $i$ at the left endpoint $a$ of the box $\Lambda$. It is readily seen that for any real $\lambda$ the symmetric operators ${}^\pm H^\Lambda - \lambda\mathbf{I}$ are invertible and we denote by ${}^\pm G_\lambda^\Lambda$ their inverses.

**Lemma VIII.4.1** *With the above notations one has:*

$$\begin{aligned} |{}^+G_\lambda^\Lambda(a,b)|^2 &= ||U_\lambda(b,a)e_1||^2 \\ |{}^-G_\lambda^\Lambda(a,b)|^2 &= ||U_\lambda(a,b)e_2||^2 \end{aligned}$$

*where* $\{e_1, e_2\}$ *is the usual basis of* $\mathbb{R}^2$.

*Proof:*
The result is a direct application of the general formula expressing the Green kernel by means of two solutions of the eigenvalue equation proved in Proposition III.2.10. ∎

**Lemma VIII.4.2** *Let $\Lambda$ be the box $[-n\,,\,n]$ and $\psi$ be a normalized eigenvector associated with the eigenvalue $\lambda$ of the operator $H^{\Lambda}$. Then for $0 < m < n$ one has:*

$$\begin{aligned} |\psi(-n)| &\leq 2|{}^{+}G_{\lambda}^{[-n,-m]}(-n,-m)| \\ |\psi(n)| &\leq 2|{}^{-}G_{\lambda}^{[m,n]}(m,n)| \end{aligned}$$

*Proof:*
If $\Lambda' \subset \Lambda$ one can consider the operator $H^{\Lambda'}$ to act on $\ell^2(\Lambda)$ if we set

$$H^{\Lambda'} = \mathbf{1}_{\Lambda'} H^{\Lambda'} \mathbf{1}_{\Lambda'}$$

where $\mathbf{1}_{\Lambda'}$ is the projection operator on $\ell^2(\Lambda')$. With this convention one can write :

$$H^{\Lambda} = (H^{[-n,-m]} - i\Gamma_{-m} + i\Gamma_{-m}) - \Gamma_{-m} - \Gamma_{-m+1} + H^{[-m+1,n]}$$

This relation gives immediately:

$$\mathbf{1}_{[-n,-m]}\psi = -{}^{+}G_{\lambda}^{[-n,-m]}(i\Gamma_{-m} - \Gamma_{-m})\psi$$

Since $||\psi|| = 1$ one obtains the first relation. The second is proved similarly. ∎

**Proposition VIII.4.3** *Let $\mathcal{I}$ be a bounded interval. Then there exists a number $\alpha_0 > 0$ such that for each $\alpha \in (0, \alpha_0)$ there exist constants $\rho_\alpha < 1$ and $C_\alpha < \infty$ satisfying:*

$$\sup_{\lambda \in \mathcal{I}} \mathbb{E}\{|{}^{\pm}G_{\lambda}^{\Lambda}(a,b)|^{\alpha}\} \leq C_\alpha \rho_\alpha^n$$

*for any box $\Lambda = [a\ b]$ of length $n$.*

*Proof:*
We already know that, under the hypothesis of this section, the upper Lyapunov exponent $\gamma(\lambda)$ is strictly positive (Proposition IV.4.25 and Theorem IV.4.15) and that the convergence of the sequence $\frac{1}{n}\mathbb{E}\{\log \frac{||U_\lambda(n)v||}{||v||}\}$ to $\gamma(\lambda)$ is uniform with respect to $(\lambda, \bar{v}) \in \mathcal{I} \otimes B$ where $B$ denotes the projective line. Taking into account the formulas relating the Green function and the propagator in Lemma VIII.4.1 the result is a simple consequence of Proposition IV.3.5 applied to the cocycle $\sigma(g, \bar{v}) = \frac{||gv||}{||v||}$. ∎

**Corollary VIII.4.4** *Let $\mathcal{I}$ be a bounded interval. Then there exist constants $\alpha_1 > 0$, $\rho > 0$ and an integer $n_1$ such that one has:*

$$\mathbb{P}\{|{}^{\pm}G_\lambda^\Lambda(a,b)| > \epsilon\} \leq e^{-n\rho}\epsilon^{-\alpha_1}$$

*for any box $\Lambda = [a,b]$ of length $n$, $\epsilon > 0$, $\lambda \in \mathcal{I}$ and $n \geq n_1$.*

*Proof:*
This is a direct consequence of Proposition VIII.4.3 together with the Chebyshev's inequality. ∎

We now precise the dependence of the propagator $U_n(\lambda)$ with respect to the parameter $\lambda$.

**Lemma VIII.4.5** *Let $g_n$ and $h_n$ be two sequences of matrices of $SL(2,,\mathbb{R})$ such that there exists a real number $\rho$ and a matrix $Y$ with $||Y|| \leq 1$ satisfying $g_k h_k^{-1} = I + \rho Y$ for all $k$. Let us set $A_n = g_n \dots g_1$ and similarly $B_n = h_n \dots h_1$, then we have:*

$$||A_n B_n^{-1} - I|| \leq C_n|\rho| \qquad \text{with} \quad C_n = 1 + \sum_{k=2}^{n}(\prod_{r=k}^{n} ||g_r||\ ||h_r||)$$

*where $I$ is the identity matrix of order two.*

*Proof:*
Write $A_{n,k} = g_n \dots g_k$ *and* $B_{n,k} = h_n \dots h_k$. Then one has:

$$\begin{aligned} A_n B_n^{-1} &= g_n \dots g_1 h_1^{-1} \dots h_n^{-1} \\ &= A_{n,2}B_{n,2}^{-1} + \rho A_{n,2} Y B_{n,2}^{-1} = \dots \\ &= I + \rho \sum_{k=2}^{n} A_{n,k} Y B_{n,k}^{-1} \end{aligned}$$

The result follows. ∎

**Corollary VIII.4.6** *Let $v$ be a unit vector in $\mathbb{R}^2$. Then one has :*

$$||U_\lambda(n)v|| \leq |\lambda - \lambda'|C_n(\lambda,\lambda') + ||U_{\lambda'}(n)v||$$

*where $C_n(\lambda,\lambda') = 1 + \sum_{k=1}^{n-1}(\prod_{r=k}^{n-1} ||g_\lambda(r)||\ ||g_{\lambda'}(r)||)$*

*Proof:*
Using the relation $g_\lambda(k)g_{\lambda'}^{-1}(k) = |\lambda - \lambda'| \begin{pmatrix} 1 & 0 \\ 0 & 0 \end{pmatrix}$ one can apply the above lemma to the sequences $g_n = g_\lambda(n)$ and $h_n = g_{\lambda'}(n)$. The inequality

$$||U_\lambda(n)v|| \leq ||U_\lambda(n)U_{\lambda'}^{-1}(n) - I|| + ||U_{\lambda'}(n)v||$$

yields the result. ∎

**Proposition VIII.4.7** *Let $\mathcal{I}$ be a compact set of energy, then there exist a finite constant $C_1$ and $\delta > 0$ such that one has:*

$$I\!\!P\{|^{\pm}G_\lambda^\Lambda(a,b)| \leq \epsilon \ \ and \ \ |^{\pm}G_{\lambda'}^\Lambda(a,b)| > \sqrt{\epsilon}\} \leq C_1^n |\lambda - \lambda'|^\delta \left(\frac{\epsilon^2}{1-\epsilon}\right)^\delta$$

*for any box $\Lambda = [a, b]$ of length $n$, $0 < \epsilon < 1$ and $\lambda, \lambda' \in \mathcal{I}$.*

*Proof:*
Using the notations of Corollary VIII.4.6 and Lemma VIII.4.1 we see that the set appearing in the left hand side of the desired inequality is contained in the set:

$$\{\ C_n(\lambda, \lambda') \geq \frac{1-\epsilon}{\epsilon^2|\lambda - \lambda'|}\ \}$$

The existence of a positive moment for the distribution of the potentials implies that $\sup_{\lambda\in\mathcal{I}} I\!\!E\{||g_\lambda(r)||^\eta\} \leq K$ where $K$ is some finite constant. The Cauchy-Schwarz inequality and the independence of the potentials imply that:

$$I\!\!E\{\prod_{r=k}^{n-1} (||g_\lambda(r)||\ ||g_{\lambda'}(r)||)^{\eta/2}\} \leq K^{n-r}$$

It follows that one has $\sup_{\lambda\in\mathcal{I}} I\!\!E\{||C_n(\lambda,\lambda')||^{\eta/2}\} < C_1^n$ and the Chebyshev's inequality yields the result. ∎

The next proposition is the essential ingredient in the proof of localization. It essentially uses the local Hölder continuity of the distribution of states proved in Theorem VI.4.8 under the hypothesis of this section.

**Proposition VIII.4.8** *For each compact interval $\mathcal{I}$ of energy, there exist constants $\alpha_2 > 0$ and $C_2 < +\infty$ such that for a box $\Lambda = [-n, n]$ one has:*

$$\begin{aligned} &I\!\!P\{\text{there exists } \lambda' \in (\lambda - \epsilon\,,\ \lambda + \epsilon) \text{ and } \psi' \in \ell^2(\Lambda),\ ||\psi'|| = 1 \\ &\text{such that } H^\Lambda\psi' = \lambda'\psi' \quad \text{and} \quad (|\psi'(-n)|^2 + |\psi'(n)|^2) \leq \epsilon^2\} \\ &\leq nC_2\epsilon^{\alpha_2} \end{aligned}$$

*for any* $0 < \epsilon < 1$ *and* $\lambda \in \mathcal{I}$

*Proof:*
Let us fix $\lambda \in \mathcal{I}$, positive integers $m, n$ and $0 < \epsilon < 1$. For each $i \in \{0 \ldots \ell\}$ let $\Lambda_i$ be the box of center $i(2n+3)$ and width $2n+1$ and let $H^i$ be the restriction of the Schrödinger operator to the box $\Lambda_i$. We remark that two such boxes are separated by two points hence the operators $H^i$ are completely decoupled. All these boxes are contained in $\widehat{\Lambda} = \cup_{i=0}^{m-1} \Lambda_i$.
Let us denote by $A_i$ the following event:
*there exists a normalized eigenfunction, say $\psi_i$ of the operator $H^i$ associated with an eigenvalue $\lambda_i$ such that:*

- $|\lambda_i - \lambda| \leq \epsilon$
- *The sum of the square of the values of $\psi_i$ at the endpoint of the box $\Lambda_i$ is less than $\epsilon^2$*

We first need to check the measurability of these sets. The eigenvalues of $H^i$ are the (simple) roots of a polynomial function with respect to $\lambda$ and with polynomial coefficients with respect to the potentials in the box $\Lambda_i$. Hence if we write them by increasing order for example, any eigenvalue is a measurable function of $\omega$. It follows that each associated eigenfunction also has measurable components since it is computed inductively from its initial value.
We remark that $A_0$ is the event whose probability appears in the Proposition and that all the events $A_i$ are independent and have the same probability. For a fixed $\omega \in \Omega$, let $J$ be the subset of integers $j \in [0 \ldots m-1]$ for which $\omega \in A_j$.
For each $j \in J$, let $\widehat{\psi}_j$ be the extension of $\psi_j$ to the whole of $\widehat{\Lambda}$, assigning the value zero to the points outside $\Lambda_j$. The support of the functions $\{\psi_j \; ; \; j \in J\}$ are disjoint, hence this family of functions is an orthonormal set in $\ell^2(\Lambda)$ satisfying

$$\begin{aligned} \|(H^{\widehat{\Lambda}} - \lambda I)\widehat{\psi}_j\| &\leq 2\epsilon \qquad \text{for } j \in J \\ < \widehat{\psi}_j \, , \, H^{\widehat{\Lambda}}\widehat{\psi}_{j'} > &= \; < H^{\widehat{\Lambda}}\widehat{\psi}_j \, , \, H^{\widehat{\Lambda}}\widehat{\psi}_{j'} >= 0 \qquad \text{for } (j, j') \in J \, , \, j \neq j' \end{aligned}$$

It follows from Temple's inequality proved in Problem I.7.1 that the operator $H^{\widehat{\Lambda}}$ has at least $|J|$ distinct eigenvalues in the interval $[\lambda - 2\epsilon \, , \, \lambda + 2\epsilon]$ and one obtains:

$$\frac{1}{m}\sum_{i=0}^{m} \mathbf{1}_{A_i}(\omega) \leq \frac{1}{m}(\# \text{ of eigenvalues of } \; H^{\Lambda} \text{ in } [\lambda - 2\epsilon \, , \, \lambda + 2\epsilon])$$

If one takes the expectation on each side and let $m$ goes to $+\infty$ one gets:

$$\mathbb{P}(A_0) \leq (2n+3)\,(N(\lambda+2\epsilon) - N(\lambda-2\epsilon))$$

where we denote by $N$ the distribution function of the integrated density of states. The local Hölder continuity of the function $N$ is expressed by:

$$(N(\lambda-2\epsilon) - N(\lambda+2\epsilon)) \leq C\epsilon^{\alpha_2}$$

and this yields the desired result. ∎

We are now in position to prove the exponential localization property. Actually we will only check that the assumptions of Theorem IX.3.2 are satisfied in our situation. Namely we will prove the following:

**Theorem VIII.4.9** *Let us denote by $\Lambda$ the box $[-n\,,\,n]$ then we have:*

**(i)** *For any energy $\lambda_0$ there exist constants $\chi > 0$, $\zeta > 0$ such that:*

$$\mathbb{P}\{\lambda_0 \notin \Sigma(H^\Lambda) \text{ and } |G^\Lambda_{\lambda_0}(0,\pm n)| \leq e^{-n\chi}\} \geq 1 - e^{\zeta\sqrt{n}}$$

*for $n$ sufficiently large.*

**(ii)** *Let $\mathcal{I}$ be a bounded interval then there exists $\tau > 0$ such that:*

$$\mathbb{P}\{d(\lambda\,,\,\Sigma(H^\Lambda)) \leq e^{-\sqrt{n}}\} \leq e^{-\tau\sqrt{n}}$$

*for $\lambda \in \mathcal{I}$ and $n$ sufficiently large.*

*Proof:*
We begin with the proof of conclusion (ii). Let $\mathcal{I}$ be a compact interval of energy and $\rho$ and $\alpha_1$ be the constants associated with the interval $\mathcal{I}$ in the Corollary VIII.4.4. Moreover let $C_1 = e^\beta$ and $\delta$ be the constants appearing in Proposition VIII.4.7. In the following we will denote by $\theta$ and $\xi$ two positive numbers that will be chosen later. Finally for any positive integer $n$ we set $m = [\xi\sqrt{n}] + 1$ where $[x]$ stands as usual for the integer part of the real number $x$.
For $k = 1, 2$ we define the sets $A_{k,\lambda}$ and $B_{k,\lambda}$ by :

$$\begin{aligned} A_{k,\lambda} &= \{|^{+}G_\lambda^{[-n,m-n]}(-n, m-n)| \leq e^{-k\theta\sqrt{n}}\} \\ B_{k,\lambda} &= \{|^{-}G_\lambda^{[n-m,+n]}(n-m, +n)| \leq e^{-k\theta\sqrt{n}}\} \end{aligned}$$

Moreover for $\lambda \in \mathcal{I}$ let us define $J(\lambda) = \{\lambda' \,;\, \lambda' \in \mathcal{I}\ |\lambda - \lambda'| \leq e^{-\sqrt{n}}\}$. In order to obtain an upper bound on the probability of the set appearing in the left hand side of (ii) we remark that this probability is no greater than the sum $a + b + c + d$ defined below. Recall that $A^c$ is the complementary set of the set $A$.

$$
\begin{aligned}
a &= \mathbb{P}\{(d(\lambda, \Sigma(H^\Lambda)) \leq e^{-\sqrt{n}}) \bigcap_{\lambda' \in J(\lambda)} (A_{1,\lambda'} \cap B_{1,\lambda'})\} \\
b &= \mathbb{P}\{A_{2,\lambda} \cap (\bigcup_{\lambda' \in J(\lambda)} A^c_{1,\lambda'})\} \\
c &= \mathbb{P}\{B_{2,\lambda} \cap (\bigcup_{\lambda' \in J(\lambda)} B^c_{1,\lambda'})\} \\
d &= \mathbb{P}\{A^c_{1,\lambda}\} + \mathbb{P}\{B^c_{1,\lambda}\}
\end{aligned}
$$

We first remark that the continuity of the Green function with respect to $\lambda$ implies its joint measurability with respect to the couple $(\omega, \lambda)$ hence the measurability of the sets defined above. If the event of probability $a$ occurs then there exists an eigenvalue $\lambda'$ of $H^\Lambda$ in $J(\lambda)$. From Lemma VIII.4.2 one can deduce that for any associated normalized eigenvector $\psi$ one has:

$$
\begin{aligned}
|\psi(-n)| &\leq |{}^+G_\lambda^{[-n,m-n]}(-n, m-n)| \leq 2e^{-\theta\sqrt{n}} \\
|\psi(+n)| &\leq |{}^-G_\lambda^{[n-m,+n]}(n-m, +n)| \leq 2e^{-\theta\sqrt{n}}
\end{aligned}
$$

If one chooses $\epsilon = \max(e^{-\sqrt{n}}, 2\sqrt{2}e^{-\theta\sqrt{n}})$ Proposition VIII.4.8 implies that one has

$$
a \leq m\ C_2 \epsilon^{\alpha_2}
$$

Hence there is no problem with the term $a$. Let us now bound the term $b$. If one chooses $\epsilon = e^{-2\theta\sqrt{n}}$ in Proposition VIII.4.7, one obtains immediately:

$$
\begin{aligned}
b &\leq C_1^m\, e^{-\delta\sqrt{n}}\, e^{-4\delta\theta\sqrt{n}} \\
&= e^{-(\delta + 4\delta\theta - \xi\beta)\sqrt{n}}
\end{aligned}
$$

and the same estimate can be proved similarly for the term $c$.
It remains now to handle the term $d$. Chebyshev's inequality together with Corollary VIII.4.4 implies that for $n$ sufficiently large one has:

$$
\begin{aligned}
\mathbb{P}\{A^c_{1,\lambda}\} &\leq e^{-m\rho}\, e^{2\theta\alpha_1\sqrt{n}} \\
&\leq e^{-(\xi\rho - 2\theta\alpha_1)\sqrt{n}}
\end{aligned}
$$

The second part of $d$ admits the same bound.

In order to conclude we have to choose $\xi$ and $\theta$ in such a way that :

$$\begin{aligned} 0 &< \delta + 4\delta\theta - \xi\beta \\ 0 &< \xi\rho - 2\theta\alpha_1 \end{aligned}$$

Obviously such a choice is always possible and this finishes the proof of the claim (ii).

We now turn our attention to the proof of (i). Since the energy $\lambda_0$ is fixed we will skip the dependence upon $\lambda_0$ in the notations. Let $\Lambda = [-n, n]$ and define the coupling operator $Q = \Gamma_1 + \Gamma_0$. Then the resolvent equation can be written:

$$G^\Lambda = G^{[0,n]} + G^\Lambda Q G^{[0,n]}$$

If we compute the value of the the Green kernel at the point $(0, n)$ one obtains :

$$G^\Lambda(0,n) = G^{[0,n]}(0,n)(1 + G^\Lambda(0,0))$$

The inequality

$$|G^\Lambda(0,0)| \leq \frac{1}{d(\lambda_0\,,\ \Sigma(H^\Lambda))}$$

added to the result (ii) implies that for $n$ large:

$$\mathbb{P}\{|G^\Lambda(0,0)| \geq e^{\sqrt{n}}\} \leq e^{-\tau\sqrt{n}}$$

The expected result will follow once we will have proved that for some positive $\alpha$ and $\beta$ one has:

$$\mathbb{P}\{\lambda_0 \notin \Sigma(H^{[0,n]}) \text{ and } |G^{[0,n]}_{\lambda_0}(0,n)| \leq e^{-n\beta}\} \geq 1 - e^{\alpha\sqrt{n}}$$

for $n$ sufficiently large. Using again the resolvent identity

$$G^{[0,n]} =^{+} G^{[0,n]} - i^{+}G^{[0,n]}\Gamma_n G^{[0,n]}$$

one obtains:

$$G^{[0,n]}(0,n) =^{+} G^{[0,n]}(0,n)\ (1 - iG^{[0,n]}(n,n))$$

This yields the desired result, using Proposition VIII.4.4 and property (ii) again. ∎

We remark that in Theorem VIII.4.9 the crucial estimate is (ii) since it implies conclusion (i) by a standard argument of the theory of products of random matrices. As we saw above , the proof of the estimate (ii) is somewhat involved for a general distribution of potentials. Assuming some regularity of this distribution, such as absolute continuity with a bounded density or Hölder continuity, such a proof is straightforward !

**Definition VIII.4.10** *A probability measure $\mu$ on $\mathbb{R}$ is said to be Hölder continuous of order $0 < \alpha \leq 1$ if there exists a finite constant $C$ such that:*

$$\mu([a\ b]) \leq C|b-a|^\alpha$$

*for any finite interval $[a,b]$. (Actually this means that the Hölder exponent is at least $\alpha$).*

**Proposition VIII.4.11** *Let us assume that the distribution of potentials has a bounded density. Then there exists a finite constant $C$ such that:*

$$\mathbb{P}(d(\lambda, \Sigma(H^\Lambda)) \leq \epsilon\} \leq nC\epsilon$$

*for any box $\Lambda = [-n, n]$ and $\epsilon > 0$.*

*Proof:*
Let us denote by $N^\Lambda(B)$ the number of eigenvalues of the operator $H^\Lambda$ lying in the Borel set $B$. Then it is easy to check that:

$$\mathbb{P}\{d(\lambda, \Sigma(H^\Lambda)) \leq \epsilon\} \leq \mathbb{E}\{N^\Lambda([\lambda-\epsilon, \lambda+\epsilon])\}.$$

Denoting as usual by $\sigma_k^\Lambda$ the spectral measure of the operator $H^\Lambda$ at the site $k$ then one has:

$$N^\Lambda(B) = \sum_{k\in\Lambda} \sigma_k^\Lambda(B).$$

For each $k \in \Lambda$ one can compute the expectation of $\sigma_k^\lambda$ first by integrating with respect to the potential at the site $k$ then Lemma VIII.1.8 implies that if the density of the distribution of potentials is bounded by $K$ one has:

$$\mathbb{E}\{\sigma_k^\Lambda(B)\} \leq K\ m(B)$$

where $m$ is the Lebesgue measure. The desired result follows from this inequality. ∎

In order to obtain an estimate of the same kind in the Hölder continuous case one needs the following lemma.

**Lemma VIII.4.12** *For $\mu$ a probability measure on $\mathbb{R}$ and $\Im m(z) > 0$, let $F(z) = \int \frac{d\mu(t)}{t-z}$ be the Stieltjes transform of $\mu$. Then one has:*

1. *For any $\epsilon > 0$ and $\lambda \in \mathbb{R}$:*

$$\mu\{[\lambda - \epsilon, \lambda + \epsilon]\} \leq 2\epsilon \, \Im m F(\lambda + i\epsilon)$$

2. *If $\mu$ is Hölder continuous of order $\alpha$ then there exists a finite constant $C$ such that*

$$\Im m F(\lambda + i\epsilon) \leq C \, \epsilon^{\alpha - 1}$$

*Proof:*
In order to obtain 1. it is enough to remark that

$$\begin{aligned} \Im m F(\lambda + i\epsilon) &= \int \frac{\epsilon d\mu(t)}{(\lambda - t)^2 + \epsilon^2} \geq \int_{\{|\lambda - t| \geq \epsilon\}} \frac{\epsilon d\mu(t)}{(\lambda - t)^2 + y^2} \\ &\geq \frac{1}{2\epsilon} \mu\{[x - \epsilon \, , \, x + \epsilon]\}. \end{aligned}$$

Introducing the sets $A_n = [\lambda + n\epsilon \, , \, \lambda + (n+1)\epsilon]$, the Proof of 2. is almost the same as for 1. if one writes:

$$\begin{aligned} \Im m F(\lambda + i\epsilon) &= \sum_{n=-\infty}^{+\infty} \int_{A_n} \frac{\epsilon d\mu(t)}{(\lambda - t)^2 + \epsilon^2} \leq \frac{1}{\epsilon} \sum_{n=-\infty}^{+\infty} \frac{1}{n^2 + 1} \mu(A_n) \\ &\leq C \, \epsilon^{\alpha - 1}. \end{aligned}$$

(Remark that for $\alpha = 1$ one recovers the classical result for a distribution with a bounded density.) ∎

**Proposition VIII.4.13** *Let us assume that the common distribution of the potentials is Hölder continuous of order $\alpha \in (0 \; 1]$. Then there exists a finite constant $C$ such that:*

$$\mathbb{P}\{d(\lambda, \Sigma(H^\Lambda)) \leq \epsilon\} \leq n \, C \, \epsilon^\alpha$$

*for any box $\Lambda = [-n \, , \, n]$ and $\epsilon > 0$.*

*Proof:*

The beginning of the proof is identical to Proposition VIII.4.11. Taking into account the relation 2. above one has only to modify slightly the end of proof of Lemma VIII.1.8 to obtain:

$$\begin{aligned}\Im m \int \frac{d\tilde{\sigma}_k^\Lambda(t)}{t-(\lambda+i\epsilon)} &\leq K(\epsilon+\Im m(b'h))^{\alpha-1} \\ &\leq K\,\epsilon^{\alpha-1}\end{aligned}$$

where $\tilde{\sigma}_k^\Lambda$ is the expectation of $\sigma_k^\Lambda$ with respect to the potential at the site $k$ and the vectors $b$ and $h$ have been introduced in the proof of Lemma VIII.1.8. Since we know that $\Im m\, b'h$ is nonnegative and $\alpha \leq 1$ one obtains the desired bound. It remains now to use the first part of Lemma VIII.4.12 to conclude the proof. ∎

**Remark VIII.4.14** For potentials with absolutely continuous or Hölder continuous distributions Lemmas VIII.4.11 and VIII.4.13 hold in any dimension. One only has to replace the integer $n$ by the volume of the box $\Lambda$. Moreover estimate (i) in Theorem VIII.4.9 will be replaced in this situation by the decay of the Green kernel obtained by Frölich & Spencer in [106] or von Dreifus in [87]. Unfortunately it is likely that the estimates (i) or (ii) in this theorem do not hold in dimension greater than one (two ?) for a general distribution of potentials. This explains why the situation is much more complex in this case.

# VIII.5 Non-Stationary Processes

This section is devoted to the discussion of the spectral properties of random Schrödinger operators associated with non-stationary processes. We will not give proofs. The interested reader may consult the notes at the end of the chapter for references of papers where complete proofs are given. The models of non-stationary random potentials which have been studied fall essentially into two classes:

- potentials obtained from a stationary process by multiplication by a function decaying at infinity
- random perturbation of a deterministic potential.

Note that both classes include potentials $V(t)$ with different asymptotics when $t \to -\infty$ and $t \to +\infty$.

## VIII.5.1 The Discrete Case

The following results are borrowed from Simon [295], Delyon [68] and Delyon, Simon & Souillard in [76]. For the sake of simplicity, we do not try to give minimal sets of assumptions. The interested reader can find them in the original works quoted above.

### Multiplicative perturbations

Let us first consider the case of a discrete, one dimensional Schrödinger operator $H$ with a potential of the form

$$V(\omega, n) = \beta a(n)\xi_n(\omega)$$

where $\beta$ is some positive real number, $\{\xi_n;\ n \in \mathbb{Z}\}$ is an i.i.d. sequence of real random variables and $a(n)$ is a real sequence. One assumes that this sequence $a(n)$ behaves as some power of $n$, that is that there exist two constants $C_1$ and $C_2$ such that

$$C_1|n|^{-\alpha} \leq |a(n)| \leq C_2|n|^{-\alpha}, \qquad n \neq 0$$

Moreover we assume that the common distribution of the random variables $\xi_n$ has a bounded density, a finite moment of order strictly greater than 2 and a zero mean.

The next theorem summarizes the essential results:

**Theorem VIII.5.1** *The following properties hold almost surely:*

1. *If $0 \leq \alpha < 1/2$, the spectrum of $H$ is pure point and for any eigenvalue $\lambda$ and associated eigenfunction $\psi_\lambda$ there exist constants $C_1, C_1', C_2, C_2'$ such that:*

$$C_1 \exp(-C_1'|n|^{1-2\alpha}) \leq |\tilde{\psi}_\lambda(n)| \leq C_2 \exp(-C_2'|n|^{1-2\alpha})$$

2. *If $\alpha = 1/2$ then there exists a constant $\beta_1$ such that the spectrum of $H$ is pure point if the coupling constant $\beta$ is greater than $\beta_1$. Furthermore, for any compact $K$ contained in $(-2, +2)$ there exists a constant $\beta_2$ such that the spectrum of $H$ is purely continuous in $K$ for $\beta < \beta_2$.*
3. *If $\alpha > 1/2$ then the spectrum of $H$ is purely continuous in $[-2, +2]$.*

*Proof:*

We only give a brief sketch of the arguments. To prove the properties related to the continuous spectrum, one has to prove a suitable lower bound on the solutions of the eigenvalue equations insuring that no solution can be an eigenvector. For a fixed energy $\lambda$ it is not too difficult to get this lower bound for almost all $\omega$, using an upper bound on the growth of the products of transfer matrices. However one needs to obtain these bounds for $\lambda$ in a given interval, for almost all $\omega$. This difficulty is solved by using an idea borrowed from the study of regularity of stochastic processes.

The pure point property of the spectrum is obtained via the classical proof of localization presented in the section "Operator Theory". The only change is that one obtains products of operators rather than powers of operators since the distribution of probability defining the Laplace transform is different at each site. Nevertheless it can be proved for the classical Laplace transform or for the operators introduced by Kunz & Souillard (see Problem VIII.6.6 and Remark VIII.3.9) that there exist constants $C_1$ and $C_2$ such that:

$$\begin{aligned} ||T_{1,\lambda,n}T_{1,\lambda,n+1}||_2 &\leq 1 - C_1\lambda^2|n|^{-2\alpha} \\ ||T_{2,\lambda,n}||_\infty &\leq \frac{C_2}{\beta|a(n)|} \end{aligned}$$

Here we have added the subscript $n$ to indicate at which site the Laplace operator is defined. It is now easy to conclude, using the criteria of localization of the operator proof. ■

The above assumption about the zero mean of the distribution of potentials is not essential since the proof can accomodate an additive perturbation, but the existence of moments is crucial as shown by the example of a Cauchy distribution for which one has localization for any $\alpha < 1$. In this case a direct computation yields stronger estimates on the norm of the Laplace transform than in the general case and obviously the lower bounds obtained in the first part of the proof fail.

### Multiplicative perturbations

The next continuous time model, called the Krönig-Penney model, can easily be reduced to the situation of a discrete Schrödinger operator. Namely one considers the operator:

$$[H(\omega)\psi](t) = -\ddot{\psi}(t) - \beta t + \sum_{n=-\infty}^{+\infty} \xi_n(\omega)\delta(t-n)$$

where $\beta$ is some positive real number, $\{\xi_n;\ n \in \mathbb{Z}\}$ is an i.i.d. sequence of real valued random variables. Moreover we assume that the common distribution of the random variables $\xi_n$ has a bounded density, a finite moment of order strictly greater than 2 and a mean zero.

**Theorem VIII.5.2** *The following properties hold almost surely:*

1. *If $0 < \beta < \beta_1$, the spectrum of $H$ is pure point and for any eigenvalue $\lambda$ and associated eigenfunction $\psi_\lambda$ there exist constants $C_1, C_1', C_2, C_2'$ such that:*

$$|\psi_\lambda(n)| \leq \begin{cases} C_1 \exp(-\frac{C_1'}{\beta}\log(n)) & \text{if } n > 0 \\ C_2 \exp(-C_1'\beta^{1/2}|n|^{3/2}) & \text{if } n < 0 \end{cases}$$

*The second bound is obtained under the additional hypothesis that the Fourier transform of the distribution of $\xi_n$ be exponentially decreasing at infinity.*

2. *If $\beta > \beta_2$ then the spectrum of $H$ is purely continuous.*

*Proof:*
The proof can be carried out in the same way as in the former theorem. Integrating a solution of the eigenvalue equation between integral values of the time one sees immediately that this equation is equivalent to a recursion relation of the form:

$$\begin{pmatrix} \psi(n+1) \\ \psi(n) \end{pmatrix} = \begin{pmatrix} a(n)\xi_n + b(n) & c(n) \\ 1 & 0 \end{pmatrix} \begin{pmatrix} \psi(n) \\ \psi(n-1) \end{pmatrix}$$

where $a(n)$, $b(n)$, $c(n)$ are some real nonrandom sequences. The important fact is that $a(n)$ behaves like $n^{-1/2}$ at infinity. More precisely it is equivalent to $(\beta n)^{-1/2}$. Hence we are in the situation 2. of Theorem VIII.5.1 and one can repeat the argument. ∎

This last model exhibits power-localized extended states and it remains to study the exact nature of the continuous spectrum in both models. The results of Kotani given below partially answer this question in the continuous case and for multiplicative perturbations.

## VIII.5.2 The Continuous Case

The discussion of the continuous case will use the random potential function studied by the Russian School, namely a process of the form $\{F(X(t)); t \in \mathbb{R}\}$ where $\{X(t); t \in \mathbb{R}\}$ is the stationary Markov process generated by the Laplace-Beltrami operator $L$ of a compact connected Riemannian manifold $M$ and $F : M \to \mathbb{R}$ is a non-flattening Morse function in the sense that there exists an integer $n_0 \geq 1$ such that for all $x \in M$ there exists an integer $k = k(x), 1 \leq k \leq n_0$ for which $D^k F(x) \neq 0$. We will assume as usual that $\inf_{x \in M} F(x) = 0, \sup_{x \in M} F(x) = 1$ and that the process is defined on a complete probability space $(\Omega, \mathcal{F}, \mathbb{P})$

We first study a class of random Schrödinger operators with different behavior on the positive and negative axes. The proofs of the following results can be found in Carmona[43].

**Theorem VIII.5.3** *Let $\{X(t); t \geq 0\}$ be the stationary process of Brownian motion on a compact connected Riemannian manifold, let $F$ be a real valued Morse function with range $[-1, +1]$ and $g(t)$ be a real valued function defined on $t \leq 0$ integrable on $(-\infty, 0)$. Then we define the potential function:*

$$\begin{aligned} V(t, \omega) &= g(t) Y(t, \omega) && t < 0 \\ &= F(X(t, \omega)) && t \geq 0 \end{aligned}$$

*where $Y(t)$ is any measurable stochastic process with bounded sample paths bounded below by $-1$ which is stochastically independent of the process $X(t)$. Then there exists a set of full probability on which the following properties hold:*

1. *The spectrum of $H$ is equal to $[-1, \infty)$.*
2. *In $(0, \infty)$ this spectrum is purely absolutely continuous.*
3. *In $[-1, 0]$ the spectrum is pure point with exponentially decaying eigenfunctions.*

*Proof:*
If we fix $\omega$ in a set of probability 1 on which the assumptions of the theorem are satisfied, the potential function is integrable near $-\infty$. Hence the spectrum contains $(0, \infty)$ and is purely absolutely continuous there. Moreover the potential function is bounded below by $-1$ so that the spectrum of $H$

is contained in $[-1,\infty)$. The ergodic properties of the process $X(t)$ show that for each $\lambda \in [-1,+1]$, for each of its neighborhood $N_\lambda$ and for each $L>0$, there exists at least one interval $I$ in $[0,\infty)$ of length greater than $L$ and such that $F(X(t)) \in N_\lambda$ for all $t \in I$, and by a classical argument this implies that the spectrum contains $[-1,+1]$. It only remains to check the positivity of the Lyapunov exponent in the positive direction, and for $\lambda < 0$ in the negative direction, to conclude. ∎

The following theorem deals with a one-sided perturbation of a periodic potential.

**Theorem VIII.5.4** *Let $\{X(t); t \geq 0\}$ be the stationary process of Brownian motion on a compact connected Riemannian manifold, let $F$ be a real valued Morse function with range $[-\epsilon,+\epsilon]$ for some positive $\epsilon$. Let $g(t)$ be a real valued periodic function. Then we define the potential function:*

$$V(t,\omega) = \begin{cases} g(t) & \text{whenever } t < 0 \\ g(t) + F(X(t,\omega)) & \text{otherwise} \end{cases}$$

*Then there exists a set of full probability on which the following properties hold:*

1. *The spectrum of $H$ is equal to $\Sigma(H) = \cup_n [a_n - \epsilon, b_n + \epsilon]$ where $\Sigma_1 = \cup_n [a_n, b_n]$ is the spectrum of the nonrandom Schrödinger operator $H_1$ with potential $g(t)$.*
2. *In $\Sigma_1$ this spectrum is purely absolutely continuous.*
3. *In $\Sigma(H) \setminus \Sigma_1$ the spectrum is pure point with exponentially decaying eigenfunctions.*

*Proof:*
Let $\lambda = \lambda' + \epsilon'$ with $\lambda' \in \Sigma_1$ and $|\epsilon'| < \epsilon$. Since $\lambda'$ is in the essential spectrum of $H_1$, for each integer $n$ we can find a smooth normalized function with compact support $\psi_n$ such that:

$$\| -\ddot{\psi}_n + (g - \lambda')\psi_n \| \leq \frac{1}{n}$$

The ergodic properties of the process $X(t)$ are such that $\mathbb{P}$-almost surely, for each integer $n > 1$ there exists in $[0,\infty)$ an interval of length greater than the diameter of the support of $\psi_n$ plus twice the period of $g(t)$ on which we have:

$$|F(X(t)) - \epsilon'| \leq \frac{1}{n}$$

This means that we can always translate $\psi_n$ in order to allow us to assume that $\psi_n$ is supported in a region where the above condition is satisfied. Now we have:

$$\begin{aligned} \| - \ddot{\psi}_n + (V - \lambda)\psi_n\| &\leq \| - \ddot{\psi}_n + (g - \lambda)\psi_n\| \\ &+ \|(F(X(_t)) - \epsilon')\psi_n\| \leq \frac{2}{n} \end{aligned}$$

This proves 1. The statements 2. and 3. follow directly from the results of Chapter VII. ∎

**Multiplicative perturbations**

Let us consider first the case of a random potential $\{V(t,w); t \in \mathbb{R}, w \in \Omega\}$ of the form:

$$V(t,\omega) = a(t)F(X(t,\omega))$$

where $a : \mathbb{R} \longrightarrow \mathbb{R}$ is a bounded below continuous function. Theorem III.4.1 tells us that the random Schrödinger operator $H(\omega) = -d^2/dt^2 + V(t,\omega)$ has absolutely continuous spectrum in $(0,\infty)$ whenever the function $a(t)$ is integrable near $+\infty$ or near $-\infty$. On the other hand we know that we have exponential localization for $a(t) \equiv 1$. Intuitively one should expect a transition from exponential localization to absolutely continuous spectrum by increasing the rate of decay of $a(t)$ at infinity.
In the preceding section, this problem was studied in the lattice case. These results have been precised by Kotani and Ushiroya who extended them to the continuous case in [202]. We now present the major steps in this last work.
To avoid the problems created by different asymptotics of the coupling $a(t)$ near $\infty$ and $+\infty$ and to make our life easier, we assume from now on that

- $a(t)$ is even in the sense that $a(t) = a(-t)$
- $\lim_{|t|\to\infty} a(t) = 0$
- $a(t) < a(s)$ whenever $0 \leq s < t$.

The second assumption implies that the spectrum of $H(\omega)$ is discrete in $(-\infty, 0)$. We concentrate on $\Sigma(H(\omega)) \cap [0,\infty) = [0,\infty)$. The results are given in terms of the function $A$ and $\gamma$ defined by:

$$A(t) = |\int_0^t a(s)^2 ds|, \quad t \in \mathbb{R}$$

and:

$$\begin{aligned}\gamma(\lambda) &= \frac{1}{4\lambda}\Re e\int_M F(x)[L+2iI\sqrt{\lambda}]^{-1}F(x)dx \\ &= \int_0^\infty \frac{\xi}{4\lambda(\xi^2+4\lambda)}\sigma_F(d\xi),\end{aligned}$$

for $\lambda > 0$, where $L$ is the Laplace-Beltrami operator on the manifold $M$ and where $\sigma_F(d\xi) =< E^{(-L)}(d\xi)F, F >$ is the spectral measure associated with the element $F$ of $L^2(M, dx)$ by the spectral resolution of the identity $E^{(-L)}$ of the (nonnegative) selfadjoint operator $-L$. Note that $\gamma(\lambda) > 0$ unless $F$ is constant.

The spectral characteristics of $H(\omega)$ on $[0,\infty)$ can then be described in the following way:

**a)** if $\lim_{|t|\to\infty} A(t) = \infty$ then with probability one the spectrum is pure point in $[0, \lambda_0]$ and singular continuous in $[\lambda_0, \infty)$ where $\lambda_0$ is defined by:

$$\lambda_0 = \sup\{\lambda \geq 0; \int_{\mathbb{R}} e^{-2\gamma(\lambda)A(t)}dt < +\infty\}$$

**b)** on the other hand, if $\int_M F(x)dx = 0$, if $A(t)$ is bounded and if the function $a$ satisfies $\int_{\mathbb{R}} a(t)^2|t|^\delta dt < +\infty$ for some positive $\delta$, then the spectrum of $H(\omega)$ is absolutely continuous on $[0,\infty)$ with probability one and for each $\lambda > 0$ there exists a solution $\varphi_\lambda$ of the eigenvalue equation such that:

$$\begin{cases} \varphi_\lambda(t) &= e^{i\sqrt{\lambda}t} + o(1) \\ \dot\varphi_\lambda(t) &= i\sqrt{\lambda}e^{i\sqrt{\lambda}t} + o(1)\end{cases}$$

the $o(1)'s$ being uniform in $\lambda$ restricted to compact sets.

Notice that the existence of the plane wave solution proves that the (absolutely continuous) spectrum is of multiplicity two since $\varphi_\lambda$ and its complex conjugate $\overline{\varphi}_\lambda$ are linearly independent. On the other hand the spectral nultiplicity of the pure point spectrum occuring in $[0, \lambda_o]$ in case a) above is one because of the usual Wronskian argument and the fact that one can prove that the generalized eigenfunctions satisfy:

$$\lim_{|t|\to\infty} \frac{1}{A(t)} \log\sqrt{|\psi_\lambda(t)|^2 + |\dot\psi_\lambda(t)|^2} = \gamma(\lambda).$$

Note also that a complete characterization can be given in the case

$$a(t) \sim \frac{1}{(1+|t|)^\alpha}, \quad |t| \to \infty$$

Indeed in this case one has

- case a) with $\lambda_o = \infty$ (i.e. the spectrum is pure point in $[0, \infty)$) in the case $0 < \alpha < 1/2$
- case a) with $0 < \lambda_o < +\infty$ (i.e., pure point spectrum in $[0, \lambda_o]$ and singular continuous spectrum in $(\lambda_o, \infty)$) when $\alpha = 1/2$
- case b), i.e., multiplicity two absolutely continuous spectrum (with plane waves eigenfunctions) when $\alpha > 1/2$.

**Additive Perturbations**

We first introduce the notations which we use in this subsubsection:

- $p(t)$ is an integrable function with $\lim_{|t| \to \infty} p(t) = 0$
- $\{X(t); t \geq 0\}$ is the stationary process of Brownian motion of a compact connected Riemannian manifold $M$ and $F : M \longrightarrow \mathbb{R}$ is a Morse function such that $\inf F = 0$ and $\sup F = 1$.

For each $\omega$ in the probability space $\Omega$ on which the Brownian motion is defined we consider the potential function

$$V(t, \omega) = F(X(t, \omega)) + p(t)$$

For each $\lambda$, $\{(X(t), \vartheta_\lambda(t)); t \geq 0\}$ is a Markov process which is not homogeneous if the function $p(t)$ is not identically 0.
We use the notation $dx$ for the normalized Riemannian volume element of $M$, the notations $\mathbb{P}_{s,x}$ and $\mathbb{E}_{s,x}$ for the probability and expectation for the Brownian motion conditioned to be at $x$ at time $s$, and we restrict the notations $\mathbb{P}$ and $\mathbb{E}$ to the case of the stationary Brownian motion (i.e.,

$$\mathbb{P} = \int_M \mathbb{P}_{0,x} dx$$

We will use the following notation for the transition operators of the Markov processes $(X(t), \vartheta_\lambda(t))$

$$T_\lambda^{s,t}(f)(x, \vartheta) = \mathbb{E}_{s,x}\{f(X(t), U_\lambda(t,s)\vartheta)\}$$

for every function $f$ on $M \times \mathbb{T}$ for which the last expectation makes sense.

It is shown in Problem VIII.6.11 and in Problem VIII.6.12 that these operators have nice regularity properties.

For each realization $\omega$ of the Brownian motion process and for each $\alpha \in \mathbb{T}$ we denote by $H_\alpha^+(\omega)$ the self-adjoint operator on $L^2([0,\infty), dt)$ which coincides on $C^\infty$ functions with compact supports in $(0,\infty)$ with the formal differential operator:

$$-\frac{d^2}{dt^2} + V(t)$$

and with boundary condition at 0 given by:

$$y(0)\cos\alpha - y'(0)\sin\alpha = 0,$$

then for each $\alpha \in \mathbb{T}$ we have, with probability one:

- $[0,\infty) \subset \Sigma(H_\alpha^+)$
- $\Sigma(H_\alpha^+) \cap [0,\infty) = \Sigma_{pp}(H_\alpha^+) \cap [0,\infty)$
- The eigenfunctions corresponding to the (dense) pure point spectrum in $[0,\infty)$ decay exponentially, the rate of exponential fall off being given by the Lyapunov exponent of the unperturbed problem.

Let us emphasize that the spectrum of $H_\alpha^+$ could have a negative part if the function $p(t)$ takes negative values. Typically we expect this part of the spectrum to be made of isolated eigenvalues with exponentially decaying corresponding eigenfunctions. This can be proven under some restrictions on $p(t)$ by using the Weyl's criteria for essential spectra.
It is elegantly shown in [193] that for any stationary non-deterministic bounded stochastic process $\{V(t); t > 0\}$ the operator $H_\alpha^+$ has dense pure point spectrum with exponentially localized eigenfunctions with probability one, for almost every (with respect to Lebesgue's measure $d\alpha$) choice of the boundary condition $\alpha$ (See also Proposition VIII.1.1). Moreover some sufficient condition is given to get rid of the "almost every boundary condition" assumption. Unfortunately it seems very difficult to perturb such a general potential and still be able to control quantities like the Lyapunov exponent and show that the localization phenomenon is preserved.
We believe that the above result should be true without the assumption of $p(t)$ going to zero at infinity. In fact, Brossard showed in [37] that if the random potential is of the form

$$V(t) = p(t) + 2\epsilon F(X(t)) - \epsilon$$

with $p(t)$ continuous and periodic, then the spectrum is almost surely equal to the set $\Sigma = \bigcup_{n\geq 0}[a_n - \epsilon, b_n + \epsilon]$ where $\Sigma_0 = \bigcup_{n\geq 0}[a_n, b_n]$ is the spectrum of the operator $H_0 + p(t)$ and exponential localization holds almost surely in the sense that $E = E_{pp}$ almost surely and the eigenfunctions are exponentially decaying. This result can be proved in a slightly easier way than before. Indeed the nonstationarity created by the deterministic function $p(t)$ is not too bad in the present situation because of the periodicity. We needed $p(t)$ to be asymptotically negligible in order to get uniform control on the propagator. In the present situation this uniformity reduces to an interval of length the period and the boundedness of $p(t)$ is enough. It is interesting to compare this result to a similar one proved by Carmona in [43]) when $V(t) = 0$ if $t \geq 0$ and $V(t) = p(t)$ if $t < 0$, then the spectrum is still almost surely equal to $\Sigma$ as a set, exponential localization still occurs in $\Sigma \setminus \Sigma_0$, but the spectrum is now absolutely continuous in $\Sigma_0$. In other words the disorder on one side of the system only cannot affect the conduction bands. The proof is a trivial consequence of Theorem III.4.1. We summarize the results discussed above in a theorem, the proof of which is left to the reader.

**Theorem VIII.5.5** *Let $p : R \to R$ be a bounded continuous periodic function, let $\epsilon > 0$ and let us denote by $\Sigma_0 = \bigcup_{n\geq 0}[a_n, b_n]$ the spectrum of the unique self-adjoint extension in $L^2(\mathbb{R}, dt)$ of the operator defined for $f \in C_c^\infty(\mathbb{R})$ by $-\ddot{f}(t) + p(t)f(t)$. Let $\{F(X(t)); t \in \mathbb{R}\}$ be the random potential of the Russian school and let $\epsilon : \mathbb{R} \to [0, \infty)$ be a function to be chosen later. The random Schrödinger operator $H_\epsilon(\omega)$ is defined by:*

$$H_\epsilon(\omega) = -\frac{d^2}{dt^2} + p(t) + 2\epsilon F(X(t, \omega)) - \epsilon$$

1. *Assume that $\epsilon(t) \equiv \epsilon > 0$. Then the spectrum of $H_\epsilon(\omega)$ is almost surely equal to $\Sigma_\epsilon = \bigcup_{n\geq 0}[a_n - \epsilon, b_n + \epsilon]$ and exponential localization holds almost surely everywhere on the spectrum.*
2. *Assume now that $\epsilon(t) \equiv 0$ for $t < 0$ and $\epsilon(t) \equiv \epsilon > 0$ for $t \geq 0$. Then the spectrum of $H_\epsilon(\omega)$ is still almost surely equal to the same nonrandom set $\Sigma_\epsilon$ and exponential localization still holds almost surely in $\Sigma_\epsilon \setminus \Sigma_0$. Moreover the spectrum is now absolutely continuous in $\Sigma_0$.*

We close this subsection with a brief discussion of a model with an appealing physical interpretation. We choose the function $p(t)$ to be of the form $p(t) = Ct$ for some constant $C \geq 0$, (or more generally $p(t)$ a continuous odd real valued function on $\mathbb{R}$ which is equivalent to $t^\alpha$ when $t \to +\infty$). The random potential $V(t,\omega) = p(t) + F(X(t,\omega))$ can be viewed as the potential for a disordered system in a constant electric field. As we already pointed out in Chapter III this problem is meaningless in the lattice case. We know that exponential localization holds in the absence of the electric field (i.e., for $C = 0$). On the other hand, the deterministic potential $p(t)$ is not asymptotically negligible when the field is turned on (i.e., when $C \neq 0$). The results discussed in Chapter III show that the spectrum is the whole line $\mathbb{R}$ and that the latter should be absolutely continuous when the potential function $t \hookrightarrow V(t)$ is twice continuously differentiable with bounded derivatives. This assumption is obviously not satisfied by the potential at hand for the Brownian paths $t \hookrightarrow X(t)$ are highly nondifferentiable. We nevertheless believe that the result still holds. An example of a random potential $\{V(t,\omega); t \in \mathbb{R}, \omega \in \Omega\}$ satisfying almost surely this smoothness asumption and for which exponential localization holds is given in [27]. It is made from a sequence of similar smooth nonoverlapping wells the depths of which are i.i.d. with continuous common density with compact support. See Problem VIII.6.9.

**Remark VIII.5.6 Aside: Spherically Symmetric Random Potentials**

We now come to the case of potential functions $\{V(x,\omega); x \in \mathbb{R}^d, \omega \in \Omega\}$ where $V(x,\omega)$ is of the form $V(|x|,\omega)$ and $\{V(t,\omega); t \geq 0, \omega \in \Omega\}$ is a random potential of the Russian school type. Recall the notations introduced in RemarkII.2.5. For each integer $k \geq 1$ the operator $\tilde{H}_k$ can be written in the form:

$$\tilde{H}_k = -\frac{d^2}{dt^2} + F(X(t)) + p_k(t) \quad \text{where} \quad p_k(t) = \left[\frac{(d-1)(d-3)}{4} + \kappa_k\right]\frac{1}{t^2}$$

So, except for the singularity of $p(t)$ at $t = 0$ we are in the situation investigated in the previous section. Unfortunately this singularity forces the study of, the operator $\tilde{H}_k$ on intervals of the form $[a,b]$ with $a > 0$, with boundary conditions at both end $a$ and $b$ and then, the limits $a \to 0$ and $b \to \infty$ have to be controlled. The limit $b \to \infty$ was handled (for a fixed) in the previous section. The limit $a \to 0$ is more subtle.
If the dimension $d$ is larger than 3 then:

$$\frac{(d-1)(d-3)}{4} + \kappa_k \leq \frac{3}{4} \qquad \text{(VIII.2)}$$

and $\tilde{H}_k$ is always in the limit point case at 0. The limit $a \to 0$ is handled in the same way. The limit $a \to -\infty$ is controlled in the usual case of the whole line and since the generalized eigenfunctions have to vanish at 0 one can conclude that the spectrum of $\tilde{H}_k$ is pure point with eigenfunctions exponentially decaying at $+\infty$. Since this is true with probability one for each $k \geq 1$ the conclusion remains true for the full operator $H$ on $L^2(\mathbb{R}^d, dx)$.
If the dimension is 2 or 3, (VIII.2) is satisfied only for the large values of $k$ (in which case the conclusion on the almost sure spectral properties of $\tilde{H}_k$ remains valid) while for the smaller values of $k$, the operator $\tilde{H}_k$ is not in the limit point case at 0 and so, it is not essentially self adjoint on the space of smooth functions with compact supports contained in $(0, \infty)$. Eventhough the definition of $\tilde{H}_k$ needs to be supplemented with a boundary condition at 0 to define a self-adjoint operator on $L^2([0, \infty), dt)$, the self-adjoint operator $\tilde{H}_k$ obtained from the direct sum decomposition of the operator $H$ on $L^2(\mathbb{R}^d, dx)$ corresponds to only one of these boundary conditions. A simple computation shows that the generalized eigenfunctions of $H_k$ have to vanish at zero so that the Dirichlet boundary condition (corresponding to $\alpha = 0$ in the notation of the previous section) is the one. One can then control the limit $a \to 0$ as before and obtain the same conclusion.

## VIII.6 Problems

**Problem VIII.6.1** *In the one dimensional discrete case it was proved in Chapter III that the Green's function of the operator $H$ can be expressed by means of the Weyl functions $m^{\pm}(\lambda)$ for $\Im m(\lambda) > 0$ by the formula:*

$$G_\lambda(0,0) = \frac{-1}{m_+(\lambda) + m_-(\lambda) + \lambda - V(0)}$$

**1** *Using the spectral theorem prove that one can write :*

$$\Im m \int \frac{d\sigma_0(t)}{t - \lambda} = \frac{\beta}{(V(0) - \alpha)^2 + \beta^2}$$

*where $\alpha$ and $\beta$ are real functions independent of the potential $V(0)$and where the function $\beta$ is strictly positive.*

**2** *Using the de la Vallée Poussin theorem prove that if the law of $V_0$ has a density bounded by a constant $C$ then the averaged measure $\sigma_0$ with* *respect to the law of $V_0$ is absolutely continuous with a density also bounded by the constant $C$.*

**Problem VIII.6.2** *Let $X = (X_1, X_2)$ be a Gaussian vector with $X_1 \in \mathbb{R}^p$ and $X_2 \in \mathbb{R}^q$ and let us denote by $K_{i,j}$ ; $i, j = 1, 2$ the covariance matrix of the vectors $X_i$ and $X_j$.*
*Prove that if $X_2$ is not degenerated then the conditional law of $X_1$ given $X_2$ is Gaussian with covariance matrix $K_{1,1} - K_{1,2}K_{2,2}^{-1}K_{2,1}$.*
*Let $\{V(n)\,; n \in \mathbb{Z}\}$ be a stationary real valued Gaussian process with zero mean and covariance function $C(n) = \mathbb{E}\{V(0)V(n)\}$ that we assume to be "uniformly elliptic". This means that there exists a positive real $\alpha$ such that for any $n$ and real $(a_1, a_2, \dots a_n)$ one has:*

$$\mathrm{Var}\big(a_1V(1) + a_2V(2) + \dots + a_nV(n)\big) \geq \alpha(a_1^2 + a_2^2 + \dots + a_n^2)$$

*Let $\mu$ be the "spectral" measure on the torus $\mathbf{T}$ i.e. $\mu$ is the unique measure with Fourier coefficient of order $n$ given by $C(-n)$. It is known that the process $V(n)$ is ergodic iff $\mu$ is a continuous measure and that this process is nondeterministic iff the logarithm of the density of the absolutely continuous part of $\mu$ is integrable on $\mathbf{T}$ (See [131] page 76).*

**1** *Let us denote by $K_n$ the covariance matrix of $n$ successive potentials. Prove that the conditional law of $V(0)$ given $V(1), \dots, V(n)$ is a Gaussian law with covariance $\rho(n) = \dfrac{\det(K_{n+1})}{\det(K_n)}$*

**2** *Prove that if the sequence $\rho(n)$ has a strictly positive lower bound then the conditional law of $V(0)$ given the $\sigma$-algebra $\mathcal{F}^*$ generated by the potentials of nonzero index has a density bounded by a constant.*

**3** *Prove that if the process $V(n)$ is ergodic and nondeterministic then one has exponential localization for the associated Schrödinger Operator. Hint: Use the formula :*

$$\lim_{n\to\infty} \frac{\det(K_{n+1})}{\det(K_n)} = \exp\Big(\frac{1}{2\pi}\int_{-\pi}^{\pi} \log f(\alpha)\, d\alpha\Big)$$

*where $f$ is the density of the absolutely continuous component of $\mu$ (See [132] page 66).*

The definitions and notations used in the next three problems can be found in Section 2.

**Problem VIII.6.3** *Let c be a nonzero real number.*

1. *Prove that:*

$$\sigma^c(\{\lambda\}) = \lim_{\epsilon\downarrow 0} \epsilon\, \Im m F_c(\lambda + i\epsilon) \quad \textit{and} \quad \lim_{\epsilon\downarrow 0} \epsilon\, \Re e F_c(\lambda + i\epsilon) = 0$$

2. *Assume that* $\sigma^c(\{\lambda\}) \neq 0$. *Using formula VIII.1 prove that one has* $F_0(\lambda + i0) = -c^{-1}$. *Moreover using the monotone convergence theorem prove that*

$$\lim_{\epsilon\downarrow 0} \epsilon^{-1} \Im m\, F_0(\lambda + i\epsilon) = B(\lambda)$$

3. *Assume that* $F_0(\lambda + i0) = -c^{-1}$ *and* $B(\lambda) < \infty$. *Prove that:*

$$\lim_{\epsilon\downarrow 0} \frac{F_0(\lambda + i\epsilon)}{\epsilon F_c(\lambda + i\epsilon)} = \frac{i}{c\sigma^c(\{\lambda\})}$$

*Conclude that* $c^2\sigma^c(\{\lambda\}) = B^{-1}(\lambda) \neq 0$.

**Problem VIII.6.4** *Let z be a complex number with* $\Im m(z) > 0$.

1. *Using the Poisson formula and relation VIII.1, prove that*

$$\tilde{F}(z) = \int \frac{d\tilde{\sigma}(t)}{t - z} = \frac{F_0(z)}{1 - iF_0(z)} \qquad \text{(VIII.3)}$$

2. *Using that* $\Im m(F_0(z)) > 0$ *show that* $\Im m(\tilde{F}(z)) \leq 1$ *and conclude that* $\tilde{\sigma}$ *is absolutely continuous.*
3. *Taking the limit when* $\epsilon \to 0$ *for* $z = \lambda + i\epsilon$ *in VIII.3 prove that:*

$$\{\lambda\ ;\ \Im m\tilde{F}(\lambda + i0) = 0\} = \{\lambda\ ;\ F_0(\lambda + i0) = 0\}$$

*Conclude that* $\tilde{\sigma}$ *is equivalent to the Lebesgue measure.*

**Problem VIII.6.5** *This problem uses heavily the results about boundary values of analytic functions stated in Chapter I.*

1. *Using the de La Vallée Poussin theorem prove that* $\sigma^c_{s.c.}$ *is supported by the set*

$$\{\lambda\ ;\ B(\lambda) = \infty\} \cup \{\lambda\ ;\ B(\lambda) < \infty, F_0(\lambda + i0) = -c^{-1}\}$$

*Using Problem VIII.6.3 the second set is at most countable, hence has* $\sigma^c_{s.c.}$ *measure 0. Conclude that* $\sigma^c_{s.c.}$ *is supported by* $\{\lambda\ ;\ B(\lambda) = \infty\}$.

2. *Using VIII.1, prove that* $\sigma^c_{s.c.}$ *is supported on the complement of* $X$ *and conclude that it is supported by* $Z$.
3. *Remark that by Problem VIII.6.3 one has* $\sigma^c_{p.p.}$ *is supported on* $Y$.
4. *Prove that:*

$$\{\lambda \,;\Im m\, F_c(\lambda + i0) > 0\} = \{\lambda \,;\Im m\, F_0(\lambda + i0) > 0\}$$

*and conclude that* $X$ *supports* $\sigma^c_{a.c.}$

**Problem VIII.6.6** *Let* $\Lambda$ *be the box* $[0, b]$ *and* $p(n)$ *be the solution of* $H\psi = \lambda\psi$ *with the boundary conditions* $\psi(0) = 1$ *and* $\psi(-1) = 0$. *We denote by* $\sigma^\Lambda_{0,n}$ *the spectral measure of the operator* $H^\Lambda$ *at the couple of sites* $(0, n)$ *with* $n \leq b$.

1. *Prove that*

$$\sum_{k\in\Lambda} p^2(k) = p(b)\frac{d}{d\lambda}p(b+1) - p(b+1)\frac{d}{d\lambda}p(b)$$

2. *Prove that for a continuous function* $f$ *one has:*

$$\begin{aligned}\sigma^\Lambda_{0,n}(f) &= \sum_{\{\lambda\,;\,p(b+1)=0\}} f(\lambda)\frac{p(n)}{|p(b)\frac{d}{d\lambda}p(b+1)|}\\ &= \lim_{\epsilon\searrow 0}\frac{1}{2\epsilon}\int_{\{\lambda\,;\,|p(b+1)|\leq\epsilon\}} f(\lambda)\frac{p(n)}{|p(b)|}\,d\lambda\end{aligned}$$

3. *We now suppose that the potentials* $V(0), \ldots, V(b)$ *form an independent sequence of real random variables whose distribution has a density with respect to the Lebesgue measure. We assume that this density is the same at the sites* $0 \ldots b-1$ *and is denoted by* $r$. *Moreover the density* $w$ *of* $V(b)$ *is assumed to be continuous. Denoting by* $\mu_b$ *the distribution of* $V(b)$ *prove that*

$$\lim_{\epsilon\searrow 0}\frac{1}{2\epsilon}\mu_b\{|p(b+1)| \leq \epsilon\} = \frac{1}{|p(b)|}w(\lambda + \frac{p(b-1)}{p(b)})$$

*Using the change of variables* $V(0), \ldots V(b-1) \hookrightarrow p(1), \ldots p(b)$ *prove that with the convention* $p_0 = 1, p_{-1} = 0$ *one has:*

$$\mathbb{E}\{\sigma^\Lambda_{0,n}(f)\} = \int f(\lambda)\frac{p_n}{p_b^2}w(\lambda + \frac{p_{b-1}}{p_b})\prod_{k=1}^{b}\frac{dp_k}{|p_{k-1}|}r(\lambda + \frac{p_{k-2}+p_k}{p_{k-1}})\,d\lambda$$

4. *Using the new change of variable*

$$p_1 \dots p_b \hookrightarrow y_1 = \frac{p_1}{p_0} \dots y_b = \frac{p_b}{p_{b-1}}$$

*prove that with the convention* $\frac{1}{y_0} = 0$ *one has:*

$$\mathbb{E}\{\sigma_{0,n}^{\Lambda}(f)\} = \int f(\lambda) \qquad \frac{1}{(|y_1| \dots |y_n|)(y_{n+1} \dots y_b)^2} w(\lambda + \frac{1}{y_b})$$
$$\prod_{k=1}^{b} r(\lambda + \frac{1}{y_{k-1}} + y_k)\, dy_1 \dots dy_b\, d\lambda$$

5. *Denoting by* $K_{\delta,\lambda}$ *the operators introduced in Remark VIII.3.9 prove that for a bounded interval* $\mathcal{I}$ *one has:*

$$\mathbb{E}\{|\sigma_{0,n}^{\Lambda}|(\mathcal{I})\} = \int_{\mathcal{I}} r(y) \frac{1}{|y|} (K_{1,\lambda}^{n-1} K_{2,\lambda}^{b-n} v_\lambda)(y)\, dy\, d\lambda$$

*where* $v_\lambda(y) = w(\lambda + \frac{1}{y})$.
*One can remark that the last integration in* $y$ *would give the value of* $K_{\lambda,1} f$ *at "infinity". Unfortunately these operators act on the real line rather than on the projective line hence we cannot express this last integration as the action of* $K_{1,\lambda}$.

6. *Remark that one has:*

$$|\sigma_{0,n}^{+}|(\mathcal{I}) = \int_{\mathcal{I}} |p(\lambda, n)|\, d\sigma_{0,0}^{+}(\lambda)$$

*and conclude that if* $w$ *and* $r$ *are bounded, a sufficient condition of exponential localization on* $\mathcal{I}$ *for the Schrödinger operator on the positive axis is that there exist constants* $C$ *and* $\rho < 1$ *such that:*

$$||K_{1,\lambda}^{n}||_2 \leq C\rho^n \quad \text{and} \quad ||K_{2,\lambda}^{n}||_2 \leq C$$

*for* $\lambda \in \mathcal{I}$ *and* $n = 0, 1, \dots$.

7. *Rewrite all the above computations for a box* $\Lambda = [a, b]$ *and the spectral measure* $\sigma_{m,n}^{\Lambda}$ *for* $m$ *and* $n$ *in the box* $\Lambda$ *and prove that for* $m \leq n$ *one has:*

$$\mathbb{E}\{|\sigma_{m,n}^{\Lambda}|(\mathcal{I})\} = \int_{\mathcal{I}} r(y) \frac{1}{|y|} (K_{0,\lambda}^{m} K_{1,\lambda}^{n-m} K_{2,\lambda}^{b-n} v_\lambda)(y)\, dy\, d\lambda$$

*where* $v_\lambda(y) = w(\lambda + \frac{1}{y})$.

*Using the criterion of localization for the strip proved in Proposition III.5.22 conclude that the bounds on the norms of the operators $K_{\delta,\lambda}$ given in 6. are also sufficient for localization on the whole lattice.*

**Problem VIII.6.7** *Let*

$$U_\lambda(t) = \begin{pmatrix} \cos(t\sqrt{\lambda}) & \frac{1}{\sqrt{\lambda}}\sin(t\sqrt{\lambda}) \\ -\sqrt{\lambda}\sin(t\sqrt{\lambda}) & \cos(t\sqrt{\lambda}) \end{pmatrix}$$

*be the propagator of the free Laplacian on the real line introduced in Chapter III and let us set $T_\lambda = U_\lambda(\sqrt{\lambda})$ for $\lambda \in \mathcal{I} = [0\ ,\ 1]$.*

**1** *Prove that the mapping $\lambda \hookrightarrow T_\lambda$ from $\mathcal{I}$ to the set of matrices of order two is continuous, that the eigenvalues of $T_\lambda$ are of modulus one and that $T_\lambda$ is diagonalizable for $\lambda \in \mathcal{I}$.*

**2** *Prove that for each $\lambda \in \mathcal{I}$ the sequence $T_\lambda^n$ is bounded and compute an upper bound.*

**3** *Prove that there does not exist a uniform bound of the sequence $T_\lambda^n$ for $\lambda \in \mathcal{I}$.*

**Problem VIII.6.8** *Let $v$ be a vector in $\mathbb{R}^2$ with coordinates $v_1$ and $v_2$. One defines the function $\bar{v} \hookrightarrow \cos^2(\bar{v})$ by the formula:*

$$\cos^2(\bar{v}) = \frac{v_1^2}{v_1^2 + v_2^2}$$

*Let us set $N = \mathbb{E}\{\sigma_0\}$ where $\sigma_0$ is the spectral measure at the site 0 of the discrete random Schrödinger operator associated with an i.i.d. potential with an absolutely continuous distribution.*
*Using the notations of the Proposition VIII.3.10 and following the lines of its proof show that $N$ has a density with respect to the Lebesgue measure given by:*

$$\frac{dN}{d\lambda} = \frac{1}{\pi}\int \cos^2(\bar{v})\phi_\lambda(\bar{v})\breve{\phi}_\lambda(\bar{v})\, dm(\bar{v})$$

**Problem VIII.6.9** *Let $\{X(n)\,;\, n \in \mathbb{Z}\}$ be an i.i.d. sequence of real random variables whose distribution is not singular with respect to the Lebesgue measure. We consider the family of potentials $\{V(t)\,;\, t \in \mathbb{R}\}$ defined by:*

$$V(t) = \sum_{n=-\infty}^{+\infty} \phi(t-n)X_n$$

*where $\phi$ is a smooth function with nonvoid compact support in $[0\,,\,1]$. It follows that for $0 \le t \le 1$ then one has $V(t) = X(0)\phi(t)$.*

1. *Let us assume first that the law of $X(0)$ is absolutely continuous with a bounded density with compact support. Prove that one can apply Theorem VIII.1.15 in this situation with $c = 1$ and $\Omega_0 = \Omega$.*
2. *Using the notion of $\Omega_0$ conditional expectation prove in the same way as in Theorem VIII.1.9 that one has exponential localization in the general case of a nonsingular distribution of potentials.*

*Note that one can choose the potential well $\phi$ so that $t \hookrightarrow V(t,\omega)$ is twice continuously differentiable with bounded derivatives for almost all $\omega \in \Omega$.*

**Problem VIII.6.10** *Let $a(n)$ be an i.i.d. sequence of bounded positive random variables whose distribution has a bounded density and let us consider the "Jacobi operator" defined by :*

$$[H\psi](n) = a(n)\psi(n+1) + a(n-1)\psi(n-1)$$

1. *If as usual we set $\sigma_{m,n}(B) =< E_B\delta_m, \delta_n >$ prove that:*

$$\lambda\sigma_{n,n} = a(n-1)\sigma_{n-1,n} + a(n)\sigma_{n+1,n}$$

2. *Adapt the proof of the lemma VIII.1.8 to the case of a non diagonal spectral measure, that is : Let $X$ be a real symmetric matrix with entries $x_{i,j}$ and $m$ be a probability measure with a bounded density, then the averaged measure $\int \sigma^X_{i,j}\, dm(x_{i,j})$ also has a bounded density.*
3. *Prove that the Lyapunov exponent associated with $H$ is strictly positive.*
   *Hint: consider the Markov chain $X(n) = (a(n), a(n-1))$ and use the results proved in Chapter IV.*
4. *Using exactly the same proof as in Theorem VIII.1.9 conclude that one has exponential localization for $H$.*

**Problem VIII.6.11** *1. Show the existence of conditional probabilities*

$$\mathbb{P}^{t,(x',\vartheta')}_{\lambda,s,(x,\vartheta)}$$

*that the process $(X(t), \vartheta_\lambda(t))$ is in the state $(x', \vartheta')$ at time $t$ knowing that is was in the state $(x, \vartheta)$ at time $s \le t$.*

*2. Show that, for each fixed $\lambda$ and $s \le t$, the map:*

$$((x,\vartheta),(x',\vartheta')) \hookrightarrow \mathbb{P}^{t,(x',\vartheta')}_{\lambda,s,(x,\vartheta)}$$

*is continuous for the weak topology on the space of probability masures on path space.*

*3. If $\lambda$ and $s \le t$ are fixed, show that for each continuous function $h : [s,t] \times M \times \mathbb{T} \to \mathbb{R}$ the function*

$$((x,\vartheta),(x',\vartheta')) \hookrightarrow \mathbb{E}^{t,(x',\vartheta')}_{\lambda,s,(x,\vartheta)} \left\{ e^{-\int_s^t h(u,X(u),\vartheta(u))du} \right\}$$

*is jointly continuous.*

**Problem VIII.6.12** *Recall that for any real numbers $\lambda$ and $\delta$ and for each $s$ and $t$ such that $0 < s < t$ the Laplace transform associated with the Markov process $(X(t), \vartheta_\lambda(t)$ is defined by:*

$$T^{s,t}_{\delta\lambda} f(x,\vartheta) = \mathbb{E}_{s,x}\{f(X(t), \theta_\lambda(t,s)) |U_\lambda(t,s)\vartheta|^\delta\}$$

*for each function $f$ on $M \times \mathbb{T}$. As before, we used the notation $\mathbb{E}_{s,x}$ for the expectation over the Brownian path starting from $x$ at time $s$.*

1. *Each of the above operators has a bounded jointly continuous kernel. In particular, they map boundedly $L^1(dx, d\vartheta)$ into $L^\infty(dx, d\vartheta)$, they are compact on the Banach space $C(M \times \mathbb{T})$ of continuous functions on the compact space $M \times \mathbb{T}$, and they are Hilbert-Schmidt on $L^2(dx, d\vartheta)$.*
2. *Under the above conditions (actually, $p(t)$ locally integrable is enough) show that for each compact interval $\mathcal{I}$ one has:*

$$\sup_{x \in M, t \ge 1, \lambda \in \mathcal{I}, \vartheta \in \mathbb{T}} \mathbb{E}_{0,x}\{|U_\lambda(t,0)\vartheta|^{-2}$$

*is a finite constant depending only on $\mathcal{I}$ and $\int_0^1 |p(t)| dt$.*

**Problem VIII.6.13** *Let $\{F(X(t,\omega)); t \in \mathbb{R}, \omega \in \Omega\}$ be the random potential of the Russian school and let $\{q(t,\omega); t > 0, \omega \in \Omega\}$ be a measurable stochastic process defined on the same complete probability space $(\Omega, \mathcal{F}, \mathbb{P})$. We assume that for almost all $\omega \in \Omega$, the sample path $t \hookrightarrow q(t,\omega)$ is a bounded function bounded below by $-1$ and we assume that the two stochastic processes $\{X(t); t > 0\}$ and $\{q(t); t > 0\}$ are independent. We also consider a continuous (coupling constant) function $a : [0,\infty) \to [0,1]$ which satisfies $\int_0^\infty a(t)dt < +\infty$. We study the spectral characteristics of the random Schrödinger operator $H(\omega) = -d^2/dt^2 + V(t,\omega)$ on $L^2(\mathbb{R}, dt)$ with:*

$$V(t,\omega) = \begin{cases} a(t)q(-t,\omega) & if\, t < 0 \\ 2F(X(t,\omega)) - 1 & if\, t \geq 0. \end{cases}$$

1. *Show that the spectrum of $H(\omega)$ is almost surely equal to $[-1,\infty)$.*
2. *Show that each (fixed) $\lambda \geq -1$ is almost surely not an eigenvalue of $H(\omega)$.*
3. *Use theoremIII.4.1 to show that the spectrum is absolutely continuous in $(0,\infty)$.*
4. *Use the proof of localization presented in this chapter to show that there is exponential localization in $[-1,0]$.*
5. *Check that the independence assumption can be replaced by $\{q(t); t > 0\}$ and $\{X_t; t > 0\}$ are conditionally independent given $X_0$.*

# VIII.7 Notes and Complements

It is fair to say that the wave of interest in the spectral properties of the random Schrödinger operators and especially the pure point nature of the spectrum for these systems, has been initiated (among the mathematically inclined mathematical physicists and some probabilists) by the works of the Russian school in the early and mid seventies. The curiosity culminated with the publication of the short note [127] by Golsheid, Molcanov and Pastur and the subsequent longer article of Molcanov [244]. To popularize the subject, an extensive study of the discrete case was then provided by Kunz and Souillard in [204] and the one-dimensional continuous case was reworked by Carmona in [42]. Many papers were written on the subject, see for example those of Royer [278], Kunz and Souillard [205], Carmona [43], [44], Brossard [37], Delyon, Kunz and Souillard [70], Lacroix [207], [208], Kotani [193], [195], ... but the most significant breakthrough was the fundamental work of Spencer and Frölich which we will present in the next chapter and from which the latest developements followed, in the one-dimensional case as well as the multidimensional one.

The so called "Kotani's trick" exposed in the first section is certainly the easiest way to prove localization for general ergodic systems. It can be used when:

- one can "vary smoothly" the value of the potential in a finite box.
- the lowest nonegative Lyapunov exponent is positive.

We added Lemma VIII.1.8 to the discrete case in order to give an easy way to extend the one-dimensional arguments to the strip. In the continuous case we have followed closely the discussion of Kotani and Simon in [200]. The computations of the averaged spectral measures in the strip performed in Chapter III allows us to extend their proof to this situation.

In a recent work N. Minami [240] gives a nice application of the ideas of Kotani to one dimensional systems by replacing the nonrandom Oseledec's theorem by the Gilbert and Pearson characterization of the various spectra in terms of the existence or nonexistence of subordinate solutions. For a given bounded interval $\mathcal{I}$ let us denote by $\mathcal{F}^*$ the $\sigma$-algebra generated by the potentials lying outside $\mathcal{I}$. Let us assume as in Section 1 that the conditional distribution of $\sigma(\omega)$ given $\mathcal{F}^*$ is $\mathbb{P}$-almost surely absolutely continuous on $\mathcal{I}$. Then, using the results described in Chapter III one can give some rather precise information on the nature of the Lebesgue decomposition of $\sigma$ on $\mathcal{I}$. More precisely,

*if for $m$ almost all $\lambda \in \mathcal{I}$ one has:*

$$\mathbb{P}\{H(\omega)u = \lambda u \quad \textit{has two solutions } u_+ \textit{ and } u_- \textit{ such that } u_+ \in L^2(0,+\infty),\ u_- \in L^2(-\infty,0)\} = 1$$

*then $\sigma(\omega)$ is pure point on $\mathcal{I}$ for $\mathbb{P}$-almost all $\omega$.*

This approach to localization was singled out by Carmona in [42]. But proving that the above probability is actually equal to 1 is usually very technical, especially in the continuous case. Kotani pushed further these ideas and he gave sufficient conditions for localization in terms of the regularity of the conditionned spectral measure, see [193], [195]. The present abstract form has nice extensions to the other types of spectra which can be proved similarly using the deterministic results of Chapter III.

*If for $m$ almost all $\lambda \in \mathcal{I}$ one has:*

$$\mathbb{P}\{H(\omega)u = \lambda u \quad \textit{has no solution subordinate at } +\infty \textit{ or has no solution subordinate at } -\infty\} = 1$$

*Then $\sigma(\omega)$ is equivalent to the Lebesgue measure on $\mathcal{I}$ for $\mathbb{P}$almost all $\omega$.*

*If for $m$ almost all $\lambda \in \mathcal{I}$ one has:*

$$\mathbb{P}\{H(\omega)u = \lambda u \quad \text{has two solutions } u_+ \text{ and } u_- \text{ which are subordinate respectively at } +\infty \text{ and } -\infty \text{ but such that } u_+ \notin L^2(0,+\infty),\ u_- \notin L^2(-\infty,0)\} = 1$$

*Then $\sigma(\omega)$ is purely singular continuous on $\mathcal{I}$ for $\mathbb{P}$almost all $\omega$.*

The same results can be proved in the discrete model. The author gives two applications to non stationary models already discussed in Section 5 where the random potential is multiplied by a function decreasing at infinity and he recovers the absolute continuity of the spectrum.

Section 2 is borrowed from Simon and Wolff [304] where they essentially obtain the results of Section 1 in the discrete case. The notion of "infinite" rank one perturbation defined in this paper allows us to handle the strip case. An extension of this theory to the continuous case was given by Howland [152]. Not surprisingly, his result is the same as in the previous section, namely :
Let us consider a second potential process $\epsilon(t,\omega')$ on some probability space $\Omega'$ such that:

- the laws of the processes $V(t)$ and $V(t) + \epsilon(t)$, once appropriately redefined on the same path space are not mutually singular,
- the process $\epsilon(t)$ vanishes identically for $t$ outside some compact set independent of $\omega$.
- The distribution of $\epsilon$ is absolutely continuous

Then the spectrum of $H$ is almost surely pure point. It is also possible to state these assumptions on a set of positive $\mathbb{P}$-measure.

Our proof of localization in Section 3 named "Operator Theory" goes back to the original work of Golsheid, Molcanov and Pastur [127]. This proof has been streamlined by Carmona in [42], [44]. In the discrete case, the proof has been given by Kunz and Souillard [204]. Royer [278] used the same approach to give still another twist to the proof of the continuous case. Quite similar techniques, together with a systematic use of averaged spectral measures have been used by Lacroix in the discrete case [206], Carmona in the continuous case [43] and extended to the strip by Lacroix in

[208], [209]. This leads to a systematic study of the Fourier-Laplace transform on boundaries of semisimple Lie groups. Not surprisingly we chose to present this last approach... We should say in our defense that it gives an unified treatment of both the continuous and discrete models. The pure point spectrum property is then obtained as a direct consequence of some bounds on the spectral radius of the Fourier-Laplace operators proved in Chapter IV. The links with the $L^2$ theory of "pseudo Laplace transform" developed by Delyon, Kunz and Souillard in [71] are then obvious since they prove exactly the same kind of estimates (but for slightly different operators). Note that all these proofs can accomodate a periodic perturbation of the potential as shown by the work of Brossard [37].

The proof of localization for general potentials given in Section 4 is directly taken from the paper of Carmona, Klein and Martinelli [50]. We merely added the systematic use of imaginary boundary conditions for the Schrödinger operator in a box. This simplifies most of the proofs. We believe that a simpler way of proving localization for general potentials certainly exists since this one refers to the multidimensional analysis developed in the next chapter! But up to now, no attempt has been successful...A generalization of these ideas to the strip by Klein, Lacroix and Speis can be found in [184].

# IX

# LOCALIZATION IN ANY DIMENSION

Contents :



Let $\beta$ be a nonzero real number and let $H(\omega)$ be the random Schrödinger operator on $\mathbb{Z}^d$ defined by

$$H(\omega)\psi(x) = H_0\psi(x) + \beta V(x,\omega)\psi(x)$$

where $\{V(x) \; ; \; x \in \mathbb{Z}^d\}$ is an i.i.d. family of potentials defined on some probability space $(\Omega, \mathbb{P})$. Such a model proposed by Anderson in [8], is usually refered to as the "Anderson model". It has been shown in the preceding chapters that the behavior at infinity of the solutions of the "eigenvalue equation"

$$H_0\psi(x) = (\lambda - \beta V(x,\omega))\psi(x)$$

is crucial in the study of spectral properties of the operator $H(\omega)$. The behavior at infinity of the Green's function $G(\lambda, x, y)$ which satisfies the above

equation for any $x \neq y$ is also of primordial interest. The links between the exponential growth of the solutions of the eigenvalue equation and the exponential decay of the Green's function have already been pointed out in the one dimensional case.
There are (at least) two methods of constructing a proof of localization in the Anderson model. The first method, already developed in chapter VIII Sections 1 & 2, is essentially two-fold:

1. For a fixed value of the energy $\lambda$, one studies the behavior at infinity of the solutions of the eigenvalue equation or of the Green's function on a set $\Omega_\lambda \subset \Omega$ of full $\mathbb{P}$ measure.
2. In order to study the $\mathbb{P}$almost sure spectral properties of $H(\omega)$ the crucial problem is then to obtain $\mathbb{P}$ almost sure properties on a whole interval of energy. Unfortunately the set $\Omega_\lambda$ depends heavily on the value of the energy $\lambda$ and it is hopeless to prove that the intersection of the sets $\Omega_\lambda$ on an interval of energy is of full exterior probability. Looking at the eigenvalue equation, one remarks that it is the same to wriggle either the energy or the potential values. Hence when the distribution of the potential at one site allows such translations (i.e. essentially when it is absolutely continuous) it is possible to show that the above $\mathbb{P}$-almost sure results at a fixed energy yield $\mathbb{P}$-almost sure properties on a whole interval of energy.

In the one-dimensional case the first step has been achieved by proving the existence and positivity of the Lyapunov exponent. This leads to the important notion of hyperbolic behavior. The second step was carried out either using the spectral theory of the Laplace transform on groups or by means of the Kotani's trick or the Simon & Wolff's method.

In the multidimensional case one can repeat the same scheme but the first step now results from the Fröhlich & Spencer's analysis of the decay of the Green's function that will be presented in the first Section of this Chapter. For the second step, the procedure used in Section 2 is a remake of the one dimensional proof.

Martinelli & Scoppola [231] then Fröhlich, Martinelli, Scoppola and Spencer [104] have initiated a quite different method which was later simplified and extended by Dreifus and Klein[88]. They prove directly some $\mathbb{P}$-almost sure properties of decay at infinity of the Green's function in boxes on a whole interval of energy. This yields a proof of localization without reference to the smoothness of the distribution of the potential at one site.

Unfortunately it seems very difficult to check the necessary assumptions without assuming any regularity properties of this distribution. Nevertheless making an extensive use of the theory of random matrices we showed in Chapter VIII Section 4 that such a procedure works in the one dimensional case, for any non constant potential with a finite moment.

# IX.1 Exponential Decay of the Green's Function at Fixed Energy

The results proved in this section are due to Fröhlich and Spencer[105]. We follow the simpler proof of these results given by Von Dreifus in his thesis [87].

For a vector $x = (x_1, x_2, \ldots, x_d) \in \mathbb{Z}^d$ one considers the supremum norm $|x| = \sup_{i=1\ldots d}(|x_i|)$ for which the balls are actually boxes. For the reader's convenience we recall below the definitions of the interior and exterior boundary of a box already used in Chapter II.

**Definition IX.1.1** *Let $\ell$ be a positive real number.*

- *The box $\Lambda_\ell(x)$ centered at the point $x$ and of size $\ell$ is defined as:*

$$\Lambda_\ell(x) = \{y \in \mathbb{Z}^d \; ; \; |x - y| \leq \frac{\ell}{2}\}$$

  *We denote by $\Lambda_\ell$ a box of size $\ell$ centered at the origin and the symbol $\Lambda$ denotes a box without reference to its center or size.*
- *The boundary $\partial\Lambda$ of the box $\Lambda$ is the subset of $\mathbb{Z}^d \times \mathbb{Z}^d$ defined by:*

$$\partial\Lambda = \{(u, v) \; ; \; |u - v| = 1 \; , \; u \in \Lambda \text{ and } v \notin \Lambda\}$$

- *One defines the interior boundary $\partial\Lambda^{in}$ by:*

$$\partial\Lambda^{in} = \{u \in \Lambda \; ; \; \text{there exists } \; v \notin \Lambda \text{ such that } (u, v) \in \partial\Lambda\}$$

- *One defines the exterior boundary $\partial\Lambda^{out}$ by:*

$$\partial\Lambda^{out} = \{v \notin \Lambda \; ; \; \text{there exists } \; u \in \Lambda \text{ such that } (u, v) \in \partial\Lambda\}$$

- *One denotes by $\Gamma_\Lambda$ the boundary operator of the box $\Lambda$ defined by the matrix elements:*

$$\Gamma_\Lambda(x,y) = \begin{cases} 1 & \textit{if } (x,y) \textit{ or } (y,x) \textit{ is in } \partial\Lambda \\ 0 & \textit{otherwise} \end{cases}$$

Let $|A|$ denote the volume (i.e. the number of points) of a finite subset $A \subset \mathbb{Z}^d$. It is easily seen that there exists a constant $c_d$ depending only on the dimension $d$ such that for a box $\Lambda$ of size $\ell$ one has:

$$|\Lambda| \leq c_d \ell^d, \quad |\partial\Lambda^{in}| \leq c_d \ell^{d-1}, \quad |\partial\Lambda^{out}| \leq c_d \ell^{d-1}$$

Let $A \subset \mathbb{Z}^d$ and $\Im m(z) \neq 0$, then the Green's kernel $G^A(z,x,y)$ can be extended to $\mathbb{Z}^d$ if one sets $G^A(z,x,y) = 0$ if $x$ or $y$ are not in $A$.

**Lemma IX.1.2** *Let $\Lambda \subset \Lambda'$ be two boxes with no common boundary and let $B$ be the complementary set of $\Lambda$ in $\Lambda'$. Then one has the resolvent equations:*

$$G^{\Lambda'} = G^\Lambda + G^B + (G^B + G^\Lambda)\Gamma_\Lambda G^{\Lambda'} = G^\Lambda + G^B + G^{\Lambda'}\Gamma_\Lambda(G^B + G^\Lambda)$$

*As a consequence one gets:*

$$(i)\; G^{\Lambda'}(z,x,y) = G^\Lambda(z,x,y) + \sum_{(u,v)\in\partial\Lambda} G^\Lambda(z,x,u)G^{\Lambda'}(z,v,y) \quad \textit{if } x \in \Lambda$$

$$(ii) G^{\Lambda'}(z,x,y) = G^\Lambda(z,x,y) + \sum_{(u,v)\in\partial\Lambda} G^{\Lambda'}(z,x,v)G^\Lambda(z,u,y) \quad \textit{if } y \in \Lambda$$

*In all this section the energy $\lambda$ is kept fixed. Hence we skip the reference to $\lambda$ in most of the definitions and constants appearing below.* In particular we use the notation $G_\epsilon(x,y)$ instead of $G(\lambda + i\epsilon, x, y)$.

**Definition IX.1.3** *Let $\gamma$ and $\ell$ be positive numbers. The box $\Lambda_\ell(x)$ is said to be a $\gamma$-"good box" if for any $\epsilon \neq 0$ one has:*

$$\sum_{y\in\partial\Lambda_\ell^{in}(x)} |G_\epsilon^{\Lambda_\ell(x)}(x,y)| \leq e^{-\gamma\ell}$$

*and it is said to be a $\gamma$-"bad box" if the above condition does not hold.*

The following Lemma shows that a big box is a good box assuming that it contains sufficiently many good little boxes. It is illustated by the next figure.

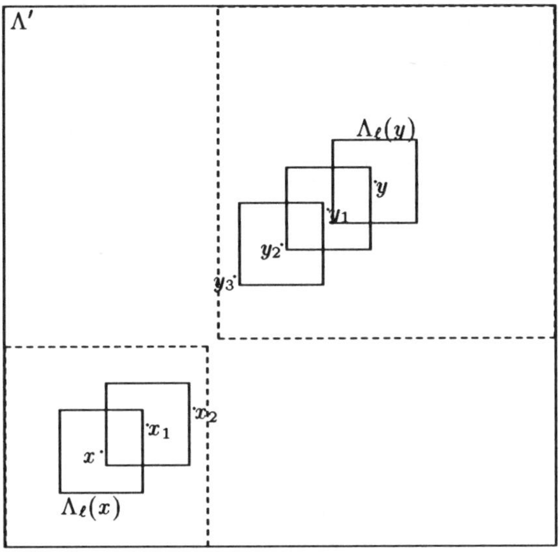

Fig. IX.1*Chained Boxes*

**Lemma IX.1.4** *Let $\Lambda'$ be a box of size $\ell' > \ell > 0$ and containing the two points $x$ and $y$. Let us assume that there exists nonnegative integers $m$ and $n$ such that:*

- *For any sequence $(x_0, x_1, \ldots, x_n)$ with $x_0 = x$ and $x_{k+1} \in \partial\Lambda_\ell^{out}(x_k)$ then $\Lambda_\ell(x_k)$ is a $\gamma$-good box contained in $\Lambda'$ for $k = 0, \ldots, n-1$.*
- *For any sequence $(y_0, y_1, \ldots, y_m)$ with $y_0 = y$ and $y_{k+1} \in \partial\Lambda_\ell^{out}(y_k)$ then $\Lambda_\ell(y_k)$ is a $\gamma$-good box contained in $\Lambda'$ for $k = 0, \ldots, m-1$.*
- $|x - y| \geq (m+n)(\ell/2+1)$

*then for any $\epsilon \neq 0$ there exists $(u, v) \in \Lambda'$ such that:*

$$|G_\epsilon^{\Lambda'}(x, y)| \leq e^{-(m+n)\gamma\ell} |G_\epsilon^{\Lambda'}(u, v)|$$

*Proof:*
It is easily seen that one has:

$$|x_n - x| \leq n(\ell/2+1) \quad \text{and} \quad |y_m - y| \leq m(\ell/2+1)$$

hence the condition $|x - y| \geq (m+n)(\ell/2+1)$ entails that no box $\Lambda_\ell(x_k)$, $k = 0 \ldots n-1$ can contain the point $y$ and that no box $\Lambda_\ell(y_k)$, $k = 0 \ldots m-1$ can contain the point $x_n$. If $n \geq 1$ we apply the resolvent equation (i) in Lemma IX.1.2 :

$$G_\epsilon^{\Lambda'}(x, y) = \sum_{(u,v) \in \partial\Lambda_\ell(x_0)} G_\epsilon^{\Lambda_\ell(x_0)}(x, u) G_\epsilon^{\Lambda'}(v, y)$$

Let $x_1$ be a point in $\partial\Lambda_\ell^{out}(x_0)$ for which the function $v \hookrightarrow |G_\epsilon^{\Lambda'}(v,y)|$ attains its maximum value. Then one has:

$$\begin{aligned} |G_\epsilon^{\Lambda'}(x,y)| &\leq |G_\epsilon^{\Lambda'}(x_1,y)| \sum_{u\in\partial\Lambda_\ell^{in}(x_0)} |G_\epsilon^{\Lambda_\ell(x_0)}(x_0,u)| \\ &\leq e^{-\gamma\ell}|G_\epsilon^{\Lambda'}(x_1,y)| \end{aligned}$$

Iterating the process to compute $G_\epsilon^{\Lambda'}(x_1,y)\dots G_\epsilon^{\Lambda'}(x_{n-1},y)$ this yields:

$$|G_\epsilon^{\Lambda'}(x,y)| \leq e^{-n\gamma\ell}|G_\epsilon^{\Lambda'}(x_n,y)|$$

If $m \geq 1$ we apply the second form of the resolvent equation (ii) in Lemma IX.1.2 to compute $G_\epsilon^{\Lambda'}(x_n,y)$ and we obtain:

$$G_\epsilon^{\Lambda'}(x_n,y) = \sum_{(u,v)\in\partial\Lambda_\ell(y_0)} G_\epsilon^{\Lambda'}(x_n,v)G_\epsilon^{\Lambda_\ell(y_0)}(u,y_0)$$

Choosing a point $y_1 \in \partial\Lambda_\ell^{out}(y_0)$ for which the function $v \hookrightarrow |G_\epsilon^{\Lambda'}(x_n,v)|$ attains its maximum value we iterate the above process $m$ times and we get:

$$|G_\epsilon^{\Lambda'}(x_n,y)| \leq e^{-m\gamma\ell}|G_\epsilon^{\Lambda'}(x_n,y_m)|$$

The desired result now follows. ∎

The next proposition is a direct application of this lemma where one chooses the box $\Lambda'$ to be of size $2\ell'$, the point $x$ at the origine and the point $y$ on the boundary of $\Lambda_{\ell'}$.

**Proposition IX.1.5** *Let $\ell' > \ell > 1$ and let us assume that the box $\Lambda_{2\ell'}$ does not contain two overlapping $\gamma$-bad boxes of size $\ell$. Then for any $\epsilon \neq 0$ and any point $y \in \partial\Lambda_{\ell'}^{out} \cup \partial\Lambda_{\ell'}^{in}$ there exist $(u,v) \in \Lambda_{2\ell'}$ such that:*

$$|G_\epsilon^{\Lambda_{2\ell'}}(0,y)| \leq e^{-\gamma\frac{\ell}{\ell+2}(\ell'-8\ell)}|G_\epsilon^{\Lambda_{2\ell'}}(u,v)|$$

*Proof:*
There is only something to prove in the case $\ell' > 8\ell$. Actually we want to show that it is possible to accumulate sufficiently many good boxes in $\Lambda' = \Lambda_{2\ell'}$ according to the process explained in Lemma IX.1.4. More precisely choosing $x = 0$, we need that the integers $m$ and $n$ obeying the conditions of this lemma also satisfy the relation $m + n \geq \frac{1}{\ell+2}(\ell' - 8\ell)$. Let $d$ be a positive integer and let $[x]$ denotes the integral part of a real

number $x$. It is readily seen that starting the construction at the point $u$ one can accumulate at least $\left[\frac{2d}{\ell+2}\right]$ "chained" boxes of size $\ell$ in the box $\Lambda_{2d}(u)$ without reaching the boundary of this box.
The hypothesis implies that if there exists one $\gamma$-bad box then any non-overlapping box contained in $\Lambda'$ is a $\gamma$-good box. Hence if there is no $\gamma$-bad box or if the distance from the center of the bad box to 0 or $y$ is no greater than $\ell/2$ the result is obvious: one begins the construction of chained boxes at the origin or at the point $y$.
Otherwise let $d$ be the norm of the center of the bad box. From the above remark one obtains that:

$$n \geq \left[\frac{2d-\ell}{\ell+2}\right] \quad \text{and} \quad m \geq \left[\frac{|y|-2d-\ell}{\ell+2}\right]$$

The result follows if we remark that $[x] \geq x-1$ and $|y| = [\ell'/2]$ or $[\ell'/2]+1$.
∎

**Definition IX.1.6** *Let $\gamma$ be a positive number and let $\xi = 2d+4$. A box $\Lambda$ of size $\ell$ is said to be $\gamma$-adapted if:*

$$\mathbb{P}\{\Lambda \text{ is a } \gamma\text{-good box }\} \geq 1-\ell^{-\xi}$$

Notice that the translation invariance of the probability measure $\mathbb{P}$ implies that this definition only involves the size of the box and not the location of its center.

Throughout this section we will consider the following set of statements:

[$H_1$ ] **There exist $\gamma_0 > 0$ and $\ell_0$ "sufficiently large" such that the box of size $\ell_0$ is $\gamma_0$-adapted.**

[$H_2$ ] **There exists a real $0 < \tau \leq 1$ and a constant $C_w$ such that one has:**

$$\mathbb{P}\{\text{dist}(\lambda, \Sigma(H^\Lambda)) \leq \eta\} \leq C_w|\Lambda|\eta^\tau$$

**for any box $\Lambda$ and any $\eta > 0$. In this case the distribution of potentials is said to obey a Wegner's estimate.**

Hypothesis $[H_1]$ will sometimes be refered to as $[H_1](\gamma_0, \ell_0)$ and the minimal value of $\ell_0$ (as a function of $\gamma_0$) will be precised in Proposition IX.1.13.

It is well known that the decay of Green's functions gives an indication of the strength of the coupling between regions far apart. Therefore when

$[H_1]$ holds for a sufficiently large $\ell_0$ this means that with high probability, sufficiently large blocks are weakly coupled for a given energy.
Hypothesis $[H_2]$ is the crucial ingredient to show that tunneling at large distance is very unlikely. We say that an energy $\Lambda$ is "resonant" in some box $\Lambda$ if it is "very close" to the spectrum of $H^\Lambda$. It is readily seen that $[H_2]$ implies that for any $\chi > 0$ one has:

$$\mathbb{P}\{\text{dist}(\lambda, \Sigma(H_\ell^\Lambda)) \leq e^{-\ell}\} \leq \ell^{-\chi}$$

for $\ell$ sufficiently large. Using the independence, this implies that for regions far apart, the probability that a given energy be simultaneously "resonant" is very small.

**Remark IX.1.7** If the distribution of potentials obeys a Wegner estimate one obtains immediately the following bound on the norm of the Green's function:

$$\mathbb{P}\{\sup_{\epsilon\neq 0} \|G_\epsilon^\Lambda\| \leq \eta\} \geq 1 - C_w|\Lambda|\eta^{-\tau}$$

We saw in Proposition VIII.4.13 that assumption $[H_2]$ holds uniformly with respect to the energy as long as the distribution of the potentials is Hölder continuous of some order $0 < \tau \leq 1$. Moreover the following proposition shows that for any $\gamma_0$ and $\ell_0$, assumption $[H_1](\gamma_0, \ell_0)$ holds if the energy or the disorder are large enough.

**Proposition IX.1.8** *Let $\ell$ be a given positive number and $\Lambda$ be a box of size $\ell$. Then for any $\eta > 0$ one has:*

$$\lim_{|\lambda|\to\infty} \mathbb{P}\{\sup_{\epsilon\neq 0} \|G^\Lambda(\lambda + i\epsilon)\| \leq \eta\} = 1$$

*If we now assume that the distribution of the potential at one site is Hölder continuous then one has:*

$$\lim_{|\beta|\to\infty} \mathbb{P}\{\sup_{\epsilon\neq 0} \|G^\Lambda(\lambda + i\epsilon)\| \leq \eta\} = 1$$

*uniformly with respect to $\lambda \in \mathbb{R}$.*

*Proof:*
Note that $\|H_0^\Lambda\| \leq \|H_0\| \leq 2d$.
Let us assume that $|\beta V(x)| \leq |\lambda| - 2d - \eta^{-1}$ for any $x \in \Lambda$. Then the

spectrum of $H^{\Lambda}$ is contained in the interval $[-|\lambda|+\eta^{-1}, |\lambda|-\eta^{-1}]$. This implies that for any $\epsilon \in \mathbb{R}$ one has $||G^{\Lambda}(\lambda+i\epsilon)|| \leq \eta$
Moreover the independence of the potentials yields:

$$\mathbb{P}\{\cap_{x\in\Lambda}\{|\beta V(x)| \leq |\lambda| - 2d - \eta{-}1\}\} = \left(\mathbb{P}\{|\beta V(0)| \leq |\lambda| - 2d - \eta^{-1}\}\right)^{|\Lambda|}$$

This last term goes to 1 when $|\lambda| \to \infty$ and we get the first part of the proposition.

On the other hand let us assume that $|\beta V(x) - \lambda| \geq 2d + \eta^{-1}$ for any $x \in \Lambda$. It follows that the distance from $\lambda$ to the spectrum of $H^{\Lambda}$ is greater than $\eta^{-1}$ hence that for any $\epsilon \in \mathbb{R}$ one has $||G^{\Lambda}(\lambda+i\epsilon)|| \leq \eta$.
Moreover:

$$\begin{aligned}\mathbb{P}\{|\beta V(x) - \lambda| &\leq 2d + \eta^{-1} \text{ for some } x \in \Lambda\} \\ &\leq |\Lambda|\,\mathbb{P}\{\beta V(0) \in [\lambda - 2d - \eta^{-1}, \lambda + 2d + \eta^{-1}]\} \\ &\leq c_d \ell^d\, C \left(\frac{2d+\eta^{-1}}{|\beta|}\right)^{\alpha}\end{aligned}$$

where the constants $C$ and $\alpha$ are given by the Definition VIII.4.10. This last term goes to 0 when $|\beta| \to \infty$ and the proof is complete. One can remark that this last property does not require the independence of the potentials. ■

This proposition yields immediately:

**Corollary IX.1.9** *Let $\ell_0$ and $\gamma_0$ be positive numbers. Then there exist $\lambda_0$ and $\beta_0$ such that $[H_1](\gamma_0, \ell_0)$ holds in the following situation:*

1. *The energy is large i.e. $|\lambda| \geq \lambda_0$*
2. *The distribution of potentials is Hölder continuous and the disorder is large i.e. $|\beta| \geq \beta_0$. In this case the result holds uniformly with respect to the energy.*

*Proof:*
Let $\Lambda_0$ be the box of size $\ell_0$, then one has:

$$\{||G^{\Lambda_0}(\lambda+i\epsilon)|| \leq \eta(c_d\ell_0^{d-1})^{-1}\} \subset \{\sum_{y\in\partial\Lambda_0^{in}} |G_{\epsilon}^{\Lambda_0}(0,y)| \leq \eta\}$$

and for $\eta = \exp(-\gamma_0\ell_0)$ one can choose $\lambda$ or $\beta$ sufficiently large in order that the probability of the left-hand set in the above inclusion be greater that $1 - \ell_0^{-\xi}$. ■

One can ask what is the exact meaning of such a result. Let us assume for example that $\beta V$ is distributed according to the uniform law on the interval $[-1,1]$. Taking into account the above inequalities one finds that

$$\lambda_0 = 2d + e^{\gamma_0 \ell_0}(c_d \ell_0^{d-1}) + (1 - \ell_0^{-2d-4})^{\frac{1}{c_d \ell_0^d}}$$

It is easily seen that $\lambda_0$ is greater that $2d+1$. But in this case we know that the spectrum of $H(\omega)$ is almost surely equal to $[-2d-1, 2d+1]$ hence this result only applies to values of the energy outside the spectrum !! Thus the result at large energy is only meaningful for unbounded potentials.

On the other hand, assuming the distribution of the potential at one site has a density bounded by some constant $C$, one finds that

$$\beta_0 = c_d \ell_0^{3d+4}\ C(2d + e^{\gamma_0 \ell_0} c_d \ell_0^{d-1})$$

hence $\beta_0$ increases exponentially fast with respect to $\ell_0$.

We saw in Chapter VIII Section 4. (Theorem VIII.4.9) that for any energy and any distribution of potential which is not constant and has a positive moment then $[H_2]$ holds and $[H_1](\gamma_0, \ell_0)$ is true for some positive $\gamma_0$ and $\ell_0$ sufficiently large. Actually $[H_1]$ is essentially a consequence of the positivity of the Lyapunov exponent and the proof of $[H_2]$ was the difficult part. At large disorder or large energy one can easily be convinced of the positivity of the exponent since one can write:

$$\frac{1}{n}\log(\|U_\lambda(n)\|) = \log(|\lambda|) + \frac{1}{n}\log(\|\prod_{k=0}^{n-1}\begin{pmatrix} -1+\beta V(k)/\lambda & -1/\lambda \\ 1/\lambda & 0 \end{pmatrix}\|)$$

or

$$\frac{1}{n}\log(\|U_\lambda(n)\|) = \log(|\beta|) + \frac{1}{n}\log(\|\prod_{k=0}^{n-1}\begin{pmatrix} V(k)-\lambda/\beta & -1/\beta \\ 1/\beta & 0 \end{pmatrix}\|)$$

Let us call "scaled matrices" the matrices appearing in the above products. In the first case, i.e. at large energy $\lambda$, the scaled matrices are "almost equal" to the constant $\begin{pmatrix} -1 & 0 \\ 0 & 0 \end{pmatrix}$. In the second case, i.e. at large disorder $\beta$ they are "almost equal" to $\begin{pmatrix} V(k) & 0 \\ 0 & 0 \end{pmatrix}$. Each limit case is degenerated hence we cannot claim that the Lyapunov exponent of the scaled matrices products converge to the Lyapunov exponent of the limit case but a careful analysis shows that the $1/n^{th}$ power of the norm of the scaled products has a positive lower bound hence the Lyapunov exponent of the initial problem goes to $+\infty$ when $|\lambda| \to \infty$ or $|\beta| \to \infty$.

## IX.1.1 Decay of the Green's Function in Boxes

Given $[H_2]$ and Lemma IX.1.5 the strategy to improve the exponential decay of the Green's function with respect to the size of a box is clear:

1. Accumulate sufficiently many "good " boxes of size $\ell_0$ in a large box with a large probability using hypothesis $[H_1](\gamma_0, \ell_0)$
2. Bound the remaining term in the resolvent expansion with a large probability using hypothesis $[H_2]$.

This section is devoted to the implementation of this strategy. The next theorem, whose proof relies on some ideas developed in percolation theory, is the central result of this section. It essentially says that there exists a scale such that if one starts with a sufficiently large $\gamma_0$-adapted box then the scaled boxes are "almost" $\gamma_0$-adapted.

**Theorem IX.1.10** *Let $\gamma_0 > 0$ be given and assume that $[H_2]$ holds. Then there exists*

- *a scale $1 < \alpha < 2$ only depending on the dimension $d$*
- *a minimal length $\ell^*$ depending on $\gamma_0$, the dimension $d$ and the Wegner's constants*

*such that*
*if $[H_1](\gamma_0, \ell_0)$ holds for some $\ell_0 \geq \ell^*$ then the boxes of size $\ell_n = (\ell_0)^{\alpha^n}$ are $5\gamma_0/6$-adapted for $n = 0, 1, \ldots$*

This theorem results immediately from the following two propositions. The first proposition establishes the induction step and the second gives the appropriate choice of $\alpha$ and $\ell^*$.

**Proposition IX.1.11** *Let $\ell' > \ell > 1$ and assume that $[H_2]$ holds. If*

$$\mathbb{P}\{\Lambda_\ell \text{ is a } \gamma\text{-good box }\} \geq 1 - \eta$$

*then for any positive $\delta$ one has*

$$\mathbb{P}\{\Lambda_{\ell'} \text{ is a } \gamma'\text{-good box }\} \geq 1 - \eta'$$

*with:*

$$\begin{aligned}\gamma' &= \gamma\frac{\ell}{\ell+2} - \frac{1}{\ell'}(\gamma\frac{8\ell^2}{\ell+2} + \log(\delta|\partial\Lambda_{\ell'}^{in}|(1+\delta|\partial\Lambda_{\ell'}^{out}|)))\\ \eta' &= |\Lambda_{2\ell'+1}|^2\eta^2 + 2C_w\delta^{-\tau}|\Lambda_{2\ell'+1}|\end{aligned}$$

*Proof:*
Let us suppose that $\Lambda_{2\ell'}$ does not contain two non overlapping bad boxes of size $\ell$ and that $|G_\epsilon^{\Lambda_{2\ell'}}(u,v)| \leq \delta$ for any $u, v \in \Lambda_{2\ell'}$. From $[H_2]$ and our hypothesis this happens with a probability at least

$$1 - (C_w\delta^{-\tau}|\Lambda_{2\ell'}| + \eta^2|\Lambda_{2\ell'}|^2) = 1 - P$$

where we used the independence of the potentials in two non overlapping boxes. It follows from Proposition IX.1.5 that in this situation one has

$$\sup_{\epsilon\neq 0} |G_\epsilon^{\Lambda_{2\ell'}}(0,y)| \leq \delta e^{-\gamma\frac{\ell}{\ell+2}(\ell'-8\ell)} = M$$

for any $y \in \partial\Lambda_{\ell'}^{out} \cup \partial\Lambda_{\ell'}^{in}$.
Let $y \in \partial\Lambda_{\ell'}^{in}$. Applying the resolvent identity to the boxes $\Lambda_{\ell'} \subset \Lambda_{2\ell'}$ according to the figure below:

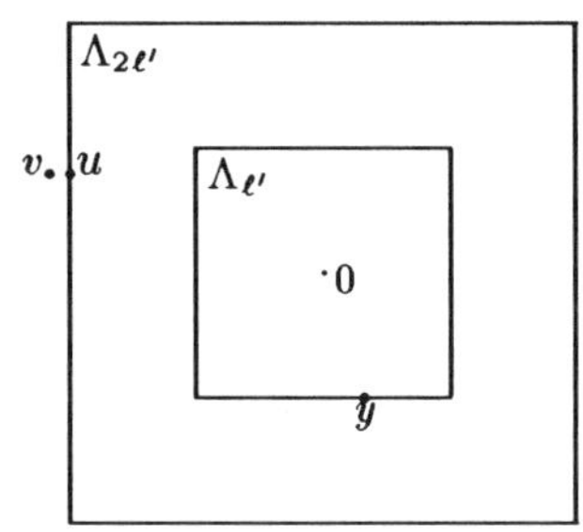


we obtain:

$$G_\epsilon^{\Lambda_{2\ell'}}(0,y) = G_\epsilon^{\Lambda'_\ell}(0,y) + \sum_{(u,v)\in\partial\Lambda_{\ell'}} G_\epsilon^{\Lambda_{2\ell'}}(0,v)G_\epsilon^{\Lambda_{\ell'}}(u,y)$$

Hence if we now suppose that $|G_\epsilon^{\Lambda_{\ell'}(0)}(.,.)|$ is bounded on $\Lambda_{\ell'}$ by some positive constant $\delta$ we get:

$$\begin{aligned}|G_\epsilon^{\Lambda'_\ell}(0,y)| &\leq |G_\epsilon^{\Lambda_{2\ell'}}(0,y)| + \delta\sum_{v\in\partial\Lambda_{\ell'}^{out}} |G_\epsilon^{\Lambda_{2\ell'}}(0,v)|\\ &\leq M(1+\delta|\partial\Lambda_{\ell'}^{out}|)\end{aligned}$$

It follows that

$$\sum_{y \in \partial \Lambda_{\ell'}^{in}} |G_\epsilon^{\Lambda'_\ell}(0,y)| \leq M |\partial \Lambda_{\ell'}^{in}| (1 + \delta |\partial \Lambda_{\ell'}^{out}|)$$

with a probability larger than $1 - (P + C_w \delta^{-\tau} |\partial \Lambda_{\ell'}|)$. Replacing $M$ and $P$ by their values one obtains the desired conclusion. ■

**Remark IX.1.12** This last proposition is only meaningful for large $\delta$ and for $\ell' >> \ell >> 1$. Otherwise one can have $\eta' > 1$ and $\gamma' < 0$ ... The strategy of the proof can be summarized as follows:

1. If the box $\Lambda_{2\ell'}$ does not contain any bad box of size $\ell$ then it is readily seen from the resolvent expansion that it is a good box.
2. The event "the box $\Lambda_{2\ell'}$ contains two non overlapping bad boxes of size $\ell$" has a probability bounded by $\eta^2 |\Lambda_{2\ell'}|^2$. The essential fact is that one can control the occurence of this event by a power of $\eta$ greater than one.
3. The difficult case is when $\Lambda_{2\ell'}$ contains overlapping bad boxes of size $\ell$. Fortunately if we are not in the second situation then the union of the intersecting bad boxes is contained in a cube of size $2\ell$ and Lemma IX.1.4 can apply.

**Proposition IX.1.13** *Let us set* $\alpha = \frac{2d+3}{2d+2}$. *For initial data* $\ell_0$, $\gamma_0$ *and* $\eta_0$ *let us consider the sequences* $\ell_n$, $\gamma_n$ *and* $\eta_n$ *defined by the recursion relations:*

$$\begin{aligned}
\ell_{n+1} &= \ell_n^\alpha \\
\gamma_{n+1} &= \gamma_n \frac{\ell_n}{\ell_n + 2} - \frac{1}{\ell_{n+1}} \left(\gamma_n \frac{8\ell_n^2}{\ell_n + 2} + \log(\delta_n |\partial \Lambda_{\ell_{n+1}}^{in}| (1 + \delta_n |\partial \Lambda_{\ell_{n+1}}^{out}|))\right) \\
\eta_{n+1} &= |\Lambda_{2\ell_{n+1}}|^2 \eta_n^2 + 2 C_w \delta_n^{-\tau} |\Lambda_{2\ell_{n+1}}|
\end{aligned}$$

*where* $\delta_n$ *is chosen such that* $2C_w \delta_n^{-\tau} |\Lambda_{2\ell_{n+1}}| = (1/2) \ell_{n+1}^{-\xi}$.

*Then there exists a minimal length* $\ell^*$ *depending on* $\gamma_0$, $d$ *and the Wegner's constants such that the above construction with initial data satisfying* $\eta_0 \geq \ell_0^{-\xi}$ *and* $\ell_0 \geq \ell^*$ *gives rise to sequences* $\gamma_n$ *and* $\eta_n$ *such that for all* $n$ *one has:*

$$\eta_n \leq \ell_n^{-\xi} \qquad \text{and} \qquad \gamma_n \geq 5\gamma_0/6$$

*Proof:*
Let us suppose that for some index $n$ one has the bound $\eta_n \leq \ell_n^{-\xi}$. It follows that

$$\eta_{n+1} \leq K_d \ell_{n+1}^{2d} \ell_n^{-2\xi} + \frac{1}{2}\ell_{n+1}^{-\xi}$$

for some constant $K_d$ only depending on the dimension $d$. Hence we will get the inequality $\eta_{n+1} \leq \ell_{n+1}^{-\xi}$ as long as :

$$K_d \ell_{n+1}^{2d} \ell_n^{-2\xi} \ell_{n+1}^{\xi} = K_d \ell_n^{\alpha(2d+\xi)-2\xi} \leq 1/2$$

With our choice of $\alpha$ one has $\alpha(2d+\xi) - 2\xi = -2$. It is now clear that if $\ell_1^*$ satisfies $K_d(\ell_1^*)^{-2} \leq 1/2$ then this inequality holds for any $\ell_n \geq \ell_0 \geq \ell_1^*$. In order to prove the second bound one remarks that

$$\gamma_n - \gamma_{n+1} = \frac{2\gamma_n}{\ell_n + 2} + \frac{1}{\ell_{n+1}}\left(\frac{8\gamma_n \ell_n^2}{\ell_n + 2} + \log(\delta_n |\partial\Lambda_{\ell_{n+1}}^{in}|(1 + \delta_n |\partial\Lambda_{\ell_{n+1}}^{out}|))\right)$$

This equality shows that the sequence $\gamma_n$ is decreasing hence we can replace $\gamma_n$ by $\gamma_0$ to obtain an upper bound on the difference $\gamma_n - \gamma_{n+1}$. If one replaces $\delta_n$ by its value we see that there exists a constant $K$ only depending on the dimension $d$ and the Wegner's constants such that

$$0 \leq \gamma_n - \gamma_{n+1} \leq \frac{2\gamma_0}{\ell_n} + \frac{1}{\ell_{n+1}}\left(8\gamma_0\ell_n + K + 2\left(\frac{\xi + d}{\tau} + d - 1\right)\log(\ell_{n+1})\right)$$

It is easily seen that for $\ell_0$ and $\alpha$ strictly greater than 1 one has the relations:

$$\sum_{n=0}^{\infty} \frac{1}{\ell_n} \leq \int_0^{\infty} \ell_0^{-\alpha^x}\, dx = \frac{1}{\log(\alpha)} \int_{\log(\ell_0)}^{\infty} \frac{e^{-t}}{t}\, dt \leq \frac{1}{\log(\alpha)\log(\ell_0)}$$

and:

$$\sum_{n=0}^{\infty} \frac{\ell_n}{\ell_{n+1}} \leq \frac{1}{(\alpha - 1)\log(\alpha)\log(\ell_0)}$$

It is now plain that there exists $\ell_2^*$ such that for $\ell_0 \geq \ell_2^*$ one has

$$\sum_{n=0}^{\infty} (\gamma_n - \gamma_{n+1}) \leq \gamma_0/6$$

This proves the proposition, if one chooses for $\ell^*$ the maximum of $\ell_1^*$ and $\ell_2^*$. ∎

Up to now we have only considered the value of the Green's function between the center and the boundary of a box. The next proposition proves the decay of the Green's function for any box and any couple of points in this box satisfying some conditions. The only relevant property is the strict positivity of the rate of decay hence we do not try to find the best possible rate as a fraction of $\gamma_0$.

**Theorem IX.1.14** *Let $\gamma_0 > 0$ be given and let assume that $[H_2]$ holds. Then there exists a minimal length $\ell^*$ such that if assumption $[H_1](\gamma_0, \ell_0)$ holds with $\ell_0 \geq \ell^*$ then one has:*

$$\mathbb{P}\{\sup_{\epsilon \neq 0} |G_\epsilon^\Lambda(x,y)| \leq e^{-\gamma_0|x-y|/8}\} \geq 1 - C_b|\Lambda||x-y|^{-d-2}$$

*for some constant $C_b$ which only depends on $\gamma_0$, the dimension $d$ and the Wegner's constants. This holds for any box $\Lambda$ and $(x,y) \in \Lambda$ such that*

$$4\,\mathrm{dist}(x, \partial\Lambda^{in}) \geq |x-y| \geq 4\ell_0$$

*Proof:*
Let $\alpha$ and $\ell^*$ be given by Theorem IX.1.10 and let $n$ be an index such that $4\ell_n \leq |x-y| \leq 4\ell_{n+1}$. We may assume that $\ell^* \geq 8$ in which case one has $\frac{\ell_n}{\ell_n+2} \geq 4/5$ for any $n$. Starting the construction at the point $x$ one can accumulate at least

$$\left[\frac{|x-y|+4}{2\ell_n+4}\right] \geq \frac{|x-y|-2\ell_n}{2\ell_n+4} \geq \frac{|x-y|}{4\ell_n+8} \geq \frac{|x-y|}{5\ell_n}$$

chained boxes of size $\ell_n$ before reaching $y$ or the boundary of $\Lambda$. Each of these boxes is a $\gamma$ good box with a probability larger than $\ell_n^{-\xi}$. Hence using the argument of Lemma IX.1.4 one obtains that:

$$|G_\epsilon^\Lambda(x,y)| \leq e^{-\gamma_0|x-y|/6}|G_\epsilon^\Lambda(v,y)|$$

for some $v \in \Lambda$ with a probability larger that $1 - |\Lambda|\ell_n^{-\xi}$.
Using a Wegner's estimate one has

$$|G_\epsilon^\Lambda(v,y)| \leq e^{(1/6)\gamma_0\ell_n} \leq e^{(1/24)\gamma_0|x-y|}$$

with a probability larger that $1 - C_w|\Lambda|e^{-(1/6)\tau\gamma_0\ell_n}$. It follows that

$$|G_\epsilon^\Lambda(x,y)| \leq e^{-(1/8)\gamma_0|x-y|}$$

with a probability at least

$$1 - |\Lambda|(C_w e^{-(1/6)\tau\gamma_0\ell_n} + \ell_n^{-\xi})$$

and this yields immediately the desired result using the relation $\ell_{n+1} \geq \ell_n^2$.
∎

### IX.1.2 Decay of the Green's Function in $\mathbb{Z}^d$

**Theorem IX.1.15** *Let $\gamma_0 > 0$ be given and assume that $[H_2]$ holds. Then there exists a minimal length $\ell^*$ such that if assumption $[H_1](\gamma_0, \ell_0)$ holds with $\ell_0 \geq \ell^*$ then one has:*

$$\mathbb{P}\{\sup_{\epsilon \neq 0} |G_\epsilon(0,x)| \leq e^{\gamma_0(\ell - |x|)/8}\} \geq 1 - C_e \ell^{-2}$$

*for $\ell \geq 8\ell_0$. The constant $C_e$ only depends on $\gamma_0$, $d$ and the Wegner's constants.*

*Proof:*
Let $\ell*$ be given by Theorem IX.1.14 and let us set $\gamma = \gamma_0/8$. For $x$ fixed and a given $\ell \geq 8\ell_0$ let us consider the increasing sequence of boxes $\Lambda_n$ centered at the origin and of size $\ell_n = 2^n(|x| + \ell)$.

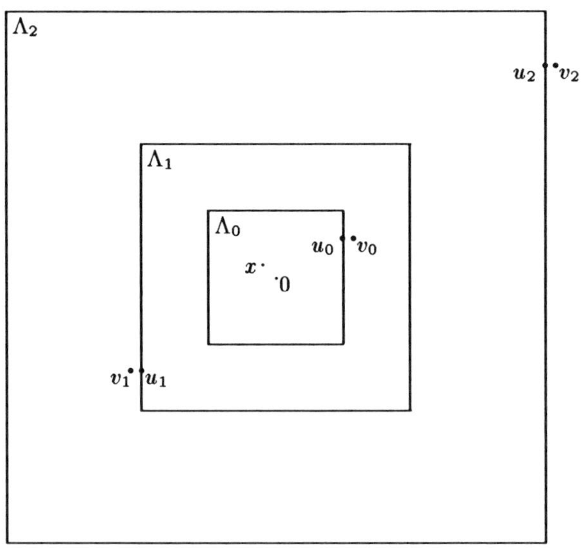


Using the above figure we first apply the resolvent identity to the box $\Lambda_0$:

$$G_\epsilon(0,x) = G_\epsilon^{\Lambda_0}(0,x) + \sum_{(u_0,v_0)\in\partial\Lambda_0} G_\epsilon(0,v_0) G_\epsilon^{\Lambda_0}(u_0,x)$$

We then compute $G_\epsilon(0, v_0)$ applying the resolvent identity to the box $\Lambda_1$.

$$G_\epsilon(0, v_0) = G_\epsilon^{\Lambda_1}(0, v_0) + \sum_{(u_1,v_1)\in\partial\Lambda_1} G_\epsilon(0, v_1)G_\epsilon^{\Lambda_1}(u_1, v_0)$$

If we iterate this procedure $n$ times we obtain:

$$\begin{aligned} G_\epsilon(0, x) &= [G_\epsilon^{\Lambda_0} + G_\epsilon^{\Lambda_1}\Gamma_0 G_\epsilon^{\Lambda_0} + G_\epsilon^{\Lambda_2}\Gamma_1 G_\epsilon^{\Lambda_1} G_\epsilon^{\Lambda_1}\Gamma_0 G_\epsilon^{\Lambda_0} + \dots \\ &+ G_\epsilon G_\epsilon^{\Lambda_n}\Gamma_n G_\epsilon^{\Lambda_{n-1}} \dots G_\epsilon^{\Lambda_1}\Gamma_0 G_\epsilon^{\Lambda_0}](0, x) \end{aligned}$$

We first bound the terms $G_\epsilon^{\Lambda_0}(0, x)$ and $G_\epsilon^{\Lambda_0}(u_0, x)$
Looking at the definition of $\Lambda_0$ we see that:

$$\text{dist}(0, \partial\Lambda_0) \geq |x| \quad \text{and} \quad 3\ \text{dist}(x, \partial\Lambda_0) \geq |x - u_0| \geq |x|$$

If we assume $|x| \geq \ell/2 \geq 4\ell_0$ Theorem IX.1.14 applies to the box $\Lambda_0$ and one has

$$|G_\epsilon^{\Lambda_0}(0, x)| \leq e^{-\gamma|x|} \quad \text{and} \quad |G_\epsilon^{\Lambda_0}(u_0, x)| \leq e^{-\gamma|x|}$$

with a probability at least:

$$1 - C_b(2|x| + \ell)^d|x|^{-\xi/2} \geq 1 - 4^d C_b|x|^{-2} \geq 1 - C'\ell^{d-\xi/2}$$

where in the second inequality we used the bound $(2|x| + \ell) \leq 4|x|$.

If $|x| \leq \ell/2$ then one has $(2|x| + \ell) \leq 2\ell$ and the Wegner's estimate yields:

$$|G_\epsilon^{\Lambda_0}(0, x)| \leq e^{\gamma\ell} \quad \text{and} \quad |G_\epsilon^{\Lambda_0}(u_0, x)| \leq e^{\gamma\ell}$$

with a probability at least

$$1 - C_w(2|x| + \ell)^d e^{-\tau\gamma\ell} \geq 1 - C''\ell^{-2}$$

It follows that in any case one has:

$$|G_\epsilon^{\Lambda_0}(0, x)| \leq e^{\gamma(\ell-|x|)} \quad \text{and} \quad |G_\epsilon^{\Lambda_0}(u_0, x)| \leq e^{\gamma(\ell-|x|)}$$

with a probability at least $1 - \bar{C}\ell^{-2}$ where the constant $\bar{C}$ only depends on $\gamma_0$, the dimension $d$ and the Wegner's constants.

We now bound the general term $|G_\epsilon^{\Lambda_n}\Gamma_n G_\epsilon^{\Lambda_{n-1}} \dots G_\epsilon^{\Lambda_1}\Gamma_0|$ in the resolvent expansion. In this expression we find the terms:

- $G_\epsilon^{\Lambda_n}(0, v_{n-1})$ with $v_{n-1} \in \partial\Lambda_{n-1}^{out}$. In this case one has:

$$\text{dist}(v_{n-1}, \partial\Lambda_n^{in}) = \frac{\ell_n}{4} \quad \text{and} \quad |v_{n-1}| = \frac{\ell_n}{4}$$

- $G_\epsilon^{\Lambda_k}(u_k, v_{n-1})$ with $v_{k-1} \in \partial\Lambda_{k-1}^{out}$ and $u_k \in \partial\Lambda_k^{in}$. In this case one has:

$$\text{dist}(v_{k-1}, \partial\Lambda_k^{in}) = \frac{\ell_k}{4} \quad \text{and} \quad 3\frac{\ell_k}{4} \geq |u_k - v_{k-1}| \geq \frac{\ell_k}{4}$$

Applying Theorem IX.1.14 to the boxes $\Lambda_k$ for $k = 1 \ldots n$ one obtains that each of these terms is bounded by $e^{-\gamma(\ell_k)/4}$ with a probability at least $1 - C_b|\Lambda_k|(\ell_k/4)^{-\xi/2}$. Hence the bound

$$\sum_{v_0 \in \partial\Lambda_0^{out}} |G_\epsilon^{\Lambda_n}\Gamma_n G_\epsilon^{\Lambda_{n-1}} \ldots G_\epsilon^{\Lambda_1}\Gamma_0|(0, v_0) \leq \prod_{k=1}^{n} c_d \ell_k^{d-1} e^{-\gamma(\ell_k)/4}$$

holds for $n = 1 \ldots m$ with a probability at least equal to:

$$1 - C_b \sum_{k=1}^{m} |\Lambda_k|(\ell_k/4)^{-\xi/2}$$

It is not difficult to check that

$$\sum_{n=1}^{\infty} |\Lambda_n|(\ell_n/4)^{-\xi/2} \leq K\ell^{d-\xi/2}$$

where $K$ only depends on $\gamma_0$, $d$ and the Wegner's constants.
Moreover for $\ell$ sufficiently large one has:

$$\sum_{n=1}^{\infty} \prod_{k=1}^{n} (c_d \ell_k^{d-1} e^{-\gamma(\ell_k)/4} \leq 1$$

and the proof is complete.

We finish this section with the celebrated theorem of Fröhlich and Spencer which is a direct consequence of the above results.

**Theorem IX.1.16** *Let us assume that the distribution of the potential at one site is Hölder continuous. For any $\gamma > 0$ there exist constants $C$, $\lambda_0$ and $\beta_0$ such that if one of the following conditions holds:*

- *Large energy i.e.* $|\lambda| \geq \lambda_0$
- *Large disorder i.e.* $|\beta| \geq \beta_0$

*then one has*

$$\mathbb{P}\{\sup_{\epsilon\neq 0}|G(\lambda+i\epsilon,0,x)|\leq e^{\gamma(\ell-|x|)}\}\geq 1-C\ell^{-2}$$

*for $\ell$ sufficiently large.*

*Proof:*
We already know that, under the Hölder continuity assumption on the distribution of the potential, hypothesis $[H_2]$ holds and that the Wegner's constants are uniform with respect to the energy. Let us choose $\gamma_0 = 8\gamma$ in Theorem IX.1.15. We know from Corollary IX.1.9 that $[H_1](\gamma_0, \ell_0)$ holds for any $\ell_0$, $\gamma_0$ assuming the disorder or the energy is sufficiently large. This yields the result. ∎

**Corollary IX.1.17** *Let assume that the distribution of potential is Hölder continuous. For any $\gamma > 0$ there exist constants $\lambda_0$ and $\beta_0$ such that if one of the following condition holds:*

- *Large energy i.e. $|\lambda| \geq \lambda_0$*
- *Large disorder i.e. $|\beta| \geq \beta_0$*

*then there exists a finite $C(\omega,\lambda)$ such that:*

$$\sup_{\epsilon\neq 0}|G(\lambda+i\epsilon,0,x)|\leq C(\omega,\lambda)e^{-\gamma|x|}\quad \mathbb{P}-a.e.$$

*Proof:*
Let us set for $n = 2, 3, \ldots$

$$\Omega(n,\lambda)=\{\omega\ ;\ \sup_{\epsilon\neq 0}|G(\lambda+i\epsilon,0,x)|\leq e^{\gamma(n-|x|)}\quad x\in\mathbb{Z}^d\}$$

then from Theorem IX.1.16 one has $\mathbb{P}(\Omega(n,\lambda)) \geq 1-Cn^{-2}$ for $n$ sufficiently large.
The sequence $\Omega(n,\lambda)$ is increasing in $n$ thus the Borel-Cantelli lemma implies that for $\mathbb{P}$almost all $\omega$ (depending on $\lambda$ !) there exists an integer $n(\omega,\lambda)$ such that $\omega \in \Omega(n(\omega,\lambda),\lambda)$. This yields the result if one sets $C(\omega,\lambda) = \exp(\gamma n(\omega,\lambda))$. ∎

**Remark IX.1.18** This corollary is certainly not completely satisfactory since it only says that for a given rate $\gamma$ of decay it is always possible to choose the energy or the disorder so large that the Green's function will admit this rate of decay. In the case of a bounded potential the spectrum is also bounded hence, at large energy, we are perhaps proving an obvious result i.e. outside of the spectrum the Green's function decays exponentially fast ! Actually the following question is of crucial importance to apply Theorem IX.1.15 on some domain of energy:
*For a given energy $\lambda$ is it possible to find a positive rate of decay $\gamma(\lambda)$ such that $[H_1](\gamma(\lambda), \ell_0)$ holds for some $\ell_0$ sufficiently large ?*
This question has received a positive answer in the one dimensional case, where $\gamma(\lambda)$ is chosen as the Lyapunov exponent. The question is still open in many dimensions and probably one has to take into account the dimension of space $d$ since it is believed that the answer is positive for any energy in the two dimensional case and negative for small energies when $d \geq 3$.

The assumptions done in this section are not always necessary and it is possible to relax them. Moreover a careful analysis shows that we have not chosen the best constants. In particular:

1. One can choose $\xi > 2d$ rather than $\xi = 2d + 4$ in the definition of a good box. (and even $\xi > d$ with another construction...)
2. Assuming $[H_1]$ and $[H_2]$ the independence of the potentials is not a crucial hypothesis. We only used this fact in Proposition IX.1.11 to bound the probability of occurence of disjoint bad boxes.
   *Let us define for positive numbers $\ell$ and $\gamma$*
   $$\Omega_x = \{\omega \ ; \ \Lambda_\ell(x) \ \textit{is a bad } \gamma\textit{-box} \} \quad \textit{and} \quad I\!P\{\Omega_x\} = \chi$$
   *then the independence implies that for $x$ and $y$ with dist$(x,y) > \ell$ one has $I\!P\{\Omega_x \cap \Omega_y\} = \chi^2$.*
   *Actually a weaker mixing condition of the form $I\!P\{\Omega_x \cap \Omega_y\} \leq \chi^{1+\rho}$ with $\rho > 0$ is sufficient to carry out the proof of Theorem IX.1.10. Dreifus and Klein used this property for Gaussian processes and for boxes far apart in [89]. Unfortunately a Wegner's estimate seems very difficult to obtain under this weak form of independence.*
3. A careful analysis of the proof also shows that one can replace our Wegner's estimate by a weaker condition of the form:
   *There exists a (suitable) $q > 0$ such that for any $\eta > 0$ one has:*
   $$I\!P\{\text{dist}(\lambda, \Sigma(H_\ell^\Lambda)) \leq \eta\} \leq |\log(\eta)|^{-q}$$

*for $\ell$ sufficiently large.*

# IX.2 Localization for A.C. Potentials

We denote by $\mathcal{B}$ the Borel sigma-algebra of $\mathbb{R}$, by $\mathcal{I}$ an interval of $\mathbb{R}$ and by $m$ the Lebesgue measure restricted to $\mathcal{I}$.
The property of exponential decay of the Green's function used in conjunction with the absolute continuity of the distribution of potentials yields a proof of localization quite similar to the one dimensional case. We present below two kinds of approaches which respectively refer to the pointwise theory and to the perturbation theory developed in Chapter VIII. The first proof is adapted from Delyon, Levy and Souillard [72] and the second is borrowed from Simon and Wolff [304].

## IX.2.1 Pointwise Theory

For any $\epsilon > 0$ the Green's function

$$G(\lambda + i\epsilon, x, y) = \int \frac{d\sigma_{x,y}(t)}{t - (\lambda + i\epsilon)}$$

is the difference of two Herglotz functions in the upper half-plane $\Pi_+$. It follows that the limit:

$$\lim_{\epsilon \downarrow 0} G(\lambda + i\epsilon, x, y) = G(\lambda, x, y)$$

exists for Lebesgue almost all $\lambda$ and any $(x, y) \in \mathbb{Z}^d$. This above limit (whenever it exists) denoted by $G(\lambda, x, y)$ is called the "pseudo Green's function". It coincides with the usual Green's function outside of the spectrum of $H$. It is readily seen that the function $G_x(y) = G(\lambda, y, x)$ is a solution of the equation $(H - \lambda)G_x = \delta_x$ and that $G_x(y) = G_y(x)$ since these relations are true for any $\lambda$ with $\Im m\, \lambda \neq 0$.

**Theorem IX.2.1** *Assume that*

1. *There exists a finite function $C(\omega, \lambda)$, a positive function $\gamma(\lambda)$ on $\mathcal{I}$ and a measurable subset $W \subset \Omega \times \mathcal{I}$ of full $\mathbb{P} \otimes m$ measure such that*

*for* $(\omega, \lambda) \in W$ *the "pseudo Green's function" exists and:*

$$|G(\lambda, 0, x)| \leq C(\omega, \lambda)\; e^{-\gamma(\lambda)|x|}$$

*2. The distribution of the potential at one site has a bounded density.*

*Then for* $\mathbb{P}$*almost all* $\omega$ *one has:*

- *The spectrum of* $H(\omega)$ *is pure point on* $\mathcal{I}$*.*
- *An eigenvalue* $\lambda$ *of* $H(\omega)$ *belonging to* $\mathcal{I}$ *is simple and the (unique) eigenfunction* $\psi(x,\omega)$ *satisfies:*

$$\limsup_{|x|\to\infty} \frac{1}{|x|} \log(|\psi(x, \omega)|) \leq -\gamma(\lambda)$$

*Proof:*
The proof follows closely the arguments developed in Proposition VIII.1.12 and Theorem VIII.1.14. Recall that we denote by $\mathcal{P}$ the set of polynomially bounded functions on $\mathbb{Z}^d$ and let us define the subset:

$W_x(c) = \{(\omega, \lambda)$; *there exists at most a function* $\psi \in \mathcal{P}$ *(up to constant factors) solution of the equation* $(H(\omega) + c\delta_x)\psi = \lambda\psi$ *and this solution (if it exists) is such that* $\limsup_{|x|\to\infty} \frac{1}{|x|} \log(|\psi(x)|) \leq -\gamma(\lambda)$ $\}$

Choose $(\omega, \lambda) \in W$ and a real number $c$ and a non-identically vanishing polynomially bounded solution of the equation $(H(\omega) + c\delta_x)\psi = \lambda\psi$. For any site $y$ the function $G_y$ exists and is exponentially decaying hence the following dot products are well defined:

$$\begin{aligned} < G_y\,,\, (H + c\delta_x)\psi > &= \lambda < G_y\,,\, \psi > \\ &= < (H + c\delta_x)G_y\,,\, \psi > \\ &= < HG_y + cG_y(x)\delta_x\,,\, \psi > \\ &= < \lambda G_y + \delta_y + cG_y(x)\delta_x\,,\, \psi > \end{aligned}$$

This yields the relation $\psi(y) = -cG_y(x)\psi(x) = -cG_x(y)$. It follows that the function $\psi$ is unique and is exponentially decaying with rate $\gamma(\lambda)$.
The set $\tilde{W} = \cap_{c\in\mathbb{R}} W_x(c)$ contains $W$, is contained in $W_x(0)$ and does not depend upon the potentials at the site $x$. It follows from Proposition VIII.1.13 that for $\mathbb{P}$almost all $\omega$, the restriction to the interval $\mathcal{I}$ of the spectral measure $\sigma_x(\omega)$ is pseudo-carried by the $\omega$ section of $W_x(0)$. Since we already know that for $\sigma_x(\omega)$ almost all $\lambda$ there exists a polynomially bounded generalized eigenfunction of $H(\omega)$ we can conclude that this

spectral measure is pure point. Repeating this argument for any $x \in \mathbb{Z}^d$ we obtain the desired result. ∎

In the one dimensional case let us choose $(\omega, \lambda)$ such that

$$\lim_{n\to\pm\infty} \frac{1}{|n|} \log \|U_\lambda(n,\omega)\| = \gamma(\lambda)$$
$$\lim_{n\to\pm\infty} \frac{1}{|n|} \log \|g_\lambda(n,\omega)\| = 0$$

We saw in Proposition VIII.1.12 that if $\gamma(\lambda) > 0$ and if $\lambda$ is not an eigenvalue of $H(\omega)$ then there exists a unique polynomially bounded solution $G_x$ of the equation $(H(\omega) - \delta_x)\psi = \lambda\psi$. Moreover this solution is exponentially decaying. Conversely it is readily seen from the Wronskian property that when this equation has an exponentially decaying solution one has $\gamma(\lambda) > 0$. Hence in the one dimensional case the positivity of the Lyapunov exponent and the exponential decay of the "pesudo Green's function" are equivalent. One can notice that this has nothing to do with randomness.
In the multidimensional case there is no simple analog of Oseledec's theorem and the exponential decay of the "almost Green's function" is a consequence of the Frölich and Spencer's result. Note that the so called "Combes-Thomas' argument" yields immediately the exponential decay of the Green's function outside the spectrum (see Problem III.6.8).

**Corollary IX.2.2** *Assume that the potential distribution has a bounded density. Then there exist constants $\lambda_0$ and $\beta_0$ such that if $|\beta| \geq \beta_0$ or if $\mathcal{I}$ is contained in $|\lambda| \geq \Lambda_0$ the conclusion of Theorem IX.2.1 hold.*

*Proof:*
We have already seen that for each $\omega$ the function $G(\lambda, x, y)$ exists for $m$ almost all $\lambda$. Let $\gamma$ be a positive number. For each $\lambda \in \mathcal{I}$, Corollary IX.1.17 asserts that the limit $G(\lambda, x, y)$ (if it exists) satisfies the required bound for $\mathbb{P}$almost all $\omega$. The result then follows from Fubini's theorem. ∎

## IX.2.2 Perturbation Theory

The theory of rank-one perturbations of self-adjoint operators developed in Section 2 of Chapter VIII can also be used to obtain a proof of localization in any dimension for smooth potentials. The exponential decay of the

Green's function is there replaced by the weaker property of $\ell^2$ summability of $x \hookrightarrow G(\lambda + i\epsilon, 0, x)$ uniformly with respect to $\epsilon \neq 0$ .
Let us first recall some notations of Chapter VIII. For $H$ a selfadjoint operator on the separable Hilbert space $\mathcal{H}$ we denote by $Pf = < f, \phi > \phi$ the projection operator on the normalized function $\phi$. For any real number $c$ we set $H_c = H + cP$ and $\sigma^c$ is the spectral measure of $H_c$ computed on the vector $\phi$. For a real $\lambda$ and a complex $z$ with $\Im m(z) > 0$ one defines the functions $B(\lambda)$ and $F_c(\lambda)$ by:

$$B(\lambda) = \int \frac{d\sigma^0(t)}{(t-\lambda)^2} \qquad F_c(z) = \int \frac{d\sigma^c(t)}{t-z}$$

We first relate the Green'function to the function $B$. Let $\{\phi_n \ ; \ n \in J\}$ be an orthonormal basis of $\mathcal{H}$ indexed by the countable set $J$ and containing the function $\phi \equiv \phi_0$. For a complex $z$ with $\Im mz \neq 0$ one defines the Green's kernel

$$G(z, n, m) = < \phi_n \ , \ (H - zI)^{-1} \phi_m >$$

**Proposition IX.2.3** *Let $\lambda$ be a real number, then:*

$$B(\lambda) = \lim_{\epsilon \downarrow 0} \sum_{n \in J} |G(\lambda + i\epsilon, 0, n)|^2$$

*Proof:*
Using the spectral theorem one can write:

$$\begin{aligned} \int \frac{d\sigma^0(t)}{|\lambda + i\epsilon - t|^2} &= < \phi_0 \ , \ |H - (\lambda + i\epsilon)I|^{-2} \phi_0 > \\ &= < (H - (\lambda + i\epsilon)I)^{-1} \phi_0 \ , \ (H - (\lambda + i\epsilon)I)^{-1} \phi_0 > \\ &= \sum_{n \in J} |G(\lambda + i\epsilon, 0, n)|^2 \end{aligned}$$

Lebesgue's theorem yields

$$B(\lambda) = \lim_{\epsilon \downarrow 0} \int \frac{d\sigma^0(t)}{|\lambda + i\epsilon - t|^2}$$

and this finishes the proof. ∎

We now turn our attention to the case of random Schödinger operators in $\mathbb{Z}^d$. Adding the parameter $\omega$ to take into account the disorder, it is readily seen from the measurability properties stated in Chapter V that the function $G_\omega(\lambda + i\epsilon, x, y)$ is jointly measurable with respect to the product sigma-algebra $\mathcal{F} \otimes \mathcal{B}$ for any $\epsilon \neq 0$ and $(x, y) \in \mathbb{Z}^d$. Let $\mathcal{I}$ be an open interval of $\mathbb{R}$ and let us define the measurable set $W$ by:

$$W = \{(\omega, \lambda) \; ; \; \lim_{\epsilon \downarrow 0} \sum_{x \in \mathbb{Z}^d} |G(\lambda + i\epsilon, 0, x)|^2 < \infty \; , \; \lambda \in \mathcal{I}\}$$

Then $W$ is an $\mathcal{F} \otimes \mathcal{B}$ measurable subset of $\Omega \times \mathbb{R}$. We now suppose that the Green's kernel is written in the orthonormal basis $\{\delta_x \; ; \; x \in \mathbb{Z}^d\}$.

**Theorem IX.2.4** *Let $H(\omega)$ be the random Schrödinger operator associated with an i.i.d. family of potentials in $\mathbb{Z}^d$ with an absolutely continuous distribution. Let $m$ be the Lebesgue measure on $\mathcal{I}$ and consider the two statements:*

*(a) $W$ is of full $\mathbb{P} \otimes m$ measure.*
*(b) For $\mathbb{P}$-almost all $\omega$ the spectrum of $H(\omega)$ is pure point in $\mathcal{I}$.*

*Then (a) implies (b) and if the distribution of potentials is equivalent to the Lebesgue measure then (b) implies (a).*

*Proof:*
The proof is exactly the same as in Theorem VIII.2.7. Let us define

$$\tilde{W} = \{(\omega, c) \; ; \; \sigma_0^c(\omega) \text{ is pure point on } \mathcal{I}\}$$

One checks immediately that $\tilde{W}$ is a measurable subset of $\Omega \times \mathbb{R}$. The theorem VIII.2.4 can be "localized" on an open interval, i.e. the two following properties are equivalent:

(i) *For $m$ almost all $c$ the spectral measure $\sigma^c$ is pure point on $\mathcal{I}$*
and
(ii) *For $m$ almost all $\lambda \in \mathcal{I}$ one has $B(\lambda) < \infty$*

From the Fubini's theorem and Proposition IX.2.3 one gets that (a) is equivalent to the fact that $\tilde{W}(\mathcal{I})$ is of full $\mathbb{P} \otimes m$ measure. Moreover $\sigma_0^c(\omega) = \sigma_0(\omega + c)$ where $\omega + c$ has the same coordinates as $\omega$ except for the coordinate of index 0 which is equal to $\omega_0 + c$. We have already remarked in Theorem VIII.2.7 that for a nonnegative measurable function $F(c, x, y)$ on the measurable space $(\mathbb{R} \times \mathbb{R} \times \mathcal{Y})$ such that $F(c, x, y) = F(0, x + c, y)$ one has:

$$\int F(c, x, y) \, dm(c) \otimes d\mu_x(x) \otimes d\mu_y(y) = \int F(0, x, y) \, dm(x) \otimes d\mu_y(y)$$

This implies that $F(0,x,y)$ is $m \otimes d\mu_y$ negligible if and only if $F(c,x,y)$ is $m \otimes \mu_x \otimes \mu_y$ negligible. Choosing $\mu_x$ as the law of $V(0)$, $\mu_y$ as the joint distribution of the potentials outside the site $x$ and $F$ as the indicator function of the complementary set of $W$ one obtains the conclusion for the spectral measure computed at the site 0. The translation invariance yields the desired result. ∎

In the one dimensional case if one assumes $\mathbb{E}\{\log(1+|V(0)|\} < \infty$ then the Lyapunov exponent exists and is positive. It follows from Proposition VIII.2.5 and Oseledec's theorem that the statement (a) of Theorem IX.2.4 is true for any interval $\mathcal{I}$ in one dimension. In the multidimensional case, statement (a) is a consequence of the exponential decay of the Green's function.

**Corollary IX.2.5** *Let us assume that the distribution of potentials is absolutely continuous and Hölder continuous. Then there exist constants $\lambda_0$ and $\beta_0$ such that if $|\beta| \geq \beta_0$ or if $\mathcal{I}$ is contained in $|\lambda| \geq \Lambda_0$ the spectrum of $H(\omega)$ is $\mathbb{P}$almost surely pure point on $\mathcal{I}$.*

*Proof:*
Let $\gamma$ be a positive number. Using Corollary IX.1.17, Fubini's theorem implies the existence of a measurable subset $W \subset \Omega \times \mathcal{I}$ of full $\mathbb{P} \otimes m$ measure such that for $(\omega, \lambda) \in W$ one has:

$$|G(\lambda + i\epsilon, 0, x)| \leq C(\omega, \lambda)\, e^{-\gamma|x|}$$

for any $\epsilon \neq 0$.
This immediately implies that for $(\omega, \lambda) \in W$ one has:

$$\sum_{x \in \mathbb{Z}^d} |G(\lambda + i\epsilon, 0, x)|^2 \leq C^2(\omega, \lambda) \sum_{n=0}^{\infty} n^{2d} e^{-2n\gamma} < \infty$$

and it follows that the condition (a) of Theorem IX.2.4 is fulfilled hence the spectrum is pure point. ∎

In general we assume that the distribution of potentials has a density in some $L^p$ for $p > 1$. It follows that it is Hölder continuous hence it obeys a Wegner type estimate.

# IX.3 A Direct Proof of Localization

As we explained in the introduction, this last section is devoted to a direct proof of localization. The strategy is to control the energy dependence in the various quantities involved in the multiscale analysis. Obviously, we reintroduce the dependence in the energy $\lambda$ in our notation.
For the purpose of the present discussion the definition of a *good box* has to be modified in the following way:

**Definition IX.3.1** *A box $\Lambda = \Lambda_\ell(x)$ will be said to be $(m, \lambda)$-regular if $\lambda \notin \Sigma(H^\Lambda)$ and if $\Lambda$ is a good box in the sense that:*

$$G^\Lambda(\lambda, x, \partial\Lambda) \leq e^{-m\ell}.$$

*where we use the notation;*

$$G^\Lambda(\lambda, x, \partial\Lambda) = \sum_{y \in \partial\Lambda^{in}} |G^\Lambda(\lambda, x, y)|$$

Obviously, any $(m, \lambda)$-regular box is also $(m', \lambda)$-regular if $m' < m$.

The main result of this section is the following.

**Theorem IX.3.2** *Let $\lambda_0 \in I\!R$ and let us assume the existence of $\ell_o > 0$ for which:*
*[$H_0$]. there exist $m_0 > 0$ and $p > d$ such that:*

$$I\!P\{\Lambda_{\ell_0} \text{ is } (m_0, \lambda_0) - \text{regular}\} \geq 1 - \frac{1}{\ell_o^p},$$

*[$H_2$]. there exist $\eta > 0$ and $q > 4p + 6d$ such that:*

$$I\!P\{\text{ dist } (\lambda, \Sigma(H^\Lambda)) < e^{-\sqrt{\ell}}\} \leq \frac{1}{\ell^q}$$

*for all $\lambda \in (\lambda_0 - \eta, \lambda_0 + \eta)$ and all $\ell > \ell_0$. Then, for any $m \in (0, m_0)$ there exists a positive constant $b$ such that, if $\ell_0 > b$ then one can find a neighborhood $\mathcal{I}$ of $\lambda_0$ such that the operator $H$ has (dense) pure point spectrum in $\mathcal{I}$ with probability one, the eigenfunctions decaying exponentially with a rate not less than $m$.*

The assumptions $[H_0]$ and $[H_2]$ are very similar to the assumptions $[H_1]$ and $[H_2]$ of the first part of the chapter. Let us emphasize the difference to shed some light on their nature. The present notion of regular box is different from the notion of adapted box (recall Definition IX.1.6) in the sense that we use real energies $\lambda$ and that, as a consequence, we have to assume that these energies are not in the spectra of the operators $H^\Lambda$'s. But as long as we keep $\lambda$ away from the spectrum of the operator $H^\Lambda$ the estimations will be similar for they will most of the time depend only upon the distance from the spectrum.

## IX.3.1 Examples

Before engaging in the technical steps of the proof, we give a few examples of models to which this theorem applies. This will help us understand the significance of the assumptions.

Let $\mu$ be the distribution of the potential at any fixed site.

We first concentrate on the assumption $[H_0]$.

It is very easy to see that this assumption is not restrictive if one is ready to work in the *low energy* or *high disorder* regime. We state the results without proofs because the latter are essentially the same as the proofs leading to Corollary IX.1.9.

1. **low energy:** for any $\ell_0 > 0$, $m_0 > 0$ and $p > d$ there exists $\lambda_0$ such that:

$$\mathbb{P}\{\Lambda_{\ell_0}(0) \text{ is } (m_0, \lambda) - \text{ regular for all } |\lambda| \geq \lambda_0\} \geq 1 - \frac{1}{\ell^p}.$$

The terminology *low energy* is certainly not very appropriate. It simply means that we are considering energies far away from the spectrum of the free Hamiltonian $H_0$. In the continuous case, $\Sigma(H_0) = [0, \infty)$, hence we are considering very negative energies and *low energies* is justified. On the other hand, the spectrum of $H_0$ is bounded in the lattice case and *low energies* covers now very large positive energies!!

As before the notion of high disorder is built into the model by using a large coupling constant in front of the potential. In this respect, we shall find the following notation convenient. If $\mu$ is the distribution of a real valued random variable $X$, then for each $\beta > 0$ we denote by $\mu_\beta$ the distribution of the random variable $\beta X$.

2. **high disorder:** for any $\ell_0 > 0$, $m_0 > 0$, $p > d$ and $\eta > 0$ there exists

$\beta_0 > 0$ such that:

$$\mathbb{P}\{\Lambda_{\ell_0}(0) \text{ is } (m_0, \lambda) - \text{ regular for all } \lambda \in [\lambda_0 - \eta, \lambda_0 + \eta]\} \geq 1 - \frac{1}{\ell_0^p}$$

for all $\lambda_0 \in \mathbb{R}$ and $\beta > \beta_0$.

**Remark IX.3.3** The above result remains true for some convex combinations of the form $\mu = (1-\epsilon)\tilde{\mu}_\beta + \epsilon\nu$ for some $\epsilon \in (0,1)$, a Hölder continuous probability measure $\tilde{\mu}$ and an arbitrary probability measure $\nu$. The proof given originally by Carmona, Martinelli and Klein in [50] is for this more general form of the result.

We now switch to the investigation of the hypothesis $[H_2]$.

1. The first example concerns potential distributions $\mu$ which have a **bounded density**. The proof is a simple consequence of Wegner's theorem. Indeed, if $\mu$ is absolutely continuous and if its density is bounded by a constant $c$, then, for any $\epsilon > 0$, $\ell > 0$ and $\lambda \in \mathbb{R}$ one has:

$$\mathbb{P}\{ \operatorname{dist}(\lambda, \Sigma(H^{\Lambda_\ell})) \leq \epsilon\} \leq \frac{2}{c}(\ell+1)^d \epsilon$$

2. The second example concerns probability measures $\mu$ which are **Hölder continuous**. If we assume that the measure $\mu$ satisfies:

$$\mu([a,b]) \leq \delta |b-a|^\alpha$$

for some $\alpha$ in $(0,1)$, some $\delta > 0$ and all $a < b$ in $\mathbb{R}$, then:

$$\mathbb{P}\{ \operatorname{dist}(\lambda, \Sigma(H^{\Lambda_\ell})) \leq \epsilon\} \leq \frac{2\delta}{1-\alpha}(\ell+1)^d \epsilon^\alpha$$

for each box $\Lambda$, $\lambda \in \mathbb{R}$ and $\epsilon > 0$ as we proved in Proposition VIII.4.13. This result was proven by Carmona, Martinelli and Klein in [50].

**Remark IX.3.4** As before, the above result remains true for some convex combinations of the form $\mu = (1-\epsilon)\tilde{\mu}_\beta + \epsilon\nu$.

Anyway, the assumption $[H_2]$ is satisfied in either case.

### IX.3.2 The Proof

Because of the resolvent equation:

$$[H^{\Lambda_{\ell_0}} - \lambda I]^{-1} = [H^{\Lambda_{\ell_0}} - \lambda_0 I]^{-1} + (\lambda - \lambda_0)[H^{\Lambda_{\ell_0}} - \lambda I]^{-1}[H^{\Lambda_{\ell_0}} - \lambda_0 I]^{-1},$$

one has: $\left(d(\lambda, \Sigma(H^{\Lambda_{\ell_0}})) \geq e^{-\sqrt{\ell_0}} \text{ and } |\lambda - \lambda_0| \leq e^{-\sqrt{\ell_0}/2}\right)$ implies $d(\lambda, \Sigma(H^{\Lambda_{\ell_0}})) \geq \frac{1}{2} e^{-\sqrt{\ell_0}}$. If moreover the box $\Lambda_{\ell_0}$ is $(m_0, \ell_0)$-regular, we also have:

$$G^{\Lambda_{\ell_0}}(\lambda, 0, y)| \leq e^{-m_0 \ell_0/2} + 2|\lambda - \lambda_0| e^{2\sqrt{\ell_0}}.$$

Now for any $p' \in (d, p)$ and any $m_0' \in (0, m_0)$ one can choose $\delta > 0$ small enough so that one has in fact proved that the hypotheses $[H_0]$ and $[H_2]$ imply the following:

$[H_1]$ for large enough $\ell_0$ one has:

$$\mathbb{P}\{\forall \lambda \in (\lambda_0 - \delta, \lambda_0 + \delta),\ \Lambda_{\ell_0} \text{ is } (m_0', \lambda) - \text{ regular}\} \geq 1 - \frac{1}{\ell^{p'}}.$$

Consequently, instead of proving Theorem IX.3.2 under the assumptions $[H_0]$ and $[H_2]$, it is enough to prove it under the assumptions $[H_1]$ and $[H_2]$. Since the energy $\lambda$ is now varying in an interval, the probabilistic estimate of the multiscale analysis given in the first part of the chapter has to be modified. Henceforth, we introduce, for each interval $\mathcal{I} \subset \mathbb{R}$ the assumptions $[H_1(\mathcal{I})]$ and $[H_2(\mathcal{I})]$. They depend also on the choice of the real constants $p > 2d$ and $m_0 > 0$, and of the integer $\ell_0 > 0$:

$[\tilde{H}_1(\mathcal{I})]$ for all $x, y \in \mathbb{Z}^d$ such that $|x - y| \geq \ell_0$,

$$\mathbb{P}\{\forall \lambda \in \mathcal{I}, \ \text{ either } \Lambda_{\ell_0}(x) \text{ or } \Lambda_{\ell_0}(y) \text{ is } (m_0, \lambda) - \text{regular }\} \geq 1 - \frac{1}{\ell_0^{2p}}$$

$[\tilde{H}_2(\mathcal{I})]$ for some $q > 4p + 6d$, all energies $\lambda$ such that $d(\lambda, \mathcal{I}) \leq e^{-\sqrt{\ell}}$ and all $\ell > \ell_0$ we have:

$$\mathbb{P}\{ \text{ dist } (\lambda, \Sigma(H^{\Lambda})) < e^{-\sqrt{\ell}}\} \leq \frac{1}{\ell^q}.$$

The localization is proved as usual by proving that the generalized eigenfunctions, which are polynomially bounded according to the results of Chapter II, are actually exponentially decaying. We do not want to make restrictive regularity assumptions on the distribution $\mu$ of the potential at a given site because we want to prove localization for singular potentials (and in particular for random potentials taking only finitely many values) . Under these constraints, Kotani's trick is not available and the proof has to be more difficult. Here is the substitute we use:

**Proposition IX.3.5** *Let us assume that $\mathcal{I} \subset \mathbb{R}$ is an interval and that $m > 0$, $\ell_0 > 0$ and $p > d$ are real numbers such that for every integer $k \geq 0$ and $x, y \in \mathbb{Z}^d$ such that $|x - y| > \ell_k$ one has:*

$$\mathbb{P}\{\forall\lambda \in \mathcal{I},\ \textit{either } \Lambda_{\ell_k}(x) \textit{ or } \Lambda_{\ell_k}(y) \textit{ is } (m,\lambda)-\textit{ regular }\} \geq 1 - \frac{1}{\ell_k^{2p}} \quad \text{(IX.1)}$$

*where the integers $\ell_k$ are defined by $\ell_{k+1} = \ell_k^\alpha$, $k = 0, 1, \ldots$ with $\alpha = 2p/(p + 2d)$. Then, all the generalized eigenfunctions of $H$ corresponding to eigenvalues in $\mathcal{I}$ are exponentially decaying with rate at least $m$.*

*Proof:*
For $x_0 \in \mathbb{Z}^d$ we define a sequence $\{A_k(x_0);\ k \geq 1\}$ domains by $A_{k+1}(x_0) = \Lambda_{2r\ell_{k+1}}(x_0)\backslash\Lambda_{2\ell_k}(x_0)$ for $k \geq 0$, where $r \geq 1$ is an integer to be chosen later. We also define $E_k(x_0)$ as the event that there exists an energy $\lambda \in \mathcal{I}$ and $x \in A_{k+1}(x_0)$ such that $\Lambda_{\ell_k}(x_0)$ and $\Lambda_{\ell_k}(x)$ are both $(m, \lambda)$-singular. Then our assumption (IX.1) implies that:

$$\begin{aligned}\sum_{k=0}^{\infty} \mathbb{P}\{E_k(x_0)\} &\leq \sum_{k=0}^{\infty} \frac{(2r\ell_{k+1} + 1)^d}{\ell_k^{2p}} \\ &\leq \sum_{k=0}^{\infty} \frac{(2r + 1)^d}{\ell_k^{2p-\alpha d}} < \infty\end{aligned}$$

and we conclude from the first Borel-Cantelli lemma that, with probability one, for any $x_0 \in \mathbb{Z}^d$ the event $E_k(x_0)$ occurs only finitely many times. Now let us assume that $\lambda \in \mathcal{I}$ and that $\psi$ is a polynomially bounded function such that $H\psi = \lambda\psi$. We need to prove that $\psi$ decays exponentially with rate at least $m$. Let $x_0$ be chosen so that $\psi(x_0) \neq 0$. Notice that the Poisson formula:

$$\psi(x_0) = \sum_{(y,y')\in\partial\Lambda_{\ell_k}(x_0)} G^{\Lambda_{\ell_k}(x_0)}(\lambda, x_0, y)\psi(y')$$

gives $\psi(x_0)$ in terms of the boundary values of $\psi$ whenever $\lambda \notin \Sigma(H^{\Lambda_{\ell_k}(x_0)})$. Moreover, if the box $\Lambda_{\ell_k}(x_0)$ is $(m, \lambda)$-regular we also have:

$$|\psi(x_0)| \leq c\ell_k^{n+d-1}e^{-m\ell_k/2}$$

for some integer $n$ independent of $k$. This implies that $\Lambda_{\ell_k}(x_0)$ is $(m, \lambda)$-singular for $k$ large enough for the contrary would imply that $\psi(x_0) = 0$. Since we also know that $E_k(x_0)$ does not occur for large $k$, we conclude that the box $\Lambda_{\ell_k}(x)$ must be $(m, \lambda)$-regular for all $x \in A_{k+1}(x_0)$. Now, let us choose a sequence of domains $\tilde{A}_{k+1}(x_0)$ such that:

$$x \in \tilde{A}_{k+1}(x_0) \qquad \Rightarrow \qquad d(x, \partial A_{k+1}(x_0)) \geq \rho|x - x_0|$$

for some $\rho \in (0, 1)$. One can choose for example:

$$\tilde{A}_{k+1}(x_0) = \Lambda_{2r\ell_{k+1}/(1+\rho)}(x_0) \backslash \Lambda_{2\ell_k/(1-\rho)}(x_0).$$

With this choice, if $|x - x_0| \geq \ell_0/(1 - \rho)$ we have

$$\begin{aligned} |\psi(x)| &\leq c\ell_k^{d-1} e^{-m\ell_k/2} \max_{y' \in \partial\Lambda^{out}} |\psi(y')| \\ &= c\ell_k^{d-1} e^{-m\ell_k/2} |\psi(y')| \end{aligned}$$

for some $y' \in \partial\Lambda^{out}$ to which one can apply the same estimate. Repeating this procedure $N$ times, and we know that we can be sure that $N \geq \rho|x - x_0|/(1 + \ell_k/2)$, one gets:

$$\begin{aligned} |\psi(x)| &\leq (c\ell_k^{d-1} e^{-m\ell_k/2})^N c(1 + |x_0| + r\ell_{k+1})^n \\ &= e^{-\epsilon\rho|x - x_0|} \end{aligned}$$

for any given $\epsilon \in (0, 1)$ provided one takes $|x - x_0|$ large enough. This gives the desired result once we choose the constants with some modicum of care. ∎

Let us summarize the situation. We proved that the hypotheses of Proposition IX.3.5 are sufficient for exponential localization with probability one in the given interval $\mathcal{I}$. The crucial estimate IX.1 has to be true at all the scales $\ell_k$. We noticed that it was essentially always at some initial scale $\ell_0$ provided the energy was low enough or the disorder large enough. The idea is then to try to propagate the probability estimate from scale to scale without loosing the control of the probability of the potential configurations for which such a propagation is possible. The following proposition gives the machinery needed to estimate the probability of the appropriate set of configurations at all scales.

**Proposition IX.3.6** *Let us assume that assumptions [$H_1$] and [$H_2$] hold for some $\ell_0 > 0$, $m_0 > 0$ and $p > 2d$. Then, for any given $m \in (0, m_0)$ the hypothesis of Proposition IX.3.5 is satisfied if $\ell_0$ is large enough.*

We have many reasons not to attempt to give a proof. Two of them can be admitted publicly. The main ideas of the multiscale analysis have been discussed in great detail in the first part of the chapter and the fact that such a result is true should not be surprising anymore. Moreover, the various steps of the proof are rather lengthy and very technical and we do not believe that the book should end in this way. After all, the reader who got to that point deserves better!

In any case, a complete proof can be found in the work [88] of von Dreifus and Klein on which we based the present cavalier discussion of the *bare hand* approach to the proof of localization.

## IX.3.3 Extensions

1. It is possible to extend the multiscale analysis to some continuous models on $\mathbb{R}^d$. This was done by Holden and Martinelli in [150]. They consider a random potential $\{V(x);\ x \in \mathbb{R}^d\}$ of the form:

$$V(x,\omega) = \sum_{\mathbf{n}\in \mathbb{Z}^d} \xi_{\mathbf{n}}(\omega)\mathbf{1}_{C_{\mathbf{n}}}(x)$$

where the $C_{\mathbf{n}}$ are the cubes of the partition of $\mathbb{R}^d$ which we used already in Chapter V, and where the $\xi_{\mathbf{n}}$ are i.i.d. bounded random variables with a bounded density. They proved the exponential decay of the Green's function at fixed energy, and as a consequence the weak form of localization given by the absence of diffusion. The technicalities are more difficult than in the lattice case. More importantly, it seems difficult to control the Wegner type estimate and push the proof further to prove the strong form of localization, namely pure point spectrum.

2. One of the many advantages of the above proof of localization is that it is possible to trace down the few places where the independence assumption is used. A close look shows that the equality $\mathbb{P}\{A \text{ and } B\} = \mathbb{P}\{A\}\mathbb{P}\{B\}$ is only used for events $A \in \mathcal{F}_\Lambda$ and $B \in \mathcal{F}_{\Lambda'}$ which are identical up to a shift in the space variable (and this implies that $\mathbb{P}\{A\} = \mathbb{P}\{B\}$), and when the distance between the regions $\Lambda$ and $\Lambda'$ is large. This means that it is presumably possible to weaken the independence assumption into some sort of mixing condition giving a decorrelation estimate for the $\sigma$-fields $\mathcal{F}_\Lambda$ and $\mathcal{F}_{\Lambda'}$. In fact an estimate of the form:

$$\mathbb{P}\{A \text{ and } B\} \leq \mathbb{P}\{A\}^{1+\epsilon}$$

for some $\epsilon > 0$ and at large distances should be enough. Such an estimate

holds for mean zero stationary Gaussian processes $\{V(x);\ x \in \mathbb{Z}^d\}$ as long as the covariance $C(x,y) = \mathbb{E}\{V(x)V(y)\}$ satisfies the following conditions:
i). there exist positive constants $c$ and $q$ such that:

$$\forall x, y \in \mathbb{Z}^d, \qquad |C(x,y)| \leq \frac{c}{|x-y|^q},$$

ii). there exists a constant $\delta > 0$ such that $C \geq \delta I$ as operators on $\mathcal{H} = \ell^2(\mathbb{Z}^d)$.
The simplest example of a covariance satisfying the above conditions is given by the kernel of the inverse of the three dimensional free Hamiltonian. Indeed, one can take $C(x,y) = H_0^{-1}(x,y)$ where $H_0$ is the nonnegative free Hamiltonian introduced in Example I.1.3 In this case, one can use $q = 1$. This result is due to von Dreifus and Klein. The details can be found in [89]. The authors show that it is actually possible to prove localization for dependent potentials in this way. They also show that the argument is not restricted to Gaussian potentials and that it can be applied to a class of Gibbs random fields at high temperature. The multiscale analysis is more involved, the estimates are very technical and the exponential decay of the eigenfunctions is lost in the process.

3. Another advantage of the approach to localization which we presented is the possibility of extension to the so-called off-diagonal disorder models. The Hamiltonian $H$ is of the form:

$$[H\psi](x) = \sum_b J(b)(\psi(x) - \psi(y))$$

where the sum is over the unoriented nearest neighbor bonds $b = \{x,y\}$ and the coefficients $J(b)$ are nonnegative i.i.d. random variables with a first moment and a bounded density, say $\rho$. Under the following extra restriction on the density:

$$\sup_{J>0}(J \vee J^2)\rho(J) < \infty,$$

Faris extended in [98] the multiscale analysis of this section to prove exponential decay of the Green's function at high energy.

# IX.4 Problems

Warning: Problem IX.4.1 is elementary, Problem IX.4.2 is technical while Problem IX.4.3 reproduce some classical facts about Gaussian processes.

**Problem IX.4.1** *Let $F$ be a strictly increasing continuous distribution function and let $\{X_n;\ n \geq 1\}$ a sequence of real valued independent random variables with common distribution $F$. For each integer $n \geq 1$ we denote by $X_{(1)} < X_{(2)} < \cdots < X_{(n-1)} < X_{(n)}$ the order statistics forming a nondecreasing rearrangement of $X_1, \cdots, X_n$. Note that we do not have ties because we assume that $F$ is continuous.*

*1. Compute the distribution function $F_{X_{(n)}}(x) = \mathbb{P}\{X_{(n)} \leq x\}$ and show that $F_{X_{(n-1)}}(x) = \mathbb{P}\{X_{(n-1)} \leq x\}$ is given by $F_{X_{(n-1)}}(x) = g_n(F(x))$ where the function $g$ is defined by $g_n(p) = p^{n-1}[n-(n-1)p]$ for $p \in [0,1]$.*

*2. Show that $\{X_{(n-1)};\ n \geq 2\}$ is a nondecreasing sequence and that:*

$$\lim_{n\to\infty} X_{(n-1)} = \lim_{n\to\infty} X_{(n)} = +\infty \quad \mathbb{P}- \text{ almost surely.}$$

*3. For each fixed $n \geq 3$ graph the function $g_n$ and show that it has a unique fixed point, say $\hat{p}_n$ in $(0,1)$.*

*4. $n$ remains fixed and we construct the sequence $\{Z_{n,k};\ k \geq 0\}$ by the following induction procedure. One sets $Z_{n,0} = X_{(n)}$ and:*

$$Z_{n,k} = \max\{Z_{n,k-1,1}, \cdots, Z_{n,k-1,n}\}$$

*where $Z_{n,k-1,1}, \cdots, Z_{n,k-1,n}$ are independent random variables with the same distribution as $Z_{n,k-1}$.*

*4.1. Prove that $F_{Z_{n,k}}(x) = F(x)^{n^{k+1}}$*

*4.2. Conclude that the sequence $\{Z_{n,k};\ k \geq 0\}$ converges in probability to $+\infty$, i.e. for all $x > 0$ one has $\lim_{k\to\infty} \mathbb{P}\{Z_{n,k} \leq x\} = 0$.*

*5. $n \geq 3$ is still fixed and we construct the sequence $\{Y_{n,k};\ k \geq 0\}$ in the following way: one sets $Y_{n,0} = X_{(n-1)}$ and $Y_{n,k} = Y_{n,k-1,(n-1}\}$, i.e. the $(n-1)$-th order statistics of $Y_{n,k-1,1}, \cdots, Y_{n,k-1,n}$ where $Y_{n,k-1,1}, \cdots, Y_{n,k-1,n}$ are independent random variables with the same distribution as $Y_{n,k-1}$.*

*5.1. Prove that $F_{Y_{n,k}}(x) = g_n\left(F_{Y_{n,k-1}}(x)\right)$.*

*5.2. Use part 3. to compute $\lim_{k\to\infty} F_{Y_{n,k}}(x)$ as a function of $x$.*

*5.3. Prove that the sequence $\{Y_{n,k};\ k \geq 0\}$ converges in probability to the unique root, say $\hat{x}_n$ of the equation $F(x) = \hat{p}_n$.*

*6. Do the explicit computations in the case $n = 3$ and*

*6.1. $F(x) = (1 - e^{-\lambda x})\mathbf{1}_{[0,\infty)}(x)$, for some $\lambda > 0$,*

*6.3. $F(x) = (2\pi)^{-1/2} \int_{-\infty}^{x} e^{(y-m)^2/2}\, dy$ for some $m \in \mathbb{R}$.*

**Problem IX.4.2** *Let us consider a family $\{J(b);\ b \in B\}$ of nonegative i.i.d. random variables indexed by the set $B$ of unoriented nearest neighbor* *bonds* $b = \{x, y\}$ *of the lattice* $\mathbb{Z}^d$. *We assume that the common distribution of the random variables* $J(b)$ *has a bounded density, say* $\rho$ *which satisfies:*

$$\sup_{J>0}(J \vee J^2)\rho(J) < \infty.$$

*We then consider the random self-adjoint operator* $H(\omega)$ *on* $\mathcal{H} = \ell^2(\mathbb{Z}^d)$ *defined by:*

$$[H(\omega)\psi](x) = \sum_{b=\{x,y\}\in B} J(b,\omega)(\psi(x) - \psi(y))$$

*For each (finite) subset* $\Lambda$ *of* $\mathbb{Z}^d$, *we define the operator* $H^\Lambda(\omega)$ *acting on* $\ell^2(\Lambda)$ *by restricting the sum to the bonds with both ends in* $\Lambda$. *We denote by* $\lambda_j^\Lambda$ *the eigenvalues of* $H^\Lambda$ *and by* $\psi_j^\Lambda$ *the corresponding normalized eigenfunctions.*

*1. Let us fix a bounded energy interval* $\mathcal{I}$ *and show the existence of a positive constant* $c$ *such that:*

$$\mathbb{E}\{< \delta_x, (H^\Lambda)^2 \mathbf{1}_\mathcal{I}(H^\Lambda)\delta_x > |\tilde{\mathcal{F}}_x\} \le c|\mathcal{I}|$$

*for every* $x \in \Lambda$ *where the conditioning in the above expectation is aver all the bonds* $J(b)$ *such that* $x \notin b$.
*Hint: check that:*

$$< \delta_x, (H^\Lambda)^2 \mathbf{1}_\mathcal{I}(H^\Lambda)\delta_x >= \sum_{\lambda_j \in \mathcal{I}} < H_\Lambda \delta_x, \psi_j >^2$$

*and that* $< H_\Lambda\delta_x, \psi_j >= \sum_b J(b)(\psi_j(x) - \psi_j(y))$. *Use these facts and the formula* $\partial\lambda_j/\partial J(b) = (\psi_j(x) - \psi_j(y))^2$ *to perform the appropriate change of variables and estimate the integral:*

$$\int \sum_{\lambda_j \in \mathcal{I}} \frac{\partial \lambda_j}{\partial J(b)}\, dJ(b)$$

*by* $2|\mathcal{I}|$.

*2. Take the expectation of the estimate of part 1. and sum over* $x \in \Lambda$ *in order to get:*

$$\mathbb{E}\{trace\ [(H^\Lambda)^2 \mathbf{1}_\mathcal{I}(H^\Lambda)]\} \le c|\Lambda|\,|\mathcal{I}|,$$

*choose* $\mathcal{I} = (\lambda - 1/M, \lambda + 1/M)$ *and conclude that:*

$$(\lambda - \frac{1}{M})^2 \mathbb{P}\{\text{ at least one eigenvalue of } H^\Lambda \text{ in } \mathcal{I}\} \le \frac{2c|\Lambda|}{M}$$

*3. Conclude that for every $\lambda > 0$ there exist constants $c > 0$ and $M_0 > 0$ such that the Wegner type estimate:*

$$\mathbb{P}\{\|[H^\Lambda - \lambda I]^{-1}\| \geq M\} \leq \frac{c|\Lambda|}{M}$$

*holds for all the finite boxes $\Lambda$ and all the real numbers $M > M_0$.*

**Problem IX.4.3** *Let $C(x,y)$ be real numbers which we regard as the matrix elements of an operator $C$ on $\mathcal{H} = \ell^2(\mathbb{Z}^d)$*
*1. Let us assume that $C$ is a nonnegative operator and prove that there exists a real valued mean zero Gaussian process $\{V(x);\, x \in \mathbb{Z}^d\}$ such that:*

$$C(x,y) = \mathbb{E}\{V(x)V(y)\}.$$

*Prove also that the process is stationary if and only if the covariance $C$ can be written in the form:*

$$C(x,y) = c(x-y)$$

*where the function $c$ is the Fourier transform of a nonnegative measure $\nu$ on the d-dimensional torus and that the process is ergodic if and only if the measure $\nu$ is continuous (i.e. has no point mass).*
*2. Prove that $C = [H_0 + mI]^{-1}$ where $H_0$ is the nonnegative free Hamiltonian introduced as Example I.1.3 in the begining of the Chapter I and $m \geq 0$, satisfies the assumptions of part 1. and that it also satisfies the assumptions i) and ii) of the second extension mentionned in the text if we also choose $m = 0$ and $d = 3$.*

# IX.5 Notes and Complements

Multidimensional localization (i.e. dense pure point spectrum with exponentially decaying eigenfunctions) remained a puzzling mathematical problem for a long time. The first breakthrough was the fundamental work of Fröhlich and Spencer [105]. They proved exponential decay of the Green's function for large disorder and for fixed energy. The motivation and the results were in terms of the average distance traveled in time $t$ by a wave packet $\psi$ which is localized around the origin at time $t = 0$. This average distance is defined as:

$$r(t) = \mathbb{E}\{\int_{\mathbb{R}^d} |x|^2\, |[e^{-itH}\psi_0](x)|^2\, dx\}^{1/2}$$

in the continuous case and:

$$r(t) = \mathbb{E}\{ \sum_{x \in \mathbb{Z}^d} |x|^2 \, |[e^{-itH}\psi_0](x)|^2\}^{1/2}$$

in the lattice case. For $V = 0$ and fr periodic potentials one has $r(t) \sim t$ which is understood as infinite conductivity. In three or more dimensions and for small disorder, this averaged distance should behave like $r(t) \sim \sqrt{t}$ which is typical of diffusin. Anderson conjectured that localization should be seen at large disorder by proving that $\sup_t r(t) < \infty$. The result of Fröhlich and Spencer does not imply this strong form of localization. It nevertheless shows that the behavior of $r(t)$ is different. Indeed, it implies that, for a wave packet of energy localized around $\lambda$ at time $t = 0$, $t^{-1}r_\lambda(t)$ converges to 0 as $t \to \infty$. This is a weaker form of localization. It is an easy consequence of the decay of the Green function at fixed energy. The original proof of the latter was very technical. In particular, it is extremely difficult to follow all the combinatorial estimates. The proof we give is self contained and much shorter. It is due to von Dreifus [87] and Spencer [310]. It took more than one year after the publication of [105] to realize that there was more than a weak form of absence of diffusion and that Fröhlich and Spencer's paper was containing information concerning the spectral type of the operators. This was first done in a note [231] by Martinelli and Scoppola who remarked that the exponential decay of the Green function implies the absence of absolutely continuous spectrum. We devoted the last part of Chapter VII to this result and we discussed the proof in the Notes and Complements. A few months later, three (independent) proofs of the existence of dense pure point spectrum with exponentially decaying eigenfunctions were announced. [72], [304] and [104]. The proof of Delyon, Lévy and Souillard [72] uses the polynomial growth of the generalized eigenfunctions and supplemented Fröhlich and Spencer's estimate with what we called Kotani's trick in the preceding chapter. Simon and Wolff [304] elaborated on the integration of the spectral measures with respect to a parameter and supplemented the Fröhlich and Spencer estimate with a fine analysis of rank one perturbations of self-adjoint operators. This proof does not rely on the previous knowledge that the growth of the generalized eigenfunctions is controlled by polynomials. Both approaches are given in detail in the first half of the present chapter. They are both elegant and relatively short (once the exponential decay of the Green's function is known). Unfortunately, they require the existence of a density *somewhere* in the distribution of the potential, and they cannot be used to prove localization for singular potentials such as the Bernouilli one for this reason.

This drawback is not shared by the third approach. Fröhlich, Martinelli, Scoppola and Spencer used in [104] some ideas developed by Jona-Lasinio, Martinelli and Scoppola in the study of a hierarchical toy model [161] to push the original proof of Fröhlich and Spencer to give localization. The crucial difference is the probability estimate. It is not done at fixed energy anymore. It is simultaneous for the energy restricted to a fixed interval. The proof of localization which we gave in the second half of the present chapter is essentially the proof of [161] in the sense that the spirit of the probability estimate and the use of the polynomial growth of the generalized eigenfunctions to conclude the pure point nature of the spectrum are taken from [161]. The details are much simpler because we followed the simplified proof given by von Dreifus and Klein in [88]. They implement in the setting of localization the ideas developed by von Dreifus and Spencer to estimate the Green's function at fixed energy. The extension to the case of off-diagonal disorder is due to Faris, [98].

The understanding of the localization in the multidimensional case seems to be limited to the lattice case. Holden and Martinelli extended in [150] the Fröhlich and Spencer result on the decay of the Green's function to some continuous multidimensional models. The main drawback is the artificial requirement that the potential be constant on the cells of a tiling of the space $\mathbb{R}^d$. It would be more interesting to prove this type of result for random potentials of the Poisson type, i.e. of the form:

$$V(x,\omega) = \sum \chi(x - x_n(\omega))$$

where the $x_n(\omega)$'s are points in $\mathbb{R}^d$ distributed according to a stationary Poisson process and where the potential well $\chi$ is a function with compact support (or decaying fast enough at infinity). The main difficulty seems to be the lack of control of the possible resonances. More precisely, the various steps of the multiscale analysis could be extended if one could control the probability that a given energy is close to the spectrum of the operator in large boxes, in other words if one had some sort of Wegner type estimate. Assuming that the potential is piecewise constant is one way to recreate the lattice structure and to make possible the proof of a weak form of such an estimate.

Faris gave in [98] an interpretation of the propagation of the probability estimates in the multiscale analysis in terms of order statistics. He noticed that, at stage $k$ of the analysis one is handling $N_k$ independent copies of the same random variable. Indeed, the crucial probability estimate can be interpreted as an exponential bound for the tail of the probability distribution of a *normalized* logarithm of the norm of the Green's function in a box. The $N_k - 1$-th largest of these random variables gives the leading order of the probability estimate for the next stage while the largest one had to be controlled as a correction term. Indeed, one has to trade off the best possible decay of the Green's function for configurations of the potential which ***avoid bad boxes***. We summarized the main steps of his abstraction in Problem IX.4.1 on order statistics. Problem IX.4.2 on off-diagonal disorder is taken from [99]. This extension of the Wegner estimate is also due to Faris. Finally, Problem IX.4.3 reviews some of the classical facts about stationary Gaussian processes. We prepare the reader for the application to the localization result for dependent potentials given by von Dreifus and Klein in [89].

For the sake of completeness we mention a multidimensional localization result for a class of unbounded potentials on $\mathbb{Z}^d$. Let $V$ be a real valued periodic function of period one on $\mathbb{R}$ which is meromorphic in some strip $\{z : |\Im m\, z| < R\}$ for some $R > 0$. Let us also assume that $V$ has exactly one pole in the interval $[0, 1)$ and no pole outside the real axis. Then Bellissard, Lima and Scoppola proved in [18] that the operator $H = \epsilon H_0 + V(\vartheta + \mathbf{n} \cdot \alpha)$ has, for almost every $\theta$, pure point spectrum with exponentially decaying eigenfunctions. The assumption on the frequencies $\alpha$ is that it is a Diophantine multi-integer in the sense that:

$$\forall \mathbf{m}, \mathbf{n} \in \mathbb{Z}^d, \qquad |\mathbf{n} \cdot \alpha + \mathbf{m}| \geq \frac{\gamma}{|\mathbf{n}|^\sigma}$$

for some $\gamma > 0$ and $\sigma > d$. As usual, the Diophantine condition is crucial for the spectrum is expected to be singular continuous otherwise because of the results of Avron and Simon [14]. See also [102] and [103]. This class of examples can be viewed as a generalization of some of the models studied first by Sarnack in [284]. Long range resonances and tunneling are not present in such models and the analysis is consequently easier.

A discussion of the results on localization would not be complete if we did not discuss the two dimensional case. After several controversial argumentations it seems that the physics community agrees on the the fact that one should have localization for all disorders. This means that one should be able to prove that the spectrum is (dense) pure point in all cases, as in one dimension. The only conjecture which is not unanimously agreed upon is the exponential decay of the eigenfunctions. Many people believe that the localization phenomenon may be weaker for small disorder and that the eigenfunctions may decay only polynomially. Settling this question is, together with proving the existence of absolutely continuous spectrum in dimension three and more, one of the most important remaining challenges of the field. The next step should be the fine analysis of the transition from absolutely continuous spectrum to pure point spectrum. There are many conjectures concerning the ***phase diagram*** giving the nature of the spectrum as a function of the energy and the coupling constant. But mathematical proofs of these results may still be a long way ...

# BIBLIOGRAPHY


[1] R. Abou-Chacra, P.W. Anderson and D.J. Thouless (1973): A self-consistent theory of localization. J. Phys. C Solid State Phys. **6**, 1734-1752.

[2] R. Abou-Chacra and D.J. Thouless (1974): Selfconsistent theory of localization:II. Localization near the band edges. J. Phys. C Solid State Phys. **7**, 65-75.

[3] M. Aizenman and and B. Simon (1982): Brownian motion and Harnack inequality for Schrödinger operators. Comm. Pure Appl. Math. **35**, 209-271.

[4] M. A. Akcoglu and U. Krengel (1981): Egodic theorems for superadditive processes. J. Reine Angew. Math. **323**, 53-67.

[5] N. I. Akhiezer and I. M. Glazman (1978): Theory of linear operators in Hilbert space I, II. Frederick Ungar Publish. Co. 3rd printing.

[6] S. Albeverio, R. Hoegh-Krohn, W. Kirsch and F. Martinelli (1982): The spectrum of the three dimensional Krönig-Penney model with random point defects. Adv. Appl. Math. **3**, 435-440.

[7] G. André and S. Aubry (1980): Analyticity breaking and the Anderson localization in incommensurate lattices. Ann. Israel Phys. Soc. **3**, 133-164.

[8] Ph. Anderson (1958): Absences of diffusion in certain random lattices. Physical Review **109**, 1492-1505.

[9] L. Arnold, G. Papanicolaou and G. Wishtutz (1989): Asymptotic analysis of the Lyapunov exponent and rotation number of the random oscillator and applications. SIAM J. Applied Math. (to appear).

[10] N. Aronszajn (1957): On a problem of Weyl in the theory of singular Sturm-Liouville equations. Am. J. Math. **79**, 597-610.

[11] F. V. Atkinson (1964): Discrete and continuous boundary problems. Mathematics in science and engineering Vol **8**, Academic Press, New York.

[12] J. Avron and B. Simon (1981): Transient and recurrent spectrum. J. Funct. Anal. **43**, 1-31.

[13] J. Avron and B. Simon (1982): Almost periodic Schrödinger operators, I. Limit periodic potentials. Comm. Math. Phys. **82**, 101-120.

[14] J. Avron and B. Simon (1982): Singular continuous spectrum for a class of almost periodic Jacobi matrices. Bull. Amer. Math. Soc. **6**, 81-85.

[15] J. Avron and B. Simon (1983): Almost periodic Schrödinger operators, II. The integrated density of states. Duke Math. J. **50**, 369-391.

[16] J. Avron, W. Craig and B. Simon (1983): Large Coupling Behavior of the Lyapunov Exponent of Tight Binding One-dimensional Random Systems. (preprint)

[17] J. Bellissard, D. R. Grempel, F. Martinelli and E. Scoppola (1989): Localization of electrons with spin-orbit or magnetic interactions in a two dimensional disordered crystal. Phys. Rev. Lett. (to appear)

[18] J. Bellissard, R. Lima and E. Scoppola (1983): Localization in $\nu$-dimensional incommensurate structures. Comm. Math. Phys. **88**, 465-477.

[19] J. Bellissard, R. Lima and D. Testard (1983): A metal-insulator transition for the almost Mathieu model. Comm. Math. Phys. **88**, 207-234.

[20] J. Bellissard, B. Jochum, E. Scoppola, D. Testard (1989): Spectral properties of one dimensional quasi-crystals. (preprint)

[21] J. Bellissard and B. Simon (1982): Cantor spectrum for the almost Mathieu equation J. Functional Anal. **48**, 408-419.

[22] J. Bellissard (1988): The strange properties of the Thue-Morse potential. (preprint)

[23] M. Ben-Artzi (1985): Remarks on Schrödinger operators with an electric field and a deterministic potential. J. Math. Anal. Appl. **109**, 333-339.

[24] M. M. Benderskii and L. A. Pastur (1969): Calculation of the average number of states in a model problem. Soviet Phys. JETP **30**, 158-162.

[25] M. M. Benderskii and L. A. Pastur (1970): On the spectrum of the one dimensional Schrödinger equation with a random potential. Math. Sb. **82**, 245-256.

[26] C. Bennewitz (1989): Spectral Asymptotics for Sturm-Liouville Equations. Proc. London Math. Soc. **59**, 294-338.

[27] F. Bentosela, R. Carmona, P. Duclos, B. Simon, B. Souillard and R. Weder (1983): Schrödinger operators with an electric field and random or determinsitic potential. Comm. Math. Phys. **88**, 387-397.

[28] J. M. Berezanskii (1968): Expansion in eigenfunctions of self-adjoint operators. Transl. Math. Monographs Vol. 17. Amer. Math. Soc. Providence, R. I.

[29] R. M. Blumenthal and R. K. Getoor (1968): Markov Processes and Potential Theory. Academic Press, New York, N.Y.

[30] R. E. Borland (1963): The nature of the electronic states in disordered one-dimensional systems. Proc. Royal Soc. London ser. A, **A274**, 529-545

[31] P. Bougerol (1988): Théorèmes limites pour les systèmes linéaires à coefficients markoviens. Probab. Th. Rel. Fields, vol **78**, 193-221

[32] P. Bougerol (1988): Comparaison des exposants de Lyapounov des processus markoviens multiplicatifs. Ann. Inst. Henri Poincaré **24**, 439-489

[33] P. Bougerol and J. Lacroix (1985): Products of random matrices with applications to random schrödinger operators. Birkhauser, Boston, Mass.

[34] A. Bovier and A. Klein (1988): The weak disorder expansion of the invariant measure for the one dimensional Anderson model. J. Stat. Phys. **51**, 501-517.

[35] A. Bovier, M. Campanino, M. Klein and F. Perez (1988): Smoothness of the density of states in the Anderson model at high disorder. Comm. Math. phys. **114**, 439-461.

[36] L. Breiman (1988): Probability. Addison Wesley. Reading Massachussets.

[37] J. Brossard (1983): Perturbations Aléatoires de Potentiels Périodiques. (preprint)

[38] A. Brunel, D. Revuz (1974): Quelques applications probabilistes de la quasi-compacité. Ann. Inst. H. Poincaré **10**, 301-337.

[39] M. Campanino and A. Klein (1986): A supersymmetric transfer matrix and differentiability of the density of states in the one dimensional Anderson model. Comm. Math. Phys. **104**, 227-241.

[40] M. Campanino and F. Perez (1989): Singularity of the density of states for one dimensional chains with random couplings. Comm. Math. Phys.

[41] R. Carmona (1979): Regularity Properties of Schrödinger and Dirichlet Semigroups. J. Functional Anal. **33**, 259-296.

[42] R. Carmona (1982), Exponential localization in one dimensional disordered systems. Duke Math. J. **49**, 191-213.

[43] R. Carmona (1983), One dimensional Schrödinger operators with random or deterministic potentials: new spectral types. J. Functional Anal. **51**, 229-258.

[44] R. Carmona (1984): One dimensional Schrödinger operators with Random Potentials. Physica **124A**, 131-188.

[45] R. Carmona (1984): Random Schrödinger operators, in Ecole d'Eté de probabilités de Saint Flour. Lect. Notes in Math. **1070**

[46] R. Carmona (1986): Exercises for a book on random potentials. IMA Random Media **7**, 45-65 Springer Verlag.

[47] R. Carmona (1990): Lévy Processes and Schrödinger Operators. (to appear).

[48] R. Carmona (1989): Path Integrals for relativistic Schrödinger operators. In Proc. Northern Summer School in Mathematical Physics. Aarhus 1988. Lect. Notes in Physics **345**, 65-92.

[49] R. Carmona, W. Chen-Masters and B. Simon (1989): Relativistic Schrödinger operators: Decay of the eigenfunctions. J. Functional Anal. (to appear)

[50] R. Carmona, A. Klein and F. Martinelli (1987): Anderson localization for Bernouilli and other singular potentials. Comm. Math. Phys. **108**, 41-66.

[51] R. Carmona and S. Kotani (1987): Inverse spectral theory for random Jacobi matrices. J. Statist. Phys. **46**, 1091-1114.

[52] A. Casher and J. L. Lebowitz (1971): Heat flow in regular and disordered harmonic chains. J. Math. Phys. **12**, 1701-1711.

[53] V. Chulaevsky (1981): On perturbations of a Schrödinger operator with periodic potential. Russ. Math. Surveys **36**, (5) 143-144.

[54] V. Chulaevsky and F. Delyon (1989): Purely absolutely continuous spectrum for almost Mathieu operators. J. Statist. Phys. (to appear)

[55] V. Chulaevsky and Ya. G. Sinai (1989): Anderson Localization for the 1-D Discrete Schrödinger operator with Two-Frequency Potential. (preprint)

[56] E. A. Coddington and N. Levinson (1955): Theory of ordinary differential equations. McGraw Hill, New York.

[57] J.E. Cohen and C. Newman (1984): The stability of large random matrices and their products. Ann. Proba. **12**, 283-310.

[58] F. Constantinescu, J. Fröhlich and T. Spencer (1983): Analiticity of the density of states and replica method for random Schrödinger operators on a lattice J. Stat. Phys. **34**, 571-596.

[59] R. Courant and D. Hilbert (1953): Methods of Mathematical Physics. Interscience Publ,. New York.

[60] W. Craig (1983): Pure point spectrum for discrete Schrödinger operators. Comm. Math. Phys. **88**, 113-131.

[61] W. Craig (1990): The trace formula for Schrödinger operators on the line. Comm. Math. Phys. (to appear)

[62] W. Craig and B. Simon (1983): Subharmonicity of the Ljapunov Index. Duke Math. J. **50**, 551-559.

[63] W. Craig and B. Simon (1983): Log-Hölder Continuity of the Integrated Density of States for Stochastic Jacobi Matrices. Comm. Math. Phys. **90**, 207-218.

[64] H. L. Cycon, R. G. Froese, W. Kirsch and B. Simon (1987): Schrödinger operators with Applications to Quantum Mechanics and Global Geometry. Springer-Verlag. New York, N.Y.

[65] P. Deift and E. Trubowitz (1979): Inverse scattering on the line. Comm. Pure Appl.Math. **32**, 121-251.

[66] P. Deift and B. Simon (1983): Almost Periodic Schrödinger Operators III, the Absolute Continuous Spectrum. Comm. Math. Phys. **90**, 389-411.

[67] R.R. del Rio Castillo (1988): On conditions which guarantee the stability of the absolutely continuous spectrum of ordinary differential operators. Quart. J. Math. Oxford **39**, 269-280.

[68] F. Delyon (1985): Apparition of purely singular continuous spectrum in a class of random Schrödinger operators. J. Statist. Phys. **40**, 621-630.

[69] F. Delyon (1987): Absence of localization in the almost Mathieu equation. J. Phys. A **20** L 21

[70] F. Delyon and P. Foulon (1986): Adiabatic Invariants and asymptotic Behavior of Lyapunov Exponents of the Schrödinger Equation. J. Statist. Phys. **45**, 41-47.

[71] F. Delyon, H. Kunz and B. Souillard (1983): One dimensional wave equations in random media. J. Phys. **A16**, 25-42.

[72] F. Delyon, Y. Lévy and B. Souillard (1985): Anderson localization for multidimensional systems at large disorder or low energy. Comm. Math. Phys. **100**, 463-470.

[73] F. Delyon, Y. Lévy and B. Souillard (1985): Approach à la Borland to multidimensional localization. Phys. Rev. Lett. **55**, 618-621.

[74] F. Delyon, Y. Lévy and B. Souillard (1985): Anderson localization for one and quasi-one dimensional systems. J. Stat. Phys. **41**, 375-388.

[75] F. Delyon and D. Petritis (1986): Absence of Localization in a Class of Schrödinger Operators with Quasiperiodic Potentials. Comm. Math. Phys. **103**, 441-444.

[76] F. Delyon, B. Simon and B. Souillard (1984): From power law localized to extended states in disordered systems. Phys. Rev. Lett. **52**, 2187-3189.

[77] F. Delyon and B. Souillard (1983): The rotation number for finite difference operators and its properties. Comm. Math. phys. **89**, 415-427.

[78] F. Delyon and B. Souillard (1984): Remark on the continuity of the density of states of ergodic finite difference operators. Comm. Math. Phys. **94**, 289-291

[79] B. Derrida and H.J. Hilhorts (1983): Singular behaviour of certain infinite products of random $2 \times 2$ matrices. J. Phys. A. Math. Ge. **16**, 2641-2654.

[80] B. Derrida and E. Gardner (1984): Lyapunov exponent of the one dimensional Anderson model:weak disorder expansion. J. de Physique. **45**, 1283-1295.

[81] Y. Derriennic (1975): Sur le théorème ergodique sous additif. C. R. Acad. Sci. Paris ser A 282, 985-988.

[82] Y. Derriennic and Y. Guivarch (1974): Théorème de renouvellement pour les groupes non moyennables. C. R. Acad. Sc. Paris **277**, 613-615.

[83] E. I. Dinaburg and Ya. G. Sinai (1976): The one dimensional Schrödinger equation with a quasi-periodic potential. Funct. Anal. Appl. **9**, 279-289.

[84] W. Donoghue (1965): On the perturbation of the spectra. Comm. Pure Appl. Math. **18**, 559-579.

[85] W. F. Donoghue Jr. (1974): Monotone matrix functions and analytic continuation. Springer Verlag, New York.

[86] M. Donsker and S.R.S. Varadhan (1975): Asymptotics for the Wiener saussage. Comm. Pure Appl. Math. **28**, 525-565.

[87] H. von Dreifus (1987): On the effects of randomness in ferromagnetic models and Schrödinger operators. Ph.D. New York Univ.

[88] H. von Dreifus and A. Klein (1989): A new proof of localization in the Anderson tight binding model. Comm. Math. Phys. **124**, 285-299.

[89] H. von Dreifus and A. Klein (1990): Anderson Localization for Correlated Gaussian Potentials. (preprint)

[90] J. Droese and W. Kirsch (1986): The effect of boundary conditions on the density of states of random Schrödinger operators. *Stoch. Proc. Appl.* **23**, 169-175.

[91] A.A. Dubrovin, V.B. Matveev and S.P. Novikov (1976): Nonlinear equations of Kortweg de Vries type, finite zone linear operators and Abelian varieties. Russian Math. Surveys **31**, 59-146.

[92] N. Dunford and J. T. Schwarz (1963): Linear operators II. John Wiley and Sons, New York.

[93] E.B. Dynkin (1978): Sufficient Statistics and Extreme Points. Ann. Proba. **6**, 705-730.

[94] M.S.P. Eastham (1970): Theory of Ordinary Differential Equations. van Nostrand Reinhold Co. London.

[95] M.S.P. Eastham (1973): The Spectral Theory of Periodic Differential Equations. Scottish Academic Press, London.

[96] H. English, W. Kirsch, M. Schroder and B. Simon (1990): Random Hamiltonians Ergodic in all but one Direction. Comm. Math. Phys. (to appear)

[97] W. G. Faris (1975): Self Adjoint Operators. Lect. Notes in Math. No. **433**, Springer Verlag, New York.

[98] W.G. Faris (1989): Localization with Off-diagonal Disorder. (preprint)

[99] W.G. Faris (1989): Order Statistics and Renormalization. (preprint)

[100] A. Figotin and L. A. Pastur (1984): An exactly solvable model of a multi-dimensional incommensurate structure. Comm. Math. Phys. **95**, 401-425.

[101] A. Figotin and L. A. Pastur (1990): Spectral properties of disordered systems in the one body approximation. Springer Verlag (to appear)

[102] S. Fishman, D. Grempel and R. Prange (1982): Chaos, quantum recurrence and Anderson localization. Phys. Rev. Lett. **49**, 509-977.

[103] S. Fishman, D. Grempel and R. Prange (1984): Localization in a d-dimensional incommensurate structure. Phys. Rev. **B329**, 4272-42.

[104] J. Fröhlich, F. Martinelli, E. Scoppola and T. Spencer (1985): A constructive proof of localization in Anderson tight binding model. Comm. Math. Phys. **101**, 21-46.

[105] J. Fröhlich and T. Spencer (1983): Absence of diffusion in the Anderson tight binding model for large disorder or low energy. Comm. Math. Phys. **88**, 151-184.

[106] J. Fröhlich and T. Spencer (1984): Mathematical theory of Anderson localization. Physica ser. A **124**, 303-310.

[107] J. Fröhlich, T. Spencer and C. E. Wayne (1986): Localization in disordered nonlinear dynamical systems. J. Statist. Phys. **42**, 247-274.

[108] J. Fröhlich, T. Spencer and P. Wittwer (1989): Localization for a Class of One Dimensional Quasi-Periodic Schrödinger Operators. (preprint)

[109] P. A. Fuhrmann (1981): Linear Systems and Operators in Hilbert Space. McGraw Hill, New York.

[110] M. Fukushima (1974): On the spectral distribution of a disordered system and the range of a random walk. Osaka J. Math. **11**, 73-85.

[111] M. Fukushima (1980): Dirichlet Forms and Markov Processes. North Holland, New York, N.Y.

[112] M. Fukushima (1981): On asymptotics of spectra of Schrödinger operators. In "Aspects Statistiques et Aspects Physiques des Processus Gaussiens," ed. C.N.R.S.

[113] M. Fukushima, H. Nagai and S. Nakao (1975): On the asymptotic property of spectra of a random difference operator. Proc. Japan Acad. **51**, 100-102.

[114] M. Fukushima and S. Nakao (1977): On spectra of the Schrödinger operator with a white noise potential. Z. Wahrsch. verw. Geb. **37**, 267-274.

[115] H. Furstenberg (1963): Non commuting random products. Trans. Amer. Soc. **108**, 377-428.

[116] H. Furstenberg and H. Kesten (1960): Products of random matrices. Ann. Math. Stat. **31**, 457-469.

[117] H. Furstenberg, Y. Kifer (1983): Random matrix products and measures on projective spaces. Israël j. of Math. **46**, 12-32.

[118] H. Fustenberg, I. Tzkoni (1971): Spherical functions and integral geometry. Israël J. of Math. **10**, 327-338.

[119] F. Gesztezy, H. Holden, B. Simon and Z. Zhao (1990): On the Toda and Kac-van Moerbeke Systems. (preprint)

[120] D.J. Gilbert (1989): On Subordinacy and Analysis of Schrödinger Operators with two Singular Endpoints. Proc. Royal Soc. Edinburgh **112A**, 213-229.

[121] D.J. Gilbert and D.B. Pearson (1987): On Subordinacy and Analysis of the Spectrum of One-Dimensional Schrödinger Operators. J. Math. Anal. Appl. **128**, 30-56.

[122] C. Glaffig (1988): Smoothness of the integrated density of states on discrete strip lattices, i.i.d. and non i.i.d. cases. J. Functional Anal. (to appear).

[123] I. Ja. Goldsheid (1980): Asymptotic properties of the product of random matrices depending on a parameter. In "Multicomponent Systems", advances in Probability, **8**, 239-283.

[124] I. Ja. Golsheid (1980): The structure of the spectrum of the Schrödinger random difference operator. Sov. Math. Dokl. **22**, 670-674.

[125] I. Ja. Golsheid (1990): Sturm-Liouville Formula for Schrödinger Operators. (preprint)

[126] I. Ja. Golsheid and S. A. Molcanov (1976): On Mott's Problem. Soviet Math. Dokl. **17**, 1369-1373.

[127] I. Ja. Golsheid, S. A. Molcanov and L. A. Pastur (1977): A pure point spectrum of the one dimensional Schrödinger operator. Funct. Anal. Appl. **11**, 1-10.

[128] I. Ja. Goldsheid and A. G. Margulis (1987): A condition for simplicity of the spectrum of Lyapunov exponents Sov. Math. Dokl. **35** 309-313.

[129] A. Ya. Gordon (1976): On the point spectrum of the one dimensional Schrödinger operator. Usp. Math. Nauk. **31**, 257-.

[130] S.A. Gredeskul, I.M. Lifshitz and L. A. Pastur (1988): Introduction to the theory of disordered systems. Wiley.

[131] U. Grenander and M. Rosenblatt (1957): Statistical Analysis of Stationary Time Series. Chelsea Publ. Co. New York N.Y.

[132] U. Grenander and G. Szegö (1958): Toeplitz Forms and their Applications. Chelsea Publ. Co. New York N.Y.

[133] L. N. Grenkova, S. A. Molcanoy and Ju. N. Sudarev (1983), On the basic states of the one Dimensional Disordered Structures. Comm. Math. Phys. **90**, 101-123.

[134] Y. Guivarch (1980): Quelques propriétés asymptotiques des produits de matrices aléatoires. In Ecole d'Eté de Probabilités de Saint-Flour VIII.Lect. Notes in Math. **774**, 177-249.

[135] Y. Guivarch (1984): Exposants caractéristiques des produits de matrices aléatoires en dépendance markovienne. Probability measures on Groups. Lect. Notes in Math. **1064**

[136] Y. Guivarc'h and A. Raugi (1986): Products of random matrices : Convergence theorems. Contemporary Math. **50**, 31-53.

[137] Y. Guivarch and A. Raugi (1987): Frontière de Furstenberg, propriétés de contraction et théorèmes de convergence. Z. für Wahrscheinlichkeitstheorie verw. Geb. **69**, 187-242.

[138] Y. Guivarch and A. Raugi (1989): Propriétés de contraction d'un semi groupe de matrices inversibles et coefficients de Lyapunov d'un produit de matrices aléatoires independantes. Israel J. Math. **65**, 165-196.

[139] P. R. Halmos (1951), Introduction to Hilbert Space and the Theory of Spectral Multiplicity. Chelsea Publ. Co. New York.

[140] B. Halperin (1967): Properties of a particle in a one dimensional random potential. Adv. Chem. Phys. **13**, 123-177.

[141] A.B. Harris (1973): Rigorous bounds on the integrated density of states of a three dimensional random alloy. Phys. Rev. **B8**, 3661-3664.

[142] B. Helffer and J. Sjostrand (1984): Multiple wells in the semiclassical limit. I. Comm. Partial Diff. Equat. **9**, 337-408.

[143] B. Helffer and J. Sjostrand (1985): Puits multiples en limite semiclassique II: Interaction moléculare-symétries perturbation. Ann. Inst. H. Poincaré ser. A. **42**, 127-212.

[144] H. Hennion (1984): Loi des grands nombres et perturbations pour des produits réductibles de matrices aléatoires indépendantes. Wahrs. Verw. Gebiete **67**, 265-278.

[145] I. Herbst and J. Howland (1981): The Stark ladder and other one dimensional external field problems. Comm. Math. Phys. **80**, 23-42.

[146] Herbst and A. Sloan (1978): Perturbations of translation invariant positivity preserving semigroups in $L^2(\mathbf{R}^n)$. Trans. Amer. Math. Soc. **236**, 325-360.

[147] M. Herman (1983): Une méthode pour minorer les exposants de Ljapunov et quelques exemples montrant le caractère local d'un théorème d'Arnold et de Moser sur le tore de dimension 2. Comment. Math. Helvetici **58**, 453-502.

[148] K. Hewitt and A. Ross (1963): Abstract Harmonic Analysis 1. Springer Verlag, Berlin, Heidelberg, New York.

[149] K. Hoffman (1965): Banach Spaces of Analytic Functions. Prentice Hall, Englewood Cliffs, N.J.

[150] H. Holden and F. Martinelli (1984): A remark on the absence of diffusion near the bottom of the spectrum for a random Schrödinger operator on $L^2(R^\nu)$. Comm. Math. phys. **93**, 197-217.

[151] L. Hörmander (1967): Hypoelliptic differential equations of second order. Acta Mathematica **119**, 147-171.

[152] J. Howland (1989): A Localization Theorem for one-dimensional Schrödinger Operators. Preprint University of Virginia.

[153] L. K. Hua (1963): Harmonic analysis of several complex variables in the classical domains. A.M.S. Providence.

[154] K. Ichihara and H. Kunita (1974): A classification of the Second Order Degenerate Elliptic Operators and its Probabilistic Characterization. Zeit fur Wahrch. Verw Gebete, vol **30**, 235-254.

[155] K. Ishii and H. Matsuda (1970): Localization of normal modes and energy transpaort in the disordered harmonic chain. Suppl Progr. Theor. Phys. **45**, 56-86.

[156] K. Ishii (1973): Localization of eigenstates and transport phenomena in the one dimensional disordered system. Supp. Progr. Theor. Phys. **53**, 77-138.

[157] R. Johnson and J. Moser (1982): The rotation number for almost periodic potentials. Comm. Math. Phys. **84**, 403-438.erratum: **90** (1983) 317-318.

[158] R. Jhonston and H. Kunz (1983): The conductance of a disordered wire. J. Phys. C **16**, 3895-3912

[159] G. Jona-Lasinio, F. Martinelli and E. Scoppola (1981): The semiclassical limit of quantum mechanics: a qualitative theory via stochastic mechanics. Phys. Rep. **77**, 313-327.

[160] G. Jona-Lasinio, F. Martinelli and E. Scoppola (1981): New approach to the semi-classical limit in quantum mechanics I: Multiple tunneling in one dimension. Comm. Math. Phys. **80**, 223-254

[161] G. Jona-Lasinio, F. Martinelli and E. Scoppola (1985): Multiple tunnelings in d-dimensions: a quantum particle in a hierarchical potential. Ann. Inst. H. Poincaré ser. A **42**, 73-108.

[162] I.S. Kac (1963): On the multiplicity of the spectrum of a second order differential operator. Soviet Math. Dokl. 1035-1038.

[163] M. Kac and P. van Moerbeke (1975): On some periodic Toda lattices. Proc. Nat. Acad. Sci. USA **72**, 1627-1629.

[164] M. Kac and P. van Moerbeke (1975): A complete solution of the periodic Toda problem. Proc. Nat. Acad. Sci. USA **72**, 2879-2880.

[165] O. Kallenberg (1976), Random Measures. Academic Press, London.

[166] Karube (1958): On the local cross section in locally compact groups. J. Math. Soc. Japan **10**, 343-347.

[167] T. Kato (1966): Perturbation Theory for Linear Operators. Springer Verlag, New York.

[168] Y. Katznelson and B. Weiss (1982): A Simple proof of some ergodic theorems. Israel J. Math. **42**, 291-296.

[169] Y. Kifer (1986): Ergodic Theory of Random Transformations. Birkhauser, Boston, Basel, Stuttgart.

[170] Y. Kifer (1989): Random Perturbations of Dynamical Systems. Birkhauser, Boston.

[171] J.F.C. Kingman (1976): Subadditive processes. In Ecole d'Eté de Probabilités de Saint-Flour V, Lect. Notes in Math. **539**, 167-223.

[172] W. Kirsch (1985): On a class of random Schrödinger opertors. Adv. Appl. Math **6**, 177-187.

[173] W. Kirsch (1985): Random Schrödinger Operators and the Density of States. In Lect. Notes in Phys. **1109**, 68-102.

[174] W. Kirsch (1990): Random Schrödinger Operators: a Course. in Proc. Northern Summer School in Mathematical Physics. Aarhus 1988. Lect. Notes in Physics. **345**, 264-370.

[175] W. Kirsch, S. Kotani and B. Simon (1985): Absence of absolutely continuous spectrum for some one dimensional random but deterministic potentials. Ann. Inst. Henri Poincaré **42**, 383-406.

[176] W. Kirsch and F. Martinelli (1982): On the ergodic properties of the spectrum of general random operators. J. Reine Angew. Math. **334**, 141-156.

[177] W. Kirsch and F. Martinelli (1982): On the spectrum of Schrödinger operators with a random potential. Com. Math. Phys. **85**, 329-350.

[178] W. Kirsch and F. Martinelli (1982): On the density of states of Schrödinger operators with a random potential J. Phys. **A 15**, 2139-2156.

[179] W. Kirsch and F. Martinelli (1983): Large deviations and the Lifschitz singularity of the integrated density of states of random Hamiltonians. Comm. Math. Phys. **89**, 27-40.

[180] W. Kirsch and F. Martinelli (1983): On the essential self adjointness of stochastic Schrödinger operators. Duke Math. J. **50**, 2155-1260. erratum ibidem.

[181] W. Kirsch and L.A. Pastur (1989): On the asymptotic behavior of the interband light absorption coefficient on the fluctuation spectrum. (preprint)

[182] W. Kirsch and B. Simon (1986): Lifschitz tails for periodic plus random potentials. J. Statist. Phys. **42**, 799-808.

[183] A. Klein, J. Lacroix and A. Speis (1989): Regularity of the density of states in the Anderson model on a strip for potentials with singular continuous distributions. J. Statist. Phys. **57**, 65-88

[184] A. Klein, J. Lacroix and A. Speis (1989): Localization for the Anderson model on a strip for singular potentials. J. Functional Anal. (to appear)

[185] A. Klein, F. Martinelli and F. Perez (1986): A rigorous replica trick approach to Anderson localization in one dimension. Comm. Math. Phys. **106**, 623-633.

[186] A. Klein and F. Perez (1985): On the density of states for random potential in the presence of a uniform magnetic field. Nucl. Phys **B251**, 199-211.

[187] A. Klein and A. Speis (1988): Smoothness of the density of states in the Anderson model on an one-dimensional strip. Ann. of Phys. **183**, 352-398.

[188] A Klein and A. Speis (1989): Regularity of the invariant measure and of the density of states in the one dimensional Anderson model. J. Functional Anal. **88**, 211-227.

[189] S. Kotani (1976): On asymptotic behavior of the spectra of a one dimensional Hamiltonian with a certain random coefficient. Publ. Res. Inst. Math. Sci. Kyoto Univ. **12**, 447-492.

[190] S. Kotani (1983): Ljapunov indices determine absolute continuous spectra of stationary one dimensional Schrödinger operators. in Proc. Taneguchi Intern. Symp. on Stochastic Analysis. Katata and Kyoto (1982), ed. K. Ito. North Holland, 225-247.

[191] S. Kotani (1983): Limit Theorems of Hypoelliptic Diffusions Processes. Probability theory and Mathematical Statistics. Lect. Notes in Math, No **1021**, Springer Verlag, New York.

[192] S. Kotani (1985): Support theorems for random Schrödinger operators. Comm. Math. Phys. **97**, 443-452

[193] S. Kotani (1984): Lyapunov exponents and spectra for one-dimensional random Schrödinger operators. Proc. Conf. on Random Matrices and their Applications. Contemporary Math. Amer. Math. Soc. Providence R. I.

[194] S. Kotani (1985): On the inverse problem for random Schrödinger operators. Contemporary Math. **41**, 267-280. Amer. Math. Soc. Providence R. I.

[195] S. Kotani (1987): One-dimensional Schrödinger operators and Herglotz functions. in Proc. Taneguchi Intern. Symp. on Stochastic Processes and Math. Physics. Katata and Kyoto (1985), ed. K. Ito. North Holland, 219-250.

[196] S. Kotani (1987): Link between periodic potentials and random potentials in one-dimensional Schrödinger operators. Proc. Intern. Conf. on Diff. Equations and Math. Physics.

[197] S. Kotani (1987): Absolutely continuous spectrum of one-dimensional random Schrödinger operators and Hamiltonian systems. Proc. Fifth Japan-USSR Proba. Symp. Lect. Notes in Math.(to appear)

[198] S. Kotani (1989): Jacobi Matrices with Random Potentials Taking Finitely Many Values. (preprint)

[199] S. Kotani and M. Krishna (1988): Almost Periodicity of some Random Potentials. J. Functional Anal. **78**, 390-405.

[200] S. Kotani and B. Simon (1987): Localization in general one dimensional systems, II. Comm. Math. Phys. **112**, 103-120

[201] S. Kotani and B. Simon (1988): Stochastic Schrödinger Matrices on the Strip. Commun. Math. Phys. **119**, 403-429.

[202] S. Kotani and N. Ushiroya (1989): One-dimensional Schrödinger Operators with Random Decaying Potentials. Commun. Math. Phys. **115**, 247-266.

[203] M. Krishna (1989): Anderson model with decaying randomness: Existence of extended states. (preprint)

[204] H. Kunz and B. Souillard (1980): Sur le spectre des opérateurs aux différences finies aléatoires. Comm. Math. Phys. **78**, 201-246.

[205] H. Kunz and B. Souillard (1983): The localization transition on the Bethe lattice. J. Phys. (Paris) Lett. **44**, 411-414.

[206] J. Lacroix (1982): Problèmes probabilistes liés a l'étude des opérateurs aux différences aléatoires. Ann. Inst. Elie Cartan **7**, Marches aléatoires et processus stochastiques sur les groupes de Lie, 80-95.

[207] J. Lacroix (1983): Singularité du spectre de l'opérateur de Schrödinger aléatoire dans un ruban ou un demi ruban. Ann. Inst. H. Poincaré ser **A38**, 385-399.

[208] J. Lacroix (1984): Localisation pour l'opérateur de Schrödinger aléatoire dans un ruban. Ann. Inst. H. Poincaré ser **A40**, 97-116.

[209] J. Lacroix (1984): The Schrödinger operator in a strip. Probability measures on groups VII, Proceedings Oberwolfach, Lect. Notes in Math. **1064** 280-297.

[210] J. Lacroix (1984): Computation of the sum of positive Lyapunov exponents in a strip. Lect. Notes in Math. **1186** 280-297.

[211] F. Ledrappier et G. Royer (1980): Croissance exponentielle de certains produits aléatoires de matrices. C. R. Acad. Sci. Paris ser. **A290**, 513-514.

[212] F. Ledrappier (1984): Quelques propriétés des exposants caractéristiques. Ecole d'Eté de Probabilités de Saint Flour. Lect. Notes in Math. **1097**.

[213] F. Ledrappier (1985): Positivity of the exponent for stationary sequences of matrices (preprint).

[214] E. Lepage (1982): Théorèmes limites pour les produits de matrices aléatoires. Lect. Notes in Math. **1064**, 258-303.

[215] E. Le Page (1983): Répartition d'état d'un opérateur de Schrödinger aléatoire. Probability measures on groups VII, Proceedings Oberwolfach, Lect. Notes in Math. **1064**, 309-367.

[216] E. Le Page (1989): Régularité du plus grand exposant caractéristique de matrices aléatoires indépendantes et applications. Ann. Inst. Henri Poincaré ser. B, **25**, 109-142

[217] B.M. Levitan (1983): Approximation of infinite zone potentials by finite zone potentials. Math. USSR Izvet. **20**.

[218] B. M. Levitan (1984): An inverse problem for the Sturm-Liouville operators in the case of finite-zone and infinite-zone potentials. Trans. Moscow Math. Soc. **1**, 1-34.

[219] B. M. Levitan (1985): On the closure of the set of finite-zone potentials. Math. USSR Sbornik **51**, 67-89.

[220] B. M. Levitan and I. S. Sargsjan (1976): Introduction to Spectral Theory. Transl. Math. Monographs, **39**, Amer. Math. Soc. Providence.

[221] B. M. Levitan and V.V. Zhikov (1978): Almost periodic functions and differential equations. Cambridge Univ. Press New York, N.Y.

[222] E.H. Lieb and D.C. Mattis (1966): Mathematical Physics in One Dimension: Exactly Solvable Models of Interacting Particles. Academic Press, New York, N.Y.

[223] I. M. Lifshitz (1965): Energy spectrum structure and quantum states of disordered condensed systems. Sov. Phys. Usp. **7**, 549-573.

[224] P. Loyd (1969): Exactly solvable model of electronic states in a three dimensional disordered Hamiltonian: non-existence of localized states. J. Phys. **C. 2**, 1717-1725.

[225] J.M. Luck (1988): Cantor spectra and scaling of gap widths in deterministic aperiodic systems. Preprint C.E.A. Saclay.

[226] L.A. Maloziumov (1988): Thouless formula and absence of absolutely continuous spectrum for Schrödinger operators with perturbed random potentials. Vestnik Mosk. Univ. ser.1, Math. Mech. **6**, 3-6.

[227] P. March and A. S. Sznitman (1987): Some connections between excursion theory and the discrete Schrödinger equation with random potential. Theor. Proba. and Related Fields **75**, 11-53.

[228] V.A. Marchenko and I.V. Ostrovsky (1975): A spectral property of the Hill operator. Math. USSR Sbornik **26**, 493-554.

[229] F. Martinelli (1985): The wave equation in random domains: Localization of the normal modes in the small frequency region. Ann. Inst. H. Poincaré ser. A **43**, 22-97.

[230] F. Martinelli and L. Micheli (1986): On the large coupling constant behavior of the Lyapunov exponent of a binary alloy. j. Statist. Phys. **48**, 1-18.

[231] F. Martinelli and E. Scoppola (1985): Remark on the absence of absolutely continuous spectrum for d-dimensional Schrödinger operator with random potential for large disorder or low energy. Comm. Math. Phys. **97**, 465-471.

[232] F. Martinelli and E. Scoppola (1987): Introduction to the mathematical theory of Anderson Localization. La Rivista del Nuovo Cimento **10** (3)

[233] V. P. Maslov and S.A. Molcanov (1989): Spectral properties of the rare scatterers and counterexamples to the hypotheses of Schrödinger and Stekloff. (preprint)

[234] R. Mayer (1987): Bounds on the density of states of random Schrödinger opertors. J. Statist. Phys. **48**, 425-447.

[235] H.P. McKean and P. van Moerbecke (1975): The spectrum of the Hill's equation. Inventiones Math. **30**, 217-274.

[236] H.P. McKean and E. Trubowitz (1976): Hill's operator and hyperelliptic function theory in the presence of infinitely many branch points. Comm. Pure and Appl. Math. **29**, 143-226.

[237] G. A. Mezincescu (1986): Internal Lifschitz Singularities of Disordered Finite Difference Schrödinger operators. Comm. Math. Phys. **103**, 167-176.

[238] G. A. Mezincescu (1987): Lifschitz Singularities for Periodic Operators Plus Random Potentials. J. Statist. Phys. **49**,1181-1190.

[239] N. Minami (1986): An extension of Kotani's theorem to random generalized Sturm Liouville operators. Comm. Math. Phys. **103**, 387-402.

[240] N. Minami (1989): Some Applications of a Theorem of Gilbert and Pearson to Random Schrödinger Operators. (preprint)

[241] S. A. Molcanov (1978): The structure of eigenfunctions of one dimensional unordered structures. Math. U.S.S.R. Izvestija **12** 69-101.

[242] S. A. Molcanov and H. Seidel (1982): Spectral Properties of the General Sturm-Liouville Equation with random Coefficients. Math. Nachr. **109**, 57-78.

[243] S. A. Molcanov and V.A. Chulaevskii (1984): Structure of the spectrum of a lacunary limit periodic Schrödinger operator. Funct. Anal. Appl. **18**, 343-344.

[244] S. A. Molcanov (1981): The Local Structure of the Spectrum of the One Dimensional Schrödinger Operator. Comm. Math. Phys. **78**, 429-446.

[245] J. Moser (1981): An example of a Schrödinger equation with almost periodic potential and nowhere dense spectrum. Comm. Math. Helvetici **56**, 198-224.

[246] J. Moser (1981): Integrable Hamiltonian Systems and Spectral Theory. Fermi Lectures. Scuola Normale Superiore, Pisa.

[247] J. Moser and J. Pöschel (1984): An extension of a result by Dinaburg and Sinai on quasi-peiodic potentials. Comm. Math. Helvetici **59**, 139-85.

[248] N. F. Mott (1981): The minimum metallic conductivity in three dimensions. Philos. Mag. **B44**, 265-77.

[249] N. F. Mott and W. D. Twose (1961): The theory of impurity conduction. Adv. Phys. **10**, 107-163.

[250] Y. Nagaoka (1985): Anderson localization. Prog. Theory phys. Suppl. **No. 84**.

[251] S. Nakao (1977): On the spectral distribution of the Schrödinger operator with a random potential. Japan J. Math. **3**, 11-139.

[252] U. Neri (1971): Singular integrals. Lect. Notes in Math. **200**, Springer Verlag, New York, N.Y.

[253] Th. M. Nieuwenhuizen and J. M. Luck (1985): Singular behavior of the density of states and the Ljapunov coefficient in binary random harmonic chains. J. Stat. Phys. **41**, 745-771.

[254] S.P. Novikov (1974): The periodic problem for the Korteweg de Vries equation. Funct. Anal. Appl. **8**, 236-246.

[255] O'Connor (1975): Disordered harmonic chain. Comm. Math. Phys. **45**, 63-77.

[256] H. Okura (1979): On the spectral distribution of certain integro-differential operators with random potential. Osaka J. Math. **16**, 633-666.

[257] V. Oseledec (1968): A multiplicative ergodic theorem, Ljapunov characteristic numbers for dynamical systems. Trans. Moscow Math. Soc. **19**, 197-231.

[258] V.G. Papanicolaou (1989): Short Time Asymptotics for the Trace of One and Multidimensional Schrödinger Semigroups. Proc. Amer. Math. Soc. **107**, 927-935.

[259] L. A. Pastur (1972): On the distribution of the eigenvalues of the Schrödinger equation with a random potential. Funct. Anal. Appl. **6**, 163-165.

[260] L. A. Pastur (1973): Spectra of random self adjoint operators. Russ. Math. Surveys **28**, 1-67.

[261] L. A. Pastur (1974): On the spectrum of the random Jacobi matrices and the Schrödinger equation on the whole axis with random potential. (preprint), Kharkov, (russian).

[262] L. A. Pastur (1977): Behavior of some Wiener integrals as $t \to \infty$ and the density of states of Schrödinger equation with a random potential. Theor. Math. Phys. **32**, 88-95.

[263] L. A. Pastur (1980): Spectral properties of disordered systems in the one body approximation. Comm. Math. Phys. **75**, 167-196.

[264] L. A. Pastur (1984): On the spectral theory of one-dimensional Schrödinger operator with limit periodic potential. Doklady Acad. Nauk. USSR **279**, 1050-1054.

[265] L.A. Pastur and B.A. Tkachenko (1984): On the spectral theory of one dimensional Schrödinger operator with limit-periodic potential. Soviet Math. Dokl. **279**, 1050-1054.

[266] D. B. Pearson (1978): Singular continuous measures in scattering theory. Comm. Math. Phys. **60**, 13-36.

[267] S.C Port and C.J. Stone (1978): Brownian Motion and Classical Potential Theory. Academic Press, New York, N.Y.

[268] J. Pöschel (1983): Examples of discrete Schrödinger operators with pure point spectrum. Comm. Math. Phys. **88**, 447-463.

[269] J. Pöschel and E. Trubowitz (1987): Inverse Spectral Theory. Academic Press, Boston, Ma.

[270] M. Raghunatan (1979): A proof of Oseledec's multiplicative ergodic theorem. Israel J. Math **32**, 356-362.

[271] M. Reed and B. Simon (1975): Methods of Modern Mathematical Physics II. Fourier II. Fourier Analysis, Self Adjointness. Academic Press, New York.

[272] M. Reed and B. Simon (1978): Methods of Modern Mathematical Physics IV. Analysis of Operators. Academic Press, New York.

[273] M. Reed and B. Simon (1979): Methods of Modern Mathematical Physics III. Scattering Theory. Academic Press, New York.

[274] M. Reed and B. Simon (1981): Methods of Modern Mathematical Physics I. Functional Analysis. Academic Press, New York. 2nd printing.

[275] VA. Ja. Reznikova (1981): The central limit theorem for the spectrum of the one dimensional Schrödinger operator. J. Stat. Phys. **5**, 291-308.

[276] M. Romiero and W. Wreszinski (1979): On the Lifshitz singularity and the tailing in the density of states for random lattice systems. J. Stat. Phys. **21**, 169-179.

[277] G. Royer (1980): Croissance exponentielle de produits de matrices aléatoires. Ann. Inst. H. Poincaré ser. B, **16**, 49-62.

[278] G. Royer (1982): Etude des opérateurs de Schrödinger à potentiel aléatoire en dimension un. Bull. Soc. Math. France **110**, 27-48.

[279] W. Rudin (1973): Functional Analysis. McGraw Hill, New York.

[280] W. Rudin (1974): Real and Complex Analysis. 2nd ed. McGraw Hill, New York.

[281] D. Ruelle (1969): A remark on bound states in potential scattering theory. Nuovo Cimento **A 61**, 655-662.

[282] D. Ruelle (1979): Ergodic Theory of Differentiable Dynamical Systems. Publ. Math. IHES **50**, 275-306.

[283] H. Rüssmann (1980):On the one-dimensional Schrödinger equation with a quasi-periodic potential. Ann. New York Acad. Sci. **357**, 90-107.

[284] P. Sarnak (1982): Spectral behavior of quasi periodic potentials. Comm. Math. Phys. **84**, 377-401.

[285] Schneider W. (1984): Rigorous scaling laws for Dyson measure. Bibos Symposium on Stochastic Processes. (preprint)

[286] M. A. Shubin (1982): The Density of States of Self Adjoint Elliptic Operators with Almost Periodic Coefficients. Amer. Math. Soc. Transl. **118**, 307-339.

[287] B. Simon (1978): Classical boundary conditions as a tool in quantum physics. Adv. In Math **30**, 268-281.

[288] B. Simon (1979): Trace Ideals and their Applications. Cambridge Univ. Press, Cambridge.

[289] B. Simon (1979): Functional Integration and Quantum Physics. Academic Press, New York, N.Y.

[290] B. Simon (1980): Brownian Motion, $L^p$ Properties of Schrödinger Operators and the localization of Binding. J. Functional Anal. **35**, 215-229.

[291] B. Simon (1981): Large Time Behavior of the $L^p$ Norm of Schrödinger Semigroups. J. Functional Anal. **40**, 66-83.

[292] B. Simon (1981): Spectrum and Continuum eigenfunctions of Schrödinger Operators. J. Functional Anal. **42**, 347-355.

[293] B. Simon (1982): Almost Periodic Schrödinger Operators: A Review. Adv. Appl. Math. **3**, 463-490.

[294] B. Simon (1982): Schrödinger Semigroups. Bull. Amer. Math. Soc. **7**, 447-526.

[295] B. Simon (1982): Some Jacobi matrices with decaying potential and dense point spectrum. Comm. math. Phys. **87**, 253-258.

[296] B. Simon (1983): Equality of the density of states in a wide class of tight-binding Lorentzian random models. Phys. Rev. **B 27**, 3859-3860.

[297] B. Simon (1983): Instantons, double wells and large deviations. Bull. Amer. Math. Soc. **8**, 323-326.

[298] B. Simon (1983): Kotani theory for one dimensional stochastic Jacobi matrices. Comm. Math. Phys. **89**, 227-234.

[299] B. Simon (1985): Localization in general one-dimensional random system, I. Jacobi Matrices. Comm. Math. Phys. **102**, 327-336.

[300] B. Simon (1985): Lifschitz Tails for the Anderson Model. J. Statist. Phys. **38**, 65-76.

[301] B. Simon (1985): Almost periodic Schrödinger operators, IV. The Maryland model. Ann. Phys. **159**, 157-183.

[302] B. Simon (1987): Internal Lifschitz Tails. J. Statist. Phys. **46**, 911-918.

[303] B. Simon and M. Taylor (1985): Harmonic analysis on $SL(2, R)$ and smoothness of the density of states in the Anderson model. Comm. Math. Phys. **101**, 1-19.

[304] B. Simon and T. Wolff (1986): Singular continuous spectrum under rank one perturbation and localization for random Hamiltonians. Comm. Pure Appl. Math. **39**, 75-90.

[305] B. Simon and T. Spencer (1989): Trace Class Perturbations and the Absence of Absolutely Continuous Spectra. Comm. Math. Phys. (to appear)

[306] Ya. G. Sinai (1987): Anderson localization for one-dimensional difference Schrödinger operator with quasiperiodic potential. J. Statist. Phys. **46**, 861-909.

[307] I. M. Slivnyak (1956): Spectrum of the Schrödinger Operator with Random Potential. Zh. Vychist. Mat. mat. Fiz. **6**, 1104-1108.

[308] E. Sorets (1989): Positive Lyapunov Exponents for Schrödinger Operators with quasi-periodic potentials. (preprint)

[309] T. Spencer (1985): The Schrödinger equation with a random potential, a mathematical review. In Les Houches Summer School of Critical Phenomena 1984. ed. K. Osterwalder and R. Stora.

[310] T. Spencer (1988): Localization for random and quasi-periodic potentials. J. Statist. Phys. **51**, 1009-1019.

[311] T. Spencer (1990): Ergodic Schrödinger Operators. in a volume dedicated to J. Moser. (to appear)

[312] J.M. Steele (1989): Kingman's subadditive ergodic Theorem. Ann. Inst. Henri Poincaré , ser. B **25**, No 93-98.

[313] S. Surace Jr (1987): The Schrödinger equation with a quasiperiodic potential. PhD N.Y. Univ.

[314] A. Sütö (1987): The Spectrum of a Quasiperiodic Schrödinger Operator. Comm. Math. Phys. **111**, 409-415.

[315] A. S. Sznitman (1988): Lifschitz Tail and Wiener Saussage on Hyperbolic Space. (preprint, May 1988)

[316] A. S. Sznitman (1989): Lifschitz Tail on Hyperbolic Space: Neumann Condition. (preprint, June 1988)

[317] A. S. Sznitman (1988/89): Lifschitz Tail and Wiener Saussage, I. (preprint Dec. 1988), II (preprint March 1989)

[318] M. Thompson (1983): The State Density for Second Order Ordinary Differential Equations with White Noise Potential. Bollettino U.M.I. **2-B**, 283-296.

[319] D. J. Thouless (1972): A relation between the density of states and the range of localization for one dimensional random systems. J. Phys. C. **5**, 77-81.

[320] D. J. Thouless (1974): Electrons in disordered systems and the theory of localization. Phys. Rep. **13**, 93-.

[321] Toda (1989): Theory of Nonlinear Lattice. 2nd Edition Springer Verlag, New York, N.Y.

[322] E. Trubowitz (1977): The inverse problem for periodic potentials. Comm. Pure Appl. Math. **30**, 321-337.

[323] V.N. Tutubalin (1965): On limit theorems for products of Random Matrices. Theor. Proba. Appl. **10**, 15-27.

[324] V.N. Tutubalin (1969): Some Theorems of the type of the strong law of large numbers. Theor. Proba. Appl. **14**, 313-319.

[325] P. van Moerbeke (1976): The spectrum of Jacobi matrices. Inventionesa Math. **37**, 45-81.

[326] T. Verheggen (1979): Transmission coefficient and heat conduction of a harmonic chain with random masses. Comm. Math. Phys. **68**, 69-82.

[327] A. D. Virtser (1979): On products of random matrices and operators. Theory Proba. Appl. **24**, 367-377.

[328] A. D. Virtser (1983): On the simplicity of the spectrum of the Ljapunov characteristic indices of a product of random matrices. Theory Proba. Appl. **26**, 122-136.

[329] F. Wegner (1981): Bounds on the density of States in Disordered Systems. Z. Phys. B. Condensed Matter. **44**, 9-15.

[330] H. von Weizsäcker (1983): Exchanging the order of taking suprema and countable intersection of sigma algebras. Ann. Inst. H. Poincaré **19**, 91-100.

[331] C.H. Wilcox (19840: Sound Propagation in Stratified Fields. Springer Verlag N.Y.

[332] Y. Yoshioka (1973): On the singularity of the spectral measures of a semi-infinite random system. Proc. Japan Acad. **49**, 665-668.

[333] S. Ẑhitomirskaya (1990): Hierarchical Scaling of the Spectrum for the 1D Quasi Periodic Schrödinger Operators with Binary Potentials. Uspekhi Matem. Nauk SSSR (to appear)

# NOTATION INDEX

# SUBJECT INDEX

**Probability and Its Applications**

*Editors*



*Probability and Its Applications* includes all aspects of probability theory and stochastic processes, as well as their connections with and applications to other areas such as mathematical statistics and statistical physics. The series will publish research-level monographs and advanced graduate textbooks in all of these areas. It acts as a companion series to *Progress in Probability*, a context for conference proceedings, seminars, and workshops.

We encourage preparation of manuscripts in LaTEX or AMS TEX for delivery in camera-ready copy, which leads to rapid publication, or in electronic form for interfacing with laser printers or typesetters.

Proposals should be sent directly to the editors or to: Birkhäuser Boston, 675 Massachusetts Avenue, Suite 601, Cambridge, MA 02139, U.S.A.

*Series Titles*

R.K. GETOOR. *Excessive Measures*
K.L. CHUNG/R.J. WILLIAMS. *Introduction to Stochastic Integration*, 2nd Edition
R. CARMONA/J. LACROIX. *Spectral Theory of Random Schrödinger Operators*